Graduate Texts in Physics

Graduate Texts in Physics publishes core learning/teaching material for graduate- and advanced-level undergraduate courses on topics of current and emerging fields within physics, both pure and applied. These textbooks serve students at the MS- or PhD-level and their instructors as comprehensive sources of principles, definitions, derivations, experiments and applications (as relevant) for their mastery and teaching, respectively. International in scope and relevance, the textbooks correspond to course syllabi sufficiently to serve as required reading. Their didactic style, comprehensiveness and coverage of fundamental material also make them suitable as introductions or references for scientists entering, or requiring timely knowledge of, a research field.

More information about this series at http://www.springer.com/series/8431

A. J. Berlinsky · A. B. Harris

Statistical Mechanics

An Introductory Graduate Course

A. J. Berlinsky
Brockhouse Institute for Materials Research
Department of Physics and Astronomy
McMaster University
Hamilton, ON, Canada

A. B. Harris
Department of Physics and Astronomy
University of Pennsylvania
Philadelphia, PA, USA

ISSN 1868-4513 ISSN 1868-4521 (electronic)
Graduate Texts in Physics
ISBN 978-3-030-28189-2 ISBN 978-3-030-28187-8 (eBook)
https://doi.org/10.1007/978-3-030-28187-8

This Springer imprint is published by the registered company Springer Nature Switzerland AG
The registered company address is: Gewerbestrasse 11, 6330 Cham, Switzerland

Preface

This book is designed to be used as a text for a year-long introductory course on statistical mechanics at the graduate level. It is introductory in the sense that it starts at the beginning. However, as a practical matter, most of the students who use it will have had undergraduate courses in thermodynamics and statistical mechanics. A reasonable familiarity with quantum mechanics is assumed, which may be found in the book *Principles of Quantum Mechanics* by R. Shankar which contains lucid presentations of all the necessary material.

Most of the content of this book is based on graduate courses that we have taught either at the University of British Columbia, at McMaster University, or at the University of Pennsylvania for which we used various existing texts. In these courses, we found ourselves picking and choosing from various available texts until we found the mix of topics and styles that worked for us. In adding to the rather long list of existing texts on this subject, we were motivated by several criteria which we felt existing texts did not appropriately satisfy. Here, we briefly enumerate those aspects and the points of view which we have attempted to incorporate in the present text.

(1) Statistical mechanics provides the bridge between the macroscopic world that we experience in the laboratory and on a day-to-day basis and the microscopic world of atoms and molecules for which quantum mechanics provides the correct description. Our approach is to begin by examining the questions that arise in the macroscopic world, to which statistical mechanics will ultimately provide answers, and explain how these are described by thermodynamics and statistical mechanics. Although the laws of thermodynamics were inferred long before the invention of statistical mechanics and quantum mechanics, through a lengthy process of empirical observation and logical inference, the meaning of the laws quickly becomes clear when viewed through the lens of statistical mechanics. This is the common perspective of both the elementary undergraduate text, *States of Matter*, by David Goodstein, and the profound and timeless graduate text, *Statistical Physics*, by Landau and Lifshitz.

(2) With regard to Landau and Lifshitz, which covers most if not all of what was known about statistical mechanics prior to the invention of the renormalization group in a concise and rigorous manner, one might well wonder whether any other text is necessary. However, Landau and Lifshitz is hardly an introductory text, even at the graduate level, and furthermore it does not contain homework problems suitable for a graduate course. Indeed, in our view, one strength of the present text is the array of homework problems which deal with interesting physical models and situations, including some problems of the type one encounters in "real" research. (Problem 5.12 arose in one of our Ph.D. theses!) Nevertheless, we recommend Landau and Lifshitz as an important reference which figures prominently in Part III of this text.

(3) A unifying thread in all of physics is the concept of scaling. In the simplest guise what this means is that before turning the crank in a calculation, one should be aware of what the essential variables will be. Even within mean-field theory, one can see so-called "data collapse" which results when the equation of state relating p, V, and T asymptotically close to the liquid–gas critical point becomes a relation between two appropriately scaled variables. In various problems, we have tried, by example, to be alert for such scaling and to point it out. (The Gruneisen relation in Sect. 5.7 is an example of such scaling.) Of course, scaling is a key aspect of renormalization group theory and in the analysis of Monte Carlo data.

(4) Mean-field theory, although not exact, is almost always the first line of attack in statistical problems. Accordingly, we have devoted several chapters to the various approaches to mean-field solutions of statistical problems. We also show that this approach is useful, not only for traditional temperature-dependent statistical mechanics problems but also for certain random problems not usually addressed in thermal statistical mechanics. We also introduce several heuristic arguments which give either exact or nearly exact results and serve to illustrate how successful theory is actually done. Examples include the Ginsburg criterion, the Harris criterion, the Imry-Ma argument, and the Flory estimate for polymers.

Considerable effort has been made to treat classical and quantum statistical mechanics problems on an equal footing. For this reason, the section on mean-field theory includes not only Ising models and lattice gases but also Hartree–Fock theory for normal and superfluid quantum gases. Later on, exact solutions are derived for classical Ising models and for the (one-dimensional) Ising model in a transverse field, the simplest quantum spin model.

It is also worth mentioning what is not included in this book. The introductory chapters make it clear that the book is about equilibrium systems. This means that the very interesting subjects of transport and other topics in nonequilibrium thermodynamics and statistical mechanics are not included. Furthermore, the choice of topics has been strongly influenced by research that we have been involved in throughout our careers. This includes both hard and soft condensed matter physics with an emphasis on magnetic systems or systems that are analogous to magnetic systems. Our first joint projects were on the orientational properties of ortho-H_2 molecules in alloys of ortho- and para-hydrogen. There are sections on liquid crystals and on polymers, but these subjects are presented mainly for the purpose of illustrating methods in statistical mechanics and are not treated in depth. There is somewhat more emphasis on systems with randomness. What are completely missing are subjects related to biological systems which are popular and important but which lie outside our expertise.

The material in this book is suitable for a two-semester course. At McMaster, the book would be covered in two one-semester courses, Graduate Statistical Mechanics and Advanced Statistical Mechanics. A reasonable objective might be to cover Chaps. 1–12 in the first semester and as much as possible of the remainder of the book in the second semester.

The book consists of four parts. Part I, Preliminaries, starts with a brief survey of statistical mechanics problems and then takes a closer look at a variety of phase diagrams, which are one way of representing the results of solving specific statistical mechanics problems. In Chap. 3, we give a brief review of thermodynamics emphasizing topics, such as Legendre transformations, and approaches, such as variational theorems, which are useful in the development of statistical mechanics and mean-field theory.

Part II presents the basic formalism with simple applications. In Chap. 4, the canonical ($p \sim exp[-E/(kT)]$) distribution is derived, and simple applications are given in Chap. 5. In Chap. 6, the grand canonical ($p \sim exp[-(E - \mu N)/(kT))$ distribution is derived and is then used, in Chap. 7, to treat noninteracting quantum gases.

Part III treats mean-field theory. Since this is the simplest nontrivial theory of interacting systems, we analyze it from several different perspectives. In Chap. 8, we derive it by neglecting correlated fluctuations in the Hamiltonian. In Chap. 9, we use a variational principle to obtain the mean-field density matrix. These approaches lead to a study of Landau expansions to analyze phase transitions. In Chap. 10, we extend the treatment of Landau expansions to treat a variety of cases in which one must consider more than one equivalent- or inequivalent-coupled order parameters.

Chapters 11 and 12 show how the variational mean-field approach can be applied to quantum problems of interacting Bose and Fermi particles. This is conventionally referred to as the Hartree–Fock approximation, and we use it to treat the problems of weakly repulsive bosons that are superfluid at low temperatures and, in Chap. 12, the problem of fermions with attractive interactions that become superconducting.

In Chap. 13, we examine the extent to which spatial correlations can be described in the context of mean-field theory, beginning by deriving the Ornstein–Zernike form of the spin–spin correlation function. We formulate the discussion in terms of scaling behavior and relations between critical exponents and round out that discussion by introducing Kadanoff's length-scaling hypothesis. We also examine where mean-field theory breaks down due to spatial correlations as described by the Ginsberg criterion. We extend the discussion by considering the Gaussian model, which we show can be derived from a partition function for discrete spins, using the Hubbard–Stratonovich transformation from a lattice theory to a field theory and then reexamine the criteria for the breakdown of mean-field theory from a field-theoretic perspective.

In Chap. 14, we show that mean-field theory for nonthermal problems, outside the usual purview of statistical mechanics, can be identified with exact solution on a recursive lattice (the Cayley tree).

Part IV concerns various ways in which fluctuations not included in mean-field theory can be taken into account. In Chap. 15, we discuss several examples of exact mappings of nonthermal problems (percolation, self-avoiding walks, and quenched randomness) onto models with canonical distribution characterized by a temperature, T, which can then be analyzed using all the machinery of statistical mechanics. In Chap. 16, we discuss the use of series expansions which enable one to effectively include fluctuations not treated within mean-field theory. Here, applications to the nonideal gas and to interacting spin systems are presented.

The only exact solutions beyond mean-field theory are presented in Chap. 17 where we study the one-dimensional Ising model in a transverse field and its analog, the two-dimensional Ising model. Historically Onsager's solution of the latter model was the first example to prove unambiguously that nonanalyticity at a phase transition could arise from the partition function, which, at least for finite size systems, is an analytic function of the temperature. In Chap. 18, the powerful numerical approach, Monte Carlo sampling (including the histogram method), is discussed, and Monte Carlo results are used to illustrate the method of finite size scaling. The penultimate two Chaps. 19 and 20 are devoted to a brief exposition of the renormalization group (RG). Chapter 19 introduces the subject in terms of real space RG transformations, which are pedagogically useful for defining recursion relations, flow diagrams, and the extraction of critical exponents, while Chap. 20 describes Ken Wilson's epsilon expansion and applies it to a number of standard problems. Chapter 20 also includes a more detailed treatment of the renormalization of the momentum dependence of coupling constants and the calculation of η than is normally found in texts. Chapter 21, provides an overview of Kosterlitz–Thouless physics for a variety of systems and further illustrates the renormalization group approach in a somewhat different context.

Although it is probably not possible to include all or even most of the chapters of Part IV in a 1-year course, several of the chapters could also be used as the basis for term projects for presentation to the class and/or submitted as reports.

We benefitted greatly from having Uwe Tauber as a reviewer for this book. Uwe provided extensive feedback and many useful suggestions and caught numerous typos. In addition we acknowledge helpful advice from colleagues at the University of Pennsylvania, C. Alcock on astrophysics, T. C. Lubensky on condensed matter physics, and M. Cohen for all sorts of questions as well as Amnon Aharony of Tel Aviv and Ben Gurion Universities for his advice and assistance. At McMaster, we are particularly grateful to Sung-Sik Lee for providing us with a detailed derivation of the calculation of η which appears in Sect. 20.4.4 and for a careful reading of several chapters. Finally, AJB would like to thank Catherine Kallin for her ongoing advice, patience, and encouragement.

Hamilton, ON, Canada A. J. Berlinsky
Philadelphia, PA, USA A. B. Harris

Contents

Part I Preliminaries

About the Authors

A. J. Berlinsky received his Ph.D. from the University of Pennsylvania in 1972. He was a post-doc at the University of British Columbia (UBC) and the University of Amsterdam before joining the faculty of UBC in 1977. In 1986, he moved to McMaster University where he is now Emeritus Professor of Physics. He also served as Academic Program Director and as Founding Director of Perimeter Scholars International at the Perimeter Institute for Theoretical Physics in Waterloo, Ontario, from 2008 to 2014 and as Associate Director of the Kavli Institute for Theoretical Physics in Santa Barbara, California from 2014 to 2016. He was an Alfred P. Sloan Foundation Fellow and he is a Fellow of the American Physical Society.

A. B. Harris received his Ph.D. from Harvard in 1962. He was a post-doc at Duke University and at the Atomic Energy Research Establishment at Harwell in the UK. He joined the faculty of the University of Pennsylvania in 1962, where he is now Professor of Physics Emeritus. He was an Alfred P. Sloan and John Simon Guggenheim Fellow, and he is a Fellow of the American Physical Society. In 2007, he was awarded the Lars Onsager Prize of the American Physical Society, "For his many contributions to the statistical physics of random systems, including the formulation of the Harris criterion, which has led to numerous insights into a variety of disordered systems."

Part I
Preliminaries

Chapter 1
Introduction

1.1 The Role of Statistical Mechanics

The way that most of us learned physics involved first acquiring the mathematical language and methods that are used to describe physics and then learning to calculate the motions of particles and fields, first using classical physics and then later quantum mechanics. Now we want to apply what we have learned to problems involving many degrees of freedom—solids, liquids, gases, polymers, plasmas, stars, galaxies and interstellar matter, nuclei, and a complex world of subnuclear particles and fields. Statistical mechanics provides a bridge between the dynamics of particles and their collective behavior. It is basically the study of the properties of interacting many-body systems which have in common the fact that they involve very large numbers of degrees of freedom.

Whatever the nature of the constituent degrees of freedom, statistical mechanics provides a general framework for studying their properties as functions of energy, volume, and number of particles for closed systems or as a function of control parameters, such as temperature, T, pressure, P, and chemical potential, μ, for systems that can exchange energy, volume, or particles with a reservoir, as well as how the properties of these systems respond to the application of external static and uniform fields. *In general, we will use the fact that the number of degrees of freedom is very large and study time- and space-averaged properties.*

What are the properties? We begin with somewhat abstract (but useful) definitions and then proceed to examples. We are interested in quantities which are *densities*, such as the number density of particles or the magnetization. Statistical mechanics will be set up so that these densities are first derivatives of the relevant *free energy* with respect to the *fields* that couple to them. Densities represent the solution of the many-body system to its own internal interactions and to the environment and fields that we apply. Much of Statistical Mechanics involves calculating free energies which can then be differentiated to yield the observable quantities of interest. The free energy can also be used to calculate *susceptibilities* which measure the change in

A. J. Berlinsky and A. B. Harris, *Statistical Mechanics*, Graduate Texts in Physics,
https://doi.org/10.1007/978-3-030-28187-8_1

density induced by a change in applied field and which hence are second derivatives of the free energy, often in the limit of zero applied field.

In general, the densities and susceptibilities of a system vary smoothly as functions of the control parameters. However, these properties can change abruptly as the system moves from one *phase* to another. We will see that the free energy is a nonanalytic function at certain surfaces in the space of fields that we control. These surfaces, phase boundaries, separate different phases.

1.2 Examples of Interacting Many-Body Systems

Before launching into a formal discussion of the basic principles of statistical mechanics, it is worth looking briefly at some of the kinds of systems that statistical mechanics can describe, focusing, in particular, on examples that will be discussed further in this book. These are all interacting systems that exhibit a variety of phases. The properties of these phases, particularly the quantities that distinguish them, are studied using statistical mechanics. The following is a list of examples to keep in mind as the subject is developed.

1.2.1 Solid–Liquid–Gas

Historically, one of the most important paradigms in statistical mechanics has been the solid–liquid–gas system which describes the behavior of a collection of atoms with long-range attractive and short-range repulsive interactions. The high-temperature phase, the gas, has uniform density, as does the higher density, intermediate temperature, and liquid phase. The low-temperature phase, which is a crystalline solid, does not have uniform density. In a crystal, the atoms choose where to sit and line up in rows in certain directions. Thus, the atoms in a crystal break the rotational symmetry of the gaseous and liquid states by picking out directions in space for the rows of atoms, and they also break translational symmetry by choosing where to sit within these rows. What remains is a discrete symmetry for translations by multiples of a lattice spacing, and, in general, a group of discrete rotations and reflections. These are much lower symmetries than the continuous translational and rotational symmetry of the gas and liquid. Symmetry plays an important role in statistical mechanics in helping to understand the relationships among different phases of matter.

1.2.2 Electron Liquid

The electron liquid is another important example of an interacting many-body system. Here, the interactions are the long-range repulsive Coulomb interaction and the interaction with a positive neutralizing background of ionic charge. In the simplest model, the ionic charge is taken to be uniform and rigid. From the point of view of phases, the electron liquid is somewhat unsatisfying. The low-temperature state is described by Landau's Fermi liquid theory which draws a one-to-one correspondence between the excitations of the Fermi liquid and the states of a noninteracting Fermi gas. As the temperature increases, the electron liquid evolves continuously into a nondegenerate Fermi gas. Transitions can occur at low density and temperature to a "Wigner crystal" or charge density wave (CDW) state, and, if there is any hint of attractive interactions, to a superconducting ground state. The stability of the Fermi liquid at low temperature against other competing ground states is the subject of continuing research.

1.2.3 Classical Spins

Although the solid–liquid–gas system is the classic paradigm, the most frequently encountered model in statistical mechanics is one or another variety of interacting magnetic "spins." By far the simplest of these is the Ising spin which can point "up" or "down" or, equivalently, take on the values ± 1. Typically, one studies spins on a lattice where the lattice serves to define which spins interact most strongly with which others. Having spins on a lattice simplifies the counting of states which is a large part of statistical mechanics.

Ising spins with interactions that favor parallel alignment are the simplest model of ferromagnetism. Spins whose energetics favor pairwise antiparallel states are the basis for models of antiferromagnetism. Other kinds of spins abound. XY spins are two-dimensional unit vectors; classical Heisenberg spins are three-component, fixed-length vectors. q-state Potts variables are generalizations of the Ising spin to q equivalent states. A spin-like variable with discrete values 0 and ± 1 can be used to model ^{3}He-^{4}He mixtures or an Ising model with vacancies. Quantum spins with $S = 1/2$ are described by SU(2) spinors.

For essentially all of these models, one can define a "ferromagnetic" pair interaction which is minimized when all spins are in the same state, as well as other pairwise and higher order interactions which stabilize more complicated ground states. Similarly one can define a (magnetic) "field" which selectively lowers the energy of one single spin state with respect to the others. We will see later that a ferromagnetic system may or may not have a phase transition to a ferromagnetically ordered state at $T_c > 0$, depending on the spatial dimensionality of the system and the nature of the spins. If it does have such a transition, then the temperature-field phase diagram has a phase boundary at zero field for $T < T_c$. This boundary separates different ground states with different values (orientations) of the spin.

1.2.4 Superfluids

Liquid ground states are special, and when a substance remains liquid down to $T = 0$, rather than freezing into a solid, other kinds of instabilities can occur. The two atomic substances that remain liquid under their own vapor pressure are the heliums, ^{3}He and ^{4}He. At low T both become superfluid, but they do this in very different ways. Even more special are gases that remain stable down to $T = 0$. Such gases are the subject of intense research, now that it is known how to confine, cool, and control them using laser beams. A kind of superfluid transition known as Bose–Einstein Condensation (BEC) has been observed in trapped bosonic gases, and, presumably, in the not-too-distant future, a superfluid transition will be observed in trapped Fermi gases. The superfluid transition in liquid ^{4}He and the BEC transition both arise, in somewhat different ways, from the bosonic nature of their constituent atoms, while the superfluid transitions in ^{3}He and the expected transitions in trapped atomic Fermi gases are analogous to the phenomenon of superconductivity in metals.

1.2.5 Superconductors

The main qualitative difference between superfluidity in atomic systems and superconductivity in metals is that, in superconductors, the superfluid (the electron condensate) is charged. This implies a δ-function response in the uniform electrical conductivity at zero frequency, and a correspondingly large low-frequency diamagnetic response. The quantity that distinguishes superfluids and superconductors from their respective normal phases is the fact that a certain quantum field operator (the creation operator for zero-momentum bosons or for "Cooper pairs" of fermions) has a nonzero thermal average in the superfluid state. This thermal average is a complex number. Its magnitude is a measure of the density of superfluid and gradients of its phase describe supercurrents.

1.2.6 Quantum Spins

In classical spin systems, although the spin variables may take on only discrete values, they all commute with each other, and it is straightforward to write down the energy in terms of the values of all the spins. For quantum spins, the task of writing down the energies and states of the system requires diagonalizing a large Hamiltonian matrix, impossibly large in the case of anything approaching a many-body system. At the same time, quantum mechanics makes possible new kinds of states, in addition to those found for classical spin systems. Quantum spins can pair into singlet states and effectively disappear from view. In quantum antiferromagnets, the basic interaction which makes neighboring spins want to be antiparallel also favors singlet pairing. Thus, there is a competition between the Néel state in which spins alternate between up and down from site to site and a spin liquid state in which the average spin on every site is zero. This has been an active area of research since the 1930s when Bethe solved the spin 1/2 antiferromagnetic chain. It remained an important area in the 1980s and 1990s when surprisingly a different behavior was found for integer and half-integer spin chains and when numerical techniques were developed which allowed essentially exact solutions to just about any one-dimensional quantum problem.

1.2.7 Liquid Crystals, Polymers, Copolymers

The study of systems of large molecules and of macromolecules, such as polymers, copolymers, and DNA, is a major subarea of statistical mechanics because of the rich variety of phases exhibited by these materials. Liquid crystals have phases which are intermediate between liquid and crystal. For example, nematic phases, in which the molecules pick out a preferred direction, break rotational but not translational symmetry. Smectic phases are layered, i.e., crystalline in one direction. In the case of polymers, the constituents are long flexible chains. Thermal effects would lead to conformations of these chains resembling a random walk, with the end-to-end distance being proportional to the square root of the number of monomer elements. Interaction effects, on the other hand, might cause the chains to collapse and aggregate, if they are attractive, or to stretch out if they are repulsive. Di-block copolymers are made from two different kinds of polymer chains, a block of A monomers and a block of B monomers, connected at a point to form an A-B di-block. Attraction of like units leads to layers which may have intrinsic curvature because of the different volumes of the two chain ends. Copolymers exhibit a rich variety of structural phases. Similar considerations apply to lipid bilayers in biological membranes.

1.2.8 Quenched Randomness

A large class of problems involves geometrical randomness. For instance, the statistics of self-avoiding walks on a regular lattice may be taken as a model for conformations of linear polymers. When the only constraint is hard-core interactions between different parts of the walk, this problem does not involve the temperature and therefore seems out of place in a course on statistical mechanics. Another example involves the dilution of a magnetic system in which magnetic ions are randomly replaced by nonmagnetic ions. If diffusion does not occur, we are dealing with a system which is not in thermal equilibrium, and this type of dilution is called "quenched." The distribution function for the size of magnetic clusters then depends on the concentration of magnetic ions, but there is no obvious way to describe this phenomenon via a Hamiltonian and to relate it to a thermodynamic system. Similarly, the properties of random resistor networks or the flow of fluids in random networks of pores seem far removed from Hamiltonian dynamics. Finally, if there exist random interactions between spins in, say, an Ising model, the correct mode of averaging when the randomness is frozen is distinct from the thermal averaging one performs for a thermodynamic system.

However, in all these cases, the triumph of modern statistical mechanics is to construct a mapping between these nonthermodynamic models and various, possibly esoteric, limits of more familiar thermodynamic models. For example, the statistics of long self-avoiding walks is related to the critical properties of an n-component vector spin system when the limit $n \to 0$ is taken. Likewise, percolation statistics are obtained from the thermal properties of a q-state Potts model in the limit $q \to 1$. Fortuitously, the treatment of critical behavior via the renormalization group is readily carried out for arbitrary unspecified values of parameters like n and q. In this way, the apparatus of statistical mechanics can be applied to a wide range of problems involving nonthermodynamic randomness.

1.2.9 Cosmology and Astrophysics

Statistical mechanics has long played a crucial role in theoretical approaches to cosmology and astrophysics. One early application was that of Chandrasekhar to the theory of white dwarf stars. Here, the temperature is high enough that the light elements of such stars are completely ionized but is low enough that the resulting electron gas is effectively at low temperature. Then, the energy of this system consists of the gravitational potential energy (which tends to compress the star) and the relativistic quantum zero-point energy (which tends to expand the star). Thus, the theory incorporates the constant G, Newton's constant of gravitation, h Planck's constant for quantum effects, and c from the relativistic kinetic energy.

1.3 Challenges

The physics that makes it into textbooks is generally "old stuff," solutions to problems that were challenging many years ago but which are now part of the body of common knowledge. This tends to mitigate the excitement associated with what were once great discoveries, and similarly, since the material is several steps removed from the forefront, the connection to what is currently "hot" and what are the great unsolved problems of the field is often less than apparent. On the other hand, statistical mechanics is an important tool in current research, and some of its more recently solved problems, the breakthroughs of the past few decades, are intimately related to things which still puzzle us today.

For example, the theory of critical phenomena was the great unsolved problem of the 50s and 60s. The problem was one of the length scales. Away from critical points, a many-body system will generally have a well-defined (maximum) length scale. If the problem can be solved on this length scale, then an effective "mean-field" theory can be written down to describe the system. For example, it is plausible that solving the problem of a cubic micron of water (or even much less) is enough to define the properties of a glass of water. What made critical phenomena intractable, until the invention of Wilson's renormalization group theory, was that the length scale diverges at a critical point. Thus, the only solution is the solution to the whole problem at all length scales. Wilson invented a way of systematically solving the short-length-scale part of the problem and then recasting the problem back into its original form. This systematic "integrating out" of shorter length scales eventually allows the long-length-scale behavior to emerge. Wilson's theory resolved a number of important mysteries such as the scaling behavior observed in the vicinity of critical points and the apparent "universality" of critical exponents. Although Wilson solved the problem of critical phenomena, there are still unsolved problems which may be connected to critical points where the application of this type of approach can lead to important new discoveries.

Quantum mechanics both simplifies and complicates statistical mechanics. On the one hand, it provides a natural prescription for state-counting which is at the root of what statistical mechanics is about. On the other hand, quantum mechanics involves non-commuting operators, so that the Hamiltonian, instead of being a simple function of the coordinates and momenta, is an operator in a Hilbert space. Since the dimensionality of this operator grows like e^N where N is, for example, the number of spins in a quantum spin system, the brute force approach to the solution of quantum problems is a very thankless task. When formulated in the language of path integrals, the time variable acts like an additional spatial dimension so that quantum critical behavior at zero temperature resembles classical behavior in one higher spatial dimension. The problem of quantum critical behavior remains an active area of both experimental and theoretical researches.

Exact solutions to model problems, both classical and quantum, play an important and distinctive role in statistical mechanics. Essentially, all 1-D classical problems can be solved exactly. Many 1-D quantum model Hamiltonians have been solved, and virtually all can be solved to a high degree of numerical accuracy using White's density matrix renormalization group (DMRG) technique. Onsager's solution to the 2-D Ising model provided a number of important clues for understanding the theory of second-order phase transitions. There is little in the way of exact results for 2-D quantum or 3-D classical models. Here, one must rely on computer simulations, such as Monte Carlo and quantum Monte Carlo, and series expansion techniques. The search for new, more efficient algorithms for computer simulations, and the evaluation of series expansions is an important area at the forefront of statistical mechanics. Overall, the study of 2-D quantum systems has accelerated significantly in recent years as a result of the desire to understand high-temperature superconductivity in the layered copper oxides. In spite of the increased effort and occasional success, there is much left to be learned in this area.

1.4 Some Key References

As mentioned in the preface, we assume a basic understanding of Quantum Mechanics at the undergraduate level. An excellent reference is the book,

R. Shankar, *Principles of Quantum Mechanics* (Plenum Press, 1980).

The authoritative source for statistical mechanics up to but not including the Renormalization Group is the classic text by Landau and Lifshitz which is concise and, although sometimes challenging, is well worth the effort to understand.

L.D. Landau, E.M Lifshitz, *Statistical Physics* Volume 5 of the Course of Theoretical Physics (Pergamon Press, 1969).

An excellent pedagogical reference on thermodynamics, which we used in preparing the chapters on basic principles, is

H.B. Callen, *Thermodynamics and an Introduction to Thermostatics*, 2nd edn. (J.W. Wiley, 1985).

A comprehensive and widely available reference on modern condensed matter physics is

N.W. Ashcroft, N.D. Mermin, *Solid State Physics*. (Brooks/Cole, 1976).

Chapter 2
Phase Diagrams

2.1 Examples of Phase Diagrams

In this chapter, we extend our discussion of the kinds of systems studied in statistical mechanics to include descriptions of their phase diagrams. Phase diagrams provide an overview of the thermodynamic behavior of an interacting many-body system. The coordinate axes are the control parameters, such as temperature, pressure or magnetic field, and the lines in phase diagrams, called phase boundaries, separate regions with distinct properties. To complete the picture, we need to know how the properties of each phase, the free energy, the densities, and the susceptibilities, vary as the control parameters are varied.

Order parameters are thermally averaged quantities (densities) which characterize ordered phases. An order parameter is nonzero in its ordered phase and zero in its disordered phase. The order parameter rises continuously from zero at a "second-order" (continuous) transition and jumps discontinuously from zero to a nonzero value at a "first-order" (discontinuous) phase transition. The terms "first order" and "second order" derive from Ehrenfest's classification of phase transitions in terms of which derivative of the free energy is discontinuous. A historical discussion of the Ehrenfest classification scheme can be found in Jaeger (1998). Unfortunately, subsequent developments have shown that this is not a particularly useful scheme, and we will see that the classification of phase transitions into universality classes, with their associated critical exponents, is a more logical classification scheme arising from the renormalization group.

2.1.1 Solid–Liquid–Gas

The canonical phase diagram is that of solid–liquid–gas, as shown in Fig. 2.1 for the specific case of water (Zemansky and Dittman 1981). The following points should be noted:

A. J. Berlinsky and A. B. Harris, *Statistical Mechanics*, Graduate Texts in Physics,
https://doi.org/10.1007/978-3-030-28187-8_2

(a) Every point (T, P) corresponds to a possible thermodynamic state of the system.
(b) Every point (T, P) corresponds to a unique phase, except that two phases can coexist on special lines and three phases can coexist at special points, such as (T_3, P_3), the triple point in Fig. 2.1.
(c) At a critical point, such as (T_c, P_c), two phases become indistinguishable.
(d) Along the liquid–gas coexistence curve, the density of the liquid, ρ_L, is greater than the density of the gas, ρ_G. The difference between these densities vanishes at the critical point as is illustrated in Fig. 2.2.
(e) The density has well-defined values for every (T, P). However, the density can be double-valued along coexistence lines and triple valued at triple points.

The above comments regarding the phase diagram drawn in the (T, P) plane should be contrasted with the structure of the corresponding diagram drawn as density versus pressure for different values of T as shown in Fig. 2.3. In the density–pressure plane, not every point corresponds to a possible value of the density. In the region containing the vertical dashed lines, the value of ρ can be thought of as defining the average density in a region where the two phases coexist. At any given point, the density will be either that of the liquid or the gas at the ends of the dashed lines, and the amount of each phase will then be determined by the value of the average density. Note that, for $T > T_c$, the density varies smoothly with pressure. At $T = T_c$, the slope of the density versus pressure curve, the compressibility, diverges at P_c, and for $T < T_c$, there is a discontinuous jump of density as the pressure is increased

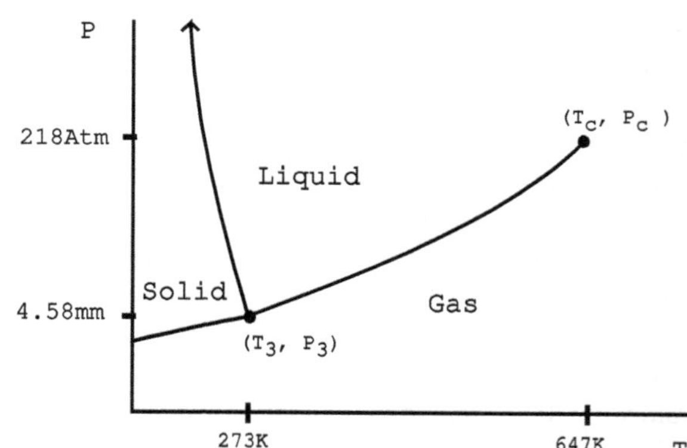

Fig. 2.1 Pressure versus temperature phase diagram of water

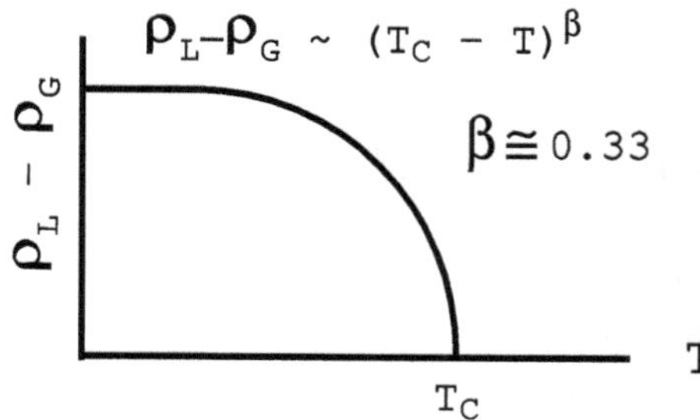

Fig. 2.2 Liquid–gas density difference, $\rho_L - \rho_G$ along the coexistence curve, plotted versus temperature

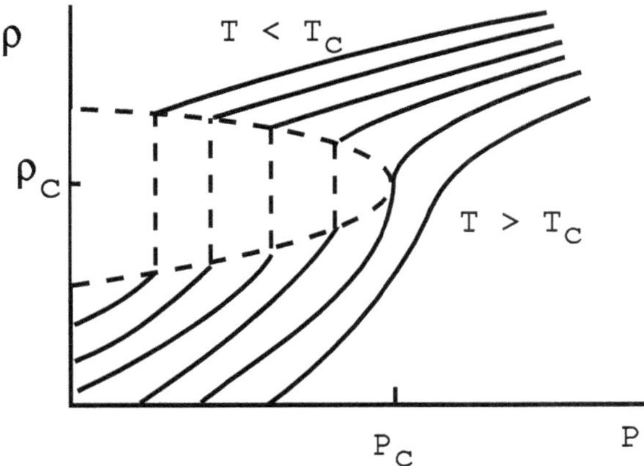

Fig. 2.3 Density of the liquid–gas system plotted versus pressure at various temperatures. The dashed lines represent a "forbidden region" where the overall density can only be obtained by having a system consisting of appropriate combinations of the two phases. A single phase having a density in the forbidden region can not occur

across the coexistence pressure at a given temperature. It is also worth mentioning that the concept of an order parameter is not particularly meaningful for the liquid–gas system. The liquid is denser than the gas, but they are both spatially uniform, and so there is no apparent order. However, there is an analogy between the liquid–gas system and the ferromagnetic system that will be discussed next, and the meaning of order in the ferromagnet is well-defined.

2.1.2 Ferromagnets

As discussed in Sect. 1.2.3, the relevant thermodynamic fields for the ferromagnet are (Kittel 1971) the magnetic field, H, and the temperature, T. A phase transition occurs at $T = T_c$ in zero field where the magnetic spins must make a decision about whether to spontaneously point up or down (for Ising spins). The presence of a nonzero magnetic field is enough to make this decision for them and thus there is no phase transition in nonzero field. This scenario is illustrated in Fig. 2.4. The phase boundary, $T < T_c$ for $H = 0$, is a coexistence line ending in a critical point. There is a striking similarity between this phase boundary and the liquid–gas coexistence line in Fig. 2.1. The main difference is that the phase diagram for the ferromagnet is more symmetric, reflecting the up–down symmetry of the magnetic system in zero field. For the liquid–gas system, this symmetry is hidden and can only be seen by analogy to the ferromagnetic system. The temperature dependence of the ferromagnetic order parameter, the magnetization jump across the temperature axis, is also strikingly similar to the jump in density across the liquid–gas coexistence line. In fact it is of great significance that the functional form of the temperature dependence close to T_c, shown in Figs. 2.2 and 2.4b, is identical for the two systems. The dependence of the magnetization on field at various temperatures is shown in Fig. 2.5. This dependence is directly analogous to the pressure dependence of the density shown in Fig. 2.3. It is observed experimentally that the critical behavior, i.e., the behavior of the order

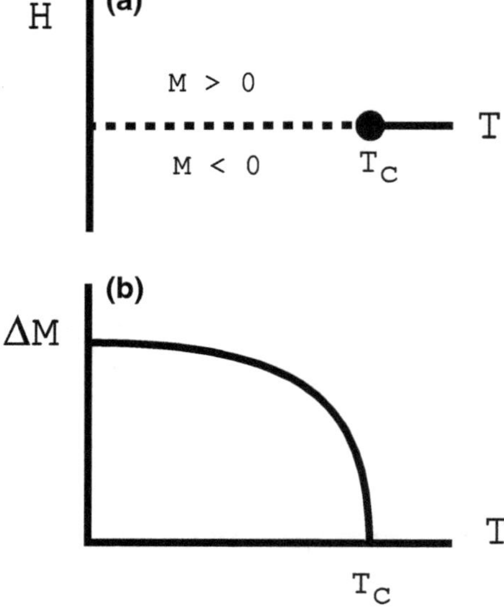

Fig. 2.4 **a** H-T phase diagram for the ferromagnet. For $T < T_c$, the magnetization jumps discontinuously as the field changes sign crossing the dashed region of the horizontal axis, whereas there is no jump crossing the solid line above T_c. The magnitude of this jump is plotted versus temperature in (**b**)

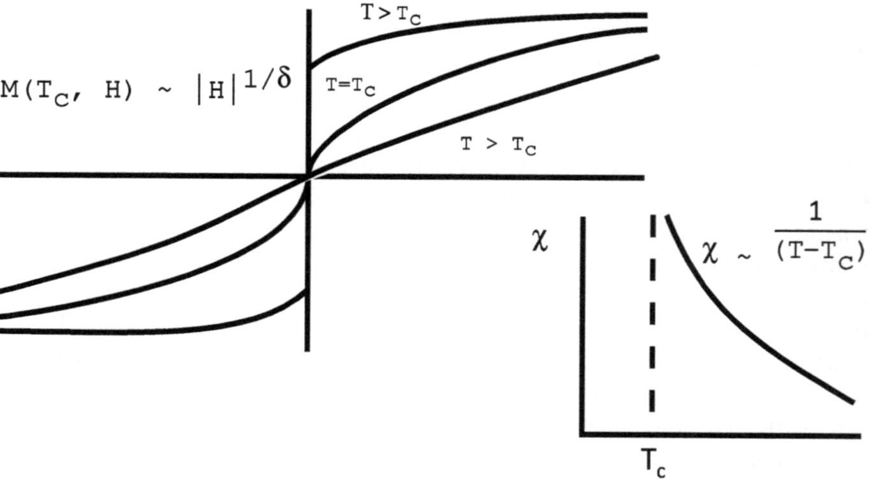

Fig. 2.5 Equilibrium magnetization versus field for the ferromagnet at various temperatures. On the bottom right, the susceptibility, $(\partial M/\partial H)_T$, is plotted for $T > T_c$

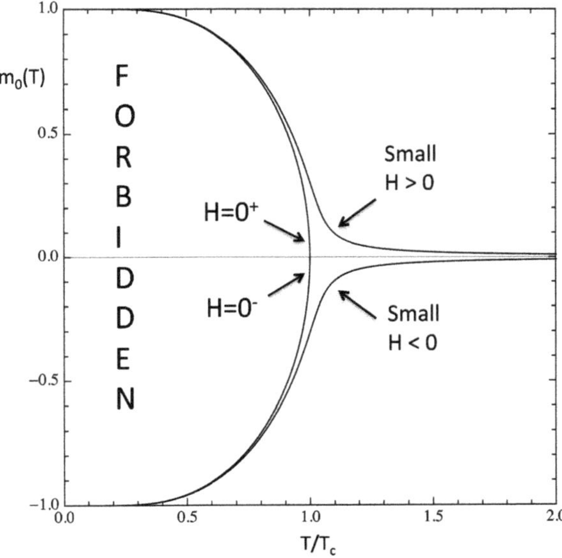

Fig. 2.6 Magnetization density m versus temperature T for the ferromagnet for the magnetic field infinitesimally positive ($H = 0^+$) or infinitesimally negative ($H = 0^-$). There is a phase transition at a critical temperature T_c (indicated by the dot) above which the magnetization is an analytic function of H at $H = 0$. For a small fields, one sees a smooth dependence on temperature, a behavior which indicates the absence of a phase transition. There is a regime of temperature in which the second derivative, d^2m/dT^2, becomes increasingly large as H is reduced toward zero

parameters and the susceptibilities close to the critical point, is identical for the magnetic and liquid–gas systems.

It is also useful to consider how the magnetization versus temperature curve evolves for small fields as the field is reduced to zero. This is shown in Fig. 2.6. One sees that the curvature near $T = T_c$ increases until, in the limit $H \to 0$, one obtains the nonanalytic curve characteristic of the phase transition. The quantity $m(H = 0^+)$, obtained as $H \to 0$ through positive values, is called the spontaneous magnetization density and is usually denoted $m_0(T)$. In analogy with the liquid–gas transition, the region with $|m| < m_0(T)$ is forbidden in the sense that one cannot have a single-phase system in this region. To obtain an average magnetization density less than $m_0(T)$ requires having more than one magnetic domain in the sample.

2.1.3 *Antiferromagnets*

The Ising Hamiltonian has the form

$$\mathcal{H} = -\frac{J}{2} \sum_{i,\delta} S_{\vec{R}_i} S_{\vec{R}_i+\vec{\delta}} - H \sum_i S_{\vec{R}_i}, \tag{2.1}$$

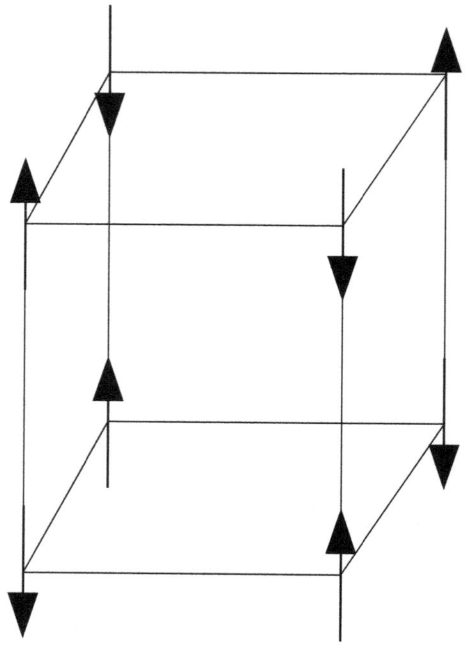

Fig. 2.7 Spins in an antiferromagnet on a cubic lattice. Ising spins can be "up" or "down" or, equivalently, they can take on values ± 1. Ferromagnetic interactions would favor parallel alignments, $\uparrow\uparrow\uparrow\uparrow$ or $\downarrow\downarrow\downarrow\downarrow$. Antiferromagnetic interactions favor arrangements, as shown here, which are locally antiparallel, $\uparrow\downarrow\uparrow\downarrow$

where $S_{\vec{R}_i} = \pm 1$, the $\vec{R}_i$ label sites on a lattice and $\vec{\delta}$ is a nearest neighbor vector. For a ferromagnet, J is positive and the Hamiltonian favors parallel spins. Here, we discuss antiferromagnets (Kittel 1971) for which J is negative and antiparallel spins are favored (Fig. 2.7). If we consider a hypercubic lattice, which is a generalized simple cubic lattice in spatial dimension d, then $\vec{\delta} = \pm\hat{x}_1, \pm\hat{x}_2, \cdots \pm \hat{x}_d$, and $\vec{R}_i + \vec{\delta}$ is a nearest neighbor of site $\vec{R}_i$. An important property of hypercubic and certain other lattices (for example the honeycomb and BCC lattices) is that they can be decomposed into two interpenetrating sublattices (A and B) which have the property that all nearest neighbors of a site on sublattice A are on sublattice B and *vice versa.* Such lattices are called "bipartite." Not all lattices are bipartite. The hexagonal (triangular) and FCC lattices are two examples of non-bipartite lattices. Antiferromagnetic order is particularly stable on bipartite lattices, since every pairwise interaction is minimized if all of the spins on one sublattice have one sign, while all the spins on the other sublattice have the opposite sign. When the lattice is not bipartite, for example, on a triangular or face-centered cubic lattice, organization into two antiferromagnetic sublattices is inhibited by geometry and the nature and possible existence of antiferromagnetic order is a subtle question. This inhibition of order by geometry is called "geometrical frustration." (Ramirez 1994 and 1996).

The ferromagnetic phase diagram is shown in Fig. 2.4a. What is the phase diagram for the antiferromagnetic case when $J < 0$? In that case the order parameter is not the average magnetization, $m = \langle S_{\vec{R}_i} \rangle$ but rather the "staggered magnetization" $m_s = \langle S_{\vec{R}_i} - S_{\vec{R}_i + \vec{\delta}} \rangle$ where $\vec{R}_i$ is restricted to be on one of the two sublattices. The phase

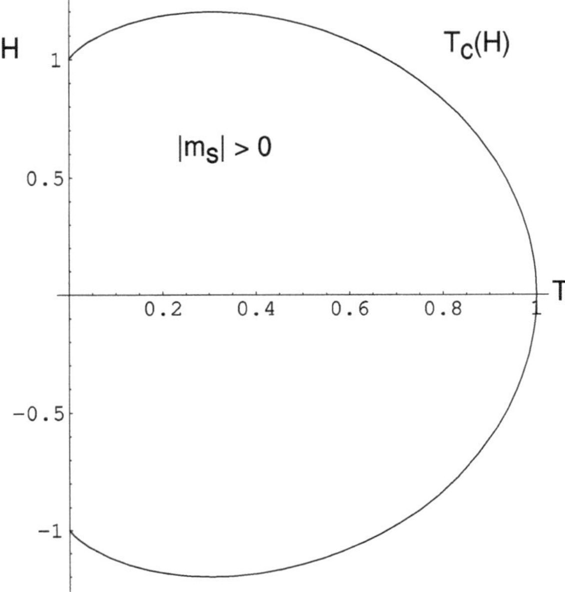

Fig. 2.8 Phase diagram for the Ising antiferromagnet in a uniform magnetic field, H. Both H and T are measured in units of Jz

diagram for the antiferromagnet consists of a region at low field and temperature in which the magnitude of the staggered magnetization is nonzero, separated by a line of continuous transitions from a high temperature, high field disordered state. The phase diagram with the phase boundary labeled $T_c(H)$ is shown in Fig. 2.8. Unlike the case for the ferromagnetic phase diagram, the sign of m_s is not specified by this phase diagram. In the region $T < T_c(H)$, m_s is nonzero, and the system spontaneously chooses its sign.

We can generalize this phase diagram by defining a field, analogous to the uniform field for a ferromagnet, which determines the sign of m_s. We define the "staggered field,"

$$H_s(\vec{R}_i) = \pm h_s, \tag{2.2}$$

where $H_s(\vec{R}_i)$ alternates in sign from one lattice site to the next. Then, the Hamiltonian for the Ising antiferromagnet becomes

$$\mathcal{H} = \frac{1}{2}|J| \sum_{i,\delta} S_{\vec{R}_i} S_{\vec{R}_i+\vec{\delta}} - H \sum_i S_{\vec{R}_i} - \sum_i H_s(\vec{R}_i) S_{\vec{R}_i}. \tag{2.3}$$

For the case where the uniform field H is zero, we can derive the H_s-T phase diagram by mapping the antiferromagnet in a staggered field onto the problem of the ferromagnet in a uniform field. We do this by a kind of coordinate transformation in which we redefine the meaning of "up" and "down" on one of the sublattices. If we

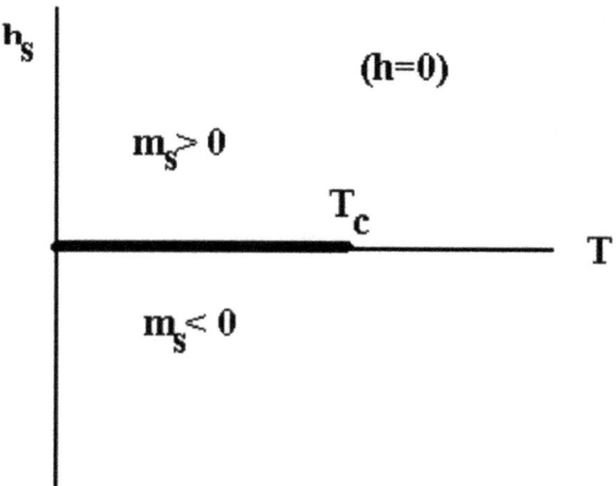

Fig. 2.9 Phase diagram for the Ising antiferromagnet in a staggered field

define an "up" spin on the B sublattice to be antiparallel to an "up" spin on the A sublattice, then the Hamiltonian becomes

$$\mathcal{H} = -\frac{1}{2}|J| \sum_{i,\delta} \tilde{S}_{\vec{R}_i} \tilde{S}_{\vec{R}_i+\vec{\delta}} - H_s \sum_i \tilde{S}_{\vec{R}_i}, \tag{2.4}$$

because $S_{\vec{R}_i} S_{\vec{R}_i+\vec{\delta}} = -\tilde{S}_{\vec{R}_i} \tilde{S}_{\vec{R}_i+\vec{\delta}}$ and $H_s(\vec{R}_i)S_{\vec{R}_i} = h_s \tilde{S}_{\vec{R}_i}$. Which is identical in form to Eq. (2.1) with $J > 0$. The corresponding phase diagram is shown in Fig. 2.9. This diagram shows that the staggered magnetization has the same sign as the staggered field. It is useful to think of the staggered field axis as a third axis, orthogonal to the temperature, and uniform field axes of Fig. 2.8. Then the region inside $T_c(H)$ for $H_s = 0$ is a surface of first-order transitions from $m_s > 0$ to $m_s < 0$

2.1.4 ^{3}He–^{4}He Mixtures

As was mentioned above in the discussion of superfluids, ^{3}He and ^{4}He remain liquid under their own vapor pressure down to $T = 0$ (Benneman and Ketterson 1975). Each becomes superfluid in its own way, as we shall now discuss. This gives rise to various interesting phase diagrams, including the phase diagram for ^{3}He–^{4}He mixtures and the P-T phase diagram of pure ^{3}He. Pure ^{4}He at 1 atm becomes superfluid at 2.18K, the temperature of the "lambda transition" which derives its name from the shape of the temperature dependence of its specific heat anomaly. The distinctive property of this superfluid phase is flow of the superfluid with no dissipation.

Since the superfluid state is a type of Bose–Einstein condensation, the temperature of the transition decreases with decreasing density of the Bose species. This is evident in Fig. 2.10 which shows the effect on the superfluid transition temperature of diluting

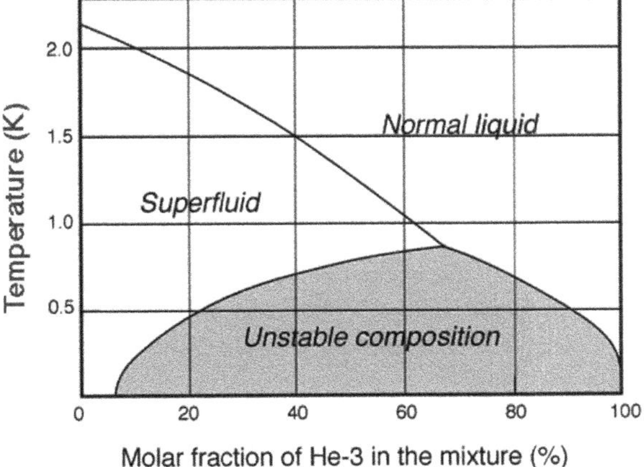

Fig. 2.10 Phase diagram of ^{3}He–^{4}He mixtures at a pressure of one atmosphere. From http://ltl.tkk.fi/research/theory/mixture.html (Courtesy of Erkki Thuneberg and Peter Berglund)

^{4}He with ^{3}He. T_λ falls below 0.9 K with the addition of about 60% ^{3}He. At this point, the λ-line ends, and the two species begin to phase separate. The phase separation transition is closely analogous to the liquid–gas transition discussed above. Both are "first order," involving a discontinuous change in density across the transition. The latent heat associated with this transition is used as the cooling mechanism in dilution refrigerators.

The phase diagram shown in Fig. 2.10 is different from those shown in Figs. 2.1, 2.4a, 2.8, and 2.9. The axes of these earlier diagrams were labeled by thermodynamic *fields*, and their topology is as described in Fig. 2.1. In Fig. 2.10, the horizontal axis is a thermodynamic *density*, as is the vertical axis in Fig. 2.3, and the same considerations apply as were discussed there. A phase diagram analogous to the $P-T$ phase diagram for the liquid–gas transition could be drawn for ^{3}He–^{4}He mixtures. In that case, the horizontal axis would be the chemical potential for ^{3}He in ^{4}He, and the λ-line would end at the beginning of the coexistence line for ^{3}He-rich and ^{3}He-poor phases. The point where a line of second-order transition changes to a first-order line is called a "tri-critical point."

At low temperature, the ^{3}He–^{4}He system phase separates into pure ^{3}He, a degenerate Fermi liquid, in the lower right-hand corner of Fig. 2.10, and a dilute mixture of about 6% ^{3}He in ^{4}He in the lower left-hand corner. Normally, one might expect the mixture to separate, at low T, into two pure phases. The fact that 6% ^{3}He dissolves into ^{4}He at $T = 0$ results from the fact that the ^{3}He atoms in ^{4}He have a much lower kinetic energy than they have in pure ^{3}He. The Fermi energy of a degenerate Fermi gas increases with the density of fermions, and this, in fact, determines the equilibrium value of 6%.

2.1.5 Pure ^{3}He

The low T end of the right-hand axis in Fig. 2.10 conceals a very low-temperature transition in pure ^{3}He to a superfluid state at about 1mK. The nature of this and other low T phases of ^{3}He are seen more clearly in the $P - T$ diagram of Fig. 2.11a. A conspicuous feature of this phase diagram is the minimum of the melting curve at around 0.3 K. Below this temperature, the melting pressure *increases* with decreasing temperature. This means that a piston containing liquid and solid in this temperature range will *cool* as the piston is compressed and liquid is converted into solid. Such an arrangement is called a Pomaranchuk cell, a convenient device for studying the physics along the melting curve. This rather unusual behavior results from the fact that, in this temperature range, the entropy of the solid is actually higher than that of the liquid. The liquid is a degenerate Fermi liquid, in which up- and down-spin states are filled up to the Fermi energy, while the solid is a paramagnetic crystal with considerable entropy in the spin system.

Lee, Richardson, and Osheroff had constructed a Pomaranchuk cell to study the low-temperature properties of ^{3}He (Osheroff et al. 1972). Late one night in 1972, Osheroff was slowly scanning the volume of the cell and monitoring the pressure, when he observed the behavior shown in Fig. 2.11b. The slope of the pressure versus volume curve changed abruptly twice, the points marked A and B in the figure. When he reversed the direction of the sweep, the features appeared again in reverse order. By observing these "glitches" and studying their properties, Lee, Richardson, and Osheroff discovered the superfluid phases of ^{3}He, for which they later were awarded the Nobel prize.

In these superfluid phases, ^{3}He atoms form Cooper pairs with total spin and orbital angular momentum both equal to 1. The two states differ in the detailed nature of this pairing. Since the spins have nuclear magnetic moments, a magnetic field will perturb the superfluid in a way that differs for the two phases. The resulting phase diagram is shown in Fig. 2.11c. Apparently, the A phase is stabilized by the application of a field, whereas, by comparison, the B phase is destabilized.

2.1.6 Percolation

The statistics of randomly occupied sites or bonds of a lattice give rise to the so-called *percolation* problem (Stauffer and Aharony 1992). First consider a regular lattice of sites, as shown in the left panel of Fig. 2.12. In the *site* percolation problem, each site is randomly occupied with probability p and is vacant with probability $1 - p$. There are therefore no correlations between the occupancies of adjacent sites. In the *bond* percolation problem, each bond is randomly occupied with probability p and is vacant with probability $1 - p$. In either case, one says that two sites belong to the same cluster if there exists a path between the two sites via bonds between occupied sites (in the site problem) or via occupied bonds (in the bond problem). As p is

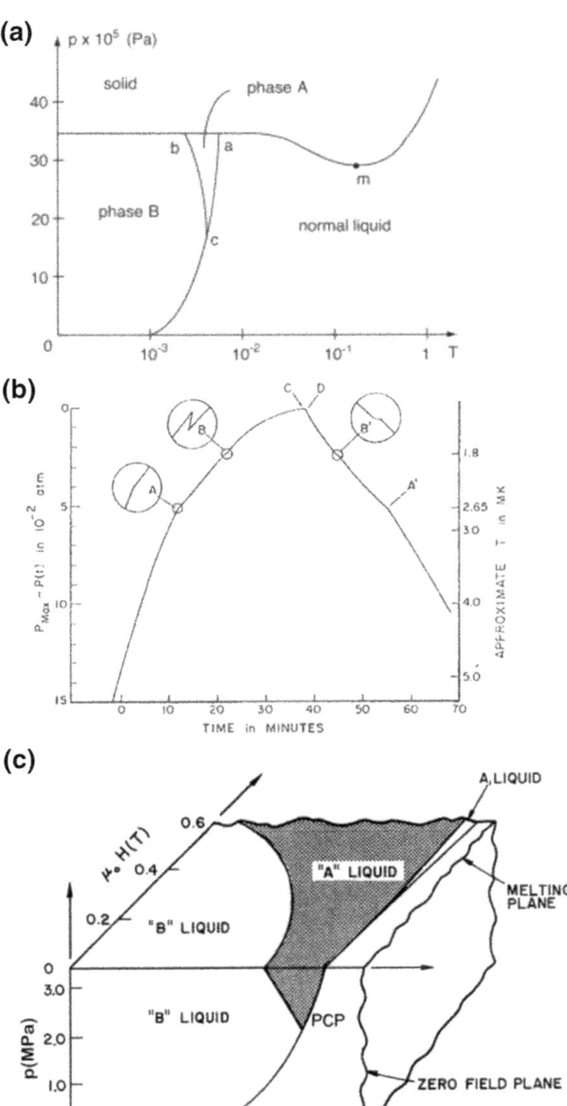

Fig. 2.11 **a** Temperature–pressure phase diagram showing the condensed phases of ^{3}He (Figure courtesy of Shouhong Wang). **b** Pressure versus time for a Pomaranchuk cell in which the volume is changing linearly with time. The two "glitches" signal transitions to the superfluid A and B states (From (Osheroff et al. 1972) with permission). **c** Pressure–temperature–magnetic field phase diagram for the normal and superfluid phases of ^{3}He (Figure from (Van Sciver 2012) with permission)

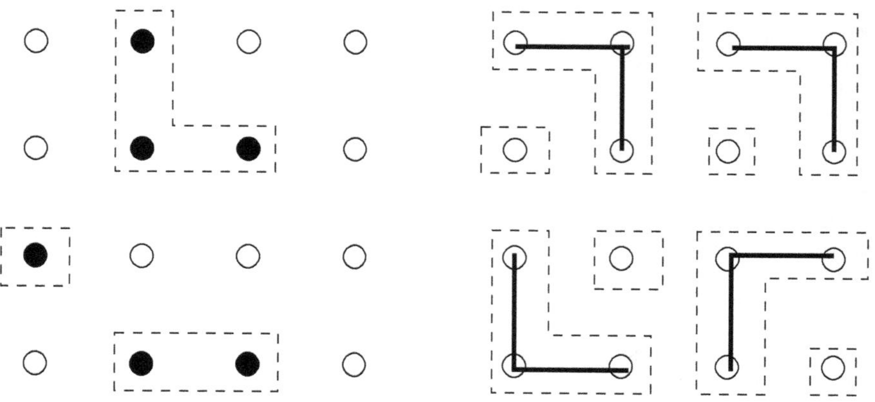

Fig. 2.12 Left: Sites of a lattice which are randomly occupied (filled circles) with probability p and are vacant (open circles) with probability $(1 - p)$. In the site problem, two neighboring sites are in the same cluster if they are both occupied. In the bond problem, two sites are in the same cluster if they are connected by an occupied bond (shown by a line). In either case, each cluster is surrounded by a dashed line

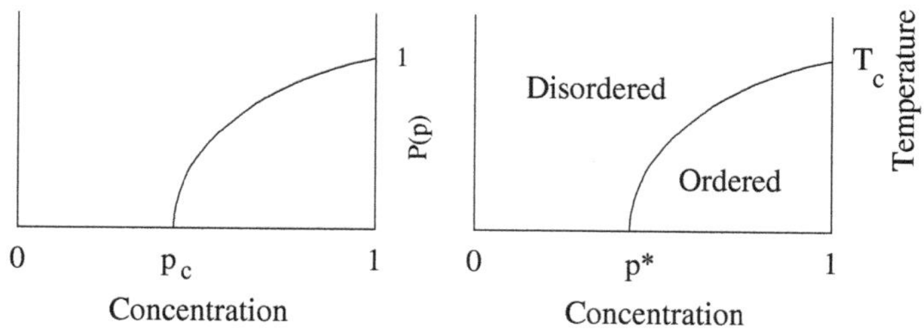

Fig. 2.13 Left: $P(p)$, the probability that a site belongs to the infinite cluster, versus concentration p. Note that $P(p) = 0$ for p less than a critical value, p_c. (For a square lattice $p_c = 1/2$ for bond percolation and $p_c \approx 0.59$ for site percolation.) Right: Phase diagram for a quenched diluted magnet. The line is $T_c(p)$ versus p. Order disappears at p* which is not necessarily the same as p_c, as discussed in the text

increased from zero, more bonds are occupied, the size of clusters increases, and one reaches a critical concentration p_c at which an infinitely large cluster appears. We will later discuss carefully what we mean by "an infinitely large" cluster. For now one may think of it as a cluster which spans the finite sample.

An important objective of percolation theory is to describe the distribution function for the size of clusters. Another quantity of interest is the percolation correlation length ξ which sets the length scale for the size of a typical cluster. The observation that ξ diverges as $p \to p_c$ suggests that the percolation problem may be analogous to a thermodynamic system like the ones discussed above. As we will see, it is possible to frame this analogy in precise mathematical terms.

There are many applications of percolation statistics to physical problems. Indeed, the model was first introduced to describe the spread of disease in orchards. The probability of the spread of disease increases as the spacing between fruit trees is decreased. To get the best yield, one would like to space the trees so that any disease that develops does not infect the entire orchard. It is apparent that the statistics of infection are the same as those discussed above for the bond model of percolation if we interpret p to be the probability that infection spread from one tree to its neighbor.

In condensed matter physics, the percolation problem describes the properties of quenched diluted systems at low temperature. Imagine replacing magnetic ions in an initially pure magnetic system by nonmagnetic ions. One can describe this system as a solid solution of the two atomic species. It is important to distinguish between two limits. In the first, called the "annealed" limit, atoms diffuse rapidly on an experimental timescale so that thermodynamic equilibrium is always achieved. This case may apply to solutions under certain conditions but is almost always the case for liquid solutions such as the He^3-He^4 mixtures discussed above. In the other limit, referred to as "quenched" dilution, atoms diffuse so slowly that the solution remains in the configuration in which it was originally prepared. Thus, the quenched system is not in thermodynamic equilibrium.

Consider properties of a quenched diluted magnetic system, whose phase diagram in zero magnetic field is shown in Fig. 2.13. For $T < T_c(1)$, the pure ($p = 1$) system exhibits long-range magnetic order. However, as p decreases the transition temperature $T_c(p)$ decreases and T_c becomes zero at some critical value $p = p^*$. If there are interactions only between nearest neighbors, then $p^* \geq p_c$ because finite clusters cannot support long-range order. In the simplest scenario considered here, $p^* = p_c$, although it is possible that certain types of fluctuations could destroy long-range order even though an infinite cluster does exist, in which case $p^* > p_c$.

From the above discussion, we conclude the existence of a critical line which separates a phase with long-range order from a disordered phase. It is interesting to discuss the critical behavior one may observe in quenched diluted magnetic systems as the critical line is approached from different directions. In such a discussion, the concept of *universality* (which states that critical properties do not depend on fine details of the system) plays a central role, as we shall see. Here, the renormalization group provides essential insight into which features of the system are relevant and which are "fine details." However, even without such high-powered formal apparatus, one can intuit some simple results. Namely, that when $p = 1$ one expects to observe pure magnetic critical phenomena as $t \to T_c(1)$. Likewise, when $T = 0$, one expects to observe the critical properties of the percolation problem. The question we will address later is what kind of critical behavior does one observe when $p < 1$ and $T > 0$.

2.2 Conclusions

We have seen in this chapter that phase diagrams can be used to characterize a great variety of interacting many-body systems and their behaviors. What remains is to learn the basic principles of statistical mechanics and how they can be used to model and calculate thermodynamic behavior, including phase diagrams. Hopefully, the rich variety of phenomena described above provide an incentive for the work required to understand this useful theoretical subject.

2.3 Exercises

1. Can the liquid–solid-phase boundary in Fig. 2.1 end in a critical point as the liquid–gas boundary does? Explain your answer.

2. Figure 2.3 shows the behavior of the density as a function of pressure for a range of temperatures near a gas–liquid critical point, such as that of water, which is labeled as (T_c, P_c) in Fig. 2.1. Draw the analogous figure for the vicinity of triple point, (T_3, P_3) in Fig. 2.1 and describe what it shows in words.

3. The problem concerns the phase diagram of liquid ^{3}He–^{4}He mixtures. In Fig. 2.10, we showed the phase diagram in the T-x_3 plane, at atmospheric pressure, where x_3 is the concentration of ^{3}He. Since x_3 is a "density," one has an inaccessible region, i.e., a region in the T-x_3 plane, labeled "unstable composition" which does not correspond to a single-phase mixture. It is instructive to represent this phase diagram when the "density" x_3 is replaced by the corresponding "field," which in this case is $\Delta\mu \equiv \mu_3 - \mu_4$, the difference between the chemical potential of the two species. Construct the phase diagram for this system in the T-$\Delta\mu$ plane for a fixed pressure of one atmosphere. Of course, without more data you cannot give a quantitatively correct diagram. So a topologically correct diagram will suffice as an answer to this exercise.

4. The figure at right shows the phase diagram for a solid S in the temperature (T)-pressure (p) plane. At zero pressure, this solid undergoes a phase transition from a hexagonal close-packed structure at low temperature to a cubic structure at higher temperature. Measurements at low pressure find a hexagonal to cubic phase boundary as shown in the figure. Do you think that this phase boundary can end at a critical point at nonzero temperature, so that by going to very high pressure one can pass from the cubic phase to the hexagonal phase without crossing a phase boundary? (Recall that this is possible in the case of the liquid and gas phases.) Briefly explain your answer.

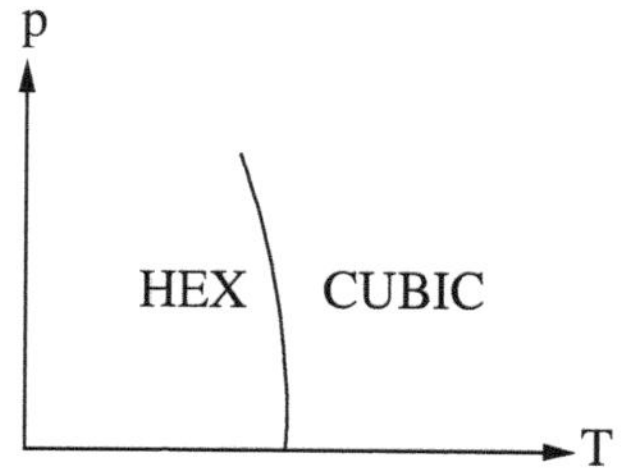

References

K.H. Benneman, J.B. Ketterson, *The Physics of Liquid and Solid Helium* (Wiley, 1975)

G. Jaeger, The ehrenfest classification of phase transitions: introduction and evolution. Arch. Hist. Exact Sci. **53**, 5181 (1998)

C. Kittel, *Introduction to Solid State Physics*, 4th edn. (Wiley, 1971)

D.D. Osheroff, R.C. Richardson, D.M. Lee, Evidence for a new phase of solid He3. Phys. Rev. Lett. **28**, 885 (1972)

A.P. Ramirez, Strongly geometrically frustrated magnets. Annu. Rev. Mater. Sci. **24**, 453, (1994)

A.P. Ramirez, Geometrical frustration in magnetism. Czech. J. Phys. **46**, 3247, (1996)

D. Stauffer, A. Aharony, *Introduction to Percolation Theory*, 2nd edn. (Taylor and Francis, 1992)

S.W. Van Sciver, 3He and refrigeration below 1 K, in *Helium Cryogenics. International Cryogenics Monograph Series* (Springer, New York, NY 2012)

M.W. Zemansky, R.H. Dittman, *Heat and Thermodynamics*, 6th edn. (McGraw-Hill, 1981)

Chapter 3
Thermodynamic Properties and Relations

3.1 Preamble

Statistical Mechanics allows us to connect the microscopic properties of material systems to the macroscopic properties that we observe and use in everyday life. It does this by providing a prescription for calculating certain functions, called thermodynamic potential functions, starting from the microscopic Hamiltonian, using the theories of statistics and classical or quantum mechanics. Exactly how to do that is the subject of Chap. 4—Basic Principles. Here we discuss what one can do with a knowledge of the thermodynamic potentials, starting from a discussion of what the natural variables are for these functions, then defining the functions along with relations between the functions and their various derivatives which correspond to observable properties of thermodynamic systems.

3.2 Laws of Thermodynamics

The laws of thermodynamics are summarized as follows.

- The zeroth law, which was named by Callen (1985), states that if systems A and B are in thermal equilibrium with system C, then A and B would already be in thermal equilibrium with each other. Note that this does not provide a metric for temperature or even that one temperature is "hotter" than another.
- The first law states that in any process (reversible or not) the change in internal energy dU is equal to the sum of the heat ΔQ added to the system plus the work ΔW done on the system. The impact of this law is that even in an irreversible process where ΔQ and ΔW are not expressible in terms of differentials of analytic functions, dU is the differential of a well-defined function of the state of the system.
- The infamous second law (in one of its many variations) introduces a scale of temperature, T (independent of the thermometric substance) and the concept of entropy, S (which is shown to increase in natural processes). Most people find that

A. J. Berlinsky and A. B. Harris, *Statistical Mechanics*, Graduate Texts in Physics,
https://doi.org/10.1007/978-3-030-28187-8_3

the concept of entropy in this formulation is unintelligible. In the simple case of gaseous systems, the result of laws one and two enable one to write, for reversible processes, that

$$dU = \Delta Q + \Delta W = TdS - PdV \;. \tag{3.1}$$

- Finally, the third law states that for a system in equilibrium, in the limit of zero temperature, the entropy can be taken to be zero.

As we will see in this book, the concept of entropy is less mysterious in statistical mechanics than in thermodynamics. Rather than starting from "Laws" of thermodynamics, we will instead infer the content of these laws from the understanding of statistical mechanics that we will derive. This will be discussed further in Sect. 3.4.

3.3 Thermodynamic Variables

Understanding the nature and number of independent variables characterizing a thermodynamic state plays a central role in statistical mechanics, even in the renormalization group (see Chaps. 19 and 20). Thermodynamics concerns itself with systems, like those discussed in the introductory chapters, which are characterized by a small number of *macroscopic* variables. Thermodynamics does not specify the number and choice of these variables. Their choice is dictated by the system under consideration. For instance, for gaseous systems one may select three independent variables out of the set, pressure P, volume V, temperature T, and number of particles N. Once three of these variables are fixed, then the remaining one is determined by what is called "the equation of state." (Similar considerations apply if the system can have a magnetization or an electric polarization.) It is sometimes useful to classify the variables as being either *extensive* or *intensive*, depending on how they scale with system size. If you divide a thermodynamic system in equilibrium into subvolumes, some variables, such as the pressure or the temperature, will be the same in each subvolume. Such variables are called *intensive*, and they are uniform across the system. Other variables, such as the volume and number of particles in a subvolume, are additive. Combining regions combines their volume and number of particles. Such variables, which scale linearly with system size, are called *extensive* variables.

From the above definition, it is clear that ratios of extensive variables are intensive. In particular, this applies to the ratio of the number of particles to the volume, which is called the *number density*, or *density* for short. As noted in Chaps. 1 and 2, it is also useful to distinguish between intensive quantities such as N/V, i.e., *densities*, and intensive variables which are external control parameters, such as pressure and temperature which we call *fields*. Along a phase boundary where two or more phases coexist, the system can have multiple values of the density, even though the pressure and temperature are uniform throughout the sample.

3.4 Thermodynamic Potential Functions

Statistical Mechanics allows us to calculate the entropy, S, of a system, which is a function of the total energy of the system, U, also called the "internal energy," its volume, V, and number of particles, N. The connection of entropy to the microscopic properties of matter falls into the realm of statistical mechanics and was provided by Boltzmann in his famous formula shown in Fig. 3.1,

$$S = k \ln W \, , \tag{3.2}$$

where k is Boltzmann's constant. If W is the total number of states accessible to a system with a given total energy, volume, and number of particles, then S is the equilibrium entropy. More generally, W could represent a sum over energetically accessible states, each weighted with its own probability. For a system in equilibrium, we will see that those probabilities are all equal and that the resulting entropy has its maximum value with respect to the weightings.

We will also see in the next chapter that the entropy defined by Boltzmann has the important property that two otherwise isolated systems in thermal contact with each other have a maximum total entropy, and hence are in equilibrium with each other, when their temperatures are equal, where the temperature, T, is defined through

$$\frac{1}{T} = \left. \frac{\partial S}{\partial U} \right|_{V,N} , \tag{3.3}$$

that is, $1/T$ is the rate at which the entropy increases with increasing energy with other system parameters held fixed. This result encompasses the Zeroth Law of Thermodynamics which says that if system A is in equilibrium with two systems B and C, then B and C are automatically in equilibrium with each other. The zeroth law implies that when systems B and C are brought into thermal contact with each other, there is no net flow of energy from one to the other. Statistical mechanics

Fig. 3.1 Ludwig Boltzmann's Tombstone. (Photo by Thomas D. Schneider)

describes this shared equilibrium by saying that the systems A, B, and C are at the same temperature and provides a meaningful definition of temperature.

For most systems, including gases, liquids, and elastic solids, W is a rapidly increasing function of energy since there are more ways to distribute larger amounts of energy. This means that the entropy, $S = k \ln W$, is also monotonically increasing, and one can invert the function $S(U, V, N)$ to obtain $U(S, V, N)$. (It also means that the temperature, defined in Eq. (3.3) is positive.) We refer to the variables, U, V, and N which determine the value of S in statistical mechanics as the "proper variables" of the thermodynamic potential function, S. Furthermore, since, as we have seen, $S(U)$ can be inverted to give $U(S)$, then the proper variables of U are S, V, and N. Holding N fixed to simplify the discussion, we can write the differential of $U(S, V)$ as

$$dU = \frac{\partial U}{\partial S}\bigg|_V dS + \frac{\partial U}{\partial V}\bigg|_S dV \equiv TdS - PdV \ , \tag{3.4}$$

where T is defined from the inverse of Eq. (3.3) and we have noted that the change in energy due to the mechanical work of compressing or expanding the system by changing its volume by an amount dV is $-PdV$. Equation (3.4) is the second law of thermodynamics.

The entropy defined by Boltzmann has another interesting and important property which is that, as $T \to 0$, the entropy of any homogeneous system approaches a constant (which we may take to be zero) independently of the values of the other thermodynamic variables. This is the Third Law of Thermodynamics, and it is very plausible and easy to understand within statistical mechanics. For instance, consider Eq. (3.2). In the zero-temperature limit $W = g$, where g is the degeneracy of the quantum ground state of the physical system of N particles. Then, the Third Law of Thermodynamics says that $(1/N) \ln g \to 0$ in the limit of infinite system size when $N \to \infty$. We will explore several consequences of the third law in the next section after we have introduced the various thermodynamic response functions.

3.5 Thermodynamic Relations

In this section, we introduce various macroscopic response functions and discuss mathematical relations between them as well as some of their applications. We also discuss how these functions must behave as $T \to 0$ in light of the third law.

3.5.1 Response Functions

The specific heat is defined as the ratio, $\Delta Q/\Delta T$, where ΔQ is the heat you must add to a system to raise its temperature by an amount ΔT. Thus, we write the specific heat at constant volume, C_V, or at constant pressure C_P as

$$C_V \equiv \left.\frac{\Delta Q}{\Delta T}\right|_V = T \left.\frac{\partial S}{\partial T}\right|_V = \left.\frac{\partial U}{\partial T}\right|_V \tag{3.5a}$$

$$C_P \equiv \left.\frac{\Delta Q}{\Delta T}\right|_P = T \left.\frac{\partial S}{\partial T}\right|_P \neq \left.\frac{\partial U}{\partial T}\right|_P . \tag{3.5b}$$

(See Exercise 3 which illustrates the fact that C_P is *not* the same as $\partial U/\partial T|_P$.) Similarly, we can write the compressibilities at constant temperature and at constant entropy as

$$\kappa_T \equiv -\frac{1}{V}\left.\frac{\partial V}{\partial P}\right|_T , \qquad \kappa_S \equiv -\frac{1}{V}\left.\frac{\partial V}{\partial P}\right|_S , \tag{3.6}$$

and the volume coefficient of thermal expansion as

$$\beta \equiv \frac{1}{V}\left.\frac{\partial V}{\partial T}\right|_P . \tag{3.7}$$

We also mention the response functions for electric and magnetic systems, the isothermal and adiabatic susceptibilities for electric and magnetic systems,

$$\chi_T \equiv \left.\frac{\partial M}{\partial H}\right|_T , \qquad \chi_S \equiv \left.\frac{\partial M}{\partial H}\right|_S \tag{3.8a}$$

$$\chi_T \equiv \left.\frac{\partial p}{\partial E}\right|_T , \qquad \chi_S \equiv \left.\frac{\partial p}{\partial E}\right|_S , \tag{3.8b}$$

where H and E are the uniform applied fields which appear in the Hamiltonian. In realistic models, where one takes into account the dipole fields of magnetic or electric moments, these "susceptibilities" will depend, not only on the composition of the system but also on the shape of the sample.

3.5.2 *Mathematical Relations*

We remind you of a few mathematical relations which are useful in thermodynamics:

$$\left.\frac{\partial y}{\partial x}\right|_{z_1,z_2,\ldots} = \left(\left.\frac{\partial x}{\partial y}\right|_{z_1,z_2,\ldots}\right)^{-1} . \tag{3.9}$$

Also (if x is a function of y and z)

$$dx = \left.\frac{\partial x}{\partial y}\right|_z dy + \left.\frac{\partial x}{\partial z}\right|_y dz , \tag{3.10}$$

so that differentiating with respect to y with x constant we get

$$0 = \left.\frac{\partial x}{\partial y}\right|_z + \left.\frac{\partial x}{\partial z}\right|_y \left.\frac{\partial z}{\partial y}\right|_x , \tag{3.11}$$

from which we obtain the mysterious relation

$$\left.\frac{\partial x}{\partial y}\right|_z = - \left.\frac{\partial x}{\partial z}\right|_y \left.\frac{\partial z}{\partial y}\right|_x . \tag{3.12}$$

From Eq. (3.10), one can also write [if x can be regarded as a function of y and z and z is in turn a function $z(y, t)$ of y and t.]

$$\left.\frac{\partial x}{\partial y}\right|_t = \left.\frac{\partial x}{\partial y}\right|_z + \left.\frac{\partial x}{\partial z}\right|_y \left.\frac{\partial z}{\partial y}\right|_t . \tag{3.13}$$

Finally, if we have

$$df(x, y) = A(x, y)dx + B(x, y)dy , \tag{3.14}$$

then

$$\left.\frac{\partial A(x, y)}{\partial y}\right|_x = \frac{\partial^2 f}{\partial y \partial x} = \frac{\partial^2 f}{\partial x \partial y} = \left.\frac{\partial B(x, y)}{\partial x}\right|_y . \tag{3.15}$$

For example

$$dU = TdS - PdV , \tag{3.16}$$

so that

$$\left.\frac{\partial T}{\partial V}\right|_S = - \left.\frac{\partial P}{\partial S}\right|_V \tag{3.17}$$

or, equivalently,

$$\left.\frac{\partial V}{\partial T}\right|_S = - \left.\frac{\partial S}{\partial P}\right|_V . \tag{3.18}$$

These type of relations are called *Maxwell relations*.

In Sects. 3.5 and 3.6, we will define the free energy, $F \equiv U - TS$, and we will show why, although U is a function of S and V, the free energy, F, is actually a function of the variables T and V. Thus, we can write

$$dF = -SdT - PdV , \tag{3.19}$$

so that

$$\left.\frac{\partial S}{\partial V}\right|_T = \left.\frac{\partial P}{\partial T}\right|_V \tag{3.20}$$

or, equivalently,

$$\left.\frac{\partial V}{\partial S}\right|_T = \left.\frac{\partial T}{\partial P}\right|_V . \tag{3.21}$$

3.5.3 Applications

For instance, using Eqs. (3.5b) and (3.13), we have

$$C_P = T \left.\frac{\partial S}{\partial T}\right|_V + T \left.\frac{\partial S}{\partial V}\right|_T \left.\frac{\partial V}{\partial T}\right|_P . \tag{3.22}$$

Using Eq. (3.20), this is

$$C_P = C_V + T \left.\frac{\partial P}{\partial T}\right|_V \left.\frac{\partial V}{\partial T}\right|_P . \tag{3.23}$$

Equation (3.12) gives

$$\left.\frac{\partial P}{\partial T}\right|_V = - \left.\frac{\partial P}{\partial V}\right|_T \left.\frac{\partial V}{\partial T}\right|_P \tag{3.24}$$

so that

$$C_P - C_V = -T \left.\frac{\partial P}{\partial V}\right|_T \left(\left.\frac{\partial V}{\partial T}\right|_P\right)^2 = VT\beta^2/\kappa_T > 0 . \tag{3.25}$$

(We will later show that κ_T is nonnegative.) The exercises will let you play with some other such relations.

3.5.4 Consequences of the Third Law

We now explore some consequences of the third law. Imagine a container of volume V filled with a homogeneous solid or quantum fluid at $T = 0$. Then, on heating this container, the entropy of its contents will be

$$S(T_f, V) = S(0, V) + \int_{0,V}^{T_f,V} \left.\frac{\partial S}{\partial T}\right|_V dT$$
$$= \int_{0,V}^{T_f,V} \frac{C_V(T, V)}{T} dT . \tag{3.26}$$

Here, we have used the third law to take $S(T = 0, V) = 0$ independent of V. Since all the entropies in this equation are finite, the specific heat must tend to zero sufficiently rapidly as $T \to 0$ to make the integral converge. For example, if S is a simple power law, T^n, at low T, then C_V must have the same T^n dependence with $n > 0$.

Another consequence of the third law is that, since $S \to 0$ as $T \to 0$ for any V, $\Delta S \to 0$ as $T \to 0$ and the Maxwell relation of Eq. (3.20) implies that

$$\lim_{T\to 0}(\partial P/\partial T)_V = 0 . \tag{3.27}$$

Furthermore, note that

$$\left.\frac{\partial V}{\partial T}\right|_P = -\left.\frac{\partial V}{\partial P}\right|_T \left.\frac{\partial P}{\partial T}\right|_V . \tag{3.28}$$

As long as the compressibility κ_T remains finite as $T \to 0$, we infer that the coefficient of thermal expansion β vanishes in the zero-temperature limit.

Some of the most important applications of the third law occur in chemistry. Consider a chemical reaction of the type

$$\mathrm{A} + \mathrm{B} \leftrightarrow \mathrm{C} , \tag{3.29}$$

where A, B, and C are compounds. (Perhaps the most famous example of such a reaction is $2H_2 + O_2 \leftrightarrow 2H_2O$.) Suppose we study this reaction at temperature T_0. The entropy of each compound (which most chemists call the "third law entropy") can be established at temperature T_0 by using

$$S(T_0, P) = S(0, P) + \int_{0,P}^{T_0,P} \left.\frac{\partial S}{\partial T}\right|_P dT$$
$$= \int_{0,P}^{T_0,P} \frac{C_P(T, P)}{T} dT . \tag{3.30}$$

(If one crosses a phase boundary, the discontinuity in crossing this phase boundary is obtained by measuring the heat of transformation ΔQ and setting $\Delta S = \Delta Q/T$.) In this way, one can predict the temperature dependence of the heat of reaction $T\Delta S$ associated with this reaction just from measurements on the individual constituents. Note the crucial role of the third law in fixing $S(0, P) = 0$ irrespective of the value of P.

It is clear from Eq. (3.30) that, at least in principle, measurements of the specific heat at constant pressure enable one to establish the entropy of a gas at some high temperature T_0 where it is well approximated by an ideal gas. As we will see later, statistical mechanical calculations show that the entropy of an ideal gas involves Planck's constant h. Thus, we come to the amazing conclusion that by using only macroscopic measurements one can determine the magnitude of h! Such is the power of the Third Law of Thermodynamics. See the end of Sect. 4.5.2 in Chap. 4 for a discussion of the method actually used for this purpose.

3.6 Thermodynamic Stability

We now examine what thermodynamics has to say about the stability of ordinary phases of matter. These results are known in chemical thermodynamics as applications of Le Chatelier's principle (Callen 1985). (A famous example of this principle in elementary physics is Lenz's Law (Giancoli 1998).) In simple language, this principle states that a system reacts to an external perturbation so as to relieve the effects of that perturbation. So if you squeeze a system, it must contract. If you heat a system, its temperature must rise. We will obtain these results explicitly in this section.

3.6.1 *Internal Energy as a Thermodynamic Potential*

Consider a system with an internal constraint characterized by a variable ξ. For instance, as shown in Fig. 3.2, our system could have an internal partition which is free to move, in which case we would expect that in equilibrium, it would adjust its position (specified by ξ) so as to equalize the pressure on its two sides. Alternatively, this constraint could simply specify the way in which the internal energy and/or volume is distributed over two halves of the system. In statistical mechanics, analogous constraints may take the form of theoretical and/or microscopic constraints. Let us write down a formula corresponding to the statement that the internal constraint of an isolated system will assume its equilibrium value in order to maximize the entropy. By "isolated," we mean that the system is contained within a thermally insulating container having rigid walls, as shown in Fig. 3.2. In this situation, we see that $dU = dV = 0$. This does not mean that dS for the system is zero, because we are considering the family of nonequilibrium states for which ξ does not assume its equilibrium value. Therefore, the statement that the entropy is maximal with respect to the constraint ξ means that the entropy is extremal, so that

$$\left.\frac{\partial S}{\partial \xi}\right|_{U,V} = 0 \tag{3.31}$$

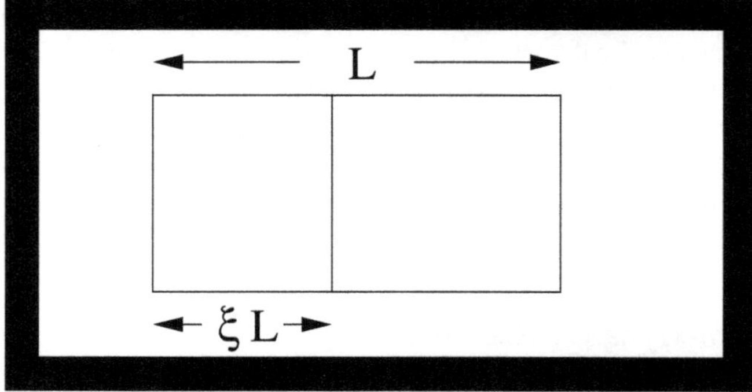

Fig. 3.2 A system isolated from the rest of the universe (as indicated by the solid wall) with a freely movable partition at a distance ξL from the left wall

and also that this extremum is a maximum, so that

$$\left.\frac{\partial^2 S}{\partial \xi^2}\right|_{U,V} < 0 \,. \tag{3.32}$$

Although we have given the specific illustration in which the parameter ξ specifies the location of an internal partition, we emphasize that many other types of internal constraints can be imagined. Note that this condition only guarantees that the system is locally stable, i.e., that it is stable with respect to infinitesimal changes in the thermodynamic variables. Indeed, it is possible to supercool a liquid to temperature below that at which a solid forms. In this case, the supercooled liquid is locally (i.e., microscopically) stable, but if sufficiently perturbed the system will solidify because it is not globally stable. As we will see later, this phenomenon is contained within simple equations of state, such as the van der Waals equation of state.

We now use the conditions of Eqs. (3.31) and (3.32) to develop a criterion involving the extremum of the internal energy U with respect to an internal constraint when the system is held at fixed volume V and entropy S. Note that *at equilibrium* [where $(\partial S/\partial \xi)_{U,V} = 0$] we have

$$\left.\frac{\partial U}{\partial \xi}\right|_{S,V} = -\left.\frac{\partial U}{\partial S}\right|_{\xi,V} \left.\frac{\partial S}{\partial \xi}\right|_{U,V} = 0 \,. \tag{3.33}$$

So with respect to variation of the constraint ξ, U is extremal at equilibrium. To see whether or not U is minimal, we consider

$$\left.\frac{\partial^2 U}{\partial \xi^2}\right|_{S,V} = -\frac{\partial}{\partial \xi}\left[\left.\frac{\partial U}{\partial S}\right|_{\xi,V} \left.\frac{\partial S}{\partial \xi}\right|_{U,V}\right]_{S,V} \,. \tag{3.34}$$

At equilibrium [where $(\partial S/\partial\xi)_{U,V} = 0$], this is

$$\left.\frac{\partial^2 U}{\partial\xi^2}\right|_{S,V} = -\left.\frac{\partial U}{\partial S}\right|_{\xi,V} \left.\frac{\partial}{\partial\xi}\left[\left.\frac{\partial S}{\partial\xi}\right]_{U,V}\right|_{S,V} = -T\left.\frac{\partial}{\partial\xi}\left[\frac{\partial S}{\partial\xi}\right]_{U,V}\right|_{S,V} . \tag{3.35}$$

Let X denote $(\partial S/\partial\xi)_{U,V}$. Then Eq. (3.35) is

$$\left.\frac{\partial^2 U}{\partial\xi^2}\right|_{S,V} = -T\left.\frac{\partial X}{\partial\xi}\right|_{S,V} . \tag{3.36}$$

Now use Eq. (3.13) with $x \to X$, $y \to \xi$, $t \to S$, and $z \to U$, with V a bystander variable. It says that

$$\left.\frac{\partial X}{\partial\xi}\right|_{S,V} = \left.\frac{\partial X}{\partial\xi}\right|_{U,V} + \left.\frac{\partial X}{\partial U}\right|_{\xi,V}\left.\frac{\partial U}{\partial\xi}\right|_{S,V} . \tag{3.37}$$

But note that at equilibrium Eq. (3.33) tells us that $(\partial U/\partial\xi)_{S,V} = 0$, so that

$$\left.\frac{\partial X}{\partial\xi}\right|_{S,V} = \left.\frac{\partial X}{\partial\xi}\right|_{U,V} = \left.\frac{\partial^2 S}{\partial\xi^2}\right|_{U,V} . \tag{3.38}$$

Thus at equilibrium [where $(\partial^2 S/d\xi^2)_{U,V} < 0$]

$$\left.\frac{\partial^2 U}{\partial\xi^2}\right|_{S,V} = -T\left.\frac{\partial^2 S}{\partial\xi^2}\right|_{U,V} > 0 . \tag{3.39}$$

Now we see from Eq. (3.33) that at fixed S and V, the first derivative of U is zero and by Eq. (3.39) that the second derivative of U with respect to the constraint is positive. Thus, *at constant entropy and volume* the internal energy U is *minimal* with respect to constraints at equilibrium. *This is why the internal energy is called a thermodynamic potential.* Thermodynamic functions, such as U, have associated extremal properties, which means that they act like potentials, in the sense that they are extremal at equilibrium when the constraint(s) are allowed to relax.

3.6.2 Stability of a Homogeneous System

We now apply the minimum principle for the internal energy to a system which is in a single *homogeneous* phase which is observed to be stable. What can we say about such a system? We consider constraining the system so that we have one mole of gas (or more generally a substance) with half a mole on each side of the partition. The system is constrained as follows: the total volume is fixed, but on the left side of the partition the volume is $(V_0 + \Delta V)/2$ and on the right side of the partition the volume is $(V_0 - \Delta V)/2$. On the left side the entropy is $(S_0 + \Delta S)/2$ and on the right side the entropy is $(S_0 - \Delta S)/2$. Up to quadratic order, the total internal energy is

$$\begin{aligned} U_{\text{tot}} &= \frac{1}{2}\Big[U(V_0 + \Delta V, S_0 + \Delta S) + U(V_0 - \Delta V, S_0 - \Delta S) \Big] \\ &= U(V_0, S_0) + \frac{1}{2} \frac{\partial^2 U}{\partial V^2}\bigg|_S (\Delta V)^2 \\ &\quad + \frac{1}{2} \frac{\partial^2 U}{\partial S^2}\bigg|_V (\Delta S)^2 + \frac{\partial^2 U}{\partial S \partial V} \Delta V \Delta S \,, \end{aligned} \tag{3.40}$$

where here U denotes the internal energy per mole. If we are dealing with a stable phase, equilibrium with the partition is what it would be without the partition, namely, $\Delta V = \Delta S = 0$. Thus the internal energy as a function of ΔV and ΔS should be minimal for $\Delta V = \Delta S = 0$. This ensures that the system is stable against small fluctuations in density. For the quadratic form in Eq. (3.40) to have its minimum there, we have to satisfy the conditions

$$a \equiv \frac{\partial^2 U}{\partial V^2}\bigg|_S \geq 0 \tag{3.41a}$$

$$b \equiv \frac{\partial^2 U}{\partial S^2}\bigg|_V \geq 0 \tag{3.41b}$$

$$c \equiv \left(\frac{\partial^2 U}{\partial V^2}\bigg|_S \right) \left(\frac{\partial^2 U}{\partial S^2}\bigg|_V \right) - \left(\frac{\partial^2 U}{\partial S \partial V} \right)^2 \geq 0 \,. \tag{3.41c}$$

We will investigate these in turn. To evaluate the derivatives remember that

$$dU = TdS - PdV. \tag{3.42}$$

We thus have

$$a = \frac{\partial}{\partial V} \frac{\partial U}{\partial V}\bigg|_S = - \frac{\partial P}{\partial V}\bigg|_S = (\kappa_S V)^{-1} \tag{3.43a}$$

$$b = \frac{\partial}{\partial S} \frac{\partial U}{\partial S}\bigg|_V = \frac{\partial T}{\partial S}\bigg|_V = \left[\frac{\partial S}{\partial T}\bigg|_V \right]^{-1} = (T/C_V) \tag{3.43b}$$

$$\begin{aligned} c &= (\kappa_S V)^{-1} (T/C_V) - \left(\frac{\partial}{\partial V} \frac{\partial U}{\partial S} \right)^2 \\ &= [T/(C_V \kappa_S V)] - [\partial T/\partial V)_S]^2 \,. \end{aligned} \tag{3.43c}$$

From these we see that thermodynamic stability requires that

$$\kappa_S \geq 0 \tag{3.44a}$$

$$C_V \geq 0 \tag{3.44b}$$

$$T/(C_V \kappa_S V) \geq [(\partial T/\partial S)_V (\partial S/\partial V)_T]^2 . \tag{3.44c}$$

The condition (3.44c) does not yield any information beyond that implied by Eqs. (3.44a) and (3.44b). (In Exercise 4b you are asked to show that the condition (3.44c) is satisfied if C_V and κ_T are both positive.) It is amazing that such bounds arise from the most harmless looking observations arising from the Laws of Thermodynamics.

3.6.3 Extremal Properties of the Free Energy

We have obtained the principle of minimum energy from the principle of maximum entropy. We now use the minimum energy principle to show that when the volume and *temperature* of the system are fixed, the actual equilibrium state is such that ξ assumes a value to minimize the free energy F, defined as

$$F \equiv U - TS . \tag{3.45}$$

To obtain this result, we consider a total system $\mathcal{T}$ whose entropy and volume is fixed. This system $\mathcal{T}$ consists of the system $\mathcal{A}$ of interest in thermal contact with a very large reservoir $\mathcal{R}$. (This construction will reappear in statistical mechanics.) The volume of each system, $\mathcal{A}$ and $\mathcal{R}$, is fixed as shown in Fig. 3.3. We characterize a constraint (which we emphasize is localized within the system $\mathcal{A}$) by the variable ξ. From the minimum energy principle, we know that equilibrium is characterized by

$$\left. \frac{\partial U_T}{\partial \xi} \right|_{S_T, V_T, V_A} = 0 , \qquad \left. \frac{\partial^2 U_T}{\partial \xi^2} \right|_{S_T, V_T, V_A} > 0 . \tag{3.46}$$

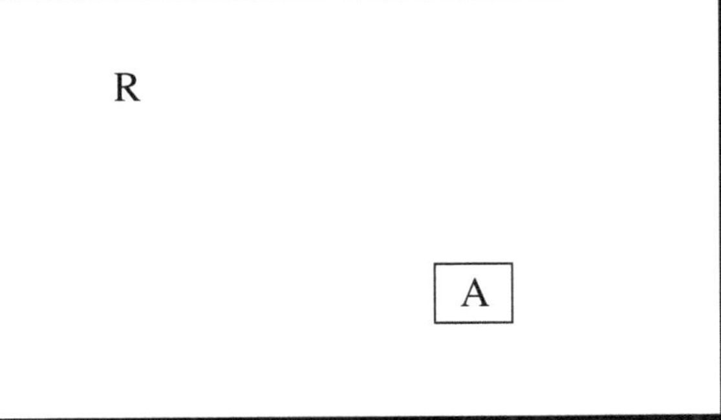

Fig. 3.3 Schematic diagram of a large reservoir $\mathcal{R}$ is thermal contact with the system of interest $\mathcal{A}$. Both systems are contained within rigid walls. The walls containing the entire system are completely thermally insulating, whereas the wall separating system $\mathcal{A}$ from the reservoir can transmit energy

Here, the subscript T labels thermodynamic functions of the total system and similarly, below, R and A refer to the reservoir and the system A, respectively. Because we are dealing with thermodynamic systems which are macroscopic, the thermodynamic potentials are additive, so that $U_T = U_R + U_A$. Thus, we write

$$\begin{aligned} 0 &= \left.\frac{\partial U_T}{\partial \xi}\right|_{S_T, V_T, V_A} = \left.\frac{\partial U_R}{\partial \xi}\right|_{S_T, V_R, V_A} + \left.\frac{\partial U_A}{\partial \xi}\right|_{S_T, V_R, V_A} \\ &= T_R \left.\frac{\partial S_R}{\partial \xi}\right|_{S_T, V_R, V_A} + \left.\frac{\partial U_A}{\partial \xi}\right|_{S_T, V_R, V_A} \\ &= -T_R \left.\frac{\partial S_A}{\partial \xi}\right|_{T_A, V_A} + \left.\frac{\partial U_A}{\partial \xi}\right|_{T_A, V_A} . \end{aligned} \tag{3.47}$$

In writing the second line, we omitted PdV terms because the volumes are fixed. In the third line, we used $dS_R = -dS_A$ and also the fact that the reservoir is very large to recharacterize the derivatives. In particular, this situation implies that the temperature of the system A is fixed to be the same as that of the very large reservoir with which it is in contact. Thus, when the constraint ξ is changed in the system A, the temperature of the reservoir (which is assumed to be vary large) remains constant and therefore T_A remains constant. Replacing T_R by T_A in Eq. (3.47), we thus have

$$0 = \left.\frac{\partial}{\partial \xi}(U_A - T_A S_A)\right|_{S_T, V_T} = \left.\frac{\partial F_A}{\partial \xi}\right|_{T_A, V_A} , \tag{3.48}$$

which shows that the free energy is indeed extremal at equilibrium when the temperature and volume are fixed.

Next we show that F is actually minimal at equilibrium. Starting from Eq. (3.47) for $\partial U_T / \partial \xi)_{S_T, V_T}$, we can write

$$0 < \left.\frac{\partial^2 U_T}{\partial \xi^2}\right|_{S_T, V_T, V_A} = -\frac{\partial T_R}{\partial \xi}\frac{\partial S_A}{\partial \xi} - T_R \frac{\partial^2 S_A}{\partial \xi^2} + \left.\frac{\partial^2 U_A}{\partial \xi^2}\right|_{V_A, T_A} . \tag{3.49}$$

Now consider the behavior of these terms as the size of the reservoir becomes very large. The first term goes to zero in this limit because a constraint in an infinitesimally small part of the total system cannot affect the temperature of the reservoir. Replacing T_R by T_A, we get the desired result:

$$0 < \left.\frac{\partial^2 F_A}{\partial \xi^2}\right|_{T_A, V_A} . \tag{3.50}$$

Equations (3.48) and (3.50) indicate that when the volume and temperature of the system are fixed, then at equilibrium the constraints assume values such as to minimize F.

One can show that similar minimum properties hold for the Gibbs function G ($dG = -SdT + VdP$), which, as we shall see shortly, describes a system whose temperature and pressure are held fixed, and for the enthalpy H ($dH = TdS + VdP$) for a system whose entropy and pressure are held fixed.

3.7 Legendre Transformations

We have shown from physical arguments that different choices of thermodynamic variables lead to different thermodynamic potential functions for which they are the proper variables, in the sense discussed for the entropy and internal energy in Sect. 3.3. For instance, when S and V are specified, the appropriate potential is the internal energy, U, while when T and V are given, the appropriate potential is the free energy, F. One can also introduce such different thermodynamic potentials from a strictly mathematical point of view, as we discuss in this section. In mathematics, the method used to transform from $U(S, V)$ to $F(T, V)$ is called a **Legendre transformation**. A particularly elegant discussion of the use of Legendre transformations in thermodynamics is given in the book *Thermodynamics* by Callen (1985) which we follow here.

Say that you know some convex function, $F(x)$. ("Convex" means that straight lines connecting any two points, $[x_1, y(x_1)]$ and $[x_2, y(x_2)]$, all lie on the same side of the curve. If the function is convex *upward*, it always has a positive second derivative. If it is convex *downward*, its second derivative is always negative. In any event, the second derivative of a convex function does not change sign as a function of x.) The stability relations for the thermodynamic potentials derived in Sect. 3.6.2 guarantee that they are convex functions of their proper variables. (e.g., the internal energy U is a convex function of S and V.) Now consider the slope $m \equiv dF/dx$ as a function of x. Our goal is as follows: we would like to construct a function which a) is a function of dF/dx and b) contains all the information present in $F(x)$.

Since $F(x)$ is convex, we can uniquely express x as a function of the slope $m = dF/dx$. That is, we can write $x = x(m)$. Then one might think that it would be possible to represent F as a function of m as $F[x(m)]$. For this function to satisfy our needs, it is necessary to show that we can use it to reconstruct the original function $F(x)$. To see why this does not work, consider the convex function $F(x) = (x + \alpha)^2$. Then $m = 2(x + \alpha)$ which can be inverted to give $x = (m/2) - \alpha$. Then $F[x(m)] = m^2/4$. Since this function is independent of α, it is clearly not possible to recover the original function just from $F[x(m)]$. A more sophisticated approach is therefore required.

Corresponding to the convex function $F(x)$, we define its Legendre transform $G(m)$ to be the y-intercept of the tangent line to F at $x = x(m)$. This is illustrated in Fig. 3.4.

Thus we write

$$G = F(x) - x\frac{dF}{dx} . \tag{3.51}$$

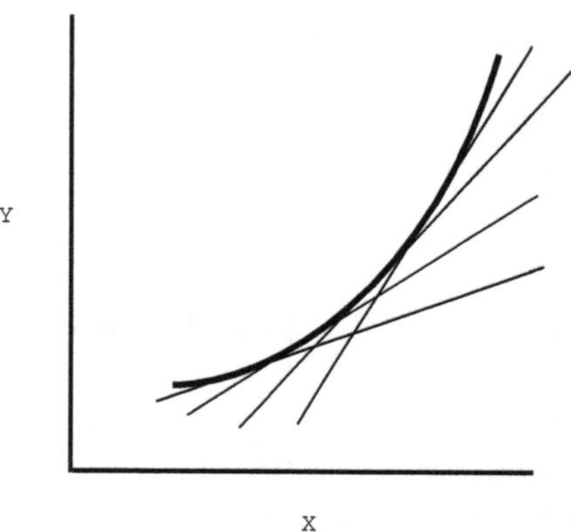

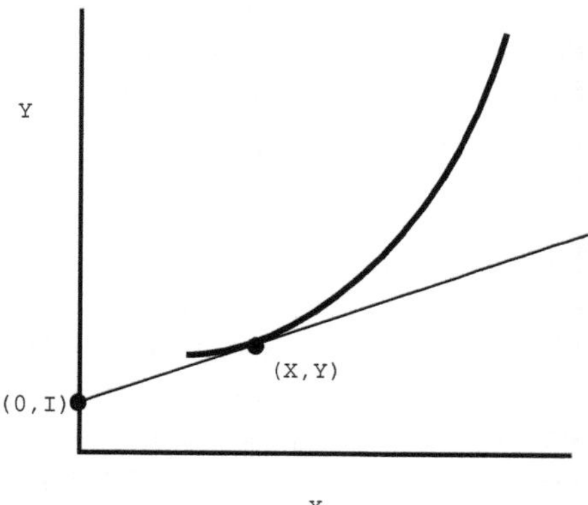

Fig. 3.4 Defining a curve by its tangents

This is the general structure of the Legendre transformation. We now show that G may be considered to be a function of $m = dF/dx$. Recall that as a result of the convexity of $F(x)$, we may consider x to be a function of m: $x = x(m)$, so that

$$G(m) = F[x(m)] - mx(m) . \tag{3.52}$$

We have thus constructed a unique function $G(m)$ which is a function of $m = dF/dx$. Furthermore, as we now show, the function $G(m)$ contains all the information originally in $F(x)$, because given $G(m)$ we can work backward to construct $F(x)$. To do this, we construct $\Phi(x)$, the Legendre transform of $G(m)$:

$$\Phi(x) = G(m) - m\frac{dG}{dm} . \tag{3.53}$$

We now show that this Φ is indeed a function of x and also that it is the desired $F(x)$. From Eq. (3.52), we have

$$\begin{aligned}\frac{dG}{dm} &= \frac{dF}{dx}\frac{dx}{dm} - m\frac{dx}{dm} - x \\ &= -x \ ,\end{aligned} \tag{3.54}$$

so that Eq. (3.53) is

$$\Phi = G(m) + mx \ . \tag{3.55}$$

Then from Eq. (3.52)

$$\Phi(x) = G[m(x)] + m(x)x = F[x(m)] - mx(m) + m(x)x = F(x) \ . \tag{3.56}$$

Thus, we have reconstructed $F(x)$ from $G(m)$.

To show that $G(m)$ is a convex function of m, it suffices to show that $d^2G/dm^2 = -dx/dm$ never changes sign. However, since F is a convex function of x, we know that $d^2F/dx^2 = dm/dx$ (which is the inverse of dx/dm) never changes sign. Therefore, $G(m)$ is a convex function of m.

It should be noted that although we can reconstruct $F(x)$ from a knowledge of $G(m)$, we cannot recover that information if all we have is $G(x)$. (It is this difficulty which prevents one from getting all the thermodynamic functions if one only has a thermodynamic potential in terms of the *incorrect* variables.) Assuming we know $G(x)$, we should determine $F(x)$ as the solution to the differential equation

$$\frac{dF(x)}{dx} = m(x) = \frac{F(x) - G(x)}{x} \ . \tag{3.57}$$

It is clear that if $F(x)$ solves this equation, then $F(x) + \alpha x$ also solves this equation. Since we can not fix the value of α, we cannot uniquely reconstruct $F(x)$ solely from the knowledge of $G(x)$.

We now apply this procedure to the dependence of $U(S, V)$ on S. First, we note that $U(S)$ is a convex function of S, as we have seen in Eq. (3.41b). For fixed V, we can express S as a unique function of $(\partial U/\partial S)_V \equiv T$. Then, we define the Legendre transformed function F (which is called the **Helmholtz Free Energy** (or simply the free energy). We write

$$F = U - \left.\frac{\partial U}{\partial S}\right|_V S = U - TS \ . \tag{3.58}$$

For fixed V, this is a function of T because

$$F = U[S(T, V), V] - TS(T, V) \equiv F(T, V) \ . \tag{3.59}$$

Then

$$dF = dU - TdS - SdT = -SdT - PdV \ , \tag{3.60}$$

so that

$$S = -\left.\frac{\partial F}{\partial T}\right|_V , \qquad P = -\left.\frac{\partial F}{\partial V}\right|_T . \tag{3.61}$$

As we will see, the formalism of statistical mechanics provides a convenient framework within which one can calculate this free energy, $F(T, V)$.

Similarly, we introduce the function $G(T, P)$, which is called the **Gibbs Free Energy** (or simply, the Gibbs function) by making a Legendre transformation to the variables T and $P \equiv -\partial F/\partial V|_T$ with $V = V(T, P)$. Thus, we write

$$G(T, P) = F - \left.\frac{\partial F}{\partial V}\right|_T V = F[T, V(T, P)] + PV(T, P) \ . \tag{3.62}$$

Then

$$\begin{aligned} dG =& \quad dF + PdV + VdP \\ =& -SdT - PdV + PdV + VdP \\ =& -SdT + VdP \ . \end{aligned} \tag{3.63}$$

Note that G is a function only of thermodynamic fields which are intensive variables.

One can similarly define the **Enthalpy**

$$H(S, P) = U - \left.\frac{\partial U}{\partial V}\right|_S V = U + PV \tag{3.64}$$

$$\begin{aligned} dH &= TdS - PdV + PdV + VdP \\ &= TdS + VdP \end{aligned} \tag{3.65}$$

which is useful, for example, for describing the cooling power of cryogenic liquids.

The extensive property or "additivity" of all of these thermodynamic potential functions implies that they can each be written as N, the number of particles, times a function of intensive variables:

$$U(S, V) = N\, u(S/N, V/N), \tag{3.66a}$$
$$F(T, V) = N\, f(T, V/N), \tag{3.66b}$$
$$H(S, P) = N\, h(S/N, P), \tag{3.66c}$$
$$G(T, P) = N\, g(T, P). \tag{3.66d}$$

It is important to realize that all the thermodynamic functions can be obtained from any one of these potentials *providing* the potential is given in terms of the correct variables. (We have already seen this restriction for the Legendre transformation.) For instance, for F the correct variables are T and V, for G the correct variables are T and P, and so forth. In Exercise 16, you are asked to show explicitly that a knowledge of the internal energy U in terms of incorrect variables does not determine all the thermodynamic functions.

3.8 N as a Thermodynamic Variable

If we allow explicitly for variation of the number of particles, then the internal energy becomes a function, not only of the entropy and the volume, but also of the number of particles, $U = U(S, V, N)$, and

$$dU = TdS - PdV + \mu dN, \tag{3.67}$$

where we have introduced the chemical potential via

$$\mu \equiv \left.\frac{\partial U}{\partial N}\right|_{S,V}. \tag{3.68}$$

The other thermodynamic potentials can also be generalized to the case of variable numbers of particles as follows:

$$dF = -SdT - PdV + \mu dN \tag{3.69}$$
$$dH = \quad TdS + VdP + \mu dN \tag{3.70}$$
$$dG = -SdT + VdP + \mu dN \; . \tag{3.71}$$

So

$$\mu = \left.\frac{\partial F}{\partial N}\right|_{T,V} = \left.\frac{\partial H}{\partial N}\right|_{S,P} = \left.\frac{\partial G}{\partial N}\right|_{T,P}. \tag{3.72}$$

If we increase N at fixed T and P, we are simply rescaling the size of the system. Then, G must be linear in N and we may write

$$G(T, P, N) = N\, g(T, P) = N\, \mu(T, P), \tag{3.73}$$

so that the chemical potential is the Gibbs free energy per particle for a system at fixed temperature and pressure. In this section, we will develop some insight into the physical interpretation of μ.

As we have mentioned the free energy, $F(T, V, N)$, is the thermodynamic potential most often calculated by the methods of statistical mechanics. A useful alternative is to calculate the thermodynamic potential Ω (called the **Grand Potential**), which is the function of T, V, and μ which is obtained as the Legendre transform of F. Thus, we write

$$\Omega = F - N\left.\frac{\partial F}{\partial N}\right|_{T,V} = F - N\mu\,. \tag{3.74}$$

One can show that Ω is a convex function of N, so that Eq. (3.68) can be inverted to give N as a function of μ. Thus, Ω is indeed a function of T, V, and μ for which

$$d\Omega = -SdT - PdV - Nd\mu\,. \tag{3.75}$$

Thus, a knowledge of Ω as a function of its proper variables (T, V, and μ) allows us to calculate the thermodynamic properties via

$$S = -\left.\frac{\partial \Omega}{\partial T}\right|_{V,\mu}, \qquad P = -\left.\frac{\partial \Omega}{\partial V}\right|_{T,\mu}, \qquad N = -\left.\frac{\partial \Omega}{\partial \mu}\right|_{T,V}. \tag{3.76}$$

Also one has

$$\Omega = F - \mu N = F - G = -PV\,. \tag{3.77}$$

Equation (3.77) gives the equation of state directly, but note that it gives the equation of state as a function of T, V, and μ. To get the equation of state as a function of T, V, and N, we still need to use Eq. (3.68) to express μ as a function of N, T, and V.

3.8.1 *Two-Phase Coexistence and $\partial P/\partial T$ Along the Melting Curve*

The thermodynamic potential $G(T, P, N)$ is a minimum with respect to internal constraints at fixed T, P, and N. We use this fact to analyze a system consisting of two phases (e.g., liquid and gas) of a single substance in equilibrium. As a result, we will derive the famous Clausius–Clapeyron equation for two-phase coexistence and, as a byproduct, we will develop more insight into the meaning of the chemical potential.

We now minimize the Gibbs function for a system in which the liquid and gas phases coexist. We take the constraint to be the number of particles in the liquid, N_ℓ. As a function of this constraint, the Gibbs function is

$$\begin{aligned} G(T, P, N_\mathrm{t}) &= N_\ell \mu_\ell(T, P) + N_\mathrm{g}\mu_\mathrm{g}(T, P) \\ &= N_\ell \mu_\ell(T, P) + [N_\mathrm{t} - N_\ell]\mu_\mathrm{g}(T, P)\,, \end{aligned} \tag{3.78}$$

where the subscripts "t," "ℓ," and "g" indicate total, liquid, and gas, respectively. Because of the linear dependence on the constraint parameter N_ℓ, G can be extremal and the two phases can coexist, with $0 < N_\ell < N_\mathrm{t}$, only if

$$\mu_\ell(T, P) = \mu_\mathrm{g}(T, P)\,. \tag{3.79}$$

When this condition is fulfilled, G is independent of the fraction of particles in each phase. Otherwise, when the condition is not fulfilled, G is lower for one of the single phases than for the two-phase coexistence. This means that although originally P and T could be considered to be *independent* variables, the condition of phase equilibrium implies a functional relation between P and T: $P = P(T)$. This function defines the coexistence line in the P–T plane. Obviously, Eq. (3.79) has the very intuitive interpretation that as a molecule goes from one phase to the other, its chemical potential does not have to change. This is the microscopic interpretation of equilibrium.

In exactly the same way, for a liquid (ℓ) and solid (s) to coexist, we have the analog of Eq. (3.79):

$$\mu_\ell(T, P) = \mu_s(T, P)\ . \tag{3.80}$$

The line in $T - P$ space along which the liquid and solid phases coexist is called the melting curve. Moving along this line,

$$\frac{1}{N_\ell} G_\ell(T + \delta T, P + \delta P, N_\ell) = \frac{1}{N_s} G_s(T + \delta T, P + \delta P, N_s). \tag{3.81}$$

Expanding to first order in the small quantities δT and δP, we find that

$$\left(\underbrace{\frac{1}{N_\ell}\frac{\partial G_\ell}{\partial T}}_{-s_\ell} - \underbrace{\frac{1}{N_s}\frac{\partial G_s}{\partial T}}_{-s_s} \right) \delta T = \left(\underbrace{\frac{1}{N_s}\frac{\partial G_s}{\partial P}}_{v_s} - \underbrace{\frac{1}{N_\ell}\frac{\partial G_\ell}{\partial P}}_{v_\ell} \right) \delta P, \tag{3.82}$$

where (v_ℓ, v_s) and (s_ℓ, s_s) are respectively the volume and entropy per particle (or if you like, per mole) in the liquid and solid states. Then, we have the slope of the melting curve expressed as

$$\frac{dP}{dT} = \frac{s_\ell - s_s}{v_\ell - v_s}, \tag{3.83}$$

which is known as *the Clausius–Clapeyron Equation*. We write $s_\ell - s_s = L_{\text{fus}}/T$, where L_{fus} is the latent heat of fusion per particle (or if you like, per mole), so that

$$\frac{dP}{dT} = \frac{L_{\text{fus}}}{T(v_\ell - v_s)}\ . \tag{3.84}$$

Consider the sign of the right-hand side. Usually, $v_s < v_l$ so that the denominator is positive. In this case, the sign of dP/dT is determined by the sign of the entropy difference per particle of the two phases. Almost always the latent heat is positive, so that normally dP/dT is positive. However, as we have seen in Sect. 2.1.5, at temperatures below the minimum of the melting curve of ^{3}He, $dP/dT < 0$ implying that the latent heat of fusion is negative. This means that the molar entropy of the solid is greater than that of the liquid. This unusual phenomenon can be attributed to the fact that the entropy of the nuclear spins is greater in the solid than in the liquid phase. Another familiar case where dP/dT is negative is that of H_2O, which expands when it freezes. In that case $v_\ell < v_s$ and dP/dT is negative, as shown in Fig. 2.1.

3.8.2 *Physical Interpretation of the Chemical Potential*

Here, we discuss the physical interpretation of μ. Consider the situation at zero temperature. There $S = 0$ is a constant, so that $\mu = (\partial U/\partial N)_{S,V}$ is simply the energy needed to add a particle to the system. For Fermi systems, this would be the energy of the lowest unoccupied single-particle energy level (if we neglect interactions between particles). For Bose systems, μ would be the energy of the single-particle ground state.

Consider a fixed value of the pressure. We know that

$$\left.\frac{\partial \mu}{\partial T}\right|_P = \frac{1}{N}\left.\frac{\partial G}{\partial T}\right|_P = -\frac{S}{N} < 0\,. \tag{3.85}$$

Since S increases with temperature, this shows that, even for Fermions whose chemical potential is positive at low temperature, at high enough temperature, μ will become negative.

We showed that stability of a single-phase system requires the internal energy U to be a convex function of S and V. Had we generalized that discussion to allow N to vary, we would have concluded that U was also a convex function of N. Furthermore, if one applies this discussion to a system at fixed T and V, the convexity of F as a function of N implies that

$$\left.\frac{\partial^2 F}{\partial N^2}\right|_{T,V} = \left.\frac{\partial \mu}{\partial N}\right|_{T,V} > 0\,. \tag{3.86}$$

Equivalently, one may say that $dN/d\mu|_{T,V}$ is positive. Now consider the behavior of N as a function of μ at fixed V and T for the liquid–gas system. Increasing μ has the effect of adding particles to the system, but the rate at which this happens depends on where the system is in its phase diagram. In Fig. 3.5, qualitative results are shown for two values of T: one for $T < T_c$ and the other for $T > T_c$, where T_c is the liquid–gas critical temperature. For $T > T_c$, no phase boundary is crossed, and N is an analytic function of μ. For $T < T_c$ the liquid–gas phase boundary is crossed; the liquid condenses; and the system becomes less compressible. At the phase boundary, the chemical potentials of the two phases are equal, whereas the density jumps discontinuously leading to the curve shown in the left-hand panel of the figure.

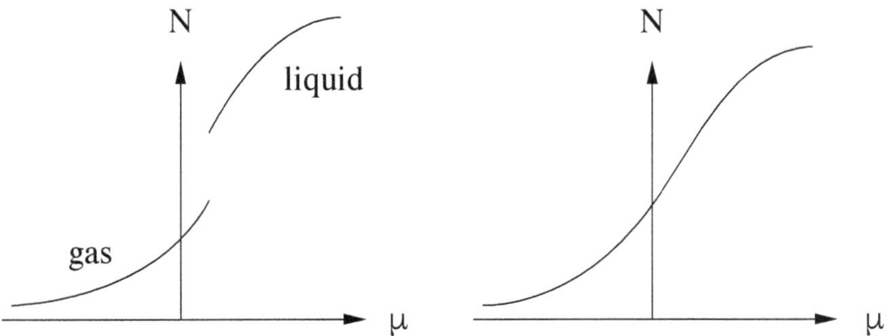

Fig. 3.5 N versus μ for; a liquid–gas system at fixed T and V. Left: $T < T_c$ and Right: $T > T_c$. One could make similar plots for N/V as a function of μ for fixed P

3.8.3 General Structure of Phase Transitions

We now restate in general terms some of the conclusions concerning phase diagrams resulting from the discussion in Chap. 2. There, we saw in Fig. 2.3, for instance, that a phase transition can give rise to a "forbidden" region in the phase diagram if one of the axes is the density. This is why it is useful to distinguish between thermodynamic fields, by which we mean thermodynamic variables which are experimentally controllable, and thermodynamic densities which respond to these fields. In this context, the pressure is a "field" and the molar volume or density is a "density," since one can experimentally control the pressure to any predetermined value, whereas the density is not so controllable when the system is not in a single-phase region. In the two-phase region, the density can assume two values, one of the liquid and the other of the gas. Densities intermediate between these two values are forbidden. (See the left panel of Fig. 3.6.) One cannot have a homogeneous system with a uniform density within this forbidden range of temperature and density. Instead, the system will separate into regions of liquid and gas.

In this same sense, the temperature is a "field" and the entropy is a "density:" In principle at least, one can control the temperature to any desired value, but one cannot maintain a homogeneous phase with a molar entropy intermediate between that of the liquid and solid, as is illustrated in the right panel of Fig. 3.6. Likewise, it is clear that the chemical potential is a "field" and the particle number or particle density is a "density." One can adjust the chemical potential to any desired value by putting the system in contact with a particle reservoir. At the phase transition, the chemical potential of the liquid and gas phases become equal, but the mass density is discontinuous there. Since the density is discontinuous at the liquid–gas transition, a value of density intermediate between that of the two phases cannot exist for a single-phase sample. (See Fig. 3.5.)

As noted in Chap. 2, the same can be said for the phase diagram of a ferromagnet. For the ferromagnet, you can control the external magnetic field H and the temperature T. In contrast, for $T < T_c$, the magnetic moment of a single-phase system

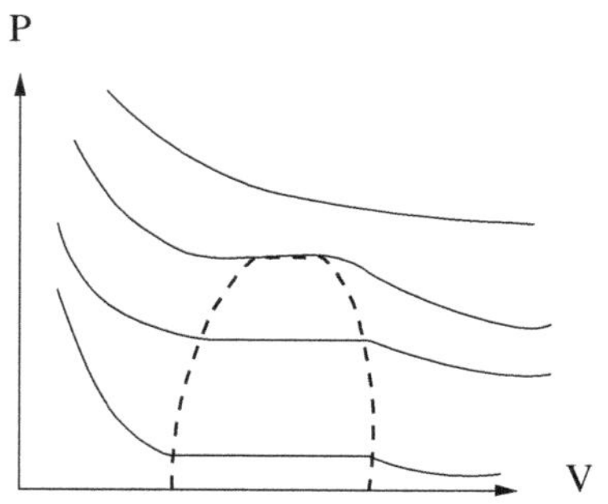

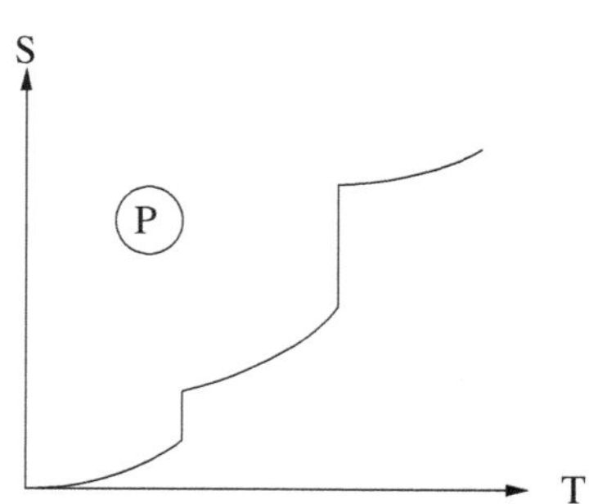

Fig. 3.6 Left: isotherms in the P–V plane which show a "forbidden" region (inside the dashed curve). Right: Molar entropy S versus temperature T (for fixed pressure P). This graph indicates that for a single-phase system, the "field" variable T can assume any desired value, but the entropy has discontinuities and cannot be adjusted to have any desired value for a single-phase system

cannot assume a value whose magnitude is less than that of the spontaneous moment (i.e., that for $H \to 0^+$) at the temperature T. We might also discuss whether or not one can control the entropy to any desired value. Suppose $T < T_c$. Now consider the entropy as a function of magnetic field H. The entropy is zero when the moments are all aligned in the field, i.e., for $H = \pm\infty$. As H is reduced to zero, we attain the largest value of the entropy, $S(T, H = 0)$. Larger values of the entropy (i. e., a phase with less order than that of the ordered phase for $H \to 0$) are inaccessible. So although the temperature can be arbitrarily adjusted (and therefore is a "field" in this terminology), the entropy (which is a "density") has an inaccessible regime.

These remarks have some impact on numerical simulations. One can imagine simulating the liquid–gas transition by treating a box of fixed volume V containing a fixed number N of particles and varying the temperature. In principle, the system will pass through the two-phase region in which the numerics will become unstable because of the formation of two phases within the sample. As a result, if one simulates a liquid at successively higher temperatures, the liquid phase may not disappear at the true thermodynamic transition temperature, but may continue to exist as a metastable superheated liquid. In contrast, if one carries out the simulations for a fixed volume but allows the temperature and chemical potential to be control parameters, one will always obtain a single-phase system. In particular, one might fix the chemical potential and vary the temperature as the control parameter. In principle, one thereby always obtains a single phase.

3.9 Multicomponent Systems

So far, in this chapter, we have considered only pure systems composed of a single substance, for example, pure He or pure H_2O. Now let us broaden our field of view to include systems which consist of N_i particles of species "i" for $i = 1, 2, \ldots$. Then, we write

$$dU = TdS - PdV + \sum_i \mu_i dN_i \,, \tag{3.87}$$

to express the fact that the internal energy can change because we add particles of species "i" to the system. From Eq. (3.87), we deduce that μ_i is the increase in internal energy per added particle of species "i", *at fixed entropy and volume.* Similarly, the other potentials are defined by

$$dF = -SdT - PdV + \sum_i \mu_i dN_i \,, \tag{3.88a}$$

$$dG = -SdT + VdP + \sum_i \mu_i dN_i \,, \tag{3.88b}$$

$$dH = \quad TdS + VdP + \sum_i \mu_i dN_i \,. \tag{3.88c}$$

Thus, Eq. (3.88a), for example, says that μ_i is the increase in the free energy, F, *at fixed temperature and volume* per added particle of species "i". Perhaps, the most important of these relations is Eq. (3.88b) which says that μ_i is the increase in Gibbs free energy, G, *at fixed temperature and pressure* per added particle of species "i".

3.9.1 Gibbs–Duhem Relation

Look again at Eq. (3.88b) and consider a process in which we simply increase all the extensive variables (volume, number of particles of species "i," thermodynamic potentials) by a scale factor, $d\lambda$. This change of scale will *not* cause the intensive variables T, P, μ_i to vary. But it will cause G to change to $G(1 + d\lambda)$, and N_i to change to $N_i(1 + d\lambda)$, so that (with $dT = dP = 0$) we have

$$dG = Gd\lambda = -SdT + VdP + \sum_i \mu_i dN_i = \sum_i \mu_i N_i d\lambda \,, \tag{3.89}$$

so that

$$G = \sum_i \mu_i N_i \,. \tag{3.90}$$

Also, since $dG = \sum_i (\mu_i dN_i + N_i d\mu_i)$, we have the Gibbs–Duhem relation

$$\sum_i N_i \, d\mu_i = -S\, dT + V dP \,. \tag{3.91}$$

3.9.2 Gibbs Phase Rule

Equations (3.79) and (3.80) for liquid–gas and liquid–solid equilibrium are simple illustrations of the Gibbs "phase rule," which is

$$f = c - \phi + 2 , \tag{3.92}$$

where f is the number of free intensive variables which may independently be fixed, c is the number of different substances, and ϕ is the number of phases (solid, liquid, gas, etc.) present. For a single substance ($c = 1$), we have seen that when $\phi = 2$ only $f = 1$ intensive variable can be fixed. When $\phi = 3$, there are NO free variables. Thus, the triple point (where solid, liquid, and gas coexist) is an isolated point in the phase diagram. For H_2O, the triple point occurs at $T = 273.16$ K (defined) and at $P = 610$ Pascals. (1 atm $\approx 10^5$ Pascals.) As another application of the phase rule, look at Fig. 2.10, the phase diagram of ^{3}He-^{4}He mixtures at one atm. Here $c = 2$, so that $f = 4 - \phi$. Outside the region of the phase-separated mixture, one has $\phi = 1$ and one has three independent intensive variables, T, p (which was fixed at atmospheric pressure for this figure), and $n_3/(n_3 + n_4)$. In the two-phase region, $\phi = 2$ and only two intensive variables are independent. We may take these to be T and p. Notice that when these two variables are fixed, then the value of ^{3}He concentration, $n_3/(n_3 + n_4)$, in each phase is fixed.

For an introductory discussion of the Gibbs phase rule, see Chap. 16 of *Heat and Thermodynamics* by Zemansky and Dittman (1981).

3.10 Exercises

1. In this problem, you are to consider the work done on an ideal gas in moving a confining piston at speed u, as shown in Fig. 3.7.

(**a**) Apply the same kind of kinetic theory arguments as are used in elementary courses to derive the ideal gas law. Assume that the piston is infinitely massive in comparison to a gas molecule. Also assume that the gas molecules collide elastically with the piston (and thereby have an increased kinetic energy after the collision if the piston is moving inward with speed u). Then show that in the limit $u \to 0$,

$$\Delta W \equiv F dx \to -P dV .$$

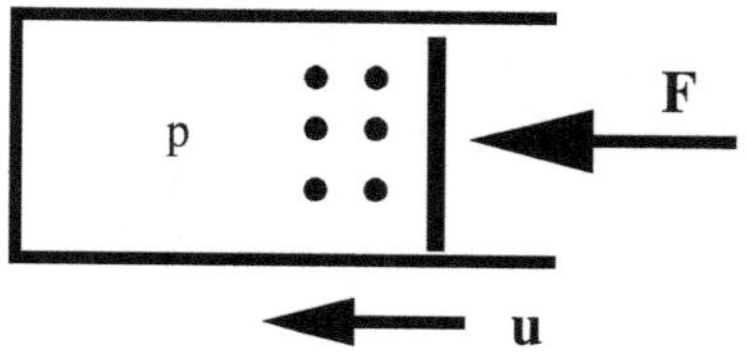

Fig. 3.7 A gas enclosed by a piston moving slowly with speed u

Must the term quadratic in u increase ΔW, must it decrease ΔW, or is the sign of its effect system-dependent?

2. This problem concerns an ideal gas for which

$$PV = NkT \ , \qquad\qquad U = U(T) \ .$$

The object of this problem is to show that work and heat are not state functions. That means that you can construct two different paths between the initial state and the final state such that the work done along the paths are different, but the internal energies at the end points are the same. For this problem, take the initial state to have pressure P_i and volume V_i and the final state pressure $P_i/2$ and volume $2V_i$. Calculate dU, ΔW, and ΔQ for a quasistatic isothermal process which takes the system from its initial state to its final state.

Now calculate dU, ΔW, and ΔQ for a different path traversing quasistatically equilibrium states. This second path consists of two segments. In one segment, heat is added to the system which is kept at constant volume. In the other segment, the piston (containing the gas) is moved quasistatically without allowing any flow of heat in or out of the system to take place. (This is an adiabatic segment.) Give the values of dU, ΔW, and ΔQ for each of the two segments.

3. Develop an expression in terms of thermodynamic response functions for the quantity $C_P - \partial U/\partial T|_P$, which, of course, is nonzero.

4. Derive some famous thermodynamic equalities:

(**a**) $\kappa_S/\kappa_T = C_V/C_P$. What does this say about the slope of adiabats as compared to the slope of isotherms in the P–V plane?

(**b**) Show that the condition of Eq. (3.44c) is true if C_V and κ_T are both positive.

(**c**) Consider a magnetic system with P and V replaced by M and H, where M is the total magnetic moment and H is the *applied* field. Recall that $dU = TdS + \mu_0 H dM$. Here, the analogs of the compressibility are the isothermal and adiabatic (magnetic) susceptibilities, which are defined as

$$\chi_T = \frac{1}{V}\frac{\partial M}{\partial H}\bigg|_T \ , \quad \chi_S = \frac{1}{V}\frac{\partial M}{\partial H}\bigg|_S \ .$$

Show that

$$C_H = C_M + \left(\frac{\mu_0 T}{V\chi_T}\right)\left(\frac{\partial M}{\partial T}\bigg|_H\right)^2$$

and

$$\chi_T/\chi_S = C_H/C_M \ ,$$

where C_M is the specific heat at constant magnetization $T\partial S/\partial T)_M$ and C_H is the similar specific heat at constant external field, H.

5. This problem concerns the chemical potential for a system consisting of a single homogeneous ideal gas for which

$$PV = NkT , \qquad U \equiv U(T) . \tag{3.93}$$

Show that the chemical potential μ can be written in the form

$$\mu(T, P) = kT\,[\phi(T) + \ln P] ,$$

and relate the function $\phi(T)$ to $U(T)$ and/or temperature derivatives of $U(T)$. (Your answer may contain an unknown constant of integration.) Give μ for an *monatomic* ideal gas. Hint: one way to do this is to construct $S(T, P)$ by integrating dS from the point (T_0, p_0) to the point (T, P).

6. The virial expansion for a low-density gas of molecules can be terminated at first order, so that the equation of state is

$$P = \frac{NkT}{V}\left[1 + \frac{N}{V}B(T)\right] ,$$

where N is the number of molecules in the gas. The heat capacity at constant volume has corresponding corrections to its ideal gas value and can be written as

$$C_V = C_{V,\mathrm{ideal}} - \frac{N^2 k}{V} C(T) ,$$

where the subscript "ideal" refers to the value the quantity would have if B vanished.

(**a**) Find the form that $C(T)$ must take in order that the expression for C_V is thermodynamically consistent with the equation of state.

(**b**) Find the leading (in powers of N/V) correction to C_P.

(**c**) Show that

$$\begin{aligned} F(T, V) &= F_{\mathrm{ideal}}(T, V) + \frac{N^2 kT}{V} B(T) \\ U(T, V) &= U_{\mathrm{ideal}}(T, V) - \frac{N^2 kT^2}{V}\frac{d}{dT} B(T) \\ S(T, V) &= S_{\mathrm{ideal}}(T, V) - \frac{N^2 k}{V}\frac{d}{dT} T B(T) . \end{aligned}$$

7. For a rubber band, one finds experimentally that

$$\begin{aligned} \left[\frac{\partial J}{\partial L}\right]_T &= \frac{at}{L_0}\left[1 + 2\left(\frac{L_0}{L}\right)^3\right] \equiv f(T, L) \\ \left[\frac{\partial J}{\partial T}\right]_L &= \frac{aL}{L_0}\left[1 - \left(\frac{L_0}{L}\right)^3\right] \equiv g(T, l) , \end{aligned}$$

where J is the tension, $a = 0.01$ Newtons/K, and $L_0 = 0.5$ m is the length of the rubber band when no tension is applied.

(**a**) Compute $\partial L/\partial T)_J$ and discuss its physical meaning.

(**b**) Show that

$$f(T, L)dL + g(T, L)dT$$

is an exact differential and find the equation of state, $J = J(L, T)$.

(**c**) Assume that the heat capacity at constant length is $C_L = 1.0$ joule/K. Find the work necessary to stretch the band reversibly and adiabatically from the initial length L_0 at temperature 300 K to a final length of 1 m. What is the change in temperature in this process?

(**d**) Suppose the rubber band initially has length 1 m and is at temperature 300K. While thermally isolated, it is then allowed to snap back freely to length L_0. After this process, the rubber band is in contact with a heat bath at temperature 300K. How much heat is exchanged between the heat bath and the rubber band? Find the change in entropy of the rubber band.

8. We know that for an ideal gas $U(T, V)$ is independent of V. What about the converse? Suppose $\partial U/\partial V)_T = 0$. What can you say then about the form of the equation of state, $P = P(T, V)$?

9. Show that

$$V\left(\frac{\partial P}{\partial T}\right)_\mu = S \qquad \text{and} \qquad V\left(\frac{\partial P}{\partial \mu}\right)_T = N\,.$$

Express μ as a function of P and T for an ideal gas and thereby illustrate these relations (Fig. 3.8).

10. Assume that the specific heat of water at constant pressure is 4.2 joules/(mole K) independent of pressure and temperature.

(**a**) One kilogram of water initial at a temperature of 0 °C is brought into contact with a large heat reservoir whose temperature is maintained at 100 °C. During this process, the water is maintained at constant pressure. What has been the change in entropy of the heat reservoir? What has been the change in entropy of the water? What has been the change in entropy of the entire system. Here and below the entire system is taken to include all reservoirs.

(**b**) If the water had been heated from 0 to 100 °C by first allowing it to come to equilibrium with a reservoir at 50 °C and then later with a reservoir at 100 °C, what would have been the change in entropy of the entire system?

(**c**) Show how (in principle) the water may be heated from 0 to 100 °C with no change in the entropy of the entire system.

(**d**) In the above, we never said whether the heat flow took place infinitesimally slowly or not. Does this matter?

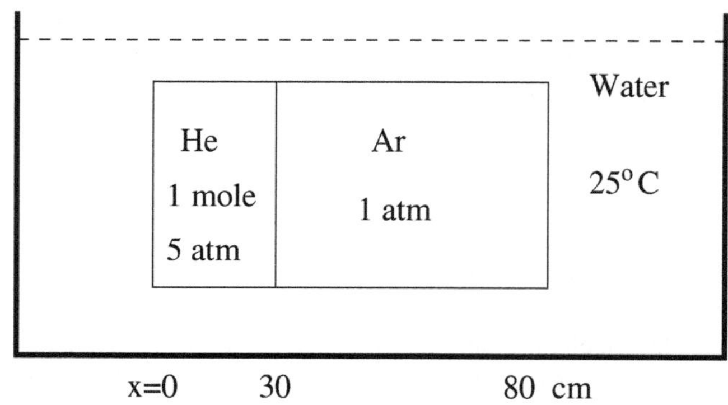

Fig. 3.8 A container surrounded by water for Exercise 11

11. As shown in Fig. 3.8, a cylindrical container of constant volume, 80 cm long, is separated into two compartments by a thin, rigid piston, originally clamped in position 30 cm from the left end. The left compartment is filled with one mole of helium gas at a pressure of 5 atm. The right compartment is filled with argon gas at 1 atm pressure. The cylinder is submerged in 1 liter of water, and the entire system is enclosed in a perfectly insulating container whose volume is fixed. Initially, the entire system is in thermal equilibrium at a temperature 25 °C. Assume the heat capacities of the cylinder and piston are zero. Also, for parts a, b, and c assume the gases are ideal. When the piston is unclamped, a new equilibrium is ultimately reached with the piston in a new position.

(**a**) What is the increase in temperature of the water?

(**b**) How far from the left end of the cylinder will the piston come to rest? Find the final equilibrium pressure.

(**c**) What is the increase in the entropy of the entire system?

(**d**) Repeat the above three parts if both helium and argon are assumed to have deviations from the ideal gas law of the form

$$PV = NkT\left[1 + \frac{aN}{VT}\right],$$

where a is a constant which gives rise to small corrections in the thermodynamic properties of the gases, as was treated in exercise #8. Work to first order in a, but do *not* assume that a for helium is the same as a for argon.

12. This problem illustrates the Legendre transformation and the notation is that of Sect. 3.7. Take $F(x) = ax^2 + bx + c$, with a, b, and c positive.

(**a**) Is this function convex?

(**b**) Construct $x(m)$.

(**c**) Construct $G(m)$.

(**d**) Is $G(m)$ convex?

Fig. 3.9 Function $F(x)$ for Exercise 13

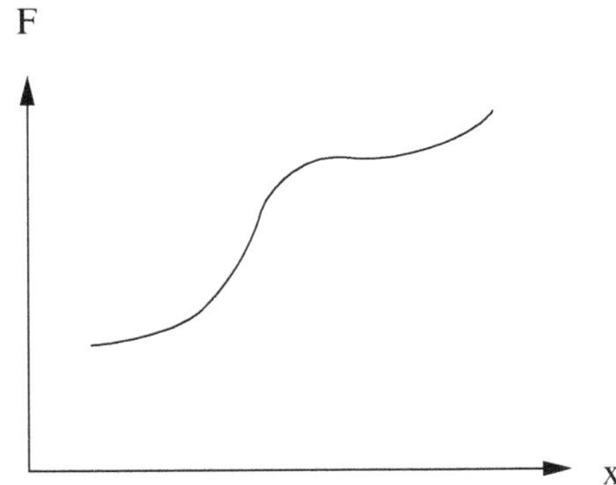

(**e**) Using a knowledge of $G(m)$ (and the fact that $dG/dm = -x$) reconstruct $F(x)$.

(**f**) Assume that you are given not $G(m)$, but rather $G(x)$. Display the family of functions $F(x)$ which give rise to this $G(x)$.

13. We introduced the Legendre transformation for a *convex* function $F(x)$. Suppose that $F(x)$ is the nonconvex function shown in Fig. 3.9. Make a graphical construction of the type of Fig. 3.4 which shows that although the y-intercept is a unique function of x, this relation cannot be inverted to give x as a unique function of the y-intercept.

14. Show that the transformation from a Lagrangian as a function of $\{q_i\}$ and $\{\dot{q}_i\}$ to a Hamiltonian as a function of $\{q_i\}$ and $\{p_i\}$ has the form of a Legendre transformation in several variables (Goldstein 1951).

15. A vertical cylinder, completely isolated from its surroundings, contains N molecules of an ideal gas and is closed off by a piston of mass M and area A. The acceleration of gravity is g. The specific heat at constant volume C_V is independent of temperature. The heat capacities of the piston and cylinder are negligibly small. Initially, the piston is clamped in a position such that the gas has a volume V_0 and is in equilibrium at temperature T_0. The piston is now released and, after some oscillations, comes to rest in a final equilibrium situation corresponding to a larger volume of gas.

(**a**) State whether the final temperature is higher than, lower than, or the same as T_0.

(**b**) State whether the entropy of the gas increases, decreases, or remains the same.

(**c**) Calculate the final temperature of the gas in terms of T_0, V_0, and the other parameters mentioned in the statement of the problem.

(**d**) Calculate the final entropy of the system in terms of T_0, V_0, and the other parameters mentioned in the statement of the problem.

(**e**) Has the total entropy of the system increased in this process?

16. (**a**) Show that a knowledge of the free energy F as a function of V and T enables a determination of all the other thermodynamic potentials, $U(S, V)$, $G(P, T)$, and $H(S, P)$.

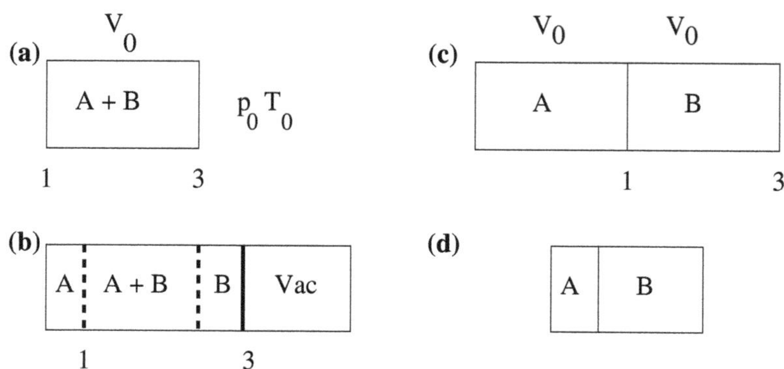

Fig. 3.10 A reversible process to separate a mixture of ideal gases. In panels a, b, and c, the numbers "1" and "3" show the location of the movable walls

(**b**) Show that a knowledge of the equation of state $P = P(V, T)$ does not enable you to calculate all the thermodynamic functions.

(**c**) Suppose that in addition to the equation of state you also know C_p as a function of T for a single value of the pressure. How do you now calculate all the thermodynamic functions?

(**d**) Show that a knowledge of $U(T, V)$ does not enable you to calculate all the thermodynamic functions.

17. In this problem, we seek to calculate the thermodynamic functions for a mixture of *ideal* gases: N_A particles of gas A and N_B particles of gas B, for which

$$PV = (N_A + N_B)kT \ , \qquad U = N_A u_A + N_B u_B \ , \tag{3.94}$$

where u_i is the internal energy per particle of species i. Since the equation of state does not allow us to calculate the entropy, we must develop an expression for it. To do so, we consider the process shown in Fig. 3.10. Panel a shows the system whose entropy we wish to calculate. Panel b shows an intermediate state of a process in which we separate the gas into its two pure subsystems. Here, we have a semi-permeable wall labeled "1" which is locked to an impermeable wall "3" so that the volume between walls "1" and "3" s always equal to V_0. The wall "1" passes only molecules of gas A, whereas wall "2" (which does not move) passes only molecules of gas B. All the walls conduct heat and the whole system is completely isolated from its surroundings. The walls are slowly moved to the right until the gases are separated, each in a volume V_0, as is shown in panel c. (The pressure in each gas is, of course, less than it was in panel a.) Now each gas is compressed so that it is at the same pressure as in panel a and the temperature of each gas is fixed to remain at T_0 (which we can do because these gases are ideal).

(**a**) Which configuration a or d is expected to have the higher entropy?

(**b**) Show that the temperature remains constant, while the movable walls are slowed moved from their initial position in panel a to that of panel c (via intermediate positions as in panel b).

(**c**) By analyzing the separation process, show that

$$U_a(P, T) = U_d(P, T)$$

so that

$$U_{AB}(T) = U_A(T) + U_B(T) \,,$$

where $U_{AB}(T)$ is the internal energy of the A-B mixture in panel a and U_A (and U_B) is the internal energy of the A subsystem (and B subsystem) in panels c and d. Show that

$$S_a(P, T) = S_d(P, T) + \Delta S \,,$$

where S_a is the entropy of the system in panel a and S_d that of the system in panel d and

$$\Delta S = -(N_A + N_B)k[x_A \ln x_A + x_B \ln x_B] \,, \tag{3.95}$$

where $x_A = N_A/(N_A + N_B)$ and $x_B = 1 - x_A$. Thus

$$S_{AB}(P, T) = S_A(P, T) + S_B(P, T) + \Delta S$$

or

$$S_{AB}(V, T) = S_A(V, T) + S_B(V, T) \,.$$

(**d**) Perhaps A and B are the different isotopes of the same atom. Can we take the limit of Eq. (3.95) when the two isotopes become identical? Comment on this equation when A and B are actually indistinguishable.

18. Consider a system at its liquid–gas transition. The temperature is T_0 and the pressure is P_0. Develop an expression for the specific heat at constant volume, $C_V(T_0, x_g)$ in terms of the properties of the liquid and gas phases separately, where x_g is the fraction of particles which are in the gas phase.

19. We know that the chemical potential μ is the energy needed to add a particle to a system when S and V are fixed. A quantity one has a better intuition about is the energy needed to add a particle to a system at constant T and V. This quantity $\bar{\mu}$. We define $\bar{\mu} \equiv \partial U/\partial N)_{T,V}$. Show that

$$N\bar{\mu} = U + V\left[P - T\left(\frac{\partial P}{\partial T}\right)_V\right] ,$$

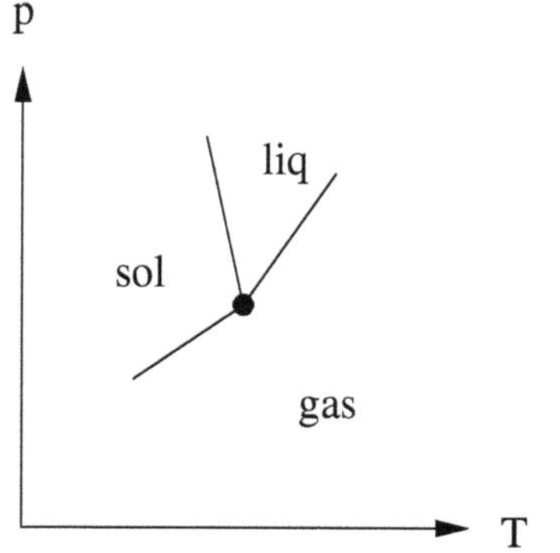

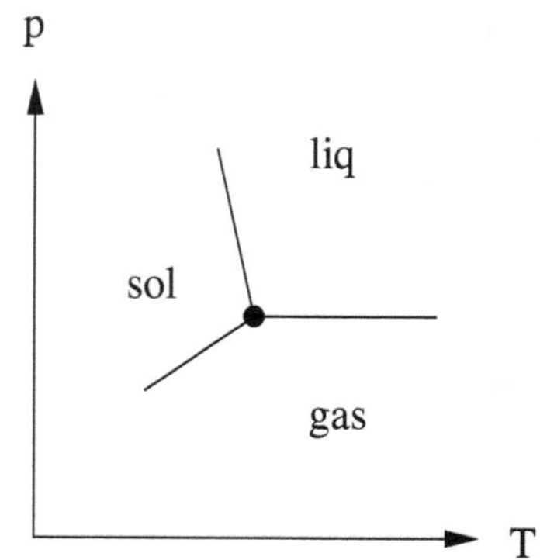

Fig. 3.11 Hypothetical phase diagrams for a pure substance

Fig. 3.12 Equation of state of a fluid with an instability

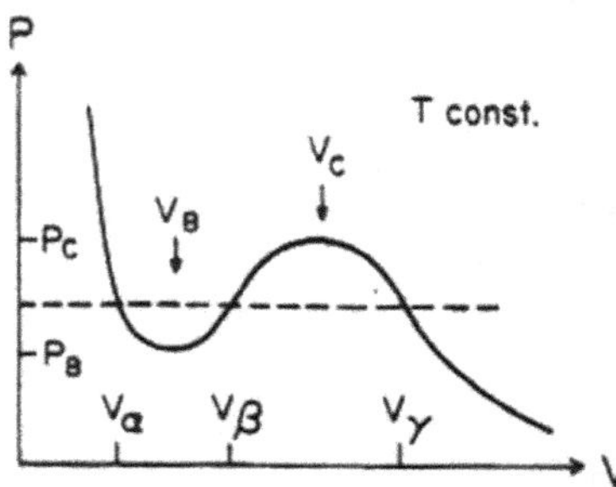

where, of course, $(\partial P/\partial T)_V$ means $(\partial P/\partial T)_{N,V}$. Thus, for an ideal gas $\bar{\mu} = U/N$, as you might expect.

20. Consider the region of the phase diagram near the triple point of a pure substance, as shown in Fig. 3.11 for a substance like water which expands when it freezes. Three first-order lines meet at the triple point, TP. The slope of the phase boundary separating the gas phase from the condensed phases, dP/dT has a discontinuous jump at the TP. Is there a thermodynamic constraint on the sign of the jump? In other words, are diagrams (a) and (b) both possible, or do thermodynamic arguments eliminate one of these possible phase diagrams?

21. From statistical mechanics, the equation of state of a fluid is found to be of the form shown in Fig. 3.12.

(**a**) Over what range of pressure is the fluid microscopically unstable? Does that mean that it is stable everywhere else?

(**b**) As shown in Fig. 3.12, for pressures in the range, $P_B \leq P \leq P_C$ there are three possible volumes V_α, V_β, and V_γ, corresponding to the three points of intersection of the equation of state with the lines of constant pressure. These three solutions correspond to three different phases, α, β, andγ. Express the Gibbs function, G, in terms of the relative fractions (x_α, x_β, and x_γ) of the fluid in each phase. Now

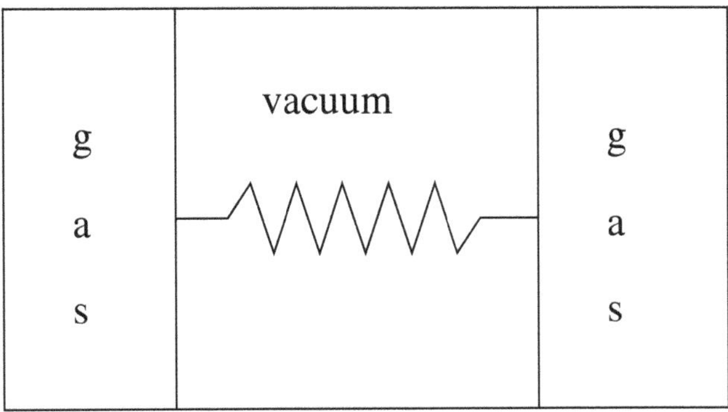

Fig. 3.13 Two samples of a gas in a three-compartment container

minimize the Gibbs function to find the x's. Show that your result can be obtained from the isotherm in the diagram by an "equal area" construction.

(c) Is the fluid stable for $V < V_B$ and/or $V > V_C$? What does the equal area construction tell you about where supercooling or superheating can take place?

(d) Draw a graph of the free energy, F, versus volume according to the isotherm of the diagram above. By considering the possibility of coexistence of more than one phase, show how the curve for the true free energy can be obtained from the one corresponding to the isotherm of the diagram shown above.

(e) Show that the construction obtained in part (d) is equivalent to the equal area rule of part (b).

(f) Draw a graph of the Gibbs function, G, versus pressure: (i) according to the isotherm of the diagram above, and (ii) according to your results of part (b). What principle tells you which curve actually occurs.

22. The system shown in Fig. 3.13 consists of a gas in a container which consists of three compartments: the left and right ones contain gas, the center one is evacuated but its walls are kept apart by a spring which obeys Hooke's law. The cross-sectional area of the compartments is A, and the spring constant is κ. The total volume of this system is fixed, but the size of the center compartment assumes its equilibrium value. Obtain an expression for the specific heat of this system in terms of the temperature T, the spring constant κ, and the following properties of the gas: C_V, $\kappa_T \equiv V^{-1}\partial V/\partial P)_T$, and $\beta_P \equiv V^{-1}\partial V/\partial T)_P$.

References

H.B. Callen, *Thermodynamics and an Introduction to Thermostatics*, 2nd edn. (J. W. Wiley, 1985)
D.C. Giancoli, *Physics*, 5th edn. (Prentice Hall, 1998)
H. Goldstein, *Classical Mechanics* (Addison-Wesley, 1951)
M. W. Zemansky, R. H. Dittman, *Heat and Thermodynamics*, 6th edn. (McGraw-Hill, 1981)

Part II
Basic Formalism

Chapter 4
Basic Principles

4.1 Introduction

The objective of statistical mechanics is to calculate the measurable properties of macroscopic systems, starting from a microscopic Hamiltonian. These systems might be isolated or nearly isolated, stand-alone systems, or they might be in contact with some kind of environment with which they can interact in various ways. If the system can exchange energy with its environment, then the environment is called a "heat bath," a "thermal reservoir," or a "refrigerator." If exchange of particles can occur, then the environment is called a "particle bath." The system might exchange volume with its environment via a piston, and there are many other analogous possibilities. We will see later how statistical mechanics deals with coupling between subsystems. For the moment, however, we focus on a system which is isolated or nearly isolated in a sense that will be discussed below.

As was stated in the first paragraph of this book, statistical mechanics provides the bridge between the dynamics of particles and their collective behavior. For atomic-scale particles, the dynamics are described by quantum mechanics. Thus we will use the theory of quantum mechanics as the base on which to build our theory. Is it really necessary to start from quantum mechanics? Certainly, it is true that the properties that we eventually want to calculate are macroscopic, and hence one might think that they could be derived from classical mechanics. To a large extent they can, and starting from classical mechanics is a traditional approach to the derivation of statistical mechanics theory. However, there are a few awkward logical gaps in this approach, which need to be "fixed up" by quantum mechanics, such as the measure for counting states which is required for calculating the entropy, and there are, of course, topics, such as the theory of quantum fluids, in which quantum mechanics plays a central role and magnetism, whose very existence violates classical mechanics.

We base our discussion on quantum mechanics because we believe it to be a complete theory of the low-energy properties of matter. For most if not all of the phenomena that we experience in our everyday lives, nonrelativistic quantum mechanics is essentially a "Theory of Everything." Furthermore, there is a well-defined

A. J. Berlinsky and A. B. Harris, *Statistical Mechanics*, Graduate Texts in Physics,
https://doi.org/10.1007/978-3-030-28187-8_4

prescription for going from quantum mechanics to the classical limit when classical mechanics is all that is needed. By starting with quantum mechanics, we obtain a theory which applies to both limits.

Next we must ask the question: What do we want quantum mechanics to do for us in deriving statistical mechanics? The answer is that we want to use it to calculate the properties that we associate with macroscopic systems in equilibrium—their equation of state (e.g., the relation between pressure, temperature, and volume for a gas), their heat capacity, their magnetization as a function of magnetic field and temperature, etc. By "equilibrium", we mean a fully relaxed state, in which the system has forgotten about any special initial conditions that can relax in microscopic times. In quantum mechanics, measurable quantities are calculated as averages of operators, and so we will need to learn how to calculate averages of operators that correspond to the thermodynamic properties of macroscopic systems.

Consider such a quantum mechanical operator, call it $\mathcal{O}$ (for operator), which is some function of the coordinates and momenta of the particles (atoms, molecules, electrons, ...) that make up the system. For the moment, we do not need to say what this operator is. Whatever it is, its exact quantum mechanical average is just its expectation value with respect to the wave function of the system.

All of the information about a specific state of the system is contained in the wave function, which can be written, in complete generality, as $\Psi(\{\vec{r}_i\}, t)$, where $\{\vec{r}_i\}$, $i = 1, \ldots, N$ are the particle coordinates, and Ψ obeys the time-dependent Schrodinger equation

$$\mathcal{H}\Psi = i\hbar \frac{\partial \Psi}{\partial t}. \tag{4.1}$$

For a truly isolated many-body system of interacting particles, $\mathcal{H}$ has no explicit time dependence, and $\Psi(\{\vec{r}_i\}, t)$ can be written as

$$\Psi(\{\vec{r}_i\}, t) = \sum_n a_n \psi_n(\{\vec{r}_i\}) e^{-iE_n t/\hbar}, \tag{4.2}$$

where the a_n are complex constants, determined by the initial conditions, and the ψ_n satisfy the time-independent Schrodinger equation

$$\mathcal{H}\psi_n = E_n \psi_n. \tag{4.3}$$

The state ψ_n is called a "stationary state" since, if the system is prepared in such a state, it will remain there with the simple time dependence $e^{-iE_n t/\hbar}$. The spectrum of energies E_n is discrete, for systems confined to a finite volume of space, and is bounded from below. That is, there is a lowest or "ground state" energy, E_0, such that

$$E_0 \leq E_n, \quad \forall\, n. \tag{4.4}$$

Note that, although ψ_n has this very simple dependence, the overall time dependence of $\Psi(\{\vec{r}_i\}, t)$ can be very complicated, depending on the initial conditions which determine the $\{a_n\}$.

When we say that "all of the information ... is contained in the wave function," we mean that every well-defined observable quantity has associated with it a particular operator, $\mathcal{O}^\alpha$, and that the value of that quantity is given by its expectation value

$$\begin{aligned}\langle \mathcal{O}^\alpha(t)\rangle &= \int d\{\vec{r}_i\}\Psi^*(\{\vec{r}_i\},t)\mathcal{O}^\alpha\Psi(\{\vec{r}_i\},t)\\ &= \sum_{m,n} a_m^* a_n \mathcal{O}^\alpha_{m,n} e^{-i(E_n-E_m)t/\hbar}, \end{aligned} \tag{4.5}$$

where

$$\mathcal{O}^\alpha_{m,n} = \int d\{\vec{r}_i\}\psi_m^*\mathcal{O}^\alpha\psi_n. \tag{4.6}$$

Thus to derive the expectation value of any observable quantity for an arbitrary state, $\Psi(\{\vec{r}_i\}, t)$ of a time-independent Hamiltonian, we would need to know the $\{\psi_n, E_n\}$ and the $\{a_n\}$ which define this state. Note that the complete set of observable operators, $\{\mathcal{O}^\alpha\}$, contains operators which are sensitive to the many microscopic internal degrees of freedom of the system, all of which are in principle observable, in addition to global, macroscopic observables.

Another way of expressing the above result, Eq. (4.5), is to write the average of $\mathcal{O}^\alpha$ in terms of a quantum mechanical "density matrix," $\rho(t)$ for the state $\Psi(\{\vec{r}_i\}, t)$, where the matrix elements of $\rho(t)$ have the values

$$\rho_{n,m}(t) \equiv a_m^* a_n e^{-i(E_n-E_m)t/\hbar}. \tag{4.7}$$

(Note the transposition of indices.) The operator $\rho(t)$ contains exactly the same information as is contained in the wave function, $\Psi(\{\vec{r}_i\}, t)$, so that knowing one is equivalent to knowing the other. Its diagonal matrix elements, $\rho_{n,n}(t)$, are all positive, and $\mathrm{Tr}\,[\rho(t)] = 1$ because $\Psi(\{\vec{r}_i\}, t)$ is normalized. This kind of density matrix is often referred to as the density matrix for a "pure state," i.e., the state $\Psi(\{\vec{r}_i\}, t)$. In terms of the operator $\rho(t)$, the average of the operator $\mathcal{O}^\alpha$ may be written as

$$\begin{aligned}\langle \mathcal{O}^\alpha(t)\rangle &= \sum_{m,n} \rho_{n,m}(t)\mathcal{O}^\alpha_{m,n}\\ &= \mathrm{Tr}\left[\rho(t)\mathcal{O}^\alpha\right], \end{aligned} \tag{4.8}$$

where the last expression follows because the previous line contains the trace of a product of matrices (because of the way that $\rho_{n,m}(t)$ was defined in Eq. (4.7)).

The density matrix $\rho(t)$ contains a great deal of microscopic information, far more than we can manage and indeed, as we will discuss, far more than we need to describe the thermodynamic properties of a system. At the same time, the wave function, $\Psi(\{\vec{r}_i\}, t)$ and the density matrix $\rho(t)$ are impossible to calculate for more than a few interacting particles. Furthermore, even if we could calculate them, the time dependence of the off-diagonal elements of $\rho_{m,n}(t)$ is typically very fast on the scale of macroscopic observation times. When performing a macroscopic measurement, one typically takes the measurement over a time interval Δt, which is very long compared to the timescales for atomic motions. Matrix elements between states with energies E_1 and E_2 such that $|E_1 - E_2|\Delta t/\hbar \gg 1$ oscillate rapidly on the experimental time scale and therefore such matrix elements effectively average to zero for thermodynamic timescales. (The case of exact degeneracy, $E_1 = E_2$, is discussed below.) This argument suggests that for thermodynamic purposes we should instead consider a density matrix $\bar{\rho}$ that is diagonal with respect to energy and has time-independent matrix elements.

We now consider the form of the density matrix, $\bar{\rho}$, for a system in equilibrium. We argued above that the off-diagonal elements of the density matrix corresponding to two different energies in Eq. (4.7) oscillate rapidly and effectively average to zero. However, even when E_m is exactly equal to E_n, the off-diagonal matrix elements of the density matrix average to zero for statistical mechanical averages. The reason for this is that it is extremely difficult, in fact essentially impossible, to isolate a system from weak, time-dependent external fields which induce transitions between nearly degenerate stationary states, causing the amplitudes to fluctuate and, most importantly, resetting the phases of the a_n. This difficulty of isolating a system completely from its environment is related to the phenomenon of "quantum decoherence" which makes it so difficult to build a quantum computer. We would argue that the main effect of very weak interactions of the system with its environment is to randomly reset the phases of the complex coefficients, a_m, while leaving the amplitudes relatively unaffected. If the phase of a_m varies randomly in time, uniformly over 0 to 2π, then the time average (indicated by an overbar) of the density matrix is

$$\overline{a_m^*(t)a_n(t)e^{-i(E_n-E_m)t/\hbar}} = \bar{\rho}_{n,n}\delta_{m,n} \, , \tag{4.9}$$

where $\delta_{m,n}$ is a Kronecker δ. Equation (4.9) says that the density matrix of a thermodynamic system is diagonal in the energy representation. Furthermore, we will argue below that $\bar{\rho}_{n,n}$ should have the same value for all states with the same energy E_n.

Before going further let us summarize the common properties of the two kinds of density matrices, $\rho(t)$ and $\bar{\rho}$, defined in Eqs. (4.7) and (4.9), both of which we will refer to as ρ. They each have matrix elements ρ_{nm} where n and m range over all the states of the system, and they are both Hermitian:

$$\rho_{nm} = \rho_{mn}^* \tag{4.10}$$

and normalized:

$$\mathrm{Tr}\rho \equiv \sum_n \rho_{nn} = 1 \, . \tag{4.11}$$

Since they are Hermitian, we can write each of them in the representation in which it is diagonal. In that basis

$$\rho = \begin{bmatrix} p_1 & 0 & 0 & 0 & \dots \\ 0 & p_2 & 0 & 0 & \dots \\ 0 & 0 & p_3 & 0 & \dots \\ 0 & 0 & 0 & p_4 & \dots \\ \dots & \dots & \dots & \dots & \dots \end{bmatrix}, \tag{4.12}$$

and we can interpret p_1 as being the probability that the system is in state $|1\rangle$, p_2 the probability that it is in state $|2\rangle$, and so forth *in the basis in which ρ is diagonal.* For the equilibrium density matrix, $\bar{\rho}$, that basis is simply the energy basis, and $\rho_{11} = p_1, \rho_{22} = p_2$, etc.

For the pure state density matrix, $\rho(t)$, derived from the quantum state, $\Psi(\{\vec{r}_i\}, t)$ which is a linear combination of energy eigenstates, the situation is a bit more subtle. This density matrix has a single eigenvalue equal to one, corresponding to the fact that the system is in state $\Psi(\{\vec{r}_i\}, t)$ with certainty, while all other eigenvalues are zero, meaning that the probability of the system being in any state orthogonal to $\Psi(\{\vec{r}_i\}, t)$ is zero. We illustrate this result below for a single spin 1/2.

This discussion also implies an implicit restriction on the density matrix, namely, that all its eigenvalues, p_k, should lie in the interval $[0, 1]$. When only a single $p_k = 1$ and all the rest are zero, the system is in a pure state. When p_k is nonzero for more than one state k, the system is not with certainty in a single quantum state. In that case, we say that the density matrix describes a "mixed state.". In particular, the equilibrium density matrix describes a mixed state in which the system occupies a set of energy eigenstates, ψ_1, ψ_2, etc. with respective probabilities $p_1, p_2, etc.$

To illustrate how these two cases differ, we consider a spin system with two states $|1\rangle$ and $|2\rangle$, such that $S_z|1\rangle = \frac{1}{2}|1\rangle$ and $S_z|2\rangle = -\frac{1}{2}|2\rangle$. If the system is in a pure state, which in full generality we can write as $\alpha|1\rangle + \beta|2\rangle$, where α and β are complex coefficients which could be functions of time, and $|\alpha|^2 + |\beta|^2 = 1$, then

$$\boldsymbol{\rho} = \begin{bmatrix} \alpha\alpha^* & \alpha\beta^* \\ \alpha^*\beta & \beta\beta^* \end{bmatrix}. \tag{4.13}$$

When diagonalized $\boldsymbol{\rho}$ has eigenvalues 1 and 0, corresponding, respectively, to the state $\alpha|1\rangle + \beta|2\rangle$ occurring with probability 1 and the state that is its orthogonal partner, $\beta^*|1\rangle - \alpha^*|2\rangle$, occurring with probability 0. One may calculate the average of the vector spin, $\vec{S} = \frac{1}{2}\vec{\sigma}$ where $\vec{\sigma}$ is the vector of Pauli matrices. This operator will have an average magnitude 1/2 and will point in a direction determined by the magnitude and phase of the ratio α/β. The point here is that if a spin 1/2 is in a single quantum state, the expectation value of the spin vector will always have magnitude 1/2, but will be oriented in a direction determined by the coefficients α and β. In contrast, suppose the system is in state $|1\rangle$ with probability $(1 + \epsilon)/2$ and is in state $|2\rangle$ with probability $(1 - \epsilon)/2$, so that

$$\rho = \begin{bmatrix} (1+\epsilon)/2 & 0 \\ 0 & (1-\epsilon)/2 \end{bmatrix}. \tag{4.14}$$

Then we have

$$\langle \vec{S} \rangle = \frac{1}{2}\mathrm{Tr}\left[\rho\vec{\sigma}\right] = (0, 0, \epsilon/2) . \tag{4.15}$$

Unless $\epsilon = \pm 1$, this result cannot be described by a single quantum state. Thus, we see that the density matrix can be used to describe mixed states which are not describable by a single wave function. This is an essential new ingredient that occurs in statistical mechanics but not in ordinary quantum mechanics.

Although our discussions are aimed at describing time-independent phenomena, it is also possible to study particles that are effectively isolated from their environments. These might be ions in an ion trap, or nuclear spins in an insulating solid that are very weakly coupled to their environment at low temperatures. In such cases, it is possible to prepare these isolated particles in a well-defined quantum state that will evolve under the influence of the system's internal Hamiltonian in a coherent way. The density matrix corresponding to this state will have nonzero off-diagonal elements which evolve according to Schrodinger's equation and oscillate in time. Eventually, of course, even these systems will feel the effects of their environments, and the off-diagonal, coherent terms will decay to zero. The difference between such nearly perfectly isolated systems and a typical thermodynamic system is that the thermodynamic system relaxes, for all practical purposes, instantaneously, while the more perfectly isolated system allows time for a series of measurements to be made, before it relaxes. These "nearly perfectly isolated systems" are exactly what is required to construct multi-qubit registers for a quantum computer.

It is also worth mentioning that there is a direct analogy between the relaxation times $\tau_{m,n}$ of the off-diagonal matrix elements of the density matrix and the so-called "transverse relaxation time," T_2 in nuclear magnetic resonance (NMR). One of the most common operations in NMR experiments involves using radio-frequency electromagnetic pulses to prepare the nuclear spins in a state that oscillates at a frequency ν_0 proportional to the nuclear Zeeman splitting. In this state, the total nuclear magnetic moment perpendicular to the applied magnetic field rotates about the direction of the applied field at the frequency ν_0. Because the nuclear spins interact with their environment, the amplitude of this oscillating transverse moment relaxes in a time T_2. If T_2 is much longer than the observation time, then the spins appear to move coherently. However, if the observation time is much larger than T_2, one is in the regime in which Eq. (4.9) is satisfied. In contrast to the case of T_2 in NMR, for most systems the relaxation is fast and the timescale in which Eq. (4.9) is not satisfied is very short.

4.2 Density Matrix for a System with Fixed Energy

Next, we consider the simplest case of the statistical density matrix for a system in which the energy E is fixed except for a very small uncertainty ϵ. In this case, $\rho_{n,n}$ is zero unless $|E - E_n| \leq \epsilon/2$. We will give two complementary "derivations" of the appropriate form for the density matrix of a system whose energy is fixed. The first relies only on macroscopic reasoning. The second invokes microscopic reversibility.

4.2.1 Macroscopic Argument

We consider the relative probabilities, i.e., the elements of the density matrix, for the group of states on the energy shell, i.e., states whose energy is within $\pm\epsilon/2$ of the fixed value of energy E. Recall from our discussions of thermodynamics that a macroscopic state of a liquid or gas is specified by fixing three macroscopic variables, one choice for which is the number of particles N, the total energy E, and the volume V. Any other parameter must then be a function of these three variables. We argue this on the same basis as one would argue in thermodynamics: Imagine having several cylinders of helium gas, all of which contain the same number of particles, N, have the same volume, V, and the same energy, E. Then many years of experiments have shown that each such cylinder of gas will have identical macroscopic properties. If we introduce another parameter, e.g., the pressure, P, then P will be the same for all the cylinders, and indeed we can make experimental determinations of P as a function of N, V, and E. It is therefore impossible to imagine differentiating between these cylinders on the basis of any additional parameter.

The crucial point here is that the distribution of probabilities within the shell of energy cannot have any relevant (see below for what "relevant" means) dependence on any parameter additional to those already fixed. It is important to realize that this argument applies to properties *in the thermodynamic limit*. In particular, when we say that the properties of such a system depend only on three macroscopic variables, we only refer to macroscopic properties, like the pressure, the specific heat, the thermal expansion coefficient, etc. What these properties have in common is that they depend on an average over all particles in the system. In contrast, if one looks at the system on a microscopic scale at short timescales, then the properties become time-dependent, and the results of different measurements of the same quantity on supposedly equivalent systems will in fact no longer be identical. A more precise statement about additional parameters would be the following. If the distribution does depend on such a parameter, it must do so in a way that becomes irrelevant for macroscopic properties in the thermodynamic limit.

In summary, the only distributions over states on the energy shell which are allowed are *parameterless* distributions. The only *parameterless* distribution we can think of is the one which weights all states on the energy shell equally. This ansatz is called the "assumption of equal *a priori* probabilities," and it is based on the thermodynamic justification described above.

We have couched this discussion in terms of a liquid or gas, but it applies equally to more complicated cases where instead of, say V, one might have to consider other or additional variables. For instance, for a single crystal, one might have to specify the strain tensor. If the gas or liquid is in a magnetic field, H, then in addition to V, one would have to specify the value of H. Indeed, in the presence of a magnetic field, directions perpendicular and parallel to the field are no longer equivalent, and a complete description of such a system would require a strain tensor.

To fix the value of density matrix elements $\rho_{n,n}$, we need to know how many states are in the energy range under consideration. For that purpose, we consider the distribution of energy levels in a many-body system. In the thermodynamic limit, the spacing between energy levels becomes arbitrarily small. In that case, we can define a smoothed density of states, $\Omega(E)$ such that the number of states having energy between E and $E + dE$ is $\Omega(E)dE$, provided that dE, although an infinitesimal, is nevertheless large compared to the spacing between energy levels. This condition is fulfilled in the thermodynamic limit. Since there are $\epsilon\Omega(E)$ states in the energy range considered, we write

$$\begin{aligned} \rho_{n,n} &= \frac{1}{\epsilon\Omega(E)} && \text{for } |E - E_n| \le \epsilon/2 \\ &= 0 && \text{for } |E - E_n| > \epsilon/2 . \end{aligned} \tag{4.16}$$

This distribution is known as the *microcanonical* distribution, where "micro" indicates that the energy is restricted to the very small range, $\pm\epsilon/2$. The above discussion is incomplete because it does not address the behavior of $\Omega(E)$ in the thermodynamic limit. In this limit, as we will see shortly, the appropriate quantity to consider is not $\Omega(E)$, but rather $(1/N)\ln\Omega(E)$ which is well defined in the thermodynamic limit.

4.2.2 Microscopic Argument

Having argued for the principle of equal *a priori* probabilities on thermodynamic grounds, we now further justify it starting from the quantum mechanical "principle of microscopic reversibility." Consider $\mathcal{H}_{\rm int}$ which describes the effect of the interaction of the system with its environment. The principle says that if $\mathcal{H}_{\rm int}$ is a time-dependent perturbation and $p_{m,n}$ is the probability that, if the system is initially in state m, it will make a transition to state n, then

$$p_{m,n} = p_{n,m}. \tag{4.17}$$

The justification for this is based on Fermi's Golden Rule and the fact that the perturbation, $\mathcal{H}_{\rm int}$, is Hermitian.

$$p_{m,n} \sim |\langle n|\mathcal{H}_{\rm int}|m\rangle|^2 = |\langle m|\mathcal{H}_{\rm int}|n\rangle|^2 \sim p_{n,m}. \tag{4.18}$$

Note that these $p_{m,n}$ are proportional to the microscopic relaxation rates, the $1/\tau_{m,n}$. If $p_{m,n} = p_{n,m}$ then, for example, although it may take the system a long time to get into state n (if $p_{m,n}$ is small for all m), it will take an equally long time to get out. Then, on average, the system will spend equally long in each accessible state and hence one should weight each energetically accessible state equally, thus confirming Eq. (4.16). (See Exercise 12.)

All of this assumes that state n is accessible on the timescale of the measurement. In the language used above, this means that $\tau_{m,n}$, the timescale over which the two states equilibrate, is short compared to the measuring time. If it is not, we must revise our notion of the possible states of the system or simply keep in mind that the system is out of equilibrium with respect to states of type m and n. Experimentally all variations are possible. It often happens that equilibration among energetically accessible states is essentially instantaneous on the timescale of a measurement (although this means that one might consider doing the measurement much more quickly and more frequently). Alternatively, there are situations in which some energetically equivalent states are never reached, because of vanishing matrix elements or insurmountable energy barriers.

A pedagogically interesting case is the equilibration of the rotational energy levels of molecular hydrogen (H_2) (van Kranendonk 1983). The two proton spins of an H_2 molecule can couple to form total nuclear spin $I = 0$ or $I = 1$. Molecules with $I = 0$ are called para-hydrogen, and those with $I = 1$ are called ortho-hydrogen. The nuclear coordinates also appear in the rotational wave function of the molecule. Since the H_2 molecule has a small moment of inertia (for a molecule), its rotational energy levels are widely spaced, with a characteristic energy of order 100 K. Since the total wave function must be antisymmetric under interchange of the two protons due to Fermi statistics, there is a coupling between the nuclear spin and the molecular rotation quantum number J. The rule is that para-hydrogen can only exist in even-J angular momentum states, and ortho-hydrogen can only be in odd-J angular momentum states. Transitions between rotational states for the same species are relatively fast, and so, for most practical purposes, the submanifolds of ortho- and para-states are each in equilibrium. However, because the rate of transitions that simultaneously change both rotational and spin quantum numbers is extremely small, the ortho concentration, which we denote by x, need not assume its temperature-dependent equilibrium value, $x(T)$. Different scenarios are possible. If we start with H_2 gas equilibrated at room temperature or above, x assumes the value $x(T) \approx 3/4$. This value corresponds to the fraction of nuclear spin states which have $I = 1$. This mixture is called "normal" hydrogen. When this gas is cooled to low temperatures, into the liquid or solid states where the value of $x(T)$ is much smaller, the ortho concentration takes many hours to equilibrate to the lower temperature value of $x(T)$. It is also possible to enrich the ortho concentration to nearly 100%, by separating ortho from para, or to catalyze the transition to nearly pure para-hydrogen at low T.

When the system is out of equilibrium, it is still possible to think of the ortho- and para-subsystems as being each in their own equilibrium state subject to the overall ortho concentration being fixed at x. This picture assumes that x varies slowly enough that the system is otherwise always in thermal equilibrium. As the reader will realize, the properties of such a mixture of ortho- and para-hydrogen depends not only on the parameters N, V, and T, but also on the value of x. The additional parameter x appears because, although the assumption of equal *a priori* probabilities is valid within the ortho- and para-manifolds separately, this assumption is violated for the relative concentrations of ortho and para H_2 on an experimental timescale.

4.2.3 Density of States of a Monatomic Ideal Gas

To illustrate the above discussion, we give an approximate calculation of the many-particle density of states $\Omega(E)$ of a low-density gas of independent particles. We start from the formula for the energy levels of a single particle of mass m in a cubic box of volume V whose side has length L:

$$E_{n_1,n_2,n_3} = \frac{h^2(n_1^2 + n_2^2 + n_3^2)}{8mL^2} , \tag{4.19}$$

where the n_α's are the positive integers and h is Planck's constant. We have assumed rigid-wall boundary conditions. As you are asked to show in Exercise 3, similar results are obtained using periodic boundary conditions. We assume that we have N identical such particles in the box. Then, the energy is given in terms of $3N$ quantum numbers as

$$E_{\{n\}} = \frac{h^2}{8mL^2} \sum_{k=1}^{3N} n_k^2 . \tag{4.20}$$

In order to obtain $\Omega(E)$, we start by calculating $\Lambda(E)$, the number of energy levels having energy less than E. Then $\Omega(E) = d\Lambda/dE$. To obtain $\Lambda(E)$, we write Eq. (4.20) as

$$\sum_{k=1}^{3N} n_k^2 = \frac{E}{E_0} \equiv R^2 , \tag{4.21}$$

where $E_0 = h^2/(8mL^2)$. Then the number of states which have energy less than or equal to E is very nearly the same as the volume of that part of the sphere of radius R in $3N$ spatial dimensions which has all coordinates positive (because the n_k's were restricted to be positive integers.) Thus

$$\Lambda(E) = V_{3N}\left(\sqrt{E/E_0}\right) 2^{-3N} , \tag{4.22}$$

where $V_d(r)$ is the volume of a sphere of radius r in d spatial dimensions and is given by

$$V_d(r) = \frac{r^d \pi^{d/2}}{(d/2)!} . \tag{4.23}$$

We then get

$$\Omega(E) = \frac{1}{(\frac{3N}{2} - 1)!E} \left(\frac{2m\pi E L^2}{h^2} \right)^{3N/2} . \tag{4.24}$$

We have as yet to take proper account of the indistinguishability of the particles. We do this in an approximate way which is appropriate in the case of a low-density gas at sufficiently high energy per particle that quantum interference effects are small. In that case one simply assumes that all particles are in different quantum states most of the time. Then, in order to not overcount states where particles are interchanged, we simply divide by $N!$, so that finally we write, correct to leading order in N,

$$\begin{aligned} \ln \Omega(E) = & \frac{3N}{2} \ln\left(\frac{2m\pi E}{h^2} \right) + N \ln V \\ & - \ln[(3N/2)!] - \ln[N!] . \end{aligned} \tag{4.25}$$

Then using Stirling's approximation, $\ln[N!] \sim N \ln N - N$, we get

$$\ln \Omega(E) = \frac{3N}{2} \ln\left(\frac{4m\pi E}{3Nh^2} \right) + N \ln(V/N) + \frac{5}{2} N . \tag{4.26}$$

In this form one sees that $\ln \Omega(E)$ for the ideal gas is extensive. (Note that E/N is an intensive variable, namely, the energy per particle.) It should be clear that, had we not taken into account the indistinguishability of particles, $\ln \Omega(E)$ would be superextensive, containing an extra factor of $\ln N$.

4.3 System in Contact with an Energy Reservoir

In this section, we consider a system in contact with an energy reservoir. In the previous chapter on thermodynamics, we found that the potential which is to be minimized in this situation is the Helmholtz free energy. Since this potential is the potential for a system at fixed volume and temperature, we may anticipate that this construction will introduce the notion of temperature.

4.3.1 Two Subsystems in Thermal Contact

As noted in the previous chapter, we are often interested, not in isolated systems, but in systems that can exchange energy. Here we consider the situation in which system 1 is in thermal contact with system 2. By "thermal contact" we mean that the

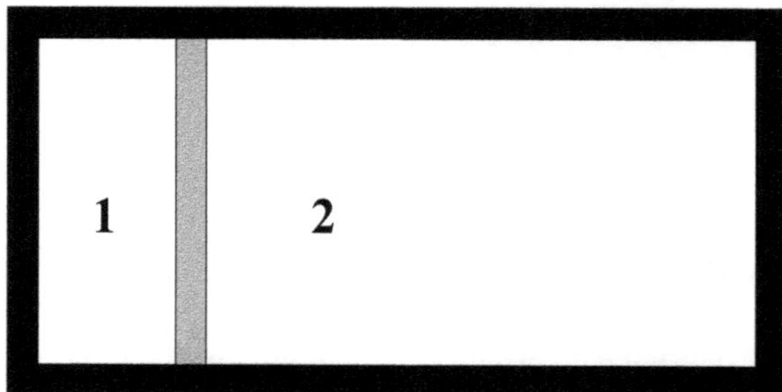

Fig. 4.1 Two subsystems, 1 and 2, in thermal contact. The wall enclosing the entire system is perfectly rigid and does not permit any exchange of energy with the rest of the universe. The wall between subsystems is fixed in position but does allow the interchange of energy. Thus the volume and number of particles of each subsystem is fixed, as is the total energy of the combined system

two systems are allowed to exchange energy, but their volumes and particle numbers are fixed. This situation is depicted in Fig. 4.1.

We can express the density of states of the combined system, 1+2, as

$$\Omega(E) = \int_0^E \Omega_1(E - E_2)\Omega_2(E_2)dE_2 \,, \tag{4.27}$$

where $\Omega_n(E_n)$, $n = 1,\ 2$, is the many-body density of states of the nth subsystem and the zero of energy is the ground state energy of each subsystem. We assume that, for the ground state energy, $E_n = 0$, $\Omega_n(0)$ is either one or, in any case, small and independent of the subsystem size, to be consistent with the Third Law of Thermodynamics.

We have seen for the ideal gas that $\Omega_n(E_n)$ for a macroscopic system varies as E_n^x where x is proportional to the number of particles in the system. This means that the spacing between energy levels becomes exponentially small for large N. This property of exponentially small spacing of energy levels for macroscopic systems is not unique to the noninteracting gas. Although their spectral weight will be moved around by interactions, the energy levels remain dense for interacting systems.

Quite generally, one expects $\Omega(E)$ to be a rapidly increasing function of E for a macroscopic system, since increasing the energy allows more choice of possible states. This is clearly true for the ideal gas, and it is also true for interacting gases, liquids, and for the vibrational many-body density of states of solids, as mentioned in the last chapter. In fact, it is generally true *except* when the spectrum of the system is bounded above. We will return to that case below and here assume that $\Omega(E)$ goes like some increasing function of E/N to a power proportional to N. This is equivalent to saying that

$$\Omega_n(E_n) \equiv e^{S_n(E_n)/k} \,, \tag{4.28}$$

where $S_n(E_n)$ is extensive in the size of the subsystem. We will soon see that $S_n(E_n)$ has all the properties of the thermodynamic entropy, and, in anticipation of that, we have included the Boltzmann constant, k (cf. Fig. (3.1) and Eq. (3.2)) in its definition. Then Eq. (4.27) can be rewritten as

$$e^{S(E)/k} = \int_0^E e^{(S_1(E-E_2)+S_2(E_2))/k} dE_2 \; . \tag{4.29}$$

The integrands of Eqs. (4.27) and (4.29) are sharply peaked functions of E_2, with a maximum at E_2^* and with essentially all of their weight within a small width, $W(E)$, around E_2^*, and so the integral is well represented as the product of the width times the maximum value at E_2^*. Taking the log of both sides gives

$$\begin{aligned} S(E) &= S_1(E - E_2^*) + S_2(E_2^*) + k \ln W(E) \; , \\ &\approx S_1(E - E_2^*) + S_2(E_2^*) \; , \end{aligned} \tag{4.30}$$

where, in the second line, we have dropped the term, $\ln W(E)$, since $\ln W(E) < \ln E$, whereas the other two terms are extensive.

The condition that the integrand is a maximum for $E_2 = E_2^*$ is

$$\left. \frac{\partial S_1(E - E_2)}{\partial E_2} \right|_{E_2^*} + \left. \frac{\partial S_2(E_2)}{\partial E_2} \right|_{E_2^*} = 0 \; . \tag{4.31}$$

(Here the partial derivatives remind us that the volume and particle number are held constant.) But since

$$\left. \frac{\partial S_1(E - E_2)}{\partial E_2} \right|_{E_2^*} = - \left. \frac{\partial S_1(E_1)}{\partial E_1} \right|_{E_1^*} , \tag{4.32}$$

where $E_1^* = E - E_2^*$, we may write Eq. (4.31) as

$$\left. \frac{\partial S_1(E_1)}{\partial E_1} \right|_{E_1^*} = \left. \frac{\partial S_2(E_2)}{\partial E_2} \right|_{E_2^*} . \tag{4.33}$$

Thus we see that, when two macroscopic systems are thermally equilibrated, the total energy is partitioned between the two systems so that the condition of Eq. (4.33) is satisfied. In fact for any system we can define a parameter β by

$$\beta \equiv \frac{1}{k} \frac{\partial S(E)}{\partial E} \tag{4.34}$$

such that, when two systems are equilibrated in thermal contact with one another, they have the same value of the parameter β. We note that, so long as $S(E)$, as defined in Eq. (4.28), is an increasing function of E, the (inverse) temperature parameter β is positive.

We mentioned above that the one case where $S(E)$ is not monotonically increasing is when the many-body density of states is zero above some maximum energy. A simple example is a macroscopic collection of two-level systems or, equivalently, N Ising spins in a magnetic field H. For that case, if we take the ground state energy to be zero for all spins up, then the maximum possible energy is $E_{max} = 2NH$ when all spins are down. This system has its maximum entropy when $E = NH$ for which $S(NH) = N \ln 2$. For $E > NH$, $S(E)$ decreases for increasing E corresponding to negative temperatures. Nevertheless, if one divides the system into two subsystems, the joint density of states in this region is strongly peaked at a value corresponding to having the same energy density in the two subsystems, when the two subsystems are at the same negative temperature. This problem is explored further in an exercise.

In the next section, we will elaborate further on the identification of $S(E) = k \ln \Omega(E)$ as the entropy and of $1/k\beta$ as the temperature T.

4.3.2 System in Contact with a Thermal Reservoir

We now consider the case where system 1 is in thermal contact with a thermal reservoir R. (Thus in Fig. 4.1 take system 2 to be R.) The reservoir R is assumed to be much larger than system 1 and we now calculate the probability P_n that system 1 is in **a particular state n** with energy E_n, given that the energy of the combined system is fixed at the value E. Since all states of the combined system with the same energy have equal *a priori* probabilities, the probability P_n is proportional to the number of states of the reservoir which have energy $E - E_n$. Thus

$$P_n \propto \Omega_R(E - E_n) , \tag{4.35}$$

where $\Omega_R(E)$ is the density of states of the reservoir. We now expand in powers of E_n. Looking back at Eq. (4.24) where we found $\Omega(E)$ for an ideal gas, if we use that form as a guide, we would conclude that an expansion of $\Omega_R(E - E_n)$ in powers of E_n gives a series in which the expansion parameter is $N_R E_n/E$, where N_R is the number of particles in the reservoir. The convergence of such an expansion is problematic. However, if we rewrite Eq. (4.35) as

$$P_n \propto e^{S_R(E-E_n)/k} , \tag{4.36}$$

and expand the exponent, then, from Eq. (4.26), an expansion of $S_R(E - E_n)$ involves the expansion parameter E_n/E, which is of order the ratio of the size of the system 1 to the size of the reservoir R. As mentioned, this ratio is assumed to be small. So not only does this expansion converge, but, in the limit of infinite reservoir size where the expansion parameter is infinitesimal, the expansion can be truncated *exactly* at first order. Thus, we have the result

$$P_n \propto \exp\left(S_R(E)/k - \frac{1}{k}\frac{\partial S_R(E)}{\partial E} E_n \right)$$
$$= \exp\left(S_R(E)/k - \beta E_n \right) . \tag{4.37}$$

Normalizing the probability, we obtain

$$P_n \equiv \rho_n = e^{-\beta E_n} / \sum_m e^{-\beta E_m} , \tag{4.38}$$

where ρ_n is the density matrix which is hence diagonal in the basis of the energy eigenstates of the system. This relation may be written in operator form as

$$\rho = e^{-\beta \mathcal{H}} / \mathrm{Tr} e^{-\beta \mathcal{H}} . \tag{4.39}$$

We define the **partition function**, Z to be

$$Z = \mathrm{Tr} e^{-\beta \mathcal{H}} , \tag{4.40}$$

so that

$$\rho = e^{-\beta \mathcal{H}} / Z . \tag{4.41}$$

Equation (4.38) is an explicit formula for the probability distribution for states of a system in equilibrium with a thermal reservoir. Note that the properties of the reservoir appear only in the scalar factor β, and the operators ρ and $\mathcal{H}$ depend only on the degrees of freedom of the system, not of the reservoir. This distribution, known as the "canonical" distribution function, forms the basis for almost all calculations in equilibrium statistical mechanics. We emphasize the generality of this result: it applies to systems independent of whether they obey classical mechanics or quantum mechanics and in the latter case whether identical particles in them obey Fermi or Bose statistics. (This is illustrated by Exercise 9 of Chap. 7.)

4.4 Thermodynamic Functions for the Canonical Distribution

We next show how these statistical quantities are related to thermodynamic properties. We consider a system with a fixed number of particles, N, volume V, and temperature parameter, β. We begin by calculating the thermal average of the entropy.

4.4.1 The Entropy

Using the principle of equal a priori probabilities for states with the same energy, we derived Eq. (4.16) for the probability that a system with an energy E_n is in a particular state n or in any other state m with the same energy, $E_m = E_n$. More precisely, we calculated the probability that the system's energy is within some very small range, $E_n - \epsilon/2 < E < E_n + \epsilon/2$. That probability is

$$\rho_n = \frac{1}{\epsilon e^{S(E_n)/k}}, \tag{4.42}$$

where we have expressed the density of states, $\Omega(E_n)$, as the exponential of the entropy for E_n. This can be inverted to give the entropy as a function of energy in terms of the probability, ρ_n, as

$$S(E_n) = -k \ln \rho_n - k \ln \epsilon \, . \tag{4.43}$$

The thermally averaged entropy for temperature $T \equiv 1/k\beta$ is

$$\langle S(E_n) \rangle = -k \langle \ln \rho_n \rangle = -k \sum_n \rho_n \ln \rho_n \, . \tag{4.44}$$

In the last step, we have dropped the $\ln \epsilon$ term which is of order unity, compared to $S(E)$ which is proportional to N. Then using the canonical probability distribution, Eq. (4.38), we can write

$$S(T) = \langle S(E_n) \rangle = k \ln Z + k\beta \sum_n E_n e^{-\beta E_n}/Z, \tag{4.45}$$

where, for the first term, we have used the fact that $\mathrm{Tr}\rho = 1$.

4.4.2 The Internal Energy and the Helmholtz Free Energy

For a system in equilibrium with a thermal reservoir at inverse temperature $k\beta$, the internal energy is simply the thermal average of the energy. That is

$$\begin{aligned} U &\equiv \langle E_n \rangle, \\ &= \sum_n E_n e^{-\beta E_n}/Z \\ &= -\frac{\partial \ln Z}{\partial \beta} \, . \end{aligned} \tag{4.46}$$

From the expression for the temperature-dependent entropy in Eq. (4.45), it is clear that, using $T = 1/k\beta$, $TS(T) = kT \ln Z + U$, so

$$-kT \ln Z = U - TS = F(T, V, N) \tag{4.47}$$

is the statistical mechanical Helmholtz free energy. This is the landmark result of statistical mechanics. The quantity $1/k\beta$ plays the role of temperature, where k is a constant relating energy and temperature, and β determines the relative probability of the system being in a given state as a function of its energy. Once units of temperature and energy have been defined, the constant k can be determined experimentally. In order to calculate the thermodynamic properties of any system described by a Hamiltonian, the strategy is to first calculate the partition function which then determines the free energy via Eq. (4.47) and provides a route for computing all thermodynamic quantities.

To explore this further, recall that, from Eq. (3.61),

$$S = -\left.\frac{\partial F}{\partial T}\right)_V , \qquad P = -\left.\frac{\partial F}{\partial V}\right)_T . \tag{4.48}$$

The first of these relations immediately reproduces Eq. (4.45). What about the second which can be written, using Eq. (4.47), as

$$\begin{aligned} P &= kT\left.\frac{\partial \ln Z}{\partial V}\right)_T , \\ &= -\frac{\sum_n (dE_n/dV)\, e^{-\beta E_n}}{\sum_n e^{-\beta E_n}} . \end{aligned} \tag{4.49}$$

How does E_n depend on the volume? We can answer this question for a gas of noninteracting particles, but that requires a more general result, which we obtain as follows. The dependence of the energy levels on volume arises from the potential energy associated with the wall. If, for instance, we have a wall at $x = L$, then we set $dE_n/dV = (1/A)dE_n/dL$, where A is the area of the wall. Note that near the wall the potential energy $V(x)$ will be given by $\phi(L - x)$, where ϕ is a function which is zero if $L - x$ is greater than a few nanometers and increases very steeply as $L - x$ goes to zero (Fig. 4.2). The Hellmann–Feynman theorem which is discussed in (Merzbacher 1998) states that.

$$dE_n/dL = \langle n|d\mathcal{H}/dL|n\rangle .$$

But because $V(x_i)$ is a function of $(x_i - L)$ we may write

$$\langle n|\frac{d\mathcal{H}}{dL}|n\rangle = -\sum_i \langle n|\frac{dV(x_i)}{dx_i}|n\rangle ,$$

where the sum is over all particles i, so that

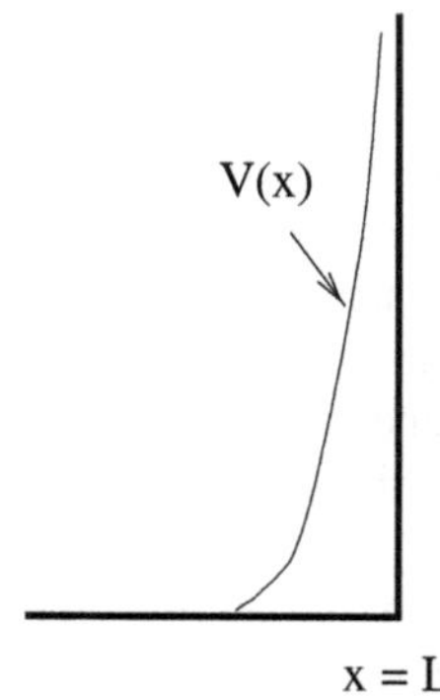

Fig. 4.2 Schematic diagram of the confining potential associated with a wall

$$dE_n/dL = -\sum_i \langle n|\frac{dV(x_i)}{dx_i}|n\rangle \ , \tag{4.50}$$

We identify $\sum_i \langle n|dV(x_i)/dx|n\rangle$ as the total force, **F** exerted on the wall at $x = L$ by all the gas molecules when the gas is in the quantum state $|n\rangle$. Taking the average over quantum states, as called for in Eq. (4.49), we obtain

$$kT\frac{\partial \ln Z}{dV}\bigg)_T = \frac{kT}{A}\frac{\partial \ln Z}{dL}\bigg)_T = P \ , \tag{4.51}$$

where P is the thermodynamic pressure.

For magnetic systems, we have the analogous results in terms of the partition function evaluated at a fixed value of the applied magnetic field H.

$$F = -kT \ln\left(\sum_n e^{-\beta E_n}\right), \quad S = -\frac{\partial F}{\partial T}\bigg)_H, \quad \mu_0 m = -\frac{\partial F}{\partial H}\bigg)_T, \tag{4.52}$$

where m is the magnetic moment of the system. To obtain the last relation in Eq. (4.52), one uses an argument analogous to that of Eq. (4.50) that

$$\frac{dE_n}{dH} = \langle n|\frac{d\mathcal{H}}{dH}|n\rangle = -\langle n|\mu_0\hat{m}|n\rangle \ , \tag{4.53}$$

where $\hat{m}$ is the magnetic moment *operator*.

4.4.2.1 The Monatomic Ideal Gas

To illustrate this formalism, we apply it to the monatomic ideal Gibbs gas. (We say Gibbs because we want to carry out the counting of states using the $1/N!$ approximation to take account of the indistinguishability of particles.) So we write

$$Z(T,V) = \frac{1}{N!}\sum_{\{n_i\}} \exp\left(-\frac{h^2}{8mL^2kT}[n_1^2 + n_2^2 + \cdots + n_{3N}^2]\right) . \tag{4.54}$$

Since the $1/N!$ counting approximation has already limited our calculation to high energy (or high temperature), we may convert the sums over the n_i's to integrals:

$$\begin{aligned} Z(T,V) &= \frac{1}{N!}\prod_{i=1,3N}\left[\int_0^\infty dn_i e^{-n_i^2 h^2/(8mL^2kT)} dn_i\right] \\ &= \frac{1}{N!}\left(\frac{2\pi m L^2 kT}{h^2}\right)^{3N/2} = \frac{1}{N!}(V/\lambda)^{3N} , \end{aligned} \tag{4.55}$$

where λ is the thermal deBroglie wavelength:

$$\lambda = \left(\frac{h^2}{2m\pi kT}\right)^{1/2} . \tag{4.56}$$

Thus

$$F(T,V) = -kT \ln Z(T,V) = kT \ln N! - NkT \ln(V/\lambda^3) . \tag{4.57}$$

Now use Stirling's approximation: $\ln N! = N \ln N - N$ to get

$$F(T,V) = -NkT \ln(v/\lambda^3) - NkT , \tag{4.58}$$

where $v \equiv V/N$ is the volume per particle. More explicitly

$$F(T,V) = -NkT\left[\ln(V/N) + \frac{3}{2}\ln[2m\pi kT/h^2] + 1\right] . \tag{4.59}$$

We thereby get the thermodynamic functions as

$$P = -\left.\frac{\partial F}{\partial V}\right)_{T,N} = NkT/V , \tag{4.60}$$

as expected. Also

$$S = -\left.\frac{\partial F}{\partial T}\right)_{V,T} = Nk\left[\ln(V/N) + \frac{3}{2}\ln[2m\pi kT/h^2] + \frac{5}{2}\right] . \tag{4.61}$$

From these we get

$$U = F + TS = (3/2)NkT \tag{4.62}$$

and

$$C_V = T \left.\frac{\partial S}{\partial T}\right)_{V,N} = (3/2)Nk \,. \tag{4.63}$$

The results for the pressure, internal energy U, and the specific heat at constant volume C_V are expected, of course. The less familiar result is that the entropy is given in terms of the system parameters and Planck's constant h and we will comment on this in a moment.

4.5 Classical Systems

4.5.1 Classical Density Matrix

We now consider the above formalism in the classical limit. The reasoning we used above, namely, that we need a parameterless distribution (since the state of a system is specified by giving the appropriate number of macroscopic variables), applies equally to classical systems. Therefore, the distribution function of an isolated system should be constant over states of the system which have the energy assigned to the system. This prescription is rather vague because it does not specify the phase space, or metric, one is supposed to invoke. One simple requirement that we could invoke to clarify this ambiguity is that the distribution function ρ at fixed energy should be time-independent. That is, if we allow the coordinates and momenta to evolve according to the classical equations of motion, we will require that the distribution function be invariant. The condition for this is well known from theoretical mechanics (Goldstein et al. 2001). Namely, the following Poisson bracket should vanish:

$$0 = \frac{d\rho}{dt} = \sum_i \left[\frac{\partial \rho}{\partial q_i}\dot{q}_i + \frac{\partial \rho}{\partial p_i}\dot{p}_i \right] = \sum_i \left[\frac{\partial \rho}{\partial q_i}\frac{\partial \mathcal{H}}{\partial p_i} - \frac{\partial \rho}{\partial p_i}\frac{\partial \mathcal{H}}{\partial q_i} \right] . \tag{4.64}$$

In writing this equation, we do not permit ρ to have any explicit dependence on time. If ρ is a function of coordinates and momenta only through the Hamiltonian, so that it is of the form

$$\rho(\{p_i\}, \{q_i\}) = F[\mathcal{H}(\{p_i\}, \{q_i\})] \,, \tag{4.65}$$

then ρ is invariant under time development governed by the Hamiltonian equations of motion. We may thus consider the classical microcanonical distribution function for which

$$F(x) = \delta(x - E) \,, \tag{4.66}$$

or the classical canonical distribution function for which

$$F(x) = \exp(-\beta x) , \tag{4.67}$$

where $\beta = 1/(kT)$. As before, the thermal average of a classical operator $\mathcal{O}$ is taken as one would expect, namely,

$$\langle \mathcal{O} \rangle = \frac{\int \prod_i dp_i \prod_i dq_i \rho(\{p_i\}, \{q_i\}) \mathcal{O}}{\int \prod_i dp_i \prod_i dq_i \rho(\{p_i\}, \{q_i\})} . \tag{4.68}$$

To obtain all the thermodynamic functions (especially the entropy), we need to give a formula for the partition function in the classical limit. It should be apparent that

$$Z = C_Z^{-1} \int d\{q_i\} d\{p_i\} e^{-\beta \mathcal{H}(\{q_i, p_i\})} , \tag{4.69}$$

where we have introduced a normalization factor C_Z.

4.5.2 Gibbs Entropy Paradox

To get a clue as to what C_Z might be, we repeat the argument first given by J. W. Gibbs long before the advent of quantum mechanics. We consider the thermodynamics of the classical ideal gas. We have

$$\begin{aligned} Z &= \frac{1}{C_Z} \int \prod_{i=1}^{3N} dp_i \int \prod_{i=1}^{3N} dq_i e^{-p_i^2/(2mkT)} \\ &= C_Z^{-1} [2m\pi kT]^{3N/2} V^N , \end{aligned} \tag{4.70}$$

from which we get

$$F = -kT \Big((3N/2) \ln[2m\pi kT] + N \ln V - \ln C_Z \Big) . \tag{4.71}$$

Correspondingly, we have

$$\begin{aligned} \frac{S}{k} &= -\frac{\partial F}{\partial (kT)} \bigg|_V \\ &= (3N/2) \ln[2m\pi kT] + N \ln V - \ln C_Z + 3N/2 \end{aligned} \tag{4.72}$$

and

$$U = F + TS = \frac{3}{2}NkT\ . \tag{4.73}$$

Also

$$P = -\left.\frac{\partial F}{\partial V}\right|_T = \frac{NkT}{V}\ , \tag{4.74}$$

which is the famous ideal gas law.

Suppose we disregard C_Z for a moment (by setting $C_Z = 1$). In that case there is a problem: if we double the size of the system (i.e., we double N and V keeping T constant), we ought to double the free energy and the entropy, but the term in $N \ln V$ indicates that, without considering C_Z, we would have

$$S(2N, 2V, T) = 2S(N, V, T) + 2N \ln 2\ , \tag{4.75}$$

for the entropy S and similarly for the free energy. This is clearly wrong: if we take two moles of a gas through a Carnot cycle, it obviously does not matter whether these two moles are in the same box or whether they are in different boxes isolated from one another. The conclusion is that entropy should be an additive function for macroscopic objects. The solution to this apparent paradox (known as the "Gibbs paradox") was noted by J. W. Gibbs (for whom the Gibbs functions and Gibbs distribution were named) even before the development of quantum mechanics. He proposed that the constant C_Z should be proportional to $N!$, say $C_Z = cN!$, where c is a constant to be determined. Then the entropy is [using Stirling's formula for $\ln(N!)$]

$$\frac{S}{k} = (3N/2)\ln[2m\pi kT] + N\ln(V/N) - \ln c + 5N/2\ . \tag{4.76}$$

Gibbs rationalized this choice for C_Z by saying that the factor of $N!$ was included in order not to overcount configurations of identical particles which are identical except for how the particles are labeled. This idea is totally unintelligible within classical mechanics, because interchanging particles does lead to classically different states. However, the inclusion of this factor of $N!$ is easily understood within quantum mechanics where it arises from the quantum mechanical treatment of identical particles. We will refer to the ideal gas when we include this factor in the partition function as a "Gibbs gas," as a reminder that the ideal gas must be treated taking account of indistinguishability of identical particles. It is noteworthy that quantum mechanics is required to explain a macroscopic property as basic as the additivity of entropy.

Since the classical partition function should be a limiting case of the quantum partition function which is dimensionless, we also need to introduce into the constant C_Z a factor which has the dimensions of $[pq]^{3N}$. This observation suggests that C_Z should be proportional to h^{3N}. Since we expect one single-particle quantum state to correspond to a volume h^3, we expect that $C_Z = h^{3N}N!$, for the Gibbs gas. With this ansatz the entropy no longer has any undetermined constant and is extensive, as we require, so that

$$\frac{S}{k} = 3N \ln[\sqrt{2m\pi kT}/h] + N\ln(V/N) + 5N/2 \,. \tag{4.77}$$

It is a remarkable fact that it is possible to determine Planck's constant (which is usually associated with microscopic phenomena) solely by the macroscopic measurements needed to fix the entropy, as embodied in Eq. (3.30).

Historical note: after Planck's work, but before the emergence of quantum mechanics, Sakur (1913) advanced theoretical arguments which suggested setting $C = (\gamma h)^{3N} N!$, for a gas, where γ is a dimensionless scale factor, probably of order unity, but not firmly fixed by theory that existed at the time. In principle, one can determine γ by measurements of the entropy, using Eq. (3.30). Actually, it is more convenient to measure the vapor pressure of an ideal gas in equilibrium with the solid at low temperature as was done by Tetrode (1912) from which he concluded that "Es erscheint plausibel, dasz z (our γ) genau 1 zu setzen ist..." (In other words, γ is exactly 1.).

With this choice for the constant C_Z, we have the free energy as given in Eq. (4.58) and the other thermodynamic functions as given thereafter.

4.5.3 Irrelevance of Classical Kinetic Energy

Normally, the integrations over momenta required for the evaluation of the partition function are straightforward Gaussian integrals. For example, for a gas of atoms one has

$$\mathcal{H} = \sum_{i=1}^{3N} \frac{p_i^2}{2m} + V(\{q_i\}) \,. \tag{4.78}$$

Then, we have

$$Z = \left(\frac{2m\pi kT}{h^2}\right)^{3N/2} \frac{1}{N!} \int \prod_{i=1}^{3N} dq_i e^{-\beta V(\{q_i\})} \,, \tag{4.79}$$

so that the kinetic energy contributes to the thermodynamics exactly as it does for an ideal gas. For classical systems, the non-ideality is entirely encoded in the potential energy. Similarly, if one wishes to evaluate the average of an operator $\mathcal{O}$ which does *not* depend on any momenta, then one has

$$\langle \mathcal{O} \rangle = \frac{\int \prod_{i=1}^{3N} dq_i e^{-\beta V(\{q_i\})} \mathcal{O}}{\int \prod_{i=1}^{3N} dq_i e^{-\beta V(\{q_i\})}} \,, \tag{4.80}$$

even when the kinetic energy depends on the coordinates, as in Exercise 9, the integrals over momenta are easy to do.

4.6 Summary

The most important statement of this chapter is that the probability that a system at temperature T whose volume and number of particles is fixed is in its nth energy eigenstate is proportional to $\exp(-\beta E_n)$, where $\beta = 1/(kT)$. This holds for classical systems as well as for quantum systems, irrespective of whether the constituent particles obey Fermi or Bose statistics. The identification of β as well as the statement that the free energy $F(T, V, N)$ is given by $F = -kT \ln Z$, where $Z \equiv \sum_n \exp(-\beta E_n)$ is made by comparing the statistical mechanical quantities with their thermodynamic counterparts. For a classical system of N identical particles (in three spatial dimensions) $Z = C_Z^{-1} \int de^{-\beta\mathcal{H}(\mathbf{p},\mathbf{q})} d^{3N} p d^{3N} q$, where $C_Z = N! h^{3n}$. The factor h expresses the fact that one quantum state corresponds to a volume h in classical phase space. Statistical mechanics provides the crucial insight into the meaning of entropy via Boltzmann's famous equation $S = k \ln W$, where W is the number of accessible states.

4.7 Exercises

1. Check that Eq. (4.23) [for the volume of a sphere in d dimensions) gives the correct answers for $d = 1, 2, 3$.

2. Suppose the reservoir consists of the ideal gas for which we calculated the logarithm of the density of states in Eq. (4.26).

(**a**) Write down the terms up to order E_n^2 in the Taylor series expansion of $\Omega_R(E - E_n)$ in powers of E_n. Display the ratio of the second term to the first term and the ratio of the third term to the second term. Thereby show that the expansion parameter is not generally very small.

(**b**) Write down the terms up to order E_n^2 in the Taylor series expansion of $\ln \Omega_R(E - E_n)$ in powers of E_n. Display the ratio of the second term to the first term and the ratio of the third term to the second term. Thereby show that the expansion parameter is of order the ratio of the size of the system to the size of the reservoir and therefore that the expansion parameter is infinitesimal for an infinitely large reservoir.

3. In deriving Eq. (4.26) (for the logarithm of the density of states of an ideal gas), we assumed rigid-wall boundary conditions, so that the wave function vanished at the walls. Suppose we had instead assumed *periodic* boundary conditions. Determine how, if at all, this would affect the result given in Eq. (4.26).

4. Carry out the integration in Eq. (4.27) exactly for $\Omega_1(E) = A_1 E^{N_1}$ and $\Omega_2(E) = A_2 E^{N_2}$. Verify Eq. (4.30) in the limit when N_1 and N_2 are assumed to be very large.
5. Expand the exponent of the exponential in the integral of Eq. (4.29) around its maximum at $E_2 = E_2^*$ to second order in $(E_2 - E_2^*)$. Considering the coefficient of $\frac{1}{2}(E_2 - E_2^*)^2$ as $1/W^2$ where W is the width of a Gaussian, relate W^2 to thermodynamic properties of the two subsystems. How does W scale with the sizes of the subsystems? Justify the neglect of the $k \ln W(E)$ term in Eq. (4.30).
6. Consider N Ising spins, $S_i = \pm 1$, in a magnetic field H, and write the energy of the spins as $E = (N - M)H$ where M is the sum of the S_i.

(**a**) Derive an exact expression for $\Omega(E, N)$.

(**b**) Using Stirling's approximation, derive an expression for $S(E, N)$. Show that $S(E, N)$ is extensive, and evaluate $S(NH, N)$. Sketch $S(E, N)$ versus E, indicating the height and position of the peak and where the entropy falls to zero.

(**c**) Consider two systems of Ising spins in thermal contact in a field H, where system 1 has N_1 spins and system 2 has N_2 spins. Derive an expression for the joint density of states, $\Omega(E - E_2, N_1)\Omega(E_2, N_2)$ and show that this function is sharply peaked as a function of E_2 at a value E_2^* that corresponds to each subsystem having the same average energy per spin. Show that this is true even in the region where more than half the spins are in their high energy state.

(**d**) Discuss how one might go about preparing a set of spins in such a negative temperature state and what would happen if such a system were put in thermal contact with a system at positive temperature, T.

7. Starting from $\Omega(E, V)$, the density of states as a function of E and V given in Eq. (4.26), obtain the pressure as a function of N, V, and T, and also the internal energy and specific heat at constant volume per particle as functions of these variables.
8. Suppose the Hamiltonian is given as a function of the external static electric field E (which is assumed to be position-independent.) How (either in classical or quantum mechanics) is the electric (dipole) moment operator related to a derivative of the Hamiltonian? Thereby relate the thermally averaged value of the dipole moment operator of the system, $\mathbf{P}$, to a derivative of a thermodynamic potential.
9. Consider the orientational properties of a system consisting of rod-like molecules. If we have N identical such molecules each localized on a lattice site in a solid, the partition function Z for the entire system is given by $Z = z^N$, where z is the single-molecule partition function. You are to calculate z. For this purpose describe each molecule by the generalized coordinates $q_1 = \theta$ and $q_2 = \phi$. To get the corresponding generalized momenta p_1 and p_2, write down the kinetic energy in terms of $\dot{q}_i$ and proceed as in theoretical mechanics. Obtain an expression for z when the molecule is in an orientational potential $V(\theta)$ in terms of an integral over the single variable θ. Notice that the phase space factor $\sin\theta$ occurs naturally and is *not* inserted by hand.
10. We have already calculated the number of microstates Ω corresponding to total energy, E, for N noninteracting and distinguishable particles confined within a box of volume V. In other words, each particle is assumed to occupy the energy levels of a particle in a box. In this problem you are to repeat the calculation for two cases:

(**a**) Each particle is in an harmonic potential with energies (relative to the ground state)

$$E(l, m, n) = (l + m + n)\hbar\omega ,$$

where l, m, and n are integers 0, 1, 2, etc.

(**b**) The gas consists of diatomic molecules which can translate and execute intramolecular vibration in which the bond length oscillates. In this model, each molecule has four quantum numbers l, m, n, j in terms of which its energies (relative to the ground state) are

$$E(l, m, n, j) = j\hbar\omega + \frac{h^2(l^2 + m^2 + n^2)}{8mL^2} ,$$

where l, m, and n are integers 0, 1, 2, etc. (In reality, diatomic molecules rotate typically with energies which are much less than that of vibration. But to avoid complicated algebra we ignore molecular rotation.)

NOTE: If you are familiar with contour integrals, it is easiest if you do this using the Laplace transform representation of the δ-function:

$$\delta(x) = \frac{1}{2\pi i}\int_{a-i\infty}^{a+i\infty} e^{xz} dz ,$$

where a (which is real) is sufficiently large that all integrations converge.

11. In classical mechanics, one writes the Hamiltonian of a system of N particles in a uniform magnetic field applied in the z-direction is

$$\mathcal{H} = \sum_{i=1}^{N} \frac{1}{2m_i}\left[\mathbf{p}_i - (q_i/c)\mathbf{A}(\mathbf{r}_i)\right]^2 + V(\{\mathbf{r}_i\}) ,$$

where q_i is the electric charge of the ith particle and we may set

$$\mathbf{A}(\mathbf{r}) = -\frac{1}{2}Hy\hat{i} + \frac{1}{2}Hx\hat{j} .$$

Also the interaction energy, V, between particles depends only on their coordinates and not on their momenta.

(**a**) Show that the classical partition function is independent of H.

(**b**) What does this imply about the thermal averaged magnetic moment of the system, assuming the validity of classical mechanics? This result that the phenomenon of magnetism is inconsistent with classical mechanics and only finds its explanation within quantum mechanics is known as the Bohr–van Leeuwen theorem (van Leeuwen 1921), (Van Vleck 1932).

12. This problem illustrates the discussion following Eq. (4.17). Let P_n be the probability that the nth quantum state is occupied. Then the time evolution of the P_n is described by what is called the "master equation,"

$$\frac{dP_n}{dt} = \sum_m \left(p_{nm} P_m - p_{mn} P_n \right) . \tag{4.81}$$

Assume that $p_{nm} = p_{mn}$ and analyze the possible solutions for the equilibrium values of the P_n's.

13. This problem concerns equilibrium for a system contained in a box with fixed rigid walls, isolated from its surroundings, and consisting of a gas with a partition (which does not allow the interchange of particles) separating the two subsystems of gas. Find the conditions for equilibrium from the maximum entropy principle when

(**a**) the partition is fixed in position, but does allow the transfer of energy.
(**b**) the partition is movable and allows the transfer of energy.
(**c**) the partition is movable but does not allow the transfer of energy.
(**d**) In case C), maximization of the entropy leads to the condition for equilibrium $p_1/T_1 = p_2/T_2$, where the subscripts refer to the two sides of the partition. This condition is strange because we would guess that we should have $p_1 = p_2$. We believe that the explanation for this implausible result is because a partition which is movable but does not allow the transfer of energy is unphysical. Try to give an argument to support this belief.

14. Suppose we change the zero of energy. That is, we set $E_n' = E_n + \Delta$, where Δ is a constant. Then we may compare the properties of the primed system for which the partition function is $Z' = \sum_n e^{-\beta E_n'}$ with those of the unprimed system for which the partition function is $Z = \sum_n e^{-\beta E_n}$.

(**a**) Display the relation between the thermodynamic potentials U, F, G, and H of the primed system to those of the unprimed system.

(**b**) How do the pressure, the specific heat, the entropy, and the compressibility of the primed system compare to those of the unprimed system?

4.8 Appendix Indistinguishability

In this appendix, we discuss briefly the counting of states of a system of three indistinguishable particles in a box. In particular, we wish to illustrate the fact that the inclusion of the factor $N!$ in Eq. (4.25) is only approximately correct.

We start by considering the family of states in which we have one particle in each of the states

$$\psi_a(\mathbf{r}) = \psi_{1,0,0}(\mathbf{r}) , \tag{4.82}$$

$$\psi_b(\mathbf{r}) = \psi_{0,0,1}(\mathbf{r}) , \tag{4.83}$$

$$\psi_c(\mathbf{r}) = \psi_{0,1,0}(\mathbf{r}) \ . \tag{4.84}$$

Classically, we consider each particle to have a number painted on it and thereby each particle is distinguished from any other particles. Then, we have six different three-body states:

$$\Psi_1 = \psi_a(\mathbf{r}_1)\psi_b(\mathbf{r}_2)\psi_c(\mathbf{r}_3) \ , \tag{4.85}$$

$$\Psi_2 = \psi_a(\mathbf{r}_3)\psi_b(\mathbf{r}_1)\psi_c(\mathbf{r}_2) \ , \tag{4.86}$$

$$\Psi_3 = \psi_a(\mathbf{r}_2)\psi_b(\mathbf{r}_3)\psi_c(\mathbf{r}_1) \ , \tag{4.87}$$

$$\Psi_4 = \psi_a(\mathbf{r}_1)\psi_b(\mathbf{r}_3)\psi_c(\mathbf{r}_2) \ , \tag{4.88}$$

$$\Psi_5 = \psi_a(\mathbf{r}_2)\psi_b(\mathbf{r}_1)\psi_c(\mathbf{r}_3) \ , \tag{4.89}$$

$$\Psi_6 = \psi_a(\mathbf{r}_3)\psi_b(\mathbf{r}_2)\psi_c(\mathbf{r}_1) \ , \tag{4.90}$$

where Ψ is the many-body wave function and we will refer to the ψ's as single-particle wave functions. The bottom line is that there are $N_C = 6$ different states if one counts "classically."

If the particles are identical, and if we follow quantum mechanics, then we can form only a single state in which one particle is in state $|a\rangle$, one in $|b\rangle$, and one in $|c\rangle$. The associated wave function is

$$\begin{aligned}\Psi = \frac{1}{\sqrt{3!}}\Big[&\psi_a(\mathbf{r}_1)\psi_b(\mathbf{r}_2)\psi_c(\mathbf{r}_3) \mp \psi_b(\mathbf{r}_1)\psi_a(\mathbf{r}_2)\psi_c(\mathbf{r}_3) \\ &+\psi_b(\mathbf{r}_1)\psi_c(\mathbf{r}_2)\psi_a(\mathbf{r}_3) \mp \psi_a(\mathbf{r}_1)\psi_c(\mathbf{r}_2)\psi_b(\mathbf{r}_3) \\ &+\psi_c(\mathbf{r}_1)\psi_a(\mathbf{r}_2)\psi_b(\mathbf{r}_3) \mp \psi_c(\mathbf{r}_1)\psi_b(\mathbf{r}_2)\psi_a(\mathbf{r}_3)\Big] \ .\end{aligned} \tag{4.91}$$

In this wave function, we have one particle in each state, but we cannot say *which* particle is *which*. It makes sense to talk about "the particle which is in state $|a\rangle$," but not the state in which particle #1 is.

We are tempted to summarize this discussion by saying that the number of quantum states for identical fermions N_F and that for identical bosons N_B are given by

$$N_B = N_F = N_C/N! \ , \tag{4.92}$$

where $N = 3$ in this illustration.

This is not the whole story. Let us now consider how many states there are in which two particles are in single-particle state $|a\rangle$ and one is in single-particle state $|b\rangle$. Classically, we can have the states $\Psi = \psi(\mathbf{r}_1)\psi(\mathbf{r}_2)\psi(\mathbf{r}_3)$ listed in Table 4.1. So for the classical case, we have

$$N_C = 3 \tag{4.93}$$

Table 4.1 Possible "Classical" States

State	$\psi(\mathbf{r}_1)$	$\psi(\mathbf{r}_2)$	$\psi(\mathbf{r}_3)$
1	ψ_a	ψ_a	ψ_b
2	ψ_a	ψ_b	ψ_a
3	ψ_b	ψ_a	ψ_a

states. How many fermion states do we have? Answer: none. We cannot put more than one particle into the same single-particle state. So $N_F = 0$. We can have exactly one bose state with the wave function

$$\Psi = \frac{1}{\sqrt{3}}\Big[\psi_a(\mathbf{r}_1)\psi_a(\mathbf{r}_2)\psi_b(\mathbf{r}_3) + \psi_a(\mathbf{r}_1)\psi_a(\mathbf{r}_3)\psi_b(\mathbf{r}_2) + \psi_a(\mathbf{r}_2)\psi_a(\mathbf{r}_3)\psi_b(\mathbf{r}_1)\Big] , \tag{4.94}$$

so that $N_B = 1$. The "divide by $N!$ rule" would give

$$N_F = N_B = N_C/3! = 1/2 . \tag{4.95}$$

Of course having a fraction of a state does not make any sense. We see that the division by $N!$ overcounts fermi states and undercounts bose states. We will see this quantitatively later on.

References

H.B. Goldstein, C.P. Poole, Jr., J.L. Safko, *Classical Mechanics*. 3rd Ed. (Pearson, 2001)

E. Merzbacher, *Quantum Mechanics*. 3rd edn. (Wiley, 1998)

O. Sackur, Die Bedeutung des elementaren Wirkungsquantums fu r die Gastheorie und die Berechnung der chemischen Konstanten, Festschrift W. Nernst zu seinem 25j ahrigen Doktorjubila um (Verlag Wilhelm Knapp, Halle a. d. S., 1912); O. Sackur, Die universelle Bedeutung des sog. elementaren Wirkungsquantums, Annalen der Physik 40, 67 (1913)

H. Tetrode, Die chemische Konstante der Gase und das elementare Wirkungsquantum, Annalen der Physik **38**, 434; *erratum*, ibid. **39**, 255 (1912)

J. van Kranendonk, *Solid Hydrogen* (Plenum, 1983)

J.H. van Leeuwen, J. de Physique, **2**(6), 361 (1921)

J.H. Van Vleck, *Electric and Magnetic Susceptibilities*, (Oxford, 1932), p. 94ff

Chapter 5
Examples

In this chapter, we will illustrate the formalism of the preceding chapter by applying it to a number of simple systems.

5.1 Noninteracting Subsystems

For noninteracting subsystems, the Hamiltonian is a sum of subsystem Hamiltonians:

$$\mathcal{H} = \sum_i \mathcal{H}_i \, . \tag{5.1}$$

We assume that there is no statistical constraint between the subsystems. Thus, if the subsystems are particles, these particles must be distinguishable. This case *includes* the case where we have indistinguishable particles which are effectively distinguishable by being localized. For example, we might have a collection of identical magnetic moments which are distinguishable by being localized at lattice sites in a solid. A particle is then distinguishable by the fact that it is the particle which is located at site 1, another is distinguishable by being located at site 2, and so forth.

For such distinguishable, noninteracting subsystems, the different $\mathcal{H}_i$ commute with each other, and the density matrix factors into a product of subsystem density matrices:

$$\rho = \frac{e^{-\beta \sum_i \mathcal{H}_i}}{\mathrm{Tr} e^{-\beta \sum_i \mathcal{H}_i}} = \frac{\prod_i e^{-\beta \mathcal{H}_i}}{\prod_i z_i} = \prod_i \rho_i \, , \tag{5.2}$$

where $z_i = \mathrm{Tr}\exp(-\beta \mathcal{H}_i)$ is the partition function for the ith subsystem and

$$\rho_i = \frac{e^{-\beta \mathcal{H}_i}}{\mathrm{Tr} e^{-\beta \mathcal{H}_i}} = \frac{e^{-\beta \mathcal{H}_i}}{z_i} \, . \tag{5.3}$$

A. J. Berlinsky and A. B. Harris, *Statistical Mechanics*, Graduate Texts in Physics,
https://doi.org/10.1007/978-3-030-28187-8_5

By the same token, the thermodynamic functions are the sum of the thermodynamic functions of the subsystems. The partition function Z of the combined system is given by the *product* of the individual subsystem partition functions:

$$Z = \prod_i z_i \,. \tag{5.4}$$

So

$$F = \sum_i F_i \,, \tag{5.5}$$

where F is the free energy of the combined system and F_i that of the ith subsystem.

5.2 Equipartition Theorem

Here, we discuss a result valid for *classical* mechanics. We start from Eq. (4.69) which allows us to write the partition function as $Z = C_Z^{-1} Z_p Z_q$, where C_Z is the appropriate normalization constant, Z_p, is the integral over the p's and Z_q that over the q's. Correspondingly, the free energy can be written as

$$F = F_p + F_q - kT \ln C_Z \,, \tag{5.6}$$

where $F_p = -kT \ln Z_p$ and $F_q = -kT \ln Z_q$. Thus the thermodynamic functions can be decomposed into contributions from F_p and those from F_q: the internal energy as $U = U_p + U_q$ and the specific heat as $C = C_p + C_q$.

Suppose the Hamiltonian is a quadratic form either in the momenta (in the case of Z_p) or in the coordinates (in the case of Z_q). For instance, suppose the potential energy is a quadratic form, so that

$$Z_q = \int_{-\infty}^{\infty} \prod_i dq_i e^{-\frac{1}{2}\sum_{ij} V_{ij} q_i q_j / kT} \,, \tag{5.7}$$

where $\mathbf{V}$ is a real symmetric matrix. Therefore, it can be diagonalized by an orthogonal matrix $\mathbf{O}$, so that

$$\tilde{\mathbf{O}}\mathbf{V}\mathbf{O} = \boldsymbol{\lambda} \,, \tag{5.8}$$

where $\boldsymbol{\lambda}_{nm} = \delta_{n,m} \lambda_n$ is a diagonal matrix (whose elements are positive to ensure stability). Thus, we introduce transformed coordinates $\mathbf{x}$ via

$$q_n = \sum_m \mathbf{O}_{n,m} x_m \,, \tag{5.9}$$

in terms of which

$$Z_q = \int_{-\infty}^{\infty} \prod_n dx_n J(x, q) e^{-\frac{1}{2}\sum_n \lambda_n x_n^2/kT} , \tag{5.10}$$

where $J(x, q)$ is the Jacobian of the transformation from the q's to the x's. This Jacobian is just the determinant of the matrix $\mathbf{O}$, so $J(x, q) = 1$. Thus

$$Z_q = \prod_n [2\pi/(\beta\lambda_n)]^{1/2} . \tag{5.11}$$

Then

$$U_q = -\left.\frac{\partial \ln Z}{\partial \beta}\right|_V = \frac{1}{2}\sum_n \frac{d \ln \beta}{d\beta} = \frac{1}{2}NkT \tag{5.12}$$

and

$$C_q = \frac{1}{2}Nk , \tag{5.13}$$

where N is the number of generalized coordinates. The same argument applies to the contribution to the thermodynamic functions from Z_p. So the general conclusion is that each quadratic degree of freedom gives a contribution to the internal energy $\frac{1}{2}kT$ and to the specific heat $\frac{1}{2}k$. In this context, the number of "degrees of freedom" is the number of variables (counting both momenta and coordinates) in the quadratic form.

We may cite some simple illustrations of this formulation. For instance, for the ideal gas, where the Hamiltonian contains only a quadratic kinetic energy, the specific heat is $3k/2$ per particle. A second example is provided by the model of a solid in which atoms are connected by quadratic springs ($V = \frac{1}{2}kx^2$) to their neighbors. In this case, there are three momenta and three coordinates per atom which contribute quadratically to the Hamiltonian. Therefore, the specific heat per atom is $3k$. Both these results are in good agreement with experiment in the appropriate range of temperature. In the case of a solid, the temperature has to be high enough that quantum effects are unimportant but low enough that anharmonic contributions to the potential energy (such as terms which are cubic and quartic in the q's) are not important. We will study quantum effects in the harmonic oscillator in a subsequent subsection.

5.3 Two-Level System

We consider a set of noninteracting spins in a uniform magnetic field, governed by the Hamiltonian

$$\mathcal{H} = -H \sum_i S_i, \quad S_i = \pm 1/2 , \tag{5.14}$$

where H is the magnetic field in energy units: $H = g\mu_B H_0$, where H_0 is the magnetic field g is approximately 2 for electrons, and $\mu_B \approx 0.927 \times 10^{-20}$ erg/gauss is the Bohr magneton. As a matrix $S_i = \begin{pmatrix} \frac{1}{2} & 0 \\ 0 & -\frac{1}{2} \end{pmatrix}$. Then the Hamiltonian for the ith spin is

$$\mathcal{H}_i = \begin{pmatrix} -H/2 & 0 \\ 0 & H/2 \end{pmatrix}, \quad e^{-\beta\mathcal{H}_i} = \begin{pmatrix} e^{\beta H/2} & 0 \\ 0 & e^{-\beta H/2} \end{pmatrix} \tag{5.15}$$

and $Z = z^N$, where z is the partition function for a single spin:

$$z = \mathrm{Tr} e^{-\beta\mathcal{H}_i} = e^{\beta H/2} + e^{-\beta H/2} = 2\cosh(\beta H/2) . \tag{5.16}$$

The thermally averaged energy per spin is given by

$$\begin{aligned} u \equiv \langle \mathcal{H}_i \rangle &= \frac{\mathrm{Tr}[\mathcal{H}_i e^{-\beta\mathcal{H}_i}]}{z} = \frac{1}{z}\mathrm{Tr}\begin{pmatrix} -(H/2)e^{\beta H/2} & 0 \\ 0 & (H/2)e^{-\beta H/2} \end{pmatrix} \\ &= -H/2\left(\frac{e^{\beta H/2} - e^{-\beta H/2}}{e^{\beta H/2} + e^{-\beta H/2}}\right) = -H/2 \tanh(\beta H/2). \end{aligned} \tag{5.17}$$

Of course there is no compelling reason to write the spin operator as a matrix, since there are no other non-commuting operators in $\mathcal{H}$. The spin operators are diagonal and could be represented by scalars. A matrix representation only becomes important for quantum spins when $\mathcal{H}$ contains different components of the spin which do not commute.

The free energy, F, obeys

$$dF = -SdT - \mu M dH \tag{5.18}$$

and the free energy per spin, $f = F/N$, of the system of $S = 1/2$ spins in a field is

$$f = -kT \ln z = -kT \ln 2 - kT \ln\cosh(\beta H/2) \tag{5.19}$$

which obeys $df = -sdT - \mu_0 m dH$. We now obtain m, the thermally averaged magnetic moment per spin. The magnetic moment operator $\hat{\mathbf{M}}$ is

$$\hat{\mathbf{M}} = g\mu_B \sum_i S_i , \tag{5.20}$$

where S_i is short-hand for S_{iz} the z-component of the spin of the ith particle. Since we are expressing the Hamiltonian in terms of the magnetic field in energy units, we correspondingly introduce the dimensionless magnetic moment operator $\mathbf{M}$ by

$$\mathbf{M} \equiv \hat{\mathbf{M}}/(g\mu_B) = \sum_i S_i \ . \tag{5.21}$$

Then the (dimensionless) magnetic moment per spin thermally averaged at temperature T (indicated by $\langle m \rangle_T$) is given by

$$\begin{aligned}
\langle m \rangle_T &= (1/N) \sum_i \langle S_i \rangle_T = \langle S_i \rangle_T = \sum_n \langle n | S_i | n \rangle p_n \\
&= \frac{\sum_{S_i=\pm 1/2} S_i e^{H S_i/(kT)}}{\sum_{S_i=\pm 1/2} e^{H S_i/(kT)}} \\
&= \frac{(1/2)e^{H/(2kT)} + (-1/2)e^{-H/(2kT)}}{e^{H/(2kT)} + e^{-H/(2kT)}} \\
&= (1/2)\tanh[H/(kT)] \ ,
\end{aligned} \tag{5.22}$$

and, the susceptibility per spin (which in our convention has the units of inverse energy) is

$$\chi = \frac{\partial m}{\partial H} = (1/4)\beta \operatorname{sech}^2(\beta H/2) \ .$$

Both $\langle m \rangle_T$ and $\chi(T)$ are shown in Fig. 5.1.

Usually the term "susceptibility" refers to the zero-field susceptibility, which, for this model, is

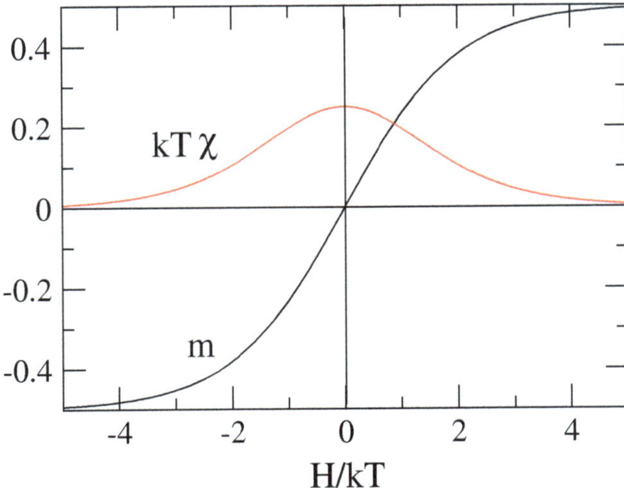

Fig. 5.1 Magnetic moment m, and susceptibility times temperature, $kT\chi$, both per spin for noninteracting $S = 1/2$ spins in a magnetic field H

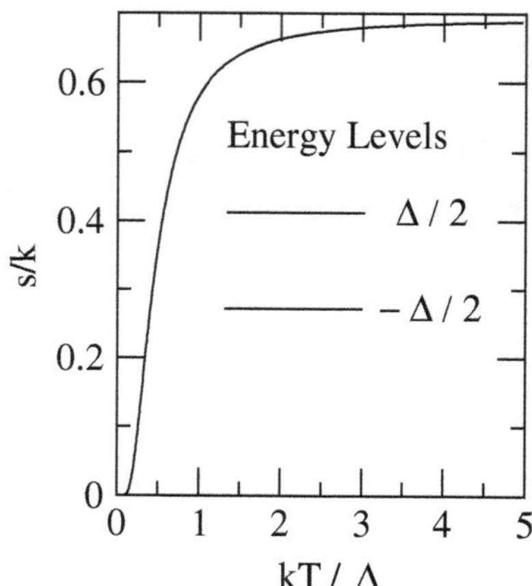

Fig. 5.2 s/k, where s is the entropy for a two-level system as a function of kT/Δ, where Δ is the separation between the two energy levels. (For the magnetic system under consideration $\Delta = H$). Note the limits $s(0) = 0$ and $s(\infty)/k = \ln 2 = 0.693$

$$\chi(H = 0) = \frac{C}{T} , \tag{5.23}$$

where $C = 1/(4k)$. This result, called **Curie's Law**, is exact for noninteracting spins. However, at sufficiently high T, even interacting spins obey Curie's law. Sufficiently high means that the thermal energy kT is large compared to the interaction energy of a spin with its neighbors.

The above results can be generalized to apply to general spin S, as is done in Exercise 1. At zero field and high temperature, one again obtains Curie's Law, Eq. (5.23), but with $C = S(S + 1)/(3k)$.

We obtain the entropy as $\mathcal{S} = (U - F)/T$. Thus, the entropy per spin, s, is given by

$$s = k\left(-(\beta H/2) \tanh(\beta H/2) + \ln 2 + \ln \cosh(\beta H/2) \right) . \tag{5.24}$$

This gives the expected results, i.e., that the entropy per spin at $T = 0$ is zero and at infinite temperature is $k \ln 2$ per spin as shown in Fig. 5.2. Most of the variation of the entropy occurs over a temperature range of order Δ, where $\Delta = H$ is the separation between energy levels. This is a special case of an important general result. If one has a system with a finite number, N_0, of energy levels, then the typical spacing of energy levels, Δ, sets the scale of the temperature range over which the entropy varies, and correspondingly the specific heat has a maximum near the characteristic temperature, $kT \approx \Delta$, Also, in general, the entropy is zero at $T = 0$ and at infinite temperature is $k \ln N_0$.

We may write the result, Eq. (5.17), for the internal energy per spin as

$$u = -(1/2)\Delta \tanh[\Delta/(2kT)] , \tag{5.25}$$

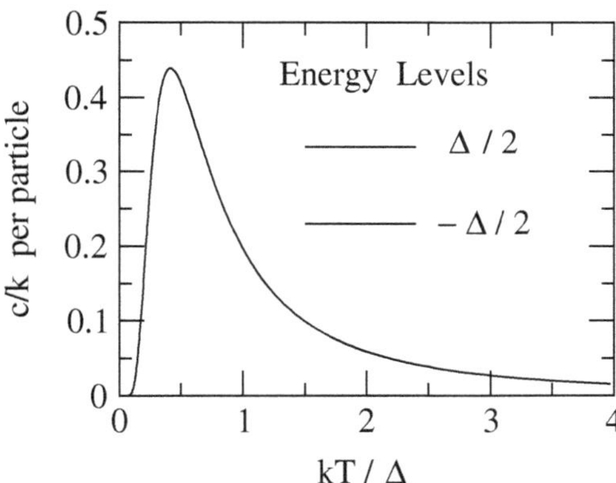

Fig. 5.3 c/k, where c is the specific heat versus kT/Δ for a two-level system

which has the same form as $\langle m \rangle_T$ shown in Fig. 5.1. The specific heat per spin for constant Δ (magnetic field) is

$$\frac{c}{k} = \frac{du}{d(kT)} = \left(\frac{\Delta}{kT}\right)^2 \frac{e^{-\Delta/(kT)}}{\left(1 + e^{-\Delta/(kT)}\right)^2} . \tag{5.26}$$

The result is shown in Fig. 5.3. This specific heat function exhibits a peak, known as a "Schottky anomaly," at around $kT = \Delta/2$. Some general properties of this graph are worth noting. First of all, at very low temperature ($kT \ll \Delta$) the specific heat goes rapidly to zero as the spins all occupy the lower level. At high temperature, the specific heat goes to zero like $1/T^2$ as the two levels become equally populated. Only when $kT \sim \Delta$ will changing the temperature significantly change the populations? So only in this regime will the specific heat be significant? Also note that, because c/k is a dimensionless function of $x = \Delta/kT$, the actual peak value of c/k is the same for nuclear spins with energy levels separated by $\Delta/k = 0.001\,\text{K}$, where it occurs in the milliKelvin range, as it is for atomic spins with $\Delta/k = 1\,\text{K}$ where it occurs around 1 K. In both cases, the maximum specific heat will be about $0.44k$ per spin.

Specific heat measurements can also be used to determine the level structure of systems with three or more levels and are particularly useful in cases where direct spectroscopy, such as magnetic resonance, is not possible (See Exercise 11).

5.4 Specific Heat—Finite-Level Scheme

Here a formula is developed for the specific heat which is especially useful for systems with a bounded energy spectrum. A bounded spectrum is one where the energy per particle is bounded. This specifically excludes cases like the harmonic oscillator where the energy levels extend to infinity.

Consider the specific heat, dU/dT, under conditions where the energies of all levels are fixed. For solids, this would mean working at fixed volume, because changing the volume changes the energy levels, and the specific heat would be C_V. Alternatively, for a magnetic system, fixing the energy levels means fixing the applied field, H. Then the specific heat would be C_H. We denote the specific heat when the energies are held constant by $C_{\{E_n\}}$. Then

$$C_{\{E_n\}} = \left.\frac{\partial U}{\partial T}\right|_{\{E_n\}} = \frac{1}{kT^2}\left[\frac{\sum_n E_n^2 e^{-\beta E_n}}{\sum_n e^{-\beta E_n}} - \left(\frac{\sum_n E_n e^{-\beta E_n}}{\sum_n e^{-\beta E_n}}\right)^2\right], \tag{5.27}$$

which we write as

$$\frac{C_{\{E_n\}}}{k} = \left(\frac{1}{kT}\right)^2 \left(\langle E_n^2\rangle_T - \langle E_n\rangle_T^2\right), \tag{5.28}$$

A simple application of this result occurs for a system with a bounded energy level spectrum in the limit of high temperatures, such that $e^{-\beta E_n} \approx 1$. Then we have

$$\begin{aligned} \frac{C_{\{E_n\}}}{k} &= \left(\frac{\langle E_n^2\rangle_\infty - \langle E_n\rangle_\infty^2}{(kT)^2}\right) \\ &= \frac{1}{(kT)^2}\left[\frac{\mathrm{Tr}\mathcal{H}^2}{\mathrm{Tr}1} - \left(\frac{\mathrm{Tr}\mathcal{H}}{\mathrm{Tr}1}\right)^2\right]. \end{aligned} \tag{5.29}$$

Two comments are in order. First we see that, in the high-temperature limit, the specific heat is proportional to $1/T^2$ with a coefficient that is the mean-square width of the energy spectrum of the system. Second, by expressing the result as a trace, which does not depend on the basis in which the Hamiltonian is calculated, this formula can be applied to cases where it is not possible to calculate the energy levels of the many-body system explicitly, but where the traces are easily performed.

5.5 Harmonic Oscillator

Here, we calculate the thermodynamic functions of a single harmonic oscillator first within classical mechanics and then within quantum mechanics. We then derive results for the thermodynamic functions per oscillator for a system of N identical but distinguishable oscillators. (Identical oscillators can be distinguished, for example, if they are labeled by different lattice sites.)

5.5.1 The Classical Oscillator

In classical mechanics, the partition function, z, for a single oscillator is

$$z = h^{-1} \int_{-\infty}^{\infty} dp \int_{-\infty}^{\infty} dq e^{-\beta \mathcal{H}(p,q)} , \tag{5.30}$$

where $\beta = 1/(k_B T)$ and, as we have seen, Planck's constant h is the appropriate normalization for the classical partition function for one generalized coordinate. Here

$$\mathcal{H} = \frac{p^2}{2m} + \frac{1}{2} k_s x^2 \equiv \frac{p^2}{2m} + \frac{1}{2} m\omega^2 q^2 , \tag{5.31}$$

where ω is the natural frequency of the oscillator: $\omega = \sqrt{k_s/m}$. Then

$$\begin{aligned} z &= h^{-1} \int_{-\infty}^{\infty} dp \int_{-\infty}^{\infty} dq e^{-\beta[p^2/(2m) + \frac{1}{2} m\omega^2 q^2]} \\ &= h^{-1} \sqrt{2m\pi kT} \sqrt{2\pi kT/(m\omega^2)} \\ &= \frac{kT}{\hbar\omega} \end{aligned} \tag{5.32}$$

and, from Eq. (4.46), the internal energy per oscillator, u, is

$$u = -\frac{d \ln z}{d\beta} = -\frac{d}{d\beta}[-\ln(\beta\hbar\omega)] = \beta^{-1} = kT . \tag{5.33}$$

Also we have the free energy per oscillator as

$$f = -kT \ln z = -kT \ln(kT/\hbar\omega) . \tag{5.34}$$

Then the entropy per oscillator, s, is given by

$$s = -\frac{df}{dT} = k \ln(kT/\hbar\omega) + k \tag{5.35}$$

and finally the specific heat per oscillator is

$$c = T\frac{ds}{dT} = \frac{du}{dT} = k . \tag{5.36}$$

There are several difficulties associated with the behavior of the classical oscillator at low temperature. The Third Law is violated in that neither the specific heat nor the entropy vanishes as $T \to 0$. In fact, the entropy not only does not go to zero, but it diverges to minus infinity in the limit of zero temperature. Why does this happen? It happens because in classical mechanics, the energetically accessible volume of phase space can become arbitrarily small as $T \to 0$, whereas in quantum mechanics the minimum number of states the system can populate is 1, as it settles into its ground state, in which case the famous equation $S = k \ln W$ gives $S = k \ln 1 = 0$. The calculation for the classical oscillator in Eqs. (5.32)–(5.36) effectively allows W to go to zero as $T \to 0$. The fact that the classical specific heat remains constant in the zero-temperature limit, also implies a violation of the Third Law, as was discussed in connection with Eq. (3.26).

5.5.2 The Quantum Oscillator

Using the energy levels of the quantum harmonic oscillator, $E_n = (n + \frac{1}{2})\hbar\omega$ where $n = 0, 1, 2, \ldots$, we write z as

$$\begin{aligned} z = \sum_n e^{-\beta E_n} = \sum_n e^{-(n+\frac{1}{2})\hbar\omega/(kT)} &= \frac{e^{-\frac{1}{2}\beta\hbar\omega}}{1 - e^{-\beta\hbar\omega}} \\ &= [2\sinh(\frac{1}{2}\beta\hbar\omega)]^{-1} , \end{aligned} \tag{5.37}$$

from which we get

$$\begin{aligned} u &= -\frac{d \ln z}{d\beta} = \frac{d}{d\beta} \ln \sinh(\frac{1}{2}\beta\hbar\omega) \\ &= \frac{\cosh(\frac{1}{2}\beta\hbar\omega)}{\sinh(\frac{1}{2}\beta\hbar\omega)}[\frac{1}{2}\hbar\omega] = \frac{1}{2}\hbar\omega \coth(\frac{1}{2}\beta\hbar\omega) \end{aligned} \tag{5.38}$$

and

$$f = -kT \ln z = kT \ln 2 + kT \ln \sinh(\frac{1}{2}\beta\hbar\omega) \tag{5.39}$$

so that

$$\begin{aligned} s &= [u - f]/T = k \ln z + (u/T) \\ &= k\left[\frac{\hbar\omega}{2kT}\coth[\frac{1}{2}\hbar\omega/(kT)] - \ln[2\sinh(\frac{1}{2}(\hbar\omega/kT)]\right] . \end{aligned} \tag{5.40}$$

This in turn leads to the result

$$\begin{aligned} c &= \frac{du}{dT} = \frac{1}{2}\hbar\omega\left[-\frac{\hbar\omega}{2kT^2}\right]\left[\frac{\sinh[\frac{1}{2}(\hbar\omega/kT)]}{\sinh[\frac{1}{2}(\hbar\omega/kT)]} - \frac{\cosh^2[\frac{1}{2}(\hbar\omega/kT)]}{\sinh^2[\frac{1}{2}(\hbar\omega/kT)]}\right. \\ &= k\left[\frac{1}{2}\frac{\hbar\omega}{kT}/\sinh[\frac{1}{2}(\hbar\omega/kT)]\right]^2 . \end{aligned} \tag{5.41}$$

Graphs of these results are shown in the accompanying figures (Figs. 5.4 and 5.5).

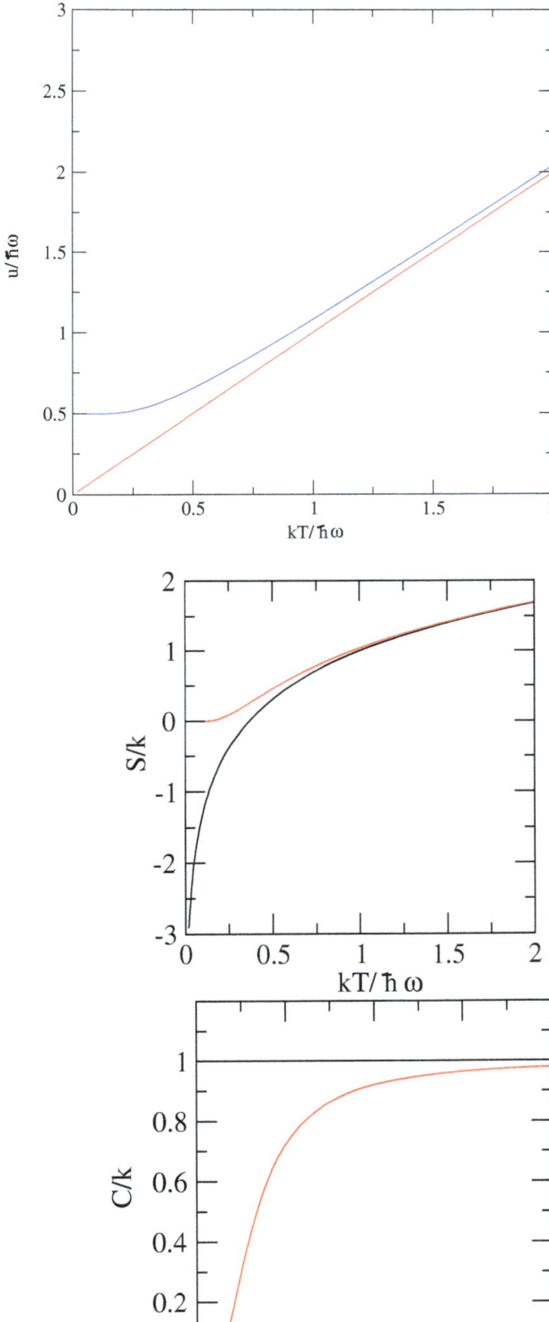

Fig. 5.4 Graphs of $u/(\hbar\omega)$ for the harmonic oscillator according to classical mechanics (lower curve) and quantum mechanics (upper curve)

Fig. 5.5 Left: s/k and right: c/k for the harmonic oscillator. The classical result is the lower curve for the entropy and the upper curve for the specific heat and the quantum result is the other curve

5.5.3 Asymptotic Limits

Here, we get the asymptotic limits of the thermodynamic functions of a quantum harmonic oscillator.

We have

$$u = \frac{1}{2}\hbar\omega\frac{\cosh[\hbar\omega/(2kT)]}{\sinh[\hbar\omega/(2kT)]}\,. \tag{5.42}$$

At high temperature, we recover the classical result:

$$u \sim \frac{1}{2}\hbar\omega\frac{1}{\hbar\omega/(2kT)} = kT\,. \tag{5.43}$$

In the zero-temperature limit

$$u = \frac{1}{2}\hbar\omega\frac{\frac{1}{2}e^{\hbar\omega/(kT)}}{\frac{1}{2}e^{\hbar\omega/(kT)}} = \frac{1}{2}\hbar\omega, \tag{5.44}$$

which is the ground state energy. For the entropy per oscillator at high temperature, we have

$$\begin{aligned} s &= \frac{\hbar\omega}{2T}\frac{1}{\frac{1}{2}\beta\hbar\omega} - k\ln[\hbar\omega/(kT)] \\ &= k + k\ln[kT/(\hbar\omega)]\,. \end{aligned} \tag{5.45}$$

To see that this result makes sense, note that, at temperature T, only states with $n\hbar\omega \le kT$ are significantly occupied. Thus, there are $W \approx kT/(\hbar\omega)$ accessible states and this argument explains that the result of Eq. (5.45) agrees with the famous equation, Entropy, $S = k\ln W$. At zero temperature

$$s = \frac{\hbar\omega}{2T} - k\ln[e^{\hbar\omega/(2kT)}] = 0\,. \tag{5.46}$$

The specific heat per particle at high temperature is

$$c = k\left[\frac{\hbar\omega/(2kT)}{\hbar\omega/(2kT)}\right] = k\,. \tag{5.47}$$

At low temperature

$$c = k\left[\frac{\hbar\omega/(2kT)}{\frac{1}{2}e^{\hbar\omega/(2kT)}}\right]^2 = k\left(\frac{\hbar\omega}{kT}\right)^2 e^{-\hbar\omega/(kT)}\,. \tag{5.48}$$

Note that the high-temperature limits coincide with the classical results, as expected. But the quantum result does not lead to any violation of the Third Law since $s \to 0$ and $c \to 0$ as $T \to 0$ (Fig. 5.5).

5.6 Free Rotator

In this section, we consider a free rotator, which describes the rotation of a symmetric diatomic molecule like N_2.

5.6.1 *Classical Rotator*

We now derive the kinetic energy of rotation for a diatomic molecule of identical atoms, each of mass m, which are separated by a distance d. If the center of the molecule is at the origin, then the atoms are located at

$$\vec{r}_{\pm} = \pm\frac{d}{2}\left(\sin\theta\cos\phi\hat{i} + \sin\theta\sin\phi\hat{j} + \cos\theta\hat{k}\right) . \tag{5.49}$$

One can use this representation to evaluate the velocity vector. One then finds the total kinetic energy K of the two atoms to be

$$K = m(d/2)^2[(\dot{\theta})^2 + \sin^2\theta(\dot{\phi})^2] . \tag{5.50}$$

The generalized momenta are

$$\begin{aligned} p_\theta &= dK/d\dot{\theta} = (md^2/2)\dot{\theta} \\ p_\phi &= dK/d\dot{\phi} = (md^2/2)\sin^2\theta\dot{\phi} \end{aligned} \tag{5.51}$$

so that

$$\begin{aligned} K &= [p_\theta^2 + (p_\phi^2)/\sin^2\theta]/(md^2) . \\ &= [p_\theta^2 + (p_\phi^2)/\sin^2\theta]/(2I) , \end{aligned} \tag{5.52}$$

where the moment of inertia is $I = 2m(d/2)^2$. The single-molecule partition function is then

$$\begin{aligned} z &= \frac{1}{2h^2}\int_{-\infty}^{\infty} dp_\theta \int_{-\infty}^{\infty} dp_\phi \int_0^\pi d\theta \int_0^{2\pi} d\phi e^{[p_\theta^2 + \sin^{-2}\theta p_\phi^2]/(2IkT)} \\ &= (I\pi kT/h^2)\int_0^\pi \sin\theta d\theta \int_0^{2\pi} d\phi = 2\pi(2I\pi kT)/h^2 . \end{aligned} \tag{5.53}$$

The factor h^{-2} is appropriate for a system of two generalized coordinates and momenta. The factor of 1/2 is included because the integration over classical states counts configurations in which the positions of the two atoms are interchanged, whereas we know that quantum mechanics takes account of indistinguishability. (This factor of 2 is asymptotically correct in the high-temperature limit.)

The free energy per molecule is

$$f = -kT \ln[4I\pi^2 kT/h^2] , \tag{5.54}$$

from which we obtain the other thermodynamic functions per molecule as

$$\begin{aligned} s &= k \ln[4I\pi^2 kT/h^2] + k \\ u &= kT \\ c &\equiv du/dT = k \, . \end{aligned} \tag{5.55}$$

As before these results conflict with the Third Law because the specific heat does not vanish and the entropy diverges to minus infinity in the zero-temperature limit.

5.6.2 *Quantum Rotator*

To analyze the analogous quantum problem, we calculate the energy levels of the quantum rotator, obtained from the eigenvalues of the Schrodinger equation:

$$-\frac{\hbar^2 (\hat{\nabla})^2}{2I}\psi = E\psi , \tag{5.56}$$

where $(\hat{\nabla})^2$ is the angular part of $r^2\nabla^2$. As is well known, the eigenfunctions of this Hamiltonian are the spherical harmonics $Y_J^m(\theta, \phi)$, where m can assume any of the $2J + 1$ values, $m = -J, -J + 1, \dots J - 1, J$ and the associated energy eigenvalue is $J(J + 1)\hbar^2/(2I)$. Next we need to consider that the two atoms in this molecule are identical. For simplicity, we assume that they are spin zero particles. Then the manifold of nuclear spin states consists of a single state. But more importantly, for integer spin atoms, the wave function must be even under interchange of particles. This requirement means that we should only allow spherical harmonics of even J. Thus we find

$$\begin{aligned} z &= \sum_{J=0,2,4\dots} \sum_{m=-J}^{J} e^{-J(J+1)\hbar^2/(2IkT)} \\ &= \sum_{J=0,2,4,\dots} (2J+1) e^{-J(J+1)\hbar^2/(2IkT)} \, . \end{aligned} \tag{5.57}$$

At high temperature, this can be approximated by an integral:

$$\begin{aligned} z &\approx \frac{1}{2}\int_0^\infty dJ(2J+1)e^{-J(J+1)\hbar^2/(2IkT)} \\ &= (IkT/\hbar^2) \,. \end{aligned} \tag{5.58}$$

(For a more accurate evaluation, one can use the Poisson summation formula, as is discussed by Pathria). Equation (5.58) agrees with the classical result in the limit of high temperature.

At very low temperature, we keep only the first two terms in the sum:

$$z \approx 1 + 5e^{-3\hbar^2/(IkT)} \,. \tag{5.59}$$

Then

$$\begin{aligned} f &= -kT\ln z = -kT\ln\left(1 + 5e^{-3\hbar^2/(IkT)}\right) \\ &\approx -5kTe^{-3\hbar^2/(IkT)} \,. \end{aligned} \tag{5.60}$$

Also

$$u = -d\ln z/d\beta \approx 15(\hbar^2/I)e^{-3\hbar^2/(IkT)} \,. \tag{5.61}$$

From these results we get

$$c/k = du/d(kT) = 45[\hbar^2/(IkT)]^2 e^{-3\hbar^2/(IkT)} \tag{5.62}$$

and

$$s/k = -df/d(kT) \approx 15\left(\frac{\hbar^2}{IkT}\right)e^{-3\hbar^2/(IkT)} \,. \tag{5.63}$$

From these results, one can see that the Third Law is satisfied. The entropy goes to zero as $T \to 0$ and, consistent with that, the specific heat goes to zero as $T \to 0$. All the thermodynamic functions have a temperature dependence at low temperature indicative of the energy gap $E(J=2) - E(J=0) = 3\hbar^2/I$ and hence involve the factor $e^{-3\hbar^2/(IkT)}$.

5.7 Grüneisen Law

In this section, we make some general observation concerning scaling properties of the energy. First, we consider systems for which all energy levels have the same dependence on volume. This situation arises in several different contexts. For example, for free nonrelativistic particles,

$$E_{l,m,n} = \frac{h^2(l^2 + m^2 + n^2)}{8mL^2} , \quad (5.64)$$

so that

$$E_\alpha(V) = e_\alpha V^{-x}, \quad (5.65)$$

where $x = 2/3$ and e_α is volume-independent. As another example, consider the relativistic limit where $E = \hbar ck$, with $k = 2\pi\sqrt{l^2 + m^2 + n^2}/L$, so that we again have Eq. (5.65), but now with $x = 1/3$. This result also applies to a system of photons in a box. In condensed matter physics, one often deals with models in which only nearest neighboring atoms are coupled. Then, the energy levels scale with this interaction which has its own special volume dependence with an associated value of x. This will be explored in a homework problem.

If we assume Eq. (5.65), then the partition function is

$$Z = \sum_{e_\alpha} e^{-\beta e_\alpha V^{-x}} , \quad (5.66)$$

so that

$$Z = G(\beta V^{-x}) , \quad (5.67)$$

where G is a function whose form we do not necessarily know. Then

$$F = -kT \ln G(\beta V^{-x}) \quad (5.68)$$

and

$$U = -\left.\frac{\partial \ln Z}{\partial \beta}\right|_V = -V^{-x} G'(\beta V^{-x})/G(\beta V^{-x}) . \quad (5.69)$$

But also

$$p = -\left.\frac{\partial F}{\partial V}\right|_T = kT[-x\beta V^{-x-1}]G'(\beta V^{-x})/G(\beta V^{-x}) . \quad (5.70)$$

Comparing these two results tells us that

$$pV = xU . \quad (5.71)$$

This implies that

$$C_V \equiv \left.\frac{\partial U}{\partial T}\right|_V = (1/x)V\left.\frac{\partial p}{\partial T}\right|_V ,$$

which is often written as

$$\left.\frac{\partial p}{\partial T}\right|_V = \gamma\frac{C_V}{V} , \tag{5.72}$$

where, in this case, $\gamma = x \equiv -d \ln E_\alpha / d \ln V$. (In condensed matter physics, this constant γ is called the Grüneisen constant).

One can express this result in more general terms if we interpret Eq. (5.64) as defining a characteristic energy $h^2/(8\,\mathrm{mL}^2)$ which defines a characteristic temperature T_0 with $kT_0 = h^2/(8\,\mathrm{mL}^2)$ for the case of noninteracting nonrelativistic particles in a box. We are then led to consider the case where the partition function is of the form $Z = Z(T/T_0)$, where T_0 is fixed by a characteristic energy in the Hamiltonian. Observe that the partition function for such a system of particles is a function of T/T_0 irrespective of what statistics due to indistinguishability we impose on the particles. Whether the particles are bosons or fermions or classically distinguishable is reflected in the *form* of the function $Z(T/T_0)$, but it does *not* change the fact that the partition function is a function of the scaled variable T/T_0. (Even if the condition that Z is a function of T/T_0 is only obeyed approximately the remarks we are about to make will still be qualitatively correct). One sees that roughly speaking T_0 is the dividing line between low and high temperatures. In the low-temperature limit, the entropy will go to zero and the energy will go to the ground state energy. In the high-temperature regime, the system will be close to that with no interactions. Indeed, if the system has a phase transition, then the critical temperature must be of order T_0. In any event, the specific heat will show a strong temperature dependence for T near T_0. One can see the truth of these statements in the case of the two-level systems, the harmonic oscillator, and the free rotator, which were considered above.

5.8 Summary

In this chapter, we have studied the thermodynamic properties of several simple noninteracting systems. A unifying theme of this chapter is that when kT is large compared to the coupling constants in the Hamiltonian, then one has the classical high-temperature regime in which typically some analog of the equipartition theorem (which says that the energy per quadratic degree of freedom is $\frac{1}{2}$ kT) applies. In the low-temperature limit, one must use quantum mechanics to describe the system in order to obtain a vanishing specific heat at zero temperature and to be consistent with the Third Law of Thermodynamics.

Two additional specific results of this chapter are as follows: (a) The partition function of a system consisting of noninteracting subsystems is given by the product over the partition functions of the individual subsystems. (b) When all energy levels have the same volume dependence ($E_\alpha = e_\alpha V^{-x}$), then $pV = xU$ and $\partial p/\partial T)_V = xC_V/V$. This result is independent of whether the particles are treated classically or obey Fermi or Bose quantum statistics.

5.9 Exercises

1. This problem concerns the thermodynamics of a system of N spins each localized to a lattice site. The Hamiltonian is then

$$\mathcal{H} = -\sum_{i=1}^{N} g\mu_B B S_{iz} ,$$

where $g = 2$, μ_B is the Bohr magneton ($\mu_B = 0.927 \times 10^{-20}$ ergs/Gauss) and $S_{iz} = -S, -S + 1 \ldots, S - 1, S$.

(a) Obtain expressions for the $S/(Nk)$, where S is the entropy, for $m \equiv M/(Ng\mu_B)$, where M is the thermal average of the total magnetic moment, and for $C/(Nk)$, where C is the specific heat (at constant B). *Note* the result for m evaluated for spin S defines a function $B_S(x)$, where $x = g\mu_B H/(kT)$ which is called the Brillouin function after the French physicist L. Brillouin.

(b) Obtain asymptotic expansions for these functions correct to order e^{-x} at low temperature and correct to order x^2 for high temperature, where $x = g\mu_B B/(kT)$.

(c) Plot these quantities (for $S = 1/2$, $S = 2$, and $S = 10$) as functions of the variable $g\mu_B B/(kT)$.

2. This problem concerns the specific heat which arises from the energy levels associated with the nuclear spins of the rare earth solid (Lu = Lutetium) at low temperature. This solid may be viewed as a collection of nuclei, each of which has nuclear spin I and thus, in a magnetic field, has $2I + 1$ equally spaced energy levels. As you know, if I is integral (as for ^{176}Lu which has $I = 7$), these nuclei are bosons, whereas if I is half-integral (as for ^{175}Lu which has $I = 7/2$), these nuclei are fermions. Now previously you derived the thermodynamic functions for similar systems. In particular, you didn't worry about whether these particles are distinguishable or not and you found that in a magnetic field H, their thermal average magnetization was given by a Brillouin function (see Exercise 1). Is it reasonable to use these results for the Lu systems referred to above? If not, what major modifications would you make in the calculations.

3. Calculate the partition function for a classical monatomic ideal gas in a uniform gravitational field in a column of cross-sectional area A. The top and bottom of the column are at heights h_1 and h_2, respectively.

(a) Find the pressure (i) at the top of the column and (ii) at the bottom of the column. Hint: the solution may be viewed as involving an application of the Hellmann–Feynman theorem.

(b) Give a physical interpretation of the magnitude of the difference between these two pressures.

(c) Estimate the pressure at the top of a 2000 m high mountain at a temperature about equal to room temperature.

4. Consider the nuclear spin system of solid nitrogen. The solid consists almost entirely of the isotope nitrogen 14, which has nuclear spin 1. Suppose these spins are subjected to electric and magnetic fields such that the three energy levels of each nuclear spin occur at $E_1 = -2\Delta/3$, $E_2 = E_3 = \Delta/3$. Calculate the nuclear spin specific heat and give its asymptotic forms for (i) $kT \ll \Delta$ and (ii) $kT \gg \Delta$.

5. This problem concerns the specific heat of the gas HCl which we treat as an ideal gas. You are to take into account electronic, vibrational, and rotational degrees of freedom, assuming these degrees of freedom do not interact with one another. The separation between H and Cl atoms in this molecule is about 13 nm. Spectroscopic studies of the vibrational levels indicate that adjacent levels are nearly equally spaced and the energy difference ΔE between adjacent vibrational levels is such that $\Delta E/(hc) = 3000\,\text{cm}^{-1}$.

(a) In what range of temperature do you expect the specific heat due to molecular rotations to be temperature-dependent?

(b) In what range of temperature do you expect the specific heat due to molecular vibrations to be temperature-dependent.

(c) Estimate the specific heat at the temperatures, $T = 1\,\text{K}$, $T = 50\,\text{K}$, and $T = 500\,\text{K}$. (In each case, assume the gas is at low enough density that it can be considered to be ideal).

6. A classical ideal gas consists of diatomic molecules. Each atom in this molecule has mass m. From measurements on the molecule in the solid phase, it is found that the interatomic potential $\phi(r)$ as a function of the interatomic separation r has the form shown in Fig. 5.6, where $\Delta \gg kT$. Assume that near the minima at $r = r_1$ and $r = r_2$ the potential takes the form

$$\phi(r) = E_0 + \frac{1}{2}k_i(r - r_i)^2\,, \qquad \text{for } (r - r_i)^2 < \Delta\,.$$

Assume that $m/m_p \gg 1$, where m_p is the proton mass. (This condition suggests that it is correct to treat the atom in the interatomic potential $\phi(r)$ classically.)

Find the thermal equilibrium value of the ratio N_1/N_2 as a function of temperature, where N_i is the number of molecules in which the mean interatomic spacing is r_i.

7. As a model of an atom with an electron which can be removed when the atom is ionized, we assume that the electron can be attached to the atom in a single bound

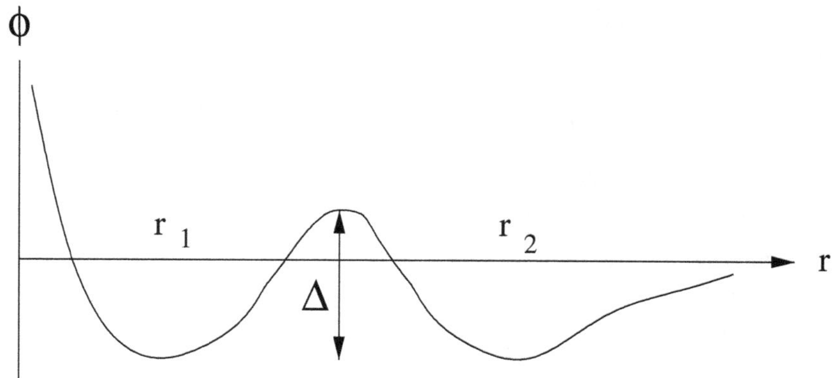

Fig. 5.6 Interatomic potential $\phi(r)$ for Exercise 6

state with energy $-3\,\text{eV}$, or in a "continuum" of states which are those of an electron in a box of volume $5\,\text{cm}^3$. Find an expression which determines the temperature to which the system must be heated in order for the probability of ionization to reach 1/2 and give an order of magnitude estimate of this temperature.

8. In this problem, you are to approximate the Hamiltonian for the nuclear spin system of solid antiferromagnetic ^{3}He by a Heisenberg model

$$\mathcal{H} = J(V) \sum_{\langle ij \rangle} \mathbf{I}_i \cdot \mathbf{I}_j \,,$$

where I_i is the nuclear spin (of magnitude 1/2) of the ith He atom and the sum is restricted to nearest neighbors in the solid. The coupling constant $J(V)$ varies as a high inverse power (maybe 5 or so) of the volume. Because the interaction is so small $J/k_B \approx 0.001\,\text{K}$, it was not convenient (in the early days of 1960) to measure the nuclear specific heat and separate this specific heat from other contributions to the specific heat. Derive a relation between the nuclear spin specific heat from the above Hamiltonian and $\partial p/\partial T)_V$ and discuss factors that make the measurement of this quantity much easier that of the specific heat. *Note* Do not attempt to evaluate the energy levels of $\mathcal{H}$.

9. A quantum mechanical harmonic oscillator is described by the Hamiltonian

$$\mathcal{H} = \frac{p^2}{2m} + \frac{1}{2} m\omega^2 x^2 \,,$$

where m is the mass, p the momentum, and x the position of the particle, and $k = m\omega^2$ is the spring constant. An impenetrable partition is clamped at $x = 0$, and the particle is placed in the half-space $x > 0$.

(a) Find all the energy levels and also the ground state wave function of the harmonic oscillator constrained to the half-space $x > 0$.

(b) Find the free energy, the internal energy, and the entropy of N such half-space oscillators in equilibrium at temperature T.

(c) Find the pressure exerted on the partition at $x = 0$ by N half-space oscillators in equilibrium at temperature T.

(d) The oscillators in part A are thermally insulated from the outside world and are in equilibrium at temperature T. The partition is removed instantaneously. When equilibrium is restored, the temperature is T_1. Relate T_1 to T and give the limiting forms for T_1 when $\hbar\omega/kT \gg 1$ and when $\hbar\omega/kT \ll 1$.

(e) Suppose the oscillators in part A are thermal insulated from the outside world and are in equilibrium at temperature T. Now the partition is quasistatically moved to $x = -\infty$ and the resulting temperature is T_2. Find T_2 in terms of T.

10. The objective of this problem is to calculate the entropy of methane (CH_4) gas as a function of temperature, for temperatures near room temperature. You are to treat this gas as an ideal gas and you are also to neglect intramolecular vibrations, since their energy is large compared to kT in the regime we consider. Thus you have to consider contributions to the entropy from the nuclear spin $S = 1/2$ of each of the four identical and indistinguishable protons, molecular rotations of the tetrahedral molecule, and their translations treated as an ideal gas. You are to construct and then evaluate the classical partition function which will provide an excellent approximation to the quantum partition function for temperatures near 300 K. From this evaluation, it is easy to get the entropy. This problem requires understanding how to normalize the classical partition function. To avoid busy work, we give the kinetic energy of rotation K in terms of the Euler angles θ, ϕ and χ:

$$K = \frac{1}{2}I\left(\dot{\theta}^2 + \dot{\phi}^2 + \dot{\chi}^2 + 2\cos\theta\dot{\phi}\dot{\chi}\right),$$

where I is the moment of inertia of the molecule about any axis, $0 \leq \theta \leq \pi$ and $0 \leq \phi \leq 2\pi$ specify the orientation of an axis fixed in the molecule and $0 \leq \chi \leq 2\pi$ describes rotations about this axis.

11. An atom (of spin 1/2) is confined in a potential well, such that its energy levels are given by

$$E(n, \sigma) = [n + (1/2)]\hbar\omega + \epsilon\sigma\ ,$$

where n assumes the values 0, 1, 2, 3, etc. and $\sigma = \pm 1/2$. Here $\hbar\omega/k = 100\,\text{K}$ and $\epsilon/k = 0.01\,\text{K}$.

(a) Estimate the entropy of this system at a temperature of 1 K.

(b) Estimate the specific heat of this system at a temperature of 1000 K.

(c) Draw a qualitatively correct graph of the specific heat versus temperature.

(d) How, if at all, would the answers to parts B and C be modified if ϵ were allowed to depend on n, so that $\epsilon_n = f_n\delta$, where $\delta/k = 0.01\,\text{K}$ and $|f_n| < 1$ for all n.

12. The specific heat is measured for a system in which at each of the N lattice sites there is a localized spin S. These spins are assumed not to interact with one another, but the value of S is not a priori known. Assume that other contributions to the specific heat (for instance, from lattice vibrations, from nuclear spin degrees of freedom, from intramolecular vibrations, etc.) have been subtracted from the experimental values, so that the tabulated values (see graph and also tables) do represent the specific heat due to these localized spins.

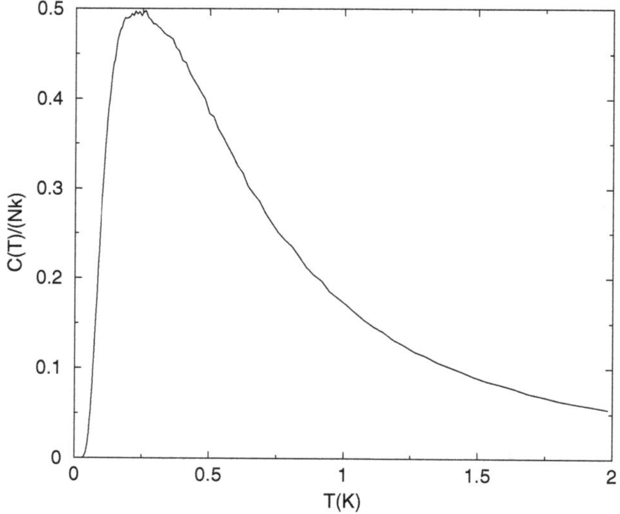

(The experimental data points have an uncertainty of order 1%). From these data points, you are to determine the energy level scheme of the localized spins. *Note* do NOT assume equally spaced levels. (The levels are NOT equally spaced.) This experimental data determines the actual energy level scheme. The graph shows the data up to $T = 2\,\text{K}$. A data table for temperatures up to $T = 12\,\text{K}$ is given below.

T(K)	$C/(Nk)$	T(K)	$C/(Nk)$	T(K)	$C/(Nk)$	T(K)	$C/(Nk)$
0.020000	0.0000008	0.099783	0.2827111	0.497836	0.3834198	2.483784	0.0360690
0.020653	0.0000015	0.103043	0.2968393	0.514099	0.3801352	2.564923	0.0343846
0.021328	0.0000025	0.106409	0.3130827	0.530893	0.3668861	2.648713	0.0322430
0.022025	0.0000043	0.109885	0.3299256	0.548236	0.3577685	2.735241	0.0300176
0.022744	0.0000071	0.113475	0.3480803	0.566146	0.3473624	2.824594	0.0284140
0.023487	0.0000118	0.117182	0.3609010	0.584640	0.3371688	2.916867	0.0265915
0.024255	0.0000188	0.121010	0.3778068	0.603739	0.3252246	3.012154	0.0253712
0.025047	0.0000297	0.124963	0.3899918	0.623462	0.3184802	3.110554	0.0237769
0.025865	0.0000462	0.129045	0.4008858	0.643829	0.3032634	3.212168	0.0221280
0.026710	0.0000706	0.133261	0.4123601	0.664861	0.2942510	3.317102	0.0209717
0.027583	0.0001057	0.137614	0.4276451	0.686581	0.2861436	3.425464	0.0196095
0.028484	0.0001576	0.142110	0.4391892	0.709009	0.2723245	3.537366	0.0186852
0.029414	0.0002293	0.146752	0.4429849	0.732171	0.2613311	3.652923	0.0173894
0.030375	0.0003327	0.151546	0.4536979	0.756089	0.2508469	3.772256	0.0165627
0.031367	0.0004677	0.156497	0.4651859	0.780789	0.2424259	3.895486	0.0152948
0.032392	0.0006576	0.161609	0.4705352	0.806296	0.2352039	4.022742	0.0145468
0.033450	0.0009153	0.166889	0.4770390	0.832635	0.2245597	4.154156	0.0136348
0.034543	0.0012582	0.172341	0.4787056	0.859836	0.2126822	4.289862	0.0128795
0.035671	0.0017057	0.177970	0.4855111	0.887924	0.2038651	4.430002	0.0120831
0.036837	0.0022771	0.183784	0.4895375	0.916931	0.1972051	4.574719	0.0112006
0.038040	0.0029982	0.189788	0.4892532	0.946885	0.1854235	4.724165	0.0105749
0.039283	0.0038953	0.195988	0.4895402	0.977817	0.1783006	4.878492	0.0099914
0.040566	0.0050322	0.202391	0.4914167	1.009760	0.1708468	5.037861	0.0093450
0.041891	0.0064886	0.209002	0.4940131	1.042747	0.1621445	5.202435	0.0087172
0.043260	0.0083117	0.215830	0.4923753	1.076811	0.1536212	5.372387	0.0082237

Values of C/Nk where C is the specific heat
The columns continue on the next page

13. The objective of this problem is to characterize adiabats for a system whose energy levels all scale the same way in terms of a parameter such as the volume V or the externally applied field H. An adiabatic process is one in which the entropy, given by $\sum_n P_n \ln P_n$, is constant. You might imagine that this quantity could remain constant by readjusting the P_n's so that the entropy functional remains constant. However, since we know that the actual P_n's which maximize the entropy are proportional to $\exp(-\beta E_n)$, we see that when all energies are modified by the same factor in an adiabatic process, the E_n' s have to remain constant.

(a) Consider a monatomic gas for which the Hamiltonian has eigenvalues given by Eq. (4.20). Use the fact that the E_n's remain constant to obtain a combination of thermodynamic variables such as p, V, or T which is constant in an adiabatic process and thereby obtain the equation of state for an adiabat of a monatomic ideal gas.

(b) Consider a paramagnet whose Hamiltonian is governed

$$\mathcal{H} = -H \sum_i S_i ,$$

T(K)	$C/(Nk)$	T(K)	$C/)Nk)$	T(K)	$C/(Nk)$	T(K)	$C/(Nk)$
0.044673	0.0103277	0.222880	0.4971782	1.111987	0.1459560	5.547890	0.0078316
0.046132	0.0130206	0.230161	0.4940729	1.148313	0.1403013	5.729126	0.0072250
0.047639	0.0158379	0.237680	0.4962512	1.185826	0.1313880	5.916283	0.0068334
0.049196	0.0193769	0.245445	0.4923761	1.224564	0.1252891	6.109554	0.0063957
0.050803	0.0233630	0.253463	0.4963020	1.264568	0.1183531	6.309138	0.0060533
0.052462	0.0280979	0.261743	0.4973703	1.305878	0.1138101	6.515243	0.0056482
0.054176	0.0339797	0.270293	0.4898964	1.348538	0.1068072	6.728080	0.0053695
0.055946	0.0401908	0.279123	0.4877608	1.392592	0.1017086	6.947871	0.0050176
0.057773	0.0474129	0.288241	0.4829935	1.438084	0.0967984	7.174841	0.0046455
0.059661	0.0550240	0.297658	0.4828484	1.485063	0.0907482	7.409226	0.0043817
0.061610	0.0640074	0.307381	0.4797770	1.533577	0.0857339	7.651268	0.0041194
0.063622	0.0733513	0.317423	0.4773291	1.583675	0.0818454	7.901217	0.0039170
0.065701	0.0836536	0.327792	0.4733523	1.635410	0.0775025	8.159331	0.0036244
0.067847	0.0954168	0.338500	0.4705285	1.688835	0.0723351	8.425876	0.0034479
0.070063	0.1067401	0.349558	0.4690031	1.744005	0.0684903	8.701130	0.0031963
0.072352	0.1205790	0.360978	0.4661604	1.800977	0.0644831	8.985375	0.0030213
0.074716	0.1356305	0.372770	0.4569451	1.859811	0.0611529	9.278906	0.0028042
0.077157	0.1496807	0.384947	0.4533222	1.920567	0.0584730	9.582026	0.0026445
0.079677	0.1635272	0.397523	0.4429092	1.983307	0.0551150	9.895047	0.0024841
0.082280	0.1791180	0.410509	0.4405149	2.048097	0.0511450	10.218295	0.0023452
0.084968	0.1978617	0.423919	0.4293444	2.115003	0.0485064	10.552102	0.0021880
0.087744	0.2144753	0.437768	0.4219385	2.184095	0.0457462	10.896815	0.0020500
0.090610	0.2303046	0.452068	0.4153853	2.255445	0.0434335	11.252788	0.0019177
0.093570	0.2478409	0.466836	0.4071684	2.329125	0.0405494	11.620389	0.0017999
0.096627	0.2651742	0.482087	0.3999674	2.405212	0.0384512	12.000000	0.0017067

where S_i assumes the values $-S$, $-S+1, \ldots, S-1, S$. Use the fact that the E_n's remain constant to obtain a combination of thermodynamic variables such as M, H, or T which is constant in an adiabatic process and thereby obtain the equation of state (giving M in terms of H and T) for this idealized paramagnet. Why is the term "adiabatic demagnetization" which is used to describe the adiabatic process to achieve low temperatures somewhat misleading?

Chapter 6
Basic Principles (Continued)

6.1 Grand Canonical Partition Function

As we have already seen, the indistinguishability of particles is a complicating factor which arises when one calculates the partition function for systems of identical particles. From a mathematical point of view, it turns out to be simpler in this case to treat a system, not with a fixed number of particles, but rather one at fixed chemical potential, μ. Physically, μ may be interpreted as the free energy for adding a particle to the system at fixed volume and temperature. μ may also be viewed as the variable conjugate to N, the number of particles, in the same way as $\beta = 1/(kT)$ is conjugate to the energy.

In this chapter, we address the enumeration of the states of a system in which the number of particles can fluctuate For this purpose, we define a partition function at fixed chemical potential which, from a physical point of view, is equivalent to considering the statistical properties of a system in contact with a large reservoir with which exchange of both energy and particles is permitted. To do this, we generalize the construction of Sect. 4.3 of the previous chapter on basic principles.

We consider a system A in contact with an infinitely large reservoir R with which it can exchange both energy and particles. The combined system has fixed volume V, fixed particle number, $N_{\rm TOT}$, and fixed energy $E_{\rm TOT}$. We know that all states of the combined system are equally probable. Consider now the probability, $P(E_n, N_A)$, that system A has N_A particles and is in its nth energy eigenstate with energy E_n. This probability is proportional to the number of states of the reservoir that are consistent with this. Thus

$$P(E_n, N_A) \propto \Omega_R(E_{\rm TOT} - E_n, N_{\rm TOT} - N_A)\,, \tag{6.1}$$

where $\Omega_R(E, N)$ is the density of states of the reservoir at energy E when it has N particles. We wish to expand Eq. (6.1) in powers of E_n and N_A. Such an expansion of Ω_R converges poorly. As before, to improve the rate of convergence, we write

A. J. Berlinsky and A. B. Harris, *Statistical Mechanics*, Graduate Texts in Physics,
https://doi.org/10.1007/978-3-030-28187-8_6

$$
\begin{aligned}
P(N_A, n) &\propto \exp\Big(\ln \Omega_R(E_{\rm TOT} - E_n, N_{\rm TOT} - N_A)\Big) \\
&= \exp\Big(S_R(E_{\rm TOT} - E_n, N_{\rm TOT} - N_A)/k\Big) ,
\end{aligned}
\tag{6.2}
$$

where S_R is the entropy of the reservoir and we used our previous identification of $k \ln \Omega$ with the entropy. Then the expansion takes the form

$$
\begin{aligned}
S(E_{\rm TOT} - E_n, N_{\rm TOT} - N_A) &= S(E_{\rm TOT}, N_{\rm TOT}) \\
&\quad - E_n \left.\frac{\partial S}{\partial E}\right)_{N,V} - N_A \left.\frac{\partial S}{\partial N}\right)_{E,V} + \dots \\
&= S(E_{\rm TOT}, N_{\rm TOT}) - E_n/T + \mu N_A/T ,
\end{aligned}
\tag{6.3}
$$

where we truncated the expansion because higher order terms are of order the ratio of the size of the system A to the size of the reservoir R. Also, we used the identification of the partial derivatives of S which follow from $dU = TdS - PdV + \mu dN$. This argument indicates that when a system of constant volume V is in contact with a thermal and particle reservoir at temperature T and chemical potential μ, then we have that

$$
P\left(E_n, N_A\right) = \frac{e^{-\beta(E_n - \mu N_A)}}{\Xi} ,
\tag{6.4}
$$

where we have introduced the grand canonical partition function Ξ as

$$
\Xi(T, V, \mu) \equiv {\rm Tr}_{E,\hat{N}} e^{-\beta(\mathcal{H} - \mu \hat{N})} = \sum_N e^{\beta \mu N} Z(T, V, N) ,
\tag{6.5}
$$

where the first equality reminds us that the trace is over both energy eigenstates and number of particles and $\hat{N}$ is the number operator. We can also define the grand canonical density matrix as

$$
\rho = \frac{e^{-\beta(\mathcal{H} - \mu \hat{N})}}{{\rm Tr}_{\hat{N},E} e^{-\beta(\mathcal{H} - \mu \hat{N})}} .
\tag{6.6}
$$

We now derive the relation between the grand canonical partition function Ξ and the thermodynamic functions. Using Eq. (6.5), we can calculate the average number of particles, $\langle \hat{N} \rangle$ directly from the grand partition function as

$$
\langle \hat{N} \rangle = -kT \frac{\partial \ln \Xi(T, V, \mu)}{\partial \mu} = -\frac{\sum_N N e^{\beta \mu N} Z(T, V, N)}{\Xi} .
\tag{6.7}
$$

So if we define the grand potential, $\Omega(T, V, \mu)$ as

$$
\Omega(T, V, \mu) = -kT \ln \Xi(T, V, \mu),
\tag{6.8}
$$

then $\Omega(T, V, \mu)$ satisfies

$$\langle \hat{N} \rangle = -\frac{\partial \Omega(T, V, \mu)}{\partial \mu}. \tag{6.9}$$

This is exactly what one would expect if $\Omega(T, V, \mu)$ is the Legendre transform of $F(T, V, N)$. That is, if

$$\Omega(T, V, \mu) = F(T, V, N) - N\left.\frac{\partial F}{\partial N}\right|_{T,V} = F(T, V, N) - \mu N. \tag{6.10}$$

This implies that

$$\begin{aligned} d\Omega &= (-SdT - PdV + \mu dN) - \mu dN - N d\mu \\ &= -SdT - PdV - N d\mu. \end{aligned} \tag{6.11}$$

The conclusion is that the statistical distribution defined by Eq. (6.5) allows us to calculate a thermodynamic potential function, $\Omega(T, V, \mu)$ which is the Legendre transform of $F(T, V, N)$ with respect to the variables N and μ. Knowledge of $\Omega(T, V, \mu)$ then enables us to compute

$$S = -\frac{\partial \Omega(T, V, \mu)}{\partial T}, \quad P = -\frac{\partial \Omega(T, V, \mu)}{\partial V}, \quad N = -\frac{\partial \Omega(T, V, \mu)}{\partial \mu}, \tag{6.12}$$

for a system in equilibrium with an energy and particle bath.

6.2 The Fixed Pressure Partition Function

The final partition function we wish to introduce is the partition function, Υ, at fixed pressure, or at fixed force, with temperature T and particle number N also fixed. Here, the trace will involve an integration over all energy eigenstates and all possible volumes. A standard way to carry out the volume integration is to place the system in a container all of whose walls are fixed except for a movable ceiling which has mass M. The pressure is then clearly $P \equiv Mg/A$, where A is the cross-sectional area of the container. In this scenario, the integration over the volume is replaced by an integration over the height, h, of the ceiling. In an exercise, the thermally averaged end-to-end length of a polymer in one dimension is calculated using a constant force partition function. The analog of the Hamiltonian for Υ at fixed pressure P, temperature, T, and particle number, N, is $\Phi = \mathcal{H} + PV$, and the "trace" is an integral over all volumes and a sum over all energy eigenstates consistent with this volume.

6.3 Grand and Fixed Pressure Partition Functions for a Classical Ideal Gas

Here, we present calculations of the thermodynamic properties of a classical ideal gas using the two partition functions which were just introduced, namely, the grand canonical partition function and the constant pressure partition function.

6.3.1 *Grand Partition Function of a Classical Ideal Gas*

The grand partition function, Ξ, is a trace over all possible values of N, the number of particles, and of energies E_n:

$$\Xi(T, V, \mu) = \mathrm{Tr}_{\hat{N},\mathcal{H}} e^{-\beta(\mathcal{H}-\mu\hat{N})} = \sum_N e^{\beta\mu\hat{N}} Z(N, T, V) . \tag{6.13}$$

We use the previous result, Eq. (4.55), for $Z(N, T, V)$:

$$Z(N, T, V) = \frac{1}{N!}\left[\frac{V}{h^3}(2m\pi kT)^{3/2}\right]^N \equiv \frac{1}{N!}\left(\frac{V}{\lambda^3}\right)^N , \tag{6.14}$$

where $\lambda = h/\sqrt{2m\pi kT}$ is the thermal de Broglie wavelength. We now perform the sum over $\hat{N}$ to get the grand potential Ω as

$$\Omega \equiv -PV = -kT \ln \Xi(T, V, \mu) = -kTe^{\beta\mu}\left(\frac{V}{\lambda^3}\right) , \tag{6.15}$$

where we have used Eq. (3.77). From this we get

$$P = -\left.\partial\Omega/\partial V\right)_{T,\mu} = kTe^{\beta\mu}/\lambda^3 . \tag{6.16}$$

For the particle number, we use Eq. (6.12)

$$N = -\left.\partial\Omega/\partial\mu\right)_{T,V} = e^{\beta\mu}(V/\lambda^3) . \tag{6.17}$$

This allows us to solve for μ in terms of N, T, and V:

$$\mu = kT \ln\left(\frac{N\lambda^3}{V}\right) . \tag{6.18}$$

Then we can write the free energy as a function of N, T, and V as

$$F = N\mu - PV = NkT \ln \left(\frac{N\lambda^3}{V} \right) - NkT \, . \tag{6.19}$$

Since this agrees with the previous result, Eq. (4.58), obtained using the canonical partition function, we have now checked that, at least for the ideal gas, the two approaches are equivalent.

6.3.2 *Constant Pressure Partition Function of a Classical Ideal Gas*

We now use the constant pressure partition function to rederive the results for the ideal gas. We consider a system in a box with fixed sides perpendicular to the x-axis at fixed separation L_x and sides perpendicular to the y-axis at fixed separation L_y. The sides perpendicular to the z axis are at $z = 0$ and at $z = L_z$, where the integration over volume is performed by integrating over L_z. The constant pressure partition function is then found by integrating over all values of L_z and over all energies with the restriction that we only include energy eigenstates for which the expectation value of the pressure agrees with the fixed value P. Thus we have

$$\Upsilon = \frac{1}{N!} \int_0^\infty dL_z \sum_\alpha e^{-\beta E_\alpha - \beta PV} \delta[P + (L_x L_y)^{-1} dE_\alpha / dL_z] \, , \tag{6.20}$$

where α labels the energies. (As defined Υ is not dimensionless. It has the same units as L_z/P. However, in the thermodynamic limit $(1/N) \ln \Upsilon$ is well defined. Alternatively, we could include an integration over the momentum of the partition and divide the result by h to get a dimensionless partition function. However, in the thermodynamic limit, we would get the same results as we get here.)

Note that we use the Hellmann–Feynman theorem to relate the pressure in an energy eigenstate to the derivative of the energy of that state with respect to L_z. Also we include a factor of $N!$ to take approximate account of the indistinguishability of quantum particles. Using the energy levels for free particles in a box,

$$\begin{aligned}
\Upsilon = \frac{1}{N!} & \int_0^\infty dL_z \int_0^\infty dn_{x1} \int_0^\infty dn_{x2} \dots \int_0^\infty dn_{xN} \\
& \times \int_0^\infty dn_{y1} \int_0^\infty dn_{y2} \dots \int_0^\infty dn_{yN} \int_0^\infty dn_{z1} \int_0^\infty dn_{z2} \dots \int_0^\infty dn_{zN} \\
& \times \exp\Bigg(- \frac{h^2}{8mkT} \Bigg[\frac{n_{x1}^2 + n_{x2}^2 \dots + n_{xN}^2}{L_x^2} \\
& + \frac{n_{y1}^2 + n_{y2}^2 \dots + n_{yN}^2}{L_y^2} + \frac{n_{z1}^2 + n_{z2}^2 \dots + n_{zN}^2}{L_z^2} \Bigg] \Bigg)
\end{aligned}$$

$$\times\delta\left(P-\frac{h^2(n_{z1}^2+n_{z2}^2\cdots+n_{zN}^2)}{4mL_xL_yL_z^3}\right)e^{-\beta PV}, \tag{6.21}$$

where we have replaced the sums over quantum numbers by integrals. Performing the integrations over the n_x's and n_y's and using the δ-function restriction, we obtain

$$\begin{aligned}\Upsilon &= \frac{1}{N!}\left(\frac{2m\pi kTL_x^2}{h^2}\right)^{N/2}\left(\frac{2m\pi kTL_y^2}{h^2}\right)^{N/2}\\ &\quad\times\int_0^\infty dL_z e^{-(3/2)\beta PL_xL_yL_z}\int_0^\infty dn_{z1}\int_0^\infty dn_{z2}\ldots\int_0^\infty dn_{zN}\\ &\quad\times\delta\left(P-\frac{h^2(n_{z1}^2+n_{z2}^2\cdots+n_{zN}^2)}{4mL_xL_yL_z^3}\right)\\ &= \frac{1}{N!}\left(\frac{2m\pi kTL_x^2}{h^2}\right)^{N/2}\left(\frac{2m\pi kTL_y^2}{h^2}\right)^{N/2}\int_0^\infty dL_z e^{-(3/2)\beta PL_xL_yL_z}\\ &\quad\times\frac{1}{2^NP}\left(\frac{4mPL_xL_yL_z^3}{h^2}\right)^{N/2}S_N,\end{aligned} \tag{6.22}$$

where

$$S_N=\int_{-\infty}^\infty d\xi_1\int_{-\infty}^\infty d\xi_2\ldots\int_{-\infty}^\infty d\xi_N\delta(1-\sum_j\xi_j^2)=\frac{\pi^{N/2}}{(\frac{N}{2}-1)!}. \tag{6.23}$$

Then

$$\Upsilon=\frac{NkT\pi^{N/2}}{3P^2L_xL_y}\frac{(3N/2)!}{N!(N/2)!}\left(\frac{2m\pi kT}{h^2}\right)^N\left(\frac{mP}{h^2}\right)^{N/2}\left(\frac{2kT}{3P}\right)^{3N/2}. \tag{6.24}$$

Using Stirling's approximation for the factorial, $N!\approx(N/e)^N$, we get (keeping only terms proportional to N)

$$\ln\Upsilon=\frac{3}{2}N\ln\left(\frac{2m\pi kT}{h^2}\right)+N\ln[kT/P], \tag{6.25}$$

so that, by comparison with Eq. (4.59) we may write

$$-kT\ln\Upsilon=F+PV=G, \tag{6.26}$$

as noted in Table 6.1. This calculation therefore demonstrates that, at least for the ideal gas, the use of the constant pressure partition function gives the same results as does the canonical partition function. In the exercises, we ask for a calculation of the thermally averaged length of a one-dimensional polymer. This calculation is most conveniently done using the one-dimensional analog of the constant pressure partition function, which in this case is a constant force partition function.

6.4 Overview of Various Partition Functions

In rather general terms, we may define the partition function at fixed variables X, Y, and Z. The general form of such a partition function, which we here call Q, is

$$Q = \mathrm{Tr} e^{-\beta\Phi} , \tag{6.27}$$

where the range of states over which the trace is to be taken depends on which partition function we are considering. When the fixed variables are N, V, and E, then we set $\Phi = 0$, the trace is over all states of N particles in the volume V, with energy equal (within ϵ) to E, and Q is the microcanonical partition function, i.e., it is just the density of states, $\Omega(E, V, N)$. When the fixed variables are N, V, and T, then we set $\Phi = \mathcal{H}$, the trace is over all states of N particles in the volume V, and Q is the canonical partition function which we have written as $Z \equiv \mathrm{Tr}\exp(-\beta\mathcal{H}) \equiv Z(T, V, N)$. By now the reader will probably agree that calculations are, generally speaking, easier in the canonical rather than the microcanonical representation.

In Table 6.1, we list the relations for various partition functions. Note that in each case we get a thermodynamic potential in terms of the appropriate variables. For instance, in the case of the microcanonical distribution we get the entropy S as a function of N, U, and V, so that we may obtain the thermodynamic functions via

$$dS = (1/T)dU + (P/T)dV - (\mu/T)dN , \tag{6.28}$$

Table 6.1 Partition functions and their relation to thermodynamics. Partial derivatives are taken with all other variables held constant. The distinction between Ω, the microcanonical density of states, and Ω the grand potential should be clear from context

	Microcanonical	Canonical	Grand canonical	Constant P
Variables	N, E, V	N, β, V	μ, β, V	N, β, P
Q	Ω	Z	Ξ	Υ
Hamiltonian	0	$\mathcal{H}$	$\mathcal{H} - \mu\hat{N}$	$\mathcal{H} + P\hat{V}$
$\frac{\partial}{\partial\beta}$ or $\frac{\partial}{\partial E}$	$\frac{\partial \ln\Omega}{\partial E} = \beta$	$\frac{\partial \ln Z}{\partial\beta} = -U$	$\frac{\partial \ln\Xi}{\partial\beta} = N\mu - U$	$\frac{\partial \ln\Upsilon}{\partial\beta} = -U - PV$
$\frac{\partial}{\partial V}$ or $\frac{\partial}{\partial P}$	$\frac{\partial \ln\Gamma}{\partial V} = \beta P$	$\frac{\partial \ln Z}{\partial V} = \beta P$	$\frac{\partial \ln\Xi}{\partial V} = \beta P$	$\frac{\partial \ln\Upsilon}{\partial P} = -\beta V$
$\frac{\partial}{\partial N}$ or $\frac{\partial}{\partial\mu}$	$\frac{\partial \ln\Gamma}{\partial N} = -\beta\mu$	$\frac{\partial \ln Z}{\partial N} = -\beta\mu$	$\frac{\partial \ln\Xi}{\partial\mu} = \beta N$	$\frac{\partial \ln\Upsilon}{\partial N} = -\beta\mu$
$-kT\ln Q$	$-TS$	$-TS + U \equiv F$	$-TS + U - N\mu \equiv \Omega = -PV$	$-TS + U + PV \equiv G$

and for the canonical distribution we get the free energy F as a function of N, T, and V, so that we may obtain the thermodynamic functions via

$$dF = -SdT - PdV + \mu dN \ . \tag{6.29}$$

6.5 Product Rule

Here, we discuss the grand partition function for a separable system, by which we mean a system whose particles can either be in subsystem A with its independent energy levels or in subsystem B with its independent energy levels. Schematically, the situation is as shown in Fig. 6.1. The grand partition of the combined system may be written as

$$\Xi_{AB} = \sum_{\{E_j,n_j\}} \sum_{\{E_k,n_k\}} e^{-\beta(E_k^{(A)} - \mu n_k^{(A)})} e^{-\beta(E_j^{(B)} - \mu n_j^{(B)})} \ , \tag{6.30}$$

where the superscripts identify the subsystem and the sum is over all states with any number of particles. Thus, for independent subsystems, we have

$$\Xi_{AB}(T, V_A, V_B, \mu) = \Xi_A(T, V_A, \mu)\Xi_B(T, V_B, \mu) \ . \tag{6.31}$$

So the two systems share the same temperature and chemical potential.

A common case of this type is when system A is a substrate on which atoms may be adsorbed and B is the gas phase above the substrate. Then the use of the grand partition function simplifies calculating what the density of particles is in the gas and what density, in atoms per unit area, is on the substrate. Since having the density of atoms on the substrate as a function of chemical potential is essentially the same as having it as a function of the pressure of the gas, the grand partition function is particularly useful in this context.

In a certain sense, Eq. (6.31) is reminiscent of Eq. (5.4). We have phrased the discussion in this chapter in terms of a first quantized formulation in which the

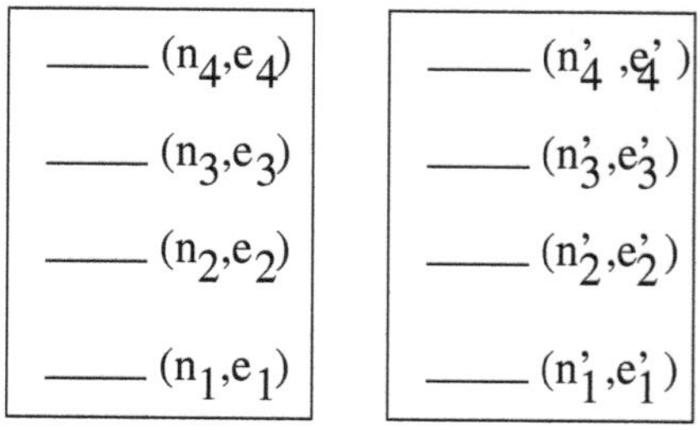

(n4,e4) (n'4,e'4)
(n3,e3) (n'3,e'3)
(n2,e2) (n'2,e'2)
(n1,e1) (n'1,e'1)

Fig. 6.1 Two independent subsystems, each with an independent spectrum of states with number of particles n_k and energy e_k

Hamiltonian is the Hamiltonian for the entire system, which consists of a volume bounded by a surface. In that formulation, the Hamiltonian does not seem to be the sum of two independent Hamiltonians. However, the situation is clearer in a formulation akin to second quantization in which we write $\mathcal{H} = \sum_\alpha n_\alpha \epsilon_\alpha$, where the ϵ_α are single-particle energy levels which can be labeled to indicate whether they correspond to wave functions localized on the surface or of particles moving in the gas. In this formulation, the Hamiltonian and also $\mathcal{H} - \mu n$ is separable and Eq. (6.31) is perfectly natural.

Indeed, as we will see in the next chapter, the product rule is useful for a system of noninteracting quantum particles. Then each single-particle state is independently populated and one can view each single-particle state as a subsystem. For single-particle state A, we have for fermions

$$\Xi_A = 1 + e^{-\beta(E_A-\mu)} , \tag{6.32}$$

since the occupation can only be zero or one, and for bosons

$$\Xi_A = \sum_{k=0}^{\infty} e^{-k\beta(E_A-\mu)} = \frac{1}{1 - e^{-\beta(E_A-\mu)}} . \tag{6.33}$$

6.6 Variational Principles

Here, we will show that the result for the microcanonical, the canonical, and grand canonical partition functions can be phrased in terms of a variational principle which is much like that developed for the ground state energy in quantum mechanics.

6.6.1 Entropy Functional

In thermodynamics, it is asserted that an isolated system evolves toward equilibrium (via spontaneous processes) and in so doing the entropy increases. In view of this statement, it should not be surprising to find that there exists a trial entropy functional which assumes a maximum value in equilibrium for an isolated system, i.e., for a system whose total energy is fixed. We define the **entropy functional**

$$\mathcal{S}\{\rho\} \equiv -k\mathrm{Tr}\rho \ln \rho = -k \sum_n P_n \ln P_n , \tag{6.34}$$

where P_n is the probability of occurrence of the nth eigenstate which has the fixed energy of the system. To be slightly more precise, the trace becomes a sum (denoted $\sum_n'$) over states with E_n within $\pm\epsilon/2$ of E, so that

$$\mathcal{S}\{P_n\} = -k \sum_n{}' P_n \ln P_n \ . \tag{6.35}$$

We now maximize $\mathcal{S}$ with respect to the P_n's subject to the constraint

$$\sum_n{}' P_n = 1 \ . \tag{6.36}$$

For this purpose, we introduce a Lagrange multiplier λ and consider the variation,

$$\delta \left[\mathcal{S} - \lambda \sum_n{}' P_n \right] = - \sum_n{}' \delta P_n \left[k \ln P_n + k + \lambda \right] \ . \tag{6.37}$$

The right-hand side of Eq. (6.37) vanishes for arbitrary variations, $\{\delta P_n\}$, if

$$P_n = e^{-1-\lambda/k} \ . \tag{6.38}$$

Since this result is independent of n, the normalization constraint gives

$$P_n = \frac{1}{\epsilon \Omega(E)}, \tag{6.39}$$

where Ω is the density of states for energy E, and therefore

$$\mathcal{S} = k \sum_n [\epsilon \Omega(E)]^{-1} \ln[\epsilon \Omega(E)] = k \ln[\epsilon \Omega(E)] \ . \tag{6.40}$$

Thus $\mathcal{S}\{\rho_n\}$ is maximized when all the ρ_n are equal and the equilibrium value of the entropy functional is given by

$$\begin{aligned} S(E) &= \max \mathcal{S} \\ &= -k \sum_n{}' [\epsilon \Omega(E)]^{-1} \ln[\epsilon \Omega(E)]^{-1} \\ &= k \ln \epsilon + k \ln \Omega(E) && (6.41) \\ &\rightarrow k \ln \Omega(E) \ . && (6.42) \end{aligned}$$

For the last step, we noted that $\ln \epsilon$ is of order unity compared to $\ln \Omega(E)$ which is proportional to N, the number of particles in the system. Thus, the variational principle for the entropy functional reproduces the famous Boltzmann result. In particular, it is appropriate that the result does not explicitly depend on the energy scale factor ϵ. Furthermore, one can think of the entropy functional for the case when all of the energetically equivalent P_n are *not* equal as some nonequilibrium entropy which will evolve in time toward the equilibrium value. This nonequilibrium entropy and its time evolution fall outside the scope of equilibrium statistical mechanics.

Both $\mathcal{S}$ and S are **additive** quantities, i.e., if the system consists of two independent subsystems "1" and "2", the total entropy of the combined system is simply the sum of the entropies of subsystems 1 and 2. Since the two subsystems are decoupled, the Hamiltonian is of the form $H = H_1 + H_2$, where H_1 and H_2 commute, and each wave function of the combined system can be written as

$$\Psi = \psi_1 \psi_2 \,. \tag{6.43}$$

It is then straightforward to show that the density matrix of the combined system is of the form

$$\rho^{(1,2)} = \rho^{(1)} \rho^{(2)} \,. \tag{6.44}$$

This then implies that the diagonal elements of the density matrix of the combined system can be written as

$$\rho^{(1,2)}(E) = \rho_m^{(1)}(E_1)\, \rho_n^{(2)}(E_2) \,, \tag{6.45}$$

where $E = E_1 + E_2$, and E_1 and E_2 are fixed.

In terms of these "ρ's", the entropy functional is

$$\begin{aligned} \mathcal{S}^{(1,2)}(E_1, E_2) &= -k \sum_{m,n}{}' \rho_m^{(1)} \rho_n^{(2)} \ln \rho_m^{(1)} \rho_n^{(2)} \\ &= -k \sum_m{}' \Big(\sum_n{}' \rho_n^{(2)}\Big) \rho_m^{(1)} \ln \rho_m^{(1)} - k \sum_n{}' \Big(\sum_m{}' \rho_m^{(1)}\Big) \rho_n^{(2)} \ln \rho_n^{(2)} \\ &= \mathcal{S}^{(1)}(E_1) + \mathcal{S}^{(2)}(E_2) \,. \end{aligned} \tag{6.46}$$

If the two systems are truly isolated from one another and cannot interact and, hence, cannot exchange energy, then the entropy is obtained by separately maximizing $\mathcal{S}^{(1)}(E_1)$ and $\mathcal{S}^{(2)}(E_2)$. Then the entropy is

$$S^{(1,2)}(E_1, E_2) = S^{(1)}(E_1) + S^{(2)}(E_2) \,. \tag{6.47}$$

In contrast, if the systems are weakly coupled and can exchange energy, then $S^{(1,2)}(E_1, E_2)$ must be maximized with respect to E_1 with $E_2 = E - E_1$. In this case of weak coupling, the variational principle gives the condition for equilibrium to be

$$\left.\frac{\partial S^{(1)}}{dE_1}\right|_V = \left.\frac{\partial S^{(2)}}{dE_2}\right|_V \,, \tag{6.48}$$

which we identify as $1/T_1 = 1/T_2$.

6.6.2 Free Energy Functional

As a practical matter, use of the variational principle for a system of fixed energy is not usually convenient. It is generally more convenient to consider systems in contact with a thermal reservoir. Accordingly, we reformulate the expression for the density matrix of a system in contact with a heat bath as a variational principle by defining the **free energy functional**

$$F(\boldsymbol{\rho}) \equiv \mathrm{Tr}[\boldsymbol{\rho}\mathcal{H}] + kT\mathrm{Tr}[\boldsymbol{\rho}\ln\boldsymbol{\rho}] \,. \tag{6.49}$$

At fixed temperature and volume, this functional is minimized (subject to the constraint that $\boldsymbol{\rho}$ have unit trace) by the equilibrium density matrix. Just to see that we are working toward a logical goal, we may check that F does give the equilibrium free energy when the density matrix assumes its equilibrium value, namely,

$$\boldsymbol{\rho} = \frac{e^{-\beta\mathcal{H}}}{\mathrm{Tr}e^{-\beta\mathcal{H}}} \equiv \frac{e^{-\beta\mathcal{H}}}{Z(T,V)} \,. \tag{6.50}$$

Then

$$\begin{aligned} F(\boldsymbol{\rho}) &= \mathrm{Tr}[\boldsymbol{\rho}\mathcal{H}] + kT\mathrm{Tr}[\boldsymbol{\rho}\ln\boldsymbol{\rho}] \\ &= U + kT\mathrm{Tr}\Big(\boldsymbol{\rho}\,[-\beta\mathcal{H} - \ln Z(T,V)]\Big) \\ &= U - U + F = F \,.\ \text{Q. E. D.} \end{aligned} \tag{6.51}$$

This evaluation indicates that we may view the term $\mathrm{Tr}\boldsymbol{\rho}\mathcal{H}$ as the trial energy and the second term on the right-hand side of Eq. (6.49) as $-T$ times the trial entropy. The role of the temperature is as one would intuitively expect: at low temperature, it is relatively important to nearly minimize the internal energy, whereas at high temperature it is important to nearly maximize the entropy.

Notice that in this variational principle we do not restrict the density matrix to be diagonal in the energy representation. So $\boldsymbol{\rho}$ is to be considered an arbitrary *Hermitian matrix*. We will show that the condition that the free energy be extremal with respect to variation of $\boldsymbol{\rho}$ leads to the condition that $\boldsymbol{\rho}$ is the equilibrium density matrix. It can also be shown (but we do not do that here) that this extremum is always a *minimum*. To incorporate the constraint on $\boldsymbol{\rho}$ of unit trace, we introduce a Lagrange multiplier λ and minimize $F(\boldsymbol{\rho}) - \lambda\mathrm{Tr}\boldsymbol{\rho}$. Because we are dealing with a trace, the variation of this functional can be written simply as if $\boldsymbol{\rho}$ were a number rather than a matrix:

$$\begin{aligned} \delta\left[F - \lambda\mathrm{Tr}\boldsymbol{\rho}\right] &= \delta\mathrm{Tr}\Big[\boldsymbol{\rho}[\mathcal{H} + kT\ln\boldsymbol{\rho} - \lambda\mathcal{I}\Big] \\ &= \mathrm{Tr}\Big(\delta\boldsymbol{\rho}\Big[\mathcal{H} + [kT - \lambda]\mathcal{I} + kT\ln\boldsymbol{\rho}\Big]\Big) = 0 \,, \end{aligned} \tag{6.52}$$

where $\mathcal{I}$ is the unit operator. The variation of F is zero for arbitrary variations of the matrix $\boldsymbol{\rho}$ if the quantity in square brackets is zero. This gives

$$\boldsymbol{\rho} \propto e^{-\beta\mathcal{H}} . \tag{6.53}$$

The constraint of unit trace then yields

$$\boldsymbol{\rho} = e^{-\beta\mathcal{H}}/\mathrm{Tr}[e^{-\beta\mathcal{H}}] , \tag{6.54}$$

which in the energy representation is simply

$$\rho_n = e^{-\beta E_n} / \sum_n e^{-\beta E_n} . \tag{6.55}$$

We may interpret the two terms on the right-hand side of Eq. (6.49) as the trial energy and trial value of $-TS$, respectively. As we shall continually see, the free energy plays a central role in statistical mechanics, and this variational principle for the free energy assumes a role analogous to that of the variational principle for the lowest eigenvalue of the Hamiltonian. Indeed, by specializing to zero temperature, this variational principle of statistical mechanics reduces to the more familiar one for the ground state energy of a quantum mechanical system. In later chapters, we will see that the variational principle obeyed by F provides a convenient route to the derivation of mean-field theories and of Landau–Ginzburg theory.

This variational principle can be extended to the case where, instead of fixed particle number, we deal with a fixed chemical potential. In that case, we have the analogous grand potential functional

$$\Omega(\boldsymbol{\rho}) \equiv \mathrm{Tr}[\boldsymbol{\rho}(\mathcal{H} - \mu\hat{N} + kT \ln \boldsymbol{\rho})] , \tag{6.56}$$

which we would minimize with respect to $\boldsymbol{\rho}$ (subject to the constraint of unit trace) to get the best possible approximation to the exact Ω.

6.7 Thermal Averages as Derivatives of the Free Energy

The free energy is useful because it can, in principle, be used to calculate thermally averaged observable properties of the system, such as order parameters, susceptibilities, and correlation functions. In this section, we will see how that is done.

6.7.1 Order Parameters

Consider a Hamiltonian of the form

$$\mathcal{H} = -\sum_\alpha h_\alpha \mathcal{O}^\alpha + \mathcal{H}_0\{O^\alpha\}, \tag{6.57}$$

where $\mathcal{H}_0$ is the value of the Hamiltonian when $h_\alpha = 0$ and is thus independent of h_α. The term $-h_\alpha \mathcal{O}^\alpha$ may already be present in the Hamiltonian or it may be added to the Hamiltonian for the express purpose of generating the result we obtain below. Note the minus sign in Eq. (6.57). With this choice of sign the thermal average $\langle \mathcal{O}^\alpha \rangle_T$ has the same sign as $h_\alpha > 0$, at least for small h_α. Then

$$\begin{aligned} \frac{\partial F}{\partial h_\alpha} &= -kT \frac{\partial}{\partial h_\alpha}\left(\ln \sum_n e^{-\beta E_n}\right) \\ &= \frac{\sum_n e^{-\beta E_n} \partial E_n / \partial h_\alpha}{\sum_n e^{-\beta E_n}} . \end{aligned} \tag{6.58}$$

But according to the Hellmann–Feynman theorem, $\partial E_n / \partial h_\alpha = -\langle n | O^\alpha | n \rangle$, in which case

$$\frac{\partial F}{\partial h_\alpha} = -\frac{\sum_n \langle n | \mathcal{O}^\alpha | n \rangle e^{-\beta E_n}}{\sum_n e^{-\beta E_n}} = -\langle O^\alpha \rangle_T , \tag{6.59}$$

or, turning this around we have

$$\langle O^\alpha \rangle_T = -\frac{\partial F}{\partial h_\alpha} . \tag{6.60}$$

Thus, all observable quantities can be expressed as first derivatives of the free energy with respect to the fields to which they are linearly coupled as in Eq. (6.57).

One example of this is for a magnetic system where we take $\mathcal{O}^\alpha = M^i$, where M^i is the ith Cartesian component of the magnetic moment and h_α is the ith component of the external field.

6.7.2 Susceptibilities of Classical Systems

Here, we discuss the susceptibilities of classical systems. In this context, "classical systems" are systems in which the operators in the Hamiltonian or those operators whose averages we wish to calculate are diagonal in the energy representation. In this classification, the Ising model is considered to be a "classical system." Correlation functions for quantum systems involve observables represented by non-commuting operators and are inherently more complicated to calculate.

For classical systems, our discussion of susceptibilities starts by considering the linear response of the system to an applied field. We define the static susceptibility as $\chi_\alpha = \partial \langle O^\alpha \rangle / \partial h_\alpha$, so that χ_α measures the response in the value of $\langle O^\alpha \rangle$ to an infinitesimal change in the field h_α. Then

$$\begin{aligned}
\chi_\alpha \equiv \frac{\partial \langle \mathcal{O}^\alpha \rangle}{\partial h_\alpha} &= \frac{\partial}{\partial h_\alpha} \left(\frac{\text{Tr}[\mathcal{O}^\alpha e^{-\beta \mathcal{H}}]}{\text{Tr} e^{-\beta \mathcal{H}}} \right) \\
&= \left(\frac{1}{\text{Tr} e^{-\beta \mathcal{H}}} \right) \frac{\partial}{\partial h_\alpha} \left[\text{Tr} \left(O^\alpha e^{-\beta \mathcal{H}} \right) \right] + \text{Tr} \left(O^\alpha e^{-\beta \mathcal{H}} \right) \frac{\partial}{\partial h_\alpha} \left(\frac{1}{\text{Tr} e^{-\beta \mathcal{H}}} \right) \\
&= \beta \frac{\text{Tr} \left[(O^\alpha)^2 e^{-\beta \mathcal{H}} \right]}{\text{Tr} e^{-\beta \mathcal{H}}} - \beta \left[\frac{\text{Tr} \left(O^\alpha e^{-\beta \mathcal{H}} \right)}{\text{Tr} e^{-\beta \mathcal{H}}} \right]^2 ,
\end{aligned} \tag{6.61}$$

so that

$$\chi_\alpha = \beta \left[\langle (O^\alpha)^2 \rangle - \langle O^\alpha \rangle^2 \right] . \tag{6.62}$$

The susceptibility is proportional to the average of the square of the deviation of the operator O^α from its mean, a positive definite quantity. To see this note that

$$\begin{aligned}
\langle (O^\alpha - \langle O^\alpha \rangle)^2 \rangle &= \langle (O^\alpha)^2 \rangle - 2 \langle O^\alpha \langle O^\alpha \rangle \rangle + \langle O^\alpha \rangle^2 \\
&= \langle (O^\alpha)^2 \rangle - \langle O^\alpha \rangle^2 .
\end{aligned} \tag{6.63}$$

More general susceptibilities $\chi_{\alpha\beta} \equiv \partial \mathcal{O}^\alpha / \partial h_\beta$ can be defined. The most common such susceptibilities are the off-diagonal elements of the magnetic or electric susceptibility tensors, such as $\chi_{xy} \equiv dm_x / dh_y$. These off-diagonal susceptibilities need not be positive.

The formalism given here applies to models in which there is a linear coupling between the field and its conjugate variable (here the magnetic moment). The situation is more complicated for Hamiltonians in which the external magnetic field enters both linearly and quadratically via the vector potential. In that case, the magnetic susceptibility may become negative (as it does for diamagnetic systems). But for simple models having only linear coupling the order-parameter susceptibility defined by Eq. (6.62) is always positive.

Similar arguments apply to the specific heat. We obtain the internal energy $U(T, V)$ as

$$U(T, V) \equiv \langle \mathcal{H} \rangle = \frac{\text{Tr}[\mathcal{H} e^{-\beta \mathcal{H}}]}{\text{Tr} e^{-\beta \mathcal{H}}} . \tag{6.64}$$

The specific heat at constant volume is

$$C_V = \left.\frac{\partial U}{\partial T}\right|_V = \left.\frac{\partial U}{\partial \beta}\right|_V \frac{d\beta}{dT} = -\frac{1}{kT^2}\left.\frac{\partial U}{\partial \beta}\right|_V$$
$$= \frac{1}{kT^2}\left(\langle \mathcal{H}^2\rangle - \langle \mathcal{H}\rangle^2\right). \quad (6.65)$$

We have therefore shown that the specific heat cannot be negative, in agreement with the same conclusion reached within thermodynamics. We also see that in the zero-temperature limit, the system is certainly in its ground state, so that $\langle \mathcal{H}^2\rangle = \langle \mathcal{H}\rangle^2$. Thus, the specific heat must go to zero as $T \to 0$, in conformity with the Third Law of Thermodynamics.

6.7.3 Correlation Functions

As we shall see throughout this book, there are many reasons why the response of the system to a spatially varying field is of interest. In this section, we will consider the *linear* response to weak inhomogeneous fields. The response could be the particle density at position $\mathbf{r}$ to the field, taken to be the potential at position $\mathbf{r}'$. Here, we consider a magnetic system where the response is the change in the magnetization (i.e., the magnetic moment per unit volume) $\delta\mathbf{m}(\mathbf{r})$ induced by the magnetic field $\mathbf{h}(\mathbf{r}')$. The most general form that such a response can take is

$$\delta m_\alpha(\mathbf{r}) \equiv \langle M_\alpha(\mathbf{r})\rangle - \langle m_\alpha(\mathbf{r})\rangle_0 = \sum_{\mathbf{r},\mathbf{r}'}\sum_\beta \chi_{\alpha,\beta}(\mathbf{r},\mathbf{r}')h_\beta(\mathbf{r}') + \mathcal{O}(h^3), \quad (6.66)$$

where $\langle \ldots \rangle_0$ is the average in zero applied field. One refers to $\chi_{\alpha,\beta}(\mathbf{r},\mathbf{r}')$ as the two-point susceptibility. If the system is homogeneous, then $\chi(\mathbf{r},\mathbf{r}') = \chi(\mathbf{r}-\mathbf{r}')$.

One sees that to evaluate the susceptibility, it is simplest to assume that the magnetic system consists of spins, $\mathbf{S}$, located at lattice sites. (To treat the continuum system requires the introduction of functional derivatives, but leads to a final result of exactly the same form as we will obtain for the lattice system.)

$$\chi_{\alpha,\beta}(\mathbf{r},\mathbf{r}') = \frac{d\langle S_\alpha(\mathbf{r})\rangle}{dh_\beta(\mathbf{r}')} = \frac{d}{dh_\beta(\mathbf{r}')}\left[\frac{\mathrm{Tr}e^{-\beta\mathcal{H}}S_\alpha(\mathbf{r})}{\mathrm{Tr}e^{-\beta\mathcal{H}}}\right], \quad (6.67)$$

where

$$\mathcal{H} = \mathcal{H}_0 - \sum_\alpha\sum_{\mathbf{r}} h_\alpha(\mathbf{r})S_\alpha(\mathbf{r}), \quad (6.68)$$

as in Eq. (6.57). Carrying out the derivatives as in Eq. (6.61), we get

$$kT\chi_{\alpha,\beta}(\mathbf{r},\mathbf{r}') = \langle S_\alpha(\mathbf{r})S_\beta(\mathbf{r}')\rangle - \langle S_\alpha(\mathbf{r})\rangle\langle S_\beta(\mathbf{r}')\rangle. \quad (6.69)$$

One can write this as

$$kT\chi_{\alpha,\beta}(\mathbf{r},\mathbf{r}') = \langle \delta S_\alpha(\mathbf{r})\delta S_\beta(\mathbf{r}')\rangle \ , \tag{6.70}$$

where $\delta S_\alpha(\mathbf{r}) \equiv S_\alpha(\mathbf{r}) - \langle S_\alpha(\mathbf{r})\rangle$. The right-hand side of Eq. (6.70) is interpreted as the two-point *correlation function*, because it measures the correlations between fluctuations in the spin at $\mathbf{r}$ to those of the spin at $\mathbf{r}'$. For an isotropic system, one often speaks of "the susceptibility," χ, by which one means the susceptibility at zero wavevector, i.e., the susceptibility with respect to a uniform field evaluated at zero field. Then one writes

$$\begin{aligned} \chi &\equiv -\frac{1}{N}\frac{\partial^2 F}{\partial h_\alpha^2} = \frac{1}{N}\sum_{\mathbf{r},\mathbf{r}'}\frac{\partial\langle S_\alpha(\mathbf{r})\rangle}{\partial h_\alpha(\mathbf{r}')} \\ &= \frac{\beta}{N}\sum_{\mathbf{r},\mathbf{r}'}\langle S_\alpha(\mathbf{r})S_\alpha(\mathbf{r}')\rangle - \langle S_\alpha(\mathbf{r})\rangle\langle S_\alpha(\mathbf{r}')\rangle \ . \end{aligned} \tag{6.71}$$

In the disordered phase, where $\langle S_\alpha(\mathbf{r})\rangle = 0$ (at zero field), this becomes

$$kT\chi = \frac{1}{N}\sum_{\mathbf{r},\mathbf{r}'}\langle S_\alpha(\mathbf{r})S_\alpha(\mathbf{r}')\rangle \ . \tag{6.72}$$

There are several simple intuitive statements one can make about the two-point susceptibility in view of what we might guess about the behavior of the spin–spin correlation function. Consider a ferromagnet. If the spin at site i is up, then the interaction favors neighboring spins being up. In the disordered phase, $\langle S_i\rangle = 0$, but if we *force* S_i to be up, its neighbors are more likely to be up. But this effect will decrease with distance. That is, we expect something like

$$\langle S_i S_j\rangle = e^{-R_{ij}/\xi} , \tag{6.73}$$

where ξ is a **correlation length**. It is clear that this correlation length must depend on temperature. At high temperature, the Boltzmann probabilities only very weakly favor neighboring spins to be parallel rather than antiparallel. The correlation length must be small. However, as the temperature is lowered, the Boltzmann factors more and more favor parallel alignment and the correlation length increases. Indeed, as the temperature approaches the critical temperature T_c below which one has spontaneous ordering, the effect of forcing spin i to be up will become long range. That is, $\xi(T) \to \infty$ as $T \to T_c$. Experimentally, it is found that, near T_c, $\xi(T) \sim |T - T_c|^{-\nu}$, where ν is a critical exponent. Then, the uniform ($\mathbf{q} = 0$) susceptibility in zero external field,

$$\chi_0 = \beta\sum_j\langle S_1 S_j\rangle \tag{6.74}$$

diverges as $\xi(T)$ diverges. Thus, Eq. (6.73) relates the divergence of the correlation length to the divergence of the susceptibility at T_c. (To obtain the correct divergence in the susceptibility one must use a more realistic representation of the correlation function in the form

$$\chi(\mathbf{r}) = \sum_r r^{-x} e^{-r/\xi} , \tag{6.75}$$

where the exponent x plays an important role in critical phenomena. This subject will be elaborated on in great detail in later chapters.)

6.8 Summary

The most important development in this chapter is the introduction of the grand canonical distribution. The grand partition function, Ξ, is given as

$$\Xi(T, V, \mu) \equiv \mathrm{Tr}_{E,\hat{N}} e^{-\beta(\mathcal{H}-\mu\hat{N})} = \sum_N e^{\beta\mu N} Z(T, V, N) , \tag{6.76}$$

where the trace is a sum over all states of the system contained in a volume V (whatever the number of particles) and $Z(N, T, V)$ is the canonical partition function for N particles at temperature T in a volume V. One can then define the grand potential function, $\Omega(T, V, \mu) = -kT \ln \Xi(T, V, \mu)$ and obtain the entropy, the pressure, and the average number of particles using Eq. (6.12). A systematic approach is to first invert the equation

$$N = -\left.\frac{\partial \Omega(T, V, \mu)}{\partial \mu}\right|_{T,V} \tag{6.77}$$

to get μ as a function of T, V, and N. One can then write the free energy as a function of these variables by

$$F(T, V, N) = N\mu(T, V, N) + \Omega(T, V, \mu(T, V, N)) , \tag{6.78}$$

from Eq. (6.10). Having the free energy as a function of its proper variables ensures that we can calculate all other thermodynamic functions. We also discussed that the grand partition function for a collection of independent subsystems is the product over grand partition functions of the individual subsystems.

A second result in this chapter is that, if one considers a Hamiltonian containing terms linear in the order parameter multiplied by an external field, the first derivative of this free energy with respect to the field gives the thermal average of the order parameter, and the second derivative gives the order-parameter susceptibility, which, for classical systems, is proportional to the mean-square fluctuations of the order parameter. Indeed if position-dependent fields are introduced, one can obtain the position-dependent susceptibility, which is the order-parameter correlation function.

Variational principles for the entropy and free energy were introduced. The latter will be extensively discussed and utilized in later chapters.

6.9 Exercises

1. Consider a model which treats the possible configurations of a linear polymer which consists of N monomer units. To simplify the problem, we make the following two assumptions. First, we only consider one-dimensional configurations of the polymer. Second, we neglect any possible hard-core interactions which occur when two or more monomer units occupy the same location. In this case, each polymer configuration can be put into a one-to-one correspondence with the set of N-step random walks in one dimension and the end-to-end displacement of the polymer will tend on average to be of order $\sqrt{N}$. To describe the statistics of such polymer configurations, you are to consider a partition function at fixed monomer number N and fixed force F in which one end of the polymer is fixed and a constant force, F, is applied to the free end of the Nth monomer unit. Use this partition function to obtain the average length of a polymer consisting of N monomers as a function of temperature T and the force F applied to the free end.
Hint: introduce a potential energy to represent the force F.

2. Use thermodynamic relations to show that if you know P as a function of μ, T, and V, you can get all the other thermodynamic functions. Does this mean that a knowledge of the equation of state, i.e., P as a function of N, T, and V suffices to get all the other thermodynamic functions?

3. A cylinder of cross-sectional area A contains N molecules which may be treated as noninteracting point masses, each of mass m. A piston of mass M rests on top of the gas. We assume that $M > Nm$, so that the gravitational potential energy of the gas is negligible compared to that of the piston and may be omitted from the Hamiltonian. Thus the Hamiltonian includes the kinetic energy of the molecules, the gravitational potential of the piston, and the translational kinetic energy of the piston (ignore any internal degrees of freedom of the piston). The whole system is weakly coupled to a heat reservoir at temperature T, which is high enough so that classical mechanics is valid. Calculate

(**a**) The heat capacity dU/dT of the system.
(**b**) The average height $\langle z \rangle$ of the piston.
(**c**) The rms fluctuation in z. That is, calculate $\Delta z = (\langle z^2 \rangle - \langle z \rangle^2)^{1/2}$.
(**d**) $P(z)$, where $P(z)dz$ is the probability that the piston is between z and $z + dz$, but don't bother to normalize $P(z)$. Also calculate $P(z, K)$, where $P(z, K)dz$ is the probability that the piston is between z and $z + dz$, given that the kinetic energy of the gas has just been measured and is found to have the value K.

4. Consider the following model of adsorption of atoms on a surface. Assume that the system consists of an ideal gas of atoms in equilibrium with adsorbed atoms. The surface consists of N_A adsorption sites which can either be vacant or occupied by a single atom which then has energy $-E_1$ or $-E_2$, respectively, where $-E_2 < -E_1 < 0$. (The zero of energy is defined to be the potential energy of an atom in the gas phase.)

(**a**) Determine the fraction of adsorption sites which are occupied as a function of T, V, and the atomic chemical potential μ.

(**b**) Determine the fraction of adsorption sites which are occupied as a function of temperature T at a fixed pressure P.

5. Consider the following model of localized electrons on a lattice of N_s sites for which the Hamiltonian is

$$\mathcal{H} = \epsilon \sum_{i\sigma} n_{i\sigma} + U \sum_{i} n_{i\uparrow} n_{i\downarrow} ,$$

where $n_{i\sigma}$, the number of electrons on site i with z-component of spin σ, is either zero or one.

(**a**) Construct the grand partition function as a function of ϵ, U, μ and $\beta \equiv 1/(kT)$.

(**b**) Determine μ (or better still the fugacity $z \equiv e^{\beta\mu}$) as a function of n, the average number of electrons per site.

(**c**) Display the electron–hole symmetry of this model. To do that show that the solution for μ when $n > 1$ can be obtained from that for $n < 1$ by a suitable transformation of parameters.

(**d**) Give the specific heat (at constant n, of course) for $U = 0$. Hint: this can be done simply.

(**e**) Determine the specific heat for large U to leading order in $e^{-\beta U}$.

6. Show that, for the grand canonical partition function, $\partial \ln \Xi / \partial \beta)_{V,\beta\mu} = -U$, where $\beta\mu$, rather than μ, is held constant, and similarly that $\partial \ln \Upsilon / \partial \beta)_{V,\beta P} = -U$, where βP, rather than P, is held constant, as indicated in Table 6.1.

7. Why does it not make any sense to introduce a partition function in which all the intensive variables β, p, and μ are fixed?

8. Qualitatively speaking, how would you expect the spin–spin correlation function to change its dependence on separation as the temperature is lowered from infinity to a temperature near the ordering transition? (Assume the system interactions between nearest neighbors are such that parallel spin configurations have lower energy than antiparallel ones.) People often speak of a "correlation volume" for such a system. What do you think this means?

Chapter 7
Noninteracting Gases

7.1 Preliminaries

For a noninteracting (ideal) gas of spin S particles in a box of volume V, the single-particle states are labeled by a wavevector $\mathbf{k}$ and a spin index σ, where σ is the eigenvalue of S_z, which assumes the $2S+1$ values, $\sigma = -S, -S+1 \ldots S$. The total energy of the system is the sum of the individual single-particle energies which, in the absence of electric and magnetic fields, are given by

$$\epsilon_{\mathbf{k},\sigma} = \frac{\hbar^2 k^2}{2m} \ . \tag{7.1}$$

Here $\mathbf{k}$, the wavevector, is quantized by the boundary conditions, but in the thermodynamic limit, it does not matter precisely what boundary conditions one chooses. It is convenient to choose periodic boundary conditions, so that for a system of volume V in a cubical box whose edges have length L ($V = L^3$), we have

$$k_\alpha = \frac{2\ell_\alpha \pi}{L} \tag{7.2}$$

for the α component ($\alpha = x, y, z$) of the wavevector, where ℓ_α ranges over integers from $-\infty$ to ∞.

The many-body Hamiltonian is thus

$$\mathcal{H} = \sum_{\mathbf{k},\sigma} \epsilon_{\mathbf{k},\sigma} n_{\mathbf{k},\sigma} \ , \tag{7.3}$$

where, for identical fermions, because of the Pauli exclusion principle, $n_{\mathbf{k},\sigma} = 0, 1$ whereas, for identical bosons, $n_{\mathbf{k},\sigma} = 0, 1, 2, \ldots, \infty$. For a system with a fixed number of particles N,

$$\sum_{\mathbf{k},\alpha} n_{\mathbf{k},\sigma} = N \ . \tag{7.4}$$

A. J. Berlinsky and A. B. Harris, *Statistical Mechanics*, Graduate Texts in Physics,
https://doi.org/10.1007/978-3-030-28187-8_7

The constraint on the total number of particles complicates the sum over states that is required to calculate the canonical partition function. (An exercise will explore this.) Accordingly, instead of working with fixed N, we evaluate the grand canonical partition function, Ξ, at chemical potential μ:

$$\Xi(T, V, \mu) = \mathrm{Tr} e^{-\beta(\mathcal{H}-\mu\hat{N})} , \tag{7.5}$$

where $\hat{N}$ is the number operator and Tr indicates a sum over values of $\hat{N}$ and over all possible energy states.

Explicitly, this may be written as

$$\begin{aligned} \Xi(T, V, \mu) &= \sum_{\{n_{\mathbf{k},\sigma}\}} \left(\prod_{\mathbf{k},\text{œ}} e^{-\beta n_{\mathbf{k},\sigma}(\epsilon_{\mathbf{k},\sigma}-\mu)} \right) \\ &= \prod_{\mathbf{k},\sigma} \left(\sum_n e^{-\beta n(\epsilon_{\mathbf{k},\sigma}-\mu)} \right) , \end{aligned} \tag{7.6}$$

where n is summed over the values 0 and 1 for fermions and over all nonnegative integers for bosons. Equation (7.6) may be viewed as an application of Eq. (6.31) where each single-particle state, labeled by $\mathbf{k}$, σ, is treated as an independent system. Henceforth, we drop explicit reference to σ as an argument in ϵ or n as long as ϵ and n are independent of σ.

For fermions,

$$\sum_{n=0}^{1} e^{-\beta(\epsilon_{\mathbf{k}}-\mu)n} = 1 + e^{-\beta(\epsilon_{\mathbf{k}}-\mu)} , \tag{7.7}$$

and **for bosons**

$$\sum_{n=0}^{\infty} e^{-\beta(\epsilon_{\mathbf{k}}-\mu)n} = \frac{1}{1 - e^{-\beta(\epsilon_{\mathbf{k}}-\mu)}} . \tag{7.8}$$

Using Eqs. (7.7) and (7.8), we obtain the grand potential from Eq. (7.6) to be

$$\Omega(T, V, \mu) \equiv -kT \ln \Xi(T, V\mu) = \mp kT \sum_{\mathbf{k},\sigma} \ln\left(1 \pm e^{-\beta(\epsilon_{\mathbf{k}}-\mu)}\right) , \tag{7.9}$$

where the upper (lower) sign is for Fermions (Bosons). Note that the probability $P_{\mathbf{k}}(n)$ that the occupation number for the single-particle state $\mathbf{k}$ is n is proportional to $\exp[\beta n(\mu - \epsilon_{\mathbf{k}})]$. Then, $\langle n_{\mathbf{k}} \rangle$, the thermally averaged number of particles in the single-particle state $\mathbf{k}$ is given by

$$\langle n_{\mathbf{k}} \rangle = \frac{\sum_n n e^{-n\beta(\epsilon_{\mathbf{k}}-\mu)}}{\sum_n e^{-n\beta(\epsilon-\mu)}} . \tag{7.10}$$

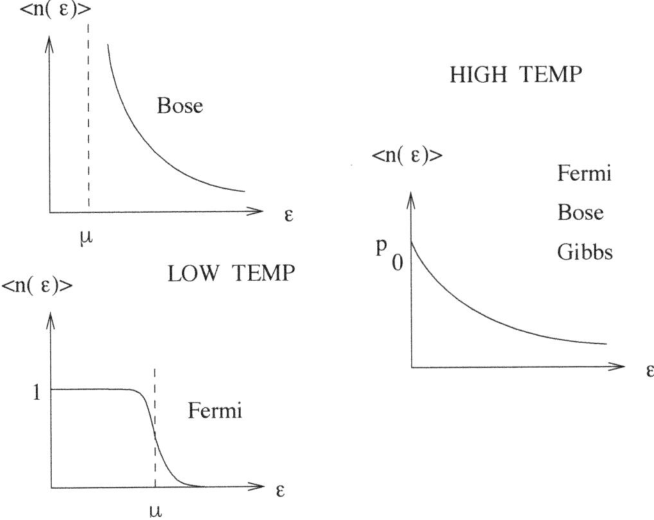

Fig. 7.1 The function $f(\epsilon) \equiv \langle n_{\mathbf{k},\sigma}\rangle_T$ vs. ϵ. In the left panel, we show the Fermi and Bose functions at very low temperature. In the right panel, we show these functions in the limit of high temperatures, where the Fermi, Bose, and Gibbs functions asymptotically coincide. Here $p(\epsilon) = p_0 e^{-\beta\epsilon}$, where $p_0 = \exp(\beta\mu) = N\lambda_D^3/V \ll 1$, where λ_D is the thermal deBroglie wavelength

For fermions, the sum runs over the values $n = 0, 1$ and we get

$$\langle n_{\mathbf{k}}\rangle = \frac{e^{-\beta(\epsilon_{\mathbf{k}}-\mu)}}{1+e^{-\beta(\epsilon-\mu)}} . \tag{7.11}$$

For bosons, the sum over n runs over all nonnegative integers and

$$\begin{aligned} \langle n_{\mathbf{k}}\rangle &= \frac{\sum_{n=0}^{\infty} n e^{-n\beta(\epsilon_{\mathbf{k}}-\mu)}}{\sum_{n=0}^{\infty} e^{-n\beta(\epsilon-\mu)}} = \frac{x/(1-x)^2}{1/(1-x)} \\ &= \frac{x}{1-x} = \frac{e^{-\beta(\epsilon_{\mathbf{k}}-\mu)}}{1-e^{-\beta(\epsilon-\mu)}} , \end{aligned} \tag{7.12}$$

where $x = \exp[\beta(\epsilon_{\mathbf{k}} - \mu)]$. We summarize these results as

$$\langle n_{\mathbf{k}}\rangle = \frac{1}{e^{\beta(\epsilon_{\mathbf{k}}-\mu)} \pm 1} . \tag{7.13}$$

These results are illustrated in Fig. 7.1.

The total number of particles is related to the chemical potential by

$$N = -\partial\Omega/\partial\mu\Big)_{T,V} = \sum_{\mathbf{k},\sigma} \frac{1}{e^{\beta(\epsilon_{\mathbf{k}}-\mu)} \pm 1} . \tag{7.14}$$

In view of Eq. (7.13), this relation is equivalent to

$$N = \sum_{\mathbf{k},\sigma} \langle n_{\mathbf{k}} \rangle = g \sum_{\mathbf{k}} \langle n_{\mathbf{k}} \rangle \,. \tag{7.15}$$

Since we have assumed here that the energy is independent of σ, the sum over σ just gives a factor $(2S+1) \equiv g$.

Similarly, the internal energy can be obtained from Eq. (7.9) using

$$\begin{aligned} U &= \Omega + TS + \mu N \,, \\ &= \Omega - T \left. \frac{\partial \Omega}{\partial T} \right)_{V,\mu} + \mu N \,, \end{aligned} \tag{7.16}$$

which might seem like a complicated way to do the calculation, except for the fact that, using Eq. (7.9),

$$- T \left. \frac{\partial \Omega}{\partial T} \right)_{V,\mu} = -\Omega + g \sum_{\mathbf{k}} \frac{(\epsilon_{\mathbf{k}} - \mu)}{e^{\beta(\epsilon_{\mathbf{k}} - \mu)} \pm 1} . \tag{7.17}$$

Combining these last two relations immediately gives the intuitively obvious result

$$U = g \sum_{\mathbf{k}} \epsilon_{\mathbf{k}} \frac{1}{e^{\beta(\epsilon_{\mathbf{k}} - \mu)} \pm 1} . \tag{7.18}$$

Next we would like to explicitly perform the $\mathbf{k}$ sum for the gas in three dimensions, but first we discuss the qualitative features we expect. Recall that we may write the single-particle energy as

$$\begin{aligned} \epsilon_{\mathbf{k}} &= \frac{\hbar^2 k^2}{2m} = \frac{h^2(\ell_x^2 + \ell_y^2 + \ell_z^2)}{2mL^2} \\ &= \frac{h^2(\ell_x^2 + \ell_y^2 + \ell_z^2)}{2m} V^{-2/3} \,. \end{aligned} \tag{7.19}$$

This means that the energy scales with $V^{-2/3}$. Thus, following the discussion in Sect. 5.7 of Chap. 5, we conclude without further calculation that

$$PV = \frac{2}{3} U \,. \tag{7.20}$$

This result is independent of statistics. It holds for fermions, bosons, or classical particles. However, it only applies in the nonrelativistic limit.

Next we want to convert the summations in Eqs. (7.15) and (7.18) into integrals. In the thermodynamic limit, the difference between adjacent single-particle energies becomes small, of order $V^{-2/3}$, and one might think that this would be sufficient to guarantee that one can replace sums over ℓ_α by their corresponding integrals. This is indeed the case, except, as we will see, for Bose systems at low temperature. We will discuss that case separately as it leads to Bose condensation. Otherwise, the sums may be converted into integrals.

This can be done by identifying the sum over integers ℓ_x, ℓ_y, and ℓ_z as integrals over these variables as we did in Eq. (4.55). However, it is instructive instead to use the following alternative prescription. We write

$$\sum_{\mathbf{k}} F(\mathbf{k}) = \frac{1}{(\Delta k_x \Delta k_y \Delta k_z)} \sum_{\mathbf{k}} F(\mathbf{k}) \Delta k_x \Delta k_y \Delta k_z \ , \tag{7.21}$$

where $\Delta k_x = (2\pi/L_x)$, where L_x is the length of the side of the confining box along the x-direction and similarly for the two other directions. The purpose for multiplying and dividing by the Δk's is to make contact with the definition of the integral in terms of the mesh size of the discrete sum. Thus

$$\sum_{\mathbf{k}} F(\mathbf{k}) = \frac{1}{(\Delta k_x \Delta k_y \Delta k_z)} \int F(\mathbf{k}) d\mathbf{k} = \frac{V}{8\pi^3} \int F(\mathbf{k}) d\mathbf{k} \ . \tag{7.22}$$

Then

$$N = \frac{gV}{8\pi^3} \int \frac{d\mathbf{k}}{e^{\beta(\epsilon_{\mathbf{k}}-\mu)} \pm 1} \tag{7.23}$$

and

$$\Omega(T, V, \mu) = \mp \frac{gVkT}{8\pi^3} \int \ln\left(1 \pm e^{-\beta(\epsilon_{\mathbf{k}}-\mu)}\right) d\mathbf{k} \ . \tag{7.24}$$

It is sometimes convenient to express results in terms of the single-particle density of states, $\rho(E)$ defined by

$$\rho(E) = \frac{1}{8\pi^3} \sum_{\sigma} \int d\mathbf{k} \delta(E - \epsilon_{\mathbf{k},\sigma}) \ . \tag{7.25}$$

This definition is such that $\rho(E)\Delta E$ is the number of single-particle energies $\epsilon_{\mathbf{k}}$ *per unit volume* in a range ΔE about E and $\rho(E)$ has the dimensions of (energy^{-1} $\times$ volume^{-1}). Here

$$\rho(E) = \frac{g}{8\pi^3} \int_0^\infty 4\pi k^2 \delta(\frac{\hbar^2 k^2}{2m} - E) dk = \frac{gE^{1/2}}{4\pi^2} \left(\frac{2m}{\hbar^2}\right)^{3/2} \ . \tag{7.26}$$

Then

$$N(T, V, \mu) = \int_0^\infty \rho(E) \frac{1}{e^{\beta(E-\mu)} \pm 1} dE \tag{7.27}$$

and

$$\Omega(T, V, \mu) = \mp VkT \int_0^\infty \rho(E) \ln\left(1 \pm e^{-\beta(E-\mu)}\right) dE \,. \tag{7.28}$$

So far, the formulas apply for a general form of $\rho(E)$. But now we specialize to the case of three dimensions, Eq. (7.19). In that case, we may integrate by parts using $\int_0^E \rho(E')dE' = (2E/3)\rho(E)$ to get

$$U = -(3/2)\Omega = V \int_0^\infty E\rho(E) \frac{1}{e^{\beta(E-\mu)} \pm 1} dE \,, \tag{7.29}$$

which agrees with Eq. (7.18). Note that Eqs. (7.20), (7.26) are specific to free particles in three dimensions, while Eq. (7.18) and the right-hand side of Eq. (7.29) (with the appropriate interpretation of V) apply for general dimension and arbitrary $\epsilon_{\mathbf{k}}$.

Finally, we make some observations about how the chemical potential depends on temperature when we invert Eq. (7.15) to get μ as a function of N, T, and V. First, in the zero-temperature limit, we can identify the value of μ. For Fermi systems, the ground state is obtained by occupying the lowest single-particle energy levels. For $T = 0$ this means that μ is equal to the energy of the highest occupied level, which is called the **Fermi energy**. For Bose systems, the ground state has all particles in the lowest energy state. For $T = 0$, this means that μ will be equal to the lowest single-particle energy, ϵ_0.

Next consider what happens at very high temperature where quantum effects become very small. What this means is that each $\langle n_{\mathbf{k}} \rangle$ is very small, in fact much less than unity. Consequently, $e^{-\beta\mu}$ must be very large. It is also significant that, when $e^{-\beta\mu}$ is large, the ± 1 in Eq. (7.14) becomes irrelevant. The conclusion is that, for fixed N and V, as the temperature increases, $\mu(N, T, V)$ will decrease and eventually become negative and large in the high-temperature limit, and that, in this limit, quantum statistics are irrelevant. An alternative way of describing the high-temperature limit is to say that this is the limit in which the *fugacity* $z \equiv e^{\beta\mu}$ is vanishingly small.

Equations (7.14) and (7.27) for $N(T, V, \mu)$ and Eqs. (7.18) and (7.29) for $U(T, V, \mu)$ emphasize how these quantities are related to the single-particle spectrum and occupation numbers $\langle n_{\mathbf{k}} \rangle$. However, it is less clear from these equations how N and U depend on their arguments. For the case of free particles in 3D, we write these in dimensionless form in terms of special functions as

$$\frac{U}{kT} = \frac{3g}{2}\left(\frac{V}{\lambda^3}\right) f_{5/2}^{\pm 1}(z) \,, \tag{7.30}$$

$$N = g\left(\frac{V}{\lambda^3}\right) f_{3/2}^{\pm 1}(z) \,, \tag{7.31}$$

where $z = \exp(\beta\mu)$ and $\lambda = h/\sqrt{2m\pi kT}$ is the thermal deBroglie wavelength. Here

$$f_\nu^a(z) = \frac{1}{\Gamma(\nu)} \int_0^\infty \frac{x^{\nu-1}dx}{z^{-1}e^x + a} . \tag{7.32}$$

Note that the results for the classical "Gibbs" gas can be obtained by setting $a = 0$ in the above results.

We point out the following consequences of the above equations. When Eq. (7.31) is solved for μ as a function of N, one sees that $z \equiv \exp(\beta\mu)$ is a function of

$$\xi \equiv gv/\lambda^3 = \frac{gv}{h^3}(2m\pi kT)^{3/2} \equiv (T/T_Q)^{3/2} , \tag{7.33}$$

where $v = V/N$ and the characteristic "quantum" temperature T_Q corresponds to the energy needed to confine a particle (in a given spin state) to the average volume per particle:

$$kT_Q = \frac{h^2}{2m\pi(gv)^{2/3}} . \tag{7.34}$$

Then Eq. (7.30) says that $pV/(NkT)$ is equal to ξ times a function of z. But since z is itself a function only of ξ, we can say that $pV/(NkT)$ is a function of the single variable ξ, or equivalently, of T/T_Q. For the ideal Fermi gas, the thermodynamic functions are usually given as a function of T/T_F, where the Fermi temperature T_F (defined below) differs from T_Q by numerical factors of order unity. Likewise, for the ideal Bose gas, the thermodynamic functions are usually given as a function of T/T_c, where T_c is the critical temperature for Bose condensation which also differs from T_Q by numerical factors of order unity. The important conclusion is that in the absence of interparticle interactions thermodynamic quantities such as $pV/(NkT)$ and the specific heat at constant volume are functions of the single reduced variable T/T_Q. This conclusion does *not* hold when interparticle interactions are non-negligible. In that more general case, $U(T, V)$ and pV/NkT are functions of the two variables T and V which cannot be expressed in terms of a single variable, as is the case here, in terms of T/T_Q.

7.2 The Noninteracting Fermi Gas

We first consider the noninteracting Fermi gas starting from Eqs. (7.30) and (7.31).

7.2.1 High Temperature

We start with the high-temperature limit. In this limit, ξ is large and $\mu/(kT)$ is large and negative. Then we can expand in powers of the fugacity $e^{\beta\mu} \equiv z \ll 1$. Keeping only the first correction, we have

$$-\frac{\Omega}{NkT} = \frac{2}{3}\frac{U}{NkT} = \frac{PV}{NkT} = \frac{4\xi}{3\sqrt{\pi}}\int_0^\infty x^{3/2}dx\left(ze^{-x} - z^2e^{-2x}\ldots\right)$$
$$= \xi\left(z - \frac{z^2}{2^{5/2}} + \cdots\right), \tag{7.35}$$

where we used $\Gamma(5/2) = 3\sqrt{\pi}/4$ and $z(\xi)$ is obtained from

$$1 = \frac{2\xi}{\sqrt{\pi}}\int_0^\infty x^{1/2}dx\left(ze^{-x} - z^2e^{-2x}\ldots\right)$$
$$= \xi\left(z - \frac{z^2}{2^{3/2}} + \cdots\right). \tag{7.36}$$

Solving Eq. (7.36) for z, we obtain correct to order ξ^{-2}

$$z = \xi^{-1}\left(1 + \frac{1}{2^{3/2}}\xi^{-1} + \cdots\right). \tag{7.37}$$

Then Eq. (7.35) is

$$\frac{PV}{NkT} = 1 + \frac{1}{2^{5/2}}\xi^{-1} + \cdots . \tag{7.38}$$

Thus we have an expansion for the pressure in powers of $\xi^{-1} = \lambda^3/(gv)$ which is proportional to the density. Such a density expansion is called a "virial expansion," and in the equation of state for the pressure P, the coefficient of $(N/V)^2$ [(which can be found from the second term on the right-hand side of Eq. (7.38)] is called the "second virial coefficient." In classical physics, the second virial coefficient results entirely from interactions between the atoms or molecules of the gas. Quantum mechanics tells us that there is also an effective interaction due to the statistics obeyed by the particles. Here we see that, as expected, the effective repulsion induced by the Pauli principle tends to increase the pressure at second order in the density. Under normal conditions (e.g., at room temperature and atmospheric pressure) quantum corrections are extremely small. Even for the most quantum gas, H_2, T_Q is of order 1 Kelvin. (See Exercise 3.) One may also view the right-hand side of Eq. (7.38) as an expansion in powers of Planck's constant h in which only terms of order h^{3k} appear, where k is an integer. The leading quantum correction is thus of order h^3.

7.2.2 Low Temperature

In the low-temperature limit, this system is referred to as a "degenerate Fermi gas." For $T = 0$, the Fermi function

$$f(\epsilon) = \frac{1}{e^{\beta(\epsilon-\mu)} + 1} \tag{7.39}$$

becomes a step function which is unity for $\epsilon < \mu$ and zero for $\epsilon > \mu$. Thus, at $T = 0$ all the single-particle states with energy less than $\mu(T = 0)$ are occupied and those with energy greater than $\mu(0)$ are unoccupied. Then the relation between the number of particles and the chemical potential is

$$\begin{aligned} N &= \frac{gV}{4\pi^2}\left(\frac{2m}{\hbar^2}\right)^{3/2}\int_0^{\mu(0)}\sqrt{\epsilon}d\epsilon \\ &= \frac{2}{3}\frac{gV}{4\pi^2}\left(\frac{2m\mu(0)}{\hbar^2}\right)^{3/2} , \end{aligned} \tag{7.40}$$

where $\mu(0)$, the chemical potential at $T = 0$, is called the Fermi energy, ϵ_F. This relation can be inverted to give

$$\mu(0) = \frac{\hbar^2}{2m}\left(\frac{6\pi^2}{g}\frac{N}{V}\right)^{2/3} \equiv \epsilon_F \equiv kT_F , \tag{7.41}$$

and we see that, as expected, T_F, the Fermi temperature, differs from the "quantum temperature" T_Q only by a numerical factor.

The energy of the system at $T = 0$ is

$$\begin{aligned} E_0 &= \frac{gV}{4\pi^2}\left(\frac{2m}{\hbar^2}\right)^{3/2}\int_0^{\epsilon_F}\epsilon^{3/2}d\epsilon \\ &= \frac{2}{5}\frac{gV}{4\pi^2}\left(\frac{2m}{\hbar^2}\right)^{3/2}\mu(0)^{5/2} \\ &= \frac{3}{5}\epsilon_F N \end{aligned} \tag{7.42}$$

and since the pressure is 2/3 the energy per unit volume, one has

$$P = \frac{2}{5}\epsilon_F N/V . \tag{7.43}$$

For $0 < kT << \epsilon_F$ the calculation of the temperature dependence of P and E is more complicated. This involves evaluating integrals of the form

$$I(T) = \int_0^\infty \frac{g(\epsilon)d\epsilon}{e^{\beta(\epsilon-\mu)} + 1} . \tag{7.44}$$

We will evaluate $I(T)$ using the Sommerfeld expansion. This approach relies on the fact that the derivative of a step function is a δ-function, and hence that the derivative of the Fermi function at low temperature, although not exactly a δ-function, is nevertheless sharply peaked near $\epsilon = \mu$. Integrating by parts, we obtain

$$I(T) = \frac{G(\epsilon)}{e^{\beta(\epsilon-\mu)}+1}\Bigg|_0^\infty + \beta\int_0^\infty \frac{G(\epsilon)d\epsilon}{\left(e^{\beta(\epsilon-\mu)/2}+e^{-\beta(\epsilon-\mu)/2}\right)^2} \tag{7.45}$$

where $G(\epsilon) = \int_0^\epsilon g(\epsilon')d\epsilon'$. For most problems of interest $G(\epsilon)e^{-\beta\epsilon} \to 0$ as $\epsilon \to \infty$, in which case the boundary terms drop out. Also, one sees that the denominator in Eq. (7.45) becomes very large when $\epsilon - \mu$ is large in magnitude compared to kT. That being the case, we may take the lower limit of integration to be $\epsilon = -\infty$. Then, in terms of the integration variable $\Delta \equiv \epsilon - \mu$, we have

$$\begin{aligned} I(T) &= \int_{-\infty}^\infty G(\mu+\Delta)\left(\frac{\beta}{\left(e^{\beta\Delta/2}+e^{-\beta\Delta/2}\right)^2}\right)d\Delta \\ &\equiv \int_{-\infty}^\infty G(\mu+\Delta)H(\Delta)d\Delta\,. \end{aligned} \tag{7.46}$$

At low temperature β is large and the function $H(\Delta)$ is very sharply peaked about $\Delta = 0$ with a width which restricts Δ to be of order kT. (Indeed, in the limit $\beta \to \infty$, $H(\Delta)$ is a representation of the delta function.) This observation suggests that by expanding G in a Taylor series in Δ we will in fact generate an expansion for $I(T)$ in powers of kT. As we will see in a moment, the expansion parameter is actually T/T_F. Since the function $H(\Delta)$ is an even function of Δ only even powers of Δ survive the integration and the expansion turns out to be one in powers of $(T/T_F)^2$. Explicitly we have

$$\begin{aligned} I(T) &= G(\mu)\int_{-\infty}^\infty \frac{dx}{(e^{x/2}+e^{-x/2})^2} \\ &\quad + \frac{(kT)^2}{2}G''(\mu)\int_{-\infty}^\infty \frac{x^2dx}{(e^{x/2}+e^{-x/2})^2} + \ldots \\ &= \int_0^\mu g(\epsilon)d\epsilon + \frac{\pi^2}{6}(kT)^2 g'(\mu) + \ldots\,. \end{aligned} \tag{7.47}$$

It is worth noting here that the Sommerfeld expansion will work poorly if there is structure on the scale of T in the density of states at the Fermi energy. This can happen, for example, in narrowband systems close to a "van Hove singularity" in the density of states. In such cases, due care must be exercised in calculating the thermal properties of the Fermionic system. The general term of this expansion is given in a convenient form in Ashcroft and Mermin (1976). There one sees that this expansion is only an *asymptotic* expansion, since eventually the high-order terms lead to a divergent series. In most cases, the first correction term in powers of T/T_F (the one we show here) is the only useful one.

We now apply the Sommerfeld expansion to the integrals in Eqs. (7.27) and (7.29). To get U we apply Eq. (7.47) with $g(\epsilon) = A\epsilon^{3/2}$ and to get N we set $g(\epsilon) = A\epsilon^{1/2}$ with $A = [gV/(4\pi^2)][2m/\hbar^2]^{3/2}$, in which case we find

$$U = \left(\frac{gV}{4\pi^2}\right)\left(\frac{2m}{\hbar^2}\right)^{\frac{3}{2}}\left(\frac{2}{5}\mu^{\frac{5}{2}}\right)\left(1 + \frac{5\pi^2}{8}\frac{(kT)^2}{\mu^2} + \cdots\right) \tag{7.48}$$

$$N = \left(\frac{gV}{4\pi^2}\right)\left(\frac{2m}{\hbar^2}\right)^{\frac{3}{2}}\left(\frac{2}{3}\mu^{\frac{3}{2}}\right)\left(1 + \frac{\pi^2}{8}\frac{(kT)^2}{\mu^2} + \cdots\right). \tag{7.49}$$

We can solve Eq. (7.49) to get μ as a function of N, T, and V as

$$\begin{aligned}\mu(T) &= \mu(T=0)\left(1 + \frac{\pi^2}{8}\frac{(kT)^2}{\mu^2} + \cdots\right)^{-2/3} \\ &= \epsilon_F\left(1 - \frac{\pi^2}{12}\frac{(kT)^2}{\epsilon_F^2}\right),\end{aligned} \tag{7.50}$$

correct to order $(kT/\epsilon_F)^2$. This gives the leading temperature dependence of μ. (As expected μ decreases as the temperature increases.)

We now determine U as a function of N, T, and V. We write U as

$$\begin{aligned}U &= E_0\left(\frac{\mu}{\epsilon_F}\right)^{5/2}\left[1 + \frac{5\pi^2}{8}\left(\frac{kT}{\mu}\right)^2\right] \\ &= E_0\left[1 - \frac{5\pi^2}{24}\left(\frac{kT}{\epsilon_F}\right)^2\right]\left[1 + \frac{5\pi^2}{8}\left(\frac{kT}{\epsilon_F}\right)^2\right] \\ &= \frac{3}{5}N\epsilon_F\left[1 + \frac{5\pi^2}{12}\left(\frac{kT}{\epsilon_F}\right)^2\right],\end{aligned} \tag{7.51}$$

correct to order $(kT)^2/\epsilon_F^2$, where we have used Eq. (7.42).

The specific heat at constant volume is then

$$C_V = \left.\frac{\partial U}{\partial T}\right|_{N,V} = Nk\frac{\pi^2}{2}\frac{kT}{\epsilon_F}. \tag{7.52}$$

This result has an intuitive explanation. We know (from the equipartition theorem) that in classical mechanics (where statistics play no role) the specific heat of translation is $(3/2)k$ per particle. If one looks at the Fermi function at low temperature (shown in Fig. 7.2), one sees that most of the states are occupied and such particles cannot change their energy. It is only the small fraction of states within kT of ϵ_F which can gain or lose energy in response to a change in the temperature of the system. If all the particles could play a role in the specific heat, the specific heat

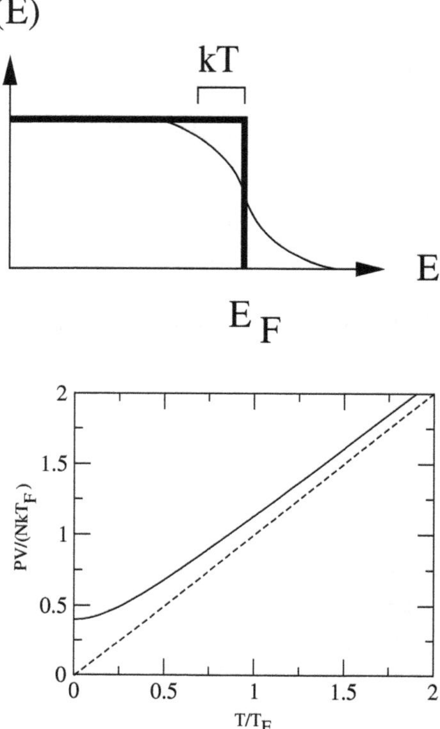

Fig. 7.2 Fermi distribution function $f(E) \equiv \langle n_{\mathbf{k},\sigma} \rangle_T$, the average number of particles in the state labeled $\mathbf{k}$, σ with $\epsilon_{\mathbf{k}} = E$, at zero temperature (heavy solid line) and at low temperature (light line). The distribution is only modified over a range of energy of order kT

Fig. 7.3 Solid line: $pV = 2U/3$ for the ideal Fermi gas. Dashed line: same for the classical monatomic Gibbs gas, to which the quantum gas asymptotes at infinite temperature. Note that the statistical repulsion of Fermions causes the pressure to be greater for the quantum system than for the analogous classical system

would be of order Nk. However, only a fraction, of order kT/ϵ_F, can play a role in the specific heat. This argument leads to the estimate $C_V \sim Nk^2T/\epsilon_F$, which is clearly the correct order of magnitude. This same logic will be used to estimate the magnetic susceptibility associated with the spin degrees of freedom.

Equations (7.30) and (7.31) are easily evaluated numerically. In Fig. 7.3, we show the pressure (which is proportional to the internal energy) of the noninteracting Fermi gas, compared to the ideal Gibbs gas, in which the particles are indistinguishable but otherwise quantum statistics are ignored. Several general comments are in order. First, we see that at a given V and T, the pressure of the Fermi gas is larger than that of the Gibbs gas, as one would expect from the statistical repulsion of Fermions. This effect becomes progressively more important as the temperature is lowered and indeed the Fermi pressure at zero temperature is nonzero, whereas that of the Gibbs gas is zero. Also, whereas $dU/dT)_V$ is constant for the Gibbs gas, the specific heat of the Fermi gas correctly (in view of the third law) goes to zero at zero temperature. The specific heat as a function of temperature is shown in Fig. 7.4.

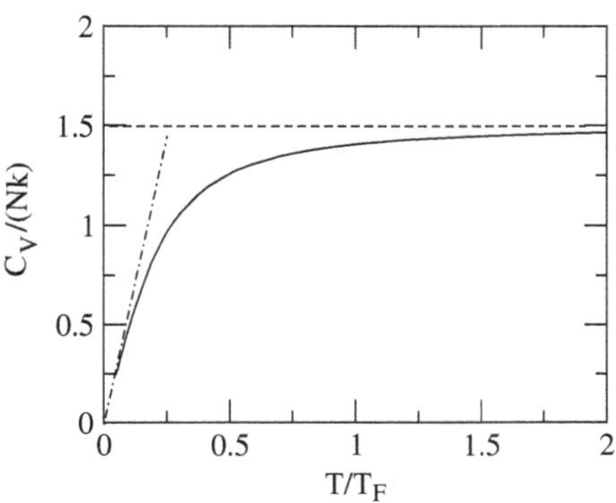

Fig. 7.4 Solid line: Specific heat of the ideal Fermi gas. The dashed line is the result for the classical Gibbs gas which the quantum result approaches in the high-temperature limit. The dash-dot line is the linear term whose slope is $\pi^2/2$

7.2.3 Spin Susceptibility at Low Temperature

Here, we consider the magnetic moment due to the spin of an electron when the electron gas is in an external magnetic field, H. This gives rise to the so-called Pauli susceptibility of the free electron gas. Note that here we are not considering the magnetic moment due to orbital motion. For most metals, the susceptibility is dominated by the spin susceptibility.

Since we consider noninteracting electrons, the energy of the system is just the sum the single-particle energies of the occupied states. The single-particle are now given by

$$\epsilon_{\mathbf{k},S} = \frac{\hbar^2 k^2}{2m} - g\mu_B HS \ , \tag{7.53}$$

where μ_B is the Bohr magneton ($\mu_B \sim 10^{-20}$ergs/Gauss), $g = 2$, and $S = \pm 1/2$ is the component of the spin of the electron in the direction of the magnetic field. Note that the energy associated with the magnetic field is normally a small perturbation. If the magnetic field is 1 Tesla (10^4 Gauss), then the energy $g\mu_B H$ is of order 10^{-16} ergs and since $k = 1.4 \times 10^{-16}$ergs/degree K, this corresponds to kT for T of order 1 Kelvin. Usually, ϵ_F is of order an electron volt in which case $T_F = \epsilon_F/k$ is of order 10^4 Kelvin. Thus, the effect of a magnetic field on the Fermi distribution is normally perturbative.

To emphasize the symmetry of this situation, let us consider the problem for which the *direction* of the magnetic field is arbitrary: $\mathbf{H} = H\hat{n}$, where $\hat{n}$ is a unit vector in some arbitrary direction. Then the single-particle energies (we still assume noninteracting electrons) are given by

$$\epsilon_{\mathbf{k},\mathbf{S}} = \frac{\hbar^2 k^2}{2m} - g\mu_B H\mathbf{S} \cdot \hat{n} \ , \tag{7.54}$$

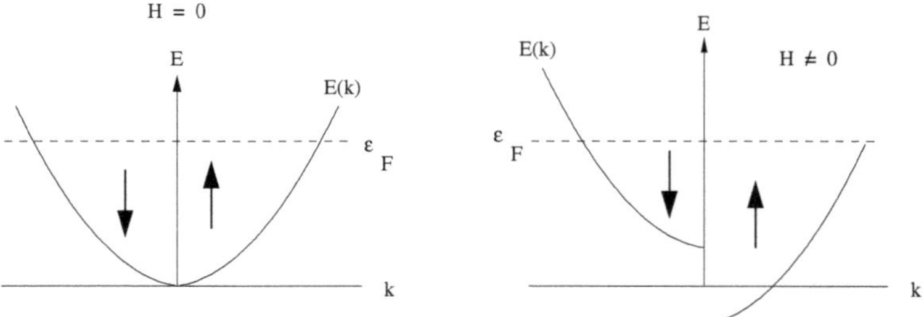

Fig. 7.5 Energy spectrum versus wavenumber k for zero field (left) and nonzero field (right). The energy of up-spin electrons is plotted for positive k and the energy of down-spin electrons is plotted for negative k. At zero temperature, all states with energy up to ϵ_F are occupied

where $\mathbf{S}$ is a vector operator. Here, we can take the "up" direction to be $\hat{n}$ and the "down" direction to be $-\hat{n}$. In that case, the eigenvalues of $\mathbf{S}\cdot\hat{n}$ are $\pm 1/2$ and we write

$$\epsilon_{\mathbf{k},S} = \frac{\hbar^2 k^2}{2m} - g\mu_B H S \,, \tag{7.55}$$

where now S assumes the values $\pm 1/2$, and

$$\epsilon_{\mathbf{k},\downarrow} - \epsilon_{\mathbf{k},\uparrow} = g\mu_B H \,. \tag{7.56}$$

We therefore have the situation shown in Fig. 7.5.

Note that the chemical potential for up-spin electrons is the *same* as for down-spin electrons. Even very weak interactions between up- and down-spin electrons or with their environment are sufficient to allow them to attain thermal equilibrium. A magnetic moment is induced by a field because more energy levels of up-spin than down-spin are occupied.

We are now going to show that the grand partition function for $H \neq 0$ can be related exactly to the grand partition function for $H = 0$ but with a shifted chemical potential. To see this we write Eq. (7.6) as

$$\begin{aligned}\Xi(T,V,\mu) &= \sum_{\{n_{\mathbf{k},\uparrow}\}}\left(\prod_{\mathbf{k}} e^{-\beta n_{\mathbf{k},\uparrow}(\epsilon_{\mathbf{k},\uparrow}-\mu)}\right)\sum_{\{n_{\mathbf{k},\downarrow}\}}\left(\prod_{\mathbf{k}} e^{-\beta n_{\mathbf{k},\downarrow}(\epsilon_{\mathbf{k},\downarrow}-\mu)}\right)\\ &= \sum_{\{n_{\mathbf{k},\uparrow}\}}\left(\prod_{\mathbf{k}} e^{-\beta n_{\mathbf{k},\uparrow}\left[\frac{\hbar^2k^2}{2m}-(\mu+\mu_B H)\right]}\right)\sum_{\{n_{\mathbf{k},\downarrow}\}}\left(\prod_{\mathbf{k}} e^{-\beta n_{\mathbf{k},\downarrow}\left[\frac{\hbar^2k^2}{2m}-(\mu-\mu_B H)\right]}\right)\end{aligned} \tag{7.57}$$

where each occupation number $n_{\mathbf{k},\sigma}$ assumes the values 0 or 1 in accordance with Fermi statistics.

So the grand partition function is a product over "up-spin" sums taken at an effective chemical potential $\mu = \mu + \mu_B H$ times another product over "down-spin" sums taken at an effective chemical potential $\mu - \mu_B H$, but we emphasize that the actual chemical potential, μ, of both spin species is the same. Expressing the grand partition function in terms of different effective chemical potentials is only a mathematical artifice. For $H = 0$, we have the product over both spin states. So the sums over a single spin direction represent the square root of the grand partition function. Thus, we may write Eq. (7.57) as

$$\Xi(\mu, T, V, H) = \left(\Xi_0(\mu + \mu_B H, T, V) \right)^{1/2} \left(\Xi_0(\mu - \mu_B H, T, V) \right)^{1/2}, \quad (7.58)$$

or

$$\begin{aligned} -kT \ln \Xi(\mu, T, V, H) &= \\ &= -1/2\Big(kT \ln \Xi_0(\mu + \mu_B H, T, V) + kT \ln \Xi_0(\mu - \mu_B H, T, V) \Big) \\ \Omega(\mu, T, V, H) &= 1/2\Big(\Omega_0(\mu + \mu_B H, T, V) + \Omega_0(\mu - \mu_B H, T, V) \Big) . \end{aligned} \quad (7.59)$$

where the subscript "0" indicates the $H = 0$ function. Thus, the grand partition function in a nonzero magnetic field is exactly related to the grand partition function in zero field.

Next we address the thermodynamics. Generalizing Eq. (6.12) to nonzero magnetic field, we can write

$$N = -\frac{\partial \Omega(\mu, T, V, H)}{\partial \mu} \Bigg)_{T,V,H} \quad (7.60)$$

which allows us to calculate μ as a function of N, T, V, and H, so that $\mu = \mu(N, T, V, H)$. Then, we can get the free energy, F, as a function of N, T, V, and H via the analog of Eq. (6.10):

$$F(N, T, V, H) = \Omega(\mu(N, T, V, H), T, V, H) + N\mu(N, T, V, H) . \quad (7.61)$$

This free energy obeys

$$dF = -SdT - PdV + \mu dN - MdH , \quad (7.62)$$

where M is the thermally averaged magnetic moment of the system.

We want to evaluate the isothermal susceptibility at zero field. We start by writing the magnetic moment as

$$M = -\frac{\partial F}{\partial H}\Bigg|_{N,T,V,H=0} . \quad (7.63)$$

Then the susceptibility at $H = 0$ is

$$\chi_0 = \left.\frac{\partial M}{\partial H}\right|_{N,T,V,H=0} = -\left.\frac{\partial^2 F}{\partial H^2}\right|_{N,T,V,H=0} . \tag{7.64}$$

From Eq. (7.61), we know that F is a function of Ω and of μ which is implicitly a function of Ω through Eq. (7.60). Furthermore, from Eq. (7.59), it is clear that $\Omega(\mu(N, T, V, H), T, V, H)$ is an even function of H. So the leading field-dependent term in the expansion of the free energy in powers of H is of order H^2. That is,

$$F(H) = F(0) - \frac{\chi_0 H^2}{2} + \mathcal{O}(H^4) , \tag{7.65}$$

and the coefficient of $-H^2/2$ in the free energy is the zero-field susceptibility χ_0.

Looking again at Eq. (7.61), we see that H enters in two ways: first as an argument of Ω and also indirectly as an argument of μ which appears in Ω and also in the second, $N\mu$, term. We now show that we need not consider explicitly the dependence of μ on H.

To see this write

$$\mu(N, T, V, H) = \mu_0 + \mu_1 H^2 + \mu_2 H^4 + \cdots \tag{7.66}$$

where the coefficients μ_k each depend on N, T, and V. Now substitute the expansion Eq. (7.66) into Eq. (7.61) to get (correct to order H^4)

$$\begin{aligned} F &= N[\mu_0 + \mu_1 H^2 + \mu_2 H^4] + \Omega(\mu_0, T, V, H] \\ &\quad + \frac{\partial \Omega}{\partial \mu}[\mu_1 H^2 + \mu_2 H^4] + 1/2 \frac{\partial^2 \Omega}{\partial \mu^2}[\mu_1 H^2]^2 + \cdots \end{aligned}$$

correct to order H^4. The terms of order H^2 vanish because

$$N + \frac{\partial \Omega}{\partial \mu} = 0 .$$

If we were working to order H^4 (to get the nonlinear susceptibility $\partial^3 M/\partial H^3$), we would have to keep track of the dependence of μ on H. But, as it is, we can replace μ everywhere by $\mu_0(N, T, V)$, the solution for μ in zero magnetic field.

As a result of the above calculations, we can write, correct to order H^2,

$$\begin{aligned} F(N, T, V, H) &= N\mu_0(N, T, V) + \frac{\Omega_0(\mu_0 + \mu_B H, T, V) + \Omega_0(\mu_0 - \mu_B H, T, V)}{2} \\ &= N\mu_0(N, T, V) + \Omega_0(\mu_0(N, T, V), T, V) + \frac{(\mu_B H)^2}{2}\frac{\partial^2 \Omega_0}{\partial \mu^2} . \end{aligned} \tag{7.67}$$

Thus we get the zero-field susceptibility in terms of quantities of the zero-field system as

$$\begin{aligned}\chi_0(N,T,V) &= -\mu_B^2 \frac{\partial^2 \Omega_0}{\partial \mu^2}\bigg|_{T,V} \\ &= \mu_B^2 \frac{\partial N}{\partial \mu}\bigg|_{T,V,H=0} .\end{aligned} \tag{7.68}$$

The above result can be used at any temperature. For the most common application, we now specialize to low temperature. To evaluate $\partial N/\partial \mu)_{T,V}$ (at $H=0$), we start from Eq. (7.49) which we write as

$$N = \frac{gV}{6\pi^2}\left(\frac{2m\mu}{\hbar^2}\right)^{3/2}\left(1+\frac{(\pi kT)^2}{8\mu^2}\right).$$

In the zero-temperature limit, then, since $N \propto \mu^{3/2}$, we may write

$$\frac{\partial N}{\partial \mu}\bigg|_{T,V} = \frac{3N}{2\mu} \tag{7.69}$$

and the spin susceptibility, which is known as the **Pauli susceptibility** for the degenerate Fermi gas, is

$$\chi = \mu_B^2 \frac{\partial N}{\partial \mu}\bigg|_{T,V} = \frac{3N\mu_B^2}{2\epsilon_F} = \mu_B^2 \rho(\epsilon_F) , \tag{7.70}$$

where the last equality holds for a general density of states. Again one can understand this result at low temperature by a simple hand-waving argument. For free spins (i.e., no Fermi statistics), we previously obtained the result

$$\chi = N\frac{g^2\mu_B^2 S(S+1)}{3kT} = N\frac{\mu_B^2}{kT} .$$

But for a Fermi gas at low temperature, $kT \ll \epsilon_F$, Fermi statistics do not permit spins outside a region in energy of order kT to respond to the external field. So we should again replace N by NkT/ϵ_F, as suggested in Fig. 7.2. This gives the correct estimate: $\chi \sim N\mu_B^2/\epsilon_F$.

7.2.4 *White Dwarfs*

It is believed that white dwarfs are composed of light elements, such as carbon and oxygen. The important observation is that the temperature at the center of the star is estimated to be of order 10^7K, which corresponds to an energy kT of order 1KeV.

(The temperature can be determined from the wavelength at the maximum intensity of radiation.) Thus all the atoms are ionized. As we will see later, ϵ_F, the Fermi energy of the electrons, is of order 1 MeV. That being the case, $kT \ll \epsilon_F$ and from the point of view of statistical mechanics the electrons in the star are essentially at zero temperature!! White dwarfs are stars with a small radius (i.e., a radius of order the radius of the earth). The radius, R, of a star can be determined from its luminosity which is given by $4\pi R^2 \sigma T^4$, where σ is the Stefan–Boltzmann radiation constant. The mass of white dwarfs are of order one solar mass. For more introductory information on white dwarfs, see (Shu 1982).

If the star has N atoms, it has NZ electrons and NA nucleons, where Z is average the atomic number and A is the average atomic mass number of the elements present in the star. Then its mass is

$$M = ANm_p \, ,$$

where m_p is the proton mass. Also its electron density is

$$n = \frac{ZN}{V} \approx \frac{ZM/Am_p}{V} = (Z/A)(\rho/m_p) \, .$$

Note that for light elements $(A/Z) \approx 2$, whether the star consists of helium or some other element(s).

First let us discuss Coulomb interactions. At low density, the Coulomb energy binds electrons and nuclei into atoms, with Z electrons bound to a nucleus with atomic number Z. However, at high enough density the electrons dissociate from the nuclei. We can take the effect of Coulomb interactions into account in an approximate way by assuming the nucleons and electrons both occupy the same spherical volume of the star. Then, we can neglect Coulomb interactions within this electrically neutral system. This same type of approximation is used to describe metals by a noninteracting gas of electrons superimposed on a uniform background of positive charge.

Next we consider the zero-point energy of the electron gas (which wants to make the star expand) and the gravitational energy of the nucleons (which wants to collapse the star). The zero-point kinetic energy of the nucleons confined to the volume of the star is negligible in comparison to the energy needed to confine electrons to this same volume. So we will consider only the gravitational potential energy of the nucleons and the zero-point kinetic energy of the electrons. The first energy involves G, whereas the second involves the constants h and c (because the highest-energy electrons are relativistic). This theory then represents a marriage of these three fundamental constants.

It is interesting to note that S. Chandrasekhar developed the beautiful theory we are about to discuss *before* he started his graduate studies in England. Revolutionary ideas are not always met with instant acclaim by the establishment, and here we have a striking example of this phenomenon. A successful career in England became impossible for Chandrasekhar, and it required great persistence on his part before his theory became accepted by the more senior scientists and his career continued at the University of Chicago. He received the Nobel prize in 1983 for his theory of white dwarfs, which is now regarded as one of the intellectual milestones in astrophysics. (Chandrasekhar 1983).

As mentioned, $kT \ll \epsilon_F$, in a white dwarf, so that the electrons are in their Fermi sea ground state. Thus, for N_e spin 1/2 electrons, we write

$$N_e = ZN = \frac{2V}{8\pi^3} \int_0^{k_F} 4\pi k^2 dk = \frac{2V}{h^3} \int_0^{p_F} 4\pi p^2 dp = \frac{8\pi V}{3h^3} p_F^3 , \quad (7.71)$$

where $p_F = \hbar k_F$ is the Fermi momentum (i.e., the momentum of an electron having the Fermi energy). This relation indicates that $p_F \propto V^{-1/3} \propto 1/R$. The kinetic energy, E_e, of the electron gas is

$$E_e = \frac{2V}{h^3} \int_0^{p_F} \epsilon(p) 4\pi p^2 dp = \frac{8\pi V}{h^3} \int_0^{p_F} \left(\sqrt{m^2c^4 + p^2c^2} - mc^2 \right) p^2 dp .$$

As we know from our previous study of the Fermi gas, there is a large zero-point pressure reflecting the repulsion due to Fermi statistics. This repulsion is overcome by the gravitational potential energy of the nucleons, an energy $-\gamma GM^2/R$, where γ is a fudge factor to take account of the inhomogeneity of the star. Thus the total energy is

$$E_{\rm TOT} = \frac{8\pi V}{h^3} \int_0^{p_F} \left(\sqrt{m^2c^4 + p^2c^2} - mc^2 \right) p^2 dp - \gamma GM^2/R .$$

The condition for equilibrium is then

$$\frac{dE_{\rm TOT}}{dR} = 0 . \quad (7.72)$$

For the differentiation, we use $dV/dR = 3V/R$ and $dp_F/dR = -p_F/R$, in which case we obtain

$$\begin{aligned} -\frac{dE_e}{dR} &= -\frac{24\pi V}{h^3 R} \int_0^{p_F} \epsilon(p) p^2 dp - \frac{8\pi V}{h^3} \left(-\frac{p_F}{R} \right) \epsilon(p_F) p_F^2 \\ &= \frac{8\pi V}{h^3 R} \int_0^{p_F} \frac{d\epsilon(p)}{dp} p^3 dp , \end{aligned} \quad (7.73)$$

where we obtained the last line by integrating by parts. Then Eq. (7.72) is

$$\gamma \frac{GM^2}{R^2} = \frac{8\pi V}{h^3 R} \int_0^{p_F} \frac{p^4 c^2 dp}{\sqrt{m^2c^4 + p^2c^2}} . \quad (7.74)$$

To get rid of the square root, set

$$p = mc \sinh\theta \, .$$

Then, we get

$$\gamma \frac{GM^2}{R^2} = \frac{8\pi V}{h^3 R} \int_0^{\theta_F} \frac{m^5 c^7 \sinh^4\theta \cosh\theta d\theta}{\sqrt{m^2c^4 + m^2c^4 \sinh^2\theta}} = \frac{8\pi V m^4 c^5}{h^3 R} \int_0^{\theta_F} \sinh^4\theta d\theta \, ,$$

where, from Eq. (7.71)

$$ZN = \frac{8\pi V m^3 c^3}{3h^3} \sinh^3\theta_F \, ,$$

so that

$$\sinh\theta_F = \left(\frac{3ZN}{8\pi V}\right)^{1/3} \left(\frac{h}{mc}\right) \equiv x \, . \tag{7.75}$$

Thus

$$\gamma \frac{GM^2}{R^2} = \frac{\pi V m^4 c^5}{h^3 R} A(x) \, , \tag{7.76}$$

where

$$A(x) = 8 \int_0^{\sinh^{-1} x} \sinh^4\theta d\theta = x(x^2+1)^{1/2}(2x^2-3) + 3\sinh^{-1} x \, ,$$

and if we use $N = M/(Am_p)$ we can write Eq. (7.75) as

$$x = \left(\frac{9\pi Z M}{4Am_p}\right)^{1/3} \frac{\hbar/(mc)}{R} \, . \tag{7.77}$$

Thus Eq. (7.76) is

$$\begin{aligned} A(x) \equiv A\left(\left\{\frac{9\pi Z M}{4Am_p}\right\}^{1/3} \frac{\hbar/mc}{R}\right) &= \frac{3\gamma h^3}{4\pi^2 m^4 c^5} \frac{GM^2}{R^4} \\ &= 6\pi\gamma \left(\frac{\hbar/mc}{R}\right)^3 \frac{GM^2/R}{mc^2} \, . \end{aligned} \tag{7.78}$$

Since we cannot solve this analytically, we first look in asymptotic regimes of small or large x. The order of magnitude values is $M = 10^{30}$ kg (the solar mass is 2×10^{30} kg) $m_p \sim 2 \times 10^{-27}$ kg, $\hbar/mc \sim 4 \times 10^{-13}$ m. So $x = 1$ when $R = 4 \times 10^6$ m, i.e., 2500 miles!! We see that large x corresponds to large mass (compared to the solar mass) and small radius (compared to the radius of the earth), whereas small x corresponds to small mass and large radius.

Notice that the mass of the star relative to the proton mass enters, the gravitational energy of the nucleons enters, and also the radius of the star in terms of the Compton wavelength of the electron enters. This theory therefore has a unique mixture of quantum mechanics, special relativity, and gravitation.

Note that since $p_F = mcx$, it means that large x is highly relativistic, whereas small x is nonrelativistic. In the latter case, $\epsilon_F = p_F^2/(2m) = 1/2(mc^2)x^2 \approx x^2/4$ MeV. So since $kT \approx 1$ KeV, we are in the zero-temperature limit unless x is extremely small.

In the limit $x \ll 1$ (but with x still large enough so that $kT \ll \epsilon_F$),

$$A(x) \approx 8 \int_0^x \theta^4 d\theta = 8/5x^5 ,$$

so that

$$\frac{M^{5/3}}{R^5} = \frac{M^2}{R^4} ,$$

i.e., $R \sim M^{-1/3}$.

In the limit $x \gg 1$, $A(x) \sim 2x^4 - 2x^2$, so that

$$2\left(\frac{\hbar/mc}{R}\right)^4 \left(\left\{\frac{9\pi ZM}{4Am_p}\right\}^{4/3}\right) - 2\left(\frac{\hbar/mc}{R}\right)^2 \left(\left\{\frac{9\pi ZM}{4Am_p}\right\}^{2/3}\right)$$
$$= \left(\frac{\hbar/mc}{R}\right)^4 6\pi\gamma\frac{GM^2}{\hbar c} ,$$

which is of the form

$$B(\tilde{R})^{-4} - C(\tilde{R})^{-2} = D(\tilde{R})^{-4} ,$$

where $\tilde{R} = Rmc/\hbar$. Thus for $D \le B$, we get the solution as

$$R = \frac{\hbar}{mc}\left\{\frac{B-D}{C}\right\}^{1/2} \tag{7.79}$$

and there is no solution if $D > B$.

Note that $B \sim M^{4/3}$, $D \sim M^2$, and $C \sim M^{2/3}$. Thus we may write Eq. (7.79) as

$$R = R_0\left(\frac{M}{M_0}\right)^{1/3}\left\{1 - \left(\frac{M}{M_0}\right)^{2/3}\right\}^{1/2} , \tag{7.80}$$

where M_0 is determined by $B(M_0) = D(M_0)$ or

$$2\left(\frac{9\pi Z M_0}{4Am_p}\right)^{4/3} = 6\pi\gamma\frac{GM_0^2}{\hbar c}\,, \tag{7.81}$$

which gives

$$M_0^{2/3} = \left(\frac{9\pi Z}{4Am_p}\right)^{4/3}\left(\frac{\hbar c}{3\pi\gamma G}\right)\,. \tag{7.82}$$

So (with $A/Z = 2$)

$$M_0 = \frac{9}{64}\left(\frac{3\pi}{\gamma^3}\right)^{1/2}\left(\frac{[\hbar c/G]^{3/2}}{m_p^2}\right) \approx 10^{30}\text{ kg}\,. \tag{7.83}$$

Thus there is a limit, called the *Chandrasekhar limit*, for the mass, above which no solution exists. This mass turns out to be about 1.44 solar masses. This prediction is verified in the sense that no white dwarfs with grater than this mass have been found. Also in Eq. (7.80)

$$R_0\left(\frac{M}{M_0}\right)^{1/3} = \frac{\hbar}{mc}\left(\frac{B}{C}\right)^{1/2} \tag{7.84}$$

so that

$$R_0 = \frac{\hbar}{mc}\left(\frac{9\pi Z}{4Am_p}\right)\left(\frac{\hbar c}{3\pi\gamma G}\right)^{1/2}\,. \tag{7.85}$$

For $\gamma = 1$, this gives $R_0 \approx 4000$ km $= 2500$ miles, which is the same order as the radius of the earth. The result of this analysis is shown in Fig. 7.6. Note the most striking aspect of this result, namely, the radius decreases as a function of mass. In contrast, when gravity is not an important factor, as for planets, the radius increases with increasing mass.

For mass larger than the Chandrasekhar limit, the white dwarf is unstable. In this case, gravitation wins and one gets a neutron star. A neutron star can be treated crudely within a model which includes the zero-point kinetic energy of neutrons competing with their gravitational energy. (One has to assume that beta decay of a neutron into a proton, an electron, and a neutrino is prevented by a small density of electrons, whose presence can otherwise be neglected. See Exercise 12).

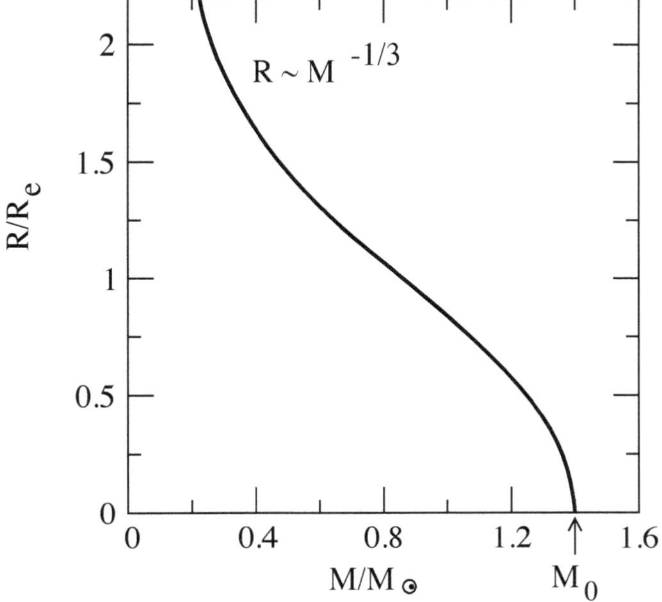

Fig. 7.6 Equation of state (radius vs. mass) of a white dwarf star. Here, M_0 is the Chandrasekhar limiting mass (3×10^{30} kg), R_e is the radius of the earth (6×10^6 m), and $M_\odot$ is the solar mass (2×10^{30} kg)

7.3 The Noninteracting Bose Gas

In this section, we will see that, although the thermodynamic properties of Bose and Fermi gases are very similar at high temperatures, they are distinctly different at low temperatures.

7.3.1 High Temperature

As mentioned earlier, the ground state of the noninteracting Bose gas is one in which all particles occupy the lowest single-particle energy eigenstate. That means that the Bose occupation number becomes singular for zero wavevector in the zero-temperature limit. In that case, it is incorrect to approximate the **k**-sums by integrals. At high temperatures, there is no such problem and the results for the Bose gas approach those for the Fermi gas for $T \to \infty$. For instance, consider Eqs. (7.23) and (7.24). There one sees that the transformation $z \to -z$ [recall that $z = \exp(\beta\mu)$] and $g \to -g$ takes one from the Fermi case to the Bose case. (This transformation does not enable us to deduce properties of the Bose-condensed phase, however.) At high temperature and low density, we can therefore use this transformation with Eq. (7.38) to obtain

$$P = \frac{NkT}{V}\left(1 - \frac{1}{2^{5/2}}\frac{N\lambda_D^3}{gV} + \dots\right). \tag{7.86}$$

This result is easily understood. For the Bose gas, particles preferentially occupy the same state. This statistical attraction causes the pressure of the noninteracting Bose gas to be less than that of the Gibbs gas. Thus for noninteracting systems at the same T and N, one has the following relation (shown in Fig. 7.3 and the left-hand panel of Fig. 7.10) between pressures of the Bose, Fermi, and Gibbs systems:

$$p_B < p_G < p_F . \tag{7.87}$$

It is also worth noting that these statistical corrections are only significant at rather high gas density. (See Exercise 3.)

7.3.2 *Low Temperature*

Now we consider the Bose gas at very low temperature where almost all the particles are in the lowest energy single-particle state, whose energy is denoted ϵ_0. The case of spin zero ($g = 1$), to which our discussion is restricted, is usually proposed as a starting point for the discussion of the properties of superfluid ^{4}He.

The discussion we are about to give can be made rigorous, but the main ideas are contained in the following simplified heuristic treatment. We start by considering the number of particles in the ground state, N_0 and the number N_1 in the lowest energy excited state, given by

$$N_0 = \frac{1}{e^{\beta(\epsilon_0-\mu)} - 1}, \tag{7.88}$$

$$N_1 = \frac{1}{e^{\beta(\epsilon_1-\mu)} - 1}. \tag{7.89}$$

Obviously, μ must be less than ϵ_0, for N_0 to be a finite positive number. Suppose $\epsilon_0 - \mu$ is of order $1/N$, so that N_0/N is nonzero in the thermodynamic limit. Since $\epsilon_1 - \epsilon_0$ is of order $L^{-2} \sim V^{-2/3}$, we see that $N_1/N_0 \to 0$ in the thermodynamic limit. Thus, the number of particles in any single-particle energy level other than the ground state is not macroscopic and it can also be shown that their contribution to the sums in Eqs. (7.14) and (7.15) may be replaced by an integral. In contrast, N_0 can become macroscopic, i.e., of order N, if $\epsilon_0 - \mu$ is of order $1/N$ in the limit of large N. This kind of discussion parallels that for any phase transition. One gets a true phase transition only in the thermodynamic limit when $N \to \infty$.

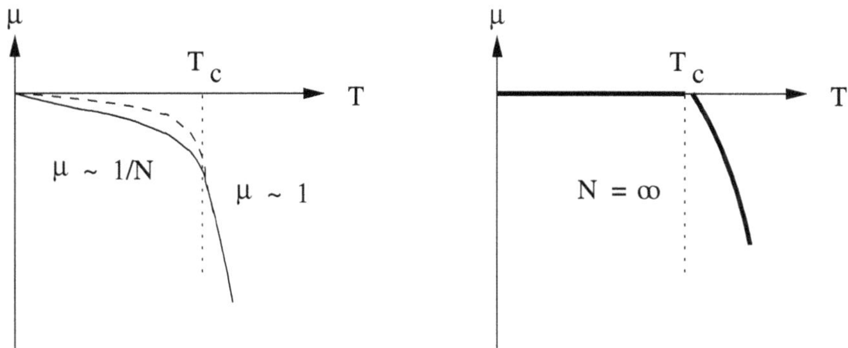

Fig. 7.7 Schematic behavior of μ versus T. Left: μ versus T for large (but finite) values of N and V, with N/V fixed. The dashed line shows the effect of approaching the thermodynamic limit by doubling the values of both N and V. In the "normal" regime for $T > T_c$ doubling N and V has only a very small effect on intensive variables such as μ. In the "Bose-condensed" regime, for $T < T_c$, since μ is proportional to $1/N$ in the thermodynamic limit, the dashed and solid curves differ by a factor of 2. Right: μ versus T in the thermodynamic limit

It is instructive to discuss the case when N is large, but not truly infinite. Then there are two regimes. In the "normal" regime, $\epsilon_0 - \mu$ is not so small as to be of order $1/N$. Then, N_0 is not macroscopic and the replacement of sums over single-particle states by integrals is permitted. $\epsilon_0 - \mu > 0$ means that in this normal regime

$$kT \gg \epsilon_0 - \mu \gg kT_Q/N \, , \tag{7.90}$$

where we have introduced the quantum temperature from Eq. (7.34) for consistency of units. In the "Bose-condensed" regime, $\epsilon_0 - \mu$ is of order $1/N$ and N_0/N is of order unity. For both regimes, we can write

$$\frac{N}{V} = \frac{N_0}{V} + \frac{1}{8\pi^3} \int_0^\infty \frac{d\mathbf{k}}{e^{\beta(\epsilon_{\mathbf{k}} - \mu)} - 1} + \mathcal{O}(V^{-1/3}) \, . \tag{7.91}$$

This equation works because in the normal regime the term N_0/V is negligible. Also the conversion from the sum to the integral introduces errors of order $V^{-1/3}$. In the Bose-condensed phase, one sets $\mu = \epsilon_0$ and N_0 is of order N so that the term N_0/V is non-negligible. This seemingly crude treatment can be substantiated by a rigorous evaluation for finite V (Pathria 1985) Henceforth, we set $\epsilon_0 = 0$.

We start by assuming that $N_0 = 0$ and ask whether the entire phase diagram can be described by the normal phase. For $N_0 = 0$, we have

$$\frac{N}{V} = \frac{1}{8\pi^3} \int_0^\infty \frac{d\mathbf{k}}{e^{\beta(\epsilon_{\mathbf{k}} - \mu)} - 1} \, . \tag{7.92}$$

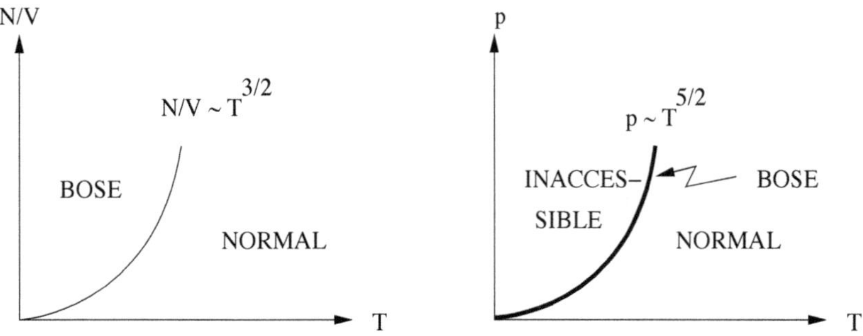

Fig. 7.8 Left: Phase diagram for the ideal Bose gas in the ρ-T plane. Right: Same in the $p - T$ plane. In each case, "Bose" refers to the Bose-condensed phase and "normal" to the normal liquid phase. In the p-T plane, the Bose-condensed phase is confined to the curve $p = p_c(T)$ and in the absence of interparticle interactions the regime $p > p_c(T)$ is inaccessible

As the temperature is reduced, μ will increase until it reaches $\mu = 0$, which we interpret as meaning that μ is negative and infinitesimal, satisfying Eq. (7.90) for $\epsilon_0 = 0$. In three dimensions, this happens at some $T = T_c > 0$. The behavior of μ as a function of T for fixed N and V is shown in Fig. (7.7), both for N large but finite in the left-hand panel and for $N \to \infty$ for fixed N/V in the right-hand panel. Since T_c is the temperature where μ first reaches zero, we can calculate it by setting $\mu = 0$ in Eq. (7.92) which gives

$$\frac{N}{V} = \frac{1}{4\pi^2}\left(\frac{2mkT_c}{\hbar^2}\right)^{3/2}\int_0^\infty \frac{\sqrt{x}dx}{e^x - 1} . \tag{7.93}$$

The integral has the value $\zeta(3/2)\sqrt{\pi}/2$, where ζ is the Riemann ζ-function. Equation (7.93) defines a curve in the density–temperature plane, as shown in the left panel of Fig. 7.8, which represents the phase boundary of the normal phase where μ is at its maximum permitted value ($\mu = 0$) and one cannot further reduce the temperature or increase the density. Eq. (7.93) gives the critical temperature, T_c, as a function of density as

$$\begin{aligned} kT_c &= \left(\frac{4\pi}{\zeta(3/2)^{2/3}}\right)\left(\frac{N}{V}\right)^{2/3}\frac{\hbar^2}{2m} = 6.625\left(\frac{N}{V}\right)^{2/3}\frac{\hbar^2}{2m} \\ &= 0.5272kT_Q . \end{aligned} \tag{7.94}$$

As expected, T_c is of order T_Q.

As we have just seen, in the normal liquid phase the temperature can not be less than T_c. However, since nothing prevents lowering the temperature below T_c, something else must happen. What happens is that the system goes into a new phase, not described by Eq (7.92). In this new phase, the single-particle ground state is macroscopically occupied and the term N_0/V in Eq. (7.91) is no longer negligible. In this Bose-condensed phase the chemical potential will satisfy

$$\mu = -A/N \ , \tag{7.95}$$

where the variable A is of order unity in the thermodynamic limit. Therefore, in the Bose-condensed phase, the grand potential is a function of T, V, and A (which plays the role of μ). However, we can equally well choose the independent variables to be T, V, and N_0, where N_0 is of order N because when Eq. (7.95) holds, then

$$N_0 = \frac{1}{e^{-\beta\mu} - 1} = \frac{kT}{-\mu} = \frac{NkT}{A} \ . \tag{7.96}$$

One can also argue for the existence of a Bose-condensed phase by considering what happens when we try to increase the density above the value on the critical line in the left panel of Fig. 7.8. This is not possible within the normal liquid phase, but in the absence of interparticle interactions there is no reason why the density cannot become arbitrarily large for nonzero values of N_0. Accordingly, there must exist a phase in which the term N_0/V is non-negligible.

Thus for $T < T_c$ the ground state is macroscopically occupied, and we can use Eq. (7.91) with $\mu = 0$, to write

$$N = N_0 + \frac{V}{4\pi^2} \left(\frac{2m}{\hbar^2} \right)^{3/2} \int_0^\infty \frac{\sqrt{\epsilon} d\epsilon}{e^{\beta\epsilon} - 1} \ . \tag{7.97}$$

Since the integral is proportional to $T^{3/2}$, we may write this as

$$N_0 = N \left[1 - \left(\frac{T}{T_c} \right)^{3/2} \right] \ . \tag{7.98}$$

The distinction between particles in the ground state and particles in excited states leads to a phenomenological "two-fluid" model (Fig. 7.9).

Since particles in the lowest energy single-particle state, $\epsilon_0 = 0$, contribute zero to the internal energy, we may write U for $T < T_c$ as

$$\begin{aligned} U &= \frac{1}{4\pi^2} \left(\frac{2m}{\hbar^2} \right)^{3/2} V \int_0^\infty \frac{\epsilon^{3/2} d\epsilon}{e^{\beta\epsilon} - 1} \\ &= \gamma N k T_c (T/T_c)^{5/2} \ , \end{aligned} \tag{7.99}$$

where $\gamma = (3/2)\zeta(5/2)/\zeta(3/2) \approx 0.770$. This gives a specific heat at constant volume, C_V, proportional to $T^{3/2}$:

$$C_V = (5/2)\gamma N k (T/T_c)^{3/2} \ . \tag{7.100}$$

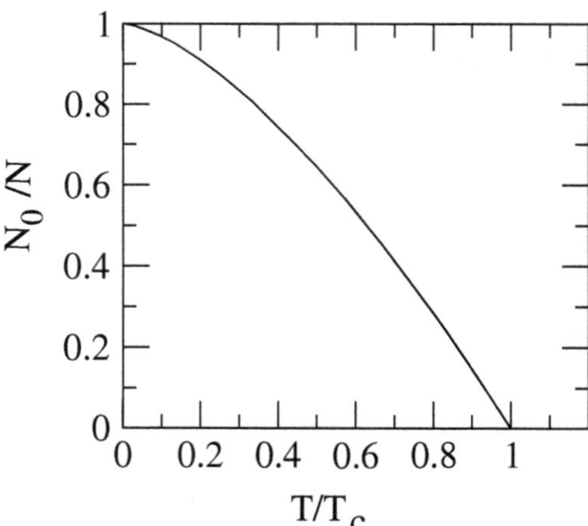

Fig. 7.9 The fraction of particles in the single-particle ground state, N_0/N, versus T/T_c for the ideal Bose gas

Note that C_V vanishes in the limit of zero temperature as required by the Third Law of Thermodynamics.

Again using $PV = (2/3)E$, we obtain the pressure as

$$\begin{aligned} P &= (NkT_c/V)(\zeta(5/2)/\zeta(3/2)(T/T_c)^{5/2} \\ &= 0.513(NkT_c/V)(T/T_c)^{5/2} . \end{aligned} \tag{7.101}$$

Writing the result in this way obscures the fact that the pressure is independent of the volume since $T_c \propto V^{-2/3}$. Substituting for T_c from Eq. (7.94) yields

$$P = 0.0851 \left(\frac{m}{\hbar^2}\right)^{3/2} (kT)^{5/2} \equiv P_c(T). \tag{7.102}$$

When the system is in the Bose-condensed phase and the volume is decreased at fixed temperature, Eq. (7.94) indicates that T_Q increases and therefore more and more particles go into the ground state, but the pressure remains constant. This means that the gas has infinite compressibility. This unphysical result is a consequence of the assumption that the Bosons are noninteracting.

Equation (7.102) implies that in the P-T plane, the Bose-condensed phase is confined to the curve $P = P_c(T)$, as is illustrated in the right panel of Fig. 7.8. For $P < P_c(T)$, one has the normal liquid phase. For $P = P_c(T)$, one has the Bose-condensed phase. An unphysical result of the infinite compressibility within this model is that the regime $P > P_c(T)$ is inaccessible. When one reduces the volume in the Bose-condensed phase, the pressure does not increase. However, when realistic interactions between particles are included, we expect that it is possible to compress the system at fixed temperature and thereby increase $P(T)$ to arbitrarily large values.

To illustrate these results, we show in Fig. 7.10 internal energy $PV = 2U/3$ in units of NkT_c as a function of T/T_c and the specific heat $C_V/(Nk)$ as a function

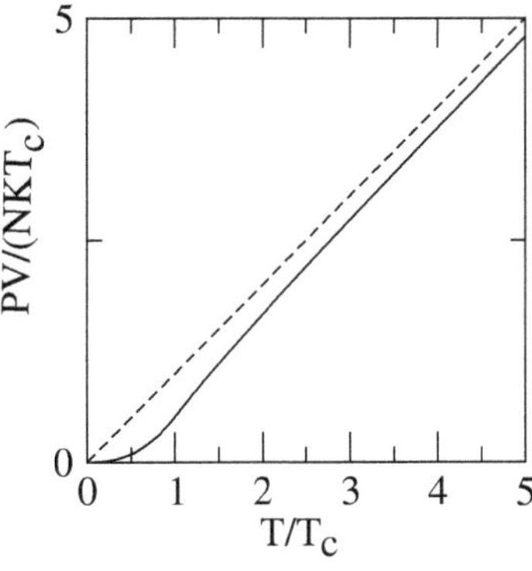

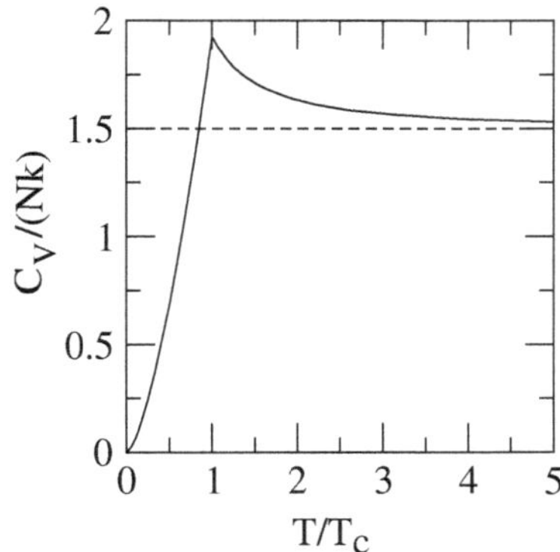

Fig. 7.10 Left: Solid line is $pV = 2U/3$ versus T for the ideal Bose gas. At the phase transition ($T/T_c = 1$), there is a discontinuity (which is not visible in this plot) in $d^2U(T)/dT^2$. The dashed line shows the result for the classical Gibbs gas. Right: Solid line is specific heat for the ideal Bose gas at fixed density (N/V) as a function of temperature. Dashed line: same for the classical Gibbs gas

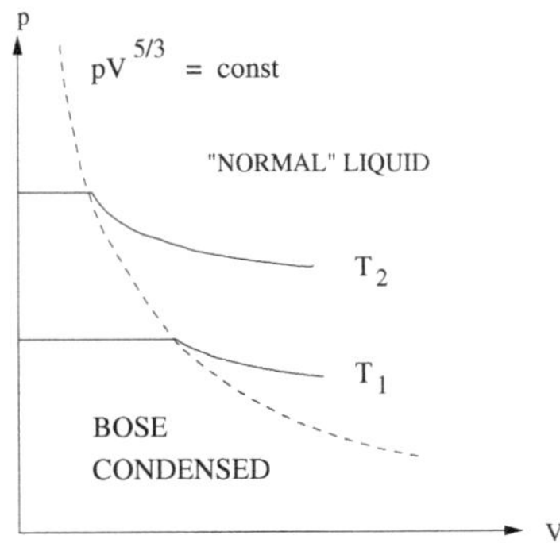

Fig. 7.11 Isotherms for the ideal Bose gas. The dashed line is the phase boundary between the "normal" liquid phase and the Bose-condensed phase

of T/T_c. Note that, due to statistical attraction between Bosons, the pressure of the quantum gas is always less than that of its classical analog. Also, note that C_V has a cusp at $T = T_c$. (See Exercise 6.) It is also instructive to plot isotherms in the $P - V$ plane, as shown in Fig. (7.11). There one sees coexistence between the normal liquid phase and the Bose-condensed phase.

7.4 Bose–Einstein Condensation, Superfluidity, and Liquid ^{4}He

Although it may initially be appealing to apply the theory of Bose condensation presented above to describe the superfluid transition in liquid ^{4}He, there are a number of important physical properties of ^{4}He which are not correctly reproduced by the noninteracting Bose liquid. For instance, the specific heat of the noninteracting gas has a finite amplitude cusp, whereas, at the superfluid transition, the specific heat of liquid helium has a divergence that is approximately logarithmic. In addition, we see that Eq. (7.94) predicts that the transition temperature should increase as the volume is decreased. This dependence on volume is not in accord with experiment for ^{4}He. Also, as we have already remarked, the above theory predicts an infinite compressibility, that is, that P is independent of V. This again is an unphysical result due to the neglect of interparticle interactions. Finally, the above theory gives no hint as to how one might calculate the critical velocity at which superfluidity breaks down. In Chap. 11, we will discuss a simplified treatment of the interacting Bose liquid which remedies most of these difficulties.

Bose–Einstein condensation (BEC) was first predicted by its two co-inventors in 1924. It was then, and remained for many years, a rather exotic prediction because it was difficult to conceive of how the appropriate conditions could be realized. The assumption in BEC is that the particles are noninteracting. The result is that, for a given density, at low enough temperature, the pressure becomes independent of density, or, in other words, the compressibility becomes infinite. Clearly, this is a system at the limit of its stability, where the slightest perturbation will lead to drastic consequences.

What kind of perturbations might occur? One obvious and inescapable perturbation results from the fact that particles, particularly atoms, interact. Interatomic interactions are typically attractive at large distances and repulsive at short distances. Thus, one might naively expect that a dilute Bose gas, close to the Bose–Einstein condensation point, would be unstable toward collapse because of the long-range attraction. As the system collapsed toward a high-density state, the short-range repulsive forces would kick in, stabilizing the system into a relatively dense, liquid state, with the average volume per atom being comparable to the volume of an atom (i.e., to the volume of the electron cloud of an atom).

The only catch to this discussion is that gas-to-liquid condensation typically occurs at temperatures much higher than that required for BEC. In fact, virtually every material, regardless of statistics, when cooled from the gas phase, first condenses into a liquid phase, and then, on further cooling, freezes into a solid. The temperature at which condensation occurs is a trade-off between the strength of the attractive potential, the size of the atomic core, and the mass of the atom. Except for the liquid heliums, ^{3}He and ^{4}He, all other materials are solid at temperatures well above where BEC might occur.

^{3}He atoms obey Fermi statistics, but ^{4}He atoms, which are Bosons, are quite relevant to this discussion. ^{4}He liquifies at atmospheric pressure at 4.2 K. This was first discovered by Kammerlingh Onnes in 1908. Reducing the pressure by pumping causes the liquid to cool by evaporation, and Onnes showed in 1910 that the liquid reaches its highest density at around 2.2 K, below which it ceases to boil. In 1923, a year before the predictions of Bose and Einstein, Dana and Onnes measured a discontinuity in the specific heat of ^{4}He, just below 2.2 K. Five years later in 1928, Keesom and Wolfke identified this discontinuity as a phase transition separating what they called He I above 2.17 K from He II below. Ten years later, Allen and Meissner and independently Kapitza showed that He II flows with zero viscosity, making it a "superfluid," in analogy to the superconductors discovered by Kammerlingh Onnes in 1911. (Kammerlingh Onnes was perhaps the most celebrated low-temperature physicist of his era and for his many seminal works he was awarded the 1913 Nobel prize in physics.)

These strange discoveries, observed in ^{4}He (but not in ^{3}He), soon raised the question of whether the transition observed at 2.17 K might be BEC. The answer to this question is clearly no, at least for the BEC described above, which is a condensation into the zero-momentum state of a dilute, noninteracting gas. Whatever was happening in ^{4}He at 2.17 K was happening in a dense, strongly interacting liquid. It is an interesting coincidence that the naive formula for T_c of Eq. (7.94) for ^{4}He atoms at the density of liquid ^{4}He gives a value close to 2 K. This will be explored further in an exercise. It is perhaps more significant that the effect of pressure on the value of T_c is to cause it to decrease rather than increase with increasing density [as predicted by Eq. (7.94)]. It seems that the repulsive interactions in this strongly correlated liquid act to suppress the tendency to superfluidity. Rather than connecting superfluidity to the model of Bose and Einstein, it is perhaps more accurate to say that this dense Bose liquid has a tendency to superfluidity, as was discussed much later by Feynman and others. Furthermore, the tendency for ^{4}He to macroscopically condense into the $q = 0$ state has been observed by neutron scattering. The condensate fraction is, however, very small, while the superfluid fraction, that is, the fraction of the fluid which flows with zero viscosity at $T = 0$ is 100%, as will be discussed in a later chapter.

But what about BEC? Can it actually happen? For many years, the belief was no. This is well-illustrated by the discussion of Landau and Lifshitz (Landau and Lifshitz 1969). Note that the word "condense" is used in this quote to connote solidification and not BEC.

> The problem of the thermodynamic properties of an "almost ideal" highly degenerate [Bose] gas ...has no direct physical significance, since gases which actually exist in Nature condense at temperatures near absolute zero. Nevertheless, because of the considerable methodological interest of this problem, it is useful to discuss it for an imaginary gas whose particles interact in such a way that condensation does not occur.

Landau and Lifshitz were not able to anticipate the almost miraculous rate of progress made in the second half of the twentieth century in the trapping and manipulation of atoms using lasers. Through various tricks, it is possible to cool these atoms to extremely low temperatures and to manipulate their atomic states so that the effective interaction between atoms in a low-density gas is repulsive. We will study the weakly repulsive Bose gas later in the chapter on Quantum Fluids. Here, it is sufficient to note that such a Bose gas does condense (in the sense of BEC), although the fraction which ends up in the $q = 0$ state is slightly depleted by interaction effects. The other consequence of the repulsive interactions is to make the compressibility finite. Indeed, it is this finite compressibility which supports a phonon-like mode, similar to those discussed in the next section, that allows the BEC to behave as a superfluid.

The experimental observation of BEC was first accomplished by Eric Cornell and Carl Wieman in a dilute gas of alkali atoms in 1995. Their accomplishment and later work by Wolfgang Ketterle were recognized by a Nobel Prize in Physics in 2001. Most of the predictions of the theory of the degenerate weakly repulsive Bose gas have since been confirmed experimentally. The history of BEC through 1996 is reviewed in Townsend et al. (1997).

7.5 Sound Waves (Phonons)

In Sect. 5.5, we discussed the statistical mechanics of a simple harmonic oscillator. Going beyond the case of a single oscillator, we note that a collection of coupled oscillators will have a set of so-called "normal modes" which, in the absence of anharmonic interactions, act like a set of independent oscillators. The thermodynamic state functions of these normal modes are simply the sum of the thermal functions of each mode.

A solid composed of atoms or molecules is like a set of coupled oscillators with many normal modes, each with its own frequency, describing oscillations in the positions of atoms in the crystal. If it is homogeneous, its lowest frequency modes, which are sound waves, can be labeled by wavevector, **k**, and, for periodic boundary conditions, the components of **k** are those given in Eq. (7.2). In a simplified picture, the allowed frequencies have the form

$$\omega(\mathbf{k}) = c|\mathbf{k}| = ck, \tag{7.103}$$

where c is the speed of sound in the material. The spectrum of $\omega(\mathbf{k})$ is bounded from below by zero, and the lowest nonzero frequency corresponds to the smallest nonzero $k = 2\pi/L$, for sample volume, L^3. This lowest excited state frequency is typically in the μKelvin range when expressed as a temperature $\hbar\omega/k_B$ for a macroscopic solid. Furthermore, since the number of modes is finite, being proportional to the number of atoms or molecules in the material, there is also a characteristic *highest* frequency corresponding to a wavevector, $k = 2\pi/a$ where a is of order the crystal lattice spacing. This highest frequency is often called the Debye frequency, Ω_D, corresponding to an energy typically in the hundreds of Kelvins.

A more complete picture would note that, in d dimension, there are d modes per atom, one longitudinal mode, involving motions along **k** and $d-1$ transverse modes with motions perpendicular to **k**. For most materials, transverse phonon frequencies have smaller sound velocities than longitudinal ones. In addition, for materials with more than one atom per unit cell, there are so-called "optical modes" whose frequencies typically lie above the acoustic part of the spectrum.

Sound waves or phonons are like bosons with a linear dispersion relation given by Eq. (7.103) and a chemical potential of zero. We have seen that having the chemical potential pinned at the bottom of the excitation spectrum is connected with Bose condensation. For the case of a superfluid, the linear excitation spectrum corresponding to a stiffness of the superfluid state allows dissipationless superflow. One might wonder then, what is it that has "condensed" for the case of the elastic solid? A solid in its ground state must choose some position in space at which to sit. However, by translational symmetry, there is no energy cost for moving the solid uniformly to a new position. (Note that uniform translation of the solid corresponds to a longitudinal mode with $k = 0$.) By choosing a particular position, the solid arbitrarily breaks this symmetry, just as a ferromagnet spontaneously breaks rotational symmetry by arbitrarily choosing a direction for all the spins to point. A similar argument can be made with regard to the transverse phonon modes and the spontaneous breaking of rotational symmetry when the crystal chooses the orientation of its crystal axes. The fact that the excitation spectrum rises continuously from zero corresponds to the fact that long-wavelength excitations look more and more uniform locally as the wavelength grows large. This behavior is an illustration of Goldstone's theorem which says that the excitation spectrum is gapless for a macroscopic system that breaks a continuous symmetry.

7.6 Summary

Without any interparticle interactions, we have seen that the Pauli principle gives rise to an effective repulsion for Fermions and an effective attraction for Bosons. These effects are completely negligible at room temperature and ambient pressure. For these noninteracting systems, the scale of energy is set by the "quantum" temperature T_Q which is defined to be of order the energy needed to confine a quantum particle in the volume per particle of a given spin state. Thus, when properly scaled, the thermodynamic functions can be expressed in terms of the single variable T/T_Q. For Fermi systems, the low-temperature properties depend only on the density of states near the Fermi energy. We also presented Chandrasehkhar's statistical theory of white dwarfs which describes the behavior of a degenerate gas of electrons and gravitational interactions involving nuclei. This theory leads to a marriage of the three fundamental constants, G, h, and c. For Bose systems, a "Bose-condensed" phase appears at low temperature in which the single-particle ground state is macroscopically occupied. To discuss Bose condensation in actual systems (such as in liquid ^{4}He), one must take interactions into account. Another example of noninteracting particles is the elementary excitations of a compressible solid which are called sound waves or phonons. These behave like a gas of relativistic bosonic particles traveling at the speed of sound with the chemical potential pinned at zero energy.

7.7 Exercises

1. For $T < 1\text{K}$, ^{3}He solidifies at pressures above about 25 atm. The liquid is well-described as an ideal degenerate Fermi gas with an effective mass $m^* = 2.5\text{M}(^3\text{He})$ and a density, along the melting curve, of $2.4 \times 10^{22}/\text{cm}^3$. The entropy of the solid is completely dominated by the entropy of the effectively noninteracting spin 1/2 nuclei, which may be treated as distinguishable because they are on a lattice. The molar volume of the solid is about 1.2 cm^3 smaller than that of the liquid. The melting pressure for $T = 0$ is $p_m(0) = 34$ atm. (1 atm.= 10^5 Nt/m^2) Calculate $p_m(T)$, the pressure along the melting curve for $0 < T < 1\text{K}$. You may need the following constants: $k_B = 1.38 \times 10^{-23}$ J/K, $N_A = 6.02 \times 10^{23}$/mole, mass of (^{3}He)= 5.01×10^{-27} kg.

2. Calculate the equation of state (relating p, V, and T) for the noninteracting Fermi and Bose gases in two dimensions. Do as much as you can exactly. Does Bose–Einstein occur in two dimensions? Explain your answer.

3a. Consider the correction to the pressure due to quantum statistics as given in Eqs. (7.38) and (7.86). Calculate the percent correction for standard temperature $T = 300\text{K}$ and pressure ($p = 1$ atm) for ^{4}He and ^{3}He gases.

(**b**) Calculate T_Q for H_2 gas at STP density, i.e., for a density when 6×10^{23} molecules occupy a volume of 22.4 liters

4(a). Calculate the Bose–Einstein transition temperature if we suppose that it applies to liquid ^{4}He, whose density is about 0.15 gr/cm^3.

(**b**) The Nobel prize for physics in 2001 was awarded for Bose condensation of trapped atoms. W. Ketterle *et al* observed Bose condensation in a sample of sodium (Na) atom at a density of 10^{12} atoms per cm^3. Estimate the transition temperature for Bose condensation. Is ^{23}Na a boson?

5. In this problem, you are to make an approximate evaluation of the *canonical* partition function Z_N^F (Z_N^B) for a quantum gas of N noninteracting Fermions (Bosons) in a box of volume V. This is to be done in the extreme low-density (high-temperature) limit where quantum statistics give small corrections to the classical ideal gas law. Express the results in terms of the classical partition function $z(\beta)$ for a *single* particle in the box of volume V. Show that, in the thermodynamic limit a correct enumeration of quantum states consistent with spin statistics yields the result that

$$Z_N(\beta) = \frac{z(\beta)^N}{N!}\left(1 \mp \frac{1}{2}N^2\frac{z(2\beta)}{z(\beta)^2} + \cdots\right).$$

Explain why we should write this as

$$\ln Z_N(\beta) = N \ln z(\beta) - \ln N! \mp \frac{1}{2}N^2\frac{z(2\beta)}{z(\beta)^2} + \cdots .$$

HINT: start by writing down exact expression for $Z(\beta)$ in terms of $z(\beta)$ for systems with a small number of particles.

6. This problem concerns the specific heat at constant volume, C_V of a Bose gas of noninteracting spinless particles.

(**a**) Show that C_V is a continuous function of temperature at T_c, where T_c is the transition temperature for Bose condensation.

(**b**) Show that dC_V/dT has a *finite* discontinuity as a function of temperature at T_c.

7a. Show that the result of Eq. (7.52) may be written as

$$C_V = \frac{\pi^2}{3} k^2 T V \rho(\epsilon_F) \, , \tag{7.104}$$

where the density of states, $\rho(E)$ was defined in Eq. (7.25). [This result holds even when $\epsilon_{\mathbf{k}} \neq \hbar^2 k^2/(2m)$.]

(**b**) Obtain an analogous formula for the spin susceptibility at zero temperature in terms of $\rho(E)$.

8. We constructed the eigenvalue spectrum of a particle in a box using periodic boundary conditions from which Eq. (7.2) followed. Calculate the effect (in the thermodynamic limit) of using different boundary conditions. Discuss.

(**a**) Antiperiodic boundary conditions. (The wave function has opposite signs on opposite sides of the cubical container of side L.)

(**b**) "Hard wall" boundary conditions which force the wave function to vanish at the walls.

9. We have said that the Fermi function which gives the probability that the single-particle state $|n\rangle$ is occupied when a noninteracting gas of Fermi particles is held at temperature T and chemical potential μ is

$$f_n = \frac{1}{e^{\beta(\epsilon_n - \mu)} + 1} \, . \tag{7.105}$$

Some people may believe the following incorrect statement.

"Equation (7.105) indicates that the states of an electron gas at temperature T are not occupied with Gibbsian probability (proportional to $e^{-\beta E}$)."

Construct the ratio $r \equiv P_b/P_a$, where P_a is the probability that the single-particle state $|a\rangle$ is occupied and the single-particle state $|b\rangle$ is unoccupied and P_b is the probability that state $|b\rangle$ is occupied and state $|a\rangle$ is unoccupied. Do you think the above statement is true?

10. In the discussion of white dwarfs, we wrote the gravitational potential energy of a sphere of radius R as $-\gamma G M^2/R$. What is the value of γ if the sphere is of uniform density?

11. Use Eq. (7.74) to show that if one treats the relativistic limit by setting the electron mass to zero, there is no regime of stability for the white dwarf. The conclusion is that even though the system may be highly relativistic, it is crucial that the lowest energy electrons are *nonrelativistic*.

12. In this problem you are to estimate the mass–radius relation for a neutron star. As we have seen, a white dwarf will become unstable if its mass increases beyond the Chandrasekhar limit. Then it may explode as a supernova and the remnant becomes a neutron star. In this case, one must consider the relations

$$p + e^- \to n + \nu\,, \qquad n \to p + e^- + \overline{\nu}\,. \tag{7.106}$$

Because the confinement energy for electrons is so large, the first reaction is favorable (and the neutrino escapes the star). The second reaction (beta decay) can only occur to a small extent, because once there are a small number of electrons and protons present, the allowed low-energy final states for beta decay are already occupied. So an approximate picture of a neutron star is that it consists of only neutrons and one must consider the balance of gravitational energy and the zero-point kinetic energy of neutrons. Noting that this scenario is essentially the same (but on a different scale) than that for white dwarfs, give an estimate for the ratio of a neutron star's mass to that of a white dwarf of the same radius. (For more on neutron stars, see Ref. (Weinberg 1972).

13. Consider two identical noninteracting spin S particles of mass m which are in a one-dimensional harmonic oscillator potential, $V(x) = (1/2)kx^2$. You are to construct the canonical partition function, $Z_2(T, V)$, for this system (a) for $S = 3$ and (b) for $S = 7/2$. Hint: take note of the results of Exercise 5.

14. This problem concerns the radiation field inside a box whose walls can absorb and reradiate photons. The Hamiltonian is taken semiclassically to be

$$\mathcal{H} = \sum_{\mathbf{k},\alpha} n_{\mathbf{k},\alpha}\epsilon_\alpha(\mathbf{k})\,,$$

where $\mathbf{k}$ is the wavevector of the photon (since boundary conditions are not usually important, you may assume that the wavevectors are determined by periodic boundary conditions), α labels the possible polarizations, and $\epsilon_\alpha(\mathbf{k})$ is the photon energy $\hbar ck$.

(**a**) The chemical potential μ of photons is taken to be zero. Why is this? (You may want to go back to the discussion of a system in thermal contact with a large reservoir.)

(**b**) Calculate the grand partition function, and from that get the internal energy $\langle\mathcal{H}\rangle_T$ and thence the pressure of the photon gas.

(**c**) Make a kinetic theory argument analogous to that used to derive the ideal gas (here using photons instead of gas atoms) to obtain a formula for $\mathcal{R}(\nu)$, the energy radiated per unit area per unit time in the frequency interval between ν and $\nu + \Delta\nu$ from an ideal radiator.

(**d**) Continuing part C: Wien's Law says that if λ_{max} is the wavelength at which $\mathcal{R}(\nu)$ is a maximum, then $\lambda_{\text{max}}T$ is a constant. Evaluate this constant. (This result is used to determine the temperature of stars.)

(**e**) The Stefan–Boltzmann radiation constant σ is defined so that

$$\int \mathcal{R}(\nu)d\nu = \sigma T^4 \ .$$

Give an expression for σ involving the fundamental constants, h, c, and k, and evaluate it numerically. (This result is used to determine the radius of stars.)

References

N.W. Ashcroft, N.D. Mermin, *Solid State Physics* (Brooks/Cole, 1976)

S. Chandrasekhar, On Stars, Their Stability, Their Evolution and Their Stability, Nobel Lecture (1983). https://www.nobelprize.org/uploads/2018/06/chandrasekhar-lecture.pdf

L.D. Landau, E.M. Lifshitz, *Statistical Physics* Volume 5 of the Course of Theoretical Physics (Pergamon Press, 1969)

R.K. Pathria, Can. J. Phys. **63**, 358 (1985)

F.H. Shu, *The Physical Universe—An Introduction to Astronomy* (University Science Books, 1982)

C. Townsend, W. Ketterle, S Stringari, *Physics World*, p. 29 (1997)

S. Weinberg, *Gravitation and Cosmology* (Wiley, Chap, 1972), p. 11

Part III
Mean Field Theory, Landau Theory

Chapter 8
Mean-Field Approximation for the Free Energy

8.1 Introduction

So far we have learned that the thermodynamic properties of a system can be calculated from the free energy of the system which is, in turn, related to the partition function. We have also learned how the partition function is defined in terms of the microscopic Hamiltonian of the system. Furthermore, for the case of particles and spins which do not interact (but which might interact with externally applied fields), we have seen how to calculate the partition function exactly, which is equivalent to solving the entire problem.

Interacting systems are a completely different matter. To begin with, there are very few exact solutions, and those that exist are complicated and not at all transparent. Furthermore, the theory which captures the correct physics of phase transitions, the renormalization group, is approximate in nature, although the "critical exponents" that it yields can approach the exact values. Series expansions and numerical simulations are both powerful methods for studying interacting systems, but these are also lengthy and complicated to implement. In this chapter, we will study a simple approximate method that can be applied to almost any interacting many-body system and which provides a simple, intuitive model of phase transitions, although it is least accurate in the vicinity of critical points. This method will be applied to a variety of problems of varying degrees of complexity. This simplest and most generally useful approximate method for the treatment of interacting many-body systems is the Mean-Field (MF) approximation which appears in various guises in virtually every area of physics. In this chapter, we show how the MF approximation arises from a neglect of the correlation between fluctuations of different particles. As a first example, we will treat the ferromagnetic Ising model. A second example will deal with liquid crystals. In the chapters which follow, we will recast the MF approximation into other more general forms which can be applied to almost any problem.

A. J. Berlinsky and A. B. Harris, *Statistical Mechanics*, Graduate Texts in Physics,
https://doi.org/10.1007/978-3-030-28187-8_8

8.2 Ferromagnetic Ising Model

We start by treating the ferromagnetic Ising model, whose Hamiltonian is written as

$$\mathcal{H} = -\frac{J}{2}\sum_{i,\delta} S_{\vec{R}_i} S_{\vec{R}_i+\vec{\delta}} - H\sum_i S_{\vec{R}_i}, \tag{8.1}$$

where H is the magnetic field in energy units, $S_{\vec{R}_i} = \pm 1$, the $\vec{R}_i$ label sites on a lattice, $\vec{\delta}$ is a vector connecting nearest neighbor sites, and $J > 0$ for ferromagnetic interactions. The number of nearest neighbors, which is also the number of different values of $\vec{\delta}$, is called z, the "coordination number." For a hypercubic lattice in d dimensions, $z = 2d$. In a more compact notation, we can write Eq. (8.1) as

$$\mathcal{H} = -J\sum_{\langle ij\rangle} S_i S_j - H\sum_i S_i \ , \tag{8.2}$$

where S_i denotes $S_{\vec{R}_i}$ and the sum over $\langle ij\rangle$ denotes a sum over pairs of nearest neighboring sites i and j. In general, if $\vec{\delta}$ takes on z values, there are $Nz/2$ nearest neighbor pairs in a lattice of N sites.

We expect the average behavior of a spin, in both the disordered and ordered phases, to be uniform. That is,

- at high temperature all spins are uniformly disordered and
- at low temperature all spins have the same nonzero average value.

So we can write

$$S_i = \langle S\rangle + \delta S_i \ , \tag{8.3}$$

where $\langle \delta S_i\rangle = 0$ and hence δS_i represents fluctuations of S_i about its mean $\langle S\rangle$ which is presumed to be the same for all sites. Then

$$\mathcal{H} = -J\sum_{\langle ij\rangle}\left[\langle S\rangle^2 + \langle S\rangle\left(\delta S_i + \delta S_j\right) + \delta S_i \delta S_j \ ,\right] - H\sum_i S_i \ . \tag{8.4}$$

In this context, the **Mean-Field (MF) Approximation** is to neglect the terms second order in the fluctuations. In other words, we neglect the fact that fluctuations on neighboring spins are not really independent of one another, but in fact are correlated. As we shall see later, as long as the spin–spin correlation length is not too large, (in particular, as long as we are not close to a continuous phase transition at which long-range order appears) this neglect of correlations inherent in the MF approximation leads to reasonable results.

Having made this approximation, the terms linear in the fluctuations can be rewritten back in terms of the original spins and the average spin $\langle S\rangle$.

$$\langle S\rangle\left(\delta S_i + \delta S_j\right) = \langle S\rangle\left(S_i + S_j\right) - 2\langle S\rangle^2 \ . \tag{8.5}$$

Substituting this back into $\mathcal{H}$ allows us to define an effective mean-field Hamiltonian, $\mathcal{H}_{\rm MF}$,

$$\mathcal{H}_{\rm MF} = \frac{JNz}{2}\langle S\rangle^2 - \sum_i (Jz\langle S\rangle + H)\, S_i$$
$$= \frac{JNz}{2}\langle S\rangle^2 - H_{\rm eff}\sum_i S_i\ , \tag{8.6}$$

and an effective field,

$$H_{\rm eff} = Jz\langle S\rangle + H\ . \tag{8.7}$$

The MF approximation has resulted in an effective noninteracting spin problem with one undetermined parameter, $\langle S\rangle$.

To determine $\langle S\rangle$, we may write the effective single-spin density matrix as

$$\rho^i_{\rm eff} = e^{\beta H_{\rm eff} S_i}/{\rm Tr}_i\, e^{\beta H_{\rm eff} S_i}\ , \tag{8.8}$$

where ${\rm Tr}_i$ indicates a trace over the states of spin i. If we define $M \equiv \langle S\rangle$ to be the magnetization (in dimensionless units), then

$$M = {\rm Tr}_i\left(S_i e^{\beta H_{\rm eff} S_i}\right)/{\rm Tr}_i\, e^{\beta H_{\rm eff} S_i}$$
$$= \tanh\beta\,(JzM + H)\ . \tag{8.9}$$

This is called a "self-consistency relation." It determines $M = \langle S\rangle$ as a function of T and H. In the rest of this section, we learn how to solve Eq. (8.9) to obtain M as a function of T and H.

8.2.1 Graphical Analysis of Self-consistency

We first analyze the self-consistency condition of Eq. (8.9) graphically. For that purpose, we introduce the variable

$$x = \beta(JzM + H)\ , \tag{8.10}$$

so that the self-consistency equation becomes

$$(kT/Jz)x - (H/Jz) = \tanh x\ . \tag{8.11}$$

When this is solved, then $M = \tanh x$. To solve Eq. (8.11), we plot both sides as a function of x and look for the intersections of the two curves. First of all, when $H = 0$ we have the situation shown in Fig. 8.1. We look for the intersection of kTx/Jz with $\tanh x$. As can be seen, when the slope of the straight line is too large there is only

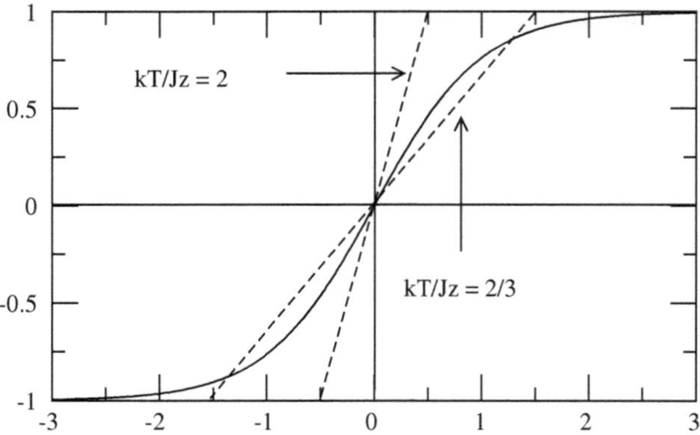

Fig. 8.1 Graphical analysis of Eq. (8.11). The solid curve is $y = \tanh x$ and the dashed line is $y = mx$, one for $m > 1$ and the other for $m < 1$

one solution, namely, $x = 0$. The slope is too large when $kT > Jz$ or $T > T_c$. For $T < T_c$, there are three solutions. The two solutions at nonzero x are equivalent and represent the two ways that the symmetry can be spontaneously broken in this Ising spin model.

The question is whether we should retain the solution at $x = 0$ or follow one of the solutions for nonzero x. To decide that we should compare the free energies of these two solutions. The free energy (per spin) is

$$f = -kT \ln \mathrm{Tr}_i \left[e^{\beta Jz\langle S\rangle S_i - \frac{1}{2}\beta Jz\langle S\rangle^2} \right] . \tag{8.12}$$

Now we want to compare the free energy when $\langle S\rangle = 0$, to that when $\langle S\rangle = S_f$, where S_f is the solution for $\langle S\rangle$ corresponding to the intersection of the curves in Fig. 8.1 for $T < T_c$. We have

$$\begin{aligned} f(S_f) &= f(0) + \int_0^{S_f} \frac{df}{d\langle S\rangle} d\langle S\rangle \\ &= f(0) + kT \int_0^{S_f} \left(Jz\beta\langle S\rangle - \beta Jz \frac{\mathrm{Tr}_i S_i e^{\beta Jz\langle S\rangle S_i}}{\mathrm{Tr}_i e^{\beta Jz\langle S\rangle S_i}} \right) d\langle S\rangle \\ f(S_f) - f(0) &= Jz \int_0^{S_f} \Big(\langle S\rangle - \tanh\big(\beta Jz\langle S\rangle\big) \Big) d\langle S\rangle \, . \end{aligned} \tag{8.13}$$

Since the integrand is negative over the entire range of integration, the free energy is lower when $\langle S\rangle = S_f$ than it is for $\langle S\rangle = 0$. In the next chapter, we will interpret these solutions in terms of potential wells for the free energy and arrive at the same conclusion in a much more intuitive way.

Now focus on the solution for positive x and consider what happens for $H > 0$. The straight line, the left-hand side of Eq. (8.11), is moved down by an amount

$-H/Jz$ and the intersection with the tanh x curve moves to larger x. Thus, the self-consistent magnetization (which is the value of y in Fig. 8.1 where the straight line intersects tanh x) increases and we thereby reproduce the behavior shown in the left panel of Fig. 2.5. So the self-consistency equation reproduces, at least qualitatively, all the phenomenology discussed in Sect. 1.2 of Chap. 2.

8.2.2 High Temperature

At high temperatures, where both M and $\beta(JzM + H)$ are small, we can expand the tanh in Eq. (8.9) to get

$$M \approx \beta(JzM + H) , \tag{8.14}$$

so that

$$M = \frac{H}{kT - Jz} = \frac{H}{k(T - T_c)} , \tag{8.15}$$

The susceptibility (at zero field H) is

$$\chi = \left.\frac{\partial M}{\partial H}\right)_T = \frac{1}{k(T - T_c)} , \tag{8.16}$$

which is called the *Curie–Weiss law*. Note that, because of interactions, χ diverges at the mean-field transition temperature T_c. In the absence of interactions, $T_c = 0$, and we recover Curie's law, Eq. (5.23).

Experimentally, it is observed that the susceptibility diverges at the ferromagnetic transition as

$$\chi \sim a\Big(T - T_c\Big)^{-\gamma}, \tag{8.17}$$

with an amplitude a. Mean-field theory predicts that $\gamma = 1$, whereas the experimental values tend to be larger than 1. This behavior is illustrated in Fig. 8.2, where the solid curve is the Curie–Weiss result, and the dashed curve indicates how fluctuations, not captured by mean-field theory, lower T_c and increase γ.

The Curie–Weiss law is useful for modeling experimental data well above T_c. When data for χ^{-1} are plotted versus temperature, they fall on a straight line at high temperature with an extrapolated intercept, the mean-field T_c, which is a measure of the strength of the interactions of a spin with its nearest neighbors. Near the mean-field T_c, the experimentally observed slope of $1/\chi$ decreases, as shown in Fig. 8.2, and $1/\chi$ intersects zero at a somewhat lower temperature indicating that MF theory gives too high a value for T_c.

The exponent γ and its partners α, β, and δ, which will be defined below, are central to the theory of continuous phase transitions which will be developed in great detail over much of the rest of this book. The notation for these exponents is standard in the phase transition literature.

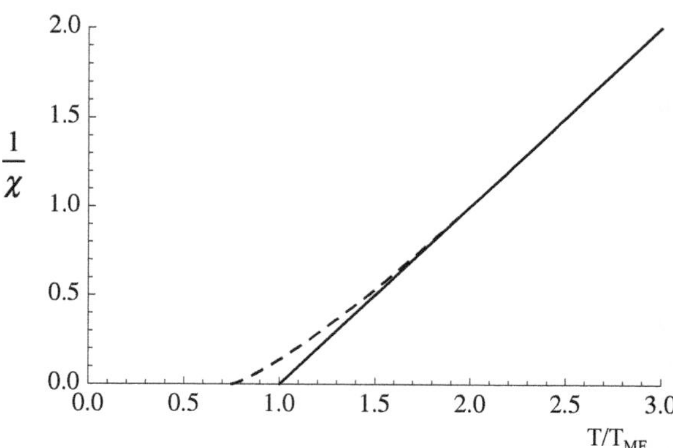

Fig. 8.2 Inverse susceptibility, $1/\chi$, plotted versus $T/T_c = T/zJ$. The solid curve is the Curie–Weiss result, while the dashed curve shows how the susceptibility is modified by fluctuations, not captured by mean-field theory, for parameters appropriate for the 3-D ferromagnetic Ising model

8.2.3 Just Below the Ordering Temperature for $H = 0$

For $H = 0$ and $T > T_c$, $M = 0$. However, just below T_c, M is nonzero but small. Therefore, to discuss the regime $T < T_c$, we expand the tanh in Eq. (8.9) to cubic order to obtain

$$M = \beta JzM - \frac{1}{3}(\beta JzM)^3 + \cdots \tag{8.18}$$

If $M \neq 0$, we can cancel one factor of M and solve for M. Then

$$\frac{1}{3}(\beta Jz)^3 M^2 = \beta Jz - 1 = \frac{1}{T}(T_c - T)$$

or

$$M = \sqrt{3}\frac{T}{T_c}\sqrt{1 - (T/T_c)}\,. \tag{8.19}$$

By choosing the positive solution for M we have implicitly treated the case of infinitesimal positive field: $H = 0^+$. For T very close to T_c, this can be written as

$$M = \sqrt{3}\left(1 - \frac{T}{T_c}\right)^{\beta}, \tag{8.20}$$

where, within mean-field theory, $\beta = 1/2$. The behavior of M as a function of T in zero field is the same as that of ΔM in Fig. 2.4.

To get the susceptibility below T_c, we return to Eq. (8.9) and take the derivative with respect to H (at $H = 0$) but with M nonzero. We get

$$\chi = \frac{\beta(Jz\chi + 1)}{\cosh^2(\beta JzM)}, \tag{8.21}$$

so that, using $\beta Jz = T_c/T$,

$$kT\chi = \frac{1}{\cosh^2(MT_c/T) - T_c/T} \,. \tag{8.22}$$

For T near T_c, we have, from Eq. (8.19), $MT_c/T = \sqrt{3(T_c - T)/T_c}$. Using $\cosh^2 x = 1 + x^2 + \mathcal{O}(x^4)$, we can write

$$\begin{aligned} \cosh^2[MT_c/T] &= 1 + [MT_c/T]^2 \\ &= 1 + 3(T_c - T)/T_c \,, \end{aligned} \tag{8.23}$$

so that

$$\begin{aligned} \cosh^2[MT_c/T] - T_c/T &= 1 - T_c/T + 3(T_c - T)/T_c \\ &\approx 2(T_c - T)/T_c \,. \end{aligned} \tag{8.24}$$

Thus, to leading order in $(T_c - T)$, we have

$$kT\chi = \frac{T_c}{2(T_c - T)} \,. \tag{8.25}$$

So approaching T_c from below we have the same value for the exponent γ as from above, but with a different amplitude. This is generally the case. Accordingly, we generalize Eq. (8.17) to

$$\chi \sim A_{\pm}|T - T_c|^{-\gamma} \,. \tag{8.26}$$

This form indicates that the exponents are assumed (and the renormalization group and experiment verify) to be equal, but with different amplitudes as T_c is approached from above (A_+) and from below (A_-).

8.2.4 At the Critical Temperature

As T approaches T_c from above, the susceptibility in zero field diverges. At $T = T_c$, for $|H/kT_c| << 1$, Eq. (8.9) becomes

$$\begin{aligned} M &= \tanh\left(M + \frac{H}{kT_c}\right) \\ M &\approx M + \frac{H}{kT_c} - \frac{1}{3}\left(M + \frac{H}{kT_c}\right)^3 + \cdots \,. \end{aligned} \tag{8.27}$$

If $H/(kT_c) << M$, which we can verify a posteriori, then

$$\frac{H}{kT_c} \approx \frac{1}{3}M^3 \tag{8.28}$$

or

$$M = \left(\frac{3H}{kT_c}\right)^{1/3} . \tag{8.29}$$

At T_c, the magnetization is observed to obey

$$M \sim |H|^{1/\delta} . \tag{8.30}$$

We have shown that mean-field theory gives $\delta = 3$. This behavior is illustrated by the M versus H curve labeled $T = T_c$ in Fig. 2.5.

8.2.5 Heat Capacity in Mean-Field Theory

Recall that the energy per spin, u, is

$$\begin{aligned} u \equiv U/N = \langle H \rangle / N &= \frac{1}{2} JzM^2 - (JzM + H)M \\ &= -\frac{1}{2} JzM^2 - HM . \end{aligned} \tag{8.31}$$

For $H = 0$ and $T < T_c$

$$u = -\frac{1}{2} Jz \frac{3T^2}{T_c^3} (T_c - T) = -\frac{3}{2} Jz \left[\left(\frac{T}{T_c}\right)^2 - \left(\frac{T}{T_c}\right)^3 \right] \tag{8.32}$$

and the specific heat per spin (at constant field), c is

$$c = \left.\frac{\partial u}{\partial T}\right)_{H=0} = \frac{3}{2} k \frac{T}{T_c} \left(\frac{3T - 2T_c}{T_c} \right) . \tag{8.33}$$

In the limit $T \to T_c^-$, $c(T_c^-)/N = (3/2)k$. Note that, for $T > T_c$, $u = 0$ so that $c = 0$. Thus, there is a specific heat "jump" at T_c. We would like to relate this to the definition of the critical exponent α such that the specific heat is given by

$$C \sim a_{\pm} |T - T_x|^{-\alpha} + \text{reg} , \tag{8.34}$$

where "reg" indicates analytic or at least less singular background terms. The discontinuity corresponds to the value $\alpha = 0$ with amplitude, $a_+ = (3/2)k$ and $a_- = 0$. Such a jump is observed in the specific heat of low-temperature superconductors. However, most other phase transitions are not mean-field-like. They typically exhibit specific heat singularities of the form $|T - T_c|^{-\alpha}$ where $0 < \alpha < 1$ corresponds to an integrable specific heat divergence, whereas $\alpha < 0$ (with $a < 0$) describes a specific heat cusp. The λ-transition in superfluid ^{4}He has a specific heat exponent divergence which is close to logarithmic in which case α is close or equal to zero.

8.3 Scaling Analysis of Mean-Field Theory

We now turn to a scaling analysis of Eq. (8.9) in order to describe the equation of state asymptotically close to the critical point at $T = T_c$ and $H = 0$. We return to the expansion of Eq. (8.9) which is

$$M = \beta(JzM + H) - \frac{1}{3}\beta^3(JzM + H)^3 + \cdots . \tag{8.35}$$

To obtain the asymptotic equation of state we drop terms which are dominated by the terms we keep, providing the dropped terms are negligible *over the entire regime* $T \approx T_c, M \approx 0$, *and* $H \approx 0$, *independent of the relative size of the variables.* According to this criterion, we drop the fifth-order terms in the expansion of the tanh function since they are always dominated by the third-order terms. The terms written in Eq. (8.35) can be rearranged in the form

$$\begin{aligned}&\frac{1}{3}(\beta Jz)^3M^3 + \beta^3(Jz)^2HM^2 + (\beta^3 JzH^2 + 1 - \beta Jz)M \\ &= \beta H - \frac{1}{3}\beta^3 H^3 .\end{aligned} \tag{8.36}$$

We want to keep only the terms which are dominant when M, H, and $|T_c - T|$ are small. Since, from Eq. (8.29), $M \sim |H|^{1/3}$, the terms in HM^2, MH^2, and H^3 are all dominated by the term βH, and hence these terms may be dropped. Also, we write $(T - T_c)/T_c \equiv t$ and work to leading order in t. Then we are left with the equation

$$\frac{1}{3}M^3 + tM - (H/Jz) = 0 . \tag{8.37}$$

To make the first two terms scale in the same way, we set $M = m\sqrt{|t|}$, so that in terms of the scaled variable $m \equiv M/|t|^{1/2}$ we have

$$\frac{1}{3}m^3 + \frac{t}{|t|}m = (H/Jz)|t|^{-3/2} , \tag{8.38}$$

or finally

$$\frac{1}{3}m^3 + \frac{t}{|t|}m = h , \tag{8.39}$$

where we have introduced the further scaled variable $h \equiv (H/Jz)/|t|^{3/2}$.

What Eq. (8.39) tells us is that, instead of being a relation between the three independent variables M, H, and T, the equation of state, close to $T = T_c$ and $H = 0$, can be expressed in terms of only two scaled variables, m and t. Given many data sets of M versus H and T, a scaling analysis involves, first, expressing T in terms

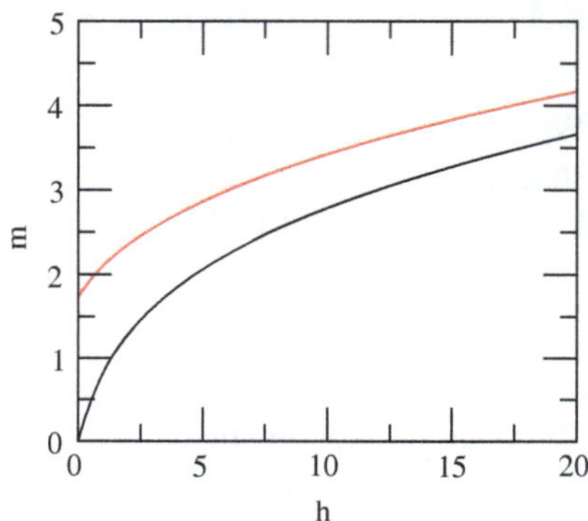

Fig. 8.3 Equation of state in terms of scaled variables $m \equiv M/\sqrt{|t|}$ and $h \equiv H/|t|^{3/2}$, where $t \equiv (T - T_c)/T_c$. Note that there are two branches to scaling curve, the upper one for $T < T_c$ and the lower one for $T > T_c$. The two branches approach one another as $h \to \infty$

of t and then using the values of t, M, and H to calculate scaled data, $m = M/|t|^{1/2}$ and $h = (H/Jz)/|t|^{3/2}$. Then the data for different t, instead of defining many curves as the case shown in Fig. 2.5, will instead fall on only two curves like the ones in Fig. 8.3. There are two curves because of the presence of the factor $t/|t|$ in Eq. (8.39) which has two branches, one for $t > 0$ and another for $t < 0$. Each branch of the function represents a relation between scaled variables. The fact that all the data falls on two curves, one for $T > T_c$ and the other for $T < T_c$, is called "data collapse."

All this assumes, of course, that the data are correctly described by mean-field theory. If the data are poorly described by mean-field theory, the plots of scaled values of m and h will instead resemble a scatter plot in the m–h plane. As it turns out, the values of the exponents found in mean-field theory are only roughly correct. Nevertheless, the general form of M versus H near T_c is well represented by the functional form of Eqs. (8.20) and (8.30), and hence collapse can still be obtained by use of the correct exponents. Even before the discovery of renormalization group theory, it was a well-known and useful strategy for experimentalists to try to represent their data to obtain data collapse, i.e., to show experimentally the existence of scaled variables even when no real theory existed. Indeed, the fact that experimental results in the critical regime could be represented to obtain data collapse was a significant stimulus in the lead up to the development of scaling theory and the renormalization group.

8.4 Further Applications of Mean-Field Theory

8.4.1 Arbitrary Bilinear Hamiltonian

The above procedure can be generalized to a Hamiltonian in which particles interact via a general bilinear interaction. Thus, we consider a Hamiltonian of the form

$$\mathcal{H} = \frac{1}{2}\sum_{i,j} \lambda_{ij} \mathcal{O}_i \mathcal{O}_j \ , \tag{8.40}$$

where the λ's are coupling constants (with $\lambda_{ij} = \lambda_{ji}$) and the operators $\mathcal{O}_i$ depend only on the degrees of freedom of the site labeled i. Following the procedure leading to Eq. (8.4), we write

$$\mathcal{O}_i = \langle \mathcal{O}_i \rangle + \delta \mathcal{O}_i \tag{8.41}$$

and we ignore second order terms in $\delta\mathcal{O}$. Thereby we obtain the mean effective field Hamiltonian as

$$\begin{aligned}\mathcal{H} &= \frac{1}{2}\sum_{i,j} \lambda_{ij}\Big(\mathcal{O}_i \langle \mathcal{O}_j\rangle + \langle \mathcal{O}_i\rangle \mathcal{O}_j - \langle \mathcal{O}_i\rangle\langle \mathcal{O}_j\rangle\Big) \\ &\equiv \sum_i \gamma_i \mathcal{O}_i - \frac{1}{2}\sum_{ij} \lambda_{ij}\langle \mathcal{O}_i\rangle\langle \mathcal{O}_j\rangle \,,\end{aligned} \tag{8.42}$$

where $\gamma_i = \sum_j \lambda_{ij}\langle \mathcal{O}_j\rangle$. Thus, this Hamiltonian describes a set of noninteracting systems, each of which is in the self-consistent effective field of its neighbors, where

$$\langle \mathcal{O}_i \rangle = \frac{\mathrm{Tr}\left(\mathcal{O}_i e^{-\beta\gamma_i \mathcal{O}_i}\right)}{\mathrm{Tr} e^{-\beta\gamma_i \mathcal{O}_i}} \,. \tag{8.43}$$

8.4.2 Vector Spins

The Ising spin is the simplest type of spin. However, there are a host of other spin-like objects that can be used to model physical systems. One can have two-component spins, called XY spins, or three-component spins which are often referred to as classical Heisenberg spins. Let us consider this latter case, where we describe the spin on site i as a three-component unit vector, $\hat{n}_i = (\sin\theta_i \sin\phi_i, \sin\theta_i \cos\phi_i, \cos\theta_i)$. The analog of summing over $S_i = \pm 1$ is to integrate over $d\Omega_i = \cos\theta_i d\theta_i d\phi_i$. The simplest interaction between these vector spins is just their scalar product, often called the classical Heisenberg interaction,

$$\mathcal{H} = -J \sum_{\langle ij \rangle} \hat{n}_i \cdot \hat{n}_j \,, \tag{8.44}$$

which, for $J > 0$, favors parallel spins. As for the Ising case, we can write the vector spin as

$$\hat{n}_i = \langle \hat{n} \rangle + \delta \hat{n}_i \,, \tag{8.45}$$

substitute this into $\mathcal{H}$, drop the terms quadratic in $\delta\hat{n}_i$, and then eliminate $\delta\hat{n}_i$ in favor of $\hat{n}_i$ and $\langle \hat{n}_i \rangle$. The result is

$$\mathcal{H}_{\rm MF} = \frac{JNz}{2}|\langle \hat{n}_i \rangle|^2 - Jz \sum_i \langle \hat{n} \rangle \cdot \hat{n}_i \,. \tag{8.46}$$

Without loss of generality, we are free to choose the ordering direction along $\pm\hat{z}$ by writing $\langle \hat{n}_i \rangle = m\hat{z}$ where $-1 < m < 1$ is a scalar. Then

$$\mathcal{H}_{\rm MF} = \frac{JNz}{2}m^2 - Jzm \sum_i n_{i,z} \,, \tag{8.47}$$

and the self-consistency condition is

$$m = \int d\Omega_i n_{i,z} e^{\beta J z m n_{i,z}} \Big/ \int d\Omega_i e^{\beta J z m n_{i,z}} \,, \tag{8.48}$$

where, as noted above Eq. (8.44), $n_{i,z} = \cos\theta_i$. After defining $x = n_{i,z} = \cos\theta_i$ and writing $\int d\Omega_i = 2\pi \int_{-1}^{1} dx$, the integrals in Eq. (8.48) are straightforward. Since the rest of this calculation is similar to that of the MF theory for the Ising model, we leave it as an exercise and move on to the more interesting case of nematic ordering of rod-like molecules in a liquid crystal.

8.4.3 Liquid Crystals

We next consider a simplified model which very crudely mimics the properties of liquid crystals. Liquid crystals are typically systems composed of long rod-like molecules which have liquid phases in which their centers-of-mass display no long-range translational order. In such a state, the orientations of the molecules may either be disordered (the isotropic liquid phase) or they may display long-range order in which the long axes of all molecules tend to be aligned parallel to one another (the "nematic" liquid phase). To simplify the calculation we describe the orientational properties of this system by a *lattice* model in which there is an orientable rod on each site of a simple cubic lattice. We assume an interaction between rods which tends to make them align along parallel axes. The main difference between this nematic ordering of molecular axes and the vector ordering that results from the interaction of Eq. (8.44) is that, even if the molecules have vector dipole moments, it is only their axes that order at the nematic transition, not their dipole moments. A liquid crystal is *not* a ferromagnet. The long-range order is not one of *direction*, but rather of the *axes*, as is illustrated in Fig. 8.4.

Accordingly, we introduce a Hamiltonian of the form

$$\mathcal{H} = -E \sum_{\langle ij \rangle} \left((\hat{n}_i \cdot \hat{n}_j)^2 - \frac{1}{3} \right) , \tag{8.49}$$

Fig. 8.4 A system of rod-like molecules which have a dipole moment shown in their isotropic (disordered) liquid phase (left) and in their nematic (ordered) liquid crystal phase (right). In the nematic phase, we depict the usual situation in which the long-range order is not completely developed—molecules tend to be aligned parallel or antiparallel, but there are still significant thermal fluctuations. Note that in either case the dipole moments show no long-range order

which is fully symmetric under the parity operation, $\hat{n}_i \to -\hat{n}_i$ for every molecule and which favors parallel ordering of axes. We have included a constant term for convenience. We can write this Hamiltonian as

$$\mathcal{H} = -E \sum_{\langle ij \rangle} \sum_{\alpha\beta} Q_{\alpha,\beta}(i) Q_{\alpha,\beta}(j) , \tag{8.50}$$

where

$$Q_{\alpha\beta}(i) \equiv n_{i,\alpha} n_{i,\beta} - (1/3)\delta_{\alpha,\beta} . \tag{8.51}$$

We then perform a mean-field calculation for the Hamiltonian of Eq. (8.50) following the procedure of Sect. 8.4.1. Note that Eq. (8.50) is not a Hamiltonian which couples scalars (as in the Ising model), or vectors (as in the Heisenberg model), but rather it couples symmetric traceless second-rank tensors. This difference reflects the difference in symmetry of the interacting entities.

Now we need to consider what, if anything, we know about $Q_{\alpha,\beta}(i)$. Except for the fact that **Q** is traceless, it is quite similar to the moment of inertia tensor. Although it is not usually done, one may characterize the orientation of a baseball bat by giving its moment of inertia in the space fixed frame. If a cylindrical rod is preferentially oriented along the z-axis, then **Q** is diagonal and we may set $Q_{zz} = (2/3)Q_0$ and $Q_{xx} = Q_{yy} = -(1/3)Q_0/2$, where Q_0 indicates the degree of preference for orientations along $\hat{z}$. For complete order, one has $Q_0 = 1$. If the rod is oriented along the x-axis then we would set $Q_{xx} = (2/3)Q_0$ and $Q_{yy} = Q_{zz} = -(1/3)Q_0$. In an exercise you are asked to give the form of **Q** for a different ordering. Since **Q** is real, symmetric, and traceless, it has five linearly independent elements. The five independent components of the real symmetric traceless second-rank tensor components $Q_{\alpha,\beta}(i)$ are isomorphic to the five complex spherical harmonics $Y_2^m(\theta_i, \phi_i)$, where θ_i and ϕ_i are the spherical angles of the unit vector $\hat{n}_i$. We want to allow for the possibility that molecules are aligned along some axis, say $\hat{n}_0$, but that components transverse to this axis are disordered. (The occurrence of order transverse to the axis is called *biaxial* order, which does not exist in the nematic liquid crystal phase.) If the quantization axis is collinear with the axis of nematic order, the absence of order transverse to the quantization axis implies that $\langle e^{im\phi}\rangle = 0$ for $m \neq 0$ and thus that

$$\begin{aligned}\langle Y_2^m(\theta_i, \phi_i)\rangle &= \delta_{m,0}\sqrt{5/(4\pi)}\langle(3\cos^2\theta_i - 1)/2\rangle \\ &\equiv \delta_{m,0}\sqrt{5/(4\pi)}A\ ,\end{aligned} \tag{8.52}$$

where the constant $A \leq 1$ describes the amplitude of long-range nematic order. For complete nematic order, $A = 1$ and for orientational disorder, as in the isotropic liquid phase, $A = 0$. When the axis of nematic ordering $\hat{n}_0$ has spherical angles θ_0 and ϕ_0, then

$$\langle Y_2^m(\theta_i, \phi_i)\rangle = AY_2^m(\theta_0, \phi_0)\ . \tag{8.53}$$

Correspondingly, in terms of Cartesian tensors this result is

$$\langle Q_{\alpha,\beta}(i)\rangle = Q_0\left(n_{0,\alpha}n_{0,\beta} - \frac{1}{3}\delta_{\alpha,\beta}\right)\ , \tag{8.54}$$

where Q_0 specifies the amplitude of the long-range orientational order and the unit vector $\hat{n}_0$ defines the axis associated with nematic order. As for the ferromagnetic Ising model, we assume this average to be independent of position, so that $\langle Q_{\alpha,\beta}(i)\rangle \equiv \langle Q_{\alpha,\beta}\rangle$.

We now follow the general formulation of Sect. 8.4.1 and write

$$Q_{\alpha,\beta}(i) = Q_0\left(n_{0,\alpha}n_{0,\beta} - \frac{1}{3}\delta_{\alpha,\beta}\right) + \delta Q_{\alpha,\beta}(i)\ . \tag{8.55}$$

When we ignore terms quadratic in δQ, we obtain the following mean-field Hamiltonian:

$$\mathcal{H}_{\rm MF} = -E\sum_{\langle ij\rangle}\sum_{\alpha,\beta}\left(Q_{\alpha,\beta}(i)\langle Q_{\alpha,\beta}\rangle + \langle Q_{\alpha,\beta}\rangle Q_{\alpha,\beta}(j) - \langle Q_{\alpha,\beta}\rangle^2\right)\ . \tag{8.56}$$

The easiest way to deal with this Hamiltonian is to fix the z-axis to lie along $\hat{n}_0$, in which case $Q_{zz}(i) = [\cos^2\theta_i - 1/3]$, $Q_{xx}(i) + Q_{yy}(i) = -Q_{zz}(i)$,

$$\langle Q_{zz}\rangle = \frac{2}{3}Q_0\,, \qquad \langle Q_{xx}\rangle = \langle Q_{yy}\rangle = -\frac{1}{3}Q_0\,, \tag{8.57}$$

and the thermal averages of the off-diagonal $Q_{\alpha\beta}$'s vanish. Then Eq. (8.56) is

$$\begin{aligned}\mathcal{H}_{\rm MF} &= -Ez\sum_i\left(Q_{zz}(i)\langle Q_{zz}\rangle + Q_{xx}\langle Q_{xx}\rangle + Q_{yy}\langle Q_{yy}\rangle - \frac{1}{2}\sum_\alpha \langle Q_{\alpha,\alpha}\rangle^2\right)\\ &= -EzQ_0\sum_i Q_{zz}(i) + \frac{1}{3}NzEQ_0^2\,,\end{aligned} \tag{8.58}$$

where z is the coordination number of the lattice and N is the total number of sites. This can be written as

$$\mathcal{H}_{\rm MF} = -EQ_0z\sum_i[\cos^2\theta_i - (1/3)] + \frac{1}{3}NzEQ_0^2\,. \tag{8.59}$$

Since $Q_0 = (3/2)\langle Q_{zz}(i)\rangle = \langle 3\cos^2\theta - 1\rangle/2$, Q_0 is determined by the self-consistency equation

$$Q_0 = \frac{\int \sin\theta d\theta e^{\beta EQ_0z\cos^2\theta}(3\cos^2\theta - 1)/2}{\int \sin\theta d\theta e^{\beta EQ_0z\cos^2\theta}}\,. \tag{8.60}$$

To solve this graphically, we proceed by analogy with the Ising model. We write this equation as

$$\frac{kTx}{zE} = F(x)\,, \tag{8.61}$$

where $x \equiv \beta EQ_0z$ and $F(x)$ is the right-hand side of Eq. (8.60). In Fig. 8.5, we plot $F(x)$ versus x and look for the intersection of this curve with the straight line $kTx/(zE)$. In terms of these variables, $F(x)$ has no explicit T dependence whereas the slope of the straight line is proportional to T.

This gives the following qualitative results, as illustrated in Fig. 8.5. The left-hand panel shows that, for high T (line A), there is only one solution to Eq. (8.61), namely, $x = 0$ or $Q_0 = 0$. This is the isotropic liquid phase. As the temperature is decreased, the slope of the straight line decreases and, at some point, a second solution appears for positive x. This happens, as indicated by the dashed line A_0 in the right-hand panel, when $F(x)/x = 0.14855$. As the slope decreases further, two solutions with nonzero values of x develop and move oppositely. In particular, for line B whose slope matches $F'(0) = 2/15 = 0.13333$ the lower of these solutions moves to $x = 0$ and the other moves to $x \approx 4.6$. All this happens because, unlike the Ising case, here $F(x)$ is initially convex upward. To see that $F(x)$ is initially convex upward, we expand the exponential in powers of x, perform the integrals, and calculate $F''(0)$.

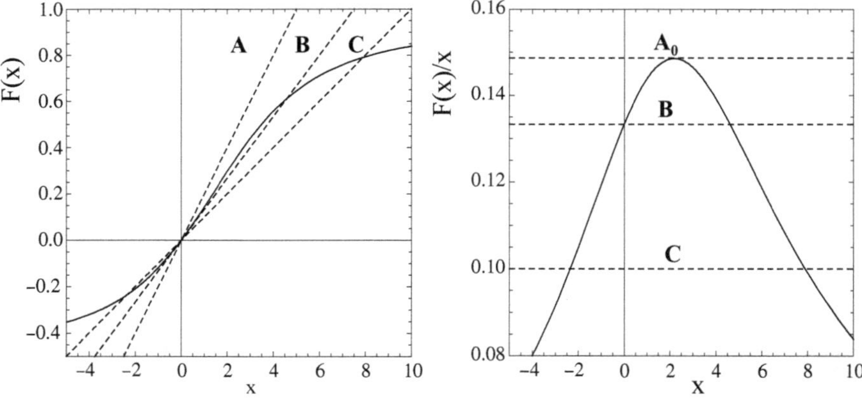

Fig. 8.5 (Left panel) Graphical solution of the self-consistent equation (8.61). The solid curve is the function, $F(x)$, and the dashed lines are (A) $3x/15$, (B) $2x/15$, and (C) $x/10$. Dashed line A intersects $F(x)$ only once, at $x = 0$. B has the same slope as $F(x)$ at $x = 0$, (cf. Eq. (8.62)), but, because $F(x)$ starts off with some upward curvature, intersects $F(x)$ twice, once at $x = 0$ and again at $x = 4.61$. Line C intersects $F(x)$ three times: once at $x = -2.35$, once at $x = 0$, and once at $x = 7.9$. (Right panel) The solid curve is F(x)/x. The dashed lines are at $A_0 = 0.14855$ which is the first value for which a second solution appears at $x \approx 2.18$. Dashed lines B and C correspond to the slopes of dashed lines B and C in the left-hand panel

$$
\begin{aligned}
F(x) &= \frac{\frac{1}{2}\int_0^\pi \sin\theta[(3/2)\cos^2\theta - (1/2)][1 + x\cos^2\theta + (1/2)x^2\cos^4\theta]}{\frac{1}{2}\int_0^\pi \sin\theta[1 + x\cos^2\theta + (1/2)x^2\cos^4\theta]} \\
&= \frac{x[(3/10) - (1/6)] + (1/2)x^2[(3/14) - (1/10)]}{1 + x(1/3) + \mathcal{O}(x^2)} \\
&= \frac{(2/15)x + (2/35)x^2}{1 + (x/3)} = (2/15)x + (4/315)x^2 \,. \qquad (8.62)
\end{aligned}
$$

So indeed $F'(0) = 2/15$ and $F''(0)$ is positive which means that $F(x)$ initially curves upward, which is why the transition cannot be continuous. Finally, for even smaller slopes, there are again three solutions, with one at negative x, as shown for the dashed line C.

Since there are multiple solutions, the question, as for the Ising model, is which solution to take. By an argument almost identical to that used for the Ising model (see Eq. (8.13)) one can write the free energy difference between the solution with no order and one with a nonzero order parameter as an integral. However, unlike the Ising case, where the integrand is negative, the integrand for the nematic case starts out positive. It is only for larger values of the order parameter that the free energy difference changes sign and a nonzero order parameter is favored.

Thus, the isotropic to nematic transition is discontinuous in that the equilibrium value of x jumps from zero to a finite value discontinuously as the temperature is lowered. The complexity of this analysis suggests that another approach that focused directly on the free energy rather than trying to analyze solutions to the self-consistent equation might be simpler and more transparent. Indeed, this is exactly how we will develop a much more intuitive analysis of this problem in the next chapter. Furthermore, focusing on the free energy will lead us to understand how the difference in symmetry between the Ising model and our simplified model for liquid crystals gives rise to a continuous transition for the former and a discontinuous transition for the latter.

8.5 Summary and Discussion

In this chapter, we have used mean-field theory to study systems which are described by an order parameter which is a scalar for the Ising model, a vector for the Heisenberg model, and a second-rank tensor for nematic liquid crystals. For a bilinear Hamiltonian of the form $\mathcal{H} = (1/2)\sum_{ij} \lambda_{ij} \mathcal{O}_i \mathcal{O}_j$, where i and j label different sites and $\mathcal{O}$ is an operator, then within the mean-field approximation we may treat this system via the effective Hamiltonian

$$\mathcal{H}_{MF} = (1/2)\sum_{ij} \lambda_{ij}[\mathcal{O}_i \langle \mathcal{O}_j \rangle_T + \langle \mathcal{O}_i \rangle_T \mathcal{O}_j - \langle \mathcal{O}_i \rangle_T \langle \mathcal{O}_j \rangle_T] , \qquad (8.63)$$

where the thermal averages $\langle \mathcal{O}_i \rangle_T$ are determined self-consistently. This effective Hamiltonian is a sum of single-site Hamiltonians, and, as such, it cannot describe the correlations between the fluctuations of $\mathcal{O}_i$ and $\mathcal{O}_j$. The mean-field approximation therefore breaks down when the $\mathcal{O}$-$\mathcal{O}$ correlation length becomes large. Near a continuous phase transition one expects to see "data collapse," which means that the equation of state, instead of being a relation between three variables, becomes a relation between two scaled variables, providing the scale factors are suitably chosen.

A word of caution is warranted about the concept of an "effective" Hamiltonian. A *microscopic* Hamiltonian has no temperature dependence. It is a quantum mechanical operator that describes the paths of microscopic particles and fields in space and time. For Hamiltonians with no explicit time dependence, the thermodynamic density matrix can be written as $\rho = \exp(-\beta\mathcal{H})/Z$ where β is the inverse temperature and $Z = \mathrm{Tr}\exp(-\beta\mathcal{H})$ and, again, $\mathcal{H}$ has no temperature dependence. However, as we have seen in discussing the variational principle for the free energy in Sect. 6.6.2, it is sensible to define a variational density matrix which is an arbitrary Hermitian operator with positive eigenvalues on the interval, [0, 1], which could, in principle, depend on many temperature-dependent parameters. A convenient form for such a trial density matrix is $\rho_{\rm trial} = \exp(-\beta\mathcal{H}_{\rm eff})/\mathrm{Tr}\exp(-\beta\mathcal{H}_{\rm eff})$ where $\mathcal{H}_{\rm eff}$ is an arbitrary Hermitian operator with dimensions of energy. It is in this context that the concept of a temperature-dependent effective Hamiltonian is sensible. The use of an effective temperature-dependent Hamiltonian is discussed further in Exercise 1.

8.6 Exercises

1. In Eq. (8.7), we derived an effective Hamiltonian $\mathcal{H}_{\rm MF}$ which is temperature dependent. One might ask whether the formalism we have presented in Chap. 4 applies when the Hamiltonian is temperature dependent. In general, the answer would be "no." According to our formalism, one should have

$$-\left.\frac{\partial \ln Z}{\partial \beta}\right)_H = \langle \mathcal{H}_{\rm MF} \rangle_T \ , \tag{8.64}$$

where $Z = {\rm Tr} e^{-\beta \mathcal{H}_{\rm MF}}$ and thermal averages are defined by

$$\langle X \rangle_T = {\rm Tr}\left[X e^{-\beta \mathcal{H}_{MF}} \right] / \ {\rm Tr}\left[e^{-\beta \mathcal{H}_{MF}} \right] \ . \tag{8.65}$$

If Eq. (8.64) is *not* fulfilled, then the question would arise as to which side (if either) of this equation should be used to calculate the internal energy. Similarly, we should have

$$\left\langle \sum_i S_i \right\rangle_T = \left\langle -\frac{d\mathcal{H}_{MF}}{dH} \right\rangle_T \ , \tag{8.66}$$

for there to be no ambiguity as to which side of this equation should be used to calculate the thermally averaged magnetic moment. This relation is also fulfilled by the MF approximation. Show that relations, Eqs. (8.64) and (8.66), are true for the MF treatment we have given for the Ising model.

2(a). This problem forms a mathematical introduction to scaling. Consider the equation

$$w^3 - xw = y \ , \tag{8.67}$$

which defines w as a function of x and y. (Assume that x and y are small, as discussed below.) Suppose that x is real positive and that we are only interested in the root for w which is near $|\sqrt{x}|$. Show that with an appropriate choice of p and r, one can write

$$w = x^p f(y/|x^r|) \ ,$$

where $f(0) = 1$. In a math course, we could ask you to prove that $f(z)$ is an analytic function for small z. Instead we will ask you to calculate the coefficients a_1 and a_2 in the expansion

$$f(z) = 1 + a_1 z + a_2 z^2 + \cdots \ .$$

This suggests, but does not prove, that $f(z)$ is analytic.

2(b). Now for some experiments. Use the solution to Eq. (8.67) in the form

$$w = 2\sqrt{x/3}\cos(\theta/3) \,, \qquad \cos\theta = \frac{1}{2} y\sqrt{27/x^3} \,,$$

to obtain $w(x, y)$ for $0.01 \leq y \leq 0.06$ for five values of x: 0.4, 0.5, 0.6, 0.7, and 0.8. On a single graph, plot the results as curves of w versus y for each of the values of x. (Use a suitable scale so that the graphs are nicely displayed.) These plots will tell you very little!!

2(c). Now plot the data differently. Plot w/x^p versus y/x^r. (If possible, use different symbols for distinguishing data for different values of x.) Observe what is called "data collapse." Knowing nothing else, what can you conclude from this data collapse?

3. Suppose the equation of state for the magnetic moment per spin of a magnet near the critical point at $H = t = 0$, where $t \equiv (T - T_c)/T_c$ is of the form

$$M^5 - atM = bH \,.$$

Define scaled variables for the magnetic moment and magnetic field so that the asymptotic equation of state involves only these two variables. Display the two branches of the scaled equation of state.

4. Generalize the mean-field result for a spin S ferromagnet for which the Hamiltonian is

$$\mathcal{H} = -J \sum_{\langle ij \rangle} S_{iz} S_{jz} - H \sum_i S_{iz} \,,$$

where $\langle ij \rangle$ indicates a sum over pairs of nearest neighboring sites and S_{iz} is the z-component of the spin on the ith site: $S_{iz} = -S, \ -S+1\,, \ldots S$.

(**a**) Obtain the self-consistent equation for $M \equiv \langle S_{iz} \rangle_T$, the thermal average of S_{iz}, in terms of a Brillouin function. (See your old homework.)

(**b**) Determine the critical exponents β and γ defined by the asymptotic forms

$$M(T, H = 0^+) \sim (T_c - T)^\beta \,, \qquad T \to T_c^-$$

$$\chi(T) \equiv \left.\frac{\partial M}{\partial H}\right)_{T,H=0} \sim (T - T_c)^{-\gamma} \,.$$

NOTE: For χ you need only consider $T > T_c$.

5. Complete the derivation of the mean-field theory for the vector spin model of Sect. 8.4.2. Starting with Eq. (8.48), do the integral and then solve the self-consistent equation numerically.

6. Give a mean-field treatment of the spin 1/2 quantum Heisenberg model, for which the Hamiltonian is

$$\mathcal{H} = -J \sum_{\langle ij \rangle} \mathbf{S}_i \cdot \mathbf{S}_j \ ,$$

where $\langle ij \rangle$ indicates a sum over pairs of nearest neighbors on a three-dimensional lattice for which each site has z nearest neighbors. Allow the spontaneous magnetization to be oriented an an arbitrary direction, consistent with the fact that the order parameter is a *vector*.

7. Consider the following model to describe the orientation of diatomic molecules adsorbed on a two-dimensional surface. For simplicity, assume the centers of the molecules are fixed on a square lattice and can assume only four orientations which are those parallel to the lattice directions. Assume the neighboring molecules have an interaction energy $-E_1$ if they are perpendicular to one another and an energy $+E_1$ if they are parallel to one another, with $E_1 > 0$. (This interaction could arise from their electric quadrupole moments.) Obtain a mean-field treatment of this problem. (Hint: this model is closely related to one of the models discussed in this chapter.) In particular (a) discuss whether or not this model has a phase transition, and if so locate its critical temperature and (b) define an order parameter and show how it is determined within mean-field theory.

8(a). This problem concerns the order parameter for liquid crystals. Give the form that $\langle \mathbf{Q} \rangle_T$ assumes if the molecules are ordered to be along (i.e., parallel or antiparallel to) an axis in the x–y plane which makes a 45° angle with the positive x- and y-axes.

8(b). Repeat the exercise of part A if the axis of ordering is in the first quadrant of the x–y plane, making an angle θ with the positive x-axis.

9. This problem concerns the equation of state connecting the variables M, H, and T is the region asymptotically close to the critical point. In the text, the scaled variables m and h are introduced in terms of which the equation of state related the *two* variables m and h. Predict the form the scaling variables m and h must assume if the values of the critical exponents β (for the magnetization) and γ (for the susceptibility) are known, but these values are *not* those of mean-field theory.

10. This problem concerns finite-size corrections to Bose condensation in three spatial dimensions. Start from

$$N = \frac{1}{\hat{\mu}} + \sum_{\mathbf{k}}{}' \frac{1}{e^{\beta \epsilon(\mathbf{k}) + \hat{\mu}} - 1} \equiv \frac{1}{\hat{\mu}} + \mathcal{S} \ , \tag{8.68}$$

where $\hat{\mu} = -\beta\mu \geq 0$. Now use the evaluation (valid for $T \approx T_c$, small $\hat{\mu}$, and large V) that

$$\mathcal{S} = AVT^{3/2} - BV\sqrt{\hat{\mu}} + CV^{2/3}\ , \tag{8.69}$$

where A, B, and C may be regarded as constants. We will study this transition for fixed particle density ρ ($\rho = N/V$). Then for fixed ρ Eq. (8.68) relates the three variables $\hat{\mu}$, V, and T. We will be concerned with finite-size corrections when V is large but not infinite.

(**a**) Determine $T_c(V = \infty)$ in terms of the parameters introduced in Eq. (8.69).

(**b**) Give the result of Eq. (8.68) in the asymptotic regime when all the three variables $\hat{\mu}$, $(T - T_c)$, and $1/V$ are infinitesimals.

(**c**) Show that the asymptotic equation of part B can be expressed in terms of the two scaled variables $(T - T_c)V^x$ and $\hat{\mu}V^y$ and give the values of the exponents x and y (This relation will be referred to as the "scaling equation.").

(**d**) Check the behavior of μ for $T > T_c$ and $T < T_c$ for infinite V according to your result for the scaling equation.

(**e**) How does your scaling equation predict N_0 will behave at $T = T_c(V = \infty)$ as V becomes large?

(**f**) For $V = \infty$, T_c is determined as the temperature at which $\partial^2\mu/\partial T^2)_\rho$ is discontinuous. For finite, but large, V we may define $T_c(V)$ to be where $\partial^3\mu/\partial T^3)_\rho$ is maximal. This condition leads to the result

$$T_c(V) = T_c(\infty) + C_1 V^\tau\ . \tag{8.70}$$

Give the value of τ, but only try to evaluate C_1 if you have nothing better to do.

(**g**) If we assume that finite-size effects involve the ratio of the system size, L, to the correlation length ξ and that for T near T_c one has $\xi \sim |T - T_c|^{-\nu}$, deduce the value of ν for Bose condensation. (There are much more direct ways to obtain this result.)

11. We have seen order parameters which are tensors of rank 0, 1, and 2 for the Ising model, the Heisenberg model, and liquid crystals, respectively. What kind of order parameter do you think would describe the orientational ordering of methane (CH_4) molecules, each of which is a regular tetrahedron? (Note that the moment of inertia tensor of a tetrahedron is diagonal.) This question can be answered in terms of nth rank tensors or spherical harmonics of degree n (James and Keenan 1959). In a more exotic vein, what kind of order parameters are needed to describe the orientational ordering of icosahedra (see Harris and Sachidanandam 1992 and Michel et al. 1992). The molecule C_{60}, known colloquially as a Buckeyball, has icosahedral symmetry, and crystalline C_{60} has an orientationally ordered phase.

References

A.B. Harris, R. Sachidanandam, Orientational ordering of icosahedra in solid C_{60}. Phys. Rev. B **46**, 4944 (1992)

H.M. James, T.A. Keenan, Theory of phase transitions in solid heavy methane. J. Chem. Phys. **31**, 12 (1959)

K.H. Michel, J.R.D. Copley, D.A. Neumann, Microscopic theory of orientational disorder and the orientational phase transition in solid C_{60}. Phys. Rev. Lett. **68**, 2929 (1992)

Chapter 9
Density Matrix Mean-Field Theory and Landau Expansions

9.1 The General Approach

In this chapter, we discuss a formulation of mean-field (MF) theory which puts the derivation in the previous chapter on a somewhat firmer footing and which often proves to be more convenient. It is based on the observation that the result of MF theory, as developed in the previous chapter, was the construction of an effective Hamiltonian which replaced interactions between particles by an effective field which was self-consistently determined. Thus, in effect, the density matrix was approximated by a direct product of single-particle density matrices. That is

$$\rho^{(N)} = \prod_{i=1}^{N} \rho_i^{(1)} , \tag{9.1}$$

where $\rho_i^{(1)}$ operates only in the space of states of the ith particle and is subject to the following constraints. It should be a Hermitian matrix satisfying $\mathrm{Tr}_i \rho(i) = 1$, where Tr_i indicates a trace only over states of the ith particle, and it should have no eigenvalues outside the interval [0,1]. It is important to remember that this is *not* the most general form that $\rho^{(N)}$ can have. In general, $\rho^{(N)}$ will be a non-separable function of the many particle coordinates. *Equation* (9.1) *is a very special form* in which the coordinates of the ith particle are subject to a static field or potential due to the other particles. From this characterization, we expect that this approximation should be similar or even equivalent to the mean-field approximation introduced in the previous chapter.

In view of this observation, we will use the variational principle for the free energy to find the best choice for the $\rho_i^{(1)}$ assuming ρ to have the form of Eq. (9.1). The strategy, therefore, is to choose a general form for $\rho_i^{(1)}$ and then vary its parameters to minimize the trial free energy of Eq. (6.49) which we write in the form

$$F_{\mathrm{tr}} = \mathrm{Tr}\Big(\mathcal{H}\rho^{(N)}\Big) + kT\,\mathrm{Tr}\left(\rho^{(N)} \ln \rho^{(N)}\right) , \tag{9.2}$$

A. J. Berlinsky and A. B. Harris, *Statistical Mechanics*, Graduate Texts in Physics,
https://doi.org/10.1007/978-3-030-28187-8_9

where $\rho^{(N)}$ is the "trial density matrix." One almost always deals with a Hamiltonian which is the sum of single-particle energies and two-body interactions and is therefore of the form

$$\mathcal{H} = \sum_i \mathcal{H}_i + \sum_{i<j} V_{ij} , \tag{9.3}$$

where $\mathcal{H}_i$ depends only on the coordinates of the ith particle and V_{ij} depends only on the coordinates of the ith and jth particle. Then the trial free energy may be written as

$$F_{\rm tr} = U_{\rm tr} - T S_{\rm tr} , \tag{9.4}$$

where the trial energy and trial entropy are, respectively,

$$U_{\rm tr} = \sum_i {\rm Tr}_i[\rho_i^{(1)} \mathcal{H}_i] + \sum_{i<j} {\rm Tr}_{ij}[\rho_i^{(1)} \rho_j^{(1)} V_{ij}] \tag{9.5}$$

and

$$S_{\rm tr} = -k \sum_i {\rm Tr}_i \rho_i^{(1)} \ln \rho_i^{(1)} , \tag{9.6}$$

where ${\rm Tr}_{ij}$ indicates a sum only over states of spins i and j.

The fundamental idea of this variational principle is that it incorporates the competition between energy and entropy. At zero temperature, the energy is minimized. As the temperature is increased, one places successively more weight on the entropy, a term which favors complete disorder, in contrast to the energy term, which favors some sort of order. Furthermore, since the entropy term disfavors order, only that type of order which is energetically favored will be consistent with the variational principle.

9.2 Order Parameters

In the previous chapter, we saw that ordered phases can be characterized by an "order parameter." In the case of the Ising model, the order parameter was a scalar, the thermally averaged magnetization at a single site. In Exercise 5 of Chap. 8, you were invited to study the Heisenberg model in which the order parameter is a vector, namely, the magnetization vector. In the case of liquid crystals, the order parameter was taken to be a second-rank tensor. Order parameters with other symmetries will be discussed in examples later on. In general, an order parameter which characterizes an ordered phase is zero in the disordered phase and is nonzero in the ordered phase. At a second-order or continuous transition, the order parameter goes smoothly to zero at the phase boundary. We will see that, in this case, an expansion in powers of the order parameter is a useful way of analyzing the mean-field behavior of the system near that boundary. This approach can be extended to the case of more than one order parameter and to weakly discontinuous (first order) transitions. In fact, the usefulness of such series expansions goes beyond that of a computational method. Expansions in powers of order parameters near phase boundaries are a way of characterizing the nature of transitions.

This method of describing phase transitions was first developed by Landau. The use of a "Landau expansion" permeates the phase transition literature. A Landau expansion can be thought of as a scenario. Different physical systems which have the same form of Landau expansion of their mean-field free energies are, in this sense, equivalent, and their phase transitions will follow the same scenario. In order to have Landau expansions of the same form, two systems must, first of all, have order parameters of the same type, e.g., scalars (as in the case of the Ising model), vectors (as in the case of the Heisenberg model), or tensors (as in the case of liquid crystals). Furthermore, the Hamiltonians which govern the two systems must respect the same symmetries. For example, the symmetry of the interactions of many models of magnetic ordering is such that, in the absence of external fields, only even powers of the order parameter appear in the expansion of the free energy.

9.3 Example: The Ising Ferromagnet

Here we again treat the Ising ferromagnet whose Hamiltonian is

$$\mathcal{H} = -J \sum_{<ij>} \sigma_{iz}\sigma_{jz} - h \sum_i \sigma_{iz} , \tag{9.7}$$

where σ_{iz} is the Pauli matrix for the z-th component of spin of the ith particle. Or more simply, σ_{iz} may be considered a scalar assuming the values ± 1. Since each spin has two states, $\rho_i^{(1)}$ is, in principle, a two-by-two matrix. However, in applying the variational principle, it is useful to keep in mind a general principle: namely, that only the type of order which is favored by the energy should be introduced into the density matrix. Here, this principle indicates that there is no point to introducing off-diagonal elements into the density matrix because these do not lower the energy. Indeed, one can go through the algebra (see Exercise 2) to show that if one starts from a density matrix with off-diagonal elements, the trial free energy is minimized when these off-diagonal elements are zero. Accordingly, we consider most general diagonal form for $\rho(i)$, which is

$$[\rho(i)]_{n,m}(a_i) = a_i \delta_{n,1}\delta_{m,1} + (1 - a_i)\delta_{n,-1}\delta_{m,-1} . \tag{9.8}$$

However, a more useful way of writing this is

$$\rho_i^{(1)} = \frac{1}{2}[\mathcal{I} + m_i \sigma_{iz}] \equiv \begin{bmatrix} \frac{1+m_i}{2} & 0 \\ 0 & \frac{1-m_i}{2} \end{bmatrix} , \tag{9.9}$$

where $\mathcal{I}$ is the unit matrix and σ_{iz} is the Pauli matrix for the z-component of the ith spin. Alternatively, one can write

$$\rho_i^{(1)} = \frac{1}{2}[1 + m_i\sigma_i] \, . \tag{9.10}$$

Then Tr_i is a sum over states with $\sigma_i = \pm 1$. This form for $\rho_i^{(1)}$ satisfies

$$\mathrm{Tr}_i \rho_i^{(1)} = \sum_{\sigma_i=\pm 1} \frac{1}{2}[1 + m_i\sigma_i] = 1 \tag{9.11}$$

$$\langle \sigma_{iz} \rangle = \mathrm{Tr}_i[\sigma_{iz}\rho_i^{(1)}] = \sum_{\sigma_i=\pm 1} \frac{1}{2}[1 + m_i\sigma_i]\sigma_i = m_i \, , \tag{9.12}$$

so that m_i is the average value of spin σ_{iz} for the trial density matrix, $\rho_i^{(1)}$. In other words, m_i is the local (at site i) order parameter. Also, using Eqs. (9.5) and (9.6), we have

$$U_{\mathrm{tr}} = -\sum_{\langle ij \rangle} J m_i m_j - h \sum_i m_i \tag{9.13}$$

and

$$\frac{S_{\mathrm{tr}}}{k} = -\sum_i \left(\frac{1+m_i}{2}\right) \ln \left(\frac{1+m_i}{2}\right) + \left(\frac{1-m_i}{2}\right) \ln \left(\frac{1-m_i}{2}\right) . \tag{9.14}$$

The strategy employed here is quite generally applicable. It is to parametrize the density matrix in terms of the average quantities (e.g., the order parameters, m_i) that the density matrix can be used to calculate.

At this point, there are two ways to proceed.

1. We can evaluate the free energy, F_N, for $\rho_N\{\sigma_i\}$ as defined above, and then minimize with respect to all of the m_i and show that the lowest free energy results when all the m_i are equal. (This is an exercise.) This approach will be generalized in Chap. 10.
2. We can "guess" that the free energy is minimized for all $m_i = m$, as we did in our initial mean-field treatment of the ferromagnet, and then solve for m as a function of h and T. This "guess" is motivated by the observation that the energy does not favor a nonuniform distribution of magnetic moments. Then m, the zero wavevector component of the magnetization, is the order parameter for this system.

For now, we will adopt the second strategy. Then the free energy per site, $\hat{f}$, is

$$\hat{f}(m) = -\frac{Jz}{2}m^2 - hm + kT\left[\left(\frac{1+m}{2}\right)\ln\left(\frac{1+m}{2}\right) + \left(\frac{1-m}{2}\right)\ln\left(\frac{1-m}{2}\right)\right]. \tag{9.15}$$

The value of m is obtained by minimizing $\hat{f}(m)$ with respect to m. Thus, we write

$$\frac{\partial \hat{f}}{\partial m} = -Jzm - h + \frac{kT}{2}\ln\frac{1+m}{1-m} = 0. \tag{9.16}$$

If we define $x = (Jzm + h)/(kT)$, then

$$\frac{1+m}{1-m} = e^{2x}$$

$$m = \tanh\left(\frac{Jzm+h}{kT}\right), \tag{9.17}$$

which is the result obtained earlier for the mean-field theory of the Ising ferromagnet and which was analyzed in detail.

9.3.1 Landau Expansion for the Ising Model for h = 0

It is instructive to consider the graphs of $\hat{f}(m)$ as a function of m for various temperatures. A qualitative analysis of the dependence of $\hat{f}$ on m is obtained by expanding $\hat{f}$ in powers of m for temperature T near T_c. To obtain this expansion, which is known as the Landau expansion, we write

$$\begin{aligned}\frac{1\pm m}{2}\ln\left(\frac{1\pm m}{2}\right) &= \frac{1\pm m}{2}\left[-\ln 2 \pm m - \frac{m^2}{2} \pm \frac{m^3}{3} - \frac{m^4}{4}\dots\right] \\ &= \frac{1}{2}\left[-\ln 2 \pm (1-\ln 2)m + \frac{m^2}{2} \mp \frac{m^3}{6} + \frac{m^4}{12} \mp \dots\right].\end{aligned} \tag{9.18}$$

The sum over + and − of these terms gives

$$\sum_{\sigma=\pm 1}\frac{1+\sigma m}{2}\ln\left[\frac{1+\sigma m}{2}\right] = -\ln 2 + \frac{m^2}{2} + \frac{m^4}{12} + \mathcal{O}(m^6). \tag{9.19}$$

Then we define $f \equiv \hat{f} + kT\ln 2$, so that

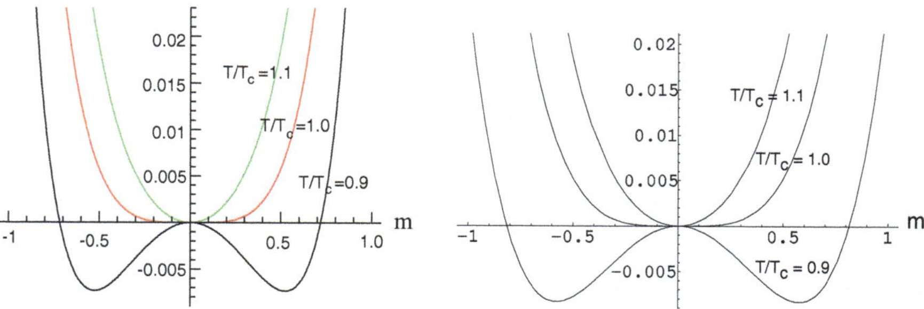

Fig. 9.1 Trial free energies, f as a function of the order parameter m for the values of temperature $t \equiv kT/Jz$ indicated. The left panel shows the full result of Eq. (9.15). The right panel shows the result according to the truncated expansion of Eq. (9.20). (Clearly, there is not much difference between the two results.)

$$f = -hm + k(T - T_c)\frac{m^2}{2} + kT\frac{m^4}{12} \,. \tag{9.20}$$

The Landau expansion to order m^4 is sufficient to illustrate essentially all of the interesting properties of the mean-field phase diagram of the Ising ferromagnet. In Fig. 9.1, we show, as a function of m, the zero-field free energy obtained using this expansion up to order m^4 for several temperatures near T_c, as compared to the full result. One sees that the Landau expansion accurately captures the qualitatively important features of the free energy.

We now comment on the symmetry and stability properties of this free energy. For the special case of $h = 0$, the free energy involves only even powers of m. This property is a consequence of the fact that the Hamiltonian is invariant under the transformation of σ_i into $-\sigma_i$, for all i. The free energy is therefore invariant under the transformation $m \to -m$. (This is an exact symmetry of the Hamiltonian, not an artifact of the mean-field approximation.) This symmetry does *not* imply that the spontaneous magnetic moment (which changes sign under the above transformation) is zero. One can see this from the curves in Fig. 9.1. Since the free energy looks like a potential well, we will discuss it in mechanical terms. For $T > T_c$, we say that the disordered phase is *stable*, meaning that it is stable with respect to formation of order (i.e., to making m be nonzero). For $T < T_c$, $m = 0$ corresponds to a local *maximum* of the free energy, and correspondingly we say that the disordered phase is unstable relative to the formation of long-range order. The free energy is minimized by either of two equivalent values of m, determined by the condition

$$\frac{1}{m}\frac{\partial f}{\partial m} = k(T - T_c) + \frac{1}{3}kTm^2 = 0. \tag{9.21}$$

The system arbitrarily selects one of these minima. Mathematically, one may cause the selection by considering the limit $h \to 0^+$ which is why one describes $m(h = 0^+)$ as the spontaneous magnetization. The fact that the magnetization in zero field is nonzero for models like the Ising model is called "spontaneous symmetry breaking." Thus, the thermodynamic state need not, and in an ordered phase, actually does not possess the full symmetry of the Hamiltonian.

Contrast this picture with the rather convoluted argument we had to give in Chap. 8, particularly Eq. (8.13), to ensure that, for $T < T_c$, we were selecting the order parameter corresponding to the lowest free energy. Had we developed a similar argument for the liquid crystal problem, it would have been much more complicated. Here the conclusion is totally trivial. The procedure used here is equivalent to that used in the previous chapter *except* that here we start from an explicit representation of the free energy as a function of m, rather than from the self-consistency equation. When there is more than one local minimum in $f(m)$ as a function of m, it is usually obvious which local minimum is a global minimum. In addition, as we will see much later, this formulation suggests that, by considering fluctuations about the minimum of the free energy, one can improve upon mean-field theory.

We now use this expansion to study the mean-field properties in the vicinity of a phase transition. Here, for $T \approx T_c$, one has

$$f = -hm + \frac{1}{2}k(T - T_c)m^2 + \frac{1}{12}kT_c m^4 \, . \tag{9.22}$$

We will consider various cases. For example, for $h \neq 0$ but small and $T > T_c$, we can neglect the m^4 term because m will be of the same order as the small parameter h. Then

$$\left.\frac{\partial f}{\partial m}\right|_T = -h + k(T - T_c)m = 0 \quad \rightarrow$$

$$m = \frac{h}{k(T - T_c)} \equiv \chi h \, , \tag{9.23}$$

the Curie–Weiss law. One sees that the coefficient of the quadratic term can be identified with the inverse susceptibility, so that we may write

$$f(h = 0) = \frac{1}{2}\chi^{-1}m^2 + \frac{1}{12}kT_c m^4 \, . \tag{9.24}$$

In our mechanical analogy, the inverse susceptibility is like a spring constant. When χ is small, the system is very stiff, or equivalently, very stable. A large χ corresponds to a very "soft" and only weakly stable system. In that case, we may visualize the possibility of large fluctuations. We will see this more clearly later on.

For $h = 0$, we set

$$\left.\frac{\partial f}{\partial m}\right|_T = k(T - T_c)m + \frac{1}{3}kT_c m^3 \, , \tag{9.25}$$

from which we conclude that $m(h = 0) = 0$ for $T > T_c$ and

$$m(h = 0^+) = \sqrt{3(T_c - T)/T_c} \, , \tag{9.26}$$

for $T < T_c$. Indeed the whole analysis of the previous chapter for T near T_c can be repeated, based on the Landau expansion.

9.4 Classical Systems: Liquid Crystals

Here we apply the variational principle to treat the lattice model of liquid crystals introduced in the previous chapter, whose Hamiltonian is

$$\mathcal{H} = -E \sum_{<ij>} \sum_{\alpha\beta} Q_{\alpha\beta}(i) Q_{\alpha\beta}(j) \, , \tag{9.27}$$

where

$$Q_{\alpha\beta}(i) = n_\alpha(i) n_\beta(i) - (1/3)\delta_{\alpha,\beta} \, , \tag{9.28}$$

with $\hat{n}_i$ a unit vector along the molecular axis. Note that both the energy and the order parameter are invariant with respect to the transformation $\hat{n}_i \to -\hat{n}_i$. First we must decide what variables the single-particle density matrix depends on. Clearly, the variables involved are the spherical angles θ_i and ϕ_i of the ith rod-like molecule. Also, Tr... means $\prod_i \int \ldots \sin\theta_i d\theta_i d\phi_i$.

A plausible ansatz for the density matrix is

$$\rho_i^{(1)}(\theta_i) = \frac{1}{4\pi} \left[\sum_n a_n P_n(\cos\theta_i) \right] , \tag{9.29}$$

independent of ϕ_i, where $\cos\theta_i \equiv \hat{n}_i \cdot \hat{n}_0$, and the variational parameters are the set of a_n's. Since we do not anticipate any ordering which distinguishes the two collinear directions, we can immediately say that $a_n = 0$ for odd n, and $a_0 = 1$ to have unit trace. A particularly simple form would be a density matrix with only the a_2 term, in which case

$$\rho_i^{(1)}(\theta_i) = \frac{1}{4\pi} \left(1 + \epsilon(3[\hat{n}_i \cdot \hat{n}_0]^2 - 1)/2\right) . \tag{9.30}$$

Of course, a compromise could be invoked whereby one allows only a_2 and a_4 to be nonzero. Alternatively, the notion of an effective Hamiltonian suggests that we could set

$$\rho_i^{(1)}(\theta_i) = C e^{\epsilon(3[\hat{n}_i \cdot \hat{n}_0]^2 - 1)/2} \, , \tag{9.31}$$

with the single variational parameter ϵ. When the degree of ordering is small Eqs. (9.30) and (9.31) are equivalent. Either one is equivalent to using Eq. (9.29) with special relations for the a_n's with $n > 2$ in terms of a_2. The equation of state will differ only slightly from one approximation to the other. Near the isotropic to nematic transition, all these approximations lead to qualitatively similar results.

For simplicity, we follow the approach based on the trial density matrix of Eq. (9.30). The condition of unit trace has already been incorporated in this form. We relate ϵ to the local order parameter:

$$\begin{aligned}\langle Q_{\alpha\beta}(i)\rangle &\equiv \langle n_{i,\alpha}n_{i,\beta} - \frac{1}{3}\delta_{\alpha\beta}\rangle \\ &= \int_0^\pi \sin\theta_i d\theta_i \int_0^{2\pi} d\phi_i [n_{i,\alpha}n_{i,\beta} - \frac{1}{3}\delta_{\alpha,\beta}]\rho_i^{(1)}(\theta_i,\phi_i) \\ &= \frac{1}{4\pi}\int_0^\pi \sin\theta_i d\theta_i \int_0^{2\pi} d\phi_i [n_{i,\alpha}n_{i\beta} - \frac{1}{3}\delta_{\alpha,\beta}] \\ &\quad\times \left[1 - \frac{\epsilon}{2} + \frac{3}{2}\epsilon \sum_{\xi\eta} n_{i\eta}n_{0,\eta}n_{i,\xi}n_{0,\xi}\right] \\ &\equiv \overline{[n_{i,\alpha}n_{i\beta} - \frac{1}{3}\delta_{\alpha,\beta}][1 - \frac{\epsilon}{2} + \frac{3}{2}\epsilon \sum_{\xi\eta} n_{i\eta}n_{0,\eta}n_{i,\xi}n_{0,\xi}]}\ ,\end{aligned} \tag{9.32}$$

where the overbar indicates a spherical average over angles.

Since the averages are independent of site indices we write

$$\overline{n_\alpha n_\beta} = (1/3)\delta_{\alpha\beta}\ . \tag{9.33}$$

Next consider $\overline{n_\alpha n_\beta n_\gamma n_\delta}$. This average vanishes unless each Cartesian index x, y, or z appears an even number of times. By symmetry, we have

$$\begin{aligned}\overline{n_x^2 n_y^2} &= \overline{n_x^2 n_z^2} = \overline{n_y^2 n_z^2} = A \\ \overline{n_x^4} &= \overline{n_y^4} = \overline{n_z^4} = B\ .\end{aligned} \tag{9.34}$$

Explicitly $\overline{n_z^4} = \overline{\cos^4\theta} = 1/5$. Thus $B = 1/5$. From $3B + 6A = \overline{(n_x^2 + n_y^2 + n_z^2)^2} = 1$ we get $A = 1/15$. These results are summarized as

$$\overline{n_\alpha n_\beta n_\gamma n_\delta} = \frac{1}{15}\left(\delta_{\alpha,\beta}\delta_{\gamma,\delta} + \delta_{\alpha,\gamma}\delta_{\beta,\delta} + \delta_{\alpha,\delta}\delta_{\beta,\gamma}\right)\ . \tag{9.35}$$

In an exercise you are asked to write down the general result for the average of a product of $2n$ components of $\hat{n}$. From Eq. (9.32), we have

$$
\begin{aligned}
\langle Q_{\alpha\beta}(i)\rangle &= \frac{3}{2}\epsilon\overline{\left([n_{i,\alpha}n_{i\beta} - \frac{1}{3}\delta_{\alpha,\beta}]\sum_{\xi\eta} n_{i\eta}n_{0,\eta}n_{i,\xi}n_{0,\xi} \right)} , \\
&= \frac{\epsilon}{10}\sum_{\xi\eta} n_{0\eta}n_{0\xi}\left(\delta_{\alpha,\beta}\delta_{\xi,\eta} + \delta_{\alpha,\xi}\delta_{\beta,\eta} \right. \\
&\quad \left. +\delta_{\alpha,\eta}\delta_{\beta,\xi} - \frac{5}{3}\delta_{\alpha,\beta}\delta_{\xi,\eta} \right) . \qquad (9.36)
\end{aligned}
$$

Thus

$$
\begin{aligned}
\langle Q_{\alpha\beta}(i)\rangle &= \frac{\epsilon}{10}\left(\delta_{\alpha,\beta} + 2n_{0\alpha}n_{0\beta} - \frac{5}{3}\delta_{\alpha,\beta} \right) \\
&= \frac{\epsilon}{15}[3n_{0,\alpha}n_{0,\beta} - \delta_{\alpha,\beta}] . \qquad (9.37)
\end{aligned}
$$

Then the trial energy is

$$
\begin{aligned}
U_{\rm tr} &= -E\sum_{\langle ij\rangle}\sum_{\alpha\beta}\langle Q_{\alpha\beta}(i)\rangle\langle Q_{\alpha\beta}(j)\rangle \\
&= -\frac{1}{2}NzE(\epsilon/15)^2\sum_{\alpha\beta}[3n_{0,\alpha}n_{0,\beta} - \delta_{\alpha,\beta}]^2 \\
&= -\frac{\epsilon^2}{75}NzE , \qquad (9.38)
\end{aligned}
$$

where z is the coordination number of the lattice. To construct the trial entropy it is useful to realize that the entropy is independent of the axis of ordering. So it is simplest to take $\hat{n}_0$ to lie along the z-axis. Then the trial entropy per molecule is

$$
\begin{aligned}
\frac{s_{\rm tr}}{k} &\equiv -{\rm Tr}\left(\rho_i^{(1)}(\theta_i)\ln\rho_i^{(1)}(\theta_i) \right) \\
&= -\frac{1}{2}\int_0^\pi \sin\theta d\theta\left(1 + \frac{\epsilon}{4\pi}[3\cos^2\theta - 1] \right) \\
&\quad \times \ln\left(\frac{1 + \frac{\epsilon}{2}[3\cos^2\theta - 1]}{4\pi} \right). \qquad (9.39)
\end{aligned}
$$

This last integral can be evaluated in closed form but the result is not very transparent. Accordingly, in order to analyze the nature of the transition we generate the Landau expansion of the free energy in powers of the order parameter ϵ. We expand $s_{\rm tr}/k$ as

$$
\frac{s_{\rm tr}}{k} = \ln(4\pi) - \frac{1}{2}\epsilon^2\overline{[P_2(\cos\theta)]^2} + \frac{1}{6}\epsilon^3\overline{[P_2(\cos\theta)]^3} - \frac{1}{12}\epsilon^4\overline{[P_2(\cos\theta)]^4} + \cdots
$$

$$= \ln(4\pi) - \frac{1}{10}\epsilon^2 + b'\epsilon^3 - c'\epsilon^4 + \cdots, \tag{9.40}$$

where b' and c' (and later b and c) are positive constants of order unity. We are therefore led to consider a free energy per site of the form

$$f = \frac{1}{2}a'(T - T_0)\sigma^2 - b\sigma^3 + c\sigma^4 , \tag{9.41}$$

with rescaled order parameter, σ, where we call the temperature at which the quadratic term changes sign T_0 (rather than T_c) because, as we will see, the phase transition does *not* occur at T_0.

For liquid crystals, the constants a', b, and c are fixed by our model. The fact that b is nonzero is a result of the fact that $\epsilon \to -\epsilon$ is not a symmetry of the system. An egg ($\langle P_2(\cos\theta)\rangle > 0$) and a pancake ($\langle P_2(\cos\theta)\rangle < 0$) are not related by symmetry (even for a chef!). The sign of b reflects the fact that positive ϵ has more phase space than negative ϵ indicating that a phase (nematic) in which rods are aligned along an axis is more common than one (discotic) in which molecules assume all orientations within a plane perpendicular to some fixed axis. However, the results for $b < 0$ can be directly deduced from those we obtain below for $b > 0$.

9.4.1 Analysis of the Landau Expansion for Liquid Crystals

As we have seen for the Ising model, to elucidate the nature of the ordering transition, it suffices to analyze the properties of the free energy given by the truncated Landau expansion. The behavior of the free energy given by Eq. (9.41) as a function of σ for positive b and various values of T is shown in Fig. (9.2).

We now give a quantitative analysis of the free energy curves. Differentiating with respect to σ we find that local extrema occur when

$$\frac{\partial f}{\partial \sigma} = a'(T - T_0)\sigma - 3b\sigma^2 + 4c\sigma^3 = 0 .$$

This gives $\sigma = 0$ or

$$\sigma^2 - \frac{3b}{4c}\sigma + \frac{a'(T - T_0)}{4c} = 0 ,$$

with roots

$$\sigma_\pm = \frac{3b}{8c} \pm \frac{1}{8c}\sqrt{9b^2 - 16a'(T - T_0)c} . \tag{9.42}$$

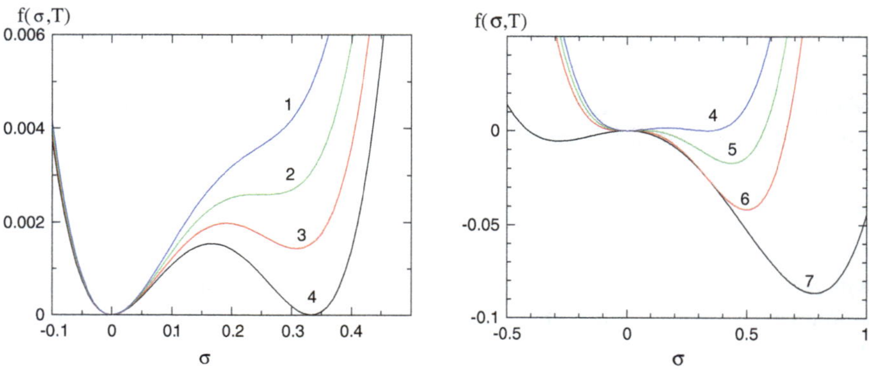

Fig. 9.2 Free energy functional, $f(\sigma, T)$ plotted versus σ for $a' = 1$, $b = 4/3$, $c = 2$ and for values of temperature $T_n = T_0 + x_n b^2/(a'c)$ with $x_1 = 0.6$, $x_2 = 9/16$, $x_3 = 153/288$, $x_4 = 1/2$, $x_5 = 0.25$, $x_6 = 0$, and $x_7 = -2$. For x_7, we plot the quantity $f(\sigma, a)/10$

These roots are real if $T < T_>$, where

$$T_> \equiv T_0 + \tfrac{9b^2}{16a'c} \ .$$

Otherwise, when $T > T_>$, as shown in Fig. 9.2, $\sigma = 0$ is the only solution. In the figure, we also show the critical curve (2) for $T = T_>$ where the two other positive roots for σ begin to be real as the temperature is lowered. As one can see in the figure, for $T < T_>$, the σ_+ root has the lower free energy. For $T_> > T > T_0$, the σ_- root is a local maximum.

A first-order (discontinuous) phase transition occurs at a temperature we denote by T_c when the minimum at $\sigma_+ \equiv \sigma_0$ and that at $\sigma = 0$ both have zero free energy. This condition, supplemented by the minimization condition, yields

$$f(\sigma_0) = \frac{1}{2}a'(T_c - T_0)\sigma_0^2 - b\sigma_0^3 + c\sigma_0^4 = 0 \tag{9.43}$$

$$f'(\sigma_0) = a'(T_c - T_0)\sigma_0 - 3b\sigma_0^2 + 4c\sigma_0^3 = 0 \ . \tag{9.44}$$

Assuming that $\sigma_0 \neq 0$ and multiplying the first equation by $4/\sigma_0$ and subtracting it from the second equation, we find that

$$\sigma_0 = \frac{a'}{b}(T_c - T_0) \ , \tag{9.45}$$

in which case, we see, from substituting back into Eq. (9.43), that

$$T_c = T_0 + \frac{b^2}{2a'c} > T_0 \tag{9.46}$$

and the discontinuity in σ is given by

$$\sigma_0 = \frac{b}{2c} \, . \tag{9.47}$$

Note that the first-order transition occurs *above* the temperature T_0 where the coefficient of σ^2 vanishes, regardless of the sign of b, and that, for b small, $T_c \approx T_0$.

There is a latent heat ℓ (per site) associated with the first-order transition

$$\ell = T_c \Delta s, \tag{9.48}$$

where $\Delta s = s(\sigma = 0) - s(\sigma = \sigma_0)$, both evaluated at $T = T_c$. From Eq. (9.41), $f = 0$ for $T > T_c$, and also $s = 0$ for $T > T_c$, since, in writing Eq. (9.41), we have dropped the $k \ln 4\pi$ term from Eq. (9.40) that does not depend on σ. For $T = T_c^-$ (where $\sigma = \sigma_0$), we have

$$s = -\left.\frac{df}{dT}\right)_{\sigma=\sigma_0} = -\frac{1}{2}a'\sigma_0^2 - \left.\left.\frac{\partial f}{\partial \sigma}\right)_T \frac{d\sigma}{dT}\right)_{\sigma=\sigma_0} = -\frac{1}{2}a'\sigma_0^2 \, , \tag{9.49}$$

where we have ignored the T-dependence of the coefficients b and c and used the fact that $df/d\sigma)_T = 0$ because the actual free energy is a minimum with respect to σ. Using Eqs. (9.47) and (9.48), we get

$$\ell = \frac{T_c a' b^2}{8c^2} \, . \tag{9.50}$$

Since this is quadratic in b, we see that the jump in the order parameter is a stronger signal of a weak first-order transition than the latent heat.

9.5 General Expansions for a Single-Order Parameter

At this point, it is instructive to step away from specific model Hamiltonians and consider the abstract question: What kinds of terms can appear in a Landau expansion for a single-order parameter σ? Consider the following expansion:

$$f = -h\sigma + \frac{1}{2}a\sigma^2 + b\sigma^3 + c\sigma^4 + d\sigma^5 + e\sigma^6 + \cdots \tag{9.51}$$

where h is a field, and a, b, c, d, and e are, in general, functions of T. To obtain qualitative results, we allow the coefficient a to be temperature dependent via

$$a = a'(T - T_0) \, , \tag{9.52}$$

where a' is temperature independent, and the other coefficients are assumed not to vanish at or near $T = T_0$ and hence are treated as temperature-independent constants near T_0.

9.5.1 Case 1: Disordered Phase at Small Field

Then, as before, we can truncate the series after the σ^2 term because σ and h are linearly related.

$$f = -h + \frac{1}{2}a\sigma^2 + \cdots , \qquad \frac{\partial f}{\partial \sigma} = 0 \rightarrow \sigma = \frac{h}{a} \tag{9.53}$$

and $\chi = \partial\sigma/\partial h$ as $h \rightarrow 0$. Therefore

$$\chi = \frac{1}{a} = \frac{1}{a'(T - T_c)} \tag{9.54}$$

and the free energy can be rewritten as

$$f = -h\sigma + \frac{1}{2}\chi^{-1}\sigma^2 + \cdots . \tag{9.55}$$

It is useful to think of the coefficient of σ^2 as the inverse susceptibility for inducing σ by an external field. In fact, when a, the coefficient of the quadratic term, becomes negative, the disordered phase with $\sigma = 0$ has become locally unstable and the system develops long-range order. In this ordered phase, the coefficient a (which is negative for $T < T_c$) is no longer identified as the susceptibility. Instead, one has an expansion about $\sigma_0(T)$, the equilibrium value of σ:

$$f = -h\sigma + \frac{1}{2}\chi^{-1}(\sigma - \sigma_0)^2 + \cdots \tag{9.56}$$

Thus, although a is negative in the ordered phase, the susceptibility is positive as is it must be.

9.5.2 Case 2: Even-Order Terms with Positive Coefficients

This is equivalent to the ferromagnet in zero field. All the odd-order terms are zero and the fourth-order term is positive. Then

$$f = \frac{1}{2}a'(T - T_c)\sigma^2 + c\sigma^4 + \cdots \tag{9.57}$$

$$\begin{aligned} \frac{\partial f}{\partial \sigma} = 0 &\rightarrow \sigma = 0 \quad (T > T_c) \\ &\rightarrow \sigma = \pm\sqrt{\frac{a'(T_c - T)}{4c}} \quad (T < T_c). \end{aligned} \tag{9.58}$$

Note that odd-order terms vanish when $f(\sigma) = f(-\sigma)$, which would typically be a consequence of a symmetry of the Hamiltonian. On the other hand, the positivity of the fourth-order coefficient does not arise from symmetry. Having a positive fourth-order term is just the simplest way of stabilizing the system. Here it is obvious that the temperature T_c at which the phase transition occurs is the same as the temperature T_0 at which the disordered phase becomes locally unstable. (Locally unstable means that the free energy is decreased when σ deviates infinitesimally from its equilibrium value.) It should also be noted that a similar phase transition can occur where the role of the temperature is played by some other control parameters such as the pressure. In that case, we would write

$$a = a'(p - p_c) \tag{9.59}$$

and one would likewise have $\sigma = \pm[a'(p_c - p)/(4c)]^{1/2}$.

9.5.3 Case 3: f Has a Nonzero σ^3 Term

We treated this generic case in connection with liquid crystals. It should be clear that the appearance of a term $b\sigma^3$ in the free energy gives rise to a first-order transition. The jump in the order parameter is proportional to b and the latent heat is proportional to b^2. When b is small and as $T \to T_c$, one may see fluctuation characteristics of a second-order transition which are eventually preempted by the first-order transition.

9.5.4 Case 4: Only Even Terms in f, But the Coefficient of σ^4 is Negative

In this case, the order parameter will become unboundedly large unless the coefficient of the next (sixth) order term is positive. Assuming T is the relevant control parameter we write the Landau expansion as

$$f = \frac{1}{2}a'(T - T_0)\sigma^2 - b\sigma^4 + c\sigma^6 + \cdots , \tag{9.60}$$

where b is assumed to be positive. Differentiating with respect to σ, we find the condition for an extremum to be

$$\frac{\partial f}{\partial \sigma} = a'(T - T_0)\sigma - 4b\sigma^3 + 6c\sigma^5 = 0 \tag{9.61}$$

or

$$\sigma^4 - \frac{2b}{3c}\sigma^2 + \frac{a'(T - T_0)}{6c} = 0\,. \tag{9.62}$$

We find the roots for σ to be

$$\sigma_\pm^2 = \frac{b}{3c} \pm \frac{1}{3c}\sqrt{b^2 - 3ca'(T - T_0)/2}\,. \tag{9.63}$$

For $T < T_> = T_0 + 2b^2/(3a'c)$, these roots are real and then the solution for σ_+ (σ_-) is a local minimum (maximum) of the free energy.

Following the same logic that was used for the case for a nonzero cubic term, the transition occurs at T_c (where $\sigma = \sigma_0$) when

$$\begin{aligned} f(\sigma_0) &= \frac{1}{2}a'(T_c - T_0)\sigma_0^2 - b\sigma_0^4 + c\sigma_0^6 = 0 \\ f'(\sigma_0) &= a'(T_c - T_0)\sigma_0 - 4b\sigma_0^3 + 6c\sigma_0^5 = 0\,, \end{aligned} \tag{9.64}$$

which gives $\sigma_0^2 = a'(T_c - T_0)/b$. Then we find that

$$T_c = T_0 + \frac{b^2}{2a'c} \tag{9.65}$$

and $\sigma_0 = \sqrt{b/(2c)}$.

In Fig. 9.3, we show curves for the free energy as a function of σ for various temperatures. Note that as the temperature is decreased T_2 is the temperature at which the metastable local minimum first appears, $T_4 = T_c$ is the thermodynamic transition temperature, and T_6 is the temperature at which the local minimum at $\sigma = 0$ first becomes unstable.

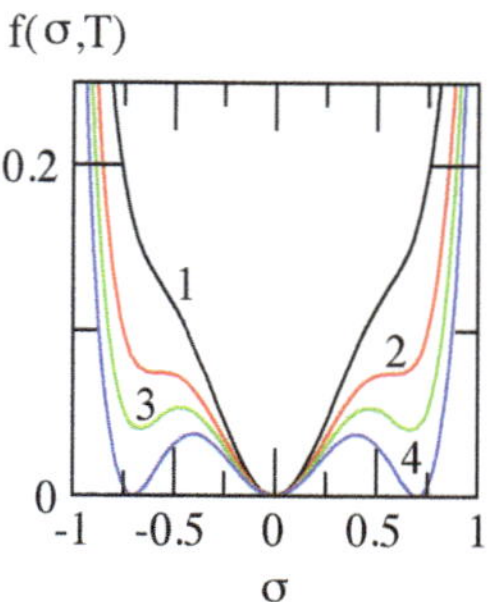

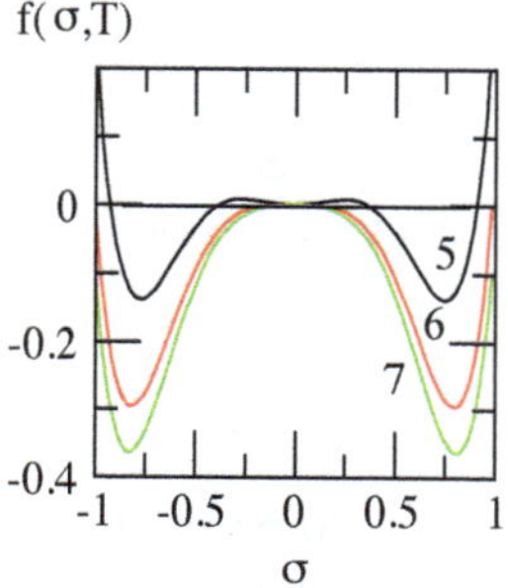

Fig. 9.3 Free energy functional, $f(\sigma, a)$ plotted versus σ for $a' = 1$, $b = 2/3$, $c = 1/2$ and for temperatures $T_n = T_0 + x_n b^2/(a'c)$, with $x_1 = 5/6$, $x_2 = 2/3$, $x_3 = 7/12$, $x_4 = 1/2$, $x_5 = 1/3$, $x_6 = 0$, and $x_7 = -0.1$

9.5.5 Case 5: Only Even Terms in f, But the Coefficient of σ^4 is Zero

Note that the transition goes from being continuous to discontinuous when the sign of the fourth-order term goes from positive to negative. What happens when the fourth-order coefficient is exactly zero? Then

$$f = \frac{1}{2}a'(T - T_0)\sigma^2 + c\sigma^6, \quad c > 0$$

$$\frac{\partial f}{\partial \sigma} = 0 \quad \rightarrow \ a'(T - T_0)\sigma + 6c\sigma^5 = 0\,, \tag{9.66}$$

so that $\sigma = 0$ for $T > T_c = T_0$ and for $T < T_c$

$$\sigma = \left[\frac{a'}{6c}(T_c - T)\right]^{1/4} . \tag{9.67}$$

Thus, the order parameter critical exponent is 1/4 instead of 1/2.

A change in sign of the fourth-order coefficient can occur if the coefficients a', T_c, b, c, etc. are functions of some parameter, say V, so that

$$f = a'(V)\Big(T - T_c(V)\Big)\sigma^2 + b(V)\sigma^4 + c(V)\sigma^6. \tag{9.68}$$

As long as $b(V)$ is positive, $T_c(V)$ defines a line of second-order transitions in the $T - V$ plane. At the “tricritical point”, (T^*, V^*), where

$$b(V^*) = 0, \quad T_c(V^*) = T_c^* \tag{9.69}$$

the transition changes from second to first order and continues along a line determined by $T_1(V)$. Another important scenario which causes the sign of b to change sign is discussed in reference to Eq. (10.63).

In any event, one does not reach this special tricritical point simply by adjusting the temperature. By suitably adjusting the temperature one can make the coefficient of σ^2 be zero. To simultaneously also have the coefficient of σ^4 be zero, requires adjusting both the temperature and a second control parameter, here called V. If we broaden the discussion to the case of nonzero field h, so that there is a term in the free energy linear in σ, then two variables must be controlled to reach the usual critical point and three variables controlled to reach the tricritical point.

9.6 Phase Transitions and Mean-Field Theory

9.6.1 Phenomenology of First-Order Transitions

Here we discuss various physical properties of such discontinuous transitions. Look at Fig. 9.2 and consider what happens as the temperature is reduced from an initial value well above T_c. It is true that the thermodynamic transition occurs when T is reduced to the value T_c. However, because there is a free energy barrier between the disordered state and the ordered state, it is possible that one can observe the metastable disordered state for some range of temperature below T_c. This phenomenon is called "supercooling." Since the barrier decreases as the temperature is lowered below T_c, the stability of the globally unstable disordered phase becomes progressively more tenuous, and at the temperature T_0 even local stability disappears.

The analogous phenomenon occurs when the temperature is increased from an initial value below T_0. In this case, the system is initially in the ordered phase and the thermodynamic phase transition occurs when the temperature is increased to the value T_c. But, as in the case of supercooling, a free energy barrier may prevent the system from achieving true equilibrium even for temperatures somewhat above T_c. Eventually, the possibility of local stability of the globally unstable ordered phase disappears when the temperature reaches the value $T_>$, above which "superheating" is obviously not possible. Thus, on rather general grounds one expects that a discontinuous transition is accompanied by the possibility of observing *supercooling* and *superheating*. This phenomenon is known as *hysteresis* and represents a unique signal of a discontinuous transition.

That a first-order transition is accompanied by hysteresis is a qualitatively correct prediction of mean-field theory. What is misleading is the size of the effect or, equivalently, the path by which the transition actually occurs. Mean-field theory assumes that the entire system has the same, spatially uniform order parameter, whereas, in real systems, there are frequent and ubiquitous temporal and spatial fluctuations, particularly near a phase transition. When a system which has been supercooled, for example, starts to undergo its transition, this will almost certainly happen in an inhomogeneous way, often by nucleation of droplets of the stable phase, which would form and then grow and coalesce. If the system was forced to undergo the transition perfectly homogeneously, the free energy barrier for the transition would be an extensive quantity and the transition would only occur when the metastable minimum became completely unstable.

It is also worth noting how fluctuations may make themselves felt near a weakly discontinuous transition, as illustrated in Fig. 9.4. First consider the zero-field susceptibility, χ, as measured in the disordered phase. As we have seen, mean-field theory predicts that $\chi = C/(T - T_0)$, where C is a constant. This form only holds for temperatures above $T = T_c$, of course. Thus, we expect the behavior as shown schematically in Fig. 9.4. Far from T_0, mean-field theory will be correct and χ will vary as $|T - T_0|^{-1}$ since the system appears to be headed toward a continuous transition at $T = T_0$. As one supercools the system and thereby approaches the instability

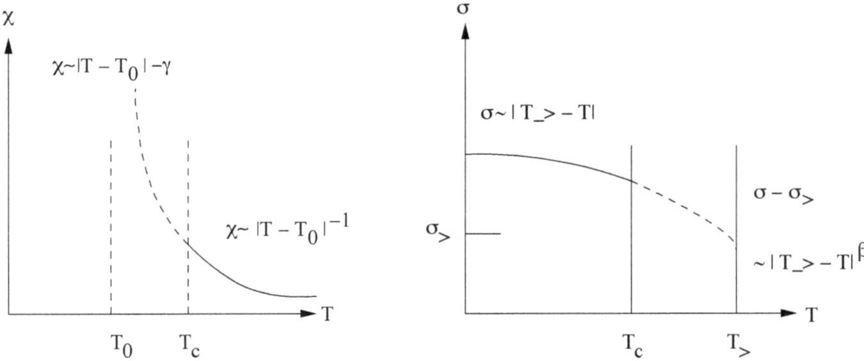

Fig. 9.4 Schematic behavior of the susceptibility (left) and the order parameter (right) near a first-order transition. The dashed part of the curves represent supercooling for $T < T_c$ and superheating for $T > T_c$

temperature T_0, one expects a crossover so that χ will begin to appear to diverge with an exponent different from $\gamma = 1$, although one never actually gets into the asymptotic regime for critical fluctuations. A similar phenomenon occurs for $T < T_c$. We write Eq. (9.42) as

$$\sigma(T) = \sigma_> + \frac{1}{2}\sqrt{\frac{a'}{c}}(T_> - T)^{1/2} , \tag{9.70}$$

where $\sigma_> = 3b/(8c)$ is the metastable value of σ at $T_>$. This equation only holds for $T < T_c$, of course. But as one superheats the system and thereby approaches the instability temperature $T_>$, we expect that fluctuations will lead to a crossover to a modified critical exponent β. Thus, if T_c and $T_>$ are not very far apart (i.e., if the transition is only weakly first order), one may be able to observe fluctuations associated with the loss of stability both of the disordered phase upon supercooling and also of the ordered phase upon superheating.

9.6.2 Limitations of Mean-Field Theory

It must be clear by now that mean-field theory and Landau theory are useful, powerful methods for describing the behavior of interacting systems which undergo phase transitions. On the other hand, it is important to be aware of the limitations of this seductively simple theory and to understand which of its predictions are reliable, which are simply qualitative, and which are clearly incorrect.

To begin with, recall the statistical mechanical definition of the free energy

$$f = -kT \ln Z , \tag{9.71}$$

where the partition function, Z, is given by

$$Z = \sum_n e^{-\beta E_n} , \tag{9.72}$$

where $\{E_n\}$ is the set of many-body energy levels of an N-particle system. When a system undergoes an equilibrium phase transition, then its free energy must in some way be singular or nonanalytic at the phase transition point. In particular, for a phase transition to occur as the temperature is varied, the free energy must necessarily be a nonanalytic function of temperature at the transition temperature. However, from the definition of the free energy as the logarithm of a sum of exponentials, with each exponent scaled by $1/T$, it would seem impossible for f to be nonanalytic in T. In fact, it is easy to see that the *only* way that f can possibly be nonanalytic is in the limit $N \to \infty$ where the sum contains an infinite number of terms, in other words, in the limit of infinite system size.

This is a fundamental statement about the nature of phase transitions, which must be satisfied by a correct theory. It applies to any kind of phase transition. For example, one class of transition which we have examined is one in which a symmetry, that exists in the disordered state, is spontaneously broken in the ordered state. For example, in the disordered state of a ferromagnet, the spins have no preferred direction. Below the ordering temperature, the system spontaneously and arbitrarily picks out a preferred direction. This cannot happen in a finite system. In a finite system, even at low temperature, the system would make rare spin-reversing excursions so that, over long periods of time, the average value of the spin would be zero. On the other hand, for an infinite or macroscopic system, the time required for a global spin reversal to occur is infinite or effectively infinite.

Mean-field theory does not model this size dependence of phase transitions. One can calculate mean-field transitions for finite as well as infinite systems and the answer is the same—the transition occurs at the mean-field transition temperature which is insensitive to system size. However, in Chap. 14, we will use the exact solution of a model on a recursive (i.e., nonperiodic) lattice to generate the mean-field result. In that formulation, a phase transition only occurs in the thermodynamic limit of infinite size. In Chap. 18, we study the Monte Carlo method for simulating classical statistical mechanical systems. There we will show data and discuss in detail how the behavior of finite systems approaches that of a critical phase transition in the limit of infinite system size.

It is important to keep in mind these limitations of mean-field theory, while recognizing how useful the theory can be, particularly in light of the great difficulty and complexity of solutions which go beyond mean-field theory. Such theories are discussed at length in Part IV of this book.

9.7 Summary

The variational principle balances minimizing energy against maximizing entropy by requiring that the trial free energy, $F_{\rm trial}$ be minimized with respect to variation of the normalized Hermitian density matrix ρ, where

$$F_{\rm trial} = {\rm Tr}\Bigl[\rho\mathcal{H} + kT\rho\ln\rho\Bigr] . \qquad (9.73)$$

Note that the temperature is the crucial parameter which dictates which of these two terms wins out. By using a variational density matrix ρ that is a product of single-particle density matrices, one develops mean-field theory in a much more systematic way than the self-consistent field truncation of the Hamiltonian discussed in the last chapter. This theory will be qualitatively reasonable for most continuous phase transitions and it can be applied to both classical and quantum systems. In introducing an order parameter, a cardinal principle is that the system only develops order of a type that lowers the energy, so only such types of order need to be considered.

In this chapter, we also studied phase transitions as described by the Landau expansion of the free energy in powers of an order parameter. A term linear in the order parameter induces some order at any temperature. When symmetry restricts the free energy to contain only even powers of the order parameter, then there will be a continuous phase transition from a high-temperature disordered state to a low-temperature ordered one, unless the sign of the fourth-order term is negative. When the Landau expansion contains a cubic term, the phase transition will be discontinuous. It was also useful to think of the Landau free energy in mechanical terms in which the coefficient of the term quadratic in the order parameter is identified as the inverse susceptibility, which is analogous to the spring constant of a harmonic potential. Then one says that the disordered phase is stable at high temperature and, as the temperature is reduced toward the transition temperature T_c, the spring constant approaches zero, the system becomes very "loose," subject to large fluctuations, and at T_c it becomes unstable against the formation of order. When the cubic term in the Landau expansion is nonzero (or when the cubic term vanishes but the coefficient of the fourth-order term is negative) the system undergoes a first-order transition, in which the system is always locally stable, making a discontinuous transition from a disordered state to a state with nonzero order. This transition is inevitably accompanied by hysteresis. Interesting modification of the critical exponents occurs when both the cubic and quartic terms vanish. (Reaching this so-called tricritical point requires adjusting three thermodynamic variables rather than two, as for the usual critical point.)

9.8 Exercises

1. Show that the free energy of the Ising model corresponding to the density matrix of Eq.(9.9) is minimized by choosing the same value m for all the m_i, as suggested below Eq. (9.14).

2. For the Ising model show that if $\rho_i^{(1)}$ is allowed to have an off-diagonal matrix element (which may be complex), the trial free energy is minimized when the density matrix is diagonal.

3. Derive the Landau expansion for the free energy of the Ising model by expanding the right-hand side of Eq. (8.13) in powers of $S_f = m$.

4. Perform a similar calculation to that of Exercise 3 for the liquid crystal problem with the mean-field Hamiltonian given by Eq. (8.59). Start by deriving an expression for the single-site free energy analogous to Eq. (8.12). Then use that to obtain the analog of Eq. (8.13) and expand the result in powers of the order parameter. (Hint: You can use the results of Eq. (8.62), extended to order x^3.)

5. Carry out mean-field theory for the Ising model in a transverse field whose Hamiltonian is

$$\mathcal{H} = -J \sum_{\langle ij \rangle} \boldsymbol{\sigma}_{iz} \boldsymbol{\sigma}_{jz} - h \sum_i \boldsymbol{\sigma}_{ix} , \tag{9.74}$$

where $\boldsymbol{\sigma}_{i\alpha}$ is the αth Pauli σ-matrix for site i.

6. Carry out mean-field theory for a q-state Potts model, whose Hamiltonian may be written in various forms, one of which is

$$\mathcal{H} = -J \sum_{\langle ij \rangle} \left(q \delta_{s_i, s_j} - 1 \right) , \tag{9.75}$$

where s_i is a "spin" variable for site i which can assume one of the q possible values $1, 2, \dots q$ and δ is the Kronecker delta. Thus, a pair of adjacent spins have energy $-J(q-1)$ if they are in the same state and energy J is they are in different states.

7. Obtain the generalization of Eq. (9.35) to the average of a product of $2p$ components of $\hat{n}$, for general p. Give a totally explicit result for $p = 3$.

8. We have discussed the case exemplified by liquid crystals in which the coefficient of σ^3 in the Landau expansion is nonzero. For such a system could one reach a special point where the coefficients of both σ^2 and σ^3 simultaneously vanish? If so, can this be done by adjusting the temperature, or how?

9. If, starting from a nematic liquid crystal, one was able to change the sign of the coefficient of σ^3 in the Landau expansion, what would Fig. 9.2 look like for the resulting system?

Chapter 10
Landau Theory for Two or More Order Parameters

10.1 Introduction

It frequently happens that more than one order parameter appears in a physical problem. Sometimes the different order parameters are "equivalent" as are the x, y, and z components of a vector spin when the interactions involving the spins are isotropic. Alternatively, it may be that different order parameters are stable in different regions of the phase diagram defined by the parameters in the Hamiltonian. In this chapter, we consider a variety of possible scenarios involving more than one order parameter from the perspective of Landau theory.

10.2 Coupling of Two Variables at Quadratic Order

10.2.1 General Remarks

It can happen that the variables that we choose to describe a physical problem couple at the quadratic level in the Landau expansion of the free energy. This coupling is analogous to the problem of coupled oscillators, and the strategy is the same, to reexpress the free energy in terms of "normal modes."

We first illustrate this situation for the case of two variables, when the free energy is of the form

$$f = \frac{a}{2}\sigma_1^2 + b\sigma_1\sigma_2 + \frac{c}{2}\sigma_2^2 + \mathcal{O}(\sigma^4). \tag{10.1}$$

For simplicity, assume that $c = a$ and from previous examples we expect that, due to the competition between energy and entropy, one has $a = \alpha(T - T_0)$. Consider the quadratic terms in the disordered state which must occur in the limit of sufficiently high temperature. Then the quadratic part of the free energy is

A. J. Berlinsky and A. B. Harris, *Statistical Mechanics*, Graduate Texts in Physics,
https://doi.org/10.1007/978-3-030-28187-8_10

$$f_2 = \frac{1}{2}(\sigma_1, \sigma_2) \begin{pmatrix} a & b \\ b & a \end{pmatrix} \begin{pmatrix} \sigma_1 \\ \sigma_2 \end{pmatrix} \tag{10.2}$$

If we define new order parameters

$$\sigma_\pm = (\sigma_1 \pm \sigma_2)/\sqrt{2} \tag{10.3}$$

then f_2 becomes

$$f_2 = \frac{1}{2}(a+b)\sigma_+^2 + \frac{1}{2}(a-b)\sigma_-^2 \tag{10.4}$$

The question is: which of the two coefficients, $a+b$ or $a-b$, first vanishes as the temperature is decreased starting from the disordered phase. If we take b to be positive, then $a-b$ vanishes at the higher temperature. This temperature, which we call T_c, is the temperature at which the disordered phase becomes unstable relative to the appearance of order *in the variable* σ_-. We have that

$$T_c = T_0 + b/a' \tag{10.5}$$

and the corresponding order parameter σ_- is called the "critical order parameter" (as contrasted to σ_+ which is a "noncritical order parameter.")

This procedure can be generalized to an arbitrarily large number of order parameters, in which case the critical order parameter is obtained from the eigenvector associated with the smallest eigenvalue of the coefficient matrix of the quadratic terms in the free energy.

10.2.2 The Ising Antiferromagnet

Next we consider a case in which the one-particle density matrix can be different on different sites—the Ising antiferromagnet on a bipartite lattice. (A bipartite lattice is one, like the simple cubic or square lattice, which can be divided into two sublattices, denoted A and B, such that each site on a given sublattice has nearest neighbors only on the *other* sublattice. A square or cubic lattice is bipartite, but the triangular lattice is *not* bipartite.) The Hamiltonian for the antiferromagnet in a uniform magnetic field h is written as

$$\mathcal{H} = J \sum_{<ij>} \sigma_i \sigma_j - h \sum_i \sigma_i \ , \tag{10.6}$$

where, as before, $\sigma_i = \pm 1$, and $J > 0$ for the antiferromagnet. Note that this Hamiltonian favors order in which spins on one sublattice are "up" and on the other sublattice are "down." Therefore, it is essential that we allow the density matrix for sites on the

A sublattice to be different from that of sites of the B sublattice. Then the density matrix of the system of N spins is of the form

$$\rho^{(N)} = \prod_{i=1}^{N/2} \rho_i^A \rho_i^B , \tag{10.7}$$

where

$$\rho_i^A = \frac{1 + m_A \sigma_i^A}{2}, \quad \rho_i^B = \frac{1 + m_B \sigma_i^B}{2}. \tag{10.8}$$

Then the free energy per site is

$$\begin{aligned} f(m_A, m_B) = {} & \frac{Jz}{2} m_A m_B - h\left(\frac{m_A + m_B}{2}\right) \\ & + \frac{kT}{2}\left[\left(\frac{1+m_A}{2}\right)\ln\left(\frac{1+m_A}{2}\right) + \left(\frac{1-m_A}{2}\right)\ln\left(\frac{1-m_A}{2}\right)\right. \\ & \left. + \left(\frac{1+m_B}{2}\right)\ln\left(\frac{1+m_B}{2}\right) + \left(\frac{1-m_B}{2}\right)\ln\left(\frac{1-m_B}{2}\right)\right] . \end{aligned} \tag{10.9}$$

This expression must be minimized with respect to *both* m_A and m_B. For the simple case, $h = 0$, the solution is $m_A = -m_B = m$ which is equivalent to the ferromagnet (cf. the discussion of Eqs. (2.2) and (2.3)).

For $h \neq 0$, we must consider a more general class of solutions in which

$$M = \frac{m_A + m_B}{2} \tag{10.10}$$

$$m_S = \frac{m_A - m_B}{2} , \tag{10.11}$$

where M is the uniform magnetization and m_S is the staggered magnetization, can both be nonzero. In terms of the original order parameters, m_A and m_B, the free energy is minimized when

$$\frac{\partial f}{\partial m_A} = \frac{Jz}{2} m_B - \frac{h}{2} + \frac{1}{4} kT \ln\left(\frac{1+m_A}{1-m_A}\right) = 0, \tag{10.12}$$

$$\frac{\partial f}{\partial m_B} = \frac{Jz}{2} m_A - \frac{h}{2} + \frac{1}{4} kT \ln\left(\frac{1+m_B}{1-m_B}\right) = 0, \tag{10.13}$$

or, adding and subtracting these two equations and transforming to the new variables,

$$JzM - h + \frac{1}{4}kT \ln \frac{(1 + M + m_S)(1 + M - m_S)}{(1 - M - m_S)(1 - M + m_S)} = 0 \qquad (10.14)$$

$$-Jzm_S + \frac{1}{4}kT \ln \frac{(1 + M + m_S)(1 - M + m_S)}{(1 - M - m_S)(1 + M - m_S)} = 0 \,. \qquad (10.15)$$

These equations are clearly satisfied by $m_S = 0$, the paramagnetic (disordered) solution. For this case, they reduce to

$$h - JzM = \frac{1}{2}kT \ln \left(\frac{1 + M}{1 - M} \right) \qquad (10.16)$$

or

$$M = \tanh \left(\frac{h - JzM}{kT} \right) , \qquad (10.17)$$

which looks exactly like the ferromagnetic equation with the sign of J reversed. It has only one solution for all values of T and h, as illustrated in Fig. 10.1.

At high temperatures, we can expand the tanh and write

$$\begin{aligned} M &\approx \frac{h - kT_c M}{kT} \\ M &= \frac{h}{k(T + T_c)} , \end{aligned} \qquad (10.18)$$

where we have substituted $kT_c = Jz$. Then

$$\chi = \frac{\partial M}{\partial h} = \frac{1}{k(T + T_c)} \,. \qquad (10.19)$$

This is the Curie–Weiss law, Eq. (8.2.2), but $1/\chi$ intercepts zero at negative temperature. Experimentally one distinguishes between magnetic systems in which the intercept of $1/\chi$, extrapolated from high temperatures, is positive or negative, calling

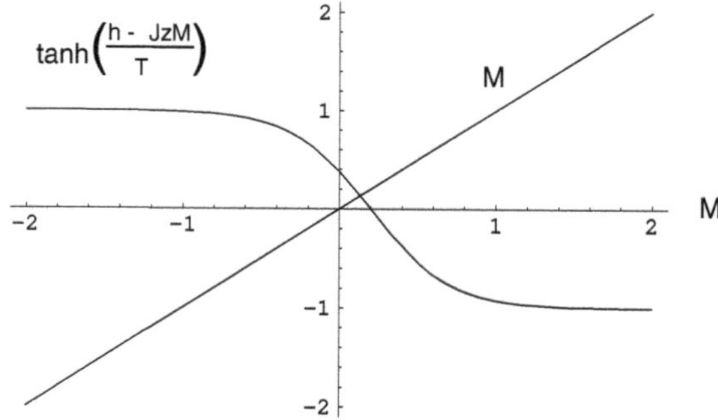

Fig. 10.1 Graphical solution of Eq. (10.17) for $h = 0.2\,\text{Jz}$ and $T = 0.5\,\text{Jz}$. Note that, regardless of the values of h and T, the solution is unique

the former ferromagnetic and the latter antiferromagnetic. In both cases, what is measured is the sign of the assumed dominant nearest neighbor pair interaction.

When does m_S become nonzero? To see this, we expand Eq. (10.15) in powers of m_S. We find

$$- Jzm_S + \frac{1}{4}kT \ln \left(1 + \frac{2m_S}{1+M} + \frac{2m_S}{1-M} + \dots \right) = 0$$

$$-Jzm_S + \frac{1}{2}kTm_S \underbrace{\left(\frac{1}{1+M} + \frac{1}{1-M}\right)}_{2/(1-M^2)} + Cm_S^3 = 0\,. \tag{10.20}$$

We assert that after some annoying algebra one finds that the constant C is positive. That being the case, we see that whether or not there is a solution with $m_S \neq 0$ depends on the sign of the quantity

$$Jz - \frac{kT}{1-M^2}\,. \tag{10.21}$$

We have a solution with $m_S \neq 0$ for temperature less than a critical value T_N, called the Néel temperature, where

$$T_N = T_0(1 - M^2)\,, \tag{10.22}$$

where we set $Jz = kT_0$. Of course, what we need to draw a phase diagram is not T_N versus M but rather T_N versus h. The relation between M and h is given by Eq. (10.17). The resulting phase diagram can be calculated using a hand calculator. The result was shown in Chap. 2 in Fig. 2.8.

10.2.3 Landau Expansion for the Antiferromagnet

Consider the Ising antiferromagnet, for which the mean-field theory was developed in Chap. 9. As we have seen, the uniform susceptibility obtained in Eq. (10.19) remains finite even at the phase transition. This is because the uniform susceptibility is the response to a uniform field, a field which does not couple to the order parameter, m_S. Accordingly, let us add to the Hamiltonian a term of the form

$$\delta\mathcal{H} = -h_S\left(\sum_{i\in A}\sigma_i - \sum_{i\in B}\sigma_i\right)\,. \tag{10.23}$$

Rather than repeat the previous calculations, it is instructive to expand the free energy in powers of m_A and m_B, using the results for the ferromagnet to avoid recalculating.

Then, transforming to average and staggered magnetizations, M and m_S, we see that the free energy per site is

$$\begin{aligned} f &= \frac{1}{2}Jz(M^2 - m_S^2) - hM - h_s m_s \\ &\quad + \frac{1}{2}kT\left(\frac{1}{2}(M+m_S)^2 + \frac{1}{12}(M+m_S)^4 + \dots \right. \\ &\quad \left. + \frac{1}{2}(M-m_S)^2 + \frac{1}{12}(M-m_S)^4 + \dots \right) \\ &= \frac{1}{2}[kT + Jz]M^2 + \frac{1}{2}[kT - Jz]m_S^2 - hM - h_S m_S \\ &\quad + Am_S^4 + BM^4 + Cm_S^2 M^2 + \dots , \end{aligned} \tag{10.24}$$

where A, B, and C (and later B' and C') are positive constants whose explicit values are not needed for this discussion. This trial free energy is a function of the two parameters M and m_S. The former is a noncritical order parameter, whereas the latter is the critical order parameter which gives rise to the phase transition. It is instructive to eliminate M (by minimizing f with respect to it) in order to obtain an effective free energy as a function of only the order parameter m_S. We will do this for small h. Minimizing with respect to M gives $M = h/[kT + Jz]$ so that

$$\begin{aligned} f &= -(1/2)h^2/[kT + Jz] + (1/2)[kT - Jz]m_S^2 - h_S m_S + Am_S^4 \\ &\quad + B'h^4 + C'm_S^2 h^2 \\ &= (1/2)[kT - Jz + C'h^2]m_S^2 - h_S m_S + Am_S^4 + \dots \\ &= (1/2)[kT - kT_c(h)]m_S^2 - h_S m_S + Am_S^4 + \dots . \end{aligned} \tag{10.25}$$

This form, which includes the lowest order effects of h, reinforces our previous analysis. We see that T_c decreases when a small uniform field is applied. Basically, this arises because inducing ferromagnetic order leaves less phase space available for the spontaneous appearance of staggered order. We see that the transition temperature is given by

$$kT_c(h) = Jz - C'h^2 + \mathcal{O}(h^4) . \tag{10.26}$$

For small values of h_S, this effective free energy gives

$$m_S = \frac{h_S}{k[T - T_c(h)]} , \tag{10.27}$$

so that we may define the *staggered* susceptibility as

$$\chi_S \equiv \left.\frac{\partial m_S}{\partial h_S}\right)_{h_S=0} = \Big(k[T - T_c(h)]\Big)^{-1} . \tag{10.28}$$

Thus, we find the order parameter susceptibility does diverge at the phase transition and the associated critical exponent γ is equal to 1 within mean-field theory.

We can use this development to see how the long-range order develops as a function of temperature, again for small h. The effective free energy (for $h_S = 0$) is

$$f = (1/2)k[T - T_c(h)]m_S^2 + Am_S^4 , \tag{10.29}$$

where the value of the constant A is not needed in the present context. The spontaneous staggered magnetization is zero for $T > T_c(h)$ and for $T < T_c(h)$ is given by

$$m_S(T) = \left([kT_c(h) - kT]/4A \right)^{1/2} . \tag{10.30}$$

So the exponent associated with the order parameter is $\beta = 1/2$ independent of h.

10.3 Landau Theory and Lattice Fourier Transforms

Landau theory and Landau expansions are particularly valuable tools for understanding the development of order that breaks the translational symmetry of the underlying lattice. We have seen how antiferromagnetic order enlarges the unit cell by distinguishing between A and B sublattices. However, the theory is much more powerful than this simple case suggests. In fact, it is capable of describing arbitrary kinds of translational symmetry, particularly when combined with the theory of *space groups* which are groups whose elements include translations, rotations, inversion, and reflection. In this section, we begin a more general study of how translational symmetry is broken by phase transitions.

The case of antiferromagnetism, which involves two variables coupled at the quadratic level, suggests that we consider a Landau free energy which at quadratic order assumes the form

$$F = 1/2 \sum_{i,j} F_2(i, j)\eta(\mathbf{r}_i)\eta(\mathbf{r}_j) , \tag{10.31}$$

where, for simplicity, the variable $\eta(\mathbf{r})$ associated with site $\mathbf{r}$ is assumed to be a scalar and $F_2(i, j)$ is a coefficient which, because of translational invariance, is a function of only $\mathbf{r}_i - \mathbf{r}_j$. This quadratic form can be diagonalized by a Fourier transformation. In accord with our interpretation that the coefficient of the quadratic term is the inverse susceptibility (also see Exercise 1), we write this quadratic form as

$$F_2 = 1/2 \sum_{\mathbf{r}_i, \mathbf{r}_j} [\chi^{-1}]_{i,j}\eta(\mathbf{r}_i)\eta(\mathbf{r}_j) . \tag{10.32}$$

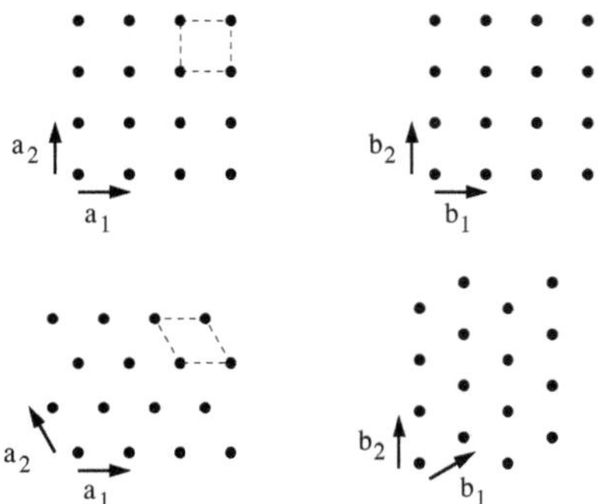

Fig. 10.2 Left panel: Square lattice (top) and hexagonal lattice (bottom). In each case, the unit cell is defined by the dashed parallelogram. Right panel: Reciprocal lattices for the square lattice (top) and hexagonal lattice (bottom). Note that the reciprocal lattice displays the same symmetry as the original lattice. In the case of the hexagonal lattice, the reciprocal lattice is obtained by a rotation of 30°

This quadratic form can be diagonalized by a complex Fourier transformation (which is a unitary transformation). The eigenvalue which first becomes zero as the temperature is lowered will define the type of order that spontaneously appears at the phase transition.

We now express $\eta(\mathbf{r})$ in a Fourier representation. Note that $\eta(\mathbf{r})$ is defined on a mesh of lattice points. For illustrative purposes, we treat the case of two dimensions, but the discussion can be extended straightforwardly to higher spatial dimension. The mesh of points is given by

$$\mathbf{r}_{m_1,m_2} = m_1\mathbf{a}_1 + m_2\mathbf{a}_2 \ , \tag{10.33}$$

where $\mathbf{a}_1$ and $\mathbf{a}_2$ are non-collinear 2-D vectors that define the "unit cell" of the lattice. For example,

For a square lattice: $\mathbf{a}_1 = (a, 0), \ \mathbf{a}_2 = (0, a)$.
For a hexagonal lattice: $\mathbf{a}_1 = (a, 0), \ \mathbf{a}_2 = (-\frac{a}{2}, \frac{a\sqrt{3}}{2})$.

These lattices are shown in Fig. 10.2.

We now define basis vectors, $\mathbf{b}_1$ and $\mathbf{b}_2$, for the **reciprocal lattice**, which satisfy

$$\mathbf{a}_i \cdot \mathbf{b}_j = 2\pi\delta_{i,j} \ . \tag{10.34}$$

For a square lattice: $\mathbf{b}_1 = (\frac{2\pi}{a}, 0), \quad \mathbf{b}_2 = (0, \frac{2\pi}{a})$.

For a hexagonal lattice: $\mathbf{b}_1 = (\frac{2\pi}{a}, \frac{2\pi}{a\sqrt{3}}), \ \mathbf{b}_2 = (0, \frac{4\pi}{a\sqrt{3}})$.

A general reciprocal lattice vector (RLV) $\mathbf{Q}$ has the form

$$\mathbf{Q} = Q_1\mathbf{b}_1 + Q_2\mathbf{b}_2 \, , \tag{10.35}$$

where Q_1 and Q_2 can be any integers. The reciprocal lattice is also shown in Fig. 10.2.

We will treat a finite lattice of points described by Eq. (10.33) having L_1 rows of sites in the direction of $\mathbf{a}_1$ and L_2 rows of sites in the direction of $\mathbf{a}_2$. Equivalently, we may impose periodic boundary conditions so that

$$\eta(\mathbf{r}) = \eta(\mathbf{r} + L_1\mathbf{a}_1) = \eta(\mathbf{r} + L_2\mathbf{a}_2) \, . \tag{10.36}$$

Any function which is defined on the above mesh of lattice points and which obeys the periodicity conditions of Eq. (10.36) can be expanded in a Fourier series of the form

$$\eta(\mathbf{r}) = \sum_{\mathbf{q}} \eta(\mathbf{q}) e^{-i\mathbf{q}\cdot\mathbf{r}} \, , \tag{10.37}$$

where the $\mathbf{r}$'s are restricted to be lattice points. Since $\eta(\mathbf{r})$ is real, we must have

$$\eta(\mathbf{q})^* = \eta(-\mathbf{q}) \, . \tag{10.38}$$

The wavevector $\mathbf{q}$ which enters the sum in Eq. (10.37) must give functions which obey the periodicity conditions of Eq. (10.36). Thus, we require that

$$e^{i\mathbf{q}\cdot(L_1\mathbf{a}_1)} = e^{i\mathbf{q}\cdot(L_2\mathbf{a}_2)} = 1 \, . \tag{10.39}$$

This condition means that $\mathbf{q}$ must be of the form

$$\mathbf{q} = (q_1/L_1)\mathbf{b}_1 + (q_2/L_2)\mathbf{b}_2 \, , \tag{10.40}$$

where q_1 and q_2 are integers. Note that

$$e^{i\mathbf{q}\cdot\mathbf{r}} \quad \text{and} \quad e^{i(\mathbf{q}+\mathbf{Q})\cdot\mathbf{r}} \tag{10.41}$$

are identical on all mesh points when $\mathbf{Q}$ is any RLV. This is illustrated in Fig. 10.3. Accordingly, the sum in Eq. (10.37) may be restricted so that it does not include two points which differ by an RLV. One way to implement this restriction is to write the $\mathbf{q}$ sum as

$$\eta(\mathbf{r}) = \sum_{q_1=0}^{L_1-1} \sum_{q_2=0}^{L_2-1} \eta(\mathbf{q}) e^{-i\mathbf{q}\cdot\mathbf{r}} \, , \tag{10.42}$$

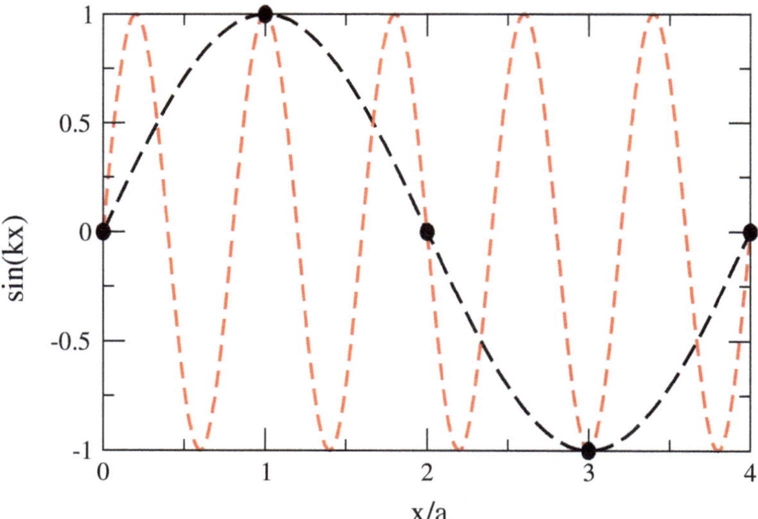

Fig. 10.3 Two waves which give the same displacements at lattice points (which are represented by filled circles) at $x_n = na$, where a is unity and the lattice points are shown by filled circles. The wavevectors differ by a reciprocal lattice vector, which, in one dimension, is $2\pi/a$

where $\mathbf{q}$ is given by Eq. (10.40). Note that this sum is over $L_1L_2 = N$ values of $\mathbf{q}$. Indeed (with appropriate normalization) the Fourier transformation is a unitary transformation from the variables $\eta(\mathbf{r}_i)$ to the $\eta(\mathbf{q})$'s. The transformation inverse to Eq. (10.37) is

$$\eta(\mathbf{q}) = \frac{1}{N}\sum_{\mathbf{r}} \eta(\mathbf{r})e^{i\mathbf{q}\cdot\mathbf{r}} , \tag{10.43}$$

where the $\mathbf{r}$'s are summed over the lattice of Eq. (10.33) with $0 \leq m_1 < L_1$ and $0 \leq m_2 < L_2$. We may verify this formulation by substituting $\eta(\mathbf{q})$ given by Eq. (10.43) into Eq. (10.37), which gives

$$\eta(\mathbf{r}_{m_1,m_2}) = \frac{1}{N}\sum_{\mathbf{q},\mathbf{r}'} e^{-i\mathbf{q}\cdot\mathbf{r}}\eta(\mathbf{r}')e^{i\mathbf{q}\cdot\mathbf{r}'} . \tag{10.44}$$

This is the relation we wish to verify. So we need to evaluate the quantity we call S, where

$$\begin{aligned} S &\equiv \frac{1}{N}\sum_{\mathbf{q}} e^{-i\mathbf{q}(\mathbf{r}-\mathbf{r}')} \\ &= \frac{1}{N}\sum_{q_1q_2} e^{-i(q_1\mathbf{b}_1/L_1+q_2\mathbf{b}_2/L_2)\cdot(\mathbf{r}-\mathbf{r}')} . \end{aligned} \tag{10.45}$$

Now we use the fact that $\mathbf{r}$ and $\mathbf{r}'$ are of the form of Eq. (10.33). So we set $\mathbf{r} = m_1\mathbf{a}_1 + m_2\mathbf{a}_2$ and $\mathbf{r}' = m_1'\mathbf{a}_1 + m_2'\mathbf{a}_2$. Then

$$\begin{aligned} S &= \frac{1}{N}\sum_{q_1 q_2} e^{i[q_1\mathbf{b}_1/L_1 + q_2\mathbf{b}_2/L_2]\cdot[(m_1'-m_1)\mathbf{a}_1 + (m_2'-m_2)\mathbf{a}_2]} \\ &= \left(\frac{1}{L_1}\sum_{q_1=0}^{L_1-1} e^{2\pi i q_1(m_1'-m_1)/L_1}\right)\left(\frac{1}{L_2}\sum_{q_2=0}^{L_2-1} e^{2\pi i q_2(m_2'-m_2)/L_2}\right) \\ &\equiv S_1 S_2 \, , \end{aligned} \tag{10.46}$$

where S_1 is the first factor in large parentheses and S_2 is the second one. Note that if $m_1' = m_1$, then $S_1 = 1$ whereas if $m_1' \neq m_1$ then

$$S_1 = \left(\frac{1 - e^{2\pi i(m_1'-m_1)}}{L_1\left[1 - e^{2\pi i(m_1'-m_1)/L_1}\right]}\right) = 0 \, . \tag{10.47}$$

Thus, $S_1 = \delta_{m_1',m_1}$ and similarly $S_2 = \delta_{m_2',m_2}$ and we have that

$$\frac{1}{N}\sum_{\mathbf{q}} e^{-i\mathbf{q}\cdot(\mathbf{r}-\mathbf{r}')} = \delta_{\mathbf{r},\mathbf{r}'} \, . \tag{10.48}$$

This relation verifies Eq. (10.43).

Had we substituted Eq. (10.37) into Eq. (10.43), we would have had to use

$$\frac{1}{N}\sum_{\mathbf{r}} e^{i\mathbf{r}\cdot(\mathbf{q}-\mathbf{q}')} = \delta_{\mathbf{q},\mathbf{q}'} \, . \tag{10.49}$$

More generally one has the relation

$$\frac{1}{N}\sum_{\mathbf{r}} e^{i(\mathbf{k}_1+\mathbf{k}_2\cdots+\mathbf{k}_n)\cdot\mathbf{r}} = \sum_{\mathbf{Q}} \delta_{\mathbf{k}_1+\mathbf{k}_2+\cdots+\mathbf{k}_n,\mathbf{Q}} \, , \tag{10.50}$$

where the sum is over all vectors $\mathbf{Q}$ of the reciprocal lattice. This relation is usually described as meaning that wavevector is conserved, but only modulo a reciprocal lattice vector. More compactly we write

$$\frac{1}{N}\sum_{\mathbf{r}} e^{i(\mathbf{k}_1+\mathbf{k}_2\cdots+\mathbf{k}_n)\cdot\mathbf{r}} = \Delta(\mathbf{k}_1 + \mathbf{k}_2 + \cdots + \mathbf{k}_n) \, , \tag{10.51}$$

where Δ is unity if its argument is an RLV and is zero otherwise. If all wavevectors are in the first BZ, nonzero values of $\mathbf{Q}$ only become relevant in the case when $n > 2$. Then, when $n = 2$ we have Eq. (10.49).

Table 10.1 Domain of Fourier variable for various domains in real space. In this table, n, m, and N are integers

Real space		Fourier space	
$-\infty < x < \infty$		$-\infty < k < \infty$	
$0 < x < L$		$k = \frac{2m\pi}{L}$	$-\infty < m < \infty$
$x = na$	$0 < n \leq N$	$k = \frac{2m\pi}{aN}$	$-\frac{N}{2} < m \leq \frac{N}{2}$

Conceptually, it is useful to compare and contrast different kinds of spatial Fourier transforms as done in Table 10.1. Fourier transforms of functions defined on a uniform infinite line have wavevectors that range over $-\infty < k < \infty$. For functions defined in finite volume, L^d, the Fourier transform involves wavevectors whose components are spaced by the finite amount, $2\pi/L$. For functions defined on a lattice, the wavevectors of a lattice Fourier transform are defined in Eq. (10.40), and form their own lattice of points in "reciprocal space."

Now we substitute the Fourier representation into Eq. (10.31) to get

$$F_2 = 1/2 \sum_{\mathbf{r}_i, \mathbf{r}_j} \sum_{\mathbf{q}, \mathbf{q}'} F_2(\mathbf{r}_{i,j}) \eta(\mathbf{q}) \eta(\mathbf{q}') e^{-i[\mathbf{q} \cdot \mathbf{r}_i + \mathbf{q}' \cdot \mathbf{r}_j]} , \tag{10.52}$$

where we have indicated that F_2 is a function of $\mathbf{r}_{i,j}$. Now set $\mathbf{r}_i = \mathbf{r}_{i,j} + \mathbf{r}_j$. Then the sums over $\mathbf{r}_i$ and $\mathbf{r}_j$ become a sum over $\mathbf{r}_{i,j}$ and a sum over $\mathbf{r}_j$. Thus, we have

$$F_2 = 1/2 \sum_{\mathbf{q}, \mathbf{q}'} \eta(\mathbf{q}) \eta(\mathbf{q}') \sum_{\mathbf{r}_{i,j}} F_2(\mathbf{r}_{i,j}) e^{-i\mathbf{q} \cdot \mathbf{r}_{i,j}} \sum_{\mathbf{r}_j} e^{-i(\mathbf{q} + \mathbf{q}') \cdot \mathbf{r}_j} . \tag{10.53}$$

We carry out the sum over $\mathbf{r}_j$ using Eq. (10.49), so that

$$F_2 = \frac{N}{2} \sum_{\mathbf{q}} F_2(\mathbf{q}) \eta(\mathbf{q}) \eta(-\mathbf{q}) , \tag{10.54}$$

where

$$F_2(\mathbf{q}) = \sum_{\mathbf{r}_{i,j}} F_2(\mathbf{r}_{i,j}) e^{-i\mathbf{q} \cdot \mathbf{r}_{i,j}} . \tag{10.55}$$

We interpret this free energy as

$$F_2 = \frac{N}{2} \sum_{\mathbf{q}} [\chi(\mathbf{q})]^{-1} \eta(\mathbf{q}) \eta(-\mathbf{q}) , \tag{10.56}$$

where $\chi(\mathbf{q})^{-1} = F_2(\mathbf{q})$. We have now transformed the free energy into diagonal form. In real space, there is a distinction between $1/\chi_{i,j}$ and $[\chi^{-1}]_{i,j}$, the i, jth

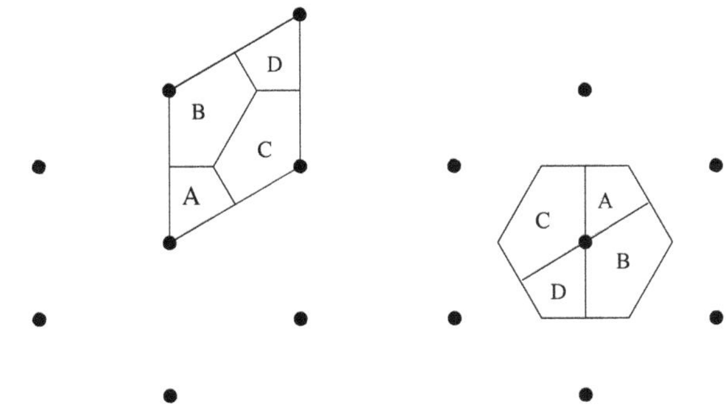

Fig. 10.4 Left: original nonsymmetric domain of summation used in Eq. (10.42). Right: symmetric domain, the first Brillouin Zone. We indicate how regions in the left-hand panel can be translated by a reciprocal lattice vector to get the domain in the right-hand panel

element of the inverse matrix. Because χ is diagonal in the $\mathbf{q}$-representation there is no analogous confusion about the inverse operation.

10.3.1 The First Brillouin Zone

In general, and in particular for the hexagonal lattice, the domain of summation of the sums in $\mathbf{q}$-space defined in Eq. (10.42) does *not* reflect the full symmetry of the system. In this context, "full symmetry" includes not only translations but also rotations, inversion, and reflections. The remedy for this is to sum over an equivalent domain (called the first Brillouin zone) which does have the symmetry of the system. We illustrate this for the hexagonal lattice. Because functions such as $\chi(\mathbf{q})$ satisfy

$$\chi(\mathbf{q}) = \chi(\mathbf{q} + \mathbf{Q}) , \tag{10.57}$$

where $\mathbf{Q}$ is any RLV, we may sum over the domain of points which is closer to the origin in $\mathbf{q}$-space than to any other RLV. This construction is shown in Fig. 10.4.

10.4 Wavevector Selection for Ferro- and Antiferromagnetism

In this section, we present a simple Landau theory analysis of a case where the free energy involves variables coupled at quadratic order. This analysis also serves to unify our treatment of the ferromagnetic and antiferromagnetic systems. We do this for a system in which $J_{ij} = J$ for nearest neighbor (i, j) pairs and is zero otherwise, but J may be positive or negative. If the J's are positive, we have a ferromagnet. If the J's are negative, we have an antiferromagnet. Then, as we have previously shown, if the order parameter m_i is the thermally averaged value of the spin at site i,

$$F = 1/2 \sum_{i,j} [kT\delta_{i,j} - J_{i,j}] m_i m_j + \mathcal{O}(m^4) , \tag{10.58}$$

which is precisely the form studied in the preceding section. So at quadratic order, we have

$$F = N/2 \sum_{\mathbf{q}} \chi(\mathbf{q})^{-1} m(\mathbf{q}) m(-\mathbf{q}) , \tag{10.59}$$

where, for nearest neighbor interactions J and using Eq. (10.55), we have

$$\chi(\mathbf{q})^{-1} = kT - 2J(\cos q_x a + \cos q_y a + \cos q_z a) . \tag{10.60}$$

In Fig. 10.5, we plot this for a sequence of temperatures as a function of aq for $q_x = q_y = q_z = q$ for $J > 0$ (in the left panel) and for $J < 0$ (in the right panel). What we see is that when the temperature is sufficiently high, all modes are stable, that is, $\chi(\mathbf{q})^{-1}$ is positive for all values of $\mathbf{q}$. As the temperature is lowered, because $\chi(\mathbf{q})$ depends on $\mathbf{q}$, an instability will develop at that wavevector for which, as a function of $\mathbf{q}$, $\chi(\mathbf{q})$ is largest. We call this "wavevector selection." We see that for the ferromagnet ($J > 0$) wavevector selection picks out $q = 0$ for the incipient ordering, whereas for the antiferromagnet ($J < 0$) wavevector selection leads to $\mathbf{q} = (\pi/a)(1, 1, 1)$.

Thus, it is not necessary to "guess" what the ordered structure will be because this Landau analysis of "wavevector selection" determines the structure of the incipient ordered phase. In the present example on a cubic lattice, because the order parameter is a scalar and the wavevector at which the instability occurs is unique for both ferro- and antiferromagnetic cases, once we know that wavevector, we immediately know the ordered structure. As we will see in later examples, the situation is more complicated when either there are simultaneous instabilities at more than one

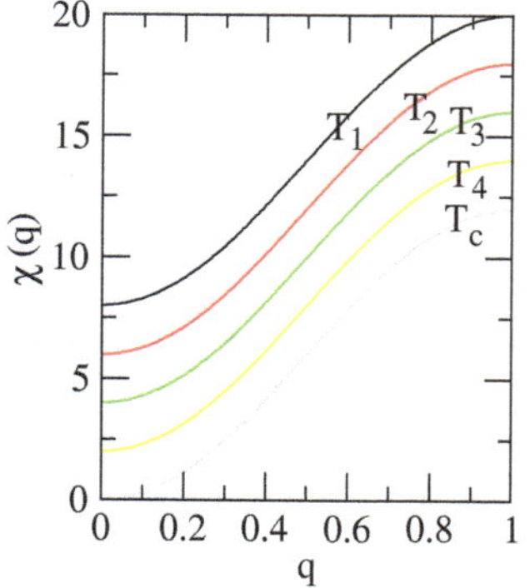

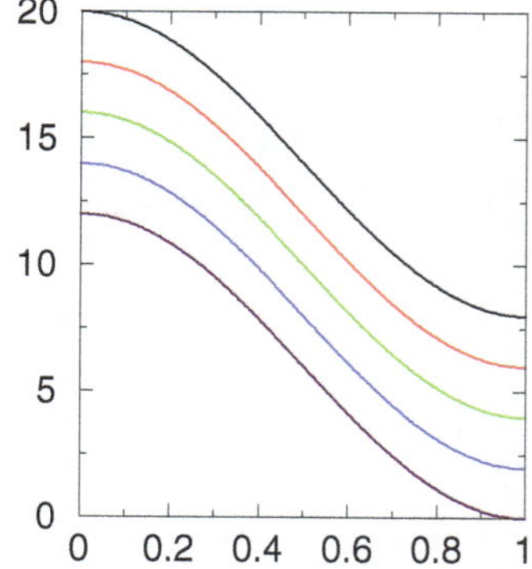

Fig. 10.5 $\chi(\mathbf{q})^{-1}$ versus q for the ferromagnet (left) and the antiferromagnet (right) for temperatures $T_1 > T_2 \cdots > T_c$. In both case, $\mathbf{q} = q(\pi/a, \pi/a, \pi/a)$

equivalent wavevector or the order parameter is more complicated than a scalar. In Sects. 10.8, 10.9 and 10.10, we will consider systems for which wavevector selection leads to states which are neither ferromagnetic nor antiferromagnetic. Wavevector selection is also the subject of Exercise 12.

10.5 Cubic Coupling

It is possible for one order parameter to couple linearly to the square of another. This leads to a type of cubic term whose effect is different from that of a cubic term involving only the critical order parameter. (In the previous chapter, the latter was seen to give rise to a discontinuous transition.) Consider the free energy

$$f = a_1\sigma_1^2 + a_2\sigma_2^2 + b\sigma_1^2\sigma_2 + c\sigma_1^4, \tag{10.61}$$

where $a_1 < a_2$ so that σ_1 is the critical order parameter, and σ_2 is called a "noncritical order parameter." In a field theory, one would "integrate out" the variable σ_2 and derive an expression for the free energy completely in terms of σ_1. In mean-field theory, the strategy is to explicitly minimize f with respect to σ_2, finding σ_2 in terms of σ_1, and express f completely in terms of σ_1.

$$\frac{\partial f}{\partial \sigma_2} = 0 \rightarrow \sigma_2 = -\frac{b\sigma_1^2}{2a_2} \tag{10.62}$$

$$f = a_1\sigma_1^2 + \left[c - \frac{b^2}{4a_2}\right]\sigma_1^4. \tag{10.63}$$

Note that if a_2 becomes too small or if b is too large, the coefficient of σ_1^4 becomes negative and the transition is driven first order by the coupling to σ_2. Also, note that the effect we find here depends on a coupling which is *linear* in the noncritical order parameter rather than by some functional dependence of c on, say, volume, as was suggested in Sect. 9.5.5. The idea is that, when the critical order parameter appears, it generates a field which induces noncritical ordering. Terms of the form $\sigma_1\sigma_2^2$ or $\sigma_1^2\sigma_2^2$ can be discarded because they are minimized for $\sigma_2 = 0$. In an exercise, the reader is asked to ascertain the effect of terms of order $\sigma_1^3\sigma_2$.

The classic example of a cubic coupling involving a noncritical order parameter driving a transition from second to first order is the Blume–Emery–Griffiths (BEG) model (Blume et al. 1971) of the superfluid transition and phase separation in ^{3}He-^{4}He mixtures. The phase diagram for this system is shown in Fig. 2.10. The system is represented by a collection of ferromagnetically interacting spins which can take on values, $S_i = 0, \pm 1$. $S_i = 0$ represents the presence of a ^{3}He atom on site i. $S_i = \pm 1$ represents a ^{4}He atom at that site. The superfluid ordering of the ^{4}He atoms corresponds to ferromagnetic ordering of the spins with unit magnitude. The Hamiltonian for this system is

$$\mathcal{H} = -J \sum_{(i,j)} S_i S_j + \Delta \sum_i S_i^2. \tag{10.64}$$

The single-site density matrix for the mixture $(^3\text{He})_x(^4\text{He})_{(1-x)}$ can be written down directly, by analogy to the Ising case. It is

$$\rho(S_i) = (1-x)\frac{1+m_i S_i}{2} S_i^2 + x(1-S_i^2). \tag{10.65}$$

From this, we can calculate the internal energy and the entropy

$$\langle \mathcal{H} \rangle / N = -Jz(1-x)^2 m^2 + \Delta(1-x) \tag{10.66}$$

$$\begin{aligned} -S/N &= (1-x)\left(\frac{1+m}{2}\right)\ln\left[(1-x)\left(\frac{1+m}{2}\right)\right] \\ &+ x\ln x + (1-x)\left(\frac{1-m}{2}\right)\ln\left[(1-x)\left(\frac{1-m}{2}\right)\right]. \end{aligned} \tag{10.67}$$

The strategy for generating the relevant Landau expansion is to expand $F = \langle \mathcal{H} \rangle - TS$ in powers of m and δ where $\delta = x - x_0$ and x_0 minimizes F for $m = 0$. The expansion will include terms of order m^2 and δ^2, a cubic term of order $m^2\delta$ and a term of order m^4. Minimizing explicitly with respect to δ results in an expression for the free energy of the form of Eq. (10.63). With a bit more algebra, it is possible to show that the transition becomes first order for $x_0 = 2/3$. This calculation is left as an exercise.

10.6 Vector Order Parameters

Instead of the scalar order parameter of the Ising model, we now consider the vector order parameter of the n-component Heisenberg model. We will construct the Landau free energy from symmetry considerations. We know that the free energy must be constructed from invariants involving the vector order parameter $\mathbf{M}$, where $\mathbf{M}$ is the magnetization vector. We assume that the free energy is invariant when all magnetic moments are uniformly rotated about an arbitrary axis. This observation rules out terms involving odd powers of $\mathbf{M}$. For the free energy to be rotationally invariant, it must be of the form

$$F = \frac{1}{2}aM^2 + \frac{1}{4}uM^4 + \ldots \tag{10.68}$$

In Cartesian notation this is

$$F = \frac{a}{2}[M_1^2 + M_2^2 \cdots + M_n^2] + \frac{u}{4}[M_1^2 + M_2^2 \cdots + M_n^2]^2 + \mathcal{O}(M^6) \quad (10.69)$$

and we assume that $a = a'(T - T_c)$.

In the ordered phase, $|\mathbf{M}|$ assumes a value determined by the interplay of the quadratic and quartic terms, but the direction of $\mathbf{M}$ is completely arbitrary because F is a function of $M^2 \equiv \sum_n M_n^2$. However, so far we have not taken account of the fact that the system is embedded in a crystal lattice which inevitably causes the free energy to be different when $\mathbf{M}$ is oriented along the coordinate axes than when it is oriented along a body-diagonal direction. This breaking of full rotational symmetry leads to a term in the free energy of the form

$$F_A = \frac{1}{4}v \sum_\alpha M_\alpha^4 \,. \quad (10.70)$$

The term, F_A, is called "cubic anisotropy." Here "cubic" refers to the symmetry of the cubic lattice and not to the order in the Landau expansion which, from Eq. (10.70), is clearly quartic. Then the free energy is of the form

$$F = \frac{1}{2}a|\mathbf{M}|^2 + \frac{1}{4}u|\mathbf{M}|^4 + \frac{1}{4}v[M_1^4 + M_2^4 \cdots + M_n^4] \,. \quad (10.71)$$

We do not need to include in the free energy a term $v' \sum_{\alpha \neq \beta} M_\alpha^2 M_\beta^2$. Such a term can be expressed as a linear combination of the two other fourth-order terms which we have already included.

For $v > 0$, the order parameter configuration which minimizes the free energy must minimize $\sum_i M_i^4$ for fixed M^2. This implies that

$$\mathbf{M} = M(\pm\frac{1}{\sqrt{n}}, \pm\frac{1}{\sqrt{n}}, \cdots \pm \frac{1}{\sqrt{n}}), \quad (10.72)$$

i.e., that $\mathbf{M}$ points toward a corner of a hypercube. Then

$$f = \frac{1}{2}aM^2 + \frac{1}{4}\left[u + (v/n)\right] M^4. \quad (10.73)$$

For $v < 0$, the order parameter must *maximize* $\sum_i M_i^4$ for fixed M^2. Then

$$\mathbf{M} = M(1, 0, \ldots, 0) \quad (10.74)$$

or $\mathbf{M}$ is parallel to any one of the other equivalent edges of the hypercube, and

$$f = \frac{1}{2}aM^2 + \frac{1}{4}(u + v)\, M^4 \,. \quad (10.75)$$

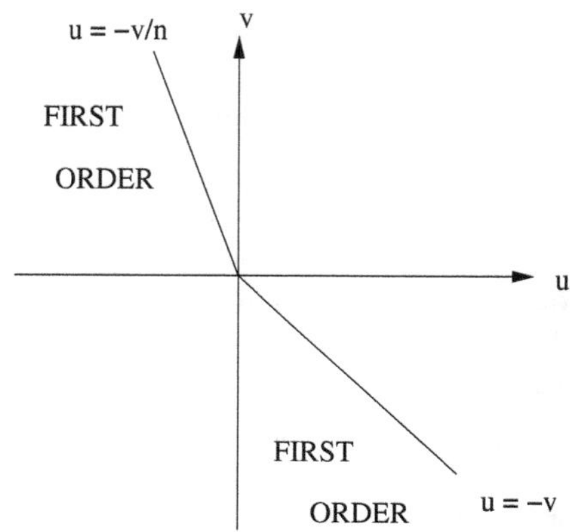

Fig. 10.6 Phase diagram for the n-vector model with cubic anisotropy according to mean-field theory. To the right of the solid lines, the ordering transition is continuous. To the left of the solid lines, the fourth-order term is negative and one has a discontinuous transition. The RG gives a somewhat different phase diagram. See (Aharony 1976)

Of course, if $u + v < 0$, the transition becomes a discontinuous one. The phase diagram within mean-field theory is given in Fig. 10.6.

10.7 Potts Models

Ising models are such a useful pedagogical tool that they tend to become addictive. Even the lattice gas, which we will study in the next section, is a kind of Ising model. In this section, we introduce a new kind of spin variable, the q-state Potts variable, and consider the mean-field theory and Landau expansion of the free energy for interacting Potts spins.

A Potts variable is a discrete variable which can have one of q distinct but equivalent values. We could call these values, $A,\ B,\ C,\ \ldots$, or we could call them red, green, blue, etc. If we represent the variable at site i as p_i, then a ferromagnetic Potts Hamiltonian may be written as

$$\mathcal{H} = -J \sum_{(i,j)} \left(q\delta_{p_i,p_j} - 1 \right) . \tag{10.76}$$

This Hamiltonian has been written so as to emphasize the equivalence of different values of p_i. The energy only depends on whether the spins on two neighboring sites are in the same ($p_i = p_j$) or different ($p_i \neq p_j$) states and not on what that state is. The normalization of Eq. (10.76), specifically the factor of q inside the sum and the constant term, has been chosen to reproduce the Ising Hamiltonian for the Ising case, $q = 2$.

A useful representation of the Potts variables is as symmetrically arranged points on the unit circle or, equivalently as the complex numbers

$$p_j = e^{i\theta_j} = e^{2\pi i n_j/q}, \quad n_j = 0, \ldots q - 1. \tag{10.77}$$

Then

$$q\delta_{p_i,p_j} = \sum_{k=0}^{q-1} e^{ik(\theta_i - \theta_j)} \tag{10.78}$$

and, in terms of these variables, the Hamiltonian is

$$\mathcal{H} = -J \sum_{(i,j)} \left[\left(\sum_k e^{ik\theta_i} e^{-ik\theta_j} \right) - 1 \right]. \tag{10.79}$$

In mean-field theory, the average $\langle \mathcal{H} \rangle$ will depend on the averages $\langle e^{\pm ik\theta_j} \rangle$ for each site j. This average is calculated using the one-particle density matrix, $\rho_j(n_j)$ as follows:

$$\langle e^{ik\theta_j} \rangle = \sum_{n=1}^{q-1} \rho_j(n) e^{ik\,2\pi n/q}. \tag{10.80}$$

Using the orthogonality relation, Eq. (10.78), and applying our usual strategy, we can express the density matrix in terms of the averages that it can be used to calculate as

$$\rho_j(n) = \frac{1}{q} \sum_{k'} \langle e^{ik'\theta_j} \rangle e^{-ik'\,2\pi n/q}. \tag{10.81}$$

You should check for yourself that the trace of this density matrix is one.

In the discussion which follows we will assume, for the case of the ferromagnetic Potts model, that the order parameters are the same at each site. Thus, we drop the subscripts on ρ. Consider the explicit form of the density matrix for specific values of q. For $q = 2$

$$\rho(n_i) = \frac{1}{2} \left[1 + m e^{-i\pi n_i} \right], \quad n_i = 0, 1, \tag{10.82}$$

where $m = \langle e^{i\pi n} \rangle$. This is exactly the density matrix for the Ising model.

For $q = 3$, the density matrix for the three-state Potts model has the form

$$\rho(n_i) = \frac{1}{3} \left[1 + a e^{-i\frac{2\pi}{3} n_i} + b e^{-i\frac{4\pi}{3} n_i} \right], \quad n_i = 0, 1, 2, \tag{10.83}$$

where

$$a = \langle e^{i\frac{2\pi}{3} n} \rangle, \quad b = \langle e^{i\frac{4\pi}{3} n} \rangle. \tag{10.84}$$

The fact that $\rho(n_i)$ is real, leads to the requirement that $b = a^*$. If we write

$$a = m e^{i\phi} \tag{10.85}$$

then the density matrix becomes

$$\rho(n_i) = \frac{1}{3}\left[1 + 2m\cos\left(\frac{2\pi}{3}n_i - \phi\right)\right]. \tag{10.86}$$

Using this expression for the density matrix, we can calculate the mean-field internal energy.

$$\begin{aligned}\langle\mathcal{H}\rangle &= -\frac{JNz}{2}\left[\underbrace{\left(\sum_k \langle e^{ik\theta_i}\rangle\langle e^{-ik\theta_j}\rangle\right)}_{1+2m^2} - 1\right] \\ &= -JNzm^2. \end{aligned} \tag{10.87}$$

The fact that $\langle\mathcal{H}\rangle$ does not depend on the phase angle ϕ raises the interesting question of what does determine ϕ. Of course, the only other possibility is the entropy.

$$\begin{aligned} S &= -\mathrm{Tr}\rho\ln\rho \\ &= -\frac{1}{3}\Big\{[1 + 2m\cos\phi]\ln[1 + 2m\cos\phi] \\ &+ [1 + 2m\cos(\phi - 2\pi/3)]\ln[1 + 2m\cos(\phi - 2\pi/3] \\ &+ [1 + 2m\cos(\phi + 2\pi/3)]\ln[1 + 2m\cos(\phi + 2\pi/3]\Big\}. \end{aligned} \tag{10.88}$$

The entropy $S(m, \phi)$ is clearly invariant under threefold rotations of the type $\phi \to \phi \pm 2\pi/3$. In fact, it is clear from the form of Eq. (10.88) that the values $\phi = 0, \pm 2\pi/3$ are extremal, and it can easily be shown, by direct calculation, that these values maximize S. Therefore, the free energy is minimized by these values of ϕ. We will see, at the end of the next section, that this distinctive behavior of the three-state Potts model can arise in systems which start out looking nothing at all like Potts models.

10.8 The Lattice Gas: An Ising-Like System

10.8.1 Disordered Phase of the Lattice Gas

As a further example of wavevector selection and coupled order parameters, we consider the lattice gas, which is often used to model the liquid–gas transition, and which can also describe various kinds of crystalline ordering. The lattice gas Hamiltonian is

$$\mathcal{H} = \frac{1}{2}\sum_{i,j} U_{i,j} n_i n_j, \tag{10.89}$$

where $U_{i,j} = U_{j,i} = U(\mathbf{R}_i - \mathbf{R}_j)$ is the interaction between particles at sites i and j, and $n_i = 0, 1$ is the occupation of site i. We now apply the variational principle for the grand potential $\Omega \equiv U - TS - \mu N$. Equation (6.56) for a system on a lattice at fixed volume is

$$\Omega(T, V) = \mathrm{Tr}\left(\boldsymbol{\rho}\left[\mathcal{H} + kT \ln \boldsymbol{\rho} - \mu N\right]\right) , \tag{10.90}$$

where N is the total number operator and μ is the chemical potential which controls the total number of particles.

A common situation is

$$\begin{aligned} U_{i,j} &= U > 0, \quad |\mathbf{R}_i - \mathbf{R}_j| = |\boldsymbol{\delta}|, \\ U_{i,j} &= 0 , \qquad\quad |\mathbf{R}_i - R_j| \neq |\boldsymbol{\delta}|, \end{aligned} \tag{10.91}$$

where $\boldsymbol{\delta}$ is a nearest neighbor vector and $U_{i,j}$ represents a repulsive nearest neighbor interaction. However, we will consider the mean-field theory for the general case of arbitrary U_{ij}.

The most general single-site density matrix for the lattice gas is

$$\rho_i(n_i) = x_i n_i + (1 - x_i)(1 - n_i) \tag{10.92}$$

which satisfies

$$\begin{aligned} \mathrm{Tr}\rho_i &= 1, \\ \langle n_i \rangle = \mathrm{Tr} n_i \rho_i &= x_i . \end{aligned} \tag{10.93}$$

In the last equation, we have used the facts that $n_i^2 = n_i$ and $n_i(1 - n_i) = 0$.

For this mean-field density matrix, we have is

$$U - \mu N \equiv \langle \mathcal{H} - \mu N \rangle = \frac{1}{2} \sum_{i,j} U_{i,j} x_i x_j - \mu \sum_i x_i , \tag{10.94}$$

and the entropy is

$$S = -\sum_i \left[x_i \ln x_i + (1 - x_i) \ln(1 - x_i)\right] . \tag{10.95}$$

We will analyze this problem using Landau theory, expanding in powers of the order parameters. What are the order parameters?

Define η_i by $x_i = x + \eta_i$ where

$$x \equiv \frac{1}{N} \sum_i x_i \rightarrow \frac{1}{N} \sum_i \eta_i = 0. \tag{10.96}$$

The η_i are the *deviations* from uniform occupancy. At high temperatures, we expect, $\eta_i = 0 \, \forall \, i$. In terms of the η_i

$$\langle \mathcal{H} - \mu n \rangle = \frac{1}{2} N x^2 \left(\sum_j U_{i,j} \right) + \frac{1}{2} \sum_{i,j} U_{i,j} \eta_i \eta_j - \mu x N \, . \tag{10.97}$$

There are no terms linear in η_i because of Eq. (10.96).

For the entropy, we need to expand the quantity

$$(x + \eta_i) \ln(x + \eta_i) + (1 - x - \eta_i) \ln(1 - x - \eta_i).$$

Again the linear terms vanish when summed over i. The result is

$$\begin{aligned} S = & -N \left[x \ln x + (1-x) \ln(1-x) \right] \\ & - \frac{1}{2} \sum_i \eta_i^2 \left(\frac{1}{x} + \frac{1}{1-x} \right) = - \frac{1}{2x(1-x)} \sum \eta_i^2 \\ & + \frac{1}{6} \sum_i \eta_i^3 \left(\frac{1}{x^2} - \frac{1}{(1-x)^2} \right) = \frac{1-2x}{6x^2(1-x)^2} \sum_i \eta_i^3 \\ & - \frac{1}{12} \sum_i \eta_i^4 \left(\frac{1}{x^3} + \frac{1}{(1-x)^3} \right) = - \frac{1 - 3x + 3x^2}{12x^3(1-x)^3} \sum_i \eta_i^4 . \end{aligned} \tag{10.98}$$

The Landau expansion for the free energy (actually the grand potential) of the lattice gas is thus

$$\Omega = F_0 + F_2 + F_3 + F_4, \tag{10.99}$$

where

$$\begin{aligned} F_0 = & \frac{1}{2} N x^2 \left(\sum_j U_{i,j} \right) - \mu x N \\ & + T N \left[x \ln x + (1-x) \ln(1-x) \right] \end{aligned} \tag{10.100}$$

$$F_2 = \frac{1}{2} \sum_{i,j} U_{i,j} \eta_i \eta_j + \frac{T}{2x(1-x)} \sum \eta_i^2 \tag{10.101}$$

$$F_3 = - \frac{(1-2x)T}{6x^2(1-x)^2} \sum_i \eta_i^3 \equiv A \sum_i \eta_i^3 \tag{10.102}$$

$$F_4 = \frac{(1 - 3x + 3x^2)T}{12x^3(1-x)^3} \sum_i \eta_i^4 \equiv B \sum_i \eta_i^4 \, . \tag{10.103}$$

It is useful to define

$$U(0) \equiv \sum_j U_{i,j}. \tag{10.104}$$

We consider first the uniform case in which all of the η_i are zero. Although simple, this case contains a great deal of useful physics. F_0 is minimized by

$$\frac{1}{N}\frac{\partial F_0}{\partial x} = xU(0) - \mu + T\ln\frac{x}{1-x} = 0 \tag{10.105}$$

or

$$\mu = xU(0) + T\ln\frac{x}{1-x}. \tag{10.106}$$

Note the log divergence of μ for $x = 0$ and $x = 1$. The quantity $\partial x/\partial\mu$ measures the compressibility of the lattice gas (See Exercise 4). From the above expression,

$$\frac{\partial x}{\partial \mu} = \frac{1}{U(0) + \frac{T}{x(1-x)}}. \tag{10.107}$$

The ability to add or subtract particles by changing the chemical potential vanishes near $x = 0$ and $x = 1$.

10.8.2 Lithium Intercalation Batteries

The Li intercalation battery is the physicist's battery. It can be understood completely in terms of physical processes. The voltage of the battery is simply related to the work required to remove a Li^+ ion from "between the layers" of the layered cathode material and return it to the Li metal anode. Such a battery is illustrated in Fig. 10.7.

As the battery is discharged, electrons flow through the external circuit, and Li^+ ions flow through the electrolyte into the cathode. In general, the voltage of the battery decreases as the layers fill with lithium. The battery is recharged by applying an external voltage which pulls electrons out of the cathode and adds electrons to the Li metal anode, while Li^+ ions flow back to the metal anode through the electrolyte.

The voltage of a lithium intercalation battery decreases as it is discharged, because it becomes less favorable, in terms of the free energy decrease, for lithium atoms to go into the layered material. In a simple theory, the free energy is lowered by two contributions, lower energy, and higher entropy. This is modeled by the dependence of $-\mu$ versus x in our mean-field formula. A comparison of the simple theory of Eqs. (10.106) and (10.107) to battery voltage data is shown in Fig. 10.8. Figure (10.8) shows that the general tendency of the voltage to decrease as the battery discharges is well represented by the uniform lattice gas theory. This is particularly the case near the fully charged or fully discharged ends where entropy dominates and the compressibility of Eq. (10.107) vanishes. However, it is clear that more is going on

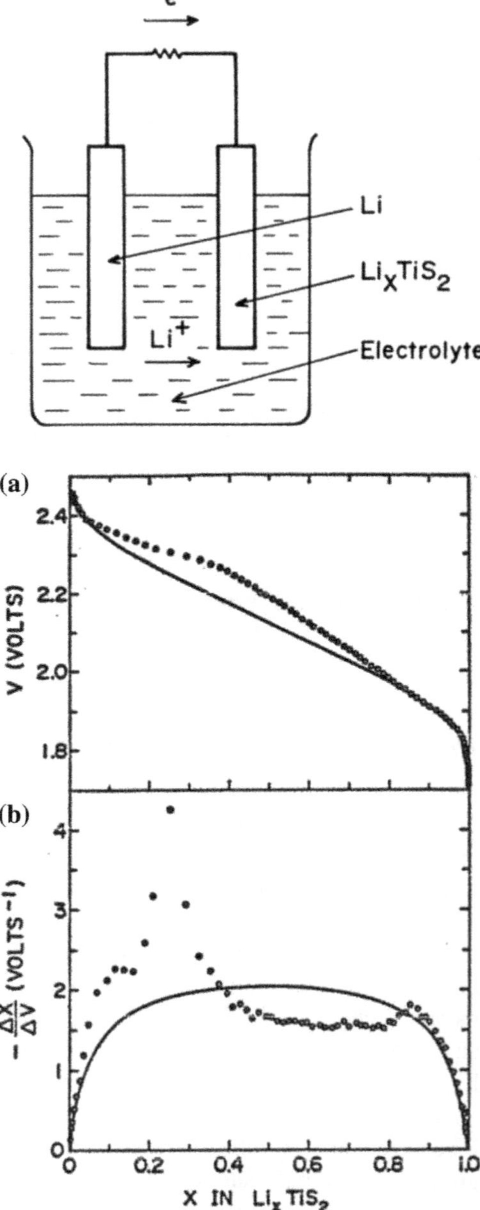

Fig. 10.7 Schematic representation of the discharge of a Li/Li_xTiS_2 intercalation battery (from J.R. Dahn, PhD. Thesis, 1982)

Fig. 10.8 **a** V(x) and **b** $-\partial x/\partial V$ for Li_xTiS_2 cells (points) after Thompson (Phys. Rev. Lett. **40**, 1511 (1978)) (Figure from Berlinsky et al. 1979). The solid lines are a fit to Eqs. (10.106) and (10.107)

at intermediate filling and that the resulting structure in the voltage curve must result from some kind of ordering effects in the interacting lattice gas of Li ions. In the next section, we examine how the ions might order on a hexagonal lattice of sites of the type that exists between TiS_2 layers.

10.9 Ordered Phase of the Lattice Gas

We now diagonalize F_2 in Eq. (10.101) using the Fourier series for η as in Eq. (10.37) or Eq. (10.42).

$$F_2 = \frac{1}{2}\sum_{i,j}\left[U_{ij} + \frac{T\delta_{i,j}}{x(1-x)}\right]\sum_{\mathbf{q},\mathbf{q}'} \eta_{\mathbf{q}} e^{i\mathbf{q}\cdot\mathbf{R}_i} \eta_{\mathbf{q}'} e^{i\mathbf{q}'\cdot\mathbf{R}_j} . \tag{10.108}$$

Replacing $\mathbf{R}_i$ by $\mathbf{R}_i = \mathbf{R}_j + \mathbf{R}_{i,j}$, the sum over $\mathbf{R}_i$ and $\mathbf{R}_j$ becomes a sum over $\mathbf{R}_{ij}$ and $\mathbf{R}_j$. Thus, we have

$$F_2 = \frac{1}{2}\sum_{\mathbf{R}_{ij}}\left[U_{ij} + \frac{T\delta_{i,j}}{x(1-x)}\right] e^{i\mathbf{q}\cdot\mathbf{R}_{ij}} \sum_{\mathbf{q},\mathbf{q}'} \eta_{\mathbf{q}}\eta_{\mathbf{q}'} \sum_{\mathbf{R}_j} e^{i(\mathbf{q}+\mathbf{q}')\cdot\mathbf{R}_j} . \tag{10.109}$$

In analogy with Eq. (10.48), we have that

$$\sum_{\mathbf{R}_j} e^{i(\mathbf{q}+\mathbf{q}')\cdot\mathbf{R}_j} = N\delta_{\mathbf{q}+\mathbf{q}',0} \tag{10.110}$$

where $\mathbf{R}_j$ is a lattice vector and $\mathbf{q}$ and $\mathbf{q}'$ are wavevectors as in Eq. (10.40). Thus, we have

$$\begin{aligned} F_2 &= \frac{N}{2}\sum_{\mathbf{q}} \chi(\mathbf{q})^{-1}|\eta(\mathbf{q})|^2 \\ &= \frac{N}{2}\sum_{\mathbf{q}} \chi(\mathbf{q})^{-1}\eta(\mathbf{q})\eta(-\mathbf{q}) \end{aligned} \tag{10.111}$$

where the wavevector-dependent inverse susceptibility is

$$\chi^{-1}(\mathbf{q}) = U(\mathbf{q}) + \frac{T}{x(1-x)}, \tag{10.112}$$

where

$$U(\mathbf{q}) = \sum_{\mathbf{R}_{i,j}} U_{i,j} e^{-i\mathbf{q}\cdot\mathbf{R}_{i,j}} . \tag{10.113}$$

Each $\eta_{\mathbf{q}}$ is a possible order parameter. We expect the $\{\eta_{\mathbf{q}}\}$ with the largest susceptibilities to be the ones that order spontaneously at the highest temperature. How does this work?

Define $\{\mathbf{q}_\alpha^*\}$ as the set of $\mathbf{q}$'s in the first Brillouin zone with the largest susceptibilities (most negative $U(\mathbf{q})$'s). The rules, due to Landau, regarding equivalent, symmetry-related $\mathbf{q}$-vectors in the first Brillouin zone are as follows:

1. A general $\mathbf{q}$ has a number of equivalent partners equal to the number of point group symmetry operations. The set of these vectors is called the "star of $\mathbf{q}$."
2. Higher symmetry $\mathbf{q}$'s will have smaller stars. For example, the star of $\mathbf{q} = (\mathbf{0}, \mathbf{0})$ contains only one vector.
3. Extrema of functions of $\mathbf{q}$ can occur at points in the Brillouin zone which have sufficiently high symmetry that the gradient of the function vanishes. In that case, the value of $\mathbf{q}$ is fixed by *symmetry* and is not affected by small changes in coupling constants that preserve the symmetry. Extrema may also occur at arbitrary points in the Brillouin zone. In that case, however, $\mathbf{q}$ will be an explicit function of all the coupling constants.

Order described by a $\mathbf{q}$-vector at a symmetry point is called *commensurate* because the translational symmetry breaking leads to a larger unit cell which is some small integer multiple of the original unit cell, and the size of the Brillouin zone is reduced proportionately. Order corresponding to a general, non-symmetry $\mathbf{q}$-point is called *incommensurate*. This discussion applies quite generally to any lattice in any dimension. Next we will consider a specific simple case.

10.9.1 Example: The Hexagonal Lattice with Repulsive Nearest Neighbor Interactions

The Fourier transform of the nearest neighbor interaction potential for the hexagonal lattice is obtained by substituting the coordinates of the six nearest neighbor sites into Eq. (10.113)

$$\begin{aligned} U(\mathbf{q}) &= 2U\left[\cos q_x a + \cos\frac{(q_x+\sqrt{3}q_y)a}{2} + \cos\frac{(q_x-\sqrt{3}q_y)a}{2}\right] \\ &= 2U\left[\cos q_x a + 2\cos\frac{q_x a}{2}\cos\frac{\sqrt{3}q_y a}{2}\right]. \end{aligned} \tag{10.114}$$

This function has the full symmetry of the hexagonal reciprocal lattice.

10.9.2 The Hexagonal Brillouin Zone

The most symmetric (Wigner–Seitz) Brillouin Zone is formed by the perpendicular bisectors of the smallest RLVs, as shown in Fig. 10.4. A general point in the BZ, such as the point $\mathbf{Q}$ in Fig. 10.9, defines 12 distinct but equivalent points. These points are generated, by 60° rotations of $\mathbf{Q}$ and by taking (Q_x, Q_y) into $(-Q_x, Q_y)$. A point on a line connecting the point Γ, $\mathbf{q} = (0, 0)$, to the center of a zone face, Y, or corner, X, defines six distinct but equivalent points. There are three distinct but equivalent points, Y, Y', and Y'', that are the centers of zone faces. The others are related by translations by a reciprocal lattice vector. Similarly, there are two distinct but equivalent zone corners, X, which may be chosen as $\pm X$, with the others being related by reciprocal lattice vectors. The zone center, the point Γ, is unique.

The function $U(\mathbf{q})$ has its minimum value, $-3U$, at the X-points, which we will denote as $\pm\mathbf{q}^* = (\pm 4\pi/3a, 0)$.

$$\chi^{-1}(\pm\mathbf{q}^*) = -3U + \frac{T}{x(1-x)} .$$

This vanishes when $T = T_c$, where

$$T_c = 3Ux(1-x) . \tag{10.115}$$

Next consider the cubic term of Eq. (10.102). Using Eq. (10.51), we write it as

$$F_3 = NA \sum_{\mathbf{q}_1,\mathbf{q}_2,\mathbf{q}_3} \eta_{\mathbf{q}_1}\eta_{\mathbf{q}_2}\eta_{\mathbf{q}_3}\Delta(\mathbf{q}_1 + \mathbf{q}_2 + \mathbf{q}_3) . \tag{10.116}$$

Similarly

$$F_4 = NB \sum_{\mathbf{q}_1,\mathbf{q}_2,\mathbf{q}_3,\mathbf{q}_4} \eta_{\mathbf{q}_1}\eta_{\mathbf{q}_2}\eta_{\mathbf{q}_3}\eta_{\mathbf{q}_4}\Delta(\mathbf{q}_1 + \mathbf{q}_2 + \mathbf{q}_3 + \mathbf{q}_4) . \tag{10.117}$$

The quadratic terms in F which involve *only* $\eta_{\mathbf{q}^*}$ and $\eta_{-\mathbf{q}^*}$ are

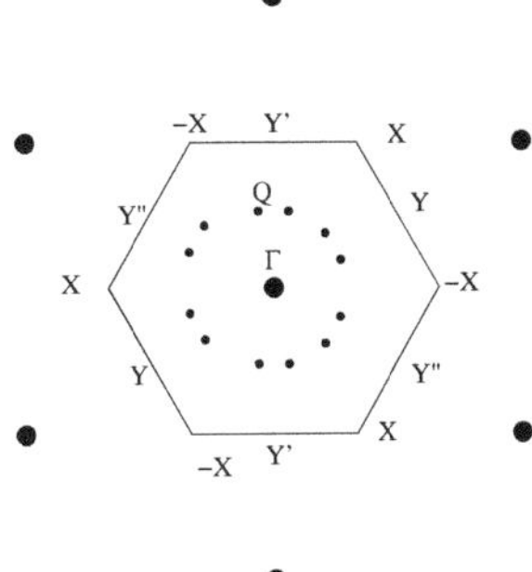

Fig. 10.9 First Brillouin zone for the hexagonal lattice. A general point (labeled **Q** generates a star consisting of 12 equivalent vectors. Note that the star of Γ contains only a single point. The star of the vector X contains X and $-X$. Note that the three vectors labeled X (or $-X$) are related by RLVs, so only one of them appears in a sum over wavevectors. Similarly, the centers of the face labeled Y (or Y' or Y'') are separated by an RLV, so the star of Y contains the three inequivalent wavevectors Y, Y', and Y''

$$\tilde{f}_2 = \tilde{F}_2/N = \frac{1}{2}\left[-3U + \frac{T}{x(1-x)}\right](\eta_{\mathbf{q}^*}\eta_{-\mathbf{q}^*} + \eta_{-\mathbf{q}^*}\eta_{\mathbf{q}^*})$$
$$= \frac{T - T_c}{x(1-x)}|\eta_{\mathbf{q}^*}|^2. \tag{10.118}$$

Next look at the cubic terms that contain only $\eta_{\pm\mathbf{q}^*}$, subject to $\mathbf{q}_1 + \mathbf{q}_2 + \mathbf{q}_3 = 0$. The condition can only be satisfied because $3\mathbf{q}^* = 2\mathbf{b}_1 - \mathbf{b}_2$ is an RLV and hence $\eta_{\mp 2\mathbf{q}^*}$ is equivalent to $\eta_{\pm\mathbf{q}^*}$. Then

$$\tilde{f}_3 = A\left[\eta_{\mathbf{q}^*}^3 + \eta_{-\mathbf{q}^*}^3\right]. \tag{10.119}$$

Note that $\eta_{\mathbf{q}^*}$ is a complex number, which we can write as

$$\eta_{\pm\mathbf{q}^*} = |\eta_{\mathbf{q}^*}|e^{\pm i\phi}. \tag{10.120}$$

Then we can write

$$\tilde{f}_3 = 2A|\eta_{\mathbf{q}^*}|^3 \cos 3\phi. \tag{10.121}$$

A similar analysis gives

$$\tilde{f}_4 = 6B|\eta_{\mathbf{q}^*}|^4. \tag{10.122}$$

So the part of the total free energy that depends only on the critical order parameter is

$$\tilde{f} = \frac{T - T_c}{x(1-x)}|\eta_{\mathbf{q}^*}|^2 + 2A|\eta_{\mathbf{q}^*}|^3 \cos 3\phi + 6B|\eta_{\mathbf{q}^*}|^4. \tag{10.123}$$

This is not actually the complete result up to fourth order in the order parameter. As we have seen in Chap. 5, one can induce additional fourth-order terms if there are cubic terms which are quadratic in the critical variables and linear in any noncritical variable. We have such cubic terms which arise when the wavevectors in Eq. (10.116) are $\mathbf{q}^*$, $-\mathbf{q}^*$ and zero. In an exercise, you are to show that this coupling leads to a renormalization of B. So we use Eq. (10.123) with the proviso that B has correction terms due to the coupling to noncritical variables.

We now minimize $\tilde{f}$ given by Eq. (10.123) with respect to $|\eta_{\mathbf{q}^*}|$ and ϕ. For $x < 1/2$, A is negative and the free energy is minimized for $\phi = 0,\ 2\pi/3,\ 4\pi/3$. For $x > 1/2$, A is positive and the free energy is minimized by $\phi = \pi/3,\ \pi,\ 5\pi/3$.

What are these ordered states? To answer that write

$$\eta_i = \eta_{\mathbf{q}^*}e^{-i\mathbf{q}^*\cdot\mathbf{R}_i} + \eta_{-\mathbf{q}^*}e^{i\mathbf{q}^*\cdot\mathbf{R}_i}$$
$$= 2|\eta_{\mathbf{q}^*}|\cos(\mathbf{q}^* \cdot \mathbf{R}_i + \phi), \tag{10.124}$$

where

$$\mathbf{q}^* = \left(\frac{4\pi}{3a}, 0\right), \quad \mathbf{R}_i = l(a, 0) + m\left(-\frac{a}{2}, \frac{\sqrt{3}a}{2}\right). \tag{10.125}$$

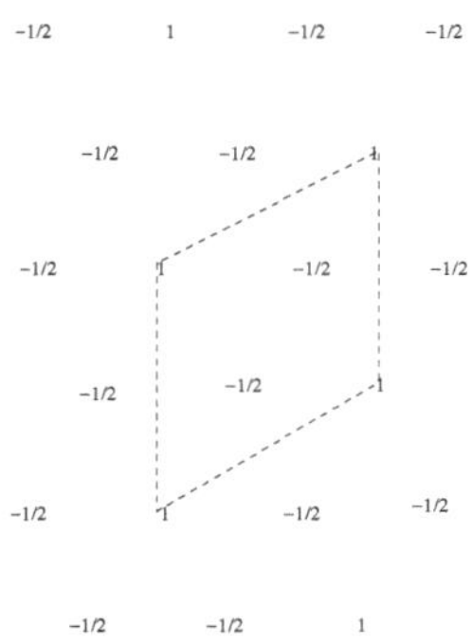

Fig. 10.10 Density variation relative to uniform density (in units of $2|\eta_{\mathbf{q}^*}|$) according to Eq. 10.126. In the ordered phase (in which the order parameter $\eta_{\mathbf{q}}$ is nonzero), the unit cell (outlined by dashed lines) contains three sites, one with enhanced density and two with depleted density. (In the disordered phase, the unit cell contains one site)

So

$$\eta(l, m) = 2|\eta_{\mathbf{q}^*}| \cos\left[2\pi\left(\frac{2l - m}{3}\right) + \phi\right]. \tag{10.126}$$

For $\phi = 0$ the cos is $+1$ when $2l - m$ is a multiple of 3, and $-1/2$ otherwise. Other values, $\phi = 2\pi/3,\ 4\pi/3$, correspond to the same state with the maxima shifted to different sublattices. For $x > 1/2$, the maxima become minima and vice versa. So $\eta(l, m)$ defines a density wave in which, for $x < 1/2$, 1/3 of the sites have enhanced density, while the other 2/3 have lower than average density. The density peaks are arranged on a $\sqrt{3} \times \sqrt{3}$ superlattice, described by Eq. (10.126). A map of the density for $x < 1/2$ is given in Fig. 10.10.

There is a direct analogy between the three degenerate values of the phase for the three-state Potts model, which was discussed in the previous section, and the three equilibrium values of the phase for the hexagonal lattice gas. In fact, the Landau expansions of the two models are quite analogous. The main difference is the x-dependence of the coefficients for the lattice gas case, which can cause the cubic term to change sign and hence stabilize the values $\phi = \pi/3, \pi, 5\pi/3$.

Returning to our theme that Landau expansions describe scenarios for how phase transitions occur, the fact that the hexagonal lattice gas and the three-state Potts model having the same Landau expansions has been used to argue that the two systems should exhibit the same critical exponents. The value of this argument is that the Potts exponents can be calculated using renormalization group techniques, while the hexagonal lattice gas can be realized experimentally. This type of equivalence is described by saying that the two systems are in the same "universality class," a concept which finds its justification within renormalization group theory.

One final word about the intercalation battery data of Fig. 10.8: the Li compressibility curve, $-dx/dV$, exhibits a prominent peak at about $x = 0.25$. This peak could correspond to the $\sqrt{3} \times \sqrt{3}$ ordering transition predicted by the lattice gas theory. As x is increased from low values, the ordering transition occurs around $x = 0.25$, which would give a peak in the compressibility, and then drops as the commensurate state fills at $x = 1/3$. However, unlike the data, the lattice gas model is symmetric around $x = 1/2$, and so the agreement with experiment is less than compelling. Further discussion and explanations for the shape of the data can be found in Dahn (1982). An intercalation system that does display the type of lattice gas ordering discussed above is described in Dahn and McKinnon (1984). More recent developments which pertain to practical Li intercalation batteries can be found in Reimers and Dahn (1992) and Chen et al. (2002).

10.10 Landau Theory of Multiferroics

A multiferroic is a material which can simultaneously exhibit at least two of the following three kinds of order: ferromagnetism (or more general spin order), ferroelectricity, and/or ferroelasticity, where the last involves a spontaneous relative displacement of ions. Such materials are of theoretical interest because they involve simultaneous breaking of disparate symmetries, particularly when they involve both magnetic and electric order. They are also interesting because of potential applications.

The particular system that we will look at exhibits spin order which is incommensurate with the periodicity of the lattice and which can induce an electric polarization, P, that is spatially uniform. The coupling which allows this to happen is actually *trilinear* in the order parameters. It is quadratic in spin-density waves of two different symmetries, with equal and opposite incommensurate wavevectors, which then couple to the polarization. The theory of this multiferroic order involves a number of new concepts which will be developed in the following two subsections and then applied to a simplified model of a multiferroic material.

10.10.1 Incommensurate Order

In Sect. 10.9, we discussed the concept of symmetry points in the Brillouin zone and argued that local maxima (as well as minima and saddle points) of the wavevector-dependent susceptibility may occur at these points where gradients of the susceptibility vanish. It was also mentioned that it is possible for the maximum susceptibility to occur at a general (non-symmetry) point in the Brillouin zone, which can give rise to an ordering that is *incommensurate* with the underlying lattice. For that case, the precise value of the wavevector of the critical order parameter will depend on both the coupling constants and, in general, on the temperature.

To illustrate how incommensurate order might come about, we consider a one-dimensional model where spins interact with first and second neighbors, for which the Hamiltonian can be written as

$$\mathcal{H} = -\sum_n \{J_1 s_n s_{n+1} + J_2 s_n s_{n+2}\}. \tag{10.127}$$

Then the analog of $U(q)$ in Eq. (10.113) is

$$J(q) = -J_1 \cos(q) - J_2 \cos(2q), \tag{10.128}$$

where $-\pi < q < \pi$, and we have set the lattice constant equal to 1. We ask for what value of q $J(q)$ is minimized. If J_1 and J_2 are both positive, it is clear that a ferromagnetic state ($q = 0$) is stabilized. For this case, both first and second neighbor interactions favor parallel spins. Keeping J_1 positive, the system remains ferromagnetic as long as $J_2/J_1 > -1/4$. The point $J_2/J_1 = -1/4$ is a kind of critical point. For $J_2/J_1 < -1/4$, $J(q)$ is minimized for $q = \pm q_0$ where $q_0 = \cos^{-1}(-J_1/4J_2)$. In the limit of $J_2/J_1 \to -\infty$, the second-neighbor antiferromagnetic coupling is dominant and $q_0 \to \pi/2$. By considering this J_1-J_2 model, we have seen that "competing" interactions which favor different periodicity orders can lead to translational symmetry breaking which is incommensurate with the underlying lattice. One consequence of this is that there is a complex phase which determines the position of the incommensurate order with respect to the underlying lattice, and this phase can vary continuously, corresponding to sliding the incommensurate wave along the direction of the incommensurate wavevector, at no cost in energy. This phase or sliding mode of an incommensurate density wave has many interesting physical consequences which are sensitive to dimensionality, disorder, and temperature. However, to discuss those would take us too far from the problem at hand.

10.10.2 NVO—a multiferroic material

The compound $Ni_3V_2O_8$, which we refer to as NVO, consists of spin 1 Ni^{2+} ions arranged in a so-called "kagomé staircase" structure shown in Fig. (10.11a). In zero applied magnetic field, these spins order in the structure (HTI) shown in Fig. (10.11b), which is an incommensurate spin-density wave with the moments aligned along the crystal a-axis, at a critical temperature $T_H \approx 9K$. At a lower temperature, $T_L \approx 6K$, the spins spontaneously develop a component of order along the crystal b-axis (LTI) as shown in Fig. (10.11c). Simultaneous with the second transition, a uniform electric polarization grows up, also oriented along the crystal b-direction, suggesting that the electric polarization requires the combination of the two magnetic orders to occur. Our goal is to understand why that is the case.

We begin by analyzing the symmetry and symmetry breaking at the high temperature transition. This case is more complicated than the situation illustrated in Fig. (10.5) for two reasons. First, there are several vector spins per unit cell, and, second, the order which develops is incommensurate. The situation is illustrated in Fig. (10.12) for the hypothetical case of two three-component spins per unit cell for a system with orthorhombic symmetry. Here we use a simple approximation to the inverse susceptibility, given by

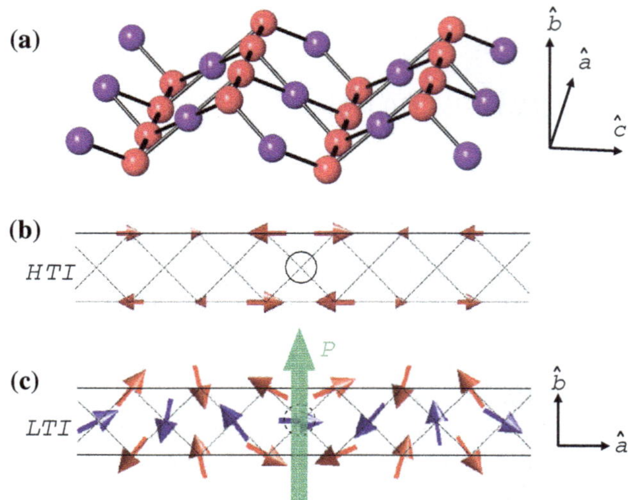

Fig. 10.11 Crystal and magnetic structures of NVO. **a** Crystal structure showing spin-1 Ni^{2+} spine sites in red and crosstie sites in blue. **b**, **c** Simplified schematic representation of the spin arrangements in the antiferromagnetic HTI and LTI phases. (Figure adapted from Lawes et al. (2005))

$$\chi_{\alpha\beta}^{-1}(\mathbf{k}) = kT\delta_{\alpha\beta} + \sum_{\mathbf{r},\boldsymbol{\delta}} J_{\alpha\beta}(\boldsymbol{\delta}) S_\alpha(\mathbf{r}) S_\beta(\mathbf{r}+\boldsymbol{\delta}) \cos(\mathbf{k}\cdot\boldsymbol{\delta}) \,. \tag{10.129}$$

For the orthorhombic case, $\chi_{\alpha\beta}^{-1}(\mathbf{k})$ is diagonal in the indices (α, β). Panel a) shows $\chi_{\alpha\alpha}(\mathbf{k})^{-1}$ for the six spin degrees of freedom per cell and suggests that spin order involving spins along the z-direction will condense at $k = 0$. Panel b) shows the temperature dependence of this lowest mode. Panel c) illustrates the case where the lowest mode is at $k = \pi$, and panel d) illustrates the case of an incommensurate

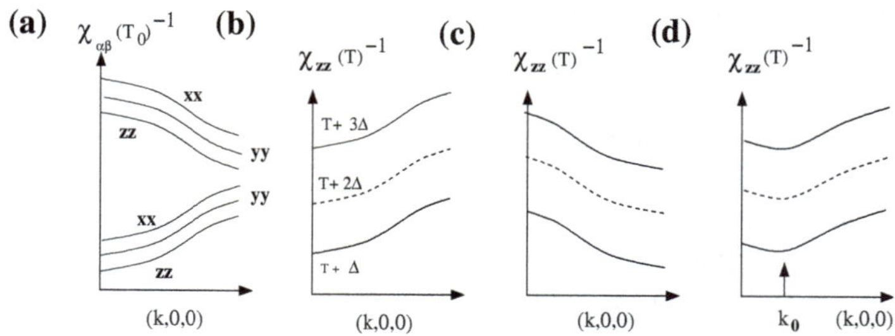

Fig. 10.12 **a** $\chi_{\alpha\alpha}(\mathbf{k})^{-1}$ for **k** along [100] at some temperature T_0 above T_c for $\alpha = x, y, z$, with two spins per unit cell, and therefore six branches to the spectrum; **b** shows the temperature dependence of $\chi_{zz}(T)^{-1}$ when the most unstable mode is ferromagnetic. Similarly, **c** shows $\chi_{zz}(T)^{-1}$ for an antiferromagnet and **d** $\chi_{zz}(T)^{-1}$ for an incommensurate magnet whose wavevector is $(k_0, 0, 0)$

ordering wavevector. Thus, this generalized Landau analysis describes not only which wavevector(s) condense, which is called “wavevector selection,” but also provides information about the nature of the spin order that arises within a unit cell.

For the case of NVO, the spins condense into an incommensurate spin-density wave state (HTI), presumably due to an interaction mechanism similar to that described in the previous subsection, choosing the a-direction for both the incommensurate wavevector and the direction of the spins. The underlying crystal lattice has orthorhombic symmetry, which means that the a, b, and c axes must be orthogonal to each other with three different lattice constants.

10.10.3 Symmetry Analysis

We discussed above the concept of the “star” of the wavevector. The star is the set of symmetry-equivalent wavevectors. For NVO, the spin-density wave is formed with incommensurate wavevector, $\vec{k}$ along $\hat{a}$. The star of this wavevector contains only $\vec{k}$ and $-\vec{k}$ which allows the construction of magnetization at each site that is real. Next, we consider the “group of the wavevector” which is the set of symmetry operations that leave $\vec{k}$ invariant.

Formally speaking, there are four symmetry operations that leave $\vec{k}$ invariant. One, which is trivial, is the identity operation which leaves all vectors invariant. If we call the $\hat{a}$, $\hat{b}$, and $\hat{c}$ directions $\hat{x}$, $\hat{y}$, and $\hat{z}$, then rotation around the $\hat{x}$-direction by 180° (C_{2x}) leaves $\vec{k}$ invariant, as do reflections that take $\hat{y}$ into -$\hat{y}$ and $\hat{z}$ into -$\hat{z}$. We call these two reflection operations σ_{xz} and σ_{xy}, where the subscripts denote the plane across which the reflection is performed.

Table (10.2) shows the effect of these symmetry operations on the components of the vectors, $\vec{r}$ and $\vec{p}$, and the pseudovector, $\vec{L}$. The spin vectors of the spin-1 Ni^{2+}

Table 10.2 Table of symmetry operations which leave x invariant. The top row lists the symmetry operations. The left-hand column contains labels for the different ways that functions can transform under symmetry operations. These are called “irreducible representations” or irreps. The right-hand column lists some objects called basis functions upon which the symmetry operations can act. For example, the irrep Γ_1 transforms like the function x which goes into itself under all the operations of the group. The functions listed are the components of the vectors, $\vec{r}$ and $\vec{p}$, and of the pseudovector, $\vec{L} = \vec{r} \times \vec{p}$. In each case, the components go into $\pm$ themselves. The entries show the actual factor

irrep	E	C_{2x}	σ_{xy}	σ_{xz}	Basis functions
Γ_1	1	1	1	1	x, p_x
Γ_2	1	1	-1	-1	L_x
Γ_3	1	-1	1	-1	y, p_y, L_z
Γ_4	1	-1	-1	1	z, p_z, L_y

ions are angular momenta, which are pseudovectors. One can think of a spin angular momentum as having the same symmetry as $\vec{L} = \vec{r} \times \vec{p}$. Then, for example, if $\vec{r}$ and $\vec{p}$ lie in the x–y plane, then $\vec{L}$ will point along $\hat{z}$. In that case, a reflection through the x–y plane leaves $\vec{r}$, $\vec{p}$, and hence $\vec{L}$ invariant. It is straightforward to work out the effect of the other symmetry operations on $\vec{L}$ by considering the effect that those operations have on $\vec{r}$ and $\vec{p}$ to construct all the entries in Table (10.2).

We should mention that a table such as Table (10.2) is called a "character table" in the language of group theory. The first row of the table lists the operations of the group, the last column lists functions which are called "basis functions for irreducible representations (irreps) of the group", and the entries, as described above, are derived from the way the basis functions transform under the operations of the group. In the first column, the names, Γ_n label the different possible types of symmetry of the irreps and their basis functions. There are many good texts on group theory which explain symmetry groups, character tables, and space groups. Two examples are the one by Tinkham (Tinkham 1964) and another by Dresselhaus et al. (Dresselhaus et al. 2008). Landau and Lifshitz also provide a concise presentation of the basics of group theory in Chap. XIII, "The Symmetry of Crystals," of their book on Statistical Physics.

We next turn to the connection between the information contained in the character table, Table 10.2, for an ordered state described by a wavevector along $\hat{a}$, and the symmetry of the spin order for such a state. Landau explained many years ago that the symmetry of the spin order that develops below T_c, under operations that leave the wavevector unchanged, must correspond to one of the irreducible representations of the group of the wavevector. For example, spin order in which all spins point along $\hat{a}$ or $\hat{x}$ corresponds to the irrep, Γ_2. (Note that the basis function L_x in Table 10.2 has the same symmetry as a spin pointed along $\hat{x}$.) Similarly, having all spins pointed along $\hat{y}$ correspond to the symmetry Γ_3 and spins pointed along $\hat{z}$ have the symmetry Γ_4.

In general, types of order corresponding to different irreps of the group of the wavevector will have different T_cs and the one with the highest T_c will develop first as the temperature is lowered. In the absence of fine-tuning, there is zero probability for order corresponding to more than one of the irreps having exactly the same T_c. However, when two different kinds of order have T_cs that are close together, there can be a sequence of transitions where first one and then the other develops. In that case, since the higher temperature order has a finite value at the lower T_c, nonlinear couplings of the two-order parameters may be important.

We describe the high-temperature HTI structure for NVO by a complex vector $\mathbf{Q_H} = Q_H e^{i\phi_H} \alpha_H^n \boldsymbol{\eta}_\mathbf{H}^n$ where Q_H is the amplitude and ϕ_H is the phase of the complex order parameter, $Q_H e^{i\phi_H}$, α_H^n describes the relative amplitude of the spin order on sublattice n, and $\boldsymbol{\eta}_\mathbf{H}^n$ is a unit polarization vector which, for the HTI phase, points predominantly along $\pm\hat{\mathbf{x}}$ for the spine sites. (α_H^n is very small on the crosstie sites.) Similarly, in the low-temperature LTI phase, the additional vector order parameter is $\mathbf{Q_L} = Q_L e^{i\phi_L} \alpha_L^n \boldsymbol{\eta}_\mathbf{L}^n$ where $Q_L e^{i\phi_L}$ is the complex order parameter for the LTI phase and $\boldsymbol{\eta}_\mathbf{L}^n$points predominantly along + or – $\hat{y}$ for the spine sites and lies in the x–y plane for the crosstie sites.

Since $\mathbf{k}$ and $-\mathbf{k}$ describe equivalent order parameters, we can write the real magnetization as a linear combination of the two. In the LTI phase for site n in unit cell $\mathbf{R}$, the order parameter is a superposition of the two incommensurate orders, $\mathbf{Q_H}$ and $\mathbf{Q_L}$,

$$\mathbf{M}(n, \mathbf{R}) = Q_H \alpha_n^H \eta_{\mathbf{H}} e^{i\mathbf{k}\cdot\mathbf{R}+i\phi_H} + Q_L \alpha_n^L \eta_{\mathbf{L}} e^{i\mathbf{k}\cdot\mathbf{R}+i\phi_L} + c.c. \,, \quad (10.130)$$

where the $+c.c.$, which involves $e^{-i\mathbf{k}\cdot\mathbf{R}}$, ensures that the magnetization is real.

Since the origin of each mode has not so far been fixed, each of the modes $\mathbf{Q_H}$ and $\mathbf{Q_L}$ can be multiplied by an overall factor of $\exp(i\mathbf{k}\cdot\Delta)$, which corresponds to shifting the origin of the incommensurate wave and is equivalent to changing the value of $(\phi_H + \phi_L)/2$. Although this overall phase is arbitrary, the *relative* phase, $\phi_H - \phi_L$, of the two-order parameters is important because it affects the energetics.

Henceforth, we will simplify the problem by dropping the index n, specializing to only one spin per unit cell. This greatly simplified model contains the essential ingredients to describe the coupling of two incommensurate spin-density waves to a uniform electric polarization. The full case of six spins per three-dimensional unit cell for NVO is treated in the literature (Harris 2007).

Imagine turning on the second ordering amplitude, Q_L, in the presence of the first amplitude, Q_H. We assume that the interactions are between fixed length spins. Then it is easier for Q_L, which corresponds to spins oriented along $\hat{y}$ on the spine sites, to develop where Q_H, which corresponds to spins ordering along $\hat{x}$, is near a node. This reasoning suggests that the phases of Q_H and Q_L, ϕ_H and ϕ_L, respectively, are related by $\phi_H - \phi_L = \pm\pi/2$.

Now we write down the lowest terms in the order parameter description of the free energy:

$$\begin{aligned} F = {} & \frac{a_H}{2}|Q_H|^2(T - T_H) + \frac{u_H}{4}|Q_H|^4 + \frac{a_L}{2}|Q_L|^2(T - T_L) + \frac{u_L}{4}|Q_L|^4 \\ & + \frac{v}{2}|Q_H|^2|Q_L|^2 + wQ_H^2Q_L^2\cos 2(\phi_H - \phi_L) \\ & + \frac{1}{2}\sum_\alpha \chi_{E_\alpha}^{-1} P_\alpha^2 + V \,. \end{aligned} \quad (10.131)$$

We have argued, on physical grounds, that Q_H and Q_L are $\pm\pi/2$ out phase. Such a relative phase is favored if $w > 0$ so that the term containing w is minimized when $\cos 2(\phi_H - \phi_L) = -1$. The polarization energy involves χ_{E_α} which is of order unity, and the last term, V, is the magnetoelectric interaction.

To induce a polarization when F is minimized, V should be linear in $\mathbf{P}$. Furthermore, the product of $\mathbf{P}$ and the Qs must be symmetric under the symmetry operations of the full space group of the crystal in the disordered phase. The fact that the polarization turns on at the LTI transition suggests that it should involve at least one power of Q_L.

If we define

$$\mathcal{Q}_X(\mathbf{k}, \mathbf{R}) = Q_X e^{i\phi_X} e^{i\mathbf{k}\cdot\mathbf{R}}, \quad X = H, L. \quad (10.132)$$

Then the leading order contribution to V is of the form

$$V = i \sum_{\mathbf{R},\alpha} \gamma_\alpha P_\alpha [\mathcal{Q}_H^*(\mathbf{k},\mathbf{R})\mathcal{Q}_L(\mathbf{k},\mathbf{R}) - \mathcal{Q}_H(\mathbf{k},\mathbf{R})\mathcal{Q}_L^*(\mathbf{k},\mathbf{R})], \quad (10.133a)$$

$$= N_{uc} \sum_\alpha \gamma_\alpha P_\alpha Q_H Q_L \sin(\phi_H - \phi_L), \quad (10.133b)$$

where N_{uc} is the number of unit cells and γ_α are real. In order to complete the argument and to show which way the polarization actually points (i.e., which γ_α are nonzero), we consider how the order parameters that enter into $P_\alpha Q_H Q_L$, for some α, transform under the symmetry operations that leave the vectors $\pm\mathbf{k}$ invariant which is a subgroup of the full space group.

From Fig. (10.11b), we know that Q_H corresponds to a mode where the spin is oriented along the $\pm\hat{a}$ or $\pm\hat{x}$ direction, and that the effect of nonzero Q_L is to introduce a component of spin along the $\pm\hat{b}$ ($\pm\hat{y}$) directions. From Table (10.2), we see that the effect of the various symmetry operations of the group of the wavevector, $\vec{k}$, on a product such as $Q_H Q_L$, which transforms like the product of pseudovectors, $L_x L_y$, is the same as their effect on the vector component, y. Thus, if the polarization $\mathbf{P}$ is along the crystal $\mathbf{b}$ ($\hat{y}$) axis, then the trilinear product, $Q_H Q_L P_y$ is invariant under those symmetry operations as is required for a term in the free energy. We also note that the term V in Eq. (10.133) is invariant under spatial inversion $\mathcal{I}$, which is also an operation of the full space group and which takes P_α into $-P_\alpha$. From Eq. (10.132), we see that $\mathcal{I}\mathcal{Q}_X = \mathcal{Q}_X^*$ which means that the square bracket in the first line, Eq. (10.133a), also goes to minus itself under inversion so that V is invariant under inversion.

The effect of the trilinear term is to lower the free energy when the three kinds of order, which includes the uniform polarization, $\mathbf{P}$, coexist. In this way, incommensurate spin order couples to uniform electric order. More information about this kind of multiferroic order can be found in (Cheong and Mustovoy 2007).

As mentioned in the second paragraph below Eq. (10.130), our discussion of the superposition of different orders has been greatly simplified by ignoring the sublattice structure of six spins per unit cell in NVO. The reason for this simplification was purely pedagogical. Landau's group theoretical symmetry-based approach is certainly capable of dealing with the full multi-sublattice system. However, the calculations are rather complicated, and, these days, they tend to be done using computer-based tools, such as ISODISTORT (Campbell et al. 2006). The main additional ingredient in a multi-sublattice system like NVO is that the sublattices transform among themselves under the operations of the group of the wavevector. Thus, for example, S_{1x}, the x-component of the spin on sublattice 1, does not transform into $\pm$ itself under group operations, but instead transforms into $\pm$ the x-component of a spin on a different sublattice. This is all worked out in a series of papers by Harris and co-workers (Harris 2007) who also developed a formalism whereby the complex phases of the amplitudes of the spins on different sublattices (which were up to then subject to guesswork) could be fixed by invoking inversion symmetry.

10.11 Summary

The key idea in this chapter, which is due to Landau, is that the term in the free energy which is quadratic in the Fourier components of the various order parameters determines the nature of the local instability of the disordered phase. When this quadratic term is positive definite, the disordered phase is locally stable against the formation of long-range order. A first-order, discontinuous phase transition occurs when there is a global instability to a state in which the order parameters are not infinitesimal. For continuous phase transitions, a local instability occurs as the temperature is lowered and wavevector selection occurs, where the instability in the quadratic form occurs at a single wavevector, or at a set (called a star) of equivalent wavevectors.

For example, for ferromagnetic interactions, wavevector zero is selected, whereas for nearest neighbor antiferromagnetic interactions on a bipartite lattice, a wavevector corresponding to the formation of a two sublattice structure is favored. In the case of vector or tensor order parameters with more than one component, higher-than-quadratic anisotropy terms can select a direction in order-parameter-component space which is favored for order. For vector order parameters in a cubic-symmetry system, fourth-order terms select either (1, 0, 0) or (1, 1, 1) as the easy axis for ordering, depending on the sign of the relevant fourth-order term in the free energy. The sign and anisotropy of the fourth-order terms can be sensitive to coupling between third-order terms which are quadratic in the critical order parameter but linear in some noncritical variable. When the fourth-order term can be adjusted to be zero, then one has a so-called tricritical point where the critical exponents assume anomalous values due to the absence of fourth-order terms in the free energy.

When the star of the critical wavevector contains more than one distinct element, then a rich variety of different orders can occur, including ones, as in the case of the lattice gas on a triangular lattice, that are equivalent to the ordered states of a Potts model, and, in the case of multiferroics, order that couples states with incommensurate magnetic order to uniform electric order. The power of Landau's theory is greatly enhanced by combining the concept of wavevector selection with the power of group-theoretical symmetry analysis.

10.12 Exercises

1. Add to the free energy of Eq. (10.31) a site-dependent field h_i which couples linearly to the order parameter. By calculating $\langle \eta(\mathbf{r}_i) \rangle$ in the presence of this field, express the nonlocal susceptibility χ_{ij} in terms of $F_2(i, j)$, thereby justifying the identification of Eq. (10.32).

2. Suppose the term $b\sigma_1^2\sigma_2$ is replaced by the term $b\sigma_1^3\sigma_2$. How would this term renormalize the free energy when σ_2 is eliminated.

3. (Blume–Emery–Griffiths Model (Blume et al. 1971)) Complete the calculation described in Chap. 5. Show that the tricritical density of ^{3}He is $x = 2/3$, and sketch the phase diagram in the $T - \Delta$ plane.

4. After Eq. (10.106) it was said that $dN/d\mu)_{T,V}$ was a "measure of the isothermal compressibility, κ_T." By considering various forms which are equivalent to $\partial V/\partial \mu)_{T,N}$, show that

$$\kappa_T = \frac{V}{N^2} \frac{\partial N}{\partial \mu}\Big)_{T,V} .$$

5. Demonstrate the equivalence of the lattice gas and Ising models. Write the lattice gas Hamiltonian in terms of Ising spin variables. Calculate the phase diagram in the $\mu - T$ plane for the lattice gas with attractive nearest neighbor interactions.

6. Consider the lattice gas with an attractive nearest neighbor interaction. Show how the phase diagram in the $P - T$ plane resembles that of the liquid–gas transition.

7. Find the renormalized value of B in Eq. (10.123) which results when the cubic coupling between the critical and noncritical variables is taken into account. Hint: proceed as in Chap. 5.

8. (Three-State Potts Model) Calculate the Landau expansion in powers of m for the three-state Potts model discussed in Eq. (9.9).

9. (Four-State Potts Model) Analyze the ferromagnetic four-state Potts model, and compare and contrast your results to the three-state case.

10. Develop mean-field theory for a classical model of diatomic molecules on a surface consisting of molecular sites labeled by i which form a square lattice. Assume that the orientational potential energy is given by

$$V(\{\theta_i\}, \{\phi_i\}) = -V \sum_i \cos 2\theta_i - J \sum_{\langle ij \rangle} \sin\theta_i \sin\theta_j \cos(\phi_i - \phi_j)$$

where the sum over $\langle ij \rangle$ is over pairs of nearest neighboring sites. Obtain the mean-field phase diagram in the V/J-$J/(kT)$ plane. The parameters V and J are not restricted to be positive.

11. Repeat problem #10, but this time do the calculations for a triangular lattice.

12. (ANNNI Model). Consider the Axial Nearest Next-nearest Neighbor Ising model for Ising spins on a simple cubic lattice. To write its Hamiltonian, we specify sites by two indices, the first of which, n, labels the plane perpendicular to the z-axis and the second of which, $\mathbf{r}$, is a vector within the plane perpendicular to the z-axis. Then the Hamiltonian is written as

$$\mathcal{H} = -J_0 \sum_n \sum_{\langle \mathbf{rr}' \rangle} \sigma(n, \mathbf{r})\sigma(n, \mathbf{r}')$$
$$+ \sum_{\mathbf{r}} \sum_n \Big(J_1 \sigma(n, \mathbf{r})\sigma(n+1, \mathbf{r}) + J_2 \sigma(n, \mathbf{r})\sigma(n+2, \mathbf{r}) \Big)$$

where $\langle \mathbf{rr}' \rangle$ indicates a sum over pairs of nearest neighbors in a given z-plane and all the J's are positive. Thus, the spins are coupled ferromagnetically within each z-plane and between planes there are antiferromagnetic first and second neighbor interactions. Construct the Landau expansion up to quadratic order in the order parameters $m(n, \mathbf{r})$ and determine the wavevector of the Fourier component of the magnetization which first becomes unstable as the temperature is lowered as a function of J_1 and J_2.

13. Derive an explicit result for $h_c(T)$, the critical value for the uniform field h as a function of T for the Ising antiferromagnet, shown in Fig. 2.8.

14. Calculate the uniform magnetic susceptibility of the Ising antiferromagnet in zero field as a function of temperature for $0 < T < T_0 = zJ$ where $|m_S| > 0$.

References

A. Aharony, Dependence of universal critical behaviour on symmetry and range of interaction, in *Phase Transitions, Critical Phenomena*, ed. by C. Domb, M. Green, vol. 6 (Academic Press, 1976)

A.J. Berlinsky, W.G. Unruh, W.R. McKinnon, R.R. Haering, Theory of lithium ordering in LixTiS2. Solid State Commun. **31**, 135–138 (1979)

M. Blume, V.J. Emery, R.B. Griffiths, Ising model for the λ transition and phase separation in He^3-He^4 mixtures. Phys. Rev. A **4**, 1071 (1971)

B.J. Campbell, H.T. Stokes, D.E. Tanner, D M. Hatch, ISODISPLACE: an internet tool for exploring structural distortions. J. Appl. Cryst. **39**, 607–614 (2006); and H.T. Stokes, D.M. Hatch, B.J. Campbell, ISODISTORT, ISOTROPY Software Suite, iso.byu.edu

Z. Chen, Z. Lu, J.R. Dahn, Staging phase transitions in LixCoO2. J. Electrochem. Soc. **149**, A1604–A1607 (2002)

S.-W. Cheong, M. Mustovoy, Multiferroics: a magnetic twist for ferroelectricity. Nat. Mater. **6**, 13 (2007)

J. Dahn, Thermodynamics and structure of LixTiS: theory and experiment. Ph.D. Thesis, UBC (1982); J.R. Dahn, D.C. Dahn, R.R. Haering, Elastic energy and staging in intercalation compounds. Solid State Commun. **42**, 179 (1982)

J.R. Dahn, W.R. McKinnon, Lithium intercalation in 2H-LixTaS2, J. Phys. **C17**, 4231 (1984)

M.S. Dresselhaus, G. Dresselhaus, A. Jorio, *Group Theory: Application to the Physics of Condensed Matter* (Springer, 2008)

A.B. Harris, Landau analysis of the symmetry of the magnetic structure and magnetoelectric interaction in multiferroics. Phys. Rev. B **76**, 054447 (2007)

G. Lawes, A.B. Harris, T. Kimura, N. Rogado, R.J. Cava, A. Aharony, O. Entin-Wohlman, T. Yildirim, M. Kenzelmann, C. Broholm, A.P. Ramirez, Magnetically driven ferroelectric order in $Ni_3V_2O_8$. Phys. Rev. Lett. **95**, 087206 (2005)

J.N. Reimers, J.R. Dahn, Electrochemical and in-Situ X-ray diffraction studies of lithium intercalation in LixCoO2. J. Electrochem. Soc. **139**, 2091 (1992)

M. Tinkham, *Group Theory and Quantum Mechanics* (McGraw-Hill, 1964)

Chapter 11
Quantum Fluids

11.1 Introduction

In Chap. 7, we treated the problem of noninteracting gases, including quantum gases where the difference between fermions and bosons leads to drastically different low-temperature properties. In the case of fermions, the ground state is one in which the lowest energy single-particle states are filled, with one fermion per state, and the thermal properties are determined by the density of states at the Fermi energy, the energy of the highest occupied states.

The behavior of noninteracting bosons is very different. In two dimensions, the chemical potential, which is negative, increases monotonically with decreasing temperature, reaching zero at $T = 0$. This means that the 2-D Bose gas has infinite compressibility at $T = 0$, unlike the Fermi gas which has a finite, nonzero compressibility. Even more remarkable is the 3-D case where the chemical potential reaches zero at the Bose–Einstein condensation (BEC) temperature, $T_c > 0$ and remains zero down to $T = 0$, as shown in the right-hand panel of Fig. 7.7.

These two contrasting behaviors of the two species of noninteracting quantum gases suggest that the effects of interactions will also be different for the two cases. Furthermore, the infinite compressibility of the noninteracting Bose gas below T_c implies an instability for attractive interactions and a singular response to weak repulsive interactions which would render the compressibility finite. Clearly, the effect of interactions on both Fermi and Bose gases is important to understand, and mean-field theory is the obvious way to begin.

Fortunately, the formalism that we have developed carries over very directly to the case of quantum gases. However, some of the phenomena which arise from the theory are new and distinctly quantum mechanical. Mean-field theory is based on the idea of a variational density matrix which is a direct product of one-particle density matrices. If $\mathcal{H}_{tr}$ is a sum of one-particle Hamiltonians, then we can write a mean-field variational (or trial) grand canonical density matrix as

$$\rho_{tr} = e^{-\beta(\mathcal{H}_{tr}-\mu\hat{n})}/\mathrm{Tr}e^{-\beta(\mathcal{H}_{tr}-\mu\hat{n})} \tag{11.1}$$

A. J. Berlinsky and A. B. Harris, *Statistical Mechanics*, Graduate Texts in Physics,
https://doi.org/10.1007/978-3-030-28187-8_11

where $\hat{n}$ is the number *operator* with average value, N, and the trace is over all possible values of $\hat{n}$. The role of $\mathcal{H}_{\rm tr}$ is to facilitate the definition of the variational density matrix, $\rho_{\rm tr}$. The parameters in $\mathcal{H}_{\rm tr}$ are the variational parameters in $\rho_{\rm tr}$. $\mathcal{H}_{\rm tr}$ is often referred to as the trial Hamiltonian. We should emphasize that the trial Hamiltonian is not the Hamiltonian of the physical system. It merely provides a convenient way of parametrizing the trial density matrix as a product of single-particle density matrices. Because it is convenient to work with the grand partition function, the density matrix includes the term $-\mu\hat{n}$ which couples the system to a particle reservoir at fixed chemical potential μ. We utilize the variational principle for the grand potential Ω as a function of the parameters in the trial Hamiltonian to determine the optimal mean-field density matrix.

The actual physical problem that we want to solve is that of a gas of interacting quantum particles. We consider the Hamiltonian

$$\mathcal{H} = \mathcal{H}_0 + \mathcal{H}_1 \tag{11.2}$$

where

$$\mathcal{H}_0 = \int d\vec{r} \sum_\alpha \Psi^+(\vec{r}\alpha)\left[\frac{|\vec{p}|^2}{2m} + U_0(\vec{r})\right]\Psi(\vec{r}\alpha) \tag{11.3}$$

$$\begin{aligned}\mathcal{H}_1 &= \frac{1}{2}\int d\vec{r}d\vec{r}\,'\, U(\vec{r}-\vec{r}\prime)\\ &\times \sum_{\alpha,\beta}\Psi^+(\vec{r}\alpha)\Psi^+(\vec{r}\,'\beta)\Psi(\vec{r}\,'\beta)\Psi(\vec{r}\alpha)\end{aligned} \tag{11.4}$$

and where, for simplicity, we have assumed a spin-independent potential and zero electromagnetic field. Here the function $U_0(\vec{r})$ is a one-body potential. In a crystal, it would represent the periodic potential of the ions. In the interest of simplicity, we will take $U_0(\vec{r})$ to be a position-independent constant and subsume it into the chemical potential μ. Note that $\mathcal{H}_1$ contains the quartic pairwise interaction governed by the interaction potential $U(\vec{r})$.

The Hamiltonian $\mathcal{H}$ applies equally to Bose and Fermi particles. The field operators $\Psi^+(\vec{r}\alpha)$ and $\Psi(\vec{r}\alpha)$ create or destroy a particle with spin α at the point $\vec{r}$. If the particles are bosons, then the field operators obey commutation relations

$$\left[\Psi^+(\vec{r}\alpha),\ \Psi(\vec{r}\,'\beta)\right] = \delta_{\alpha,\beta}\delta(\vec{r}-\vec{r}\,') \tag{11.5}$$

$$\left[\Psi(\vec{r}\alpha)\,,\ \Psi(\vec{r}\,'\beta)\right] = 0\,. \tag{11.6}$$

If the particles are fermions, then the field operators obey the analogous anticommutation relations.

$\mathcal{H}_0$ has the form of a general sum of one-particle Hamiltonians for a system of identical particles. It is not diagonal in position space because the momentum operator implicitly couples the values of the field at neighboring points in space. To reduce $\mathcal{H}_0$ to a diagonal form, we Fourier transform the field operators by writing

$$\Psi^+(\vec{r}\alpha) = \frac{1}{\sqrt{V}} \sum_{\vec{k}} a^+_{\vec{k}\alpha}\, e^{i\vec{k}\cdot\vec{r}} \,, \tag{11.7}$$

where the operators $a^+_{\vec{k}\alpha}$ and $a_{\vec{k}\alpha}$, respectively, create and destroy particles of spin α with momentum $\hbar\vec{k}$. The allowed values of the wavevector $\vec{k}$ may conveniently be determined by periodic boundary conditions with respect to the volume V. (In this connection, recall the results of Exercise 8 of Chap. 7.) These operators obey the appropriate Bose (−) or Fermi (+) commutation relations:

$$\left[a_{\vec{k}\alpha}, a^+_{\vec{k}'\beta}\right]_\mp = \delta_{\vec{k},\vec{k}'}\delta_{\alpha\beta} \,, \qquad \left[a_{\vec{k}\alpha}, a_{\vec{k}'\beta}\right]_\mp = 0 \,, \tag{11.8}$$

where $[A, B]_\mp = AB \mp BA$.

Then

$$\mathcal{H}_0 = \sum_{\vec{k},\alpha} \left(\frac{\hbar^2 k^2}{2m}\right) a^+_{\vec{k}\alpha} a_{\vec{k}\alpha} \equiv \sum_{\vec{k},\alpha} E_0(\vec{k})\hat{n}_{\vec{k}\alpha} \tag{11.9}$$

$$\mathcal{H}_1 = \frac{1}{2V} \sum_{\vec{k}_1,\vec{k}_2,\vec{q}} \sum_{\alpha,\beta} U(q) a^+_{\vec{k}_1-\vec{q},\alpha} a^+_{\vec{k}_2+\vec{q},\beta} a_{\vec{k}_2,\beta} a_{\vec{k}_1,\alpha} , \tag{11.10}$$

where we introduced the notation, $E_0(\vec{k}) \equiv \hbar^2 k^2/2m$ and $\hat{n}_{\vec{k}\alpha} \equiv a^+_{\vec{k}\alpha} a_{\vec{k}\alpha}$.

The mean-field variational approach consists of choosing a trial density matrix of the form of Eq. (11.1) in terms of a trial Hamiltonian, $\mathcal{H}_{\rm tr}$, which for the case of "normal" quantum fluids (as opposed to superfluids) is taken to have the form

$$\mathcal{H}_{\rm tr} = \sum_{\vec{k},\alpha} E_{\rm tr}(\vec{k})\hat{n}_{\vec{k}\alpha} \,. \tag{11.11}$$

Note that this Hamiltonian is diagonal in the occupation number representation. The trial parameters in $\mathcal{H}_{\rm tr}$, namely, the $E_{\rm tr}(\vec{k})$, will be determined by minimizing the mean-field trial free energy (actually the grand potential)

$$\Omega_{\rm tr} = {\rm Tr}\,[\rho_{\rm tr}(\mathcal{H}_0 + \mathcal{H}_1 - \mu\hat{n})] + kT\ {\rm Tr}\,[\rho_{\rm tr} \ln \rho_{\rm tr}] \,. \tag{11.12}$$

Note the distinction between the variational parameters $\{E_{\rm tr}(\vec{k})\}$ and the kinetic energy $E_0(\vec{k})$. The form of Eq. (11.11) is quite general for translationally invariant states of a normal quantum fluid. Nevertheless, Eq. (11.11) is not the most general quadratic form that we could have written, since we could have added terms which create or destroy pairs of particles. Such terms are important in the theory of superfluids which will be discussed later.

The entropy term in $\Omega_{\rm tr}$ in Eq. (11.12) can be written in a variety of ways. It can be expressed in terms of the variational parameters, $E_{\rm tr}(\vec{k})$. Alternatively, we may express it in terms of occupation numbers as follows. We write

$$
\begin{aligned}
-TS &= kT\ \mathrm{Tr}\,[\rho_{\mathrm{tr}} \ln \rho_{\mathrm{tr}}] \\
&= kT\ \mathrm{Tr}\left\{ \frac{e^{-\beta(\mathcal{H}_{\mathrm{tr}}-\mu\hat{n})}}{\mathrm{Tr}\ \left[e^{-\beta(\mathcal{H}_{\mathrm{tr}}-\mu\hat{n})}\right]} \right. \\
&\qquad \left. \times \left(-\beta(\mathcal{H}_{\mathrm{tr}}-\mu\hat{n}) - \ln\left[\mathrm{Tr}\ e^{-\beta(\mathcal{H}_{\mathrm{tr}}-\mu\hat{n})}\right]\right)\right\} \\
&= -\langle \mathcal{H}_{\mathrm{tr}}-\mu\hat{n}\rangle_{\mathrm{tr}} - kT \ln\left[\mathrm{Tr}\ e^{-\beta(\mathcal{H}_{\mathrm{tr}}-\mu\hat{n})}\right] , \qquad (11.13)
\end{aligned}
$$

where $\langle\ \rangle_{\mathrm{tr}}$ denotes an average with respect to the trial density matrix of Eq. (11.1). For instance

$$
\langle \mathcal{H}_0 - \mu\hat{n}\rangle_{\mathrm{tr}} = \sum_{\vec{k},\alpha}\big[E_0(\vec{k})-\mu\big]\langle \hat{n}_{\vec{k}\alpha}\rangle_{\mathrm{tr}} = \sum_{\vec{k},\alpha}\big[E_0(\vec{k})-\mu\big] n_{\vec{k}}\,. \qquad (11.14)
$$

Here we introduced the thermodynamic average of $\hat{n}_{\vec{k}\alpha}$:

$$
n_{\vec{k}} \equiv \langle \hat{n}_{\vec{k}\alpha}\rangle_{\mathrm{tr}} = \frac{1}{e^{\beta[E_{\mathrm{tr}}(\vec{k})-\mu]} \mp 1}\,. \qquad (11.15)
$$

For later use we note that

$$
\frac{1 \pm n_{\vec{k}}}{n_{\vec{k}}} = e^{\beta[E_{\mathrm{tr}}(\vec{k})-\mu]}\,. \qquad (11.16)
$$

Also,

$$
\begin{aligned}
-kT \ln\left[\mathrm{Tr}\ e^{-\beta(\mathcal{H}_{\mathrm{tr}}-\mu\hat{n})}\right] &= -kT \ln\left[\mathrm{Tr}\ e^{-\beta\sum_{\vec{k},\alpha}[E(\vec{k})-\mu]\hat{n}_{\vec{k}\alpha}}\right] \\
&= -kT \sum_{\vec{k},\alpha} \ln\left[\sum_n e^{-n\beta[E(\vec{k})-\mu]}\right] \\
&= -kT \sum_{\vec{k},\alpha} \ln[1+n_{\vec{k}}] \quad \text{for bosons}\,, \qquad (11.17) \\
&= \quad kT \sum_{\vec{k},\alpha} \ln[1-n_{\vec{k}}] \quad \text{for fermions}\,. \qquad (11.18)
\end{aligned}
$$

Putting these results together we get

$$
-TS = kT \sum_{\vec{k},\alpha}\left[n_{\vec{k}} \ln n_{\vec{k}} \mp (1 \pm n_{\vec{k}}) \ln(1 \pm n_{\vec{k}})\right]\,. \qquad (11.19)
$$

The advantage of this representation is that now we may treat the $n_{\vec{k}}$ as variational parameters.

Next we need to calculate $\langle \mathcal{H}_1 \rangle_{\rm tr}$. This involves evaluating averages of the form

$$\langle a^+_{\vec{k}_1-\vec{q},\alpha} a^+_{\vec{k}_2+\vec{q},\beta} a_{\vec{k}_2,\beta} a_{\vec{k}_1,\alpha} \rangle_{\rm tr} \ .$$

Since, in the occupation number representation, the averages are a sum of diagonal matrix elements, it is clear that, whatever the last two operators, $a_{\vec{k}_2,\beta} a_{\vec{k}_1,\alpha}$, do, the first two operators, $a^+_{\vec{k}_1-\vec{q},\alpha} a^+_{\vec{k}_2+\vec{q},\beta}$, must undo. We distinguish between two cases, $(\vec{k}_2, \beta) \neq (\vec{k}_1, \alpha)$ and $(\vec{k}_2, \beta) = (\vec{k}_1, \alpha)$. Then

$$\begin{aligned} \langle a^+_{\vec{k}_1-\vec{q},\alpha} a^+_{\vec{k}_2+\vec{q},\beta} a_{\vec{k}_2,\beta} a_{\vec{k}_1,\alpha} \rangle_{\rm tr} &= \left(1 - \delta_{\vec{k}_1,\vec{k}_2}\delta_{\alpha,\beta}\right) \\ \times \Big[\langle a^+_{\vec{k}_1-\vec{q},\alpha} a_{\vec{k}_1,\alpha}\rangle_{\rm tr} \langle a^+_{\vec{k}_2+\vec{q},\beta} a_{\vec{k}_2,\beta}\rangle_{\rm tr} &\pm \langle a^+_{\vec{k}_1-\vec{q},\alpha} a_{\vec{k}_2,\beta}\rangle_{\rm tr} \langle a^+_{\vec{k}_2+\vec{q},\beta} a_{\vec{k}_1,\alpha}\rangle_{\rm tr} \Big] \\ + \delta_{\vec{q},0}\delta_{\vec{k}_1,\vec{k}_2}\delta_{\alpha,\beta} \langle a^+_{\vec{k}_1,\alpha} a^+_{\vec{k}_1,\alpha} a_{\vec{k}_1,\alpha} a_{\vec{k}_1,\alpha}\rangle_{\rm tr} \ . \end{aligned} \tag{11.20}$$

The last average is zero for fermions, because it involves having two particles in the same state.

Evaluating the quadratic averages, we find that for bosons the last term in Eq. (11.20) is proportional to

$$\begin{aligned} C &\equiv \langle \hat{n}_{\vec{k}_1,\alpha} (\hat{n}_{\vec{k}_1,\alpha} - 1) \rangle_{\rm tr} \\ &= \frac{\sum_n e^{-n\beta[E_{\rm tr}(\vec{k}_1)-\mu]} n(n-1)}{\sum_n e^{-n\beta[E_0(\vec{k}_1)-\mu]}} \ . \end{aligned} \tag{11.21}$$

If we let x denote $\exp\{-\beta[E_{\rm tr}(\vec{k}_1) - \mu]\}$, then we have

$$\begin{aligned} C &= \sum_n n(n-1)x^n \Big/ \sum_n x^n = \frac{2x^2}{(1-x)^2} \\ &= 2n^2_{\vec{k}_1} \ . \end{aligned} \tag{11.22}$$

Thus, for bosons (as indicated by the superscript "B") Eq. (11.20) is

$$\begin{aligned} \langle a^+_{\vec{k}_1-\vec{q},\alpha} a^+_{\vec{k}_2+\vec{q},\beta} a_{\vec{k}_2,\beta} a_{\vec{k}_1,\alpha} \rangle^B_{\rm tr} &= \left(1 - \delta_{\vec{k}_1,\vec{k}_2}\delta_{\alpha,\beta}\right) \\ \times \left[n_{\vec{k}_1} n_{\vec{k}_2} \delta_{\vec{q},0} + n_{\vec{k}_1} n_{\vec{k}_2} \delta_{\vec{q},\vec{k}_1-k_2} \delta_{\alpha\beta} \right] &+ 2\delta_{\vec{k}_1,\vec{k}_2} \delta_{q,0} \delta_{\alpha\beta} n^2_{\vec{k}_1} \\ = n_{\vec{k}_1} n_{\vec{k}_2} \delta_{\vec{q},0} + n_{\vec{k}_1} n_{\vec{k}_2} \delta_{\vec{q},\vec{k}_1-k_2} \delta_{\alpha\beta} \ . \end{aligned} \tag{11.23}$$

For fermions (as indicated by the superscript "F"),

$$\langle a^+_{\vec{k}_1-\vec{q},\alpha} a^+_{\vec{k}_2+\vec{q},\beta} a_{\vec{k}_2,\beta} a_{\vec{k}_1,\alpha} \rangle^F_{\rm tr} = \left(1 - \delta_{\vec{k}_1,\vec{k}_2}\delta_{\alpha,\beta}\right)$$

$$\times \left[n_{\vec{k}_1} n_{\vec{k}_2} \delta_{\vec{q},0} - n_{\vec{k}_1} n_{\vec{k}_2} \delta_{\vec{q},\vec{k}_1 - \vec{k}_2} \delta_{\alpha,\beta} \right]$$

$$= n_{\vec{k}_1} n_{\vec{k}_2} \delta_{\vec{q},0} - n_{\vec{k}_1} n_{\vec{k}_2} \delta_{\vec{q},\vec{k}_1 - \vec{k}_2} \delta_{\alpha,\beta} \,, \tag{11.24}$$

which vanishes, as it must, for $(\vec{k}_2, \beta) = (\vec{k}_1, \alpha)$.

We may write the results of Eqs. (11.24) and (11.23) as

$$\langle a^+_{\vec{k}_1 - \vec{q},\alpha} a^+_{\vec{k}_2 + \vec{q},\beta} a_{\vec{k}_2,\beta} a_{\vec{k}_1,\alpha} \rangle_{\rm tr} = n_{\vec{k}_1} n_{\vec{k}_2} \delta_{\vec{q},0} \pm n_{\vec{k}_1} n_{\vec{k}_2} \delta_{\vec{q},\vec{k}_1 - k_2} \delta_{\alpha\beta} \,, \tag{11.25}$$

for bosons (+) and fermions (−).

This result is an example of Wick's theorem which says that the grand canonical average of a product of Fermi or Bose operators, with respect to a Hamiltonian that is quadratic in those operators, is equal to the sum of all possible products of averages of pairs of those operators (keeping track of the minus signs that result from permuting the order of fermion operators). (See also the appendix of Chap. 20 for the analogous result for Gaussian averages of c-numbers.) Thus, Wick's theorem allows us to write Eq. (11.25) without any calculation at all.

11.2 The Interacting Fermi Gas

Making use of Eqs. (11.14), (11.19), and (11.24), we can write Eq. (11.12) for the total variational free energy of the Fermi gas as

$$\begin{aligned} \Omega_{\rm tr} &= \sum_{\vec{k},\alpha} \left[E_0(\vec{k}) - \mu \right] n_{\vec{k}} \\ &+ \frac{1}{2V} \sum_{\vec{k}_1,\alpha,\vec{k}_2,\beta} \left[U(0) n_{\vec{k}_1} n_{\vec{k}_2} - U(\vec{k}_1 - \vec{k}_2) n_{\vec{k}_1} n_{\vec{k}_2} \delta_{\alpha,\beta} \right] \\ &+ kT \sum_{\vec{k},\alpha} \left[n_{\vec{k}} \ln n_{\vec{k}} + (1 - n_{\vec{k}}) \ln(1 - n_{\vec{k}}) \right] . \end{aligned} \tag{11.26}$$

Setting the variation of $\Omega_{\rm tr}$ with respect to $n_{\vec{k}}$ equal to zero gives

$$\begin{aligned} \frac{1}{2S+1} \frac{\partial \Omega_{\rm tr}}{\partial n_{\vec{k}}} &= E_0(\vec{k}) - \mu + \frac{U(0)}{V} \sum_{\vec{k}',\beta} n_{\vec{k}'} \\ &- \frac{1}{V} \sum_{\vec{k}'} U(\vec{k} - \vec{k}') n_{\vec{k}'} + kT \ln \left(\frac{n_{\vec{k}}}{1 - n_{\vec{k}}} \right) \\ &= 0 \,. \end{aligned} \tag{11.27}$$

We set

$$n_{\vec{k}} = \frac{1}{e^{\beta[E_{\rm tr}(\vec{k})-\mu]} + 1} , \qquad (11.28)$$

so that $E_{\rm tr}(\vec{k})$ is identified as the energy of a quasiparticle of wavevector $\vec{k}$. Equation (11.16) indicates that the last term in Eq. (11.27) is $-[E_{\rm tr}(\vec{k}) - \mu]$, so that

$$E_{\rm tr}(\vec{k}) = E_0(\vec{k}) + \left(U(0) - \frac{\overline{U(\vec{k})}}{2S+1} \right) n , \qquad (11.29)$$

where

$$\overline{U(\vec{k})} = \frac{2S+1}{N} \sum_{\vec{k}'} U(\vec{k} - \vec{k}') n_{\vec{k}'} \qquad (11.30)$$

and $n = N/V$ is the average density of particles since $N = \sum_{\vec{k},\alpha} n_{\vec{k}}$. Since $n_{\vec{k}'}$ depends on $E_{\rm tr}(\vec{k}')$, Eq. (11.29) is a self-consistent equation to determine the trial quasiparticle energies $E_{\rm tr}(\vec{k})$.

The following points about Eqs. (11.29) and (11.30) warrant comment:

1. The interaction term $U(0)n$ is called the “direct” or “Hartree” term. It represents the direct interaction of one fermion with the average density of other particles.

2. The second interaction term is the “exchange” or “Fock” term. It describes additional interactions among particles with the same spin quantum number. For a given potential, the sign of the exchange interaction is opposite for bosons and fermions.

3. An important model interaction is the point or δ-function interaction. The Fourier transform of this interaction is independent of wavevector, which means that $\overline{U(\vec{k})} = U(0)$ for a point interaction. Then for the case $S = 0$, hypothetical “spinless fermions” feel no effect from a point interaction (because of the exclusion principle). More generally, the interaction term is $U(0)n[2S/(2S+1)]$ for a δ-function interaction between spin S fermions. [We will see later that spin-0 bosons feel an interaction, $2U(0)n$.]

4. For the case of a Coulomb interaction, the particles will usually also interact with a neutralizing background charge (e.g., the ionic charge in a solid) which can be treated as uniform. Then the $\vec{q} = 0$ interaction energy, $U(0)$, is canceled, and the Coulomb contribution to the quasiparticle energy, $E_{\rm tr}(\vec{k})$, comes completely from the exchange term.

11.3 Fermi Liquid Theory

The most important result of the previous section is that, for the mean-field theory defined by Eqs. (11.1) and (11.11), the effect of interactions in a system of fermions is to modify the effective single-particle dispersion relation. The Hartree term shifts the chemical potential, and the Fock term, which arises from exchange, i.e., particle statistics, changes the energy–momentum relation for the single-particle states. The fermionic states described by the modified dispersion relation are called "quasiparticle states." The mean-field ground state corresponds to occupying the N lowest energy quasiparticle states, and an "excitation" corresponds to moving a quasiparticle from an occupied state to an empty one. Such excitations determine the thermal properties of interacting fermions. It turns out that, for repulsive interactions, this version of Hartree–Fock theory for fermions works remarkably well. On the other hand, for attractive interactions, the fermionic ground state is unstable against the more stable superconducting state which will be discussed in the next chapter.

A more general theory of interacting fermions, called Landau's Fermi liquid theory, was developed by Landau in the late 1950s (Landau 1957a, b, 1958). The basic idea of Fermi liquid theory is that, as interactions are turned on, the noninteracting system evolves continuously into a qualitatively similar system in which the main effect of interactions is to renormalize the single-particle dispersion relation. This means that the low-lying thermal excitations correspond to particle–hole excitations as in the noninteracting case. The quasiparticles themselves are spin 1/2 objects which carry a single electron charge. The main effect of interactions, the change in their energy–momentum relation, $E_{qp}(\vec{k}) - \mu$, can be described in terms of an effective mass, $E_{qp}(\vec{k}) - \mu \approx \hbar^2 k^2/2m^* - \mu$. Fermi liquid theory goes beyond simple Hartree–Fock by introducing a "lifetime" or (inverse) width to the quasiparticle energy. Remarkably, it can be shown that, because of the Pauli exclusion principle, the width of the quasiparticle energies goes like $(E_{qp}(\vec{k}) - \mu)^2 + (k_B T)^2$ so that, at low temperatures, quasiparticle states within $k_B T$ of the Fermi energy are well defined with very long lifetimes.

Fermi liquid theory is also useful, even in the presence of interactions that favor some form of superconductivity, for describing the high-temperature "normal" state from which superconductivity condenses. An example is the very interesting kind of superconductivity that occurs in liquid ^{3}He. The review by Leggett (1975) describes both Fermi liquid theory and the theory of superconductivity and applies them to the case of superfluid ^{3}He.

Fermi liquid theory works well for most metals as long as the Fermi energy which is of the order electronic bandwidth is large compared to the scale of electron–electron interactions. This is the case in so-called "conventional metals," such as copper, aluminum, or gold. The theory can be extended to treat systems, such as iron, where the exchange interaction leads to itinerant ferromagnetism. (The theory here might be called something like "spin-dependent, self-consistent Hartree–Fock.") However, Fermi liquid theory breaks down in narrowband systems where interaction effects dominate. An example is the so-called "heavy fermion materials" where transport is incoherent at high temperatures and becomes coherent and Fermi-liquid-like at low temperatures but with extremely large effective masses. Another case where Fermi liquid theory breaks down is in high-temperature superconductors. High T_c materials exhibit a linear resistivity at high temperatures and transport is quite incoherent. As the temperature is lowered, instead of going into a heavy fermion state, they are condensed into a high-temperature superconducting state.

We will not pursue the theory of Fermi liquids any further, which would take us too far into quantum many-body theory and condensed matter physics. We will, however, extend our mean field, Hartree–Fock theory to treat superfluids and superconductors below.

11.4 Spin-Zero Bose Gas with Short-Range Interactions

We now consider the mean-field treatment of an interacting Bose gas of spin-zero particles with short-range interactions. Later, we will treat the Bose-condensed phase. However, here we start by assuming a normal state in which no single-particle occupation numbers are macroscopic. In this case, the analog of Eq. (11.26) is

$$\Omega_{\rm tr} = \sum_{\vec{k}}[E_0(\vec{k}) - \mu]n_{\vec{k}} + \frac{1}{2V}\sum_{\vec{k},\vec{k}'}\left(U(0) + U(\vec{k}-\vec{k}')\right)n_{\vec{k}}n_{\vec{k}'}$$
$$+kT\sum_{\vec{k}}\left[n_{\vec{k}}\ln n_{\vec{k}} - (1+n_{\vec{k}})\ln(1+n_{\vec{k}})\right] , \qquad (11.31)$$

where we have used Eqs. (11.19) and (11.23). Differentiation with respect to $n_{\vec{k}}$ gives

$$-kT\ln\frac{n_{\vec{k}}}{1+n_{\vec{k}}} = E_0(\vec{k}) - \mu + \frac{1}{V}\sum_{\vec{k}'}\left(U(0) + U(\vec{k}-\vec{k}')\right)n_{\vec{k}'} . \qquad (11.32)$$

In the discussion which follows, we will always be considering potentials which are in some sense "short range." For simplicity, we consider purely repulsive, spherical potentials. If the potential is nonzero over a range b, then its Fourier transform, $U(\vec{k})$, is a constant approximately equal to $U(0)$ for $k \ll 1/b$ and falls to zero for $k \gg 1/b$. One length scale relevant to bosons at finite temperature is their thermal de Broglie wavelength, $\Lambda \equiv 2\pi\hbar/\sqrt{2\pi mkT}$. For $k \gg 1/\Lambda$, $n_{\vec{k}}$, defined in Eq. (11.15) with the minus sign, is a rapidly decreasing function of $|k|$. If the range of the potential is much smaller than the thermal de Broglie wavelength, then $U(\vec{k}) \approx U(0)$ for $0 < k < 1/b$ and $b << \Lambda$. For a low-density, degenerate atomic Bose gas, this is not a very stringent condition. The system is degenerate if the de Broglie wavelength is comparable to or larger than the interatomic spacing. The density is low if the interatomic spacing is large compared to the range of the potential. Under these conditions, it is reasonable to replace $U(\vec{k})$ by $U(0)$ in Eq. (11.32). Then, using Eq. (11.16) we identify the left-hand side of Eq. (11.32) as $E_{\rm tr}(\vec{k}) - \mu$, so that

$$E_{\rm tr}(\vec{k}) = E_0(\vec{k}) + 2U(0)n . \qquad (11.33)$$

In terms of the $n_{\vec{k}}$, the free energy becomes

$$\Omega = \sum_{\vec{k}} \left\{ E_0(\vec{k}) - \mu + 2U(0)n \right\} n_{\vec{k}} - U(0)N^2/V$$
$$+ kT \sum_{\vec{k}} \left[n_{\vec{k}} \ln n_{\vec{k}} - (1 + n_{\vec{k}}) \ln(1 + n_{\vec{k}}) \right] . \qquad (11.34)$$

To obtain this equation, we have added and subtracted $U(0)N^2/V \equiv U(0)n \sum_{\vec{k}} n_{\vec{k}}$ in order to allow us to write Ω in terms of the grand potential of the ideal Bose gas (IBG). Also, since $n_{\vec{k}}$ is given by Eq. (11.15), we identify $E_{\rm tr}(\vec{k})$ as the energy of a quasiparticle of wavevector $\vec{k}$. Explicitly we have

$$n_{\vec{k}} = \frac{1}{e^{\beta[E_0(\vec{k})+2U(0)n-\mu]} - 1} . \qquad (11.35)$$

This result shows that $n_{\vec{k}}$ may be identified as that of an IBG at chemical potential, $\tilde{\mu} \equiv \mu - 2U(0)n$. Thus, the sum of the first and last terms in Eq. (11.34) is equal to the grand potential of an IBG at chemical potential, $\tilde{\mu}$:

$$\Omega(\mu, T, V) = \Omega_{IBG}(\tilde{\mu}, T, V) - U(0)N^2/V . \qquad (11.36)$$

Note: the right-hand side of this equation is a function of μ, T, and V, because $n_{\vec{k}}$ is a function of μ, T, and V via Eq. (11.18). See also Exercise 5.

The Helmholtz free energy for a system of N particles is $F = \Omega + \mu N$ or

$$F(N, T, V) = F_{IBG}(N, T, V) + \frac{U(0)}{V} N^2 , \qquad (11.37)$$

where

$$F_{IBG}(N, T, V) = \Omega_{IBG}(\tilde{\mu}, T, V) + \tilde{\mu} N . \qquad (11.38)$$

Given F, we calculate the pressure at fixed T and N as

$$P = -\frac{\partial F}{\partial V} = P_{IBG} + U(0)n^2 . \qquad (11.39)$$

This result gives the correction to the equation of state due to interactions between particles. Note that so far we have not specified whether the interaction is repulsive or attractive. As long as the system is not Bose condensed, the effect of interactions is simply to increase or decrease the pressure as specified in Eq. (11.39), depending on the sign of $U(0)$.

11.5 The Bose Superfluid

For a noninteracting Bose gas, we saw in Chap. 7 that the phase diagram allows for the existence of a "normal" gaseous phase and a Bose-condensed phase in which the lowest energy single-particle state is macroscopically occupied. We expect that the possibility of a "normal" phase and a Bose-condensed phase might persist in the presence of weak interactions between particles. However, what actually happens will depend on the nature of the interactions. In general, most systems become solids rather than gases, liquids, or superfluids when cooled to very low temperature. The most striking exception for bosonic atoms is the case of liquid helium 4, which becomes a Bose superfluid at temperatures below about 2 K. The reason this system becomes superfluid rather than solid is because He^4 atoms are too light to be localized by the very shallow He–He interatomic potential. Quantum mechanics tells us that the potential energy required for localizing an atom is inversely proportional to the atomic mass, and for helium 4 this energy is large enough to prevent solidification at the density of liquid helium at atmospheric pressure. As the pressure is increased, the potential well due to interatomic interactions becomes deeper and quantum effects become less important. Thus, increasing the pressure eventually stabilizes the solid phase even for helium.

Helium 4 is a relatively dense liquid—not a dilute gas. For simplicity, we will consider the case of a dilute gas of spin-zero bosons, with single-particle states labeled by momentum. Our earlier calculation in Sect. 11.4 of the free energy of the spin-zero Bose gas with short-range interactions implicitly assumed that the ground state was not macroscopically occupied. We know that for the noninteracting Bose system, macroscopic occupation of the ground state leads to a distinct phase which is infinitely compressible. Such a state is unstable against attractive interactions, but might be only moderately perturbed by weak repulsive interactions when the density of atoms is sufficiently low. So we now consider how macroscopic occupation of the ground state is affected by weak, short-range repulsive interactions.

First, note that matrix elements of the operators a_0 and a_0^+ are of order $\sqrt{N_0}$, where N_0 is the number of particles in the single-particle ground state: $N_0 \equiv \langle a_0^+ a_0 \rangle$. In the supposed Bose-condensed phase, N_0 is extensive, i.e., it is proportional to the volume V. This suggests that we should display these operators explicitly in the interaction Hamiltonian. Thus, we write

$$
\begin{aligned}
\mathcal{H}_1 = \frac{1}{2V} & \sum_{\vec{k}_1,\vec{k}_2,\vec{q}} U(q) a^+_{\vec{k}_1-\vec{q}} a^+_{\vec{k}_2+\vec{q}} a_{\vec{k}_2} a_{\vec{k}_1} \\
& + \frac{a_0}{V} \sum_{\vec{k}_1,\vec{q}} U(q) a^+_{\vec{k}_1-\vec{q}} a^+_{\vec{q}} a_{\vec{k}_1} + \frac{a_0^+}{V} \sum_{\vec{k}_1,\vec{q}} U(q) a^+_{\vec{k}_1-\vec{q}} a_{-\vec{q}} a_{\vec{k}_1} \\
& + \frac{a_0^+ a_0^+}{2V} \sum_{\vec{q}} U(q) a_{\vec{q}} a_{-\vec{q}} + \frac{a_0 a_0}{2V} \sum_{\vec{q}} U(q) a^+_{-\vec{q}} a^+_{\vec{q}}
\end{aligned}
$$

$$+\frac{a_0^+ a_0}{V} \sum_{\vec{q}} [U(q) + U(0)] a_{\vec{q}}^+ a_{\vec{q}} + \frac{U(0)}{V} a_0^+ a_0^+ a_0 a_0 \,, \tag{11.40}$$

where the sums over momenta now exclude states with $k = 0$. Note that, because of momentum conservation, there is no cubic term in the operators a_0 and a_0^+. Since $a_0^+ a_0$ is macroscopic, its fluctuations are of relative order $1/V$, and consequently we replace this operator by the "c"-number N_0. To make further progress we will restrict our analysis to the case where the interactions are assumed to have only a small effect on the macroscopically occupied ground state. In other words, we assume that the density of particles *not* in the ground state is small.

Since their density is small, we can ignore the quartic and cubic interaction terms between particles with $k \neq 0$. The remaining terms involve various quadratic combinations of a_0 and a_0^+ along with an N_0^2/V term. As we shall see, the terms involving $(a_0)^2$ and $(a_0^+)^2$ are the crucial new ingredients we need to incorporate into the theory. Accordingly, we write the interaction Hamiltonian as

$$\begin{aligned}\mathcal{H}_1 = &\frac{a_0^+ a_0^+}{2V} \sum_{\vec{q}} U(q) a_{\vec{q}} a_{-\vec{q}} + \frac{a_0 a_0}{2V} \sum_{\vec{q}} U(q) a_{-\vec{q}}^+ a_{\vec{q}}^+ \\ &+\frac{N_0}{V} \sum_{\vec{q}} [U(q) + U(0)] a_{\vec{q}}^+ a_{\vec{q}} + \frac{N_0^2}{V} U(0) \,. \end{aligned} \tag{11.41}$$

The purpose of this discussion is to elucidate the correlations which must be incorporated into a good trial density matrix. From Eq. (11.41), we are led to introduce into $\mathcal{H}_{\rm tr}$ a term of the form

$$1/2 \sum_{\vec{k}} \Gamma_{\vec{k}} \left(a_{\vec{k}}^+ a_{-\vec{k}}^+ + a_{-\vec{k}} a_{\vec{k}} \right) , \tag{11.42}$$

With this term included, the variational Hamiltonian no longer conserves the number of particles. Of course, the actual spontaneous production of helium atoms would require the absurdly large energy of a few GeVs. However, creation (or annihilation) of a pair of particles with wavevectors $\vec{k}$ and $-\vec{k}$ is in reality accompanied by the concomitant annihilation (or creation) of two $q = 0$ particles, i.e., by a compensating change in N_0. Since manifolds having different numbers of $q \neq 0$ particles are distinct, it is actually not necessary to explicitly keep track of how N_0 varies as $q \neq 0$ particles are created or destroyed. That is why replacing a_0 and a_0^+ by $\sqrt{N_0}$ works. To get a complete, physically correct wave function which explicitly conserves particles, one simply multiplies each component of the wave function having n $q \neq 0$ particles by a wave function having $N - n$ particles in the $q = 0$ single-particle state.

11.5.1 Diagonalizing the Effective Hamiltonian

These arguments motivate us to consider a more general variational Hamiltonian

$$\mathcal{H}_{\rm tr} = \sum_{\vec{k}} E_{\rm tr}(\vec{k}) a_{\vec{k}}^{+} a_{\vec{k}} + 1/2 \sum_{\vec{k}} \Gamma_{\vec{k}} \left(a_{\vec{k}}^{+} a_{-\vec{k}}^{+} + a_{-\vec{k}} a_{\vec{k}} \right) . \qquad (11.43)$$

Here, in addition to $\{E_{\rm tr}(\vec{k})\}$ we have the set of variation parameters, $\{\Gamma_{\vec{k}}\}$ needed to describe the Bose-condensed phase. Note that the terms which do not conserve the number of $q \neq 0$ particles cause averages like $\langle a_{\vec{k}}^{+} a_{-\vec{k}}^{+} \rangle_{\rm tr}$ to be nonzero. Systems with such nonzero averages are said to have "off-diagonal (in the number of particles) long-range order." We shall see that nonzero averages of $\langle a_{\vec{k}}^{+} a_{-\vec{k}}^{+} \rangle_{\rm tr}$ will only occur when the system is Bose condensed, that is, when N_0, the occupation of the zero momentum state, is macroscopic. In that case, Eq. (11.43) will allow us to calculate quadratic averages of operators with $\vec{k} \neq 0$. We will still have to calculate N_0, as we did in Chap. 7, by subtracting the number of particles with $\vec{k} \neq 0$ from the total number N.

In order to calculate any averages with respect to $\rho_{\rm tr}$ we first need to diagonalize $\mathcal{H}_{\rm tr} - \mu \hat{n}$. To do this, we write $a_{\vec{k}}$ and $a_{\vec{k}}^{+}$ in terms of new Bose operators as

$$\begin{aligned} a_{\vec{k}} &= u_{\vec{k}} b_{\vec{k}} + v_{\vec{k}} b_{-\vec{k}}^{+} \\ a_{\vec{k}}^{+} &= u_{\vec{k}} b_{\vec{k}}^{+} + v_{\vec{k}} b_{-\vec{k}} , \end{aligned} \qquad (11.44)$$

where, for the purpose of this calculation, we can take $u_{\vec{k}}$ and $v_{\vec{k}}$ to be real. This transformation (which mixes particle creation and annihilation operators) is known as a "Bogoliubov transformation" because it was used by Bogoliubov in his seminal treatment of quantum fluids (Bogoliubov 1947). This type of particle nonconserving transformation was also used by Holstein and Primakoff in their remarkable early (1940) paper (Holstein and Primakoff 1940) on ferromagnetism in which they treated both exchange and dipolar interactions via a Hamiltonian of the same form as that of Eq. (11.43).

The fact that the $a_{\vec{k}}$ and $b_{\vec{k}}$ operators and their conjugates each obey Bose commutation relations implies that

$$\begin{aligned} [a_{\vec{k}}, a_{\vec{q}}^{+}] &= [u_{\vec{k}} b_{\vec{k}} + v_{\vec{k}} b_{-\vec{k}}^{+} , u_{\vec{q}} b_{\vec{q}}^{+} + v_{\vec{q}} b_{-\vec{q}}] \\ &= u_k u_q [b_{\vec{k}} , b_{\vec{q}}^{+}] + v_k v_q [b_{-\vec{k}}^{+} , b_{-\vec{q}}] \\ &= (u_{\vec{k}}^2 - v_{\vec{k}}^2) \delta_{\vec{k},\vec{q}} . \end{aligned} \qquad (11.45)$$

Thus, the assumption that $a_{\vec{k}}$ and $b_{\vec{k}}$ both obey Bose commutation relations requires that

$$u_{\vec{k}}^2 - v_{\vec{k}}^2 = 1 . \qquad (11.46)$$

We can then write

$$
\begin{aligned}
\mathcal{H}_{\rm tr} &= \sum_{\vec{k}} \big[E_{\rm tr}(\vec{k}) - \mu\big] \Big[u_{\vec{k}}^2 b_{\vec{k}}^+ b_{\vec{k}} + v_{\vec{k}}^2 b_{-\vec{k}} b_{-\vec{k}}^+ + u_{\vec{k}} v_{\vec{k}} (b_{\vec{k}}^+ b_{-\vec{k}}^+ + b_{-\vec{k}} b_{\vec{k}})\Big] \\
&\quad + \frac{1}{2}\Gamma_{\vec{k}} \Big[(u_{\vec{k}}^2 + v_{\vec{k}}^2)(b_{\vec{k}}^+ b_{-\vec{k}}^+ + b_{-\vec{k}} b_{\vec{k}}) \\
&\qquad\qquad + u_{\vec{k}} v_{\vec{k}} (b_{\vec{k}}^+ b_{\vec{k}} + b_{-\vec{k}} b_{-\vec{k}}^+ + b_{\vec{k}} b_{\vec{k}}^+ + b_{-\vec{k}}^+ b_{-\vec{k}})\Big] \\
&= \sum_{\vec{k}} \Big(\big[E_{\rm tr}(\vec{k}) - \mu\big] v_{\vec{k}}^2 + \Gamma_{\vec{k}} u_{\vec{k}} v_{\vec{k}}\Big) \\
&\quad + \sum_{\vec{k}} \Big(\big[E_{\rm tr}(\vec{k}) - \mu\big](u_{\vec{k}}^2 + v_{\vec{k}}^2) + 2\Gamma_{\vec{k}} u_{\vec{k}} v_{\vec{k}}\Big) b_{\vec{k}}^+ b_{\vec{k}} \\
&\quad + \sum_{\vec{k}} \left(\big[E_{\rm tr}(\vec{k}) - \mu\big] u_{\vec{k}} v_{\vec{k}} + \frac{1}{2}\Gamma_{\vec{k}} (u_{\vec{k}}^2 + v_{\vec{k}}^2)\right) (b_{\vec{k}}^+ b_{-\vec{k}}^+ + b_{-\vec{k}} b_{\vec{k}}) \,. \qquad (11.47)
\end{aligned}
$$

The $u_{\vec{k}}$'s and $v_{\vec{k}}$'s are chosen so that the last term above is zero and the normalization condition Eq. (11.46) is satisfied. The result is that

$$
u_{\vec{k}}^2 = \frac{[E_{\rm tr}(\vec{k}) - \mu] + \epsilon(\vec{k})}{2\epsilon(\vec{k})} \qquad (11.48)
$$

$$
v_{\vec{k}}^2 = \frac{[E_{\rm tr}(\vec{k}) - \mu] - \epsilon(\vec{k})}{2\epsilon(\vec{k})} \qquad (11.49)
$$

$$
u_{\vec{k}} v_{\vec{k}} = -\frac{\Gamma(\vec{k})}{2\epsilon(\vec{k})} \,, \qquad (11.50)
$$

where

$$
\epsilon(\vec{k}) = \sqrt{[E_{\rm tr}(\vec{k}) - \mu]^2 - \Gamma(\vec{k})^2} \,. \qquad (11.51)
$$

Substituting back into $\mathcal{H}_{\rm tr}$ we find

$$
\mathcal{H}_{\rm tr} = \sum_{\vec{k}} \left[\frac{\epsilon(\vec{k}) - E_{\rm tr}(\vec{k}) + \mu}{2} + \epsilon(\vec{k}) b_{\vec{k}}^+ b_{\vec{k}}\right] . \qquad (11.52)
$$

This means that averages with respect to $\rho_{\rm tr}$ satisfy

$$
\langle b_{\vec{k}}^+ b_{\vec{k}} \rangle_{\rm tr} = \left(e^{\beta\epsilon(\vec{k})} - 1\right)^{-1} \equiv N_{\vec{k}} \qquad (11.53)
$$

$$
\langle b_{\vec{k}}^+ b_{-\vec{k}}^+ \rangle_{\rm tr} = \langle b_{\vec{k}} b_{-\vec{k}} \rangle = 0 \,. \qquad (11.54)
$$

These relations justify interpreting the energy $\epsilon(\vec{k})$ as the quasiparticle energy and the thermal average number of quasiparticles at wavevector $\vec{k}$ is denoted $N_{\vec{k}}$.

Using these relations, we can also evaluate averages of the original field operators,

$$\begin{aligned}\langle a_{\vec{k}}^{+} a_{\vec{k}} \rangle_{\rm tr} &= \langle (u_{\vec{k}} b_{\vec{k}}^{+} + v_{\vec{k}} b_{-\vec{k}})(u_{\vec{k}} b_{\vec{k}} + v_{\vec{k}} b_{-\vec{k}}^{+}) \rangle_{\rm tr} \\ &= u_{\vec{k}}^2 \langle b_{\vec{k}}^{+} b_{\vec{k}} \rangle_{\rm tr} + v_{\vec{k}}^2 \langle b_{-\vec{k}} b_{-\vec{k}}^{+} \rangle_{\rm tr} \\ &= (u_{\vec{k}}^2 + v_{\vec{k}}^2) \langle b_{\vec{k}}^{+} b_{\vec{k}} \rangle_{\rm tr} + v_{\vec{k}}^2 \\ &= \frac{[E(\vec{k}) - \mu](N_{\vec{k}} + \frac{1}{2})}{\epsilon(\vec{k})} - \frac{1}{2} \, . \end{aligned} \tag{11.55}$$

In a similar way, it can be shown that

$$\langle a_{\vec{k}}^{+} a_{-\vec{k}}^{+} \rangle_{\rm tr} = \langle a_{\vec{k}} a_{-\vec{k}} \rangle_{\rm tr} = -\frac{\Gamma_{\vec{k}}}{\epsilon(\vec{k})} (N_k + \frac{1}{2}) \, . \tag{11.56}$$

These last two relations imply that

$$\left(\langle a_{\vec{k}}^{+} a_{\vec{k}} \rangle_{\rm tr} + \frac{1}{2}\right)^2 - \frac{1}{2} \langle a_{\vec{k}}^{+} a_{-\vec{k}}^{+} \rangle_{\rm tr}^2 - \frac{1}{2} \langle a_{-\vec{k}} a_{\vec{k}} \rangle_{\rm tr}^2 = \left(N_{\vec{k}} + \frac{1}{2}\right)^2 , \tag{11.57}$$

a constraint that relates the different possible averages for a given $\vec{k}$.

11.5.2 *Minimizing the Free Energy*

We use the identities, Eqs. (11.55) and (11.56), to evaluate the internal energy, $\mathrm{Tr}[\rho_{\rm tr}(\mathcal{H}_0 + \mathcal{H}_1)]$ as

$$\begin{aligned}\mathrm{Tr}\,[\rho_{\rm tr}(\mathcal{H}_0 + \mathcal{H}_1)] &= \sum_{\vec{k}} E_0 \langle a_{\vec{k}}^{+} a_{\vec{k}} \rangle_{\rm tr} \\ &+ \sum_{\vec{k}_1, \vec{k}_2, \vec{q}} U(\vec{q}) \langle a_{\vec{k}_1 - \vec{q}}^{+} a_{\vec{k}_2 + \vec{q}}^{+} a_{\vec{k}_2} a_{\vec{k}_1} \rangle_{\rm tr} \, . \end{aligned} \tag{11.58}$$

To evaluate this expression we need to generalize our result for the average of four operators in a Bose-condensed system, Eq. (11.23), to the case where pairs of creation or destruction operators can have nonzero average values. We proceed as follows. First we write the terms in which the number of $q \neq 0$ particles is conserved. These terms follow directly from Eq. (11.23). Then we consider the terms where the number of $q \neq 0$ particles is *not* conserved. These terms may be obtained from the first line of Eq. (11.41) by setting $\langle a_0^{+} a_0^{+} \rangle_{\rm tr} = \langle a_0 a_0 \rangle_{\rm tr} = N_0$. Thereby we obtain the result that

$$\langle \mathcal{H}_1 \rangle_{\text{tr}} = \frac{U(0)}{2V} N_0^2 + \frac{N_0}{V} \sum_{\vec{k}} \Big(U(0) + U(\vec{k}) \Big) \langle a^+_{\vec{k}} a_{\vec{k}} \rangle_{\text{tr}}$$
$$+ \frac{N_0}{2V} \sum_{\vec{k}} U(\vec{k}) \Big(\langle a^+_{\vec{k}} a^+_{-\vec{k}} \rangle_{\text{tr}} + \langle a_{\vec{k}} a_{-\vec{k}} \rangle_{\text{tr}} \Big)$$
$$+ \frac{1}{2V} \sum_{\vec{k},\vec{k}'} \Big(U(0) + U(\vec{k} - \vec{k}') \Big) \langle a^+_{\vec{k}} a_{\vec{k}} \rangle_{\text{tr}} \langle a^+_{\vec{k}'} a_{\vec{k}'} \rangle_{\text{tr}}$$
$$+ \frac{1}{2V} \sum_{\vec{k},\vec{k}'} U(\vec{k} - \vec{k}') \langle a^+_{\vec{k}} a^+_{-\vec{k}} \rangle_{\text{tr}} \langle a_{\vec{k}'} a_{-\vec{k}'} \rangle_{\text{tr}} \,. \tag{11.59}$$

As in the case of the noninteracting gas, the most convenient way to characterize the chemical potential in the Bose-condensed phase is to explicitly consider the thermodynamic limit in which the chemical potential is equal to the minimum excitation energy and replace the order $1/N$ corrections to the chemical potential by the number of particles in the ground state, N_0, as per Eq. (7.95) and the discussion that follows it. Thus, in the Bose-condensed state, the chemical potential is determined by

$$\tilde{\epsilon}(k = 0) = \sqrt{[E_{\text{tr}}(k = 0) - \mu]^2 - \Gamma^2_{k=0}} \;=\; 0 \,, \tag{11.60}$$

so that the excitation energy corresponding to the operator $b^+_{\vec{k}}$ in the limit, $\vec{k} \to 0$, is zero. Thus

$$\mu = E_{\text{tr}}(k = 0) - \Gamma(k = 0) \tag{11.61}$$

and, just as in the case of the noninteracting gas, N_0 is determined by the relation

$$N = N_0 + \sum_{\vec{k}} \langle a^+_{\vec{k}} a_{\vec{k}} \rangle_{\text{tr}} \,. \tag{11.62}$$

Now we are ready to minimize the free energy Ω_{tr} with respect to the parameters in the variational density matrix ρ_{tr}. One approach, which is very tedious, is to evaluate all of the averages in Eq. (11.59) using the density matrix expressed in terms of the operators $b^+_{\vec{k}}$ and $b_{\vec{k}}$ and then minimize with respect to the variational parameters, $E(\vec{k})$ and $\Gamma_{\vec{k}}$. A much easier and more elegant way to achieve the same result is to treat the averages $\langle a^+_{\vec{k}} a_{\vec{k}} \rangle_{\text{tr}}$, $\langle a^+_{\vec{k}} a^+_{-\vec{k}} \rangle_{\text{tr}}$, $\langle a_{-\vec{k}} a_{\vec{k}} \rangle_{\text{tr}}$, and $N_{\vec{k}}$ as independent parameters, subject to the constraint, Eq. (11.57). The constraint is handled using a set of Lagrange multipliers, $\{\lambda_{\vec{k}}\}$. Thus, the task is to minimize the function

$$\begin{aligned}\Xi_{\rm tr} &= {\rm Tr}[\rho_{\rm tr}(\mathcal{H}_0+\mathcal{H}_1)] \\ &\quad + kT\sum_{\vec{k}\neq 0}\Big[N_{\vec{k}}\log N_{\vec{k}} - \big(1+N_{\vec{k}}\big)\log\big(1+N_{\vec{k}}\big)\Big] \\ &\quad - \sum_{\vec{k}\neq 0}\lambda_{\vec{k}}\Big\{\big(\langle a^+_{\vec{k}}a_{\vec{k}}\rangle_{\rm tr}+\frac{1}{2}\big)^2 - \frac{1}{2}\langle a^+_{\vec{k}}a^+_{-\vec{k}}\rangle^2_{\rm tr} - \frac{1}{2}\langle a_{-\vec{k}}a_{\vec{k}}\rangle^2_{\rm tr} - \big(N_{\vec{k}}+\frac{1}{2}\big)^2\Big\}.\end{aligned} \tag{11.63}$$

Then differentiating fearlessly with respect to each of the averages, we obtain

$$\begin{aligned}\frac{\partial\,\Xi_{\rm tr}}{\partial\langle a^+_{\vec{k}}a_{\vec{k}}\rangle_{\rm tr}} &= E_0(\vec{k}) - \mu + \frac{N_0}{V}\Big(U(0)+U(\vec{k})\Big) \\ &\quad + \frac{1}{V}\sum_{\vec{k}'}\Big(U(0)+U(\vec{k}-\vec{k}')\Big)\langle a^+_{\vec{k}'}a_{\vec{k}'}\rangle_{\rm tr} \\ &\quad - 2\lambda_{\vec{k}}\big(\langle a^+_{\vec{k}}a_{\vec{k}}\rangle_{\rm tr}+\frac{1}{2}\big) = 0\end{aligned} \tag{11.64a}$$

$$\begin{aligned}\frac{\partial\,\Xi_{\rm tr}}{\partial\langle a^+_{\vec{k}}a^+_{-\vec{k}}\rangle_{\rm tr}} &= \frac{N_0}{2V}U(\vec{k}) + \frac{1}{2V}\sum_{\vec{k}'}U(\vec{k}-\vec{k}')\langle a_{\vec{k}'}a_{-\vec{k}'}\rangle_{\rm tr} \\ &\quad + \lambda_{\vec{k}}\langle a^+_{\vec{k}}a^+_{-\vec{k}}\rangle_{\rm tr} = 0\end{aligned} \tag{11.64b}$$

$$\begin{aligned}\frac{\partial\,\Xi_{\rm tr}}{\partial\langle a_{-\vec{k}}a_{\vec{k}}\rangle_{\rm tr}} &= \frac{N_0}{2V}U(\vec{k}) + \frac{1}{2V}\sum_{\vec{k}'}U(\vec{k}-\vec{k}')\langle a^+_{\vec{k}'}a^+_{-\vec{k}'}\rangle_{\rm tr} \\ &\quad + \lambda_{\vec{k}}\langle a_{\vec{k}}a_{-\vec{k}}\rangle_{\rm tr} = 0\end{aligned} \tag{11.64c}$$

$$\frac{\partial\,\Xi_{\rm tr}}{\partial N_{\vec{k}}} = kT\ln\frac{N_{\vec{k}}}{1+N_{\vec{k}}} + 2\lambda_{\vec{k}}\big(N_{\vec{k}}+\frac{1}{2}\big) = 0\,. \tag{11.64d}$$

Solving for the various averages, we obtain

$$\begin{aligned}\langle a^+_{\vec{k}}a_{\vec{k}}\rangle_{\rm tr}+\frac{1}{2} &= \frac{1}{2\lambda_{\vec{k}}}\Big[E_0(\vec{k}) - \mu + \frac{N_0}{V}\Big(U(0)+U(\vec{k})\Big) \\ &\quad + \frac{1}{V}\sum_{\vec{k}'}\Big(U(0)+U(\vec{k}-\vec{k}')\Big)\langle a^+_{\vec{k}'}a_{\vec{k}'}\rangle_{\rm tr}\Big]\end{aligned} \tag{11.65a}$$

$$\begin{aligned}\langle a^+_{\vec{k}}a^+_{-\vec{k}}\rangle_{\rm tr} &= \langle a_{-\vec{k}}a_{\vec{k}}\rangle_{\rm tr} \\ &= -\frac{1}{2\lambda_{\vec{k}}}\left[\frac{N_0}{V}U(\vec{k}) + \frac{1}{V}\sum_{\vec{k}'}U(\vec{k}-\vec{k}')\langle a_{\vec{k}'}a_{-\vec{k}'}\rangle_{\rm tr}\right]\end{aligned} \tag{11.65b}$$

$$N_{\vec{k}} + \frac{1}{2} = -\frac{kT}{2\lambda_{\vec{k}}} \ln \frac{N_{\vec{k}}}{1 + N_{\vec{k}}} = \frac{\epsilon(\vec{k})}{2\lambda_{\vec{k}}} . \tag{11.65c}$$

Thus, we have

$$\frac{1}{2\lambda_{\vec{k}}} = \frac{N_{\vec{k}} + \frac{1}{2}}{\epsilon(\vec{k})} . \tag{11.66}$$

If we use Eq. (11.66) in conjunction with Eq. (11.65a) and compare the result to Eq. (11.55), we can identify

$$E_{\text{tr}}(\vec{k}) = E_0(\vec{k}) + \frac{N_0}{V}\Big(U(0) + U(\vec{k})\Big) + \frac{1}{V}\sum_{\vec{k}'}\Big(U(0) + U(\vec{k} - \vec{k}')\Big)\langle a^{+}_{\vec{k}'} a_{\vec{k}'}\rangle_{\text{tr}} . \tag{11.67}$$

Likewise if we use Eq. (11.66) in conjunction with Eq. (11.65b) and compare the result to Eq. (11.56), we can identify

$$\Gamma_{\vec{k}} = \frac{N_0}{V} U(\vec{k}) + \frac{1}{V}\sum_{\vec{k}'} U(\vec{k} - \vec{k}')\langle a_{\vec{k}'} a_{-\vec{k}'}\rangle_{\text{tr}} . \tag{11.68}$$

Then μ is determined by $\mu = E_{\text{tr}}(0) - \Gamma_0$ or

$$\mu = \frac{N}{V} U(0) + \frac{1}{V}\sum_{\vec{k}'} U(\vec{k}')\left(\langle a^{+}_{\vec{k}'} a_{\vec{k}'}\rangle_{\text{tr}} - \langle a_{\vec{k}'} a_{-\vec{k}'}\rangle_{\text{tr}}\right) . \tag{11.69}$$

Substituting back into the expression for the excitation energies, we find

$$\begin{aligned}\epsilon(\vec{k}) = \Bigg\{ & \Bigg[E_0(\vec{k}) + \frac{N_0}{V} U(k) \\ & + \frac{1}{V}\sum_{\vec{k}'}\Big[U(\vec{k} - \vec{k}') - U(\vec{k}')\Big]\langle a^{+}_{\vec{k}'} a_{\vec{k}'}\rangle_{\text{tr}} + \frac{1}{V}\sum_{\vec{k}'} U(\vec{k}')\langle a_{\vec{k}'} a_{-\vec{k}'}\rangle_{\text{tr}}\Bigg]^2 \\ & - \left[\frac{N_0}{V} U(\vec{k}) + \frac{1}{V}\sum_{\vec{k}'} U(\vec{k} - \vec{k}')\langle a_{\vec{k}'} a_{-\vec{k}'}\rangle_{\text{tr}}\right]^2\Bigg\}^{1/2} \end{aligned} \tag{11.70}$$

$$= \Bigg\{\Bigg(E(\vec{k}) + \frac{1}{V}\sum_{\vec{k}'}\Big[U(\vec{k} - \vec{k}') - U(\vec{k}')\Big]\Big[\langle a^{+}_{\vec{k}'} a_{\vec{k}'}\rangle_{\text{tr}} - \langle a_{\vec{k}'} a_{-\vec{k}'}\rangle_{\text{tr}}\Big]\Bigg)$$

$$+ \left(E_0(\vec{k}) + \frac{1}{V} \sum_{\vec{k}'} \left[U(\vec{k} - \vec{k}') - U(\vec{k}') \right] \langle a^+_{\vec{k}'} a_{\vec{k}'} \rangle_{\text{tr}} \right.$$

$$\left. + \frac{2N_0}{V} U(k) + \frac{1}{V} \sum_{\vec{k}'} \left[U(\vec{k} - \vec{k}') + U(\vec{k}') \right] \langle a_{\vec{k}'} a_{-\vec{k}'} \rangle_{\text{tr}} \right) \Bigg\}^{1/2} . \quad (11.71)$$

These expressions are rather complicated, and it is useful to write them in a more compact, transparent form. To do this, we define a wavevector-dependent effective mass $m^*(\vec{k})$ and an effective potential, $\tilde{U}(\vec{k})$, as follows:

$$\frac{\hbar^2 k^2}{2m^*(\vec{k})} \equiv \frac{\hbar^2 k^2}{2m} + \frac{1}{V} \sum_{\vec{k}'} \left[U(\vec{k} - \vec{k}') - U(\vec{k}') \right] \left[\langle a^+_{\vec{k}'} a_{\vec{k}'} \rangle_{\text{tr}} - \langle a_{\vec{k}'} a_{-\vec{k}'} \rangle_{\text{tr}} \right] \quad (11.72a)$$

$$\frac{N}{V} \tilde{U}(\vec{k}) \equiv \frac{N_0}{V} U(\vec{k}) + \frac{1}{V} \sum_{\vec{k}'} U(\vec{k} - \vec{k}') \langle a_{\vec{k}'} a_{-\vec{k}'} \rangle_{\text{tr}} . \quad (11.72b)$$

We expect that, in the low-density limit for weak interactions, the effective mass, $m^*(\vec{k})$ and the effective potential, $\tilde{U}(\vec{k})$ will approach their bare values.

In terms of these renormalized functions, the excitation energies and thermal averages are given by

$$\epsilon(\vec{k}) = \sqrt{\frac{\hbar^2 k^2}{2m^*(\vec{k})} \left(\frac{\hbar^2 k^2}{2m^*(\vec{k})} + 2\tilde{U}(\vec{k}) n \right)} \quad (11.73a)$$

$$\langle a^+_{\vec{k}} a_{\vec{k}} \rangle_{\text{tr}} = \frac{\frac{\hbar^2 k^2}{2m^*(\vec{k})} + \tilde{U}(\vec{k}) n}{\epsilon(\vec{k})} \left[N_{\vec{k}} + \frac{1}{2} \right] - \frac{1}{2} \quad (11.73b)$$

$$\langle a^+_{\vec{k}} a^+_{-\vec{k}} \rangle_{\text{tr}} = -\frac{\tilde{U}(\vec{k}) n}{\epsilon(\vec{k})} (N_{\vec{k}} + \frac{1}{2}) . \quad (11.73c)$$

11.5.3 Discussion

The remainder of our treatment will be confined to the case $T = 0$. This simply means that the functions $N_{\vec{k}}$ in the above expressions can be set to zero. Equations (11.72a)–(11.73c), supplemented with the condition determining N_0, Eq. (11.62), form a complete set of self-consistent Hartree–Fock equations for the various averages as functions of $\vec{k}$.

We begin by again asking whether it is reasonable to simply set $U(\vec{k}) \approx U(0)$ in the above expressions, for, if we can, the self-consistent equations simplify considerably. Earlier in this chapter, we argued that we could ignore the momentum dependence of $U(\vec{k})$ provided that the range of U is small compared to the thermal wavelength and the average interparticle spacing. At $T = 0$, of course, the thermal wavelength is infinite. However, another length scale has appeared in the present calculation. It is basically the inverse of the wavevector at which the expression, Eq. (11.73a), crosses over from being linear to quadratic in k, which we can write as

$$\Lambda_{\rm Int} \approx \left(\frac{\hbar^2}{4mU(0)n} \right)^{1/2} . \tag{11.74}$$

Note that, for sufficiently weak interactions, this length scale will be larger than the average interparticle spacing, $n^{-1/3}$.

In the earlier finite T calculation, the momentum sums all involved factors of $n_{\vec{k}}$ which vanish exponentially with increasing k. However, in the present $T = 0$ calculation, the quantities being summed vanish either like inverse powers of k or they vanish as the potential $U(k)$ cuts off for large k. If we were to approximate $U(k)$ by a constant, then the convergence of k-sums would depend on the asymptotic power law dependence for large k of the quantity being summed. Thus, we will not replace $U(k)$ by the constant $U(0)$, but will instead proceed more cautiously.

We also note that the energy scale set by $U(0)n$ is not very physical if the real space potential, $U(r)$, has a hard core. The $k = 0$ limit of the Fourier transform of $U(r)$ is the spatial average of $U(r)$. By contrast, the potential energy scale of the dilute Bose gas must be set by something like the average of the potential function, weighted with the square of the pair wave function. Since the pair wave function vanishes in the region of a hard core, the actual value of the core energy, if it is large, can not contribute significantly to the average. We saw above in the previous section that, if terms off-diagonal in the number of excited particles are ignored, then the ground state energy is $E_0 = \frac{1}{2}U(0)nN$. This expression which is first order in $U(0)$ is like the first Born approximation to the ground state energy. The second and higher order Born approximations correct for the fact that pairs of particles avoid each other. This series is slowly converging for hard-core potentials. The result may be written in terms of the t-matrix which is the exact solution to the two-body problem. We shall see below how the t-matrix may be used to generate a more accurate Hartree–Fock-type theory, and we will discuss the subtleties associated with that approach.

11.5.4 Approximate Solution

We begin by putting the self-consistent equations in a nondimensional form. We take $U(0)n$ as the unit of energy and b, the range of the potential, as the unit of length. Then, for example

$$\frac{\epsilon(\vec{k})}{U(0)n} = \sqrt{\frac{\hbar^2(kb)^2}{2m^*b^2U(0)n} \left(\frac{\hbar^2(kb)^2}{2m^*b^2U(0)n} + 2\frac{\tilde{U}(\vec{k})}{U(0)} \right)} . \tag{11.75}$$

We define the small dimensionless constant α by

$$\alpha = \frac{2mb^2 U(0)n}{\hbar^2} \,. \tag{11.76}$$

Then writing $\vec{x} = \vec{k}b$, and $u(x) = U(k)/U(0)$, the self-consistent equations become

$$f(x) \equiv \langle a^+_{\vec{x}/b} a_{\vec{x}/b} \rangle_{\rm tr} = \frac{1}{2}\left[\frac{\frac{m}{m^*}x^2 + \alpha\tilde{u}(x)}{\sqrt{\frac{m}{m^*}x^2\left(\frac{m}{m^*}x^2 + 2\alpha\tilde{u}(x)\right)}} - 1\right] \tag{11.77}$$

$$g(x) \equiv \langle a^+_{\vec{x}/b} a^+_{-\vec{x}/b} \rangle_{\rm tr} = -\frac{1}{2}\frac{\alpha\tilde{u}(x)}{\sqrt{\frac{m}{m^*}x^2\left(\frac{m}{m^*}x^2 + 2\alpha\tilde{u}(x)\right)}} \,, \tag{11.78}$$

where

$$\frac{m}{m^*}x^2 = x^2 + \frac{\alpha}{N}\sum_{\vec{x}'}\left[u(\vec{x}-\vec{x}') - u(\vec{x}')\right] f(x') \tag{11.79}$$

$$\tilde{u}(x) = u(x)(N_0/N) + \frac{1}{N}\sum_{\vec{x}'} u(\vec{x}-\vec{x}')g(x') \tag{11.80}$$

$$N_0/N = 1 - \frac{1}{N}\sum_{\vec{x}'} f(x') \,. \tag{11.81}$$

To complete the nondimensionalization of this problem and proceed further, we need to convert the sums over $\vec{x}$ to integrals. We write

$$\frac{1}{N}\sum_{\vec{x}} = \frac{V}{(2\pi b)^3 N}\int d^3x = \frac{1}{\hat{\beta}(2\pi)^3}\int d^3x \,, \tag{11.82}$$

where $\hat{\beta} = b^3 n$ is the average number of particles in an interaction volume, b^3. Thus, the problem is specified by two small dimensionless parameters, the other, α, being the ratio of the average interaction energy per particle, $U(0)n$, to the kinetic energy of a particle with wavevector $1/b$, $\hbar^2/2mb^2$.

With the help of Eqs. (11.62) and (11.59), we can write the ground state energy as

$$E_0 = \sum_{\vec{k}} \frac{\hbar^2 k^2}{2m}\langle a^+_{\vec{k}} a_{\vec{k}} \rangle_{\rm tr} + \frac{U(0)}{2V}\overbrace{\left[N_0^2 + 2N_0(N-N_0) + (N-N_0)^2\right]}^{N^2}$$

$$+\frac{N_0}{V}\sum_{\vec{k}}U(k)\left[\langle a^+_{\vec{k}}a_{\vec{k}}\rangle_{\rm tr}+\langle a^+_{\vec{k}}a^+_{-\vec{k}}\rangle_{\rm tr}\right]$$

$$+\frac{1}{2V}\sum_{\vec{k},\vec{k}'}U(\vec{k}-\vec{k}')\left[\langle a^+_{\vec{k}}a_{\vec{k}}\rangle_{\rm tr}\langle a^+_{\vec{k}'}a_{\vec{k}'}\rangle_{\rm tr}+\langle a^+_{\vec{k}}a^+_{-\vec{k}}\rangle_{\rm tr}\langle a_{\vec{k}'}a_{-\vec{k}'}\rangle_{\rm tr}\right]. \quad (11.83)$$

Changing to dimensionless variables,

$$\frac{E_0}{N}=U(0)n\left\{\frac{1}{2}+\frac{1}{\alpha\hat{\beta}}\int\frac{d^3x}{(2\pi)^3}x^2f(x)\right.$$
$$+\frac{n_0}{n}\frac{1}{\hat{\beta}}\int\frac{d^3x}{(2\pi)^3}u(x)\left[f(x)+g(x)\right]$$
$$\left.+\frac{1}{2\hat{\beta}^2}\int\frac{d^3x}{(2\pi)^3}\int\frac{d^3x'}{(2\pi)^3}u(\vec{x}-\vec{x}')\left[f(x)f(x')+g(x)g(x')\right]\right\}. \quad (11.84)$$

To get the leading order dependence on the expansion parameters, we make a number of simplifying approximations. We approximate $u(x)$ by a unit step function, $u(x)=1$ for $|x|<c$ where c is a dimensionless momentum cutoff, and $u(x)=0$ otherwise. One would expect the constant c to be of order unity. We set $m=m^*$, $n_0=n$, and we ignore the last term, the double integral, which we will see to be higher order in the interaction. Then

$$\frac{1}{\alpha\hat{\beta}}\int\frac{d^3x}{(2\pi)^3}x^2f(x)=\frac{1}{2\pi^2\alpha\hat{\beta}}\int_0^c\frac{1}{2}\left[\frac{x^2+\alpha}{\sqrt{x^2\left(x^2+2\alpha\right)}}-1\right]x^4\,dx$$
$$=\frac{1}{4\pi^2\hat{\beta}}\left(\frac{\alpha c}{2}-\frac{4\sqrt{2}\alpha^{3/2}}{5}+\ldots\right) \quad (11.85)$$

$$\frac{1}{\hat{\beta}}\int\frac{d^3x}{(2\pi)^3}u(x)\left[f(x)+g(x)\right]=\frac{1}{2\pi^2\alpha\hat{\beta}}\int_0^c\frac{1}{2}\left[\frac{x^2}{\sqrt{x^2\left(x^2+2\alpha\right)}}-1\right]x^2\,dx$$
$$=\frac{1}{4\pi^2\hat{\beta}}\left(-\alpha c+\frac{4\sqrt{2}\alpha^{3/2}}{3}+\ldots\right), \quad (11.86)$$

where in both cases ... refers to terms that are higher order in α. Combining these results, we find

$$\frac{E_0}{N} = U(0)n\left\{\frac{1}{2} - \frac{\alpha c}{8\pi^2\hat{\beta}} + \frac{2\sqrt{2}\alpha^{3/2}}{15\pi^2\hat{\beta}} + \dots\right\}. \tag{11.87}$$

Thus, the mean-field ground state energy, $U(0)n/2$, is corrected by two kinds of terms. The first term, which is negative, depends on $\alpha c/\hat{\beta} = 2mU(0)c/\hbar b$ which is independent of the density of particles, n, but which depends on the large wavevector cutoff, c/b. This term, when multiplied by $U(0)n/2$ is, in fact, the second Born correction to the ground state energy. It is the leading correction due to the fact that the pair wave function vanishes when the two particles are within a hard-core radius b of each other. It diverges as $b \to 0$ for fixed $U(0)$. The fact that this energy is large and negative means that the particles can lower their energy considerably by staying away from each other's hard cores.

Schematically, the second Born correction is of the form

$$\Delta E = -\frac{n}{2V}\sum_{\vec{k}\neq 0}\frac{|U(k)|^2}{\hbar^2k^2/2m}, \tag{11.88}$$

which also depends linearly on the large momentum cutoff. Since the Born series is slowly converging for short-range potentials, the usual way of dealing with corrections from these kinds of short distance correlations is to everywhere replace the real potential, $U(k)$, and its leading Born-series corrections by the exact two-body t-matrix. In the low energy limit, the t-matrix may be written in terms of the so-called "scattering length", a. Thus, one makes the replacements

$$U(k) \to t(k) \to \frac{4\pi\hbar^2 a}{m} \tag{11.89}$$

independent of k, and then drops the cutoff-dependent terms. This replacement, with its associated prescription, is called a "pseudopotential."

The next leading term in the expression for the ground state energy, Eq. (11.87), is positive and proportional to $\alpha^{3/2}/\hat{\beta}$. This term is independent of the range of the potential b. Equivalently, it is independent of the large momentum cutoff, c/b. The correction to the ground state energy due to this term is

$$\begin{aligned} U(0)n\frac{2\sqrt{2}\alpha^{3/2}}{15\pi^2\hat{\beta}} &= \frac{U(0)n}{2}\frac{4\sqrt{2}}{15\pi^2}\left(\frac{2mU(0)}{\hbar^2}\right)^{3/2}n^{1/2} \\ &\to \frac{2\pi\hbar^2 a}{m}\frac{128}{15\sqrt{\pi}}(a^3n)^{1/2}. \end{aligned} \tag{11.90}$$

Thus, the total ground state energy for the weakly repulsive Bose superfluid in terms of the scattering length is

$$\frac{E_0}{N} = \frac{2\pi\hbar^2 an}{m}\left(1 + \frac{128}{15\sqrt{\pi}}(a^3n)^{1/2}\right), \tag{11.91}$$

which is a classic result, due to Bogoliubov (1947). An equally important and classic result is the depletion of the occupation of the $\vec{k} = 0$ state which is given by the convergent integral

$$\begin{aligned}\frac{n_0}{n} &= 1 - \frac{1}{\hat{\beta}}\int \frac{d^3x}{(2\pi)^3} f(x) \\ &= 1 - \frac{1}{2\pi^2\hat{\beta}}\int_0^\infty \frac{1}{2}\left[\frac{x^2+\alpha}{\sqrt{x^2\left(x^2+2\alpha\right)}} - 1\right] x^2\,dx \\ &= 1 - \frac{\sqrt{2}\alpha^{3/2}}{12\pi^2\hat{\beta}} = 1 - \frac{\sqrt{2}}{12\pi^2}\left(\frac{2mU(0)}{\hbar^2}\right)^{3/2} n^{1/2} \\ &\to 1 - \frac{8}{3\sqrt{\pi}}(a^3n)^{1/2}, \end{aligned} \tag{11.92}$$

where we have made the pseudopotential substitution in the last step.

11.5.5 Comments on This Theory

We have seen that interactions deplete the occupation of the $\vec{k} = 0$ condensate by an amount proportional to $\sqrt{a^3n}$. Furthermore, interactions increase the energy of the Bose gas at $T = 0$, in leading order, by an amount proportional to na. The next leading correction, also positive, which was generated by the non-number-conserving terms in the variational Hamiltonian, is of order $na(a^3n)^{1/2}$, and it is expected that the remainder of the, possibly divergent, perturbation expansion is an expansion in the small parameter (a^3n). The quasiparticle dispersion relation, $\epsilon_{\vec{k}}$, starts off linearly and then curves up, as for type (B) of Fig. 11.1.

The fact that it is linear at low energies is related, by an argument due to Landau (Landau and Lifshitz 1969) §67, to the property of superfluidity. This relationship is discussed in the next section. The fact that the dispersion relation curves upward

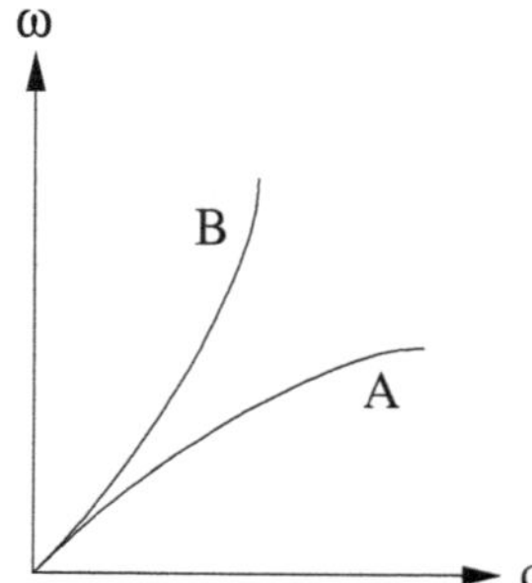

Fig. 11.1 Energy versus wavevector spectrum of the form $\hbar\omega(q) = cq + \gamma q^3$ with $\gamma < 0$ (A) and $\gamma > 0$ (B)

means that the excitations themselves have a finite lifetime, since one excitation can decay into two with the same total momentum and energy, and there are cubic terms in the Hamiltonian which will cause this to happen. (To conserve energy for a spectrum of type B, the excitations in the final state have additional energy by having equal and opposite components of wavevector transverse to the initial wavevector.)

In the early 1950s Feynman (1953a, b, 1954), Feynman and Cohen (1956) derived a theory of superfluid helium based on a variational wave function for interacting bosons. Feynman's theory goes beyond mean-field theory because it assumes a knowledge of the two-particle correlation function. Feynman's theory is one of a class of theories now called the "Single-Mode Approximation." For superfluid helium, this theory gives a stable linear dispersion relation, for the single mode, with downward curvature and a "roton minimum" at finite wavevector.

We have gone to a great deal of effort in this chapter to explain how the Hartree–Fock method and the pseudopotential (scattering length) approximation combine to give a reasonable description of the weakly interacting, low-density Bose fluid. The simplest Hartree–Fock calculation, with no off-diagonal long-range order, says that the ground state energy is proportional to the average potential times the average density. The variational calculation for the Bose-condensed state, including terms off-diagonal in the number operator (excitations from the condensate), improves the energy in two ways. It partially corrects the average potential, incorporating short-range correlations, and it captures the effects of long-range phase coherence, which modifies the excitation spectrum and corrects the ground state energy per particle by a positive amount proportional to $na(a^3n)^{1/2}$ as shown in Eq. (11.91). The fact that the low energy excitations are linear in wavevector may be thought of as the result of a superfluid "stiffness" which also gives rise to the higher order correction to E_0. Going further, one can correct both the first (mean field) term and the second, phase coherence term, by substituting the t-matrix, written in terms of the scattering length, for the potential in both terms. Although this procedure is correct, it is far more than a simple variational mean-field or Hartree–Fock result.

11.6 Superfluid Flow

Consider a normal gas of N noninteracting bosons moving in a pipe with total momentum $\vec{P}$ and energy E. Then

$$\vec{P} = \sum_{i=1}^{N} \vec{p}_i \tag{11.93}$$

$$E = \sum_{i=1}^{N} \frac{p_i^2}{2m} . \tag{11.94}$$

We can minimize the total energy for fixed total momentum with the help of a Lagrange multiplier. That is we minimize

$$\mathcal{E} = E - \vec{\lambda} \cdot \vec{P} \tag{11.95}$$

which gives

$$\vec{p}_i = m\vec{\lambda}, \quad \forall\, i, \tag{11.96}$$

so that

$$\vec{p}_i = \frac{\vec{P}}{N} \tag{11.97}$$

$$E = \frac{P^2}{2mN} \,. \tag{11.98}$$

Physically this corresponds to a macroscopic occupation of the single-particle state with momentum $\vec{P}/N$. We ask the question: How can this momentum decay due to exchange of energy and momentum with the walls?

Assume that the walls are cold, so that the Bose fluid's energy and momentum are not changed by absorbing energy and/or momentum from the pipe. However, the fluid has plenty of energy and momentum to give to the walls. Consider an elementary process in which a small amount of momentum, $\Delta\vec{p}$, with some component along $\vec{p}_i$ is transferred from one of the particles to the wall. The change of energy of the particle is $\Delta E = (\vec{p}_i - \Delta\vec{p})^2/2m - p_i^2/2m = (\Delta\vec{p})^2/2m - \vec{p}_i \cdot \Delta\vec{p}/m$. As long as $\Delta\vec{p}$ is not too large, this change in energy is clearly negative, and so the system can lower its energy and momentum by transferring ΔE and $\Delta\vec{p}$ to the walls of the pipe. Eventually, the flowing gas will slow down and stop in the pipe.

If the system is a degenerate, weakly interacting Bose gas, the situation is different since changing one of the momenta changes the interaction energy. From Eq. (11.37), the change in interaction energy due to changing the number of particles in the condensate from N to $N-1$ is $U(0)N/V$, a positive energy. The total change in energy is $\Delta E = U(0)N/V + (\Delta\vec{p})^2/2m - \vec{p}_i \cdot \Delta\vec{p}/m$ which is positive provided that $p_i^2/2m < U(0)N/V$.

However, we have seen that the energy and excitations of a degenerate weakly repulsive Bose gas are not correctly described by the theory used to derive Eq. (11.37) which predicts an energy gap, $U(0)N/V$, for excitations. Instead, the system develops off-diagonal long-range order and its excitation spectrum is linear in momentum, starting from $E = 0$. This is intermediate between the case of free particles and that of a gap for low-lying excitations. The question is: are there negative energy excitations of the flowing superfluid which can transfer momentum from the superfluid to the walls of the pipe? To answer this we need to calculate the ground state energy and low-lying excitations of a uniformly flowing superfluid.

Consider a Hamiltonian for bosons condensed in the state $\vec{q}$

$$\mathcal{H}_{\text{tr}}(\vec{q}) = \sum_{\vec{k}\geq 0}\left[\xi_{\vec{q}+\vec{k}}a^{+}_{\vec{q}+\vec{k}}a_{\vec{k}} + \xi_{\vec{q}-\vec{k}}a^{+}_{\vec{q}-\vec{k}}a_{\vec{q}-\vec{k}}\right] + \sum_{\vec{k}\geq 0}\Gamma_{\vec{k}}\left(a^{+}_{\vec{q}+\vec{k}}a^{+}_{\vec{q}-\vec{k}} + a_{\vec{q}-\vec{k}}a_{\vec{q}+\vec{k}}\right), \quad (11.99)$$

where the notation $\vec{k} \geq 0$ means that $\vec{k}\cdot\vec{q} \geq 0$ and, for $\vec{k}\cdot\vec{q} = 0$, only one of $\vec{k}$ and $-\vec{k} \neq 0$ is included. The first summation is just the usual kinetic energy separated into two parts. The second sum corresponds to excitations in which two particles are destroyed (created) in the $\vec{q}$-condensate and two particles with total momentum $2\vec{q}$ are created (destroyed). This Hamiltonian can be diagonalized in the same way that $\mathcal{H}_{\text{tr}}(0)$ was diagonalized earlier. The transformation is

$$a^{+}_{\vec{q}\pm\vec{k}} = u_k b^{+}_{\vec{q}\pm\vec{k}} + v_k b_{\vec{q}\mp\vec{k}}, \quad (11.100)$$

where $u_k^2 - v_k^2 = 1$. $\mathcal{H}_{\text{tr}}(\vec{q})$ is diagonalized by u_k, v_k which satisfy

$$2u_k v_k = \frac{-\Gamma_k}{\sqrt{\frac{1}{4}\left(\xi_{\vec{q}+\vec{k}} + \xi_{\vec{q}-\vec{k}}\right)^2 - \Gamma_k^2}}$$

$$u_k^2 + v_k^2 = \frac{\left(\xi_{\vec{q}+\vec{k}} + \xi_{\vec{q}-\vec{k}}\right)/2}{\sqrt{\frac{1}{4}\left(\xi_{\vec{q}+\vec{k}} + \xi_{\vec{q}-\vec{k}}\right)^2 - \Gamma_k^2}}. \quad (11.101)$$

The result is

$$\begin{aligned}\mathcal{H}_{\text{tr}}(0) = &\sum_{\vec{k}\geq 0}\left[\sqrt{\frac{1}{4}\left(\xi_{\vec{q}+\vec{k}} + \xi_{\vec{q}-\vec{k}}\right)^2 - \Gamma_k^2} - \frac{1}{2}\left(\xi_{\vec{q}+\vec{k}} + \xi_{\vec{q}-\vec{k}}\right)\right] \\ &\sum_{\vec{k}\geq 0}\left\{\left[\frac{1}{2}\left(\xi_{\vec{q}+\vec{k}} + \xi_{\vec{q}-\vec{k}}\right) + \sqrt{\frac{1}{4}\left(\xi_{\vec{q}+\vec{k}} + \xi_{\vec{q}-\vec{k}}\right)^2 - \Gamma_k^2}\right]b^{+}_{\vec{q}+\vec{k}}b_{\vec{q}+\vec{k}}\right. \\ &\left.\left[-\frac{1}{2}\left(\xi_{\vec{q}+\vec{k}} + \xi_{\vec{q}-\vec{k}}\right) + \sqrt{\frac{1}{4}\left(\xi_{\vec{q}+\vec{k}} + \xi_{\vec{q}-\vec{k}}\right)^2 - \Gamma_k^2}\right]b^{+}_{\vec{q}-\vec{k}}b_{\vec{q}-\vec{k}}\right\}. \quad (11.102)\end{aligned}$$

For weakly interacting bosons,

$$\xi_{\vec{q}\pm\vec{k}} = \frac{\hbar^2(\vec{q}\pm\vec{k})^2}{2m} - \mu, \quad (11.103)$$

where μ is determined by

$$\frac{\hbar^2 q^2}{2m} - \mu = \Gamma_k = \frac{4\pi\hbar^2 an}{m} \tag{11.104}$$

and we have written the interaction in terms of the scattering length a. Then the energies for excitations from the flowing condensed state are

$$\begin{aligned} \epsilon\left(\vec{q} \pm \vec{k}\right) &= \pm \frac{\hbar^2 \vec{q} \cdot \vec{k}}{m} + \sqrt{\left(\frac{\hbar^2 k^2}{2m} + \frac{4\pi\hbar^2 an}{m}\right)^2 - \left(\frac{4\pi\hbar^2 an}{m}\right)^2} \\ &= \pm \frac{\hbar^2 \vec{q} \cdot \vec{k}}{m} + \sqrt{\frac{\hbar^2 k^2}{2m}\left(\frac{\hbar^2 k^2}{2m} + \frac{8\pi\hbar^2 an}{m}\right)} . \end{aligned} \tag{11.105}$$

At this point, we can remove the restriction on $\vec{k}$ and drop the $\pm$ sign. Basically what this result says is that, if $\vec{k}$ is along $\vec{q}$, then the excitation has a higher energy than it would have if $\vec{q}$ were zero, and if $\vec{k}$ is antiparallel to $\vec{q}$ then the excitation energy is decreased. The increase or decrease in the excitation energy in a flowing gas by an amount $\hbar^2 \vec{q} \cdot \vec{k}/m$ is essentially a Doppler shift. Since both the Doppler shift and the excitation energy itself are linear in k for small k, the effect of a nonzero flow velocity, $\hbar\vec{q}/m$, is to increase (decrease) the slope of the linear dispersion relation for $\vec{k}$ parallel (antiparallel) to $\vec{q}$. For some critical value of flow velocity, $v_c = \hbar q_c/m = (\hbar/m)\sqrt{4\pi an}$, the slope of the spectrum for excitations antiparallel to $\vec{q}$ vanishes. Above this critical velocity, the flow is heavily damped since the moving fluid can easily transfer momentum and energy to the walls of the pipe.

Note that this argument is far more general than the theory of the weakly interacting Bose gas. Provided that the excitation spectrum of the superfluid is linear, sublinear, or gapped, the fluid can flow without dissipation below some critical velocity. The result of Eq. (11.105) that the excitation spectrum is shifted by $\hbar^2 \vec{q} \cdot \vec{k}/m$ is a general consequence of Galilean invariance.

Note also that it is clear from the way that the argument was formulated that the density of superfluid, under conditions where superflow can occur, $\hbar\vec{q}/m < v_c$, is exactly equal to the density of the fluid. It is not, for example, equal to n_0, the density of particles in the condensate. For a weakly interacting Bose gas near $T = 0$, these two quantities are nearly equal. However, for a strongly interacting Bose liquid, for example, for superfluid ^{4}He, the condensate fraction is only a small fraction of the total density. Nevertheless, the density of superfluid which flows without dissipation at $T = 0$ is the total density of the fluid. This follows from the argument based on Galilean invariance given above.

11.7 Summary

Here we have constructed a trial density matrix for quantum system in which we neglect interactions between excitations. We implemented this approximation by introducing a "trial Hamiltonian" $\mathcal{H} - \mu\hat{n}$, quadratic in quasiparticle excitation operators, with respect to which thermodynamic averages are defined as usual. The parameters in the trial Hamiltonian are the variational parameters in the trial density matrix, ρ, where

$$\rho = \frac{e^{-\beta(\mathcal{H}-\mu\hat{n})}}{\mathrm{Tr}e^{-\beta(\mathcal{H}-\mu\hat{n})}} . \tag{11.106}$$

For the Bose fluid with repulsive weak interactions, one may still have a phase transition from the normal liquid phase into the Bose-condensed, or superfluid, phase. In the Bose-condensed phase, there is macroscopic occupancy of the single-particle ground state, as in the noninteracting case, but, in order to properly describe the effects of interactions on the quasiparticle excitations, it is necessary to introduce "anomalous" averages $\langle a_{\vec{q}} a_{-\vec{q}} \rangle_{\mathrm{tr}}$ and $\langle a^+_{\vec{q}} a^+_{-\vec{q}} \rangle_{\mathrm{tr}}$. Although these averages appear to violate number conservation, the number of particles is actually conserved and these averages simply correspond to the creation or annihilation of excitations with the corresponding depletion of the macroscopically occupied single-particle ground state. For example, $\langle a_{\vec{q}} a_{-\vec{q}} \rangle_{\mathrm{tr}}$ really denotes $N_0^{-1} \langle a_0^+ a_0^+ a_{\vec{q}} a_{-\vec{q}} \rangle_{\mathrm{tr}}$.

For Fermi systems with weak repulsive interactions, Landau's Fermi liquid theory shows how even rather strong interactions do not change the nature of the elementary excitation spectrum, which consists of particle–hole excitations from a filled Fermi sea. The main effect of interactions is to modify the dispersion relations which can be described by an effective mass, and to generate a quasiparticle lifetime which diverges as $T \to 0$ at the Fermi surface. For weak *attractive* interactions, there is a transition to a superfluid state which is similar to what happens in the Bose case, as we will see in the next chapter.

11.8 Exercises

1. Give explicitly the steps needed to go from Eq. (11.13) to Eq. (11.19).

2. To illustrate the use of Wick's theorem discuss the perturbative treatment of the quantum Hamiltonian for a particle in a one-dimensional anharmonic potential:

$$\mathcal{H} = \frac{p^2}{2m} + \frac{k}{2}x^2 + \lambda x^4 .$$

It seems intuitively clear that a simple approximation *to get the anharmonic frequency of oscillation* is to replace the quartic term by an effective quadratic term. Schematically one writes

$$x^4 \to \langle x^2 \rangle x^2 .$$

Describe in detail how this should be done and verify that your prescription yields results which coincide with first-order perturbation theory. Note that since $\langle x^2 \rangle$

depends on the quantum number n, the oscillator energy levels are no longer equally spaced as they are for $\lambda = 0$.

3. Give explicitly the steps needed to obtain the last line of Eq. (11.13)

4. In the "equation-of-motion" method, one constructs a quasiparticle operator $b_{\mathbf{k}}^{+}$ which is a linear combination of $a_{\mathbf{k}}^{+}$ and $a_{-\mathbf{k}}$ which satisfies

$$[\mathcal{H}, b_{\mathbf{k}}^{+}]_{-} = \omega_{\mathbf{k}} b_{\mathbf{k}}^{+} .$$

Show that that approach can be used to derive the same transformation as found in Sect. 11.5.1.

5. This exercise is similar in spirit to Exercise 1 of Chap. 6. In the treatment of the interacting Fermi gas, the symbol N was introduced to denote $\sum_{\vec{k},\alpha} n_{\vec{k},\alpha}$. It seems physically obvious that this definition should be equivalent to $N = -\left.\partial\Omega/\partial\mu\right)_{TV}$. Show that these two relations for N are equivalent to one another. (Remember: μ appears implicitly in Ω in many places!).

11.9 Appendix—The Pseudopotential

This section contains a brief discussion of the low-energy pseudopotential. The main result is that for systems at low density, it is possible to include the effects of short-range correlations in the pair wave function by substituting the pseudopotential, written in terms of the s-wave scattering length, for the bare potential. However, when using this substitution, terms corresponding to multiple scattering of the pseudopotential, which typically lead to ultraviolet-divergent momentum integrals, must be excluded. This exclusion is not arbitrary, since these effects have already been included in the pseudopotential. The fact that they lead to divergent momentum integrals provides a convenient method for removing them. The discussion below illustrates how this approach can be formalized in terms of a projection operator.

If the distance, r, between the particles is larger than the range of the potential, b, then the wave function for this coordinate obeys the equation

$$\left(\nabla^2 + k^2\right)\psi = 0 , \tag{11.107}$$

where $k^2 = 2\mu E/\hbar^2$, μ is the reduced mass, and E is the energy in the center of mass frame. For $kr << 1$ the k^2 term can be neglected giving

$$\frac{1}{r^2}\frac{\partial}{\partial r} r^2 \frac{\partial \psi}{\partial r} = 0 , \tag{11.108}$$

which has solutions of the form

$$\psi(r) = A\left(1 - \frac{a}{r}\right) . \tag{11.109}$$

The net effect of the scattering potential is captured in the "scattering length", a, which determines where the wave function, extrapolated from outside the range of the potential, intersects zero.

The idea of the pseudopotential approximation is to replace the real potential, whose scattering length is a, as defined above, with a pseudopotential which is a delta function, with suitable weight, at the origin. The pseudopotential is chosen so that the wave function is identical to that of the real potential for $r > b$. For $r < b$, the wave function has the extrapolated value, given by Eq. (11.109).

For such a pseudopotential, the form of the Schrodinger Equation at low energy, close to the origin, is the same as Laplace's Equation for the electrostatic potential of a point charge,

$$\nabla^2 \phi = 4\pi e \delta(r) , \tag{11.110}$$

which has the solution

$$\phi(r) = \frac{e}{r} + \text{const.} . \tag{11.111}$$

The normalization constant, A, in Eq. (11.109) can be extracted from the wave function at small r as follows:

$$A = \frac{\partial}{\partial r}\left(r\psi(r)\right)_{r=0} . \tag{11.112}$$

Therefore, Eq. (11.109) is the solution to the equation

$$\nabla^2 \psi = 4\pi a \delta(r)\left(\frac{\partial}{\partial r} r\right)\psi . \tag{11.113}$$

The combination of the delta function, the differential operator, and the boundary condition for large r ensures that the properly normalized wave function will have the behavior described by Eq. (11.109).

We generalize Eq. (11.113) to nonzero energy and (noting that $m = 2\mu$) write the result as a Schrodinger equation of the form,

$$\left\{-\frac{\hbar^2}{2m}\left(\nabla_1^2 + \nabla_2^2\right) + \frac{4\pi\hbar^2 a}{m}\delta(\vec{r}_{12})\left(\frac{\partial}{\partial r_{12}} r_{12}\right)\right\}\psi(\vec{r}_1, \vec{r}_2) = E\psi(\vec{r}_1, \vec{r}_2) , \tag{11.114}$$

where $\vec{r}_{12} = \vec{r}_1 - \vec{r}_2$. The pseudopotential is thus

$$V(r) = \frac{4\pi\hbar^2 a}{m}\delta(\vec{r})\left(\frac{\partial}{\partial r} r\right) . \tag{11.115}$$

What is the meaning of the curious operator, $\delta(\vec{r})\left(\frac{\partial}{\partial r}r\right)$, on the right-hand side of Eq. (11.115)? In general, we expect that solutions to Eq. (11.114) can be expanded in a series of the form

$$\psi(\vec{r}_1, \vec{r}_2) = e^{i\vec{k}\cdot(\vec{r}_1+\vec{r}_2)}\left[\frac{B}{r_{12}} + \sum_{n=0}^{\infty} C_n r_{12}^n\right]. \tag{11.116}$$

The effect of $\frac{\partial}{\partial r}r$, acting on this wave function, is to annihilate the term B/r_{12} which is singular for small r_{12}. The constant term C_0 is preserved, and all of the higher order terms, $C_n r_{12}^n$, are irrelevant because of the delta function, $\delta(\vec{r}_{12})$. Thus, if it were not for the singular term, B/r_{12}, the factor, $\frac{\partial}{\partial r}r$, could simply be dropped. Keeping it has the effect of eliminating the unphysical, divergent term $\delta(\vec{r}_{12})B/r_{12}$.

When working with Fourier-transformed wave functions in momentum (k-) space, the consequence of a term such as this, which diverges at small separations, would be k-space integrals that diverge for large k.

References

N.N. Bogoliubov, On the theory of superfluidity. J. Phys. X **I**, 23 (1947)

R.P. Feynman, Atomic theory of the λ transition in helium. Phys. Rev. **91**, 1291 (1953a)

R.P. Feynman, Atomic theory of liquid helium near absolute zero. Phys. Rev. **91**, 1301 (1953b)

R.P. Feynman, Atomic theory of the two-fluid model of liquid helium. Phys. Rev. **94**, 262 (1954)

R.P. Feynman, M. Cohen, Phys. Rev. **102**, 1189 (1956)

T. Holstein, H. Primakoff, Field dependence of the intrinsic domain magnetization of a ferromagnet. Phys. Rev. **58**, 1098 (1940)

L.D. Landau, The theory of a fermi liquid. Sov. Phys. JETP **3**, 920 (1957a)

L.D. Landau, The theory of a fermi liquid. Sov. Phys. JETP **5**, 101 (1957b)

L.D. Landau, The theory of a fermi liquid. Sov. Phys. JETP **8**, 70 (1958)

L.D. Landau, E.M. Lifshitz, Statistical physics, in *Course of Theoretical Physics*, vol. 5 (Pergamon Press, 1969)

A.J. Leggett, A theoretical description of the new phases of liquid He^3. Rev. Mod. Phys. **47**, 331 (1975)

Chapter 12
Superconductivity: Hartree–Fock for Fermions with Attractive Interactions

12.1 Introduction

In the last chapter, we used mean-field theory to explore how interactions affect the behavior of quantum gases. We found that, for fermions, the simplest mean-field theory, which we called Hartree–Fock theory, gives a good qualitative description for the case of weakly repulsive interactions. In fact, we mentioned that a more general theory, called Fermi liquid theory, extends this qualitative behavior to interactions that need not be so weak. Eventually, as interactions are cranked up, a transition will occur to some other state, which might be a heavy fermion state, high-temperature superconductivity, a charge or spin density wave state, ferromagnetism, or antiferromagnetism. Precisely what happens will depend on the details of the interactions, on the crystal structure, and on the electron density. Such behaviors are called strongly correlated since they are driven by strong interactions.

The case of bosons was quite different. There, even weak repulsive interactions were sufficient to change the low-energy, low-temperature behavior of the system in a qualitative way. Not only do repulsive interactions lower the energy of the ground state because of Bose statistics, but, in addition, correlations that destroy a pair of particles in the $k = 0$ condensate and simultaneously create a $(\vec{k}, -\vec{k})$ pair, lead to a gapless linear excitation spectrum which supports superfluidity. These correlations arise from terms in the interaction potential of the form $U(\vec{k})a^{+}_{\vec{k}}a^{+}_{-\vec{k}}a_0a_0$, which connect excitations to the macroscopically occupied $k = 0$ ground state.

12.2 Fermion Pairing

Fermions have their own low-energy instability which arises from *attractive* interactions. Here, instead of $(\vec{k}, -\vec{k})$ pairs interacting with a $k = 0$ condensate, the interactions that destabilize and reorganize the ground state arise from scattering between $(\vec{k}_F, -\vec{k}_F)$ and $(\vec{k}'_F, -\vec{k}'_F)$ pairs around the Fermi surface. To understand the special

A. J. Berlinsky and A. B. Harris, *Statistical Mechanics*, Graduate Texts in Physics,
https://doi.org/10.1007/978-3-030-28187-8_12

significance of $(\vec{k}_F, -\vec{k}_F)$ pairs, we first note that since the ground state of the Fermi gas is gapless, the effect of an interaction that scatters a pair of electrons at or just below the Fermi energy to unoccupied states at or just above the Fermi energy must be treated using degenerate perturbation theory. A key question is how large is the phase space for such excitations?

If we write the interaction as

$$\mathcal{H}_1 = -\frac{1}{2\Omega_0} \sum_{\mathbf{k}_1, \mathbf{k}_2, \mathbf{q}} \sum_{\alpha, \beta} V(\mathbf{q}) a^+_{\mathbf{k}_1+\mathbf{q},\alpha} a^+_{\mathbf{k}_2-\mathbf{q},\beta} a_{\mathbf{k}_2,\beta} a_{\mathbf{k}_1,\alpha}, \tag{12.1}$$

where the operators $a_{\mathbf{k}_1,\alpha}$ and $a_{\mathbf{k}_2,\beta}$ are spin 1/2 fermion operators ($\alpha, \beta = \pm 1/2$) that obey anticommutation relations, and $V(\mathbf{q})$ is the Fourier transform of the scattering potential which is attractive if it is positive because of the $-$ sign in Eq. (12.1). Consider processes in which states $(\mathbf{k}_1, \alpha)$ and $(\mathbf{k}_2, \beta)$ lie at or just below the Fermi energy and are occupied, while states $(\mathbf{k}_1 + \mathbf{q}, \alpha)$ and $(\mathbf{k}_2 - \mathbf{q}, \beta)$, which lie at or just above the Fermi energy, are empty. The amplitude for scattering between these two pair states is $-V(\mathbf{q})/(2\Omega_0)$. The question which we posed above about the phase space for such scattering comes down to how many values of $\mathbf{q}$ are there which allow this process to be resonant so that the initial and final states all lie on the Fermi surface?

For simplicity of visualization, we consider a spherical Fermi surface, and, without loss of generality, chose $\mathbf{k}_1$ and $\mathbf{k}_2$ to be symmetrically located about the $\hat{z}$-direction in the x–z plane. Thus, we can write

$$\mathbf{k}_1 = (k_F \sin\theta, 0, k_F \cos\theta), \tag{12.2a}$$

$$\mathbf{k}_2 = (-k_F \sin\theta, 0, k_F \cos\theta). \tag{12.2b}$$

As shown in Fig. 12.1, if $\mathbf{q}$ is a vector in the x–z plane and if $\mathbf{k}_1 + \mathbf{q}$ lies on the Fermi sphere, then $\mathbf{k}_2 - \mathbf{q}$ does *not* lie on the Fermi sphere. However, if $\mathbf{q}$ is a vector in the x–y plane and if $\mathbf{k}_1 + \mathbf{q}$ lies on the Fermi sphere, then $\mathbf{k}_2 - \mathbf{q}$ also lies on the Fermi sphere. The two vectors $\mathbf{k}_1 + \mathbf{q}$ and $\mathbf{k}_2 - \mathbf{q}$ both lie on a circle on the Fermi sphere

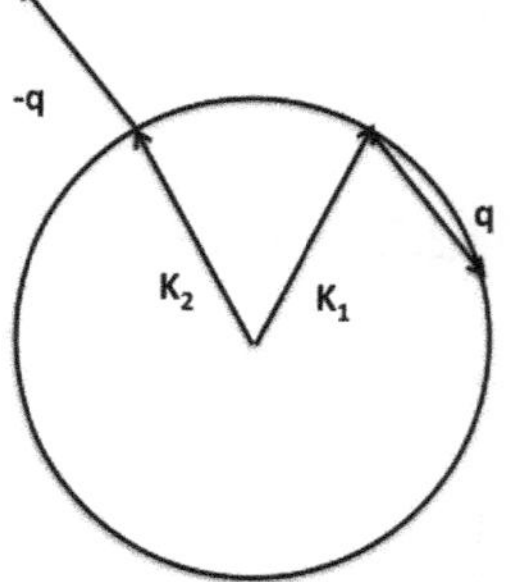

Fig. 12.1 For Fermi wavevectors $\mathbf{k}_1$ and $\mathbf{k}_2$, in the $x - -z$ plane as defined in Eq. (12.2), if $\mathbf{q}$ is such that $\mathbf{k}_1 + \mathbf{q}$ also lies on the Fermi surface, in the same plane as $\mathbf{k}_1$ and $\mathbf{k}_2$, then $\mathbf{k}_2 - \mathbf{q}$ lies well off the Fermi surface

oriented perpendicular to k_z with a radius $k_F \sin\theta$. This means that there is more phase space available for pair scattering on the Fermi sphere as $\mathbf{k}_1$ and $\mathbf{k}_2$ tip away from each other, with the largest value occurring for $\theta = \pi/2$. However, something else happens when $\theta = \pi/2$ or, equivalently, when $\mathbf{k}_2 = -\mathbf{k}_1$. The phase space for $\mathbf{q}$ grows dramatically to include the entire Fermi sphere! In other words, pair scattering can connect $(\mathbf{k}_1, -\mathbf{k}_1)$ to any other $(\mathbf{k}_1 + \mathbf{q}, -\mathbf{k}_1 - \mathbf{q})$ pair on the Fermi sphere. This means that the effect of pair scattering on states of the form $(\mathbf{k}_1, -\mathbf{k}_1)$ is singular and can have a dramatic effect on the Fermi sea. It is easy to see that this last result is not specific to spherical Fermi surfaces, but applies equally well to any system for which the single-particle electron energies obey, $E(\mathbf{k}, \alpha) = E(-\mathbf{k}, \beta)$.

The fact that a $(\mathbf{k}_F, -\mathbf{k}_F)$ pair of electrons interacts with all other such pairs around the Fermi surface was used by Cooper (1956) to show that such a pair, interacting via an attractive interaction in the presence of a filled Fermi sea, would be bound by a finite energy, and thus represented an instability of the Fermi sea to the formation of such "Cooper pairs." This somewhat artificial treatment of a single pair, quickly led Bardeen, Cooper, and Schrieffer (BCS) (Bardeen et al. 1957) to construct a proper many-body wave function of correlated Cooper pairs for which they were awarded the 1972 Nobel prize. We will provide a version of their derivation below which is similar in spirit to that of the weakly repulsive Bose gas in the last chapter.

12.3 Nature of the Attractive Interaction

Before treating the instability of the Fermi gas to attractive interactions, it is worth spending a few moments considering what could generate such an interaction, specifically for electrons in solids which is the situation that BCS considered. Electrons carry negative charge, so one might expect their interactions to be overwhelmingly repulsive. However, they also move in a compensating background of positive ions since the overall system is neutral. Interesting effects arise from the fact that the ionic background is compressible. An electron attracts ions creating a local region of slightly enhanced positive charge. However, the timescales for electronic and ionic motion are very different, with the ions being roughly two orders of magnitude slower, and so, as an electron moves through the crystal, it leaves behind a positively charged "wake" of displaced ions as shown in Fig. 12.2. It is important that the positive wake persists for a relatively long time so that a second electron surfing in that wake need not be too close to the first and thus can avoid the direct repulsive Coulomb interaction. This effect is called "retardation" and it allows a net attractive interaction to occur. Another consequence of retardation is that the attraction due to the slow response of the ions is less effective for electrons that are moving too fast, or too fast relative to each other. As a practical matter, this justifies an approximation made by BCS which is to ignore interactions between electrons in states with energies (measured with respect to the Fermi energy) larger than the natural cutoff frequency which, for this problem, is the one-phonon bandwidth.

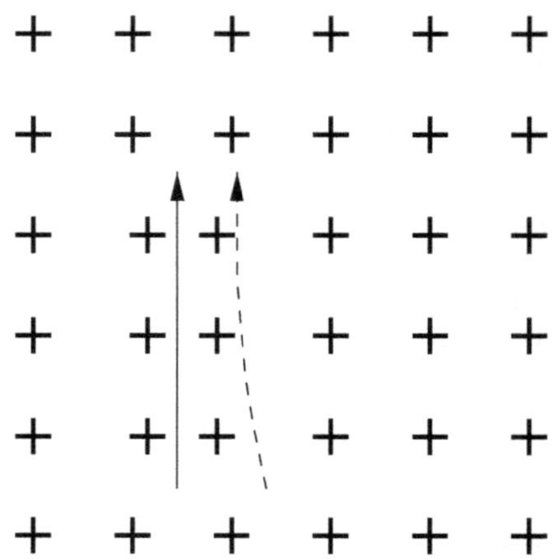

Fig. 12.2 An electron (whose path is indicated by a solid line) causes a distortion in the lattice of positive ions which can then attract a second electron

A more microscopic way of viewing the effective electron–electron interaction is as a process of "phonon exchange." The coupling of electrons to the ions allows a single electron to emit or absorb a phonon and scatter into a different energy–momentum state, conserving total energy and momentum. An effective electron–electron interaction results when two electrons exchange a virtual phonon. The details of the interaction, how it depends on the energy and momentum carried by the phonon and how the direct Coulomb interaction between the electrons is minimized, require a detailed quantum many-body calculation, but the qualitative behavior is as described in the previous paragraph.

Another topic that should be mentioned is the spin state of the Cooper pair. The wave function of a pair of electrons must be antisymmetric under the interchange of the two particles. If the potential is symmetric under spatial inversion, then, if the spatial part of the wave function is even under particle interchange, the spin part must be odd (a spin 0, singlet state) whereas if the spatial part of the wave function is antisymmetric, then the spin part must be even (a spin 1, triplet state). In the original BCS theory, it was assumed that the effective attraction due to ionic lattice distortion, the so-called phonon mechanism, would be minimized by a nodeless, spatially symmetric pair wave function which was called s-wave, by analogy to the s-state of an atom. This means that the states forming a Cooper pair are of the form, $(\mathbf{k}_1, \alpha)$ and $(-\mathbf{k}_1, -\alpha)$. Within a few years of the publication of BCS theory, Anderson and Morel (1961) and Balian and Werthamer (1963) produced theories for odd parity, triplet superconductivity which turned out to be directly applicable to the neutral fermionic superfluid, ^{3}He, where the strongly repulsive core of the He–He potential favors odd parity "p-wave" pairing with a triplet spin state. Subsequently, a number of other different kinds of space/spin pair wave functions have been discovered, most notably the spin-singlet, d-wave pairing functions of the high T_c superconductors. Certain crystalline materials are thought to exhibit triplet p-wave superconductivity analogous to ^{3}He, as well as triplet f-wave. Furthermore, certain non-centrosymmetric systems may have mixtures of singlet and triplet pairing, particularly in the presence of spin–orbit coupling.

BCS theory involved two, rather distinct but equally important discoveries. One was understanding the nature of the correlations, i.e., Cooper pairing, which arises from almost any kind of attractive electron–electron interaction, and the successful treatment of these correlations in mean-field theory, which is the subject of the next few sections. The second major discovery was the actual mechanism that drives superconductivity in essentially all superconductors that were known at that time, i.e., the phonon mechanism. This common mechanism for a broad range of superconductors accounted for the many striking similarities of superconductivity in different materials as well as the fact that superconductivity was at that time exclusively a low-temperature phenomenon. Subsequent discoveries demonstrated that the phonon mechanism is not the only way that superconductivity can arise. In particular, we now know that stronger direct electron–electron interactions, along with exchange, can lead to interactions that are magnetic in nature, but which nevertheless support different forms of superconductivity, including high-temperature superconductivity.

12.4 Mean-Field Theory for Superconductivity

To describe the statistical mechanics of the fermion pairing instability, we will proceed as in the preceding chapter. Namely, we introduce a variational (or trial) grand canonical density matrix, $\boldsymbol{\rho}_{\rm tr}$, parametrized by the trial Hamiltonian, $\mathcal{H}_{\rm tr}$ as

$$\boldsymbol{\rho}_{\rm tr} = \frac{e^{-\beta\mathcal{H}_{\rm tr}}}{{\rm Tr}e^{-\beta\mathcal{H}_{\rm tr}}} , \tag{12.3}$$

where the trace includes a sum over all possible values of $\hat{n}$ and the trial Hamiltonian (actually $\mathcal{H} - \mu\hat{n}$) is

$$\mathcal{H}_{\rm tr} = \sum_{\mathbf{k},\alpha}\Big[E_{\rm tr}(\mathbf{k}) - \mu\Big]a^+_{\mathbf{k},\alpha}a_{\mathbf{k},\alpha} + \sum_{\mathbf{k}}\Delta_{\mathbf{k}}a^+_{\mathbf{k},\uparrow}a^+_{-\mathbf{k},\downarrow} + \sum_{\mathbf{k}}\Delta^*_{\mathbf{k}}a_{-\mathbf{k},\downarrow}a_{\mathbf{k},\uparrow} , \tag{12.4}$$

where the trial parameters are $E_{\rm tr}(\mathbf{k})$ and $\Delta_{\mathbf{k}}$. Since the perturbation is relatively weak, $E_{\rm tr}(\mathbf{k})$ differs only perturbatively from the bare kinetic energy $E_0(\mathbf{k})$ (except near the Fermi surface) and Δ will be small compared to the width of the band, $E_0(\mathbf{k})$.

As in the case of Bose condensation, this Hamiltonian is not to be taken literally in that we do not actually create or destroy pairs of electrons. Since for the grand canonical ensemble the system is assumed to be in contact with a particle bath, pair creation, or annihilation is implicitly accompanied by annihilation or creation of a pair of particles each having an energy equal to the chemical potential. We include only terms that create up-spin–down-spin pairs which lead to spin-singlet pairing in the variational Hamiltonian. For p-wave pairing as in superfluid ^{3}He, one would instead include terms that describe triplet pairing.

As before, the next step is to diagonalize $\mathcal{H}_{\rm tr}$. Then the fermionic states and excitation energies of the diagonal Hamiltonian will allow us to calculate the probability distribution of occupied states as well as all averages of products of the original operators, $a^+_{\mathbf{k},\alpha}$, etc. via Wick's theorem. Looking at Eq. (12.4), we see that $a^+_{\mathbf{k},\uparrow}$ multiplies a linear combination of $a_{\mathbf{k},\uparrow}$ and $a^+_{-\mathbf{k},\downarrow}$. This implies that "normal mode" operators will involve linear combinations of the form

$$\gamma^+_{\mathbf{k},\uparrow} = u_{\mathbf{k}} a^+_{\mathbf{k},\uparrow} + v_{\mathbf{k}} a_{-\mathbf{k},\downarrow} \tag{12.5a}$$

$$\gamma_{-\mathbf{k},\downarrow} = u'_{\mathbf{k}} a_{-\mathbf{k},\downarrow} + v'_{\mathbf{k}} a^+_{\mathbf{k},\uparrow} \ . \tag{12.5b}$$

Note that this diagonalization process is similar to that for bosons and uses a number-nonconserving (Bogoliubov) transformation. We may impose the requirement that $u_{\mathbf{k}}$ and $u'_{\mathbf{k}}$ are positive real. Then the requirement that the transformed operators $\gamma_{\mathbf{k},\alpha}$ obey the same anticommutation relations as the a's yields $u'_{\mathbf{k}} = u_{\mathbf{k}}$ and $v'_{\mathbf{k}} = -v^*_{\mathbf{k}}$, where

$$u^2_{\mathbf{k}} + |v_{\mathbf{k}}|^2 = 1 \ . \tag{12.6}$$

The inverse transformation is

$$a_{\mathbf{k}\uparrow} = u_{\mathbf{k}} \gamma_{\mathbf{k}\uparrow} - v^*_{\mathbf{k}} \gamma^+_{-\mathbf{k}\downarrow} \tag{12.7a}$$

$$a^+_{-\mathbf{k}\downarrow} = v_{\mathbf{k}} \gamma_{\mathbf{k}\uparrow} + u_{\mathbf{k}} \gamma^+_{-\mathbf{k}\downarrow} \ . \tag{12.7b}$$

Substituting into $\mathcal{H}_{\rm tr}$ we find

$$\begin{aligned}
\mathcal{H}_{\rm tr} = \sum_{\mathbf{k}} (E_{\rm tr}(\mathbf{k}) - \mu) \Big\{ & \left(u_{\mathbf{k}} \gamma^+_{\mathbf{k},\uparrow} - v_{\mathbf{k}} \gamma_{-\mathbf{k},\downarrow}\right) \left(u_{\mathbf{k}} \gamma_{\mathbf{k},\uparrow} - v^*_{\mathbf{k}} \gamma^+_{-\mathbf{k},\downarrow}\right) \\
& + \left(u_{\mathbf{k}} \gamma^+_{-\mathbf{k},\downarrow} + v_{\mathbf{k}} \gamma_{\mathbf{k},\uparrow}\right) \left(u_{\mathbf{k}} \gamma_{-\mathbf{k},\downarrow} + v^*_{\mathbf{k}} \gamma^+_{\mathbf{k},\uparrow}\right) \Big\} \\
& + \sum_{\mathbf{k}} \Delta_{\mathbf{k}} \left(u_{\mathbf{k}} \gamma^+_{\mathbf{k},\uparrow} - v_{\mathbf{k}} \gamma_{-\mathbf{k},\downarrow}\right) \left(u_{\mathbf{k}} \gamma^+_{-\mathbf{k},\downarrow} + v_{\mathbf{k}} \gamma_{\mathbf{k},\uparrow}\right) \\
& + \sum_{\mathbf{k}} \Delta^*_{\mathbf{k}} \left(u_{\mathbf{k}} \gamma_{-\mathbf{k},\downarrow} + v^*_{\mathbf{k}} \gamma^+_{\mathbf{k},\uparrow}\right) \left(u_{\mathbf{k}} \gamma_{\mathbf{k},\uparrow} - v^*_{\mathbf{k}} \gamma^+_{-\mathbf{k},\downarrow}\right) \ .
\end{aligned} \tag{12.8}$$

In the second line, we replaced $\mathbf{k}$ by $-\mathbf{k}$. We require the coefficient of $\gamma^+_{\mathbf{k},\uparrow}\gamma^+_{-\mathbf{k},\downarrow}$ to vanish, so that

$$-2\,(E_{\rm tr}(\mathbf{k})-\mu)\,u_{\mathbf{k}}v^*_{\mathbf{k}}+\Delta_{\mathbf{k}}u^2_{\mathbf{k}}-\Delta^*_{\mathbf{k}}v^{*2}_{\mathbf{k}}=0\,. \tag{12.9}$$

If we write $\Delta_{\mathbf{k}}=|\Delta_{\mathbf{k}}|e^{i\phi_{\mathbf{k}}}$, then the solution to Eq. (12.9) is of the form

$$u_{\mathbf{k}}=|u_{\mathbf{k}}| \tag{12.10a}$$

$$v_{\mathbf{k}}=|v_{\mathbf{k}}|\,e^{-i\phi_{\mathbf{k}}}, \tag{12.10b}$$

and Eq. (12.9) yields

$$-2\,(E_{\rm tr}(\mathbf{k})-\mu)\,u_{\mathbf{k}}|v_{\mathbf{k}}|+|\Delta_{\mathbf{k}}|(u^2_{\mathbf{k}}-|v_{\mathbf{k}}|^2)=0\,. \tag{12.11}$$

The equation is satisfied if

$$u_{\mathbf{k}}=\cos\theta_{\mathbf{k}}\,,\qquad |v_{\mathbf{k}}|=\sin\theta_{\mathbf{k}}\,, \tag{12.12}$$

where $\theta_{\mathbf{k}}$ is determined by

$$\sin 2\theta_{\mathbf{k}}=2|u_{\mathbf{k}}||v_{\mathbf{k}}|=\frac{|\Delta_{\mathbf{k}}|}{\epsilon_{\mathbf{k}}} \tag{12.13}$$

$$\cos 2\theta_{\mathbf{k}}=|u_{\mathbf{k}}|^2-|v_{\mathbf{k}}|^2=\frac{E_{\rm tr}(\mathbf{k})-\mu}{\epsilon_{\mathbf{k}}}\,, \tag{12.14}$$

where

$$\epsilon_{\mathbf{k}}=\sqrt{(E_{\rm tr}(\mathbf{k})-\mu)^2+|\Delta_{\mathbf{k}}|^2}\,. \tag{12.15}$$

The remaining terms in $\mathcal{H}_{\rm tr}$ can be evaluated in terms of these $u_{\mathbf{k}}$'s and $v_{\mathbf{k}}$'s. After some algebra, we obtain an expression similar to that for the Bose case,

$$\mathcal{H}_{\rm tr}=\sum_{\mathbf{k}}\{E_{\rm tr}(\mathbf{k})-\mu-\epsilon_{\mathbf{k}}\}+\sum_{\mathbf{k},\alpha}\epsilon_{\mathbf{k}}\gamma^+_{\mathbf{k},\alpha}\gamma_{\mathbf{k},\alpha}\,. \tag{12.16}$$

In view of this diagonalized form, $\gamma^+_{\mathbf{k},\alpha}$ is said to create "quasiparticles" and $\epsilon_{\mathbf{k}}$ is the quasiparticle energy.

Then we have

$$\langle\gamma^+_{\mathbf{k},\alpha}\gamma_{\mathbf{k},\alpha}\rangle_{\rm tr}\equiv{\rm Tr}\left[\gamma^+_{\mathbf{k},\alpha}\gamma_{\mathbf{k},\alpha}\boldsymbol{\rho}_{\rm tr}\right]=\frac{1}{e^{\beta\epsilon_k}+1}\equiv f_k, \tag{12.17}$$

and the averages of $a^+_{\mathbf{k},\alpha}$ and $a_{\mathbf{k},\alpha}$ can be expressed as

$$\langle a_{\mathbf{k},\alpha}^{+} a_{\mathbf{k},\alpha} \rangle_{\text{tr}} = \frac{(E_{\text{tr}}(\mathbf{k}) - \mu)\left(f_k - \frac{1}{2}\right)}{\epsilon_k} + \frac{1}{2} \tag{12.18a}$$

$$\langle a_{-\mathbf{k},\downarrow} a_{\mathbf{k},\uparrow} \rangle_{\text{tr}} = \frac{\Delta_{\mathbf{k}}\left(f_k - \frac{1}{2}\right)}{\epsilon_k} . \tag{12.18b}$$

These results give $\mathcal{H}_{\text{tr}}$ and the various grand canonical averages in terms of the variational parameters $E_{\text{tr}}(\mathbf{k})$ and $\Delta_{\mathbf{k}}$, and in the next section we will minimize the free energy with respect to these parameters. The transformation from the a's to the γ's is referred to as a transformation from particles to quasiparticles.

We will see in a moment that $E_{\text{tr}}(\mathbf{k})$ is essentially identical to $E_0(\mathbf{k})$ except for energies very near the Fermi energy. Therefore, we anticipate that the energy dependence of the transformation coefficients u and v should be as shown in Fig. 12.3, where $|v_k|^2$ and $|u_k|^2$ are plotted versus $(E - \mu)/\Delta$, and $\Delta_{\mathbf{k}}$ is assumed to be small compared to the Fermi energy. For energies well below the Fermi energy μ (so that $E_{\text{tr}}(\mathbf{k}) - \mu$ is large and negative), $\cos 2\theta \approx -1$, so that $u = 0$ and $v = 1$. In this limit, the quasiparticle creation operator $\gamma_{\mathbf{k}}^{+}$ is essentially the particle destruction operator, $a_{\mathbf{k}}$. The quasiparticle energy far from the Fermi energy is simply the magnitude of $E_0(\mathbf{k}) - \mu$. However, near the Fermi energy, the quasiparticle energy exhibits a gap equal to $\Delta(\mathbf{k})$. (Many observable quantities involve creation of two quasiparticles in which case the "gap" or thermal activation energy observed experimentally would be some average over the Fermi surface of $2\Delta(\mathbf{k})$.)

Using the above results one can address the question of how the original single-particle states contribute to the ground state many-body wave function. This is equivalent to calculating the quantity $\sum_{\alpha} \langle a_{\mathbf{k},\alpha}^{+} a_{\mathbf{k},\alpha} \rangle_{\text{tr}}$ at $T = 0$, which describes which of the original $(\mathbf{k}, \alpha)$ states are occupied at $T = 0$. A straightforward calculation shows that this is equal to $2|v_{\mathbf{k}}|^2$. The quantity $|v_{\mathbf{k}}|^2$ is shown in the left-hand panel of Fig. 12.3. It resembles a Fermi distribution for a finite temperature $\text{T} = \Delta$.

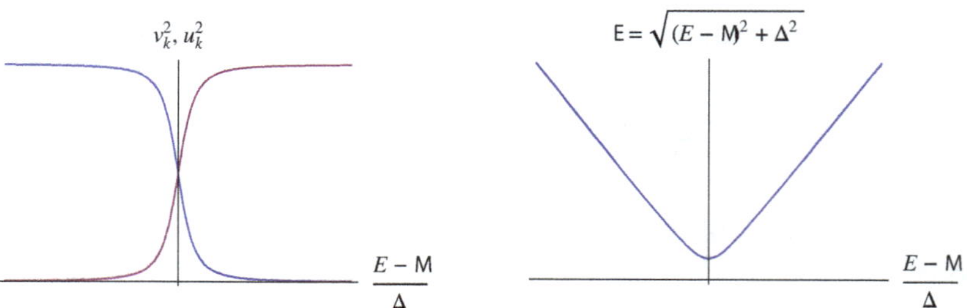

Fig. 12.3 Left: The coefficients $|v_k|^2$ and $|u_k|^2$ according to Eqs. (12.12–12.14). Right: Quasiparticle energy ϵ versus the particle energy measured from the Fermi energy in units of Δ from Eq. (12.15)

12.5 Minimizing the Free Energy

As we did for quantum fluids in the preceding chapter, we again consider the variational density matrix $\boldsymbol{\rho}_{\rm tr}$ for the interacting Fermi gas to depend parametrically on all the averages that it can be used to generate, namely, $\langle a^+_{\mathbf{k},\alpha} a_{\mathbf{k},\alpha}\rangle$, $\langle a^+_{\mathbf{k},\uparrow} a^+_{-\mathbf{k},\downarrow}\rangle$, $\langle a_{-\mathbf{k},\downarrow} a_{\mathbf{k},\uparrow}\rangle$, and $f_{\mathbf{k}}$. The trial free energy (actually the trial grand potential) is

$$\begin{aligned}\Omega_{\rm tr} &= \sum_{\mathbf{k},\alpha}\left(\frac{\hbar^2k^2}{2m}-\mu\right)\langle a^+_{\mathbf{k},\alpha} a_{\mathbf{k},\alpha}\rangle_{\rm tr}\\ &-\frac{1}{2\Omega_0}\sum_{\mathbf{k}_1,\mathbf{k}_2,\mathbf{q}}\sum_{\alpha,\beta} V(\mathbf{q})\langle a^+_{\mathbf{k}_1+\mathbf{q},\alpha} a^+_{\mathbf{k}_2-\mathbf{q},\beta} a_{\mathbf{k}_2,\beta} a_{\mathbf{k}_1,\alpha}\rangle_{\rm tr}\\ &+kT\sum_{\mathbf{k},\alpha}\big[f_{\mathbf{k}}\ln f_{\mathbf{k}}+(1-f_{\mathbf{k}})\ln(1-f_{\mathbf{k}})\big]\,,\end{aligned} \tag{12.19}$$

where we used the expression for the entropy from Eq. (11.19), and we have replaced $E_0(\mathbf{k})$ by $\hbar^2k^2/2m$. As before, we use Wick's theorem to evaluate the average of the four-operator term as a sum over products of all possible averages of operators taken in pairs. Thereby, we get

$$\begin{aligned}\Omega_{\rm tr} &= \sum_{\mathbf{k},\alpha}\left(\frac{\hbar^2k^2}{2m}-\mu\right)\langle a^+_{\mathbf{k},\alpha} a_{\mathbf{k},\alpha}\rangle_{\rm tr}\\ &-\frac{1}{2\Omega_0}\sum_{\mathbf{k},\mathbf{k}'}\sum_{\alpha,\beta}\Big(V(0)-V(\mathbf{k}-\mathbf{k}')\delta_{\alpha,\beta}\Big)\langle a^+_{\mathbf{k},\alpha} a_{\mathbf{k},\alpha}\rangle_{\rm tr}\langle a^+_{\mathbf{k}',\beta} a_{\mathbf{k}',\beta}\rangle_{\rm tr}\\ &-\frac{1}{2\Omega_0}\sum_{\mathbf{k},\mathbf{k}'}\sum_{\alpha} V(\mathbf{k}-\mathbf{k}')\langle a^+_{\mathbf{k},\alpha} a^+_{-\mathbf{k},-\alpha}\rangle_{\rm tr}\langle a_{-\mathbf{k}',-\alpha} a_{\mathbf{k}',\alpha}\rangle_{\rm tr}\\ &+kT\sum_{\mathbf{k},\alpha}\big[f_{\mathbf{k}}\ln f_{\mathbf{k}}+(1-f_{\mathbf{k}})\ln(1-f_{\mathbf{k}})\big]\,.\end{aligned} \tag{12.20}$$

The trial parameters we seek to determine, $E_{\rm tr}(\mathbf{k})$ and $\Delta_{\mathbf{k}}$, appear implicitly in $f_{\mathbf{k}}$ and the averages of products of particle creation and destruction operators.

The most convenient way to perform the minimization with respect to the trial parameters is to minimize $\Omega_{\rm tr}$ with respect to variation of the various averages. However, in so doing we must take account of the fact that these averages are not all independent of one another. Indeed, from Eqs. (12.15), (12.18a), and (12.18b), we have the constraint

$$\left[\langle a^+_{\mathbf{k},\alpha} a_{\mathbf{k},\alpha}\rangle_{\rm tr}-\frac{1}{2}\right]^2+\left|\langle a_{-\mathbf{k},-\alpha} a_{\mathbf{k},\alpha}\rangle_{\rm tr}\right|^2=\left[f_k-\frac{1}{2}\right]^2. \tag{12.21}$$

Accordingly, we introduce Lagrange parameters $\lambda_{\mathbf{k},\alpha}$ and now have to minimize the effective potential $\Xi_{\rm tr}$ given by

$$
\begin{aligned}
\Xi_{\text{tr}} = &\sum_{\mathbf{k},\alpha}\left(\frac{\hbar^2k^2}{2m}-\mu\right)\langle a^+_{\mathbf{k},\alpha}a_{\mathbf{k},\alpha}\rangle_{\text{tr}} \\
&-\frac{1}{2\Omega_0}\sum_{\mathbf{k},\mathbf{k}'}\sum_{\alpha,\beta}\Big(V(0)-V(\mathbf{k}-\mathbf{k}')\delta_{\alpha,\beta}\Big)\langle a^+_{\mathbf{k},\alpha}a_{\mathbf{k},\alpha}\rangle_{\text{tr}}\langle a^+_{\mathbf{k}',\beta}a_{\mathbf{k}',\beta}\rangle_{\text{tr}} \\
&-\frac{1}{2\Omega_0}\sum_{\mathbf{k},\mathbf{k}'}\sum_{\alpha}V(\mathbf{k}-\mathbf{k}')\langle a^+_{\mathbf{k},\alpha}a^+_{-\mathbf{k},-\alpha}\rangle_{\text{tr}}\langle a_{-\mathbf{k}',-\alpha}a_{\mathbf{k}',\alpha}\rangle_{\text{tr}} \\
&+kT\sum_{\mathbf{k},\alpha}\big[f_{\mathbf{k}}\ln f_{\mathbf{k}}+(1-f_{\mathbf{k}})\ln(1-f_{\mathbf{k}})\big] \\
&+\sum_{\mathbf{k},\alpha}\lambda_{\mathbf{k},\alpha}\left\{\big[\langle a^+_{\mathbf{k},\alpha}a_{\mathbf{k},\alpha}\rangle_{\text{tr}}-\tfrac{1}{2}\big]^2+\big|\langle a_{-\mathbf{k},-\alpha}a_{\mathbf{k},\alpha}\rangle_{\text{tr}}\big|^2-\left[f_k-\tfrac{1}{2}\right]^2\right\}.
\end{aligned}
\tag{12.22}
$$

This is minimized by

$$
\begin{aligned}
\frac{\partial\,\Xi_{\text{tr}}}{\partial\langle a^+_{\mathbf{k},\alpha}a_{\mathbf{k},\alpha}\rangle_{\text{tr}}} = &\left(\frac{\hbar^2k^2}{2m}-\mu\right)-\frac{1}{\Omega_0}\sum_{\mathbf{k}'}\Big(V(0)-V(\mathbf{k}-\mathbf{k}')\Big)\langle a^+_{\mathbf{k}',\alpha}a_{\mathbf{k}',\alpha}\rangle_{\text{tr}} \\
&+2\lambda_{\mathbf{k},\alpha}\big[\langle a^+_{\mathbf{k},\alpha}a_{\mathbf{k},\alpha}\rangle_{\text{tr}}-\tfrac{1}{2}\big]=0\,.
\end{aligned}
\tag{12.23a}
$$

$$
\begin{aligned}
\frac{\partial\,\Xi_{\text{tr}}}{\partial\langle a^+_{\mathbf{k},\uparrow}a^+_{-\mathbf{k},\downarrow}\rangle_{\text{tr}}} = &-\frac{1}{\Omega_0}\sum_{\mathbf{k}'}V(\mathbf{k}-\mathbf{k}')\langle a_{-\mathbf{k}',\downarrow}a_{\mathbf{k}',\uparrow}\rangle_{\text{tr}} \\
&+2\lambda_{\mathbf{k},\alpha}\langle a_{-\mathbf{k},\downarrow}a_{\mathbf{k},\uparrow}\rangle_{\text{tr}}=0\,.
\end{aligned}
\tag{12.23b}
$$

$$
\frac{\partial\,\Xi_{\text{tr}}}{\partial f_{\mathbf{k}}}=2kT\ln\left(\frac{f_{\mathbf{k}}}{1-f_{\mathbf{k}}}\right)-4\lambda_{\mathbf{k},\alpha}\left(f_{\mathbf{k}}-\frac{1}{2}\right)=0\,. \tag{12.23c}
$$

These three equations, along with the constraint equation, determine the equilibrium values of the averages of pairs of fermion operators, as well as the Lagrange parameters $\lambda_{\mathbf{k}\alpha}$ and the equilibrium Fermi distribution functions $f_{\mathbf{k}}$, which, in turn, determine the variational parameters $E_{\text{tr}}(\mathbf{k})$ and $\Delta_{\mathbf{k}}$. We denote the minimized values by the subscript, "eq". However, for simplicity of notation, we will use the same symbols for the equilibrium gap, $\Delta_{\mathbf{k}}$ and for the equilibrium Fermi function, $f_{\mathbf{k}}$ as for the variational ones, and indicate explicitly which is meant in the text.

Combining Eqs. (12.17) and (12.23c), we get

$$
\lambda_{\mathbf{k},\alpha}=-\frac{\epsilon_{\mathbf{k}}}{2f_{\mathbf{k}}-1}\,. \tag{12.24}
$$

Now we rewrite Eq. (12.23a) using Eqs. (12.18a) and (12.24) to get

$$
\begin{aligned}
E_{\rm eq}(\mathbf{k}) &= \left(\frac{\hbar^2 k^2}{2m}\right) - \frac{1}{\Omega_0}\sum_{\mathbf{k}'}\Big(V(0) - V(\mathbf{k}-\mathbf{k}')\Big)\langle a^+_{\mathbf{k}',\alpha} a_{\mathbf{k}',\alpha}\rangle_{\rm eq}\,, \\
&= \left(\frac{\hbar^2 k^2}{2m}\right) - \frac{1}{\Omega_0}\sum_{\mathbf{k}'}\Big(V(0) - V(\mathbf{k}-\mathbf{k}')\Big)\Bigg(\frac{\left(E_{\rm eq}(\mathbf{k}') - \mu\right)\left(f_{\mathbf{k}'} - \frac{1}{2}\right)}{\epsilon_{\mathbf{k}'}} + \frac{1}{2}\Bigg)\,,
\end{aligned}
\tag{12.25}
$$

where, in the last line, we have used Eq (12.18a) evaluated for the equilibrium values of the various functions.

Next we rewrite Eq. (12.23b) using Eqs. (12.18b) and (12.24) to get

$$
\Delta_{\mathbf{k}} = -\frac{1}{\Omega_0}\sum_{\mathbf{k}'} V(\mathbf{k}-\mathbf{k}')\langle a_{-\mathbf{k}',\downarrow} a_{\mathbf{k}',\uparrow}\rangle_{\rm eq}\,. \tag{12.26}
$$

Then, using Eq. (12.18b) on the rhs, we obtain the self-consistent equation for $\Delta_{\mathbf{k}}$ from Eq. (12.26) as

$$
\Delta_{\mathbf{k}} = -\frac{1}{\Omega_0}\sum_{\mathbf{k}'} V(\mathbf{k}-\mathbf{k}')\frac{\Delta_{\mathbf{k}'}}{\epsilon_{\mathbf{k}'}}\left(f_{\mathbf{k}'} - \frac{1}{2}\right)\,. \tag{12.27}
$$

The $\Delta_{\mathbf{k}}$ determined by this equation is called the "self-consistent gap."

The Fermi factor has the value

$$
f_{\mathbf{k}'} - \frac{1}{2} = \frac{1}{e^{\beta\epsilon_{\mathbf{k}'}} + 1} - \frac{1}{2} = -\frac{1}{2}\tanh\frac{\beta\epsilon_{\mathbf{k}'}}{2}\,, \tag{12.28}
$$

so that

$$
\Delta_{\mathbf{k}} = \frac{1}{\Omega_0}\sum_{\mathbf{k}'} V(\mathbf{k}-\mathbf{k}')\frac{\Delta_{\mathbf{k}'}}{2\epsilon_{\mathbf{k}'}}\tanh\frac{\beta\epsilon_{\mathbf{k}'}}{2}\,. \tag{12.29}
$$

Equations (12.25) and (12.29) are both self-consistent equations in the sense that they determine the equilibrium values, $E_{\rm eq}(\mathbf{k})$ and $\Delta_{\mathbf{k}}$, in terms of functions which themselves depend on $E_{\rm eq}(\mathbf{k})$ and $\Delta_{\mathbf{k}}$. In particular, $f_{\mathbf{k}'}$ and $\epsilon_{\mathbf{k}'}$ are functions of $E_{\rm eq}(\mathbf{k}')$ and $\Delta_{\mathbf{k}'}$.

12.6 Solution to Self-consistent Equations

It is useful to consider the nature of the solution to Eqs. (12.25) and (12.29). In particular, we expect these equations to define not only a low-temperature condensed phase, which describes superconductivity but also, of course, the higher temperature normal metal phase. Equation (12.25), which describes how the interaction renormalizes the effective kinetic energy of the electrons, has two main consequences. The first is to shift the chemical potential which is a parameter that is ultimately constrained by the density of conduction electrons. The second is to renormalize the electron effective mass. To simplify the following pedagogical discussion, we will assume that the effective mass is essentially the same in the normal and superconducting phases. This means that we solve Eq. (12.25) to determine the effective mass, which we call m^* for the normal state, and hold that fixed in the superconducting phase, which amounts to reducing the variational freedom, since only $\Delta_{\mathbf{k}}$ is then varied. However, we expect $\Delta_{\mathbf{k}}$ to contain most of the effects of condensation into the superconducting phase.

We see that Eq. (12.29) is indeed satisfied by $\Delta_{\mathbf{k}} = 0$ and this solution must describe the normal metal phase. For the case of the Ising model, there is a transition from a high-temperature disordered phase to a low-temperature ordered phase at a temperature comparable to that of the spin–spin interaction that drives the order. In the case of electronic systems, if the attractive interaction was on an electronic scale of order eV's, then the ordering temperature might be expected to be of order an eV (i.e., 10^4K). However, for this problem, the electron–electron interaction is mediated by the slow ionic motion described above, and the potential $V(\mathbf{q})$ is only effective acting on electron states within a phonon energy of the Fermi energy, where the typical phonon energy is of the order of the Debye energy, $\hbar\omega_D \approx 10^2$K. In fact, we will see that the energy gap, $\Delta_{\mathbf{k}}$, that arises a low temperatures is typically much smaller than $\hbar\omega_D$, as is the ordering temperature which is comparable to $\Delta_{\mathbf{k}}(T = 0)$.

In any event, the structure of Eq. (12.29) is just what we want. If $\Delta_{\mathbf{k}}$ is zero, then the "anomalous average" $\langle a_{-\mathbf{k}\downarrow} a_{\mathbf{k}\uparrow} \rangle$ vanishes and we have a normal metal phase. But we know that such self-consistent equations can also have a low-temperature nonzero solution for $\Delta_{\mathbf{k}}$ and we therefore expect that $\Delta_{\mathbf{k}}$ will play the role of an order parameter which is nonzero in the phase that describes superconductivity.

In order to solve Eq. (12.29) for $\Delta_{\mathbf{k}}$, it is necessary to make some simplifying assumptions about the form of the potential. We approximate $V(\mathbf{k} - \mathbf{k}')$ by the so-called separable form,

$$V(\mathbf{k} - \mathbf{k}') \to V f(k) f(k'), \tag{12.30}$$

where $f(k) = 1$ for $\mathbf{k}$ on the Fermi surface, and $f(k) \to 0$ beyond some cutoff distance in k-space from the Fermi surface. This seemingly artificial form accommodates two cases of general interest. For the case of interaction via the phonon mechanism treated by Bardeen, Cooper and Schrieffer (BCS), the phonon can carry any momentum, while its energy is typically small compared to electronic energy scales. Then the initial and final states of electrons scattering by exchanging a phonon will all be within $\hbar\omega_D$ of the Fermi energy, and hence are confined to a thin shell around the Fermi surface. The second case of interest is that of short-range electron–electron scattering in a one-band model. In this case, scattering is equally likely between all states in the single band. The k-sum ranges over the Brillouin zone, and $f(k) = 1$ for all scattering processes. These two limits are referred to as "weak coupling" and "strong coupling," respectively. Here we will treat only the case of weak coupling.

For the interaction of Eq. (12.30), $\Delta_{\mathbf{k}}$ will be of the form

$$\Delta_{\mathbf{k}} = \Delta f(k) . \tag{12.31}$$

Then one factor of $\Delta f(k)$, which is assumed to be nonzero, can be canceled from each side of Eq. (12.29) which becomes

$$1 = V \frac{1}{\Omega_0} \sum_{\mathbf{k}'} f(\mathbf{k}')^2 \frac{\tanh \frac{\beta \epsilon_{\mathbf{k}'}}{2}}{2\epsilon_{\mathbf{k}'}} . \tag{12.32}$$

For weak-coupling superconductors, it is convenient to linearize the electron band energy around the chemical potential $\mu = E_F = \hbar^2 k_F^2/2m^*$. Then

$$\begin{aligned} \frac{\hbar^2 k^2}{2m^*} - \frac{\hbar^2 k_F^2}{2m^*} &= \frac{\hbar^2}{2m^*}(k + k_F)(k - k_F) \\ &\approx \underbrace{\frac{\hbar k_F}{m^*}}_{v_F} \hbar(k - k_F) . \end{aligned} \tag{12.33}$$

Note that, in terms of v_F, $\hbar v_F k_F = 2E_F$.

12.6.1 The Energy Gap at $T = 0$

For $T = 0$ the tanh equals 1. To simplify the integrals and make them analytically tractable, we consider the weak-coupling limit. Thus, we take $f(k) = 1$ for $\hbar v_F|\mathbf{k} - \mathbf{k}_F| < \hbar\omega_D$ where $\mathbf{k}_F$ is the Fermi wavevector along the direction of $\mathbf{k}$, and $f(k) = 0$ otherwise. Then the self-consistency condition for the gap at $T = 0$ becomes

$$\begin{aligned} 1 &\approx \frac{V k_F^2}{2\pi^2} \int_{k_-}^{k_+} \frac{dk}{\sqrt{|\Delta|^2 + (\hbar v_F k)^2}} \\ &= \frac{V k_F^2}{2\pi^2 \hbar v_F} \int_{-\hbar\omega_D}^{\hbar\omega_D} \frac{d\epsilon}{\sqrt{|\Delta|^2 + \epsilon^2}} \\ &\approx \frac{V k_F^2}{\pi^2 \hbar v_F} \ln\left(\frac{2\hbar\omega_D}{\Delta}\right) , \end{aligned} \tag{12.34}$$

where $k_\pm = \pm\omega_D/v_F$ and $\Delta \equiv \Delta(T = 0)$.

If we define a dimensionless coupling constant

$$g = \frac{V k_F^2}{\pi^2 \hbar v_F} = \frac{V k_F^3}{2\pi^2 E_F} , \tag{12.35}$$

(recalling that that V has the units of energy $\times$ length^3) then

$$\Delta = 2\hbar\omega_D e^{-1/g} . \tag{12.36}$$

Note that Δ is a nonanalytic function of g, and that the gap is nonzero, in principle, even for infinitesimal g. This nonanalyticity indicates that this instability will not be found in any finite order of perturbation theory. Since the theory only involves states near the Fermi surface, it is useful to generalize the above result by expressing the coupling constant g in terms of the density of states (per unit volume) at the Fermi level, given in Eq. (7.26). Then we have the general result

$$g = \rho(E_F)V . \tag{12.37}$$

12.6.2 Solution for the Transition Temperature

For nonzero temperature, the self-consistency relation is

$$1 = \frac{g}{2}\int_{-\hbar\omega_D}^{\hbar\omega_D} \frac{\tanh\frac{\beta}{2}\sqrt{|\Delta(T)|^2+\epsilon^2}}{\sqrt{|\Delta(T)|^2+\epsilon^2}} d\epsilon . \tag{12.38}$$

This nonlinear equation can be solved numerically for the order parameter, $\Delta(T)$. At $T = T_c$, $\Delta = 0$, and

$$\begin{aligned} 1 &= \frac{g}{2}\int_{-\hbar\omega_D}^{\hbar\omega_D} \frac{\tanh\frac{\epsilon}{2T_c}}{\epsilon} d\epsilon \\ &= g\int_0^{\frac{\hbar\omega_D}{2T_c}} \frac{\tanh x}{x} dx \\ &= g\left(\tanh\frac{\hbar\omega_D}{2T_c}\ln\frac{\hbar\omega_D}{2T_c} - \int_0^{\frac{\hbar\omega_D}{2T_c}} \ln x \,\mathrm{sech}^2 x dx\right) . \end{aligned} \tag{12.39}$$

For weak coupling, $g \ll 1$, we shall see that $\hbar\omega_D/T_c \gg 1$, and so we can set this quantity equal to ∞ in the tanh and in the integral. The result is

$$1 \approx g\left(\ln\frac{\hbar\omega_D}{2T_c} + 0.81878\right) . \tag{12.40}$$

Solving for T_c gives

$$T_c = 1.134\hbar\omega_D e^{-1/g} , \tag{12.41}$$

justifying the assumption that $T_c \ll \hbar\omega_D$ for $g << 1$. Comparing this result to Eq. (12.36), we obtain the famous weak-coupling relation

$$2\Delta = 3.53 T_c \,. \tag{12.42}$$

12.7 Free Energy of a BCS Superconductor

To conclude this topic, we evaluate the equilibrium free energy for the variational mean-field theory described above. The equilibrium free energy for fixed volume and chemical potential is given by Eq. (12.20), evaluated for the values of thermal averages which minimize the full Eq. (12.22) which includes the constraint, Eq. (12.21). To begin, we write the equilibrium value of Eq. (12.20) as

$$\begin{aligned}\Omega_{\rm eq} &= \sum_{\mathbf{k},\alpha}\left(\frac{\hbar^2 k^2}{2m^*} - \mu\right)\langle a^+_{\mathbf{k},\alpha} a_{\mathbf{k},\alpha}\rangle_{\rm eq} \\ &\quad - \frac{1}{2\Omega_0}\sum_{\mathbf{k},\mathbf{k}'}\sum_{\alpha} V(\mathbf{k}-\mathbf{k}')\langle a^+_{\mathbf{k},\alpha} a^+_{-\mathbf{k},-\alpha}\rangle_{\rm eq}\langle a_{-\mathbf{k}',-\alpha} a_{\mathbf{k}',\alpha}\rangle_{\rm eq} \\ &\quad + kT\sum_{\mathbf{k},\alpha}\big[f_{\mathbf{k}}\ln f_{\mathbf{k}} + (1-f_{\mathbf{k}})\ln(1-f_{\mathbf{k}})\big]\,, \end{aligned} \tag{12.43}$$

where we have incorporated the Hartree–Fock terms into an effective mass, and the various averages and $f_{\mathbf{k}}$ are the ones that minimize Eq. (12.22). Then, using Eq. (12.26) we can rewrite this as

$$\begin{aligned}\Omega_{\rm eq} &= \sum_{\mathbf{k},\alpha}\left(\frac{\hbar^2 k^2}{2m^*} - \mu\right)\langle a^+_{\mathbf{k},\alpha} a_{\mathbf{k},\alpha}\rangle_{\rm eq} - TS_{\rm eq} \\ &\quad + \frac{1}{2}\sum_{\mathbf{k}}\Delta_{\mathbf{k}}\langle a^+_{\mathbf{k},\uparrow} a^+_{-\mathbf{k},\downarrow}\rangle_{\rm eq} + \frac{1}{2}\sum_{\mathbf{k}}\Delta^*_{\mathbf{k}}\langle a_{-\mathbf{k},\downarrow} a_{\mathbf{k},\uparrow}\rangle_{\rm eq} \end{aligned} \tag{12.44}$$

$$\begin{aligned} &= \langle \mathcal{H}_{\rm tr}\rangle_{\rm eq} - TS_{\rm eq} \\ &\quad - \frac{1}{2}\sum_{\mathbf{k}}\Delta_{\mathbf{k}}\langle a^+_{\mathbf{k},\uparrow} a^+_{-\mathbf{k},\downarrow}\rangle_{\rm eq} - \frac{1}{2}\sum_{\mathbf{k}}\Delta^*_{\mathbf{k}}\langle a_{-\mathbf{k},\downarrow} a_{\mathbf{k},\uparrow}\rangle_{\rm eq}, \end{aligned} \tag{12.45}$$

where we have added and subtracted a term equal to the second line of Eq. (12.44) in order to express the result in terms of $\mathcal{H}_{\rm tr}$. The quantity, $S_{\rm eq}$ in Eq. (12.44) is the mean-field entropy evaluated for the equilibrium parameters, i.e., using the equilibrium $f_{\mathbf{k}}$'s, while $\mathcal{H}_{\rm tr}$ is the mean field or trial Hamiltonian from Eq. (12.4) which was diagonalized above with the result given in Eq. (12.16). Its average $\langle \mathcal{H}_{\rm tr}\rangle_{\rm eq}$ is the mean-field internal energy, and $\langle \mathcal{H}_{\rm tr}\rangle_{\rm eq} - TS_{\rm eq}$ is the associated free energy. Since, except for an additive constant, $\mathcal{H}_{\rm tr}$ is quadratic in fermion operators, we know, from Eqs. (7.9) and (12.16), the exact result for this free energy,

$$\langle \mathcal{H}_{\mathrm{tr}} \rangle_{\mathrm{eq}} - T S_{\mathrm{eq}} = \sum_{\mathbf{k}} \left\{ \frac{\hbar^2 k^2}{2m^*} - \mu - \epsilon_{\mathbf{k}} \right\} - 2T \sum_{\mathbf{k}} \ln \left(1 + e^{-\beta \epsilon_{\mathbf{k}}} \right) . \tag{12.46}$$

It is worthwhile to emphasize here that the quantity Ω_{eq} is the free energy calculated variationally, using the "real" Hamiltonian and the mean-field density matrix. Combining the result (12.46) with the last line of Eq. (12.45) which can be rewritten using Eq. (12.18b) Ω_{eq} is given by

$$\Omega_{\mathrm{eq}}(\Delta, T) = \sum_{\mathbf{k}} \left\{ \frac{\hbar^2 k^2}{2m^*} - \mu - \epsilon_{\mathbf{k}} \right\} - \sum_{\mathbf{k}} \frac{|\Delta_{\mathbf{k}}|^2 \left(f_k - \frac{1}{2} \right)}{\epsilon_k} - 2T \sum_{\mathbf{k}} \ln \left(1 + e^{-\beta \epsilon_{\mathbf{k}}} \right). \tag{12.47}$$

The quantity $\delta E_{0,sc} = \Omega_{\mathrm{eq}}(\Delta, 0) - \Omega_{\mathrm{eq}}(0, 0)$ represents the superconducting condensation energy in the ground state, referenced to the ground state energy of the normal Fermi gas.

$$\begin{aligned} \delta E_{0,sc} = & \sum_{\mathbf{k}} \left\{ \left| \frac{\hbar^2 k^2}{2m^*} - \mu \right| - \sqrt{\left(\frac{\hbar^2 k^2}{2m^*} - \mu \right)^2 + |\Delta|^2} \right\} \\ & - \sum_{\mathbf{k}} \frac{|\Delta_{\mathbf{k}}|^2 \left(f_k - \frac{1}{2} \right)}{\epsilon_k}. \end{aligned} \tag{12.48}$$

The last term in Eq. (12.48) can be rewritten, using Eq. (12.32), as

$$\begin{aligned} - \sum_{\mathbf{k}} \frac{|\Delta_{\mathbf{k}}|^2 \left(f_k - \frac{1}{2} \right)}{\epsilon_k} &= \sum_{\mathbf{k}} \frac{|\Delta_{\mathbf{k}}|^2}{2\epsilon_k} \tanh \frac{\beta \epsilon_{\mathbf{k}'}}{2} \\ &= |\Delta|^2 \sum_{\mathbf{k}} \frac{f(\mathbf{k})^2}{2\epsilon_k} \tanh \frac{\beta \epsilon_{\mathbf{k}'}}{2} \\ &= \frac{\Delta^2}{V} \Omega_0. \end{aligned} \tag{12.49}$$

Then, following the treatment of Eqs. (12.32)–(12.34), we can write

$$\begin{aligned} \delta E_{0,sc} &= \frac{2\,\Omega_0 k_F^2}{(2\pi)^2 \hbar v_F} \int_{-\hbar\omega_D}^{\hbar\omega_D} d\epsilon \left\{ |\epsilon| - \sqrt{\epsilon^2 + \Delta^2} \right\} + \frac{\Delta^2}{V} \Omega_0 \\ &\approx \frac{\Omega_0 g}{V} \left\{ (\hbar\omega_D)^2 - \left[(\hbar\omega_D)^2 + \frac{1}{2}\Delta^2 + \Delta^2 \underbrace{\ln \left(\frac{2\hbar\omega_D}{\Delta} \right)}_{\frac{1}{g}} \right] \right\} + \frac{\Delta^2}{V} \Omega_0 \\ &= -\frac{g\Delta^2}{2V} \Omega_0. \end{aligned} \tag{12.50}$$

The approximations in the second line above include expanding $\sqrt{(\hbar\omega_D)^2+\Delta^2}$ to order Δ^2 and neglecting Δ^2 compared to $(\hbar\omega_D)^2$ in the numerator of the logarithm.

There has been considerable cancelation among the various terms in deriving $\delta E_{0,sc}$. In the normal Fermi gas, interactions shift the energy by an amount of order $-VN^2/\Omega_0$. At first, it appeared that the correction to the energy due to the nonzero value of Δ was of order $-\Delta^2\Omega_0/V$. However, the actual lowering of the ground state energy is smaller by an extra factor of $g=\rho(E_F)V \ll 1$. Using this representation, the condensation energy is

$$\delta E_{0,sc} = -\frac{1}{2}\Delta^2\Omega_0\rho(E_F) \sim -\frac{\Delta^2}{E_F}N \,. \tag{12.51}$$

This very small energy shift (usually $\Delta < 1$ meV and $E_F > 1$ eV) is the condensation energy of the superconducting state.

12.8 Anderson Spin Model

About a year after BCS published their landmark theory, P. W. Anderson, then at Bell Labs, offered an alternative way of looking at the BCS ground state and its excitations which is particularly well suited for a statistical mechanics text and, in particular, to our discussion of mean-field theory (Anderson 1958).

Anderson started from what BCS had called the "reduced Hamiltonian." (Indeed he described his theory as an "Improved Treatment of the B.C.S. Reduced Hamiltonian.")

$$\begin{aligned}\mathcal{H}_{RED} &= \sum_{\mathbf{k}} E(\mathbf{k})\left(n_{\mathbf{k}\uparrow}+n_{-\mathbf{k}\downarrow}\right)\\ &\quad -\frac{1}{\Omega_0}\sum_{\mathbf{k},\mathbf{q}} V(\mathbf{k}-\mathbf{q})a^+_{\mathbf{k},\uparrow}a^+_{-\mathbf{k},\downarrow}a_{-\mathbf{q},\downarrow}a_{\mathbf{q},\uparrow}\,.\end{aligned} \tag{12.52}$$

Then referring the energy to the chemical potential and dropping an irrelevant constant, $\mathcal{H}_{RED}$ may be written as

$$\begin{aligned}\mathcal{H}_{RED} &= -\sum_{\mathbf{k}} (E(\mathbf{k})-\mu)\left(1-n_{\mathbf{k}\uparrow}-n_{-\mathbf{k}\downarrow}\right)\\ &\quad -\frac{1}{\Omega_0}\sum_{\mathbf{k},\mathbf{q}} V(\mathbf{k}-\mathbf{q})a^+_{\mathbf{k},\uparrow}a^+_{-\mathbf{k},\downarrow}a_{-\mathbf{k},\downarrow}a_{\mathbf{k},\uparrow}\,.\end{aligned} \tag{12.53}$$

This $\mathcal{H}_{RED}$ may be viewed as the Hamiltonian for a system of ($\mathbf{k}\uparrow$, $-\mathbf{k}\downarrow$) pair states which may be either occupied or empty. For a particular ($\mathbf{k}\uparrow$, $-\mathbf{k}\downarrow$), there are two possibilities:

$$\begin{pmatrix} 0 \\ 1 \end{pmatrix} \quad \rightarrow \text{ The pair state is full.} \tag{12.54a}$$

$$\begin{pmatrix} 1 \\ 0 \end{pmatrix} \quad \rightarrow \text{ The pair state is empty.} \tag{12.54b}$$

In this basis, the operators appearing in $\mathcal{H}_{RED}$ may be represented by 2×2 matrices,

$$1 - n_{\mathbf{k}\uparrow} - n_{-\mathbf{k}\downarrow} = \begin{pmatrix} 1 & 0 \\ 0 & -1 \end{pmatrix} \tag{12.55a}$$

$$a^+_{\mathbf{k},\uparrow} a^+_{-\mathbf{k},\downarrow} = \begin{pmatrix} 0 & 0 \\ 1 & 0 \end{pmatrix} \quad \text{empty} \rightarrow \text{full,} \tag{12.55b}$$

$$a_{-\mathbf{k},\downarrow} a_{\mathbf{k},\uparrow} = \begin{pmatrix} 0 & 1 \\ 0 & 0 \end{pmatrix} \quad \text{full} \rightarrow \text{empty.} \tag{12.55c}$$

These matrices are the Pauli operators σ_z, σ^-, and σ^+, one for each ($\mathbf{k} \uparrow, -\mathbf{k} \downarrow$)) pair state. The Pauli operators are, in turn, related to the spin $\frac{1}{2}$ operators:

$$s_z = \frac{1}{2} \sigma_z \tag{12.56a}$$

$$s^{\pm} = s_x \pm i s_y = \sigma^{\pm}. \tag{12.56b}$$

What does it mean to restrict the Hilbert space to the manifold of states in which $\mathbf{k} \uparrow$ and $-\mathbf{k} \downarrow$ are both either full or empty? It means that, when we calculate the partition function, states in which $\mathbf{k} \uparrow$ is full while $-\mathbf{k} \downarrow$ is empty and vice versa will not be summed over. Note that the interaction term in $\mathcal{H}_{RED}$ does not connect states with different numbers of broken pairs. In general, states with broken pairs will have higher energies than states in which all electrons are paired. Thus, by keeping only paired states, we are keeping the states which mix together to form a low-energy ground state for $\mathcal{H}_{RED}$.

Within this manifold and using Eq. (12.56), $\mathcal{H}_{RED}$ has the form

$$\begin{aligned} \mathcal{H}_{RED} &= -\sum_{\mathbf{k}} 2\,(E(\mathbf{k}) - \mu)\, s_{z\mathbf{k}} \\ &\quad - \frac{1}{\Omega_0} \sum_{\mathbf{k},\mathbf{q}} V(\mathbf{k} - \mathbf{q}) \left(s_{x\mathbf{k}} s_{x\mathbf{q}} + s_{y\mathbf{k}} s_{y\mathbf{q}} \right), \end{aligned} \tag{12.57}$$

where the index, $\mathbf{k}$, stands for ($\mathbf{k} \uparrow, -\mathbf{k} \downarrow$)).

It is fairly easy to see what the ground state of this system will be. First consider the case $V(\mathbf{k} - \mathbf{q}) = 0$. Then the first term in $\mathcal{H}_{RED}$ is minimized when all pair states inside the Fermi surface are occupied and all states outside are empty. Thus, the ground state is basically a filled Fermi sea. In terms of spins, all the spins for **k** inside the Fermi surface are up, and all the ones outside the Fermi surface are down.

What happens when $V(\mathbf{k}-\mathbf{q}) \neq 0$? In that case, the interactions favor spins pointing in the x–y plane, while the kinetic energy favors spins pointing along $\pm\hat{z}$. For states close to the Fermi surface, the kinetic energy, which is like a Zeeman energy, is small, and the interaction can have a large effect, tipping the spins away from the z-axis. To see this mathematically, consider the local field acting on spin $\mathbf{s_k}$. It is

$$\mathbf{H_k} = 2\,(E(\mathbf{k})-\mu)\,\hat{z} + \frac{2}{\Omega_0}\sum_{\mathbf{q}} V(\mathbf{k}-\mathbf{q})\langle \mathbf{s}_{\perp,\mathbf{q}}\rangle, \tag{12.58}$$

where $\langle \mathbf{s}_{\perp,\mathbf{q}}\rangle$ is the expectation of $\mathbf{s}_{\perp,\mathbf{q}}$ in the state which minimizes $\langle \mathcal{H}_{RED}\rangle$. [Note that the 2 in front of the interaction term is the standard result for a mean field which arises from pairwise interactions.] If $V(\mathbf{k}-\mathbf{q})$ is weak, then only spins with $\mathbf{q}$ near the Fermi surface will have nonzero average transverse components.

We can make the same approximation for $V(\mathbf{k}-\mathbf{q})$ as was made in Eq. (12.30). In particular, we take the function $f(q)$ to be 1 if $|E(\mathbf{q})-\mu| < \hbar\omega_D$ and 0 otherwise. The restriction of the $\mathbf{q}$ sum to this region around the Fermi surface will be indicated by a prime over the sum. Then

$$\mathbf{H_k} = 2\,(E(\mathbf{k})-\mu)\,\hat{z} + \frac{2V}{\Omega_0}\sum_{\mathbf{q}}{}'\langle \mathbf{s}_{\perp,\mathbf{q}}\rangle. \tag{12.59}$$

We assume that the ground state is a direct product state in which the spinor describing spin **k** is

$$\begin{pmatrix} \sin\frac{\theta_\mathbf{k}}{2} \\ \cos\frac{\theta_\mathbf{k}}{2} \end{pmatrix}. \tag{12.60}$$

This spinor describes a state in which the average spin tips along x, making an angle of $\theta_\mathbf{k}$ with the $-z$-axis. Then

$$\langle s_{z\mathbf{k}}\rangle = \frac{1}{2}\left(\sin^2\frac{\theta_\mathbf{k}}{2} - \cos^2\frac{\theta_\mathbf{k}}{2}\right) = -\frac{1}{2}\cos\theta_\mathbf{k} \tag{12.61a}$$

$$\langle s_{x\mathbf{k}}\rangle = \sin\frac{\theta_\mathbf{k}}{2}\cos\frac{\theta_\mathbf{k}}{2} = \frac{1}{2}\sin\theta_\mathbf{k} \tag{12.61b}$$

$$\langle s_{y\mathbf{k}}\rangle = 0. \tag{12.61c}$$

The direction of $\langle \mathbf{s_k}\rangle$ will be parallel to the local field, $\mathbf{H_k}$, so

$$\tan\theta_\mathbf{k} = \frac{\frac{V}{\Omega_0}\sum_{\mathbf{q}}'\sin\theta_\mathbf{q}}{2\,(E(\mathbf{k})-\mu)}. \tag{12.62}$$

Then

$$\cos\theta_{\mathbf{k}} = \frac{(E(\mathbf{k})-\mu)}{\sqrt{(E(\mathbf{k})-\mu)^2 + \frac{1}{4}\left(\frac{V}{\Omega_0}\sum_{\mathbf{q}}' \sin\theta_{\mathbf{q}}\right)^2}} \quad (12.63a)$$

$$\sin\theta_{\mathbf{k}} = \frac{\frac{V}{2\Omega_0}\sum_{\mathbf{q}}' \sin\theta_{\mathbf{q}}}{\sqrt{(E(\mathbf{k})-\mu)^2 + \frac{1}{4}\left(\frac{V}{\Omega_0}\sum_{\mathbf{q}}' \sin\theta_{\mathbf{q}}\right)^2}}. \quad (12.63b)$$

We can identify the **q**-sum with the BCS energy gap Δ

$$\Delta = \frac{V}{2\Omega_0}\sum_{\mathbf{q}}{}' \sin\theta_{\mathbf{q}} = \frac{V}{2\Omega_0}\sum_{\mathbf{q}}{}' \frac{\Delta}{\sqrt{(E(\mathbf{k})-\mu)^2 + \Delta^2}}. \quad (12.64)$$

Then Δ is the region in energy of $E(\mathbf{k}) - \mu$ over which the spins rotate from $\theta_{\mathbf{k}} = 0$ to $\theta_{\mathbf{k}} = \pi$, and the self-consistent equation for Δ is the same as derived earlier for $T = 0$ in Sect. 12.6.1.

To make the connection to BCS quasiparticles, we note that the ground state of Anderson's spin model is one in which each spin is oriented along the local field. Excitations from this mean-field state correspond to flipping a single spin. Since each spin couples to every other spin in the shell around the Fermi surface, flipping one spin hardly perturbs the ground state at all. The energy to flip the spin is $|\mathbf{H}_{\mathbf{k}}|$ which we interpret as the energy to create a pair of quasiparticle excitations. This energy is

$$2\epsilon(\mathbf{k}) = 2\sqrt{(E(\mathbf{k})-\mu)^2 + \Delta^2}. \quad (12.65)$$

In fact, the situation is somewhat more complicated than this as Anderson showed. Just flipping a spin does not generate an eigenstate of the coupled spin Hamiltonian. The true eigenstates are collective modes, like spin waves. We argued above that flipping one spin hardly perturbs the local field on any other spin, because each spin is coupled to a macroscopic number of other spins in the shell around the Fermi surface. In a more complete treatment, Anderson showed that all except for one of the collective modes have energies very similar to those of the single-spin-flip modes. The one mode which is different has zero energy. This is the mode which corresponds to a uniform rotation of all spins about the z-direction.

To see why this mode has zero energy, we note that the choice of ground state in which all spins lie in the x–z plane was arbitrary. The choice could have been the y–z plane or any other plane which includes the z-axis, since the Hamiltonian is invariant under rotations about z. Thus, a uniform rotation about z does not change the ground state energy, and hence an excitation which causes such an infinitesimal rotation has zero energy. In fact, one can go further and show that there are a set of excitations which are continuous deformations of this zero-energy mode which have energies which grow continuously from zero. These are ones in which the ordering plane varies periodically in space with wavevector $\mathbf{q}$. Anderson showed that such excitations have energies which vary linearly with $|\mathbf{q}|$, much like phonons. Just as for phonons, this is an example of a Goldstone mode, a mode which arises in principle whenever the ground state breaks a continuous symmetry of the Hamiltonian.

Unfortunately, the existence of such a low-lying mode would seem to negate one of the key features of the BCS superconducting state, namely, the existence of an energy gap which makes the ground state so robust. Having a linear collective mode would make a superconductor more like a Bose superfluid, which we have seen also has a linear mode. Anderson went on to show that the distinctive feature of a BCS superconductor, the gap, is preserved by the long-range Coulomb interaction between electrons. What he showed was that the linear collective mode of a superconductor describes fluctuating inhomogeneous charge distributions. Because of the Coulomb interaction, these charge fluctuations oscillate at the plasma frequency which, for a typical metal, is much larger than the superconducting energy gap.

So, in this sense, it may seem somewhat fortuitous that BCS arrived at the correct result by only calculating the single-Cooper-pair excitation spectrum while neglecting the long-range Coulomb interaction. They were saved by the fact that the Coulomb interaction strongly suppresses charge fluctuations, and hence the part of the collective spectrum corresponding to charge fluctuations is pushed to high frequency, well above the gap.

The raising of the Goldstone modes to high frequency by long-range interactions is also important in elementary particle physics where it is referred to as the "Higgs mechanism," or, more properly, as the Anderson–Higgs mechanism. It arises in the theory of the spontaneous breaking of the symmetry between the weak and electromagnetic interactions which is analogous to the metal-to-superconductor transition. The same problem arises there as occurs in the BCS theory, namely, that the low-lying Goldstone mode is not observed, and again the problem is resolved by the effect of long-range interactions.

Finally, we note that the Higgs mechanism should not be confused with the "Higgs particle" or "Higgs Boson," which was first observed at the Large Hadron Collider at CERN in 2012. The Higgs particle is the elementary particle analog of the longitudinal collective mode, i.e., to oscillations of the Anderson spin in the x–z plane, which we have seen occurs at energies of order Δ. In 2013, the Nobel Prize in Physics was awarded to François Englert and Peter Higgs for their theoretical discovery of the Higgs Boson in the context of elementary particle physics.

12.9 Suggestions for Further Reading

There is a vast literature on the subject of superconductivity—a subject of which this chapter has barely scratched the surface. If we were to attempt to suggest a list of useful references, it would quickly grow out of control. And so, we mention only two which we have found particularly useful. The first is the classic text by de

Gennes (1999) which is in many ways close in spirit to the present text, particularly in its self-consistent field treatment of Bogoliubov-de Gennes theory. The second is the review article by Leggett (1975) which, in addition to its treatment of triplet superconductivity, also includes a useful review of Landau's Fermi liquid theory.

12.10 Summary

The BCS model for superconductivity presented here is analogous to that for a quantum Bose fluid in that the key process involves pairs of particles with zero total momentum. In the case of fermions with attractive interactions, $(\mathbf{k}\uparrow, -\mathbf{k}\downarrow)$ singlet pairs of electrons scatter among the manifold of such states at the Fermi energy. It is the enormous phase space for scattering among these states that leads to the superconducting instability. As for the Bose liquid, we do not have to explicitly keep track of particles at the Fermi energy and, as a result, the variational Hamiltonian, which is quadratic and which defines the variational density matrix does not explicitly conserve particle number. This Hamiltonian is diagonalized by a Bogoliubov (particle nonconserving) transformation and the problem of fermions interacting via an attractive interaction is then treated using our standard approach of variational mean-field theory.

12.11 Exercises

1. Fill in the details of the discussion of Fig. (12.1) by sketching how the vectors $\mathbf{k}_1+\mathbf{q}$ and $\mathbf{k}_2-\mathbf{q}$ can both lie on the Fermi surface, provided that $\mathbf{q}$ lies in the x–y plane. Explain why and how the phase space for resonant scattering varies with θ and why and how special the value $\theta=\pi/2$ is.
2. Use the equations of motion method (see Exercise 4 of Chap. 11) applied to the Hamiltonian $\mathcal{H}_{\rm tr}$ to find the transformation to quasiparticle operators.
3. What is the effect of the quasiparticle transformation applied to a normal metal for which $\Delta_{\mathbf{k}}$ vanishes?
4. Explain how you would derive the Landau expansion for a superconductor and carry the calculation as far as you can.
5. This exercise concerns the Bogoliubov transformation of Eq. (12.5b). Show that if $u_{\mathbf{k}}$ and $u'_{\mathbf{k}}$ are both positive real and the operators obey anticommutation relations that $u'_{\mathbf{k}}=u_{\mathbf{k}}$, $v'_{\mathbf{k}}=v_{\mathbf{k}}$, and $u_{\mathbf{k}}^2+|v_{\mathbf{k}}|^2=1$.

References

P.W. Anderson, Random-phase approximation in the theory of superconductivity. Phys. Rev. **112**, 1900 (1958)

P.W. Anderson, P. Morel, Generalized Bardeen–Cooper–Schrieffer states and the proposed low-temperature phase of liquid He^3. Phys. Rev. **123**, 1911 (1961)

R. Balian, N.L. Werthamer, Superconductivity with pairs in a relative p wave. Phys. Rev. B **131**, 1553 (1963)

J. Bardeen, L. Cooper, J.R. Schrieffer, Theory of superconductivity. Phys. Rev. **108**, 1175 (1957)

L. Cooper, Bound electron pairs in a degenerate Fermi gas. Phys. Rev. **104**, 1189 (1956)

P.G. de Gennes, *Superconductivity of Metals and Alloys (Advanced Books Classics)* (Westview Press, 1999)

A.J. Leggett, A theoretical description of the new phases of He^3. Rev. Mod. Phys. **47**, 331 (1975)

Chapter 13
Qualitative Discussion of Fluctuations

13.1 Spatial Correlations Within Mean-Field Theory

At first glance, it might seem that mean-field theory has nothing to say about correlations between spins on different sites, since these were the first things dropped in its derivation. To confirm this, we calculate the spin–spin correlation function, using the mean field density matrix. For Ising spins we have

$$\begin{aligned}\langle S_{\mathbf{R}_i} S_{\mathbf{R}_i+\mathbf{r}}\rangle &= \mathrm{Tr}\Bigg[\Bigg(\prod_j \rho_j\Bigg) S_{\mathbf{R}_i} S_{\mathbf{R}_i+\mathbf{r}}\Bigg] \\ &= \langle S_{\mathbf{R}_i}\rangle \langle S_{\mathbf{R}_i+\mathbf{r}}\rangle \\ &= m^2 \end{aligned} \tag{13.1}$$

for $\mathbf{r} \neq 0$, where $m = 0$ for $T > T_c$ and m is nonzero for $T < T_c$. So there are zero spatial correlations and the spin–spin *correlation function* $C(\mathbf{r})$ is zero:

$$C(\mathbf{r}) \equiv \langle S_{\mathbf{R}} S_{\mathbf{R}+\mathbf{r}}\rangle - \langle S_{\mathbf{R}}\rangle\langle S_{\mathbf{R}+\mathbf{r}}\rangle = 0 \, . \tag{13.2}$$

However, it is intuitively clear that as the temperature is reduced, the range of spatial correlations ought to increase and probably diverge at the critical temperature at which long-range order first appears.

To see this we now consider the zero field susceptibility in the disordered phase. Although we do this for the ferromagnetic Ising model, the results we will find have close analogs in the order-parameter susceptibility of most systems which exhibit continuous phase transitions. As we have seen several times before (e.g., see Eq. (10.58)), the trial free energy in zero magnetic field can be written as

$$F = \frac{1}{2}\sum_{ij} F_2(i, j) m_i m_j + \mathcal{O}(m^4) \, , \tag{13.3}$$

A. J. Berlinsky and A. B. Harris, *Statistical Mechanics*, Graduate Texts in Physics,
https://doi.org/10.1007/978-3-030-28187-8_13

where m_i is identified as $m_i = \langle S_i \rangle$ and

$$F_2(i, j) = kT\delta_{i,j} - J_{i,j} , \tag{13.4}$$

with $J_{ij} = J > 0$ for nearest neighbors and $J_{ij} = 0$ otherwise. If we now include the effect of a *site-dependent* magnetic field, h_i, we get

$$F = \frac{1}{2}\sum_{ij} F_2(i, j)m_i m_j - \sum_i h_i m_i + \mathcal{O}(m^4) . \tag{13.5}$$

Minimization indicates that $\langle S_i \rangle \equiv m_i$ is determined by

$$\sum_j F_2(i, j)m_j = h_i , \tag{13.6}$$

so that

$$m_i = \sum_j \chi_{ij} h_j , \tag{13.7}$$

where

$$\chi_{ij} = \left[F_2^{-1}\right]_{ij} . \tag{13.8}$$

In Fourier representation we write Eq. (13.6) as

$$F_2(\mathbf{q})m(\mathbf{q}) = h(\mathbf{q}) , \tag{13.9}$$

where

$$h(\mathbf{q}) = \sum_i e^{i\mathbf{q}\cdot\mathbf{r}_i} h_i , \qquad F_2(\mathbf{q}) = \sum_i e^{i\mathbf{q}\cdot(\mathbf{r}_i - \mathbf{r}_j)} F_2(i, j) , \tag{13.10}$$

and Eq. (13.7) as

$$m(\mathbf{q}) = \chi(\mathbf{q})h(\mathbf{q}) , \tag{13.11}$$

so that

$$\chi(\mathbf{q}) = 1/F_2(\mathbf{q}) , \tag{13.12}$$

and

$$\chi_{ij} = \frac{1}{N}\sum_{\mathbf{q}} \chi(\mathbf{q})e^{i\mathbf{q}\cdot(\mathbf{r}_j - \mathbf{r}_i)} = \frac{1}{N}\sum_{\mathbf{q}} \frac{e^{i\mathbf{q}\cdot(\mathbf{r}_j - \mathbf{r}_i)}}{F_2(\mathbf{q})} . \tag{13.13}$$

If $\mathcal{H}$ is the zero field Hamiltonian we may write

$$\langle S_i \rangle = \frac{\mathrm{Tr} S_i e^{-\beta\mathcal{H}+\beta\sum_i h_i S_i}}{\mathrm{Tr} e^{-\beta\mathcal{H}+\beta\sum_i h_i S_i}} , \tag{13.14}$$

so that

$$\begin{aligned} kT\chi_{ij} &\equiv \frac{\partial \langle S_i \rangle_T}{\partial h_j} \\ &= \langle S_i S_j \rangle_T - \langle S_i \rangle_T \langle S_j \rangle_T . \end{aligned} \tag{13.15}$$

Thus in the disordered phase at zero field, where all $\langle S_i \rangle_T = 0$, we have

$$kT\chi_{ij} = \langle S_i S_j \rangle_T \tag{13.16}$$

from which it follows that the spin–spin correlation function in the disordered phase is

$$\begin{aligned} \mathcal{C}(\mathbf{r}_{ij}) = \langle S_i S_j \rangle_T &= \frac{kT}{N} \sum_{\mathbf{q}} \chi(\mathbf{q}) e^{i\mathbf{q}\cdot(\mathbf{r}_j - \mathbf{r}_i)} \\ &= \frac{kT}{N} \sum_{\mathbf{q}} \frac{e^{i\mathbf{q}\cdot(\mathbf{r}_j - \mathbf{r}_i)}}{F_2(\mathbf{q})} . \end{aligned} \tag{13.17}$$

To summarize this development, we identify the matrix of coefficients of the quadratic term in the Landau free energy as the inverse susceptibility matrix. Inversion of this matrix then gives (apart from the factor kT) the spin–spin correlation function, $\mathcal{C}(\mathbf{r})$.

The largest contribution to the sum in Eq. (13.17) comes from $q \approx 0$. We can write

$$J(\mathbf{q}) = J \sum_{\delta} e^{-i\mathbf{q}\cdot\delta} = J \sum_{\delta} \Big[1 - i\mathbf{q}\cdot\delta - \frac{1}{2}(\mathbf{q}\cdot\delta)^2 + \ldots \Big] , \tag{13.18}$$

where δ is summed over all nearest neighbors of a site. We will do the calculation for a hypercubic lattice in d dimensions. Then

$$\sum_{\delta} 1 = 2d = z, \quad \sum_{\delta} \delta = 0, \quad \sum_{\delta} (\mathbf{q}\cdot\delta)^2 = 2q^2 , \tag{13.19}$$

so that

$$J(\mathbf{q}) = 2dJ - Jq^2 \equiv kT_c - Jq^2 \tag{13.20}$$

and

$$\mathcal{C}(\mathbf{r}) = \frac{(T/T_c)}{(2\pi)^d} \int_{B.Z.} d^d\mathbf{q} \, \frac{e^{-i\mathbf{q}\cdot\mathbf{r}}}{t + \hat{J}q^2} + \mathcal{O}(q^4) \,, \tag{13.21}$$

where $t \equiv (T - T_c)/T_c$ and $\hat{J} = J/(kT_c)$. This result for the correlation function is known as the Ornstein–Zernicke form. We evaluate the correlation function in the critical regime ($t \ll 1$, $r/a \gg 1$). In so doing we extend the integration over all $\mathbf{q}$. (This is permissible as long as the integral converges at large q, as will be the case as long as $d < 5$). Also we set the prefactor, $T/T_c = 1$. Then

$$\mathcal{C}(\mathbf{r}) = \left(\frac{1}{2\pi}\right)^d \int_0^\infty \frac{q^{d-1}dq}{t + \hat{J}q^2} \int d\Omega_q e^{-i\mathbf{q}\cdot\mathbf{r}} \,, \tag{13.22}$$

where $d\Omega$ indicates an integration over all angles of a d-dimensional vector. We can manipulate this into a dimensionless integral by setting $\mathbf{q} = \sqrt{t/\hat{J}}\,\mathbf{k}$, in which case we have

$$\mathcal{C}(\mathbf{r}) = \frac{1}{\hat{J}r^{d-2}} \Phi\left(\mathbf{r}\sqrt{t/\hat{J}}\right) \,, \tag{13.23}$$

where

$$\Phi(\mathbf{s}) = \frac{s^{d-2}}{(2\pi)^d} \int_0^\infty \frac{k^{d-1}dk}{1 + k^2} \int d\Omega_k e^{-i\mathbf{k}\cdot\mathbf{s}} \,. \tag{13.24}$$

Note that the surface area of a d-dimensional unit sphere can be obtained from the result quoted in Eq. (4.23), *viz.*

$$\Omega_d = \frac{2\pi^{d/2}}{(\frac{d}{2} - 1)!} \,, \tag{13.25}$$

so that

$$\Phi(\mathbf{s}) = \frac{2s^{d-2}\pi^{\frac{d}{2}}}{(\frac{d}{2} - 1)!(2\pi)^d} \int_0^\infty \frac{k^{d-1}dk}{1 + k^2} \left\langle e^{-i\mathbf{k}\cdot\mathbf{s}} \right\rangle_\Omega \,, \tag{13.26}$$

where $\langle\ \rangle_\Omega$ indicates an average over angles. To deal with the angular average we take the direction of $\mathbf{s}$ as the polar axis, so that $\mathbf{k}\cdot\mathbf{s} = ks\cos\theta$. In d dimensions the element of surface is proportional to $\sin^{d-2}\theta$, so that

$$\begin{aligned}\left\langle e^{-i\mathbf{k}\cdot\mathbf{s}}\right\rangle_{\Omega} &= \frac{\int_0^{\pi}\sin^{d-2}\theta e^{-iks\cos\theta}d\theta}{\int_0^{\pi}\sin^{d-2}d\theta} \\ &= \frac{\sqrt{\pi}\left(\frac{2}{ks}\right)^{\frac{d}{2}-1}\Gamma\left(\frac{d-1}{2}\right)J_{\frac{d}{2}-1}(ks)}{\sqrt{\pi}\Gamma[(d-1)/2]/\Gamma(d/2)} \\ &= 2^{\frac{d}{2}-1}\Gamma(d/2)J_{\frac{d}{2}-1}(ks)(ks)^{1-\frac{d}{2}}\,, \end{aligned} \tag{13.27}$$

where $J_\alpha(x)$ is a Bessel function of the first kind, and $\Gamma(y)$ is a gamma function ($\Gamma(n) = (n-1)!$). Thus

$$\Phi(\mathbf{s}) = \frac{s^{\frac{d}{2}-1}}{(2\pi)^{\frac{d}{2}}}\int_0^{\infty}\frac{k^{\frac{d}{2}}J_{\frac{d}{2}-1}(ks)dk}{1+k^2}\,. \tag{13.28}$$

This last integral can be found in Gradshteyn and Ryzhik Eq. (6.565.4) (Gradshteyn and Ryzhik (1994). The result for the correlation function is

$$C(r) = \frac{1}{\hat{J}(2\pi)^{\frac{d}{2}}}\frac{\left(\sqrt{t/\hat{J}}\right)^{\frac{d}{2}-1}}{r^{\frac{d}{2}-1}}K_{\frac{d}{2}-1}\left(r\sqrt{t/\hat{J}}\right)\,. \tag{13.29}$$

Note that the Bessel function $K_n(x) \sim \sqrt{\frac{\pi}{2x}}e^{-x}$ for $x \gg 1$, so that the correlation function decays exponentially for large r.

We define the "correlation length" to be the inverse of the coefficient of r.

$$\xi = \sqrt{\frac{\hat{J}}{t}} \quad \rightarrow \quad \nu = \frac{1}{2}. \tag{13.30}$$

Thus ν, the correlation length exponent, which was first introduced below Eq. (6.73), has the value $\nu = 1/2$ in mean-field theory. Then for $r \gg \xi$ we have

$$C(r) = \sqrt{\frac{\pi}{2}}\frac{1}{\hat{J}(2\pi)^{\frac{d}{2}}}\xi^{2-d}\left(\frac{\xi}{r}\right)^{\frac{d-1}{2}}e^{-r/\xi}\,. \tag{13.31}$$

What happens at $T = T_c$ when $\xi \gg r$ even for very large r? Then the argument of the Bessel function is small and we use

$$K_n(x) \sim \frac{1}{2}\left(\frac{2}{x}\right)^n \quad (n \neq 0)\,. \tag{13.32}$$

Then

$$C(\mathbf{r}) \sim \frac{2^{\frac{d}{2}-2}}{\hat{J}(2\pi)^{\frac{d}{2}}r^{d-2}}\,. \tag{13.33}$$

Note that within mean-field theory, the parameter ξ completely determines the behavior of the correlation function apart from the power law prefactor. A form which is consistent both with mean-field theory and with a renormalization (RG) group treatment of fluctuations is

$$\mathcal{C}(\mathbf{r}) = \frac{A}{r^{d-2+\eta}} G(\mathbf{r}/\xi) , \tag{13.34}$$

where A is an amplitude of order unity and $G(\mathbf{s})$ decays exponentially at large argument. Of course, within mean-field theory one has $\eta = 0$. In the integral that defines the correlation function, the main contribution comes from $q < 1/\xi$. A better (than mean field) theory would be one that did a better job of handling long wavelength (small q) fluctuations. In such a more complete theory, η may be nonzero. Note that at T_c where the correlation length is infinite, the correlation function is a power law controlled by the exponent η.

13.2 Scaling

In the vicinity of a critical point, e.g., $(T, H) \approx (T_c, 0)$ for a ferromagnet, the dominant singular behaviors of various thermodynamic quantities follow power laws in variables measured with respect to the critical point. Accordingly, we define *critical exponents* for the singular part of the specific heat, $c_s(\tau, H)$, the spontaneous magnetization, $m(\tau, 0)$ for $\tau < 0$, the susceptibility, $\chi(\tau, H)$, the correlation function, $\mathcal{C}(r, \tau, H)$, and the correlation length, $\xi(\tau, H)$, as functions of $\tau = (T - T_c)/T_c$ and H, close to the critical point via

$$c_s(\tau, 0) \sim |\tau|^{-\alpha} \tag{13.35a}$$
$$m(\tau, 0) \sim (-\tau)^{\beta} \tag{13.35b}$$
$$\chi(\tau, 0) \sim |\tau|^{-\gamma} \tag{13.35c}$$
$$m(0, H) \sim |H|^{1/\delta} \tag{13.35d}$$
$$\mathcal{C}(\mathbf{r}, 0, 0) \sim 1/r^{d-2+\eta} \tag{13.35e}$$
$$\xi(\tau, 0) \sim |\tau|^{-\nu} . \tag{13.35f}$$

Before the invention of Renormalization Group theory, these critical exponents were thought possibly to have different values depending on whether the critical point was approached from low temperatures or from high temperatures. But, the RG has shown that critical exponents are the same in these two cases. However, the *critical amplitudes* can be different above and below T_c. Thus, for the susceptibility, for example, we write

$$\chi(\tau, 0) \sim A_+ \tau^{-\gamma} \quad \tau \to 0^+ \tag{13.36a}$$
$$\chi(\tau, 0) \sim A_- |\tau|^{-\gamma} \quad \tau \to 0^- . \tag{13.36b}$$

As we shall see, the values of the critical exponents do not depend on many details of the model (we will make this more precise later), and in this sense are *universal*. In contrast, the values of the critical temperature and amplitudes do depend on details of the model and are thus not universal. However amplitude ratios (Aharony and Hohenberg 1976) such as A_+/A_-, are universal. In this paradigm the amplitude for the spontaneous magnetization for $T > T_c$ is zero.

One of the central issues of the theory of critical phenomena is the physical meaning of these power laws. What does it mean when a function obeys a power law? Another issue is whether the different exponents are independent or related to each other. As we shall see, only two of these exponents are independent. Thus, within the RG one can calculate ν and η and then expresses all the other exponents in terms of those two. This paradigm is referred to as "two-exponent scaling".

13.2.1 Exponents and Scaling

Functions which are power laws obey scaling relations. Say, for example, that $f(\tau)$ has the property that, when τ is scaled by a factor λ, the value of the function is multiplied by a factor k. That is

$$f(\lambda\tau) = kf(\tau) \tag{13.37}$$

The solution to this equation has the form

$$f(\tau) = a\tau^{-\alpha} \tag{13.38}$$

Substituting Eq. (13.38) into Eq. (13.37) gives

$$\begin{aligned} a(\lambda\tau)^{-\alpha} &= ka\tau^{-\alpha} \\ k &= \lambda^{-\alpha} = e^{-\alpha \ln \lambda} \\ \ln k &= -\alpha \ln \lambda \\ \alpha &= -\ln k/\ln \lambda \,. \end{aligned} \tag{13.39}$$

So the scaling relation implies that $f(\tau)$ obeys a power law, and the ratio of the logarithms of the scale factors k and λ determines the exponent α.

13.2.2 Relations Between Exponents

Here we derive a relation between the correlation function exponents, ν and η and the susceptibility exponent γ. From Eq. (6.72) for the disordered phase and, using Eq. (13.34). Assuming an exponential form for the function $G(\mathbf{r}/\xi)$, we can write

$$\begin{aligned}\chi &= \beta \sum_{\mathbf{r}} C(\mathbf{r}) \\ &= \beta A \alpha_d \int_0^\infty x^{d-1} \frac{e^{-x/\xi}}{x^{d-2+\eta}} dx \\ &\sim \xi^{2-\eta} \,. \end{aligned} \tag{13.40}$$

Clearly the susceptibility diverges because the correlation length diverges at T_c. Since $\chi \sim |T - T_c|^{-\gamma}$ and $\xi \sim |T - T_c|^{-\nu}$, we find that

$$\gamma = \nu(2 - \eta) \tag{13.41}$$

a scaling law first derived by Fisher (1964). We will derive other relations later.

13.2.3 Scaling in Temperature and Field

Back in the 1960s, when it was noticed that thermodynamic functions followed power laws and that the exponents were interrelated, it was conjectured that the free energy per spin, which is a function of the temperature and field variables, τ and H, might obey a two-parameter scaling relation. In particular, if the total free energy per spin consists of a regular part plus a singular part,

$$f_{\text{total}} = f_{\text{regular}} + f_{\text{singular}} \tag{13.42}$$

and abbreviating f_{singular} to f_s, then $f_s(\tau, H)$ was conjectured to obey the scaling relation

$$f_s(\lambda^{y_T}\tau, \lambda^{y_H} H) = \lambda^d f_s(\tau, H) \,, \tag{13.43}$$

where y_T and y_H are called "thermal" and "magnetic" exponents. One of the beautiful results of the RG approach is that it provides a derivation of this result. The solution to Eq. (13.43) has the form

$$f_s(\tau, H) = |\tau|^x g\left(\frac{H}{|\tau|^{\frac{y_H}{y_T}}}\right) . \tag{13.44}$$

Using this representation we have

$$f_s(\lambda^{y_T}\tau, \lambda^{y_H} H) = \lambda^{x y_T} |\tau|^x g\left(\frac{\lambda^{y_H} H}{(\lambda^{y_T})^{\frac{y_H}{y_T}} |\tau|^{\frac{y_H}{y_T}}}\right) \tag{13.45}$$

which agrees with Eq. (13.43) provided that

$$x = \frac{d}{y_T} \, . \tag{13.46}$$

We define new exponents via

$$\Delta = \frac{y_H}{y_T} \tag{13.47a}$$

$$2 - \alpha = x = \frac{d}{y_T} \tag{13.47b}$$

so that the singular part of the free energy becomes

$$f_s(\tau, H) = |\tau|^{2-\alpha} g\left(\frac{H}{|\tau|^\Delta}\right) \, . \tag{13.48}$$

To see what this means we write it as

$$\frac{f_s(\tau, H)}{|\tau|^{2-\alpha}} = g\left(\frac{H}{|\tau|^\Delta}\right) \, . \tag{13.49}$$

In this form we see that instead of a relation between three variables f_s, τ, and H, we have a relation between two scaled variables $f_s/|\tau|^x$ and $H/|\tau|^y$. In Chap. 8 we called this "data collapse," and showed that it happened within mean-field theory. Here we assert that this is a consequence of Eq. (13.43), although the scaling exponents need not be those predicted by mean-field theory.

This form for the singular part of the free energy implies a number of relations among critical exponents. First, it is clear that the singular part of the specific heat in zero field is proportional to the second derivative of $f_s(\tau, 0)$ with respect to τ. This means that α in Eq. (13.48) is in fact the specific heat exponent. Furthermore, for $\tau < 0$, the spontaneous magnetization is

$$\begin{aligned} m &= \left.\frac{\partial f_s}{\partial H}\right|_{H=0} \\ &= |\tau|^{2-\alpha-\Delta} g'(0) \, , \end{aligned} \tag{13.50}$$

which means that

$$\beta = 2 - \alpha - \Delta \, . \tag{13.51}$$

So far, all we have done is identify the exponents α and β based on an expression for the free energy as a function of τ and H. However, now we will see that using Eq. (13.48) to calculate additional exponents will lead to nontrivial new relations among exponents. For example, the singular part of the susceptibility in zero field is the second derivative of f_s with respect to H.

$$\chi = \left.\frac{\partial^2 f_s}{\partial H^2}\right|_{H=0}$$
$$= |\tau|^{2-\alpha-2\Delta} g''(0) \tag{13.52}$$

so that

$$-\gamma = 2 - \alpha - 2\Delta \,. \tag{13.53}$$

Combining this with Eq. (13.51) gives $\Delta = \beta + \gamma$ and

$$\alpha + 2\beta + \gamma = 2 \tag{13.54}$$

a relation due to Rushbrooke (1963) (who derived it as an inequality) which is satisfied by mean-field theory, $(\alpha, \beta, \gamma) = (0, 1/2, 1)$, and by the 2D Ising model, $(\alpha, \beta, \gamma) = (0, 1/8, 7/4)$.

Finally, for general small values of τ and H,

$$m = \frac{\partial f_s}{\partial H} = |\tau|^{2-\alpha-\Delta} g'\left(\frac{H}{\tau^\Delta}\right)$$
$$\sim |\tau|^{2-\alpha-\Delta} \left|\frac{H}{\tau^\Delta}\right|^{\frac{1}{\delta}} \,. \tag{13.55}$$

The last line has the correct $|H|^{1/\delta}$ dependence for when $\tau = 0$. However, in order for all of the factors of τ to cancel out in this limit (so that m neither vanishes nor diverges) it is necessary that

$$2 - \alpha - \Delta = \frac{\Delta}{\delta} \,. \tag{13.56}$$

Eliminating Δ in favor of the other exponents, we obtain the additional relation

$$\gamma = \beta(\delta - 1) \tag{13.57}$$

which is called Widom's scaling relation (Widom 1964).

The net result of the above analysis is that, if the scaling relation, Eq. (13.43), is satisfied, then the four exponents, α, β, γ, and δ, can all be expressed in terms of two independent exponents. This, of course, begs the question of why Eq. (13.43) should be satisfied. We should also not forget the two other exponents, η and ν, which describe the critical behavior of the correlation function. In fact, we earlier derived a relation between γ, ν, η, and the dimensionality d, Eq. (13.41) due to Fisher (1964). This relation results from the definition of the susceptibility in terms of the correlation function. There is, as we shall see, one further relation coupling the thermodynamic and correlation length exponents. This relation arises from a picture that motivates Eq. (13.43) and provides a physical basis for the scaling hypothesis. At the same time, it lays the foundations for a theory for calculating all of the exponents, starting from first principles.

13.3 Kadanoff Length Scaling

Kadanoff (1966) proposed dividing a lattice of spins into a superlattice of blocks of spins as shown in Fig. 13.1. The situation that he was concerned with was that of a system close to its critical point so that the spin–spin correlation length is large, both compared to a lattice spacing, a, and compared to the linear block dimension, La. In this situation it seems plausible to divide the free energy up into intra-block and inter-block parts, much as we did in the derivation of the Gaussian model, and to define a "block spin" variable, $\mu_\alpha = \pm 1$, corresponding, for example, to the sign of the magnetization of block α.

Following Kadanoff, we work with dimensionless coupling constants $K = \beta J$ and H which now represents the magnetic field divided by the temperature. This notation is standard, for example, in RG theory, and so it will be used generally in the next several chapters. K and H are the values of the couplings for the real spins at some temperature and field which are assumed to be close to the critical point $(K, H) = (K_c, 0)$, where $K_c = J/T_c$. We also define a dimensionless correlation length, $\xi(K, H)$ which is the correlation length in units of lattice spacings, a.

Returning to the block spins, μ_α, Kadanoff conjectured that one could define effective inter-block coupling constants, K' and H', and that these should be related to the original coupling constants, K and H, by the requirement that the spin–spin correlation length should correspond to the same physical dimension for the two kinds of spins. That is,

$$\xi(K', H')La = \xi(K, H)a \ . \tag{13.58}$$

Of course, this condition is not sufficient to completely determine the relationship between (K', H') and (K, H). It must be supplemented by the condition that free energy of the system of block spins is equal to the free energy of the original spins. We write the free energy of the original spin system as $Nf(K, H)$ where N is the number of spins. The free energy of the block spins, consists of two parts, the intra-block

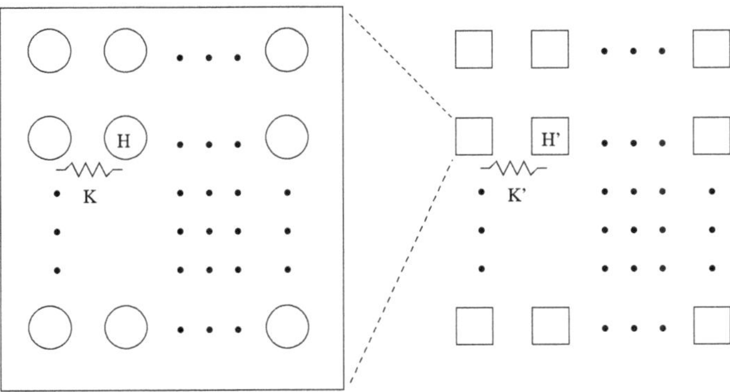

Fig. 13.1 Kadanoff length scaling

part which we write as $(N/L^d)g_{\rm block}(K, H)$, so that $g_{\rm block}(K, H)$ is the intra-block energy of a single block and, because L is finite, is never singular as a function of K and H.

For the inter-block free energy, Kadanoff made the plausible guess, $(N/L^d)f(K', H')$ where $f(K', H')$ is the free energy of a system of Ising spins with coupling constants K' and H'. Thus, K' and H' must satisfy two conditions, Eq. (13.58) and

$$Nf(K, H) = \frac{N}{L^d}g(K, H) + \frac{N}{L^d}f(K', H') . \tag{13.59}$$

Furthermore, since the function $g(K, H)$ is never singular, we can equate the singular parts of the left- and right-hand sides of Eq. (13.59) by writing

$$f_s(K', H') = L^d f_s(K, H) , \tag{13.60}$$

where, from Eq. (13.58) K' and H' also satisfy

$$\xi(K', H') = \xi(K, H)/L . \tag{13.61}$$

In order to use these equations to derive scaling relations, it is necessary to shift the origin so that K is measured with respect to K_c. To do this, we define the new variable t by

$$t = K - K_c \tag{13.62}$$

f_s and ξ are then assumed to obey

$$f_s(t', H') = L^d f_s(t, H) \tag{13.63a}$$

$$\xi(t', H') = \xi(t, H)/L , \tag{13.63b}$$

where (t, H) is a vector measuring the displacement from the critical point.

At this point Kadanoff makes the same scaling assumption that we made above in Eq. (13.43). The only difference this time is in the interpretation. The scale factor λ becomes the block size L, *representing a change in length scale*, and f_s obeys

$$f_s(L^{y_T}t, L^{y_H}H) = L^d f_s(t, H) . \tag{13.64}$$

We have already discussed the solution of such scaling relation and the relations between exponents that it implies. Thus, so far we have nothing new. However, we have one extra scaling relation involving the correlation length ξ, which can be written for zero field as

$$\xi(L^{y_T}t, 0) = \xi(t, 0)/L . \tag{13.65}$$

If we write the solution of this equation in the form $\xi(t, 0) = \xi_0|t|^{-\nu}$, then

$$\xi_0 L^{-\nu y_T}|t|^{-\nu} = \xi_0|t|^{-\nu}L^{-1}$$
$$\therefore \quad y_T = 1/\nu \,. \qquad (13.66)$$

Recalling our earlier result, Eq. (13.47b) for the thermal exponent, $2-\alpha = d/y_T$, we obtain the so-called hyperscaling relation

$$2-\alpha = d\nu \qquad (13.67)$$

due to Josephson (1966). This relation is satisfied for the 2D Ising model, where $\alpha = 0$, $d = 2$, and $\nu = 1$. In mean-field theory, where $\alpha = 0$ and $\nu = 1/2$, it is only satisfied for $d = 4$. All of the relations among exponents may be summarized as

$$2-\alpha = d\nu = 2\beta + \gamma = \beta(1+\delta) = \frac{d\gamma}{2-\eta} \,. \qquad (13.68)$$

Kadanoff's arguments seem quite plausible, if not compelling. However, in the absence of an actual calculation, showing that the effective inter-block Hamiltonian has the form of a nearest-neighbor Ising Hamiltonian and relating (K', H') to (K, H) for a given value of L, they remain little more than an inspired guess. In a later chapter, we will examine how such a calculation can be done, beginning with a very simple model that we have already solved, the classical one-dimensional Ising model.

13.4 The Ginzburg Criterion

Mean-field theory is not very accurate around critical points where correlated fluctuations make significant contributions to the free energy. On the other hand, far away from critical points, where fluctuations about the mean may be relatively small, mean-field theory should do a good job. It is reasonable to ask when to expect mean-field theory to work and when it will fail. It is even reasonable to try to answer this question in terms of mean-field theory itself by looking at where the theory becomes inconsistent. Such a criterion for the applicability of mean-field theory was formulated by Ginzburg (1961). The argument goes as follows.

For a mean-field-type theory to work just below T_c, where the order parameter is small and the correlation length is large, then it is necessary that the fluctuations of the order parameter around its mean in a correlation volume should be small compared to the average of the order parameter over this volume. If this is the case, then we can neglect the order-parameter fluctuations and estimate the free energy from the average order parameter. This may be viewed as an elaboration of the derivation of the mean-field approximation in Chap. 8. The condition may be written as

$$\left\langle \left[\int_{\Omega_\xi} d^d\vec{r} \big(m(\vec{r}) - \langle m \rangle \big) \right]^2 \right\rangle \ll \left\langle \int_{\Omega_\xi} d^d\vec{r}\, m(\vec{r}) \right\rangle^2 , \qquad (13.69)$$

where $\Omega_\xi = \alpha_d \xi^d$. The right-hand side is just $\alpha_d^2 \xi^{2d} \langle m \rangle^2$. The left-hand side is

$$\int_{\Omega_\xi} d^d\vec{r} \int_{\Omega_\xi} d^d\vec{r}\,' \Big\langle \Big(m(\vec{r}) - \langle m \rangle\Big)\Big(m(\vec{r}\,') - \langle m \rangle\Big)\Big\rangle = \alpha_d \xi^d \int_{\Omega_\xi} d^d\vec{r}\, C(r)$$
$$= \alpha_d^2 \xi^d \int_0^\xi r^{d-1} dr \frac{e^{-r/\xi}}{r^{d-2+\eta}} = \alpha_d^2 \xi^{d+2-\eta} \underbrace{\int_0^1 x^{1-\eta} e^{-x} dx}_{A} . \tag{13.70}$$

So the Ginzburg criterion is

$$A\xi^{d+2-\eta} \ll \xi^{2d} \langle m \rangle^2 . \tag{13.71}$$

If we write $\xi = \xi_0 t^{-\nu}$ and $m = m_0 t^\beta$, where $t = (T_c - T)/T_c$, then the criterion becomes

$$A m_0^{-2} \xi_0^{2-d-\eta} \ll t^{2\beta - \nu(d+\eta-2)} . \tag{13.72}$$

This can only be satisfied close to T_c if the exponent is negative. The condition that the exponent be negative yields a relation between the dimensionality and the exponents that

$$\frac{2\beta}{\nu} + (2 - \eta) < d . \tag{13.73}$$

For mean-field exponents, $\beta = \nu = 1/2$ and $\eta = 0$ this corresponds to $4 < d$. The dimension d_c at which $d_c = 2\beta/\nu + (2 - \eta)$ is called the upper critical dimension for mean-field theory. Above this dimension mean-field theory gives correct exponents and at $d = d_c$ there are usually logarithmic corrections to mean-field behavior in the critical region. Note that for the tricritical point discussed in Chap. 8, the order-parameter exponent $\beta = 1/4$ and the upper critical dimension is $d_c = 3$.

Within mean-field theory, with $\beta = \nu = 1/2$ and $\eta = 0$, the Ginzburg criterion for the temperature range in which mean field theory applies may be written as

$$t \gg \left[\frac{A}{m_0^2 \xi_0}\right]^{2/(4-d)} . \tag{13.74}$$

The quantity on the right is the width of the critical region. It depends inversely on the bare ($T = 0$) correlation length and the mean-field-specific heat jump which is proportional to m_0^2. For a system such as liquid ^{4}He near its superfluid transition, which has a short bare correlation length and a weak specific heat jump, the width of the critical region for $d = 3$ is of order 0.3. For conventional superconductors which have bare correlation lengths on the order of thousands of Å and large specific heat jumps, the critical region can be as small as 10^{-15}. Remarkably, high T_c superconductors, with their relatively short coherence lengths, quasi-two-dimensional structure, and "d-wave" gaps, have critical regions much more like superfluid helium than like conventional superconductors. It is also clear that whatever the width of the critical region in $d = 3$, the corresponding width will become larger as the dimensionality is reduced.

13.5 The Gaussian Model

Next, we consider how we might go beyond mean-field theory by including long-wavelength fluctuations of the order-parameter field into the sum over states of the partition function. In doing so we will develop a useful theoretical tool, the functional integral, and gain further insight into the Ginzburg criterion. In this section we give a heuristic derivation of the Gaussian model. In the next section we use the Hubbard–Stratonovich transformation to formally derive the field-theoretic Hamiltonian.

Consider the partition function for the Ising model.

$$Z = \mathrm{Tr}\, e^{-\beta\mathcal{H}} = \prod_i \sum_{S_i=\pm 1} e^{-\beta\mathcal{H}\{S_i\}} . \tag{13.75}$$

Of course if we could calculate this sum, that would solve the problem, but we do not know how to do that. As usual we will do it only approximately. We divide the lattice up into blocks labeled by the index B, and write Z as

$$\begin{aligned} Z &= \prod_B \prod_{i\in B} \sum_{S_i=\pm 1} e^{-\beta\mathcal{H}\{S_i\}} \\ &= \prod_B \mathrm{Tr}_{\{m_B^\alpha\}}\, e^{-\beta[F_B\{m_B^\alpha\}+F_{\mathrm{Int}}(\{m_B^\alpha\},\{m_{B'}^\alpha\})]} , \end{aligned} \tag{13.76}$$

where $\{m_B^\alpha\}$ are internal degrees of freedom of block B, F_B is the free energy of a block and F_{Int} is the interaction energy of neighboring blocks. To keep the problem tractable and also for physical reasons, we restrict ourselves to one degree of freedom per block, namely its magnetization, $N_B m_B$, where N_B is the number of spins per block and m_B is the average magnetization per site of block B. If the magnetization per site is small, then for a small isolated block, we might expect a Landau expansion to work reasonably well. Thus we approximate F_B by

$$F_B/N_B = \frac{a'}{2}(T - T_c)m_B^2 + bm_B^4 \tag{13.77}$$

and we model the interaction between blocks by

$$F_{\mathrm{Int}}/N_B = \frac{K}{2} \sum_{(B,B')} \left(m_B - m_{B'}\right)^2 , \tag{13.78}$$

where (B, B') is summed over pairs of nearest neighbor blocks. Then

$$Z = \int \prod_B dm_B \, e^{-\beta N_B \left[\sum_B \frac{a'}{2}(T-T_c)m_B^2 + bm_B^4 + \frac{K}{2}\sum_{(B,B')}\left(m_B - m_{B'}\right)^2\right]} , \tag{13.79}$$

where each m_B is integrated from -1 to 1. We can treat this as a continuum problem by making the following replacements which constitute the definition of the resulting functional integral. We write

$$m_B \to m(\vec{r}) \tag{13.80a}$$

$$\int \prod_B dm_B \to \int \mathcal{D}m(\vec{r}) \tag{13.80b}$$

$$\sum_B \to \frac{1}{V_B} \int d^d r \tag{13.80c}$$

$$\frac{1}{2}\sum_{B'}\left(m_B - m_{B'}\right)^2 \to \xi_0^2 |\vec{\nabla} m|^2 . \tag{13.80d}$$

Note that both the left- and right-hand sides of Eq. (13.80c) are equal to V/V_B, the number of blocks. We also define the density $\rho = N_B/V_B$. Then the partition function is given by the functional integral

$$Z = \int \mathcal{D}m(\vec{r}) \exp\left[-\beta\rho \int d^d r \left(\frac{a'}{2}(T - T_c)m(\vec{r})^2 + bm(\vec{r})^4 + \frac{K\xi_0^2}{2}|\vec{\nabla} m|^2\right)\right] . \tag{13.81}$$

How does one calculate such an integral? We will do so only for the case of the so-called Gaussian model in which $b = 0$ and T is assumed to be greater than T_c. Then we can work with the Fourier-transformed magnetization variables, which we write as

$$\rho \int m(\vec{r})^2 d^d r = \sum_{\vec{q}} |m(\vec{q})|^2 , \tag{13.82a}$$

$$\rho \int d^d r |\vec{\nabla} m|^2 = \sum_{\vec{q}} q^2 |m(\vec{q})|^2 . \tag{13.82b}$$

Then

$$Z = \int \prod_{\vec{q}} dm(\vec{q}) \, e^{-\frac{\beta}{2}\sum_{\vec{q}}\left[a'(T-T_c) + K\xi_0^2 q^2\right]|m(\vec{q})|^2} . \tag{13.83}$$

The $m(\vec{q})$ are complex functions, and so the meaning of the integrals needs to be explained. Because $m(\vec{r})$ is a real function, its Fourier transform, $m(\vec{q})$ satisfies, $m(-\vec{q}) = m(\vec{q})^*$. Then we can write

$$Z = \prod_{\vec{q}>0} \int dm(\vec{q})dm(-\vec{q})\, e^{-\beta \sum_{\vec{q}>0}\left[a'(T-T_c)+K\xi_0^2 q^2\right]m(\vec{q})m(-\vec{q})}\,, \tag{13.84}$$

where $\vec{q} > 0$ means that the product and the sum are over only half the space of q's, say the ones with $q_z \geq 0$. If we write $m(\pm\vec{q}) = m'_{\vec{q}} \pm i m''_{\vec{q}}$, then the integral for a single $\vec{q}$ is equal to

$$\int_{-\infty}^{\infty} dm'_{\vec{q}} \int_{-\infty}^{\infty} dm''_{\vec{q}}\, e^{-\beta\left[a'(T-T_c)+K\xi_0^2 q^2\right]\left[(m'_{\vec{q}})^2+(m''_{\vec{q}})^2\right]}$$

$$= \frac{\pi}{\beta\left[a'(T-T_c)+K\xi_0^2 q^2\right]}\,, \tag{13.85}$$

where we extended the range of integration to $\pm\infty$. Dropping the restriction on the product over $\vec{q}$, we have

$$Z = \prod_{\vec{q}} \sqrt{\frac{\pi}{\beta\left[a'(T-T_c)+K\xi_0^2 q^2\right]}}\,. \tag{13.86}$$

What can we do with this? The free energy per spin is

$$f = -\frac{T}{N}\ln Z = \frac{T}{2N}\sum_{\vec{q}} \ln \beta\left[a'(T-T_c)+K\xi_0^2 q^2\right] - \frac{T}{2}\ln \pi\,. \tag{13.87}$$

The internal energy is

$$\frac{U}{N} = \frac{\partial \beta f}{\partial \beta} = \frac{1}{2N}\sum_{\vec{q}} \left\{T - \frac{a' T^2}{a'(T-T_c)+K\xi_0^2 q^2}\right\} \tag{13.88}$$

and the specific heat is

$$\frac{C}{N} = \frac{1}{N}\frac{\partial U}{\partial T} = \frac{1}{2N}\sum_{\vec{q}} \frac{(a')^2 T^2}{[a'(T-T_c)+K\xi_0^2 q^2]^2} + \ldots\,, \tag{13.89}$$

where ... represents terms that are less singular as $T \to T_c$. We define the correlation length ξ as

$$\xi^2 = \frac{K\xi_0^2}{a'(T-T_c)}\,. \tag{13.90}$$

Then the singular part of the specific heat can be written as an integral as

$$
\begin{aligned}
C_{\text{sing}} &= \frac{T^2}{(T-T_c)^2}\frac{V/(2\pi)^d}{2N}\int\frac{d^dq}{\left[1+\xi^2q^2\right]^2} \\
&= \frac{\alpha_d}{2(2\pi)^d\rho}\frac{T^2}{(T-T_c)^2}\int_0^{q_{\max}}\frac{q^{d-1}dq}{\left[1+\xi^2q^2\right]^2}\,. \qquad (13.91)
\end{aligned}
$$

Changing to the dimensionless integration variable, $x = \xi q$, this becomes

$$
C_{\text{sing}} = \frac{\alpha_d}{2(2\pi)^d\rho}\left(\frac{Ta'}{K\xi_0^2}\right)^2\xi^{4-d}\int_0^{\xi q_{\max}}\frac{x^{d-1}dx}{\left[1+x^2\right]^2}\,. \qquad (13.92)
$$

As $\xi \to \infty$ there are three distinct possibilities:

1. $(d < 4)$ Then the integral converges and the specific heat exponent α is determined by

$$
C_{\text{sing}} \sim \xi^{4-d} \to \alpha = \frac{1}{2}(4-d)
$$

.

2. $(d = 4)$ Then

$$
C_{\text{sing}} \sim \log\,\xi q_{\max}
$$

which is only weakly singular $(\alpha = 0)$

3. $(d > 4)$ Then the integral diverges like $(q_{\max}\xi)^{d-4}$ which cancels the prefactor, and hence the specific heat does not diverge.

We can use this expression for the singular part of the specific heat due to Gaussian fluctuations, just above T_c, to derive an alternate formulation of the Ginzburg criterion. It is straightforward to show that the mean-field-specific heat jump for the Landau free energy of Eq. (13.77) is $\Delta C = (a')^2T_c/8b$. Close enough to T_c, the specific heat due to Gaussian fluctuations becomes larger than this mean-field-specific heat jump. The Gaussian specific heat has the form

$$
C_{\text{sing}} = C_0 t^{\frac{d}{2}-2},
$$

and so fluctuations become important in the critical region defined by

$$
C_0 t^{\frac{d}{2}-2} > \Delta C.
$$

Thus the width of the critical region is given by

$$
\left(\frac{C_0}{\Delta C}\right)^{\frac{2}{4-d}} < t\,. \qquad (13.93)
$$

Note that ΔC depends inversely on the coefficient, b, of the fourth-order term in the Landau expansion of the free energy which determines the magnitude of thermal fluctuations in the amplitude of the order parameter. C_0 depends inversely on the stiffness as

$$C_0 \sim (K\xi_0^2)^{-d/2}.$$

where $K\xi_0^2$ controls the magnitude of long-wavelength spatial fluctuations. It is the interplay of these two independent quantities which determines whether the critical region is large or small.

13.6 Hubbard–Stratonovich Transformation

In this section, we introduce a transformation which converts the partition function for Ising spins into a field theory where the role of the trace is replaced by a functional integral over field configurations. We then show that, under suitable conditions, this formulation reproduces mean-field theory, and we confirm our earlier estimate of the dimension above which mean-field theory becomes valid. We start from an equality involving a multivariable Gaussian integral. Let **M** be a symmetric matrix, then

$$\left(\prod_i \int_{-\infty}^{\infty} dx_i\right) e^{-\frac{1}{2}\sum_{ij} x_i [\mathbf{M}^{-1}]_{i,j} x_j} e^{\sum_i u_i x_i} = \left[\text{Det}\,(\pi \mathbf{M})\right]^{1/2} e^{\frac{1}{2}\sum_{ij} u_i M_{ij} u_j} . \tag{13.94}$$

This relation can be established by introducing new coordinates which diagonalize the matrix **M**, in which case the integrals reduce to a product of one-dimensional Gaussian integrals. We apply this relation to the ferromagnetic Ising model, by setting $M_{ij} = \beta J_{ij}$ and $u_i = S_i$, where $S_i = \pm 1$. Then we have

$$Z(T, \{H_i\}) = \text{Tr}\left[e^{\frac{1}{2}\sum_{ij} \beta J_{ij} S_i S_j + \beta \sum_i H_i S_i}\right]$$

$$= A\left[\prod_i \int_{-\infty}^{\infty} dx_i\right] \text{Tr}\left[e^{\frac{kT}{2}\sum_{ij} x_i [\mathbf{J}^{-1}]_{ij} x_j + \sum_i (x_i + \beta H_i) S_i}\right] \tag{13.95a}$$

$$= A'\left[\prod_i \int_{-\infty}^{\infty} dx_i\right]\left[e^{-\frac{kT}{2}\sum_{ij} x_i [\mathbf{J}^{-1}]_{ij} x_j + \sum_i \ln\cosh(x_i + \beta H_i)}\right], \tag{13.95b}$$

where A and A' do not qualitatively affect the properties of the model near the phase transition, and we have performed the trace over the $\{S_i\}$ which results in the ln cosh terms in the exponent.

We now need to discuss the matrix inverse of J_{ij}. We define

$$J(\mathbf{q}) = \sum_j J_{ij} e^{i\mathbf{q}\cdot(\mathbf{r}_i - \mathbf{r}_j)} \tag{13.96}$$

Then the matrix inverse of J_{ij} is given by

$$[\mathbf{J}^{-1}]_{ij} = \frac{1}{N}\sum_{\mathbf{q}} \frac{e^{-i\mathbf{q}\cdot(\mathbf{r}_i-\mathbf{r}_j)}}{J(\mathbf{q})} . \tag{13.97}$$

For a cubic lattice in d dimensions we have

$$J(\mathbf{q}) = 2J\left(\cos(q_1 a) + \cos(q_2 a) + \ldots \cos(q_d a)\right) . \tag{13.98}$$

Strictly speaking, we are not allowed to invert the matrix $\mathbf{J}$ because some of its eigenvalues are zero. That is, $J(\mathbf{q})$ vanishes, for example, when $q_1 a = q_2 a \cdots = q_n a = \pi/2$. However, we are really interested in the contributions from q near zero, because it is this region of $\mathbf{q}$-space which determines properties at long length scales. Since these degrees of freedom are expected to dominate critical phenomena, it is only the behavior of $J(\mathbf{q})$ at small q which is relevant for our discussion. It would take us too deeply into a technical discussion to properly deal with this problem, so we will only say that as long as we focus on small q, nothing bad will happen.

So we replace the matrix inverse of $\mathbf{J}$ by its form for small q which is well-behaved, and we introduce $x(\mathbf{q})$, the Fourier transform of x_i:

$$x_i = \frac{1}{N}\sum_{\mathbf{q}} x(\mathbf{q})e^{-i\mathbf{q}\cdot\mathbf{r}_i} \tag{13.99}$$

and write

$$\begin{aligned}\frac{1}{2}\sum_{ij} x_i[\mathbf{J}^{-1}]_{ij}x_j &= \frac{1}{2N}\sum_{\mathbf{q}} x(\mathbf{q})\mathbf{J}(\mathbf{q})^{-1}x(-\mathbf{q}) \\ &\to \frac{1}{2N}\sum_{\mathbf{q}} x(\mathbf{q})x(-\mathbf{q})\left(\frac{1}{zJ} + cq^2\right) , \end{aligned} \tag{13.100}$$

where c is a positive constant. In terms of a continuum theory this gives

$$\frac{1}{2}\sum_{ij} x_i[\mathbf{J}^{-1}]_{ij}x_j \to \frac{1}{2}\int d\mathbf{r}\left(\frac{1}{zJ}x(\mathbf{r})^2 + c|\nabla x(\mathbf{r})|^2\right) \tag{13.101}$$

and the partition function is

$$Z = A'\int_{-\infty}^{\infty} \mathcal{D}x(\mathbf{r})e^{-\beta\mathcal{F}(\{x(\mathbf{r})\})} , \tag{13.102}$$

where $\mathcal{D}$ indicates a functional integral, i.e., an integral over all $x(\mathbf{r})$ and

$$\mathcal{F}(\{x(\mathbf{r})\}) = \frac{1}{2}\int d\mathbf{r}\Big(-2kT \ln\cosh[x(\mathbf{r}) + \beta H(\mathbf{r})] + (kT)^2\Big(\frac{1}{zJ}x(\mathbf{r})^2 + c|\nabla x(\mathbf{r})|^2\Big)\Big) . \tag{13.103}$$

We have now converted the partition function in terms of traces over Ising spins into a functional integral over field variables $x(\mathbf{r})$ where $-\infty < x(\mathbf{r}) < \infty$. The free energy of Eq. (13.103) is related to what is called the Landau–Ginzburg–Wilson (LGW) free energy. One derives the LGW free energy by expanding the ln cosh term as a power series to order $x(\mathbf{r})^4$. This will be discussed in more detail in Chap. 20 where the LGW free energy is used to derive the ϵ-expansion.

Without the gradient term the free energy of Eq. (13.103) is a stable potential as long as $kT > zJ$, i.e., $T > T_c$. For $T < T_c$ and in a uniform field a uniform minimum occurs for

$$\frac{\delta\mathcal{F}}{\delta x} = -2kT\tanh\left(x(\mathbf{r}) + \beta H(\mathbf{r})\right) + \frac{2(kT)^2}{kT_c}x(\mathbf{r}) = 0, \tag{13.104}$$

which leads to a minimum for $x = x^*$, where

$$\frac{Tx^*}{T_c} = \tanh(x^* + \beta H) , \tag{13.105}$$

which is equivalent to the mean-field solution of Eq. (8.9). Finally, we identify the meaning of x^* by evaluating the magnetization via

$$\begin{aligned} m \equiv -\langle S_i\rangle &= \left.\frac{\partial F}{\partial H}\right|_T \\ &= -\left.\frac{\partial F}{\partial x}\right|_{T,H}\left.\frac{\partial x}{\partial H}\right|_T - \left.\frac{\partial F}{\partial H}\right|_{T,x} \\ &= -\left.\frac{\partial F}{\partial H}\right|_{T,x} = \tanh(x + \beta H) . \end{aligned} \tag{13.106}$$

Thus we see that for small x (and $H = 0$), x plays the role of m.

We now study the free energy of Eq. (13.103) when $H = 0$. We expand the ln cosh term as

$$-2kT\ln\cosh[x(\mathbf{r})] = -kTx(\mathbf{r})^2 + \mathcal{O}(x^4) . \tag{13.107}$$

Keeping only terms up to order x^2, gives the Gaussian model studied in the preceding section.

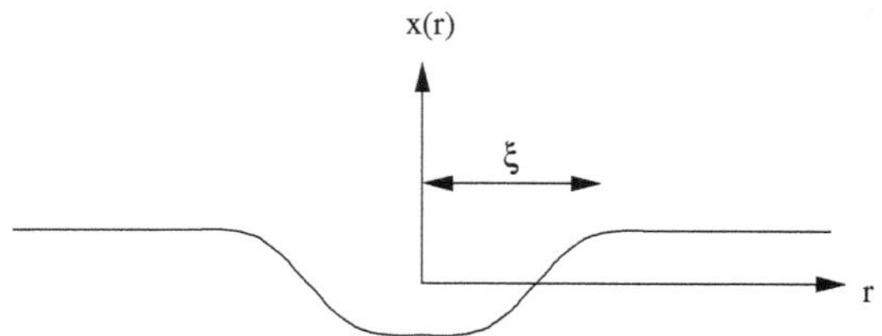

Fig. 13.2 The minimum size domain of down spins in a background of up spins. Here ξ is the correlation length, the minimum length over which $x(\mathbf{r})$ can vary significantly

More generally, we now make some heuristic estimates of the LGW free energy including the effect of the gradient squared term. In particular we ask, when does the uniform solution dominate the functional integral in Eq. (13.102)? We would argue that configurations in which $x(\mathbf{r})$ varies on a length scale less than ξ will not contribute significantly to the functional integral. This is controlled by the size of the coefficient c of the gradient squared. If the gradient term is much larger than kT, it will prevent nonuniform solutions from contributing to the functional integral and the mean-field solution will dominate. Otherwise, if the gradient term is smaller than kT, spatially nonuniform configurations have to be taken into account. We estimate this term as follows. A nonuniform solution will involve creating a domain of, say, "down" spins in an otherwise uniform background of "up" spins, as shown in Fig. 13.2. As we have said, this region must have a linear dimension of at least ξ, and therefore a volume ξ^d. At the center of this region we have a down spin and outside this region spins are up. So the gradient is estimated to be

$$|\nabla x(\mathbf{r})| \sim |m_0(T)/\xi| \, . \tag{13.108}$$

The free energy of this configuration is thus

$$\begin{aligned}\Delta F &\sim \int |\nabla x(\mathbf{r})|^2 d^d\mathbf{r} \\ &\approx m_0(T)^2 \xi^{d-2} \sim |T - T_c|^{2\beta-(d-2)\nu} \, .\end{aligned} \tag{13.109}$$

Mean-field theory will be valid if $\Delta F \gg kT$. Thus mean-field theory holds at high dimension, where the exponent is negative, i.e., for $d > d_c$, where d_c, the upper critical dimension, is given by

$$d_c = 2 + \frac{2\beta_{\rm MF}}{\nu_{\rm MF}} \, , \tag{13.110}$$

where the subscript "MF" indicates a mean-field value. In the case of the Ising or Heisenberg models, where $\beta_{\rm MF} = 1/2$ and $\nu_{\rm MF} = 1/2$, this gives $d_c = 4$. For percolation, where $\beta_{\rm MF} = 1$ and $\nu_{\rm MF} = 1/2$, this gives $d_c = 6$. In any case, when $d > d_c$, the surface area associated with a fluctuating domain is so large that the cost in free energy to create such a domain is prohibitive and mean-field theory is correct. For $d < d_c$ fluctuations are important, but this argument does not tell us how to take them into account.

13.7 Summary

In this chapter, we have mainly discussed scaling phenomena near a continuous phase transition at temperature T_c. An important function which characterizes the role of fluctuations is (in the case of spin systems) the spin–spin correlation function $C(\mathbf{r}) \equiv \langle S_{\mathbf{R}} S_{\mathbf{R}+\mathbf{r}} \rangle - \langle S_{\mathbf{R}} \rangle \langle S_{\mathbf{R}+\mathbf{r}} \rangle$. Apart from a factor of kT this quantity is equal to the two-point susceptibility at separation $\mathbf{r}$. The coefficient of the term in the Landau expansion which is quadratic in the order parameter is identified as the inverse susceptibility. Near the phase transition the correlation function usually assumes the form $C(\mathbf{r}) = Ar^{2-d-\eta}\Phi(r/\xi)$, where A is a constant and Φ is a function whose exact form is hard to determine but which, qualitatively, one expects to fall off exponentially for large r/ξ. Exponents for various response functions are defined in Eq. (14.33). One generally has "two-exponent scaling," whereby knowledge of two of these exponents determines the values of all the critical exponents. These results may be expressed in terms of the two exponents η and ν, which are the natural outputs of RG calculations.

To go beyond mean-field theory, i.e., to include fluctuations, it is convenient to develop a field theory in terms of Gaussian variables (which unlike spin variables assume values between $-\infty$ and $+\infty$). Here we obtained a Gaussian (quadratic) field theory using a simple block spin argument. This was supplemented by a more formal derivation of a field theory using the Hubbard–Stratonovich transformation from spin to Gaussian variables. We then gave two versions of the Ginzburg criterion to determine when fluctuations could be ignored asymptotically near the critical point. These arguments rely on a physical picture in which the correlation length plays a central role. The result is that fluctuations can be neglected when the spatial dimensionality d is greater than a critical value d_c, given in terms of the mean-field values of some critical exponents. For the Ising model $d_c = 4$, whereas for percolation $d_c = 6$. The field theory developed here will be treated using the RG in a later chapter.

13.8 Exercises

1. Calculate

$$\int d\Omega_d = \int d\phi \prod_{k=1}^{d-2} \sin^k \theta_k d\theta_k \equiv \alpha_d$$

Check by evaluating

$$\int dx_1 \ldots dx_d e^{-A(x_1^2+x_2^2+\ldots x_d^2)} = \int_0^\infty e^{-Ar^2} r^{d-1} dr \int d\Omega_d$$

Integrals should be done with the help of Gradshteyn and Ryzhik (1994).

2. Use a Hubbard–Stratonovich transformation to obtain the LGW free energy for a Heisenberg model in which the spins are classical n-component unit vectors. (Hint: you may want to make use of some of the integrals quoted in this chapter.)

3. The purpose of this exercise is to show that the specific heat near the critical point for small positive α does not differ very much from that for small negative α. To show this, you are to plot the specific heat in the interval between $t = -0.0002$ and $t = +0.0002$ for $\alpha = +0.05$ and $\alpha = -0.05$. Assume the specific heat to be of the form

$$C_n = A_n + B_n |t|^{-\alpha_n} \ ,$$

where $A_1 = 0$, $B_1 = 1.0$, $\alpha_1 = 0.05$, and $\alpha_2 = -0.05$. Show that by proper choice of A_2 and B_2 you can make the two specific heats C_1 and C_2 be quite similar. Did you know before completing this exercise what the sign of B_2 had to be?

13.9 Appendix—Scaling of Gaussian Variables

Here we discuss the scaling of Fourier variables. For simplicity we base our discussion on the Gaussian model. We introduce the Fourier transform on a simple cubic lattice by

$$x(\mathbf{q}) \equiv \frac{1}{N^r} \sum_i e^{i\mathbf{k}\cdot\mathbf{r}_i} x_i \ , \tag{13.111}$$

where N is the total number of sites and r is an exponent which we will leave unspecified for the moment. The inverse transformation is

$$x_i = \frac{1}{N^{1-r}} \sum_{\mathbf{q}} e^{-i\mathbf{q}\cdot\mathbf{r}_i} x(\mathbf{q}) \ . \tag{13.112}$$

If we want the transformation to be a unitary transformation (as one does if $x_{\mathbf{q}}$ is a creation or annihilation operator), then the choice $r = 1/2$ is enforced. For example, look back at Eq. (11.7). This choice is, in fact, the "standard" choice used in condensed matter physics. On the other hand, when discussing mean-field theory, we chose $r = 1$, as in Eq. (10.43). With that choice the equilibrium values of the order parameters are zero in the disordered phase and are of order unity in the ordered phase.

However, when the x_i are Gaussian variables, as one obtains using a Hubbard–Stratonovich transformation or other route to a field theory, then if one has nearest-neighbor interactions, the quadratic Hamiltonian (obtained via the Hubbard–Stratonovich transformation, for instance) is of the form

$$
\begin{aligned}
\mathcal{H}_0 &= \frac{1}{2}\sum_{ij} v_{i,j} x_i x_j \\
&= \frac{1}{2N^{1-2r}} \sum_{\mathbf{q}} v(\mathbf{q}) x(\mathbf{q}) x(-\mathbf{q}) \,, \qquad (13.113)
\end{aligned}
$$

where $v_{i,j}$ is a short-ranged function and $v(\mathbf{q})$ is its Fourier transform.

Note that all functions are defined on a mesh of points in $\mathbf{q}$-space. In $\mathbf{q}$-space, the volume per point is $\Delta\mathbf{q} = (2\pi)^d/N$. So the Hamiltonian is wavevector space is

$$
\begin{aligned}
\mathcal{H}_0 &= \frac{N^{2r}}{2} \sum_{\mathbf{q}} v(\mathbf{q}) x(\mathbf{q}) x(-\mathbf{q}) \Delta\mathbf{q} \\
&\to \frac{N^{2r}}{2} \int \frac{d\mathbf{q}}{(2\pi)^d} v(\mathbf{q}) x(\mathbf{q}) x(-\mathbf{q}) \,. \qquad (13.114)
\end{aligned}
$$

Thus the Hamiltonian density in wavevector space will be independent of N if we choose $r = 0$.

Furthermore, with this choice of r we have

$$
\begin{aligned}
C(\mathbf{q}) \equiv \langle x(\mathbf{q}) x(-\mathbf{q}) \rangle &= \frac{\mathrm{Tr} e^{-\beta\mathcal{H}_0} x(\mathbf{q}) x(-\mathbf{q})}{\mathrm{Tr} e^{-\beta\mathcal{H}_0}} \\
&= \frac{kTN}{v(\mathbf{q})} \,. \qquad (13.115)
\end{aligned}
$$

Thus we have that $C(\mathbf{q})\Delta\mathbf{q}$ is also independent of N in the thermodynamic limit.

One might well ask how the free energy becomes extensive when one goes over to the continuum formulation in Fourier space. After all, the Fourier integrals are integrations over the first Brillouin zone, so that $\mathcal{H}_0$ does not appear to be proportional to the system size. However, one has to remember that the partition function is now a functional integral over the $x(\mathbf{q})$'s, each integrated from $-\infty$ to ∞. Thus

$$
\begin{aligned}
Z &= \prod \left(\int_{-\infty}^{\infty} dx(\mathbf{q}) \right) e^{-\frac{1}{2NkT} \sum_{\mathbf{q}} v(\mathbf{q}) x(\mathbf{q}) x(-\mathbf{q})} \left| \frac{\partial x_{i=1} x_{i=2} \cdot x_{i=N}}{\partial x(\mathbf{q}_1) x(\mathbf{q}_2) \cdot x(\mathbf{q}_N)} \right| \\
&= A \prod z(\mathbf{q}) \,, \qquad (13.116)
\end{aligned}
$$

where A is a constant (which depends on N, of course) and $z(\mathbf{q})$ is the partition function for wavevector $\mathbf{q}$. Since the product has N terms, its logarithm (which gives the free energy) will be extensive. But it is very convenient that in wavevector space, the Hamiltonian density is constant in the thermodynamic limit and that scaling requires the choice $r = 0$.

References

A. Aharony, P.C. Hohenberg, Universal relations among thermodynamic critical amplitudes. Phys. Rev. B **13**, 3081 (1976)

M.E. Fisher, Correlation functions and the critical region of simple fluids. J. Math. Phys. **5**, 944 (1964)

V.L. Ginzburg, Some remarks on phase transitions of the 2nd kind and the microscopic theory of ferroelectric materials. Sov. Phys. Solid State **2**, 1824 (1961)

I.S. Gradshteyn, I.M. Ryzhik, *Table of Integrals, Series, and Products*, 5th ed (Academic Press, 1994)

B.D. Josephson, Relation between the superfluid density and order parameter for superfluid He near T_c. Phys. Lett. **21**, 608 (1966)

L.P. Kadanoff, Scaling laws for Ising models near T_c. Physics **2**, 263 (1966)

G.S. Rushbrooke, On the thermodynamics of the critical region for the Ising problem. J. Chem. Phys. **39**, 842 (1963)

B. Widom, Degree of the critical isotherm. J. Chem. Phys. **41**, 1633 (1964)

Chapter 14
The Cayley Tree

14.1 Introduction

In the previous chapter, we saw that the MF approximation becomes correct in the limit of infinite spatial dimensionality. In this limit, fluctuations in the field sensed by one spin (in the Ising model) becomes small and our neglect of correlated fluctuations seems appropriate. In this chapter, we consider the construction of exact solutions to statistical problems on the Cayley tree. As we will see in a moment, the infinite Cayley tree (which is sometimes called a "Bethe lattice") provides a realization of infinite spatial dimensionality. The Cayley tree is a lattice in the form of a tree (i.e., it has no loops) which is recursively constructed as follows. One designates a central or seed site as the zeroth generation of the lattice. The first generation of the lattice consists of z sites which are neighboring to the seed site. Each first- generation site also has z nearest neighbors: one already present in the zeroth generation and $\sigma \equiv z - 1$ new sites added in the second generation of the lattice. The third generation of sites consists of the σ new sites neighboring to the second-generation sites. There are thus z first-generation sites and $z\sigma^{(k-1)}$ kth generation sites. In Fig. 14.1 we show four generations of a Cayley tree with $z = 3$. Note that a tree with $z = 2$ is just a linear chain.

We have previously remarked that mean-field theory should become progressively more accurate as the coordination number increases. As we will see below, a more precise statement is that for systems with short-range interactions, the critical exponents become equal to their mean-field values at high spatial dimensionality. It is therefore of interest to determine the spatial dimensionality of the Cayley tree. There are at least two ways to address this question. One way is to study how the number $N(r)$ of lattice points in a spherical volume of radius r increases for large r. If we consider a tree with k generations the number of sites is

$$\begin{aligned} N(k) &= 1 + z + z\sigma + z\sigma^2 + \ldots z\sigma^{k-1} \\ &= 1 + z\frac{\sigma^k - 1}{\sigma - 1} . \end{aligned} \tag{14.1}$$

A. J. Berlinsky and A. B. Harris, *Statistical Mechanics*, Graduate Texts in Physics,
https://doi.org/10.1007/978-3-030-28187-8_14

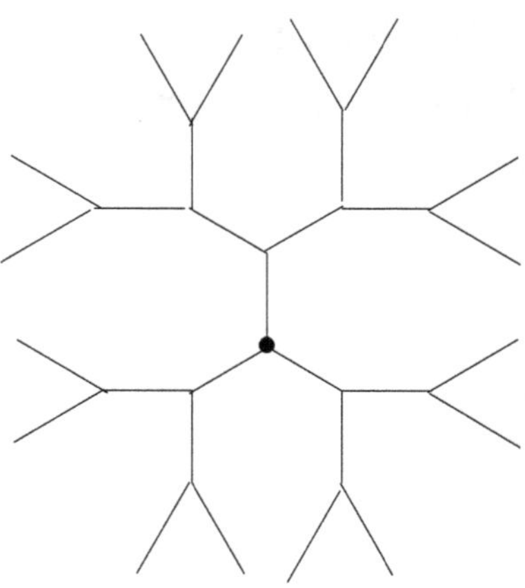

Fig. 14.1 Cayley tree of four generations with $z = 3$

We now need to relate k to a distance r. In general the k steps can be considered to form a random walk, since one should regard them as going in random directions as a path is traced from one generation to the next. In that case $r \sim \sqrt{k}$, or $k = Ar^2$, for large k, where A is a constant. Thus at large r we have

$$\ln N(r) \sim k \ln \sigma = r^2 \ln \sigma \, . \tag{14.2}$$

Then, in the limit of large r, we have

$$\begin{aligned} d &= d \ln N(r)/d \ln r = r d \ln N(r)/dr \\ &= 2r^2 \ln \sigma \; \rightarrow \infty \, . \end{aligned} \tag{14.3}$$

Thus we conclude that the Cayley tree is an infinite dimensional lattice.

The definition of dimensionality through Eq. (14.2) was introduced by the mathematician Hausdorff (1919) and can be applied to structures which have a noninteger dimensionality. This Hausdorff dimension is equivalent to what Mandelbrot (1977) has called the "fractal dimension." The spatial dimension of a smooth curve in a plane (e.g., the circumference of a circle) has dimension unity. However, if the curve is sufficiently irregular the dimension of a curve, such as a random or self-avoiding walk, has a dimension greater than one. We will later see that correlation functions at the second- order phase transition have noninteger fractal dimension.

Another measure of the dimensionality is found by the relation

$$d = \lim_{r \to \infty} \frac{r S(r)}{V(r)} \, . \tag{14.4}$$

where $S(r)$ is the surface area of a sphere of radius r. For the Cayley tree we identify $S(k)$ as the number of sites in the kth generation:

$$S(k) = z\sigma^{k-1} \, , \tag{14.5}$$

and for the volume we use $N(k)$ from Eq. (14.1), so that

$$d = \lim_{k\to\infty} \frac{r_k z \sigma^{k-1}}{1 + z\frac{\sigma^k - 1}{\sigma - 1}} \ . \tag{14.6}$$

For large k this gives

$$d = \lim_{r\to\infty} \frac{r_k(\sigma - 1)}{\sigma} = \infty \ . \tag{14.7}$$

So again we see that the Cayley tree is an infinite dimensional structure. Note that in this criterion we do not need to invent a metric for distance. All we needed to know was that $r_k \to \infty$ as $k \to \infty$.

Note that the fact that the lattice is a recursive one (without any loops) leads to the possibility of establishing recursive constructions for the partition function. Thus, we anticipate being able to construct exact solutions for various model on the Cayley tree. From the fact that the tree corresponds to infinite spatial dimension we conclude that we thereby have a means for identifying mean-field theory from such a solution. This is particularly important from models which, like percolation or quenched random spin systems, do not have any trivial connection to a canonical partition function.

14.2 Exact Solution for the Ising Model

Here we treat the ferromagnetic Ising model for which

$$\mathcal{H} = -J\sum_{\langle ij\rangle} \sigma_i\sigma_j \ , \tag{14.8}$$

where $\langle ij\rangle$ indicates that the sum is over nearest-neighbor bonds. Our first step is to write $e^{\beta J\sigma_i\sigma_j}$ as

$$\begin{aligned} e^{\beta J\sigma_i\sigma_j} &= \cosh(\beta J) + \sinh(\beta J)\sigma_i\sigma_j \\ &= \cosh(\beta J)\biggl(1 + t\sigma_i\sigma_j\biggr) \ , \end{aligned} \tag{14.9}$$

where $t \equiv \tanh(\beta J)$. Then the partition function is

$$Z = \cosh(\beta J)^{N_b}\mathrm{Tr}\biggl\{\prod_{\langle ij\rangle}\biggl(1 + t\sigma_i\sigma_j\biggr)\biggr\} \ , \tag{14.10}$$

where N_b is the number of nearest-neighbor bonds. Imagine expanding the product into its 2^{N_b} terms and taking the trace of each term. Only terms in which all σ's appear either not at all or an even number of times survive the trace. But, because the lattice has no loops, the only such term is the first term, i.e., the one independent of t. This evaluation seems to indicate that the specific heat has no singularity as a function of temperature since the prefactor of the trace is analytic. However, as we shall see, the conclusion that this system does not exhibit a phase transition is incorrect.

Instead of the partition function, let us calculate the susceptibility assuming a disordered phase. Normally one would write the susceptibility *per site* χ as

$$\chi = N^{-1} \sum_{i,j} \langle \sigma_i \sigma_j \rangle \; . \tag{14.11}$$

We do *not* do that here because such a sum unduly weights the large proportion of the sites on the surface of the tree. Instead we identify the properties of the seed site as being representative of a translationally invariant lattice in high dimension. Thus we take

$$\chi = \sum_i \langle \sigma_0 \sigma_i \rangle \; , \tag{14.12}$$

where "0" labels the seed site. We have

$$\begin{aligned} \langle \sigma_0 \sigma_i \rangle &= \frac{\mathrm{Tr} \sigma_0 \sigma_i e^{-\beta \mathcal{H}}}{\mathrm{Tr} e^{-\beta \mathcal{H}}} \\ &= \frac{\mathrm{Tr} \sigma_0 \sigma_i \prod_{\langle jk \rangle} [1 + t \sigma_j \sigma_k]}{\mathrm{Tr} \prod_{\langle jk \rangle} [1 + t \sigma_j \sigma_k]} \; . \end{aligned} \tag{14.13}$$

Here we need to say a word about taking the trace. The trace of the unit operator, $\mathcal{I}$ in the space of two states of the ith spin is

$$\mathrm{Tr} \begin{bmatrix} 1 & 0 \\ 0 & 1 \end{bmatrix} = 2 \; . \tag{14.14}$$

If we are working in the space of state of k spins, each of which has two states, the total number of states in this space is 2^k and the trace of the unit operator *in this space* is 2^k. Therefore, one sees that the value of the trace does depend on the space over which the trace is to be taken. For this reason, when expanding the partition function, it is convenient to work in terms of "normalized traces." By the normalized trace of an operator $\mathcal{O}$ we mean

$$\mathrm{tr} \mathcal{O} \equiv \frac{\mathrm{Tr} \mathcal{O}}{\mathrm{Tr} 1} \; . \tag{14.15}$$

Note the use of "tr" rather than "Tr" to indicate this normalized trace. The advantage of introducing this quantity is that normalized traces of an operator which depends only on the coordinates of one site or on a small group of sites does not depend on the number of states in the entire system. Indeed if Tr_i indicates a trace over only the states of the ith spin, then if $\mathcal{O}$ depends only on the ith spin, we may write

$$\begin{aligned} \mathrm{tr}\mathcal{O}(i) &= \frac{\mathrm{Tr}\mathcal{O}(i)}{\mathrm{Tr}\mathcal{I}} \\ &= \frac{\mathrm{Tr}_1\mathrm{Tr}_2\ldots\mathrm{Tr}_i\ldots\mathrm{Tr}_N\mathcal{O}(i)}{\mathrm{Tr}_1\mathrm{Tr}_2\ldots\mathrm{Tr}_i\ldots\mathrm{Tr}_N\mathcal{I}} \\ &= \frac{\mathrm{Tr}_i\mathcal{O}(i)}{\mathrm{Tr}_i\mathcal{I}} . \end{aligned} \tag{14.16}$$

Thus if one is dealing with a normalized trace, one can ignore sites which do not explicitly appear in the operator $\mathcal{O}$. If $\mathcal{O}$ depends on sites i, j, and k, then

$$\mathrm{tr}\mathcal{O}(i,j,k) = \frac{\mathrm{Tr}_i\mathrm{Tr}_j\mathrm{Tr}_k\mathcal{O}(i,j,k)}{\mathrm{Tr}_i\mathrm{Tr}_j\mathrm{Tr}_k\mathcal{I}} . \tag{14.17}$$

Thus we write

$$\begin{aligned} \langle\sigma_0\sigma_i\rangle &= \frac{\mathrm{Tr}\sigma_0\sigma_i\prod_{\langle jk\rangle}[1+t\sigma_j\sigma_k]}{\mathrm{Tr}\mathcal{I}} \times \frac{\mathrm{Tr}\mathcal{I}}{\mathrm{Tr}\prod_{\langle jk\rangle}[1+t\sigma_j\sigma_k]} \\ &= \frac{\mathrm{tr}\sigma_0\sigma_i\prod_{\langle jk\rangle}[1+t\sigma_j\sigma_k]}{\mathrm{tr}\prod_{\langle jk\rangle}[1+t\sigma_j\sigma_k]} . \end{aligned} \tag{14.18}$$

You should think of the restricted traces in the same way as the usual traces, except that its normalization is much more convenient, since its value does not depend on the size of the system. Also it is important to note that

$$\mathrm{tr}\mathcal{I} = 1 . \tag{14.19}$$

Now consider the expansion of the denominator in Eq. (14.18). When the product is written out, it consists of a sum of 2^{N_b} terms. Each term in this sum is a product of "bond" factors. By a bond factor, we mean $\sigma_j\sigma_k$ associated with the bond connecting nearest-neighboring sites j and k. We associate with each term a diagram, as in Fig. 14.2, as follows. For each such bond factor, we draw a heavy solid line between the two sites j and k and also place dots adjacent to sites j and k to represent the presence of spin operators at these sites. Before we take the trace, the first term has no dots, so it is just the unit operator with normalized trace unity. Succeeding terms in the expansion are represented by a collection of one or more bonds with associated dots. But notice that no matter how you arrange bonds, each diagram will have at least one site with an odd number of dots. (This is true because this lattice, unlike a "real" lattice, has no loops.) The product over σ's thus contains an odd number of operators associated with such a site. The trace of an odd power of any σ is zero. So the denominator of Eq. (14.18) is unity:

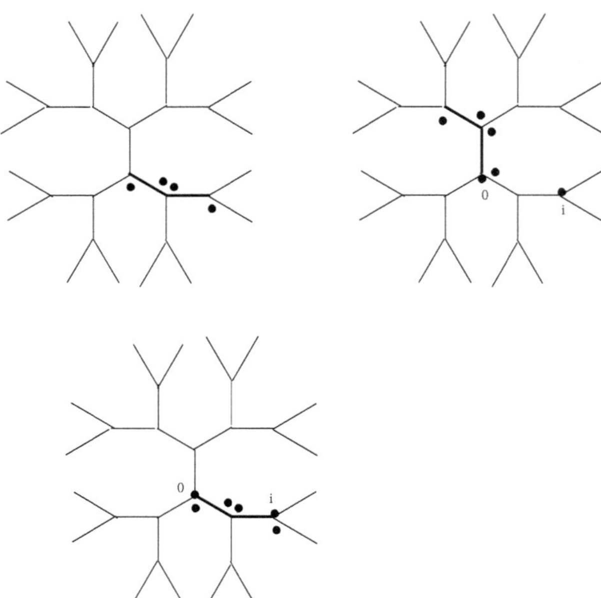

Fig. 14.2 **a** Top left: a diagram representing a term in the expansion of the denominator of Eq. (14.18). The bonds selected are represented by heavy lines. One sees that the sites at the end of the path have an odd number of dots, representing an odd number of spin operators. **b** Top right: a term in the expansion of the numerator which vanishes because the ends of the path do not coincide with the sites "0" and "i". **c** Bottom: the only nonzero term in the expansion of the numerator in which the bonds selected form a path connecting the two sites "0" and "i

$$\mathrm{tr} \prod_{\langle jk \rangle} [1 + t\sigma_j \sigma_k] = 1 \; . \tag{14.20}$$

Next consider the expansion of the numerator in Eq. (14.18) before taking the trace. The prefactor $\sigma_0 \sigma_i$ gives dots at these two sites. Then we must consider the set of 2^{N_B} diagrams one can obtain by adding to these two dots a collection of bonds with dots at each end of the bond. We note that only diagrams in which each site has an even number of bonds will survive the trace. For instance, suppose i is a second-generation site as shown in Fig. 14.2. Then only the term in the expansion of the product which has bonds which form a path connecting sites 0 and i will contribute. Thus, from $\prod_{\langle jk \rangle}(1 + t\sigma_j\sigma_k)$ the only term that survives the trace is $(t\sigma_0\sigma_1)(t\sigma_1\sigma_i)$, where 1 labels the site between sites 0 and i. So when i is a second-generation site, we have

$$\begin{aligned} \mathrm{tr}\sigma_0\sigma_i \prod_{\langle jk \rangle}(1 + t\sigma_j\sigma_k) &= \mathrm{tr}\sigma_0\sigma_i(t\sigma_0\sigma_1)(\mathrm{t}\sigma_1\sigma_i) \\ &= t^2 \end{aligned} \tag{14.21}$$

and

$$\langle \sigma_0 \sigma_i \rangle = t^2 . \tag{14.22}$$

In general, if i is a gth generation site, the only term out of the product which survives the trace is the one in which the bonds form the unique path from the seed site to site i. This path has $g(i)$ bonds, so that

$$\mathrm{Tr} \sigma_0 \sigma_i \prod_{\langle jk \rangle} [1 + t\sigma_j \sigma_k] = t^{g(i)} , \tag{14.23}$$

where $g(i)$ is the generation number associated with site i. Now, since there are z identical contributions (of t) from the first generation, $z\sigma$ identical contributions (of t^2) from the second generation, and so forth, where $\sigma = z - 1$, we have

$$\begin{aligned} \chi &= \sum_j t^{k(j)} = 1 + zt + z\sigma t^2 + z\sigma^2 t^3 + \ldots \\ &= 1 + \frac{zt}{1 - \sigma t} = \frac{1 + t}{1 - \sigma t} . \end{aligned} \tag{14.24}$$

This result shows that indeed we have a mean-field like divergence in the susceptibility ($\gamma = 1$) at a critical temperature, T_c, at which

$$\tanh(J/T_c) = 1/\sigma . \tag{14.25}$$

For large z, this coincides with our previous result that $T_c = zJ$. For finite z this version of mean-field theory gives the same results for the various critical exponents as any other version of mean-field theory, but the value of T_c depends on which version of the theory one is using.

It is instructive to calculate the spontaneous magnetization $\langle \sigma_i \rangle_T$. (To avoid spurious surface effects, this evaluation is done for the seed site.) Such a calculation raises a number of subtle issues. We will present such a calculation for percolation and leave the analogous calculation for the Ising model as a homework exercise.

14.3 Exact Solution for the Percolation Model

We consider the "bond" version of the percolation problem (Stauffer and Aharony 1992) in which a lattice of sites is connected by bonds which can either be occupied (with probability p) or unoccupied (with probability $1 - p$). One should note that the statistical average of any quantity, say, X, is given by

$$\langle X \rangle_p \equiv \sum_{\mathcal{C}} P(\mathcal{C}) X(\mathcal{C}) , \tag{14.26}$$

where $P(\mathcal{C})$ is the probability of having the configuration $\mathcal{C}$ of occupied and unoccupied bonds and $X(\mathcal{C})$ is the value of X in that configuration. In this prescription, it would seem that it is necessary to sum over all 2^N configurations, where N is the number of bonds, because each bond has two possible states: either it can be occupied or it can be unoccupied. However, if X only depends on a subset of b bonds, we only need sum over the 2^b configurations involving these bonds.

14.3.1 Susceptibility in the Disordered Phase

The position-dependent percolation susceptibility, $\chi_{i,j}$, (so-called for reasons we will see later) is defined as the probability that sites i and j are in the same connected cluster of occupied bonds. To make contact with Eq. (14.26), we define a connectivity indicator function $\nu(i, j)$ such that $\nu(i, j) = 1$ if sites i and j are connected and is zero if they are not connected. Then the two-point susceptibility is defined as

$$\chi_{i,j} = \langle \nu(i, j) \rangle_p \tag{14.27}$$

and the uniform susceptibility is defined as

$$\chi \equiv N^{-1} \sum_{i,j} \chi_{i,j} ,$$

where N is the total number of sites in the lattice. For a periodic lattice the quantity $\sum_j \chi_{ij}$ does not depend on i. For the Cayley tree we set i equal to the seed site to avoid any problems with surface effects and write

$$\chi \equiv \sum_j \chi_{0,j} . \tag{14.28}$$

Note that $\chi_{0,0} = 1$ and $\chi_{0,j}$ is the same for all sites j which are in the same generation. Consider $\chi_{0,1}$, the probability that the seed site is connected to a given first generation site. This probability is clearly p: the sites are connected if, and only if, the bond between them is occupied. Likewise $\chi_{0,2}$, the probability that the seed site is connected to a given 2nd generation site is p^2, because for these sites to be connected the two bonds separating them must be occupied. There are no indirect paths connecting two sites on a tree. Taking account of the number of sites in each generation we get that

$$\begin{aligned} \chi &= 1 + zp + z\sigma p^2 + z\sigma^2 p^3 + \cdots \\ &= 1 + \frac{zp}{1 - \sigma p} = \frac{1 + p}{1 - \sigma p} . \end{aligned}$$

We may interpret χ as being the average number of sites to which a given site is connected. This is the mean cluster size. (Some authors call this the mean square cluster size—it depends on how you count.) We see that this quantity increases with increasing p until it diverges at $p_c = 1/\sigma$, the critical concentration. It goes without saying that this result is reminiscent of that for the Ising model on a Cayley tree, namely,

$$\chi_{\mathrm{Ising}} = \frac{1+t}{1-\sigma t} ,$$

where

$$t = \tanh(J/kT) .$$

14.3.2 Percolation Probability

We now consider the probability that a given site (in our case the seed site) is in the infinite cluster. This quantity is sometimes called the percolation probability. We should define carefully what we mean by "being in the infinite cluster." The best way to look at this is to define the probability that the seed site is in the infinite cluster as

$$P_{0,\infty} \equiv 1 - P_{0,f} , \tag{14.29}$$

where $P_{0,f}$ is the probability that the seed site is in a *finite* cluster. The quantity $P_{0,f}$ can be calculated as a power series in p. For instance, the probability that the seed site is in a cluster of size 1, that is, that it is not connected to any neighbors is $(1-p)^z$. The probability that the seed site is in a cluster of size 2, is $zp(1-p)^{(2z-2)}$, because there are z way to select the second site of the cluster (one site has to be the seed site) and to isolate these two sites, we must have the bonds connected them to other sites vacant. There are $2z-2$ such bonds which have to be vacant to isolate the cluster of two sites. So, correct to order p we have

$$P_{0,f} = (1-p)^z + zp(1-p)^{2z-2} = 1 - zp + zp = 1 + \mathrm{O}(p^2) .$$

You may go to higher order in p. The result, for all orders in p, will be

$$P_{0,f} = 1.$$

This result is clearly correct: for small p there is zero probability to get an infinite cluster. So all coefficients in the expansion of $P_{0,f}$ in powers of p must vanish. Correct to order p we have

$$P_{0,\infty} = 1 - P_{0,f} = 0, \qquad \text{for } p < p_c . \tag{14.30}$$

We expect this result to hold for $p < p_c$.

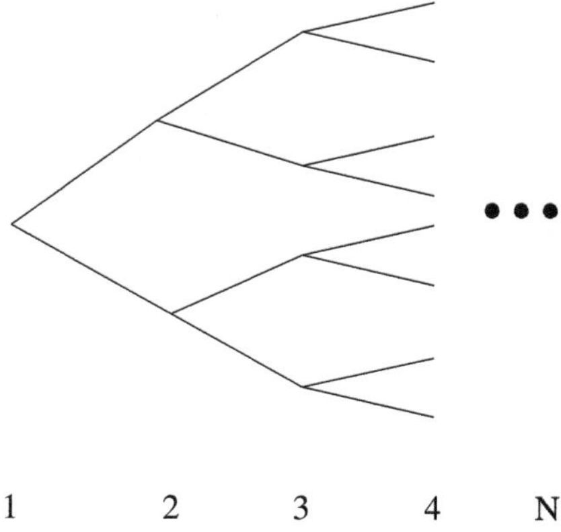

Fig. 14.3 A branch of the Cayley tree for $z = 3$ emanating from the first generation site. The integers at the bottom indicate the number of the generation

What happens for $p > p_c$? We construct a recursive equation for the probability that the branches emanating form a finite occupied cluster. We define $P_{1,f}$ to be the probability that the first generation site is not part of an infinite occupied cluster in any of the *outward* branches attached to it. (Such a *outward* branch from site i is the set of sites which can be reached from site i without going through any sites in generation lower than that of site i. Such a first-generation branch is shown in Fig. 14.3.

Note that $P_{1,f}$ is *not* the same as the probability that the first-generation site is in a finite cluster. Even if all $(z-1)$ second-generation branches attached to the first-generation site give rise to finite clusters, this site could be part of an infinite occupied cluster which is reached via the seed and then connecting to a different first-generation branch. For the central site to be in a finite cluster, it must be either isolated from each of the z first-generation sites, with probability $(1-p)$ or, if it is connected, with probability p, to this first-generation site, then the first-generation site is not part of an infinite cluster. The probability that the first-generation branch is part of a finite cluster is $P_{1,f}$. So

$$P_{0,f} = \sum_{k=0}^{z} \frac{z!}{(z-k)!k!}(1-p)^k (pP_{1,f})^{z-k} .$$

The kth term in this series corresponds to considering the case when the seed site has k unoccupied bonds and $z-k$ occupied bonds, each of which is connecting to a finite branch. Thus

$$P_{0,f} = \left((1-p) + pP_{1,f}\right)^z . \tag{14.31}$$

Clearly $P_{0,f}$ is unity if, and only if, $P_{1,f}$ is unity, as is obvious.

The above equation does not solve the problem. It relates the probability that the seed site is in a finite cluster to the slightly different quantity that each first generation outward branch is finite. A branch (see Fig. 14.3) has an origin with $z - 1 \equiv \sigma$ neighbors each of which have σ outer neighbors, etc. So we can write a recursive equation for $P_{1,f}$:

$$P_{1,f} = \sum_{k=0}^{\sigma} \frac{\sigma!}{\sigma - k!k!}(1-p)^k (pP_{2,f})^{\sigma-k} ,$$

where $P_{2,f}$ is the probability that the second-generation site has only finitely occupied branches emanating outward from it. More generally, the probability that a gth generation branch is finite is related to the probability that its $g + 1$th generation branches are finite:

$$\begin{aligned} P_{g,f} &= \sum_{k=0}^{\sigma} \frac{\sigma!}{\sigma - k!k!}(1-p)^k (pP_{g+1,f})^{\sigma-k} \\ &= \Big((1-p) + pP_{g+1,f} \Big)^{\sigma} . \end{aligned} \tag{14.32}$$

The analysis of these equations now depends critically on whether or not the tree has an infinite number of generations. If the tree has a finite number of generations, then obviously each $P_{g,f} = 1$ and there is no possibility that any site could be in an infinite cluster.

In contrast, for an infinite tree we may impose the self-consistency condition that $P_{2,f}$ is the same as $P_{1,f}$:

$$\begin{aligned} P_{1,f} &= \sum_{k=0}^{\sigma} \frac{\sigma!}{\sigma - k!k!}(1-p)^k (pP_{1,f})^{\sigma-k} \\ &= \Big((1-p) + pP_{1,f} \Big)^{\sigma} . \end{aligned} \tag{14.33}$$

Note that only in the limit of an infinite system can we admit the possibility that $P_{k,f}$ is not unity. This observation agrees with the fact that a phase transition is only possible in the infinite size limit. As we expect and as we shall see, only when p is large enough can there exist a solution to Eq. (14.33) with $P_{1,f} < 1$.

Now we analyze the self-consistency Eq. (14.33). As we have already noted, $P_{1,f} = 1$ is a solution and this must be the only solution for $p < p_c$. This equation is similar to the self-consistent equation

$$m = \tanh(Jzm/kT) \tag{14.34}$$

we had for the Ising model which also always has the analogous solution with $m = 0$. For $T > T_c$ this is the only solution. The phase transition is signaled by the emergence of a nonzero solution for the order parameter. That is also true here for percolation. Accordingly, we now try to solve Eq. (14.33) by setting

$$P_{1,f} = 1 - \epsilon \,,$$

where ϵ is an infinitesimal. Then Eq. (14.33) is

$$\begin{aligned} 1 - \epsilon &= \Big((1-p) + p - p\epsilon \Big)^{\sigma} \\ &\approx 1 - p\sigma\epsilon + \frac{1}{2}\sigma(\sigma - 1)p^2\epsilon^2 + \cdots \end{aligned}$$

This is

$$(p\sigma - 1)\epsilon = \frac{1}{2}\sigma(\sigma - 1)p^2\epsilon^2 \,.$$

Obviously, one solution which is always present has $\epsilon = 0$, which corresponds to $P_{1,f} = 1$. A second solution is only possible if it has ϵ positive because $P_{1,f}$ cannot be greater than 1. Such a solution will begin to appear when $p\sigma > 1$, or $p > p_c$, where $p_c = 1/\sigma$ is the same critical concentration at which the percolation susceptibility diverged. Thus for $p = p_c^+$ we have the solution

$$\epsilon = A(p - p_c) + \cdots \,.$$

You can go back to the percolation probability $P_{0,\infty}$ and show that it is proportional to ϵ. For instance, Eq. (14.31) gives

$$\begin{aligned} P_{0,f} &= \Big((1-p) + p - pA(p - p_c) \Big)^{z} \approx 1 - zpA(p - p_c) \\ &= 1 - A'(p - p_c) \,. \end{aligned} \tag{14.35}$$

So, we have the result,

$$P_{0,\infty} = A'(p - p_c)^{\beta} \,,$$

with $\beta = 1$. So although both percolation and the Ising model have the same value of the susceptibility exponent $\gamma = 1$, their values of β are different: $\beta = 1/2$ for the Ising model and $\beta = 1$ for percolation. These results show that these two models must be different in some fundamental sense.

14.4 Exact Solution for the Spin Glass

Here, we consider another model which has no obvious connection to a canonical partition function. This model is a random Ising model in which each nearest-neighbor interaction independently assumes the values $+J_0$ and $-J_0$. This is the simplest version of what is called a spin-glass model (Mydosh 2015). Thus

$$\mathcal{H}(\{J\}) = \sum_{\langle ij \rangle} J_{ij} S_i S_j \ , \tag{14.36}$$

where $J_{ij} = \eta_{ij} J_0$, with $\eta_{ij} = \pm 1$. We consider the case of *quenched* randomness, so that the randomness is not in thermodynamic equilibrium. In this case the way we are supposed to take averages is the following. For any random configuration of the J's we are to construct the partition function $Z(\{J\})$ and then, for this configuration of J's, the free energy $F(\{J\})$. The (averaged) free energy of the system is then found by averaging $F(\{J\})$ over the J's:

$$F(T) = \int \prod dJ_{ij} F(\{J\}) P(\{J\}) \ , \tag{14.37}$$

where $P(\{J\})$ is the probability of having the configuration $\{J\}$. Here, because we assumed a simple special distribution of J's we write

$$F(T) = \frac{1}{2^{N_B}} \sum_{\eta_{ij}=\pm 1} F(\{\eta_{ij} J_0\}) \ , \tag{14.38}$$

where N_B is the number of bonds. Note that we are performing an average of the free energy. Had we taken an average of the partition function, we could have interpreted this additional summation as part of a trace. In that case (which corresponds to annealed randomness), we could have invoked the standard canonical formulation. But averaging the free energy is *not* incorporated in a standard canonical formulation in any obvious way. The physical difference between these two averages is the following. Suppose the exchange interaction between sites i and j depends on whether or not an oxygen occupies an interstitial site between sites i and j. In our model, one might give the interpretation that J_{ij} is antiferromagnetic if the two sites have an intervening oxygen ion present and is ferromagnetic if the intervening oxygen ion is absent. This actually has some justification within a model of superexchange interactions. But that is not crucial in this discussion. What matters here is whether the oxygen ions diffuse rapidly in and out of the interstitial sites. If this diffusion is rapid, then we would expect the distribution of oxygen ions to always be a thermal equilibrium distribution and the appropriate average to take would be to average the partition function, including a factor $e^{-\beta E}$, where E is the energy of the configuration of oxygen ions. On the other hand, if the diffusion is glacially slow, so that no diffusion occurs over any experimentally possible time scale, then we are stuck with whatever distribution of ions we started with. Assuming that the starting distribution was random, we would then conclude that we should perform a "quenched" average, i.e., we should average, not the partition function, but rather we should average the free energy, because the distribution of oxygen ions is not in thermal equilibrium and does not, therefore, respond by changing as the temperature of the system is changed. Rather this distribution is temperature independent. This is the limit that we are considering here. (We will see later, however, that a trick can be used to relate quenched randomness to a limiting case of annealed randomness). Equation (14.38) implies that we should calculate the susceptibility as an average over the J's of the susceptibility calculated for each configuration of J's.

How do we expect $\langle S_i S_j \rangle$ to behave for some configuration of J's? This depends on what the J's are, of course. However, at sufficiently high temperature, this correlation function will fall off rapidly with separation. As the temperature is lowered, eventually one will enter a state with some kind of long-range order. If the J's are predominantly ferromagnetic, we would expect a transition into a ferromagnetic state. On the other hand, if the interactions are predominantly antiferromagnetic, we would expect, that there would be a transition into an antiferromagnetic state. More generally, we would expect the spin system to become rigid as the temperature is sufficiently lowered. The nature of the rigid ordering would be random (because the J's are random), but in each configuration of the J's there would be a freezing of the spins into the state favored by the particular configuration of J's. If we denote the average over J's by $[\ ]_J$, then what we are saying is that because the freezing is random

$$[\langle S_i S_j \rangle]_J \tag{14.39}$$

is zero. If there is perfect freezing then $\langle S_i S_j \rangle = \pm 1$ depending on whether the configuration of J's favors parallel alignment of the two spins i and j or antiparallel alignment of the two spins. (Note that in a random configuration, we will not be able to predict any particular dependence on separation to these ± 1's.) What to do? Since we think that freezing is reflected by the *amplitude* of the correlation function, we will define a spin-glass correlation function $f_{SG}(i, j)$ via

$$f_{SG}(i, j) \equiv [\langle S_i S_j \rangle^2]_J \tag{14.40}$$

and correspondingly, for the Cayley tree, we define the spin-glass susceptibility (as contrasted to the ordinary susceptibility) as

$$\chi_{SG} = \sum_j f_{SG}(0, j) = \sum_j [\langle S_0 S_j \rangle^2]_J \ . \tag{14.41}$$

Thus even though we expect $[\langle S_i S_j \rangle]_J = 0$, the spin-glass susceptibility can diverge if there is long-range freezing. Due to the randomness the critical temperature at which that divergence occurs ought to be much less than $kT_0 = zJ_0$.

We now evaluate the spin-glass susceptibility for a Cayley tree in order to get what the MF result should be for this model. We use the analog of Eq. (14.9) to write

$$e^{\beta J_{ij}\sigma_i\sigma_j} = \cosh(\beta J)\Big(1 + \eta_{ij} t \sigma_i \sigma_j\Big) , \tag{14.42}$$

where η_{ij} is the sign of J_{ij} and $t \equiv \tanh(\beta|J_{ij}|)$. Now compare this calculation to that for the ferromagnetic Ising model where we saw that

$$\chi = \sum_j \langle \sigma_0 \sigma_j \rangle_T \tag{14.43}$$

and that the average on the right-hand side of this equation involved only the single term which represented the interactions associated with the unique path from the seed (0) to the site j. Thus

$$\langle \sigma_0 \sigma_j \rangle_T = t^{d(j)} . \tag{14.44}$$

Here we must also include the signs associated with each random interaction:

$$\langle \sigma_0 \sigma_j \rangle_T = t^{d(j)} \prod \eta_{mn} , \tag{14.45}$$

where the product is over all the bonds involved in the unique path from the seed to site j. Note that each η_{ij} appears once and $[\eta_{ij}]_J = 0$. Only the term with no η's survives the average over the η's. So

$$\sum_j [\langle \sigma_0 \sigma_j \rangle]_J = 1 , \tag{14.46}$$

In contrast consider the spin-glass correlation function

$$\big\langle \sigma_0 \sigma_j \big\rangle_T^2 = t^{2d(j)} \prod \eta_{mn}^2 = t^{2d_{0j}} . \tag{14.47}$$

Because we have taken the square of the usual correlation function, the average over the distribution of J's is superfluous for this model where only the signs of the J's are randomized. Performing the sum over shell of neighbors of the seed site, we get

$$\begin{aligned} \chi_{\rm SG} &= 1 + zt^2 + z\sigma t^4 + z\sigma^2 t^6 + \cdots \\ &= 1 + \frac{zt^2}{1-\sigma t^2} = \frac{1+t^2}{1-\sigma t^2} . \end{aligned} \tag{14.48}$$

(This result is clearly reminiscent of that previously obtained for the Ising model and for percolation.) This gives

$$\chi_{\rm SG} \sim \frac{A}{T - T_{\rm SG}}\,, \tag{14.49}$$

where the transition temperature is given by $\tanh(kT_{\rm SG}) = 1/\sqrt{\sigma}$, which gives a much smaller critical temperature (at least for $\sigma > 1$) than for the nonrandom model. Assuming the quantity we have calculated is really the order parameter susceptibility, this calculation gives $\gamma = 1$. By a more complicated calculation, one can show that the long-range order in $\mathcal{S} \equiv [\langle S\rangle^2]_J$ varies for $T \to T_{\rm SG}^-$ as

$$\mathcal{S} \sim (T_{\rm SG} - T)\,, \tag{14.50}$$

so that $\beta = 1$. These critical exponents are the same as for percolation. Beyond mean-field theory the critical exponents for percolation and the spin-glass are no longer the same.

Of course, as with percolation, a crucial issue (which will be addressed in the next chapter) is to somehow describe these problems by a canonical partition function and identify correctly their order parameter and the field, h, conjugate to the order parameter, so that the order parameter and susceptibility can be obtained from successive derivatives of the free energy with respect to the identified field h.

14.5 General Development of the Tree Approximation

In this section, we will give a general approach (Harris 1982) which may be used to generate exact solutions on the Cayley tree and which also forms the basis for developing a perturbation expansion in powers of $1/d$, where d is the spatial dimensionality. We will now develop the so-called "tree approximation" which allows us to collect all terms in the diagrammatic expansion of the partition function which are tree-like, i.e., which have no loops. More generally, one obtains an expansion in which the first term represents a self-consistent solution for a single site and corrections involve diagrams with no free ends which can be embedded in the lattice. Since the Cayley tree does not permit any such no-free-end diagrams, the first term yields the exact solution for the Cayley tree.

14.5.1 General Formulation

To start, we assume that the Hamiltonian contains of a sum of pairwise interactions between nearest-neighboring sites and also that it is expressed in terms of classical (i.e., commuting) variables. Thus the Hamiltonian is assumed to have the form

$$\mathcal{H} = \sum_i \mathcal{H}_1(x_i) + \sum_{\langle ij\rangle} \mathcal{H}_2(x_i, x_j)\,, \tag{14.51}$$

where $\mathcal{H}_1$ describes the single-site potential (and therefore depends only on variables x_i associated with the single site i, and $\mathcal{H}_2$ describes the interaction energy between sites i and j. In particular, the term $\mathcal{H}_1$ may include the coupling to the field h conjugate to the order parameter. We assume that all sites are equivalent and that all nearest-neighbor interactions are equivalent. We may then write

$$Z = \mathrm{Tr} e^{-\beta\mathcal{H}} = \mathrm{Tr}\left\{\prod_i f_i \prod_{\langle ij\rangle} f_{ij}\right\}, \tag{14.52}$$

where $f_1 = \exp[-\beta\mathcal{H}_1(x_i)]$ and $f_{ij} = \exp[-\beta\mathcal{H}_2(x_i, x_j)]$. Since each site appears z times in the product over nearest-neighbor pairs $\langle ij\rangle$, where z is the coordination number of the lattice, we may introduce an arbitrary single-site function g_i which depends only on the variables of site i, as follows:

$$Z = \mathrm{Tr} e^{-\beta\mathcal{H}} = \mathrm{Tr}\left\{\prod_i (g_i^z f_i) \prod_{\langle ij\rangle}\left(\frac{f_{ij}}{g_i g_j}\right)\right\}. \tag{14.53}$$

We will determine g_i so as to facilitate establishing an exact solution for the Cayley tree. (One may view the factor g_i as an effective single-site potential which we determine to best approximate the pair Hamiltonian.) We therefore write

$$Z = \mathrm{Tr} e^{-\beta\mathcal{H}} = \mathrm{Tr}\left\{\prod_i (g_i^z f_i) \prod_{\langle ij\rangle}(1 + V_{ij})\right\}, \tag{14.54}$$

where

$$V_{ij} = \frac{f_{ij}}{g_i g_j} - 1\,. \tag{14.55}$$

Now we consider evaluating this partition function in a perturbation series in powers of V_{ij}. We represent each term in the expansion of Eq. (14.54) in powers of V_{ij} by a diagram. For this purpose we represent the factor V_{ij} by a bond connecting sites i and j. Then we see that each term in this perturbation series corresponds to a diagram consisting of occupied and unoccupied bonds on the lattice. We now implement the "tree" approximation in which we take advantage of the arbitrariness of the function g_i to choose its value so as to eliminate all diagrams with a "free end." By a free end we mean a site to which exactly one bond is attached. One can see that for the Cayley tree, all diagrams (apart from the diagram with no bonds at all) have at least one free end. The only diagrams that do not have a free end involve closed loops of bonds. But such diagrams cannot occur on a tree. The result is that the tree approximation gives an exact solution for the Cayley tree, at least in the absence of broken symmetry.

To eliminate diagrams with a free end, consider a free end at site k and perform the trace over the variables of this site. If the site k is a free end, it appears in only one factor in the product over bonds $\langle ij \rangle$ in Eq. (14.54). Thus the only factors which involve the variables of site k are $f_k g_k^z V_{ik}$, where i labels the only site with a bond connecting to the free end site, k. The contribution from a diagram with a free end at site k will therefore vanish providing we choose g_k so that

$$\mathrm{Tr}_k f_k g_k^z V_{ik} = 0 , \tag{14.56}$$

where Tr_k indicates a trace over only variables of site k. Using Eq. (14.55) we see that this is

$$\mathrm{Tr}_k \left(f_k g_k^z - f_k f_{ik} g_k^{z-1} / g_i \right) = 0 \tag{14.57}$$

or

$$g_i = \frac{\mathrm{Tr}_k f_{ik} f_k g_k^\sigma}{\mathrm{Tr}_k g_k^z f_k} , \tag{14.58}$$

where $\sigma = z - 1$. This is a nonlinear equation for g and as such, might seem to be hard to solve. Actually, in most cases it is easy to solve. After g is determined to eliminate diagrams with free ends, we get the tree approximation (which is exact for a tree lattice) that $Z = \mathcal{Z}^N$, where N is the total number of sites and

$$\mathcal{Z} = \mathrm{Tr}_i g_i^z f_i \tag{14.59}$$

is the single-site partition function.

14.5.2 Ising Model

As an illustration, let us apply this approach to the Ising model. Then $f_i = e^{H\sigma_i}$ and $f_{ij} = e^{J\sigma_i\sigma_j}$, where we temporarily set $\beta J = J$ and $\beta H = H$ in which case Eq. (14.58) is

$$g_i = \frac{\mathrm{Tr}_j \left(e^{H\sigma_j} g_j^\sigma e^{J\sigma_i\sigma_j} \right)}{\mathrm{Tr}_j e^{H\sigma_i} g_j^z} , \tag{14.60}$$

whose solution we may write in the form

$$g_i = A e^{B\sigma_i} . \tag{14.61}$$

By substituting this ansatz into Eq. (14.60) one finds that A and B satisfy

$$A \cosh B = \frac{\cosh J \cosh(H + \sigma B)}{A \cosh(H + zB)} \tag{14.62}$$

$$A \sinh B = \frac{\sinh J \sinh(H + \sigma B)}{A \cosh(H + zB)} . \tag{14.63}$$

Since we cannot solve these equations for arbitrary H, we expand in powers of H. We work to order H^2 and write

$$A = (\cosh J)^{1/2} \left(1 + \frac{1}{2} a H^2 + \mathcal{O}(H^4)\right) \tag{14.64}$$

$$B = bH + \mathcal{O}(H^3) , \tag{14.65}$$

where a and b are independent of H and are found to be

$$a = -t(1+t)\chi_0^2 \tag{14.66}$$

$$b = t\chi_0 , \tag{14.67}$$

where $t = \tanh(J)$ and $\chi_0 = (1 - \sigma t)^{-1}$.

Now $Z = \mathcal{Z}^N$, where

$$\begin{aligned} \mathcal{Z} &= \mathrm{Tr}_i\, f_i g_i^z = \mathrm{Tr}_i\, A^z e^{(zb+1)H\sigma_i} \\ &= 2A^z \cosh[(zb+1)H] . \end{aligned} \tag{14.68}$$

Up to order H^2 this is

$$\begin{aligned} \mathcal{Z} &= 2\cosh^{z/2} J \left(1 - \frac{1}{2} zt(1+t)\chi_0^2 H^2\right) \cosh[(zt\chi_0 + 1)H] \\ &= 2\cosh^{z/2} J \left(1 + \frac{(1+t)H^2}{2(1-\sigma t)}\right) . \end{aligned} \tag{14.69}$$

This implies that

$$kT\chi = \partial^2 \ln Z/\partial H^2\Big)_{H=0} = N \frac{1+t}{1-\sigma t} , \tag{14.70}$$

as we found before. Here, however, one can develop corrections to the result for the Cayley tree by evaluating diagrams on a hypercubic lattice which have no free ends. We will return to this point in the chapter on series expansions. Also see Exercise 4.

14.5.3 Hard-Core Dimers

We can also apply this technique to nonthermal statistical problems providing we can identify their statistical generating function with a partition function, Z of the form $Z = \mathrm{Tr}\exp(-\beta\mathcal{H})$, where $\mathcal{H}$ can then be interpreted as the Hamiltonian for the statistical problem. Clearly in such statistical problems, the first key step in applying the above formalism is to construct such a Hamiltonian. The idea that a probability generating function for a nonthermal problem can be related to a partition function will be explored in detail in a later chapter on mappings. Here, we illustrate the construction of a Hamiltonian whose partition function coincides with the grand partition function which counts the number of ways one can arrange hard-core dimers on a lattice. A hard-core dimer is an object which can occupy a nearest-neighbor bond on a lattice with the hard-core constraint that no site can be part of more than one dimer. We introduce the Hamiltonian by writing

$$e^{-\beta\mathcal{H}} = e^{x\sum_{\langle ij\rangle}\mathbf{S}_i\cdot\mathbf{S}_j} . \tag{14.71}$$

We want to arrange it so that the expansion of the partition function in powers of x counts all possible dimer configurations, where $x = \exp(\beta\mu)$, where μ is the chemical potential for dimers. (The quantity $\exp(\beta\mu)$ is known as the dimer *fugacity*.) For this scheme to work, we must prevent dimers from touching. Accordingly, we impose the following *trace rules* on the operators $\mathbf{S}_i$:

$$\mathrm{Tr}_i(\mathbf{S}_i)^n = C_n , \tag{14.72}$$

with $C_0 = C_1 = 1$ and $C_n = 0$ for $n > 1$. The reader is probably wondering how we are going to actually construct such an operator. It turns out that we do not need to explicitly display such an operator. The only property of this operator we need in order to construct the partition function is the trace rules of Eq. (14.72). We see that the fact that the trace of two or more operators at the same site vanishes, implements exactly the hard-core constraint for dimers. With these rules

$$e^{-\beta\mathcal{H}} = \prod[1 + e^{\beta\mu}\mathbf{S}_i\mathbf{S}_j] \tag{14.73}$$

and furthermore the trace rules do not allow dimers to touch. This is what we need to get the hard care dimer partition function. Thus the partition function $A = \mathrm{Tr}\exp(-\beta\mathcal{H})$ will indeed give the grand partition function for dimers as a function of their chemical potential μ:

$$Z = \sum_{\mathcal{C}} e^{\beta\mu n(\mathcal{C})} , \tag{14.74}$$

where the sum is over all configuration $\mathcal{C}$ of dimers and $n(\mathcal{C})$ is the number of dimers present in the configuration $\mathcal{C}$. From Z we can get the dimer density ρ_D as a function of the dimer chemical potential via

$$\rho_D \equiv \frac{\langle n \rangle}{N} = \frac{1}{N\beta}\frac{\partial \ln Z}{\partial \mu} = \frac{1}{N}\frac{d \ln Z}{d \ln x} . \tag{14.75}$$

Note that these "spin operators" commute with one another, so we have mapped the athermal problem of dimers on a lattice into a statistical problem involving classical spins with a given Hamiltonian. (More such mappings will be discussed in a later chapter.)

Now we consider Eq. (14.58) where, for the dimer Hamiltonian, we have $f_j = 1$ and $f_{ij} = e^{x\mathbf{S}_i\mathbf{S}_j} = 1 + x\mathbf{S}_i\mathbf{S}_j$. (We can linearize the exponential because higher order terms vanish in view of the trace rules.) For this form of f_{ij} one sees that g_i has to be of the form

$$g_i = A + B\mathbf{S}_i , \tag{14.76}$$

and we can develop equations for the constants A and B by substituting this form into Eq. (14.58):

$$A + B\mathbf{S}_i = \frac{\mathrm{Tr}_j(1 + x\mathbf{S}_i\mathbf{S}_j)(A + B\mathbf{S}_j)^\sigma}{\mathrm{Tr}_j(A + B\mathbf{S}_j)^z} . \tag{14.77}$$

Using the trace rules we may simplify the right-hand side of this equation so that

$$A + B\mathbf{S}_i = \frac{A^\sigma + \sigma A^{\sigma-1}B + \mathbf{S}_i(xA^\sigma)}{A^z + zA^\sigma B} . \tag{14.78}$$

This gives rise to the two equations

$$A = \frac{A^\sigma + \sigma A^{\sigma-1}B}{A^z + zA^\sigma B} = \frac{A + \sigma B}{A^2 + zAB} \tag{14.79}$$

$$B = \frac{xA^\sigma}{A^z + zA^\sigma B} = \frac{x}{A + zB} . \tag{14.80}$$

We may solve Eq. (14.79) for B as

$$B = \frac{A^3 - A}{\sigma - zA^2} . \tag{14.81}$$

Substituting this into Eq. (14.80) leads to

$$A^4\left(1 + z^2x\right) - A^2\left(2z\sigma x + 1\right) + x\sigma^2 = 0 . \tag{14.82}$$

Thus

$$A^2 = \frac{2xz\sigma + 1 \pm \sqrt{1+4x\sigma}}{2(1+z^2x)} \,. \tag{14.83}$$

Apart from some slightly annoying algebra, this determination of the constants A and B allows us to obtain $Z = \mathcal{Z}^N$ via

$$\begin{aligned} \mathcal{Z} &= \mathrm{Tr}_i f_i g_i^z = \mathrm{Tr}_i [A + B\mathbf{S}_i]^z = A^z + zA^\sigma B \\ &= \frac{A^z}{zA^2 - \sigma} \,, \end{aligned} \tag{14.84}$$

where A is given by Eq. (14.83). In order that $Z > 0$ for small x, we must choose the positive sign in Eq. (14.83). Furthermore, since dZ/dx has to be positive, we must retain the positive sign for all x. From Eq. (14.84), one can get the dimer density as a function of the dimer chemical potential. What we see without any calculation is that A is an analytic function of μ, so that in this approximation (and probably also on finite-dimensional lattices), this model does not have a phase transition corresponding to a transition between a gas and a liquid of dimers. For a discussion of this result and for results for some periodic lattices, see (Nagle 1966).

14.5.4 Discussion

We have said that because Eq. (14.58) eliminates diagrams with free ends, it provides an exact solution for the Cayley tree. This assertion is only true in a limited sense. It is clear that we cannot discuss the ordered phase within such a perturbative expansion.

14.6 Questions

Here are some questions we will address later.

- In what sense, if at all, is the quantity we have called the percolation (or spin-glass) susceptibility related to the response in some sort of external field?
- Is there any analog to the specific heat exponent?
- Can this statistical model of percolation or spin-glass be related to any Gibbsian probability distribution. That is, can we construct a Hamiltonian to describe percolation statistics?
- Are the different values of the critical exponents indicative of different symmetry?

14.7 Summary

The main point of this chapter is that one can often obtain the exact solution to various statistical models on the Cayley tree, which represents the solution at infinite spatial dimension. Even if it is not clear how, if at all, the model results from a partition function, such a solution should correspond to mean-field theory. Examples given here include the percolation problem and the spin-glass problem (both of which can actually be mapped into problems with canonical partition functions). Further examples of exact solutions on the Cayley tree are those for the Hubbard model (Metzner and Vollhardt 1989), the statistics of branched polymers (Harris 1982), and diffusion-limited aggregation (Vannimenus et al. 1984). In an exercise, we suggest how to obtain the tight-binding density of states on the Cayley tree.

14.8 Exercises

1. Get the spontaneous magnetization of the Ising model for $T < T_c$ in zero magnetic field. Do the following calculations for zero magnetic field:

(**a**) For the interactions of the seed site with its neighbors use

$$V \equiv \prod_i e^{-\beta J \sigma_0 \sigma_i} = C \prod_i [1 + t\sigma_0 \sigma_i] ,$$

where $t \equiv \tanh(\beta J)$ and $C = \cosh(\beta J)^{\sigma+1}$. Then we have

$$\langle X \rangle = \mathrm{Tr} X V \prod_i e^{-\beta \mathcal{H}_i} / \mathrm{Tr} V \prod_i e^{-\beta \mathcal{H}_i} .$$

Here, $\mathcal{H}_i$ is the Hamiltonian for the branch emanating from the ith first-generation site when that branch is disconnected from the rest of the tree. Use this relation to write down an exact equation which relates the magnetization of the seed site, M_0, to the magnetization that the first-generation site would have, M_1, if its branch were isolated (i.e., governed by $\mathcal{H}_i$).

(**b**) Write down an exact equation which relates M_1 to M_2 the magnetization of a second-generation site if the branch emanating from the second-generation site were isolated from the rest of the tree.

(**c**) For a finite tree what is the unique solution for M_n, the similarly defined magnetization of the nth generation site?

(**d**) Now consider the infinite tree. Set $M_1 = M_2$ to get a self-consistent equation for M_1 and thereby evaluate M_0. If you do all this correctly, your answer, for $\tau \equiv (T_c - T)/T$, when $\tau \ll 1$ will be

$$M_0 = \left(\frac{3z^2\mathcal{T}}{2(z-1)} \ln[z/(z-2)] \right)^{1/2} .$$

2. In this problem you are to obtain the density of states $\rho(E)$ for the tight binding model on a Cayley tree. To discuss this problem one introduces the Green's function $G_{ij}(E) \equiv \left([E\mathcal{I} - \mathbf{T}]^{-1} \right)_{ij}$, where $\mathcal{I}$ is the unit matrix, $T_{ij} = t$ if sites i and j are nearest neighbors and is zero otherwise. In terms of G we have the density of states $\rho(E)$ as $\rho(E) = (1/\pi)\Im G_{00}(E - i0^+)$, where "0" labels the seed site (to avoid unphysical effects of the surface of the tree.) We evaluate $G_{00}(E)$ by expanding in an infinite series in powers of the matrix $\mathbf{T}$:

$$\mathbf{G} = \mathbf{G}_0 + \mathbf{G}_0\mathbf{T}\mathbf{G}_0 + \mathbf{G}_0\mathbf{T}\mathbf{G}_0\mathbf{T}\mathbf{G}_0 + \dots , \tag{14.85}$$

where $G_0 = E^{-1}\mathcal{I}$. One can think of this series as the generating function which sums all random walks.

(**a**) Consider walks which go from the origin to the first generation site, then do arbitrary walks in that branch, return to the origin, go to the first-generation site, do arbitrary walks in that branch etc. Thereby relate $G_{00}(E)$ to $G_{11}^{(1)}(E)$ which is the generating function for all random walks which start and end at the first-generation site and take place only within the branch starting at that first-generation site.

(**b**) Similarly relate $G_{11}^{(1)}$ to $G_{22}^{(2)}$, the generating function for walks which begin and end at the same second-generation site but which never pass through the first-generation site.

(**c**) For an infinite tree equate $G_{11}^{(1)}$ and $G_{22}^{(2)}$. Obtain $G_{11}^{(1)}$ and thereby $G_{00}(E)$. Give an explicit formula for the density of states on a Cayley tree.

3. In this problem you are to construct an algebraic equation which determines the generating function for branched polymers on a Cayley tree. This generating function is given by

$$Z(K) \equiv \sum_n a_n K^n ,$$

where K is the fugacity for adding a monomer unit to the polymer and a_n is the number of branched polymers which can be constructed by adding monomers (i.e., bonds) starting from the seed site. The low order terms are

$$Z(K) = 1 + (\sigma + 1)K + [(3/2)\sigma^2 - (1/2)\sigma]K^2 + \cdots$$

The average number, N, of monomers in a polymer at fugacity K is given by $N = (K/Z)dZ(K)/dK$.

(**a**) Determine $Z(K)$ as follows:

(1) Write an expression for $Z(K)$ in terms of $Z_1(K)$, where $Z_1(K)$ is the generating function for one of the $\sigma + 1$ branches connected to the seed site when this branch is isolated.

(2) Write an expression for $Z_1(K)$ in terms of $Z_2(K)$, where $Z_2(K)$ is the generating function for a branch which starts at the second-generation site. Determine $Z_1(K)$ for an *infinite* Cayley tree by equating $Z_2(K)$ and $Z_1(K)$.

(**b**) Solve explicitly for $Z_1(K)$ for the case $\sigma = 2$. Find the critical value, K_c, of K at which N diverges. Give an explicit equation for $N(K)$ for $\sigma = 2$. Find the exponent x such that for $K \to K_c$, $N \sim |K - K_c|^{-x}$.

(**c**) Extra credit: determine K_c for arbitrary values of σ.

4. Show that Ising model susceptibility for the Cayley tree with coordination number $z \equiv 2d >> 1$ coincides with the result for a cubic lattice in d spatial dimensions at leading order in $1/d$. (This may be too hard.)

5. Consider the higher order susceptibility $\chi^{(3)}(T)$ which is defined to be $-\partial^4 F(H,T)/\partial H^4\big|_T$ evaluated at $H = 0$. In this exercise you are to make an asymptotic evaluation of $\chi^{(3)}(T)$ for $T \to T_c$ for a ferromagnetic nearest-neighbor Ising model on a Cayley tree of coordination number z. To avoid boundary issues set

$$\chi^{(3)}(T) = -\sum_{i,j,k} \frac{\partial^4 F(T,\{H_i\})}{\partial H_0 \partial H_i \partial H_j \partial H_k}\Bigg)_{\{H_i=0\}} ,$$

where "0" labels the seed site. An exact evaluation is not required. Instead obtain an asymptotic result in the form

$$\chi^{(3)}(T) \sim A|T - T_c|^{-\gamma_3} ,$$

for $T \to T_c$, where you evaluate A and γ_3 exactly. Check that your result for γ_3 agrees with the asymptotic analysis of mean-field theory in Sect. 8.3.

6. We did not explicitly evaluate Eq. 14.75 for the dimer density. Show that

$$\rho_D = \frac{z}{4(1+z^2x)}\left(1 + 2zx - r\right) ,$$

where $r = \sqrt{1 + 4\sigma x}$.

7. The Cayley tree does not have a natural d-dimensional metric. By that we mean that the distance between two points is only connected naturally to the number of steps it takes to go from one point to another and does not depend on the space in which the tree is embedded. To associate a d-dimensional metric with distance, we say that the "distance" r_{ij} between two points i and j on the tree is given by $r_{ij}^2 = d_{ij}$, where

d_{ij} is the number of steps it takes to go from site i to site j. Using this algorithm we can calculate distance-dependent quantities in the limit of infinite spatial dimension using the Cayley tree. (In the limit of infinite spatial dimensionality, every step is likely to be perpendicular to every other step, justifying the above formula for r_{ij}.)

For instance, we may define the correlation length ξ for the Ising model by

$$\xi^2 \equiv \frac{\sum_j \langle \sigma_i \sigma_j \rangle r_{ij}^2}{\sum_j \langle \sigma_i \sigma_j \rangle} . \tag{14.86}$$

Evaluate this for a Cayley tree for the Ising model IN THE DISORDERED PHASE by setting $r_{ij}^2 = d_{ij}$, and taking site i to be the seed site. If we set $\xi \sim (T - T_c)^{-\nu}$, what does this predict for the critical exponent ν?

8. Use a calculation analogous to that of problem 7 to obtain the correlation length ξ for percolation on a Cayley tree and thereby get ν for percolation in the limit of infinite spatial dimensionality.

References

A.B. Harris, Renormalized ($\frac{1}{\sigma}$) expansion for lattice animals and localization. Phys. Rev. B **26**, 337 (1982)

F. Hausdorff, Dimension und äußeres Maß. Math. Ann. **79**(157), 179 (1919)

B.B. Mandelbrot, *Fractals: Form, Chance, and Dimension*, Revised edn. (WH Freeman and Co, 1977)

W. Metzner, D. Vollhardt, Correlated lattice fermions in $d = \infty$ dimensions. Phys. Rev. Lett. **62**, 324 (1989)

J.A. Mydosh, Spin glasses: redux: an updated experimental/materials survey. Rep. Prog. Phys. **78**, 052501 (2015)

J.F. Nagle, New series-expansion method for the dimer problem. Phys. Rev. **152**, 190 (1966)

D. Stauffer, A. Aharony, *Introduction to Percolation Theory*, 2nd edn. (Taylor and Francis, 1992)

J. Vannimenus, B. Nickel, V. Hakim, Models of cluster growth on the Cayley tree. Phys. Rev. B **30**, 391 (1984)

Part IV
Beyond Mean-Field Theory

Chapter 15
Exact Mappings

In this chapter, we discuss exact mappings which relate statistical problems in which there is no reference to a temperature, to statistical mechanical models to which we can apply the various techniques of analysis discussed in this text. We have already seen one example of an exact mapping in connection with the statistics of hard-core dimers on a Cayley tree. (The Hamiltonian we derived there could be used for any lattice.) Here we consider additional examples of this approach.

15.1 q-State Potts Model

The Potts model was introduced earlier, and, in particular, the mean-field theory for the 3-state Potts model was derived in Sect. 10.7. The Hamiltonian for the q-state Potts model is a sum of nearest-neighbor pair interactions and may be written as

$$\mathcal{H} = -J \sum_{<ij>} [q\delta_{s_i,s_j} - 1] , \tag{15.1}$$

where s_i and $s_j = 1, 2, \ldots q$ label the states of Potts "spins" i and j, respectively. If the two spins are in the same state, then $\delta_{s_i,s_j} = 1$ and their interaction energy is $-(q-1)J$ whereas if the two spins are in different states, their interaction energy is J. For positive J, this is a ferromagnetic model in that spins have the lowest energy when they are in the same state. For $q = 2$, this model is exactly the Ising model in which the spins have energy $-J$ if they are both in the same state (either both up or both down) and they have energy J if they are in different states, one up and the other down.

A. J. Berlinsky and A. B. Harris, *Statistical Mechanics*, Graduate Texts in Physics,
https://doi.org/10.1007/978-3-030-28187-8_15

15.1.1 Percolation and the $q \to 1$ Limit

We will show here that in the limit when $q \to 1$, the partition function for this model can be used to generate the statistics of the percolation problem exactly (Kasteleyn and Fortuin 1969; Fortuin and Kasteleyn 1972). Write the nearest-neighbor interaction as

$$\mathcal{H}_{ij} = -J[q\delta_{s_i,s_j} - 1] . \tag{15.2}$$

Now we will need $\exp(-\beta\mathcal{H}_{ij})$, where $\beta = 1/(kT)$. Since this exponential assumes two values depending on whether or not s_i and s_j are equal, we may write

$$e^{-\beta\mathcal{H}_{ij}} = A + B\delta_{s_i,s_j} . \tag{15.3}$$

We determine A and B by forcing this relation to hold in the two cases: case 1 when $s_i = s_j$ and case 2 when $s_i \neq s_j$. In the first case we get

$$e^{\beta J(q-1)} = A + B \tag{15.4}$$

and in the second case

$$e^{-\beta J} = A \tag{15.5}$$

so that

$$B = e^{\beta J(q-1)} - e^{-\beta J} . \tag{15.6}$$

Then the partition function, Z_P, for a system of N_s sites is

$$\begin{aligned} Z_P &= \mathrm{Tr} e^{\beta J \sum_{<ij>}[q\delta_{s_i,s_j}-1]} \\ &= \sum_{\{s_i\}} \prod_{<ij>} \left\{ A + B\delta_{s_i,s_j} \right\} \\ &= \sum_{\{s_i\}} \prod_{<ij>} \left\{ \left(e^{\beta J(q-1)}\right)\left(e^{-\beta Jq} + [1 - e^{-\beta Jq}]\delta_{s_i s_j}\right) \right\} \\ &= e^{N_B \beta J(q-1)} \sum_{\{s_i\}} \prod_{<ij>} \left[(1-p) + p\delta_{s_i s_j} \right] , \end{aligned} \tag{15.7}$$

where N_B is the total number of bonds in the system and

$$p = 1 - e^{-\beta Jq} \tag{15.8}$$

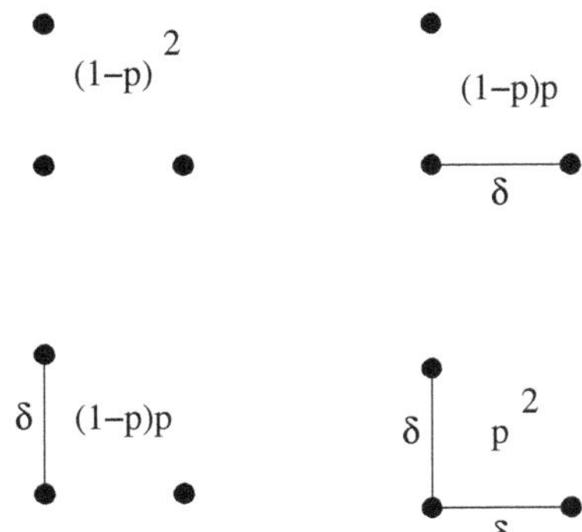

Fig. 15.1 p-dependent weight factors for various diagrams. The δ's indicate that the Potts spins connected by solid lines must be in the same state

can be identified as the probability of an occupied bond in the percolation problem (Fig. 15.1). Next we expand the product over nearest-neighboring bonds in Eq. (15.7), and identify each factor $1 - p + p\delta_{s_i,s_j}$ as the sum of two terms, the first of which, $(1 - p)$, we associate with the bond being vacant, and the second, $p\delta_{s_i,s_j}$, we associate with the bond being occupied. The product $\prod_{\langle ij \rangle}$ in Eq. (15.7) is then the sum of 2^{N_B} terms, each of which consists of a product over bonds in which for each bond either the factor $(1 - p)$ is chosen or the factor $p\delta_{s_i,s_j}$ is chosen. If $n_{\rm occ}$ ($n_{\rm vac}$) is the number of occupied (vacant) bonds, then we see that each of these 2^{N_B} terms is of the form $(1 - p)^{n_{\rm vac}} p^{n_{\rm occ}} \mathcal{P}$, where $\mathcal{P}$ is the product over occupied bonds of δ_{s_i,s_j}. We thus have a one-to-one correspondence between these 2^{N_B} terms in the expansion of the partition function of the Potts model and the 2^{N_B} configurations $\mathcal{C}$ of the percolation problem. Note that the weighting we assign in the expansion of the Potts model partition function, namely $(1 - p)^{n_{\rm vac}} p^{n_{\rm occ}}$, is exactly the same as the probability $P(\mathcal{C})$ for the configuration $\mathcal{C}$ in the percolation problem. This probability $P(\mathcal{C})$ is normalized, of course, so that

$$\sum_{\mathcal{C}} P(\mathcal{C}) = 1\,. \tag{15.9}$$

Accordingly, we may express the product over bonds in Eq. (15.7) in terms of a sum over configurations $\mathcal{C}$ of the percolation problem, by which we mean a sum over all 2^{N_B} arrangements of occupied and vacant bonds. This discussion shows that we may write Eq. (15.7) in the form

$$Z_P = e^{N_B \beta J (q-1)} \sum_{\{s_i\}} \sum_{\mathcal{C}} P(\mathcal{C}) \prod_{\rm occ} \delta_{s_i s_j}\,, \tag{15.10}$$

where the product is over occupied bonds. Note that each occupied bond forces the two sites it connects to be in the same one of the q possible states of the Potts model. The mapping is such that within each cluster of the percolation problem one has perfect correlations within the Potts model and between different percolation

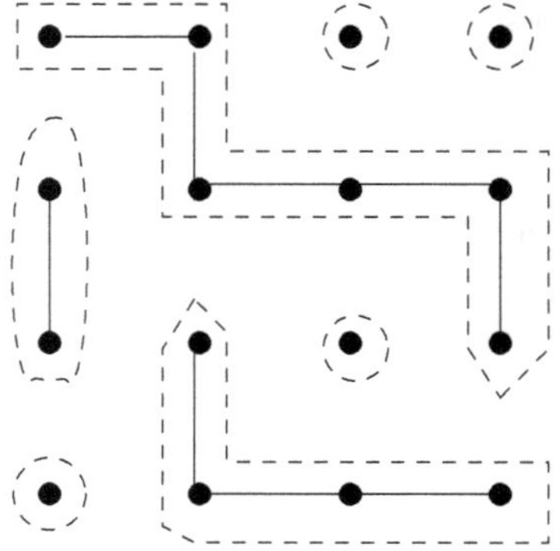

Fig. 15.2 Resolution of sites into clusters (each surrounded by a dashed line). A cluster is defined to be a group of sites connected via solid lines

clusters, the Potts model variables are totally uncorrelated. Thus, the probability of being in the same cluster will be associated with the Potts model correlation function. However, for the moment we deal only with the partition function. When we interchange the order of summations over clusters and over Potts states, we write Eq. (15.10) as

$$Z_P = e^{N_B \beta J(q-1)} \sum_{\mathcal{C}} P(\mathcal{C}) \sum_{\{s_i\}} \prod_{\text{occ}} \delta_{s_i s_j} \,. \tag{15.11}$$

Consider a diagrammatic interpretation of the product over occupied bonds, as shown in Fig. 15.2, which shows a typical configuration of occupied bonds (the solid lines). Each cluster of occupied bonds is enclosed by a dashed line. Sites which have no adjacent bonds are isolated sites which form a cluster of size one. The configuration shown has 9 occupied bonds and 15 unoccupied bonds and hence the probability that its occurs is $p^9(1-p)^{15}$. In this configuration there are 7 clusters enclosed by a dashed line. Note that the delta functions (which accompany each occupied bond) force all sites which are in the same cluster of occupied bonds to be have the same value of s_i. So instead of having to sum over an independent variable s_i for each site, this sum reduces to a sum over s_α, where s_α is the value of s_i for all sites in the cluster α. Thus

$$Z_P = e^{N_B \beta J(q-1)} \sum_{\mathcal{C}} P(\mathcal{C}) \sum_{s_{\alpha_1}} \sum_{s_{\alpha_2}} \sum_{s_{\alpha_3}} \cdots \sum_{s_{\alpha_{N_c(\mathcal{C})}}} 1 \,, \tag{15.12}$$

where we have to sum over the s's for each cluster α_1, α_2, ... $\alpha_{N_c(\mathcal{C})}$ of the $N_c(\mathcal{C})$ clusters which make up the configuration $\mathcal{C}$. Since each s_α is summed over q-states, we get

$$Z_P = e^{N_B \beta J(q-1)} \sum_{\mathcal{C}} P(\mathcal{C}) q^{N_c(\mathcal{C})} \,, \tag{15.13}$$

where $N_c(\mathcal{C})$ is the number of clusters in the configuration $\mathcal{C}$. In the configuration shown in the figure above, $N_c(\mathcal{C})$ is 7. In the limit $q \to 1$, we write

$$q^{N_c(\mathcal{C})} = [1 + (q-1)]^{N_c(\mathcal{C})} = 1 + (q-1)N_c(\mathcal{C}) + \mathcal{O}[(q-1)^2]\,. \quad (15.14)$$

Therefore, to lowest nontrivial order in $(q-1)$ we have

$$\begin{aligned} Z_P &= e^{N_B\beta J(q-1)} \sum_{\mathcal{C}} P(\mathcal{C}) \Big[1 + (q-1)N_c(\mathcal{C}) \Big] \\ &= e^{N_B\beta J(q-1)} \Big[1 + (q-1) \sum_{\mathcal{C}} P(\mathcal{C}) N_c(\mathcal{C}) \Big] , \end{aligned} \quad (15.15)$$

where we used the normalization that $\sum_{\mathcal{C}} P(\mathcal{C}) = 1$. Then we obtain

$$\ln Z_P = N_B(q-1)\beta J + (q-1)\sum_{\mathcal{C}} P(\mathcal{C})N_c(\mathcal{C}) + \mathcal{O}[(q-1)^2]\,. \quad (15.16)$$

Therefore we may define a dimensionless free energy per site f by

$$f = -\lim_{q\to 1}\Big[\ln Z_P/[N_s(q-1)] \Big] = -\frac{N_B}{N_s}\beta J - \frac{\langle N_c \rangle_P}{N_s}\,, \quad (15.17)$$

where N_s is the number of sites and

$$\langle X \rangle_P \equiv \sum_{\mathcal{C}} P(\mathcal{C}) X(\mathcal{C}) \quad (15.18)$$

is the percolation average of the quantity X. Thus, apart from the analytic term $N_B\beta J/N_s$, the free energy per site (divided by $q-1$ in the limit $q \to 1$) is the average number of clusters per site.

Equation (15.17) gives an interpretation for the specific heat index α_P of the $q \to 1$ Potts model. In the Potts model, the free energy for T near T_c goes like $|T_c - T|^{2-\alpha_P}$. According to this mapping, there will also be a singularity in the average number of clusters as a function of p for p near p_c. In light of Eq. (15.8) we have

$$kT = -\frac{Jq}{\ln(1-p)} \quad (15.19)$$

so that, as a function of p,

$$f \sim \left| \frac{Jq}{\ln(1-p_c)} - \frac{Jq}{\ln(1-p)} \right|^{2-\alpha_P} \sim |p - p_c|^{2-\alpha_P}\,. \quad (15.20)$$

Thus, using Eq. (15.17),

$$\frac{\langle N_c \rangle}{N_s} \sim |p - p_c|^{2-\alpha_P} + \text{Reg} \,, \tag{15.21}$$

where "Reg" indicates analytic contributions. Without this mapping it is doubtful that anyone would have guessed that the critical exponent associated with the average number of clusters in the percolation problem can be identified as the specific heat exponent of a statistical mechanical model.

Admittedly, the average number of clusters per site is not the most interesting property of the percolation problem. However, the fact that the free energy of the $q \to 1$-state Potts model can be identified with a property of the percolation problem leads us to hope that more interesting properties, such as the percolation probability (the probability that a site is in an infinite cluster) might also be accessible via the Potts model.

We already noted that the mapping is such that when two sites are in the same cluster, there are strong correlations between their corresponding Potts variables. Thus we are led to investigate the significance of the pair correlation function of the Potts model. For that purpose we now add a field. That is, we add to the Hamiltonian a term of the form

$$\delta\mathcal{H} = -\sum_i h[q\delta_{s_i,1} - 1] \,, \tag{15.22}$$

which gives the state "1" a lower energy than the others. We now have an additional factor in $e^{-\beta\mathcal{H}}$ of

$$\prod_i \left(e^{\beta hq\delta_{s_i,1}} e^{-\beta h} \right) = e^{\beta(q-1)N_s h} \prod_i e^{\beta hq(\delta_{s_i,1}-1)} \,. \tag{15.23}$$

Then the partition function is

$$Z_P = e^{N_B \beta J(q-1)} e^{(q-1)\beta N_s h} \sum_{\mathcal{C}} P(\mathcal{C}) \sum_{\{s_i\}} \prod_{\text{occ}} \delta_{s_i s_j} \prod_i e^{\beta qh[\delta_{s_i,1}-1]} \,. \tag{15.24}$$

Again the diagrammatic interpretation says that the sums over the s's reduce to sums over s_{α_1} for cluster α_1, s_{α_2} for cluster α_2, and so forth up to $s_{\alpha_{N_c(\mathcal{C})}}$ for cluster $\alpha_{N_c(\mathcal{C})}$. So, in analogy with Eq. (15.12), we have

$$\begin{aligned} Z_P &= e^{N_B \beta J(q-1)} e^{\beta(q-1)N_s h} \\ &\quad \times \sum_{\mathcal{C}} P(\mathcal{C}) \sum_{s_{\alpha_1}} \sum_{s_{\alpha_2}} \sum_{s_{\alpha_3}} \cdots \sum_{s_{\alpha_{N_c}(\mathcal{C})}} \prod_{k=1}^{N_c(\mathcal{C})} \left(\prod_{i \in \alpha_k} e^{\beta hq[\delta_{s_i,1}-1]} \right) \\ &= e^{N_B \beta J(q-1)} e^{\beta(q-1)N_s h} \\ &\quad \times \sum_{\mathcal{C}} P(\mathcal{C}) \sum_{s_{\alpha_1}} \sum_{s_{\alpha_2}} \sum_{s_{\alpha_3}} \cdots \sum_{s_{\alpha_{N_c}(\mathcal{C})}} \prod_{k=1}^{N_c(\mathcal{C})} e^{\beta hqn(\alpha_k)[\delta_{s_{\alpha_k},1}-1]} \,, \end{aligned} \tag{15.25}$$

where $n(\alpha_k)$ is the number of sites in the α_kth cluster. Look at the product over k. When $s_{\alpha_k} = 1$, then the factor under that product sign is unity. Otherwise, for the $(q-1)$ other values of s_{α_k} that factor is $e^{-\beta hqn(\alpha_k)}$. So

$$Z_P = e^{N_B\beta J(q-1)} e^{N_s\beta(q-1)h} \times \sum_{\mathcal{C}} P(\mathcal{C}) \prod_{k=1}^{N_c(\mathcal{C})} \left(1 + (q-1)e^{-n(\alpha_k)\beta hq}\right). \tag{15.26}$$

As a check, note that for $h = 0$ this reduces to Eq. (15.13). For $q \to 1$ this is

$$\begin{aligned} Z_P &= e^{N_B\beta J(q-1)} e^{N_s\beta(q-1)h} \sum_{\mathcal{C}} P(\mathcal{C}) \left(1 + (q-1) \sum_{k=1}^{N_c(\mathcal{C})} e^{-n(\alpha_k)\beta hq} + \mathcal{O}[(q-1)^2]\right) \\ &= e^{N_B\beta J(q-1)} e^{N_s\beta(q-1)h} \\ &\quad \times \left(1 + (q-1) \sum_{\mathcal{C}} P(\mathcal{C}) \sum_{k=1}^{N_c(\mathcal{C})} e^{-n(\alpha_k)\beta hq} + \mathcal{O}[(q-1)^2]\right). \end{aligned} \tag{15.27}$$

Then we can extract the free energy,

$$\begin{aligned} f &= -\lim_{q\to 1} \left(\ln Z/[N_s(q-1)]\right) \\ &= -(N_B/N_s)\beta J - \beta h - \frac{1}{N_s} \sum_{\mathcal{C}} P(\mathcal{C}) \sum_{\alpha=1}^{N_c(\mathcal{C})} e^{-n(\alpha_k)\beta h}. \end{aligned} \tag{15.28}$$

For $h = 0$ this reduces to Eq. (15.17).

We now evaluate the spontaneous magnetization (per site) which is

$$\begin{aligned} m &\equiv -\beta^{-1} \lim_{h\to 0^+} \left.\frac{\partial f}{\partial h}\right|_T \\ &= 1 + \frac{1}{N_s} \lim_{h\to 0^+} \sum_{\mathcal{C}} P(\mathcal{C}) \sum_{k=1}^{N_c(\mathcal{C})} [-n(\alpha_k)] e^{-\beta hn(\alpha_k)}. \end{aligned} \tag{15.29}$$

The proper order of limits is to let the system size become infinite *before* letting $h \to 0$. That is what is meant by $h \to 0^+$. If we consider a large system (say 10^6 sites), then the distribution of cluster sizes for p significantly greater than p_c will look like that shown in Fig. 15.3. The area under the entire curve is unity because any site has to belong to *some* cluster. The area under the bump at $n \sim 10^6$ is the probability that a site is in a cluster which is of order the system size. For large, but finite systems, the bump is not a delta function. To take the correct limit, we first let $N \to \infty$, and then let $h \to 0$, such that $hN \gg 1$. In this limit, for clusters which are

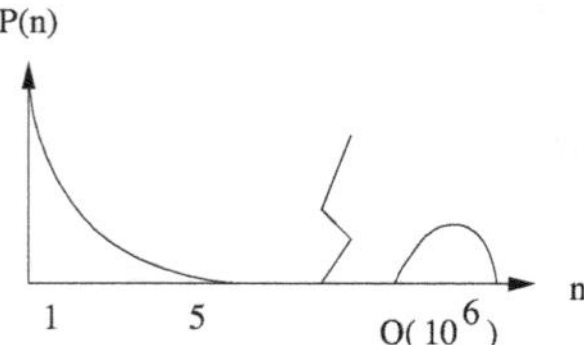

Fig. 15.3 $P(n)$ the probability that a site is in a cluster of n sites. This is a schematic result for a system of 10^6 lattice sites for p somewhat larger than p_c

of order the system size, the factor $\exp[-n(\alpha)\beta h]$ will go to zero. In this sense, the sum over cluster is limited to "finite" clusters. Thus we have

$$\begin{aligned} m &= 1 + \frac{1}{N_s} \sum_{\mathcal{C}} P(\mathcal{C})[-n_f(\mathcal{C})] \\ &= \frac{N_s - \langle n_f \rangle_p}{N_s} = P_\infty \,, \end{aligned} \tag{15.30}$$

where n_f is the number of sites in "finite" clusters and P_∞ is the probability that a given site is in "the infinite cluster." The fact that P_∞ is associated with the spontaneous magnetization of the corresponding Potts model justifies our calling it an order parameter. Note also that, from the way we calculated P_∞, it is clear that this quantity is zero for a finite system. For a finite system the sum over $n(\alpha)e^{-n(\alpha)0^+} \to n(\alpha)$ just yields the total number of sites in the system and $P_\infty = 0$. Of course, for small p all sites are in small clusters and $P_\infty = 0$. It is only for p greater than or equal to some critical value p_c that we have an infinite cluster in the thermodynamic limit. Note that when $p > p_c$ there is only a single infinite cluster. However, for $p = p_c$ more than one infinite cluster exists, at least for low enough dimension (Stauffer and Aharony 1992; Aizenman and Barsky 1987).

We will not do it here—but if you introduce a position-dependent field h_i on the ith site, then one can speak of the susceptibility $\chi(i, j) \equiv \partial^2 f/\partial h_i \partial h_j)_{h_i=0}$. One can prove (see Exercise 4) that this quantity is given by

$$\chi(i, j) = \langle \nu(i, j) \rangle_P - P_\infty^2 \,, \tag{15.31}$$

where $\nu(i, j)$ is unity if sites i and j are in the same cluster and is zero otherwise. This is exactly the quantity which we called (without justification at the time) the susceptibility. This formula is analogous to that for the Ising model

$$kT\chi(i, j) = \langle \sigma_i \sigma_j \rangle_T - \langle \sigma_i \rangle_T^2 \,. \tag{15.32}$$

15.1.2 Critical Properties of Percolation

So, what does all this prove? Suppose we can use the machinery of statistical mechanics to calculate the thermodynamic properties of the q-state Potts model as a function of q. Then what this proves is that in the limit $q \to 1$ the magnetization of the Potts model is the probability P_∞ (in the percolation problem) that a site is in the infinite cluster and the position-dependent susceptibility of the Potts model, $\chi(i, j)$, gives the probability (in the percolation problem) that sites i and j are in the same cluster. In this identification, the probability p of the percolation problem is related to the temperature (in the Potts model) by Eq. (15.8). This relation leads to an exact relation between p_c for the percolation problem to T_c for the $q \to 1$-state Potts model. For lattices with nearest-neighbor coordination number z, the mean-field theory leads to the estimate that $kT_c = zJ$. Then, Eq. (15.8) tells us that

$$p_c = 1 - e^{-J/kT_c} = 1 - e^{1/z} \sim 1/z \,. \tag{15.33}$$

This is an estimate of p_c which can be obtained by considering the probability that a walk can continue over occupied bonds. (That argument might suggest $p_c = 1/(z-1)$ which we found for the Cayley tree.) As expected, high temperature (where the Potts model is disordered) corresponds to low concentration (where there is no long-range order in the percolation problem). Also, low temperature in the Potts model corresponds to high concentration in the percolation problem. In this limit, both models display broken symmetry in that the order parameter does not go to zero as the conjugate field goes to zero. Thus, P_∞ in the percolation problem corresponds to spontaneous magnetization (spontaneous symmetry breaking) in the Potts model.

A most useful application of this mapping concerns the critical exponents of percolation. We use the customary notation for the critical exponents of the Potts model. Thus for the q-state Potts model we have the asymptotic behaviors:

$$\begin{aligned} C_h &\sim A_\pm |T - T_c|^{-\alpha_q} + \text{Reg} \\ M_0(T) &\sim B(T_c - T)^{\beta_q} \\ \chi(T) &\sim C_\pm |T - T_c|^{-\gamma_q} \\ \chi(0, \mathbf{r}) &\sim r^{-d+2-\eta_q} f(r/\xi(T)) \\ \xi(T) &\sim D_\pm |T - T_c|^{-\nu_q} \,, \end{aligned} \tag{15.34}$$

where “Reg” indicates a regular (background) contribution. Here A, B, etc. are amplitudes which may be different above (+) and below (−) T_c. Also C_h is the specific heat at constant magnetic field, M_0 is the spontaneous magnetization $M(H = 0^+, T)$, $\chi(T)$ is the uniform susceptibility, $\chi(0, \mathbf{r})$ is the two-point susceptibility, $\xi(T)$ is the correlation length, and $f(x)$ is a scaling function which is usually taken to be $f(x) \sim e^{-x}$. The above relations define the q-dependent exponents, α_q, β_q, γ_q, η_q, and ν_q, of which we may take η_q and ν_q as the two independent exponents in terms of which all other critical exponents can be expressed. For the percolation problem we similarly set

$$\begin{aligned} \langle N_c \rangle_P &\sim A'_{\pm}|p - p_c|^{-\alpha_P} + \text{Reg} \\ P_\infty(p) &\sim B'(p_c - p)^{\beta_P} \\ \langle n(\alpha) \rangle &\sim C'_{\pm}|T - T_c|^{-\gamma_P} \\ \chi(0, \mathbf{r}) &\sim r^{-d+2-\eta_P} f'(r/\xi(T)) \\ \xi(T) &\sim D'_{\pm}|T - T_c|^{-\nu_P} \ . \end{aligned} \tag{15.35}$$

Then the mapping indicates that $\alpha_P = \alpha_{q=1}$ and similarly for all the other exponents. Thus a calculation of the critical exponents of the q-state Potts model (as a function of q) allows an evaluation of the critical exponents of percolation.

It may be thought that having to calculate exponents as a function of the parameter q would lead to enormous difficulties, not the least of which is how to analytically continue a function of integers to the value $q = 1$. As it happens, in renormalization group calculations q appears in ways which are almost trivial to handle, and the extrapolation to $q = 1$ poses no problem.

15.2 Self-avoiding Walks and the n-Vector Model

Here, we show how the statistics of self-avoiding walks (SAW's) can be treated by consideration of the so-called n-vector model. The model of SAWs also describes the conformations of linear polymers, when each monomer is restricted to occupy a lattice site. The conclusion from this demonstration is that we can use whatever machinery we have to calculate properties of the n vector model for general n, and then, by taking the limit $n \to 0$ (de Gennes 1972), we obtain results for SAWs. As it happens, the renormalization group (RG) approach is ideally suited to this use because it is easy to obtain RG results for the n-vector model as an explicit function of n. These results can then be trivially continued to the limit $n \to 0$, yielding RG results for SAWs.

15.2.1 Phenomenology of SAWs

First, let us review a few facts about walks in a lattice (Slade 1994). We start with some qualitative observations. A SAW may be considered to be a variant of a random walk (RW). An RW is one in which each step is from one site to a randomly selected nearest-neighbor site, without any regard for whether the walk visits sites more than once. An RW is sometimes called "a drunkard's walk." A key relation that we will explore in a moment is that between the number of steps, N, in a walk and the end-to-end displacement $\mathbf{r}_N$ of such a walk. For a random walk the well-known result is

$$\langle r_N^2 \rangle \sim N \ , \tag{15.36}$$

where the brackets indicate an average over all equally weighted N-step random walks.

A natural question that arises is whether the fact that the walks are self-avoiding changes the relationship (15.36) in a qualitative way. To address this issue we ask whether, the last third, say, of a long RW tends to intersect the first third, say, of that RW. This is equivalent to asking whether, generically, two infinitely long random walks intersect one another. If we could consider a walk to be a straight line, then we would say that two straight lines intersect in two dimensions, but, in general, they need not intersect in higher spatial dimension, but because they twist and turn, RWs typically have some thickness. The question is, how should we characterize this? If the width remains finite as the length becomes infinite, then the object is clearly one dimensional and it is correct to view the RW as a straight line. But we know from Eq. (15.36) that the number of points in an RW increases with its linear dimension as R^2. We may identify this as the effective dimensionality of the RW. This dimension, called the fractal dimension by Mandelbrot (1977) was actually defined long ago by the mathematician Hausdorff (1919) after whom it is alternatively named. So the question whether two RWs generically intersect one another is equivalent to the question of whether two objects of fractal, or Hausdorff, dimension 2 intersect one another. If the two objects in question live in a space whose dimension d is greater than the sum of the dimensions of the two objects (here this sum is $2 + 2 = 4$), then the objects do not collide except by amazing coincidence, the same way that it would be a complete accident if two lines in three dimensional space were to intersect one another. The conclusion from this discussion is that we expect that the relation of Eq. (15.36) will also hold for SAWs when the spatial dimension is greater than 4. Therefore, the forthcoming discussion of SAWs is mainly focussed on what happens for $d < 4$.

We now discuss the mathematical characterization of SAWs. Let $c_p(\mathbf{r})$ be the number of SAWs of p steps which start at the origin and end at $\mathbf{r}$. Then, if c_p is the total number of p-step SAWs, we may write

$$c_p = \sum_{\mathbf{r}} c_p(\mathbf{r}) \, . \tag{15.37}$$

Numerics suggest that for $p \to \infty$ one has Slade (1994)

$$c_p \sim p^x \mu^p \, , \tag{15.38}$$

where μ is a constant (called the connectivity constant) and x is a critical exponent we wish to determine. Basically the factor, μ, is the typical number of options you have for the next step in a long walk, so that very crudely $\mu \sim z - 1$. In two dimensions $x = 11/32$ and in three dimensions $x \approx 1/6$. In high dimensions $x = 0$.

One can define a generating function for the set, $\{c_p\}$ as

$$\chi_{\mathrm{SAW}}(z) \equiv \sum_p c_p z^p \, , \tag{15.39}$$

so that, if $\chi_{\rm SAW}(z)$ is known, then the c_p can be derived from derivatives of $\chi_{\rm SAW}(z)$ in the limit $z \to 0$. By analogy to the grand partition function in statistical mechanics, we can interpret z in terms of a fugacity as $z = \exp(\beta\mu^*)$ where μ^* is effectively a chemical potential for adding a monomer unit. Roughly speaking z can be thought of as the probability of adding a monomer to the end of an existing polymer or of adding a step to an N-step SAW. We will also show later that $\chi_{\rm SAW}(z)$ can be interpreted as a "susceptibility."

Assuming the result of Eq. (15.38) we see that

$$\chi_{\rm SAW}(z) \sim \sum_p p^x (z\mu)^p \;, \tag{15.40}$$

which has a radius of convergence, $|z\mu| < 1$, and a singularity of the form $1/[1 - z\mu]^{x+1}$. This means that we may write the SAW susceptibility as

$$\chi_{\rm SAW}(z) = A[z_c - z]^{-(x+1)} + R(z) \;, \tag{15.41}$$

where A is some amplitude, $z_c = 1/\mu$ and $R(z)$ is a regular function (or, in any case, $R(z)$ is less singular at $z = z_c$ than the leading term). The steps leading to Eq. (15.41) are explored more thoroughly in Exercise 1.

This form of the SAW susceptibility smells like critical phenomena. If we can determine the exponent with which the SAW susceptibility diverges at the critical point where $z = z_c$, we can immediately interpret this as being $1 + x$, thereby determining the exponent x.

Perhaps, the most important critical exponent is the one which relates the average end-to-end distance r to the number of steps N. We write for large N that

$$r \sim N^{\nu_{\rm SAW}} \;, \tag{15.42}$$

and $\nu_{\rm SAW}$ is the associated critical exponent of interest. As mentioned above, $\nu_{\rm SAW} = 1/2$ for $d > 4$. We define the SAW correlation length as being the value of r as a function of the control parameter z, so that we write

$$\xi_{\rm SAW}(z)^2 \equiv \frac{\sum_{\mathbf{r}} \sum_p c_p(\mathbf{r}) r^2 z^p}{\sum_{\mathbf{r}} \sum_p c_p(\mathbf{r}) z^p} \;. \tag{15.43}$$

The object of the game is to obtain a quantity related to $\nu_{\rm SAW}$. Therefore we substitute Eq. (15.42) into Eq. (15.43) to get (for $z \sim z_c$)

$$\begin{aligned} \xi_{\rm SAW}(z)^2 &\sim \frac{\sum_{\mathbf{r}} \sum_p c_p(\mathbf{r}) p^{2\nu_{\rm SAW}} z^p}{\sum_{\mathbf{r}} \sum_p c_p(\mathbf{r}) z^p} = \frac{\sum_p c_p p^{2\nu_{\rm SAW}} z^p}{\sum_p c_p z^p} \\ &\sim \frac{\sum_p p^x \mu^p p^{2\nu_{\rm SAW}} z^p}{\sum_p p^x \mu^p z^p} \;. \end{aligned} \tag{15.44}$$

Using Eqs. (15.40) and (15.41) we find that

$$\xi_{\rm SAW}(z)^2 \sim [1 - z/z_c]^{-2\nu_{\rm SAW}} . \tag{15.45}$$

Thus if we can obtain the critical exponent of $\xi_{\rm SAW}(z)$ as z approaches the critical point at $z = z_c$, we will be able to infer the value of $\nu_{\rm SAW}$.

Even in early numerical studies of SAWs (Domb 1969), results were given for self-avoiding *polygons* consisting of N steps SAWs which return to the origin. The number, $c_N(0)$, of such self-avoiding polygons of N steps is found to obey (as $N \to \infty$)

$$c_N(0) \sim N^{-y} \mu^N , \tag{15.46}$$

where y is another critical exponent for SAWs (see Exercise 2).

15.2.2 Mapping

Now let us consider a classical spin model with the following Hamiltonian:

$$\mathcal{H} = -nJ \sum_{<ij>} \sum_{\alpha=1}^{n} S_\alpha(i) S_\alpha(j) , \tag{15.47}$$

where $\mathbf{S}$ is an n-component classical spin of unit length and which can assume any orientation on the unit sphere in n dimensions. For $n = 3$ we have the usual three dimensional spin vector whose orientation may be described by the angles θ_i and ϕ_i. For $n = 2$ we have a two-component vector which points from the origin to any point on the unit circle. Its orientation can be specified by an angle θ_i. For $n = 1$ we speak of a unit vector in one dimension. In one dimension, the orientation of a unit vector is either along the positive or along the negative axis. Thus for $n = 1$ the model reduces to an Ising model.

The objective of the discussion which follows is to show that in the limit $n \to 0$ the susceptibility of this n-vector model, which will be denoted $\chi_n(T)$ is identical to the SAW susceptibility, $\chi_{\rm SAW}(z)$, with the proper relation between the control parameters T and z. As the first step in this program, we will show that $\lim_{n\to 0} \hat{Z}_n = 1$, where $\hat{Z}_n$ is a suitably normalized partition function of the n-vector model at temperature T.

We write the partition function as

$$\begin{aligned} Z_n &= {\rm Tr} e^{-\beta\mathcal{H}} \equiv \int d\Omega_1 \int d\Omega_2 \ldots \int d\Omega_{N_s} e^{n\beta J \sum_{\langle ij\rangle} \mathbf{S}_i \cdot \mathbf{S}_j} \\ &= \Omega^{N_s} \int \frac{d\Omega_1}{\Omega} \int \frac{d\Omega_2}{\Omega} \ldots \int \frac{d\Omega_{N_s}}{\Omega} e^{n\beta J \sum_{\langle ij\rangle} \mathbf{S}_i \cdot \mathbf{S}_j} , \end{aligned} \tag{15.48}$$

where $d\Omega_k$ indicates an integral over all orientations of the unit vector $\mathbf{S}_k$, $\Omega \equiv \int d\Omega_k$ is the phase space integral for a single unit vector, and N_s is the total number of sites. It is convenient to write this as

$$Z_n = \Omega^{N_s} \hat{Z}_n , \tag{15.49}$$

where we introduce the normalized partition function $\hat{Z}_n$ via

$$\begin{aligned} \hat{Z}_n &\equiv \int \frac{d\Omega_1}{\Omega} \int \frac{d\Omega_2}{\Omega} \dots \int \frac{d\Omega_{N_s}}{\Omega} e^{n\beta J \sum_{\langle ij \rangle} \mathbf{S}_i \cdot \mathbf{S}_j} \\ &= \left\langle e^{n\beta J \sum_{\langle ij \rangle} \mathbf{S}_i \cdot \mathbf{S}_j} \right\rangle_\Omega , \end{aligned} \tag{15.50}$$

where $\langle X \rangle_\Omega$ denotes the quantity X averaged over all orientations of all spins. Note, however, that if the quantity X depends only on the orientations of spins i_1, i_2, ... i_p, then the average can be restricted to the orientations of only the spins i_1, i_2, ... i_p upon which X actually depends. Thus if $X = X_i X_j$, where X_i depends only on the orientation of spin i and X_j depends only on the orientation of spin j, then

$$\langle X_i X_j \rangle_\Omega = \langle X_i \rangle_{\Omega_i} \langle X_j \rangle_{\Omega_j} , \tag{15.51}$$

where the subscript Ω_k on the averages indicates that the average need be taken only over the orientations of spin k. We will continually apply this result in what follows.

For a single $n = 3$ component vector, this definition is such that

$$\langle X(\theta, \phi) \rangle = \frac{\int \sin\theta d\theta d\phi X(\theta, \phi)}{\int \sin\theta d\theta d\phi} = \frac{1}{4\pi} \int \sin\theta d\theta d\phi X(\theta, \phi) . \tag{15.52}$$

However, as will be seen, this angular average is best done in terms of Cartesian components rather than in terms of spherical angles. For instance, recall that we use unit length spins, so that

$$\sum_\alpha S_\alpha^2 = 1 . \tag{15.53}$$

Since all components are equivalent, we infer the angular average to be

$$\langle S_\alpha S_\beta \rangle_\Omega = \delta_{\alpha,\beta}/n . \tag{15.54}$$

We state without proof further results of averages:

$$\begin{aligned} \langle S_\alpha S_\beta S_\gamma S_\delta \rangle_\Omega &= \frac{\delta_{\alpha,\beta}\delta_{\gamma,\delta} + \delta_{\alpha,\gamma}\delta_{\beta,\delta} + \delta_{\alpha,\delta}\delta_{\beta,\gamma}}{n(n+2)} \\ \langle S_\alpha S_\beta S_\gamma S_\delta S_\epsilon S_\eta \rangle_\Omega &= \frac{\sum \delta\delta\delta}{n(n+2)(n+4)} , \end{aligned} \tag{15.55}$$

where $\sum \delta\delta\delta = \delta_{\alpha,\beta}\delta_{\gamma,\delta}\delta_{\epsilon,\eta} + \ldots$ indicates a sum over the 15 ways of contracting the 6 indices into three equal pairs of indices. Note that all nonzero averages of an arbitrary number of spin components carry one factor of $1/n$ and hence are of order $1/n$ as $n \to 0$. To have a nonzero average each component S_α must appear an even number (0, 2, 4, etc.) of times.

Now we consider the partition function, or actually the normalized partition function $\hat{Z}_n$. We now show that in the limit $n \to 0$, $\hat{Z}_n \to 1$. We have

$$\hat{Z}_n = \left\langle \left[1 + n\beta J \sum_{<ij>} \sum_{\alpha} S_\alpha(i) S_\alpha(j) \right.\right. \\ \left.\left. + \frac{(n\beta J)^2}{2} \sum_{<ij>} \sum_{<kl>} \sum_{\alpha\beta} S_\alpha(i) S_\alpha(j) S_\beta(k) S_\beta(l) + \ldots \right) \right\rangle_\Omega . \quad (15.56)$$

In evaluating this we use the fact that by Eq. (15.51) the average of a product of spin operators is equal to the product of the averages taken over each site with a spin that explicitly appears in the product. The average of 1 is 1. The average of $S_\alpha(i)$ is zero, so the term of order βJ vanishes. Look next at the term of order $(\beta J)^2$. We need

$$\langle S_\alpha(i) S_\alpha(j) S_\beta(k) S_\beta(l) \rangle_\Omega . \quad (15.57)$$

In order that components of the spin *at each site* should appear an even number of times it is necessary that the pair of indices ij and the pair kl be identical. Since the sum is over pairs of nearest neighbors, this means that the pair $\langle kl \rangle$ must be identical to the pair $\langle ij \rangle$. So the contribution, $\delta_2 \hat{Z}_n$, to $\hat{Z}_n$ which is of order $(\beta J)^2$ is

$$\begin{aligned} \delta_2 \hat{Z}_n &= \frac{(n\beta J)^2}{2} \sum_{\langle ij \rangle} \sum_{\alpha} \sum_{\beta} \langle S_\alpha(i) S_\beta(i) \rangle_{\Omega_i} \langle S_\alpha(j) S_\beta(j) \rangle_{\Omega_j} \\ &= \frac{(n\beta J)^2}{2} \sum_{\langle ij \rangle} \sum_{\alpha} \langle S_\alpha^2(i) \rangle_{\Omega_i} \langle S_\alpha^2(j) \rangle_{\Omega_j} . \end{aligned} \quad (15.58)$$

We now focus on the powers of n, writing,

$$\begin{aligned} \delta_2 \hat{Z}_n &= A n^2 \sum_{\alpha} \langle S_\alpha^2(i) \rangle_{\Omega_i} \langle S_\alpha^2(j) \rangle_{\Omega_j} \\ &= A n^2 \sum_{\alpha=1}^{n} (1/n)^2 = A n \to 0 \end{aligned} \quad (15.59)$$

in the $n \to 0$ limit.

This result suggests that all terms beyond the first one in Eq. (15.56) vanish. To see that and also for later use we will invoke a diagrammatic interpretation of the expansion. We start with a diagram of the lattice in which each site is represented by a dot. Then the term $n\beta J S_\alpha(i)S_\alpha(j)$ is represented by an interaction line joining sites i and j and at sites i and j we put the component label α of the spins which are involved and as a reminder we also associate a factor of n with this interaction line. So each Greek letter in the diagram represents a spin component and each line an interaction. With each line, we write a factor n because we want to keep track of the powers of n. In the expansion of $\exp(-\beta\mathcal{H})$ a term of order $(\beta J)^p$ will have p interaction lines, and each of these lines will have component labels at each of the sites which are connected by the interaction line in question. In principle we can have several lines joining a given pair of nearest neighboring sites.

Let us now codify the power of n (as $n \to 0$) in terms of the properties of the diagram.

(1) If a diagram contains p interaction lines, each line carries a factor of $n\beta J$, so in all this contributes a factor n^p.
(2) At each site we will have to take an angular average over the n-component spin. But, as we have seen, all such nonzero averages diverge as $1/n$ as $n \to 0$. So we have a factor $1/n$ for every site at which we have a nonzero number of spin components, irrespective of the number of operators which partake of this average (as long as this number is nonzero).
(3) At each vertex, component labels α, β, etc., must appear an *even* number of times. Each sum over a component label gives a factor of n. We can be sure that the number of such powers of n is at least as large as the number of connected components which make up the diagram. (A diagram with two disjoint polygons has two connected components.) Thus, for purposes of estimation, we attribute one power of n for each connected component of the diagram.

Let us apply this reasoning to the term we evaluated in Eq. (15.59), and which is shown in the left panel of Fig. 15.4. We have already put the two interactions on the same bond, so that spin components could appear an even number of times. For that we also need to set $\alpha = \beta$. Then rule one gives $n \times n = n^2$. Rule two gives $(1/n) \times (1/n) = n^{-2}$. Rule three says that we set $\alpha = \beta$, but we have to sum over α

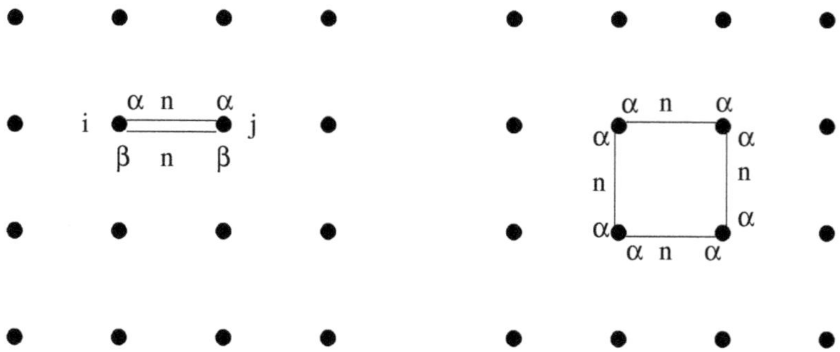

Fig. 15.4 Diagrammatic representation of terms in Eq. (15.56)

so we have from that another factor of n. All told this diagram gives $(n^2)(n^{-2})(n) = n \to 0$, as we got explicitly.

As a further example let us consider the contribution from the diagram shown in the right panel of Fig. 15.4. There rule one gives n^4 because we have four interaction lines. Rule two gives n^{-4} because we have four sites with spin operators to be averaged. We have set the component labels at sites equal in pairs to get a nonzero average. But we still have one sum over the index α which gives (according to rule 3) a factor of n. So overall, this diagram is of order $n^4(n^{-4})(n) = n \to 0$. In general, a diagram for Z will consist of a union of one or more polygons. Assume that the polygons have no sites in common. Then each polygon has k bonds (giving a factor n^k) and k sites (giving a factor n^{-k}). But the sum over the component index associated with the polygon gives a factor n. So each polygon gives a factor of n. But if polygons have sites in common, then there are fewer sites than this estimate assumes and the contribution will include more factors of n than if the polygons had no sites in common. Thus we have shown that

$$Z_n = 1 + \mathcal{O}(n) \,. \tag{15.60}$$

Next we calculate a more interesting quantity, namely the susceptibility of the n-component Heisenberg model $\chi_n(\mathbf{r})$, defined by

$$\begin{aligned} \chi_n(\mathbf{r}) &= \frac{\mathrm{Tr} \sum_\alpha S_\alpha(0) S_\alpha(\mathbf{r}) e^{-\beta \mathcal{H}}}{\mathrm{Tr} e^{-\beta \mathcal{H}}} = \frac{\left\langle \sum_\alpha S_\alpha(0) S_\alpha(\mathbf{r}) e^{-\beta \mathcal{H}} \right\rangle_\Omega}{\left\langle e^{-\beta \mathcal{H}} \right\rangle_\Omega} \\ &= \left\langle \sum_\alpha S_\alpha(0) S_\alpha(\mathbf{r}) e^{-\beta \mathcal{H}} \right\rangle_\Omega \,. \end{aligned} \tag{15.61}$$

In this case even before we have taken any account of the factor $e^{-\beta\mathcal{H}}$, we have two spin operators $S_\alpha(0)$ and $S_\alpha(\mathbf{r})$. So we start by putting these Greek indices near the site at the origin and near the site at $\mathbf{r}$ which is indicated by a cross. Then for each interaction, we add lines each of which carries a factor of n (and also βJ) and a Greek index at each of its two sites for the two spin components involved in the interaction. Since we start with one spin component at the origin and one spin component at $\mathbf{r}$, it seems clear that we should invoke a set of lines which connect these two sites. We show one such a diagram in the left-hand panel of Fig. 15.5.

Now rule one says that the 10 lines in this diagram give a factor n^{10}. We also have 11 sites, so rule 2 gives a factor n^{-11}. Finally, to get an even number of components at each site, all components must be equal to α, which is summed over and according to rule 3. This gives an additional factor of n. So the overall factor of powers of n is $(n^{10})(n^{-11})(n) = 1$. Each line carries a factor of βJ, so this diagram gives a contribution to the susceptibility $\chi(0, \mathbf{r})$ of $(\beta J)^{10}$. More generally, the susceptibility $\chi(\mathbf{r})$ will have contributions of $(\beta J)^p$ for each possible p-step SAW which connects the sites at 0 and at $\mathbf{r}$.

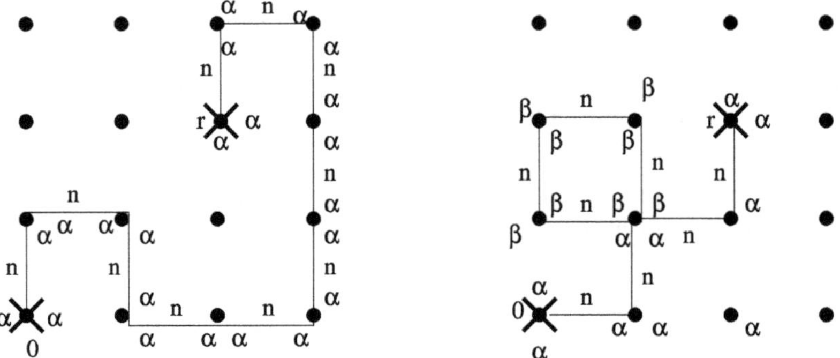

Fig. 15.5 Diagrammatic representation of terms in Eq. (15.61). The crosses indicate sites 0 and **r** which are the arguments of the susceptibility

But are there other, distinct types of diagrams that contribute to the susceptibility? Consider the diagram in the right hand panel of Fig. 15.5 which also involves an even number of spin components at each site but which is not a SAW. This diagram has 8 bonds, so rule one gives n^8. It also involves 8 sites, so rule 2 gives a factor n^{-8}. Now consider the component labels. We have assigned component labels α and β each of which occur an even number of times and therefore give a nonzero result. Since the contribution will be different depending on whether or not $\alpha = \beta$, we write the contribution C from such a diagram as

$$C = A + B\delta_{\alpha,\beta} \; . \tag{15.62}$$

For the term A the indices α and β are summed over independently and this sum gives a factor n^2. The sum over the term B requires that $\alpha = \beta$ and this sum gives a factor n. So the lowest order in n comes from contributions where all the indices of a connected components have the *same* Greek index. Thus the contribution from the diagram in the right panel of Fig. 15.5 vanishes in the $n \to 0$ limit. A generalization of this analysis indicates that a SAW in combination with any number of polygons, whether or not they intersect the SAW, will give rise to a diagram which vanishes in the $n \to 0$ limit. Thus, the diagrammatic expansion of the susceptibility, $\chi(\mathbf{r})$ counts only SAWs which begin at the origin and end at $\mathbf{r}$.

Since such an n step walk carries a factor $(\beta J)^n$, we conclude that

$$\chi_n(\mathbf{r}) = \sum_p c_p(\mathbf{r})(\beta J)^p \; . \tag{15.63}$$

That is, the two-point susceptibility of the n-vector model in the limit $n \to 0$ reproduces exactly the generating function for SAWs between these two points.

The next point to be discussed is how we should obtain the critical exponents of SAWs assuming that we know them for the n-vector model. First let us talk about the critical exponent γ. We write

$$\chi_n \equiv \sum_{\mathbf{r}} \chi_n(\mathbf{r}) = |T - T_c|^{-\gamma_n} \tag{15.64}$$

where γ_n is the critical exponent (assumed known) for the susceptibility of the n-vector model for $n \to 0$. From Eq. (15.63) we have

$$\chi_n = \sum_p c_p(\beta J)^p \sim \sum_p p^x(\mu\beta J)^p \sim |1 - \beta J\mu|^{-(1+x)} \, . \tag{15.65}$$

From this we deduce a relation between the critical temperature of the n-vector model, $T_c^{(n)}$ and the connectivity constant μ of SAWs, namely

$$kT_c^{(n)} = J\mu \, . \tag{15.66}$$

(Only in special cases do we know either of these exactly, but the relation between them is exact.) Since the connectivity constant of SAWs is easy to access numerically, this relation may tell us more about the n-vector model for $n \to 0$, than it does about SAWs. Furthermore

$$\chi_n \sim |1 - (\beta/\beta_c)|^{-(1+x)} \sim |T - T_c|^{-(1+x)} \, . \tag{15.67}$$

In view of Eq. (15.64) we identify the exponent γ_n of the n-vector model with $1 + x$, where x is the exponent in Eq. (15.38) for SAWs.

Next let us see what can be said about the correlation length. We would like to get the SAW exponent ν_{SAW}, defined by Eq. (15.42), which gives the typical end-to-end displacement of a SAW of N steps from results (like those from the RG) on the n-vector model. We may define the correlation length ξ_n for the n-vector model via

$$\xi_n^2 = \frac{\sum_{\mathbf{r}} \chi_n(\mathbf{r}) r^2}{\sum_{\mathbf{r}} \chi_n(\mathbf{r})} \sim |T - T_c|^{-\nu_n} \, , \tag{15.68}$$

where ν_n is the correlation length exponent for the n-vector model. For $\chi_n(\mathbf{r})$ we use Eq. (15.63), to get

$$\begin{aligned} \xi_n^2 &= \frac{\sum_{\mathbf{r}} \sum_p c_p(\mathbf{r})(\beta J)^p r^2}{\sum_{\mathbf{r}} \sum_p c_p(\mathbf{r})(\beta J)^p} \\ &= \frac{\sum_{\mathbf{r}} \sum_p c_p(\mathbf{r})(\beta J)^p p^{2\nu_{\mathrm{SAW}}}}{\sum_{\mathbf{r}} \sum_p c_p(\mathbf{r})(\beta J)^p} \\ &= \frac{\sum_p p^x(\beta J\mu)^p p^{2\nu_{\mathrm{SAW}}}}{\sum_p p^x(\beta J\mu)^p} \, . \end{aligned} \tag{15.69}$$

Thus, as we saw for Eq. (15.44),

$$\xi_n^2 \sim [1 - \beta J \mu]^{-2\nu_{\rm SAW}}$$
$$\sim [T - T_c]^{-2\nu_{\rm SAW}} . \qquad (15.70)$$

Thus we have shown that the ν exponent for the $n \to 0$ Heisenberg model is identical to the exponent $\nu_{\rm SAW}$ we defined for SAWs. If we anticipate the RG results for the n-vector model, we then have that $\nu_{\rm SAW} = 1/2$ for $d > 4$ and for $d = 4 - \epsilon$ we have, to order ϵ^2:

$$\nu_{n=0} = \frac{1}{2} + \frac{\epsilon}{16} + \frac{15\epsilon^2}{512} = \nu_{\rm SAW} . \qquad (15.71)$$

Obviously, with the n-component results at hand, it is a trivial calculation to get corresponding results for SAWs. The analogous result for the SAW exponent x is

$$\gamma_{n=0} - 1 = \frac{\epsilon}{8} + \frac{13\epsilon^2}{256} = x_{\rm SAW} . \qquad (15.72)$$

15.2.3 Flory's Estimate

A beautiful argument which gives a good estimate for $\nu_{\rm SAW}$ (and for which, in part, he was awarded the Nobel prize in 1974) was given by Flory (1941). For this estimate one models the free energy of N-step SAWs, assuming an energy penalty for overlaps. If their typical end-to-end displacement is r, we expect that SAWs will conspire to choose r so as to minimize the free energy for a given r. The argument assumes d spatial dimensions.

We start by estimating the energy of SAWs. This energy must come from the repulsive energy which keeps atoms in the polymer apart. Say the atoms have an energy Δ when they overlap. We may estimate the density of atoms in the volume to be N/r^d. If we put down N atoms in a volume r^d, we expect the number of overlaps to be proportional to N^2/r^d with an energy of order $(N^2/r^d)\Delta$. Note: we don't care what Δ is—we only need to understand the dependence on r and N.

Now we discuss the entropy of an N step random walk with end-to-end distance r. We will use the formula $S = k \ln W$, where W is the total number of walks of N steps which have their displacement equal to $\mathbf{r}$. This number will be $\mu^N P(\mathbf{r})$, where μ is the connectivity constant, introduced in Eq. (15.38), and $P(\mathbf{r})$ is the probability that the random walk has displacement $\mathbf{r}$. This is found from the diffusion equation, where time is replaced by number of steps:

$$\frac{\partial P(\mathbf{r}, N)}{\partial N} = D\nabla^2 P(\mathbf{r}, N) , \qquad (15.73)$$

where D is the diffusion constant (when the time for one step is unity). This equation has the solution

$$P(\mathbf{r}, N) = \frac{1}{\sqrt{4\pi DN}} e^{-r^2/(4DN)} . \tag{15.74}$$

So

$$S = k \ln W \sim kN \ln \mu - kr^2/(4DN) . \tag{15.75}$$

Thus the free energy is

$$F = U - TS = \frac{AN^2}{r^d} - BN + \frac{Cr^2}{N} , \tag{15.76}$$

where A, B, and C are constants. The key idea is that we have reasonably estimated the dependence on r and N. Now optimize with respect to r. Setting $\partial F/\partial r = 0$ gives

$$\frac{dAN^2}{r^{d+1}} = \frac{2Cr}{N} . \tag{15.77}$$

This gives

$$r \sim N^{3/(d+2)} . \tag{15.78}$$

This result is correct for $d = 2$ (the exact result is $\nu = 3/4$) and for $d = 4$, where mean-field theory begins to be correct and SAWs become like random walks with $\nu = 1/2$. For $d = 3$, the Flory result, $\nu = 3/5$ can be compared to the epsilon expansion result, $\nu_{n=0} = \nu_{\mathrm{SAW}} = 0.5918$ from Eq. (15.71) for $\epsilon = 1$. So this simple estimate (referred to as the Flory estimate) is really quite good.

15.3 Quenched Randomness

Here we discuss the effects of allowing parameters in the Hamiltonian to be random. In Sect. 15.4 of the preceding chapter, we discussed the distinction between quenched and annealed randomness. For simplicity, we will discuss quenched randomness within a spin model whose Hamiltonian is

$$\mathcal{H} = -\sum_{\langle ij \rangle} J_{ij} \mathbf{S}_i \cdot \mathbf{S}_j , \tag{15.79}$$

where $\mathbf{S}_i$ are classical n-component vectors. We now allow each J_{ij} to be an independent random variable of the form $J_{ij} = J_0 + \delta J_{ij}$. For simplicity we will assume that

$$P(J_{ij}) = \frac{1}{\sqrt{2\pi\sigma^2}} e^{-[J_{ij}-J_0]^2/(2\sigma^2)} . \tag{15.80}$$

We will assume that the randomness is quenched, so that the various random configurations do not come to thermal equilibrium but remain given by the probability distribution of $P(J_{ij})$. In that case, the properties of the system are obtained from the configurationally averaged free energy given by

$$[F]_J = \int \prod_{\langle ij \rangle} \Big(P(J_{ij}) dJ_{ij} \Big) F(\{J_{ij}\}) , \tag{15.81}$$

where $[X]_J$ denotes the value of X averaged over the distributions of the J_{ij}'s.

Properties like the specific heat C or the zero-field susceptibility χ are then obtained as their respective average over the distribution of J's as $[C]_J$ and $[\chi]_J$. Since these quantities can be obtained by suitable derivatives of the free energy, it suffices for us to calculate the configurationally averaged free energy $[F]_J \equiv -kT[\ln Z]_J$.

15.3.1 Mapping onto the n-Replica Hamiltonian

Our aim is to develop a convenient way to perform the configurational average of the logarithm of the partition function. That is, we want to evaluate

$$[F(\{J\}]_J \equiv -kT \int \ln Z(\{J\}) P(\{J\}) \prod dJ_{ij} . \tag{15.82}$$

Superficially it might seem hopeless to get analytic results for the average of a logarithm. However, we make use of what P. W. Anderson has called a "hoary trick," namely that for small n,

$$X^n = 1 + n \ln X + \mathcal{O}(n^2) . \tag{15.83}$$

To apply this idea here we introduce an n-replicated Hamiltonian, $\mathcal{H}_n$ by

$$\mathcal{H}_n = -\sum_{\langle ij \rangle} \sum_{\alpha=1}^{n} J_{ij} \mathbf{S}_i^{(\alpha)} \cdot \mathbf{S}_j^{(\alpha)} , \tag{15.84}$$

where now at each site we introduce n replicas of the original spins. As written this replica Hamiltonian simply describes n noninteracting and identical Hamiltonians. So the partition function for this replicated system, which we denote $\mathcal{Z}_n$ is given by

$$\mathcal{Z}_n(\{J\}) = Z(\{J\})^n , \tag{15.85}$$

where Z is the partition function of the system of interest. Our notation emphasizes that these quantities depend on the configuration of the random variables J_{ij}. Now consider taking the average of $\mathcal{Z}_n$ for $n \to 0$:

$$\begin{aligned} [\mathcal{Z}_n(\{J\})]_J &= [Z(\{J\})^n]_J \\ &= 1 + n[\ln Z(\{J\})]_J + \mathcal{O}(n^2) . \end{aligned} \tag{15.86}$$

Then the free energy, $\mathcal{F}_n$, of the n-replica system is

$$\begin{aligned} \mathcal{F}_n &= -kT \ln[\mathcal{Z}_n(\{J\})]_J \\ &= -kT \ln\left(1 + n[\ln Z(\{J\})]_J + \mathcal{O}(n^2)\right) \\ &= -nkT[\ln Z(\{J\})]_J + \mathcal{O}(n^2) , \end{aligned} \tag{15.87}$$

where $[\ln Z(\{J\})]_J$ is the configurationally averaged free energy whose calculation is the objective of the present discussion. What Eq. (15.87) means is that we can get the configurationally averaged free energy we want by (1) averaging the n-replicated partition function over the randomness and then (2) taking the resulting free energy per replica $\mathcal{F}/n$ in the limit $n \to 0$. (This last step eliminates the unwanted contributions of higher order in n.) The advantage of this procedure is that it is far easier to average the partition function over randomness than it is to average its logarithm.

We can carry out this program explicitly within a simple approximation. We have

$$[\mathcal{Z}_n]_J = \mathrm{Tr} \prod_{\langle ij \rangle} \int \left(P(J_{ij}) e^{\beta J_{ij} \sum_{\alpha=1}^n \mathbf{S}_i^{(\alpha)} \cdot \mathbf{S}_j^{(\alpha)}} \right) dJ_{ij} , \tag{15.88}$$

where the Tr indicates an integration over the orientations of all spins $\mathbf{S}_i^{(\alpha)}$. We assume each nearest-neighbor interaction J_{ij} is a random variable with an independent probability distribution given by Eq. (15.80). Then we can perform the Gaussian integrals over the J_{ij} which gives

$$[\mathcal{Z}]_J = \mathrm{Tr} \prod_{\langle ij \rangle} e^{-\beta \mathcal{H}_{ij}^{(n)}} , \tag{15.89}$$

where the effective Hamiltonian is

$$\begin{aligned} \mathcal{H}_{ij}^{(n)} &= -J_0 \sum_{\alpha=1}^n \mathbf{S}_i^{(\alpha)} \cdot \mathbf{S}_j^{(\alpha)} \\ &\quad - \beta\sigma^2 \sum_{\alpha=1}^n \sum_{\beta=1}^n (\mathbf{S}_i^{(\alpha)} \cdot \mathbf{S}_j^{(\alpha)})(\mathbf{S}_i^{(\beta)} \cdot \mathbf{S}_j^{(\beta)}) . \end{aligned} \tag{15.90}$$

We see from Eq. (15.90) that the effect of averaging over the randomness is to induce a spatially uniform coupling between different replicas with strength $\beta\sigma^2$.

It still remains to analyze what the effect of such replica interactions might be. The renormalization group is perfectly suited for such an analysis. However, we can also give a heuristic argument that predicts results which are consistent with renormalization group treatments.

15.3.2 The Harris Criterion

Here we present a qualitative discussion of the effect of quenched randomness (Harris 1974), which leads to new insights as well as a quantitative result for the nearest-neighbor spin Hamiltonian of the previous section. We again consider

$$\mathcal{H} = -\sum_{\langle ij \rangle} J_{ij} \mathbf{S}_i \cdot \mathbf{S}_j \ , \tag{15.91}$$

where the $\mathbf{S}_i$ are classical spins and the J_{ij}'s are quenched random variables. For the purpose of the present discussion, we assume that the random fluctuations in these coupling constants are small compared to their average value, J_0. The question then arises: Does the introduction of an infinitesimal amount of quenched randomness change the asymptotic critical behavior? Alternatively, are the critical exponents stable with respect to the introduction of infinitesimal randomness?

To answer this question we argue as follows. Imagine the system to be at some temperature T close to the transition temperature T_c. At this temperature, we may consider the system to consist of independent subvolumes with a linear dimension of order the correlation length $\xi(T)$ as shown in Fig. 15.6. Each such subvolume will actually have a slightly different value of T_c because the coupling constants in that subvolume randomly have an average value which differs from J_0. The average value of J_{ij} within a subvolume of linear dimension $\xi(T)$ (which we identify as being proportional to T_c) has a random distribution whose width, according to the central limit theorem is of order $N^{-1/2}$, where N is the number of random variables which are being averaged. In this case, for d spatial dimensions we are averaging of order $\xi(T)^d$ variables, and the width of the resulting distribution of T_c's is therefore of order $\xi^{-d/2}$. If this width of the distribution of the T_c's of the subvolumes is *less* than $|T - T_c|$, then we conclude that randomness does not affect the critical properties of the system because, as we get closer and closer to the transition, the coupling constants are averaged over sufficiently large volumes that the system looks homogeneous. Thus the condition that the critical exponents be unchanged by infinitesimal randomness is that

$$\xi(T)^{-d/2} < \frac{|T - T_c|}{T_c} \tag{15.92}$$

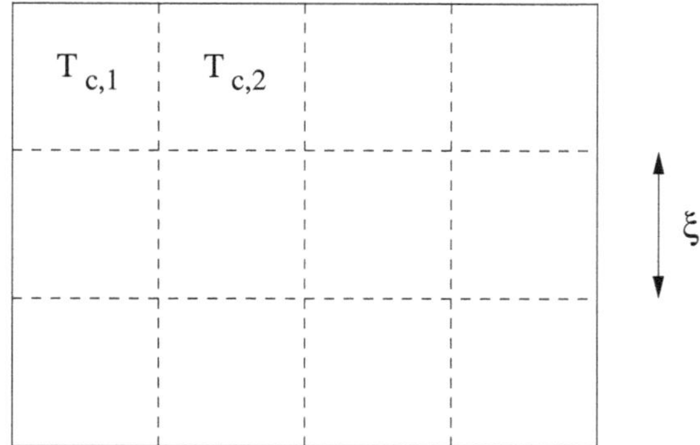

Fig. 15.6 Dividing a system into blocks, each of linear dimension ξ. Each block may be considered independent of its neighboring blocks because correlations beyond this length scale can be ignored. Since the coupling constants are random variables, the average coupling strength (which determines T_c) varies from one block to the next

or, since $\xi(T) \sim |T - T_c|^{-\nu}$,

$$\left(\frac{|T - T_c|}{T_c}\right)^{d\nu/2} < \frac{|T - T_c|}{T_c} . \tag{15.93}$$

Thus, in a random system close to T_c, we expect that $d\nu/2 > 1$ or, equivalently, that $2 - d\nu$ should be negative. If this condition (known as the Harris criterion) is fulfilled for the homogeneous system, then the introduction of infinitesimal randomness will not change the critical exponents, providing, of course, that these exponents are deduced from the behavior arbitrarily close to the critical point. (If you go far enough away from the critical point, then the apparent values of the critical exponents will indeed depend on the randomness.) If one assumes hyperscaling, Eq. (13.67), so that $\alpha = 2 - d\nu$, where α is the critical exponent for the specific heat, then one gets the oft-quoted statement of this criterion that, if the specific heat exponent of a quenched random system is negative, then disorder is irrelevant. In this simple form, the argument does not say what happens to systems which, in the absence of randomness, have a positive specific heat exponent. What one can say is that if the random system has critical behavior described by exponents, then the new exponents induced by randomness must be such that the new α is negative.

It is crucial that we have implicitly assumed that there are no long-range correlations between random couplings at different sites. Two simple examples (Harris 1974) make this clear. In the first, we imagine simply allowing all the nearest neighbor J's to be identical, with their common value taken from some distribution, $P(J)$. Then any quantity will be replaced by its average over this distribution, i.e.,

$$[X]_J = \int P(J)X(J)dJ . \tag{15.94}$$

This averaging will remove divergences in the susceptibility and singularities in the specific heat, giving a "rounded" transition. An experimentalist would refer to this as "sample inhomogeneity" which might be thought of as measuring a distribution of slightly different samples.

A somewhat different kind of rounding due to correlated disorder was found within a mathematically rigorous analysis of a two-dimensional Ising model with so-called striped randomness (McCoy and Wu 1968). To describe this model we specify the interactions as being $J(x, y)$, where x and y are coordinates of the center of the nearest neighbor bond. All interactions between nearest neighbors separated by a lattice constant in the x-direction were taken to be J_0, whereas those between nearest neighbors in the y direction were fixed to be $J(x, y) = J_0 + \epsilon F(y)$, where $F(y)$ is a random variable with a fixed width and the calculations were done in a controlled way to leading order in $\epsilon \ll 1$. In this model interactions at a given value of y are completely correlated over infinite range in the x direction as shown in Fig. 15.7. The qualitative argument for the Harris criterion, given above, assumed no such correlations. However, for this model with striped randomness, a correlation area $\xi(T)^2$ contains not $\xi(T)^2$ random variables but only $\xi(T)$ random variables. So the condition for stability with respect to randomness is now that

$$\xi(T)^{-1/2} < |T - T_c| \tag{15.95}$$

or

$$|T - T_c|^{\nu/2} < |T - T_c| \; . \tag{15.96}$$

Since $\nu = 1$ for the two-dimensional Ising model, this criterion is not fulfilled and striped randomness is expected to have a dominant effect on the critical behavior. The result that this type of randomness leads to a rounded transition is therefore consistent with the heuristic argument.

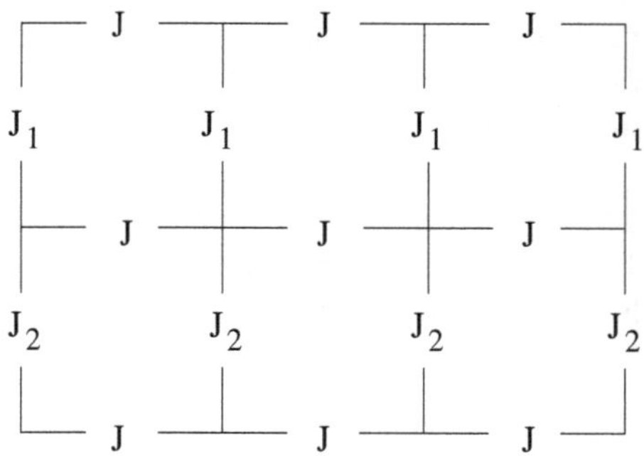

Fig. 15.7 Coupling constants for a system with quenched striped randomness. The horizontal coupling constants J are spatially uniform but the vertical coupling constants are uniform over an individual "stripe," but the values J_1, J_2, etc., are random variables

15.4 Summary

In this chapter, we have studied a number of exact mappings of athermal statistical problems onto Hamiltonian models which are subject to standard treatments of statistical mechanics. First, we showed that the cluster statistics of the bond percolation problem could be exactly reproduced by the free energy of the q-state Potts model in the limit $q \to 1$. Similarly, the statistical properties of self-avoiding walks (or equivalently the conformations of linear polymers) could be reproduced by the free energy of the n-vector model in the limit $n \to 0$. In addition, properties of quenched random systems (in which the free energy is averaged over a random distribution of coupling constants), can be calculated by considering the n-replicated Hamiltonian in the limit $n \to 0$ by using the identity $\langle Z^n \rangle \approx \langle 1 + n \ln Z \rangle + \mathcal{O}(n^2)$. When the replicated partition function is averaged over randomness, the problem is then reduced to a canonical problem in statistical mechanics.

Finally, but probably at least as important, are the qualitative arguments which allow us to develop an intuitive understanding of the effects of randomness. Thus, the Flory argument gives the approximation for the correlation length exponent for self-avoiding walks that $\nu_{\text{SAW}} \sim 3/(d+2)$. The Harris criterion indicates that the critical exponents of a system for which $2 - d\nu$ is positive (which usually means that the system has a positive specific heat exponent) are unstable with respect to randomness. Finally, (and this appears in an exercise), the Imry–Ma argument (Imry and Ma 1975) indicates that the ground state of a ferromagnet is only stable with respect to an infinitesimal random field for spatial dimension greater than 2. All these arguments can be fruitfully extended in various directions.

15.5 Exercises

1. In studying Sect. 15.2.1, the reader may have noticed that neither the SAW susceptibility of Eq. (15.39) nor the function $1/[1 - z\mu]^{x+1}$ have power series expansions of exactly the same form as the right-hand side of Eq. (15.40). In the first case, this is because the expression Eq. (15.38) for c_p, the number of SAWs of length p, is only asymptotically exact in the limit $p \to \infty$. In the case of the function $1/[1 - z\mu]^{x+1}$, the exact expansion is obtained from the binomial expansion analytically continued to a nonintegral exponent, which has different coefficients than in Eq. (15.40). Equation (15.40) can, however, be summed exactly. The result can be written in terms of the Hurwitz Lerch Phi function, which can be found, for example in Mathematica, and which is also discussed in Eq. 9.55 of Gradshteyn and Ryzhik (1994). Using Mathematica or some similar package and/or results from Gradshteyn and Ryzhik, determine the constant A in Eq. (15.41). Test this result numerically for different values of x.

2. Find the relation which gives the polygon exponent, y, for SAWs defined in Eq. (15.46), in terms of an exponent of the n-vector model in the limit $n \to 0$. Hint: evaluate the partition function for the n-vector model to first order in n.

3. Consider a spin $S = 1$ system ($S_{iz} = -1,\ 0,\ 1$) for which

$$\mathcal{H} = -J\sum_{\langle ij\rangle}(2S_{iz}^2 - 1)(2S_{jz}^2 - 1) - L\sum_i(2S_{iz}^2 - 1) . \qquad (15.97)$$

where the first sum is over pairs of nearest neighboring sites on a periodic lattice $\mathcal{L}$ in two or more spatial dimensions. The partition function Z is defined as usual:

$$Z = \mathrm{Tr} e^{-\beta\mathcal{H}} .$$

Let $Z_I(H, T)$ denote the partition function of the ferromagnetic Ising model whose Hamiltonian $\mathcal{H}_I$ is

$$\mathcal{H}_I = -J\sum_{\langle ij\rangle}\sigma_i\sigma_j - H\sum_i\sigma_i , \qquad (15.98)$$

where $\sigma_i = \pm 1$.

(**a**) Observe that the operators in Eq. (15.97) assume the values ± 1. Use this fact to show that there is an EXACT relation between Z and $Z_I(H, T)$ which involves the Ising model (on the same lattice $\mathcal{L}$) in a temperature-dependent field:

$$Z = A(T)Z_I(H(T), T) ,$$

where you should determine both $A(T)$ (which is a simple known function of T) and $H(T)$. Suppose the exact value of the transition temperature of the Ising model on the lattice $\mathcal{L}$ is known. The phase diagram of the Ising model in the field temperature (H-T) plane is as shown at right. In particular the only lack of analyticity occurs on crossing the $H = 0$ axis for $T < T_c$.

(**b**) Using this knowledge of the phase diagram of the Ising model, determine the range of values of L for which the spin 1 model of Eq. (15.97) has a phase transition as a function of temperature (Fig. 15.8).

4. Derive Eq. (15.31). Hint: start from the generalization of Eq. (15.24) to allow site-dependent fields h_i.

5. Calculate the mean-field values of the critical exponents β and γ for the q-state Potts model using a trial free energy for the optimal single particle density matrix. Hint: introduce an order parameter such that state 1 has an enhanced occupation and the other states are less occupied.

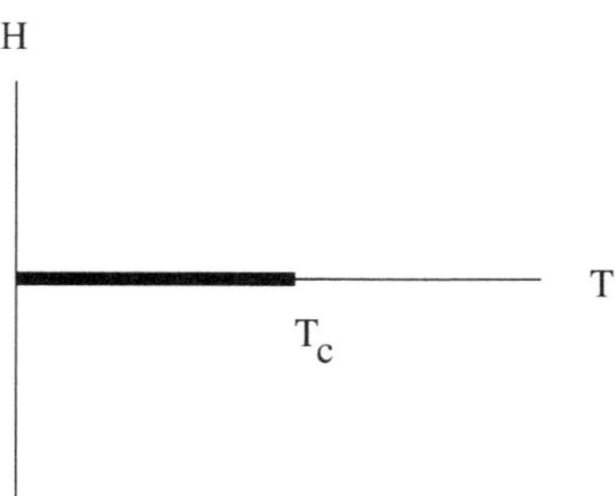

Fig. 15.8 Phase diagram of the Ising model in the (H)-T plane

6. Consider the Ising model in a random field.

(**a**) Construct the replica Hamiltonian that describes this model, in which each site independently experiences a random field $(-h_i\sigma_{iz})$ with distribution function $P(h_i)$. (In other words there are no correlations between the field acting on one site and that acting on another site.) For simplicity take

$$P(h) = \frac{1}{\sqrt{2\pi\sigma}} e^{-h^2/(2\sigma^2)} . \tag{15.99}$$

(**b**) Reproduce the Imry–Ma argument (Imry and Ma 1975) whereby one discusses, for a system in d spatial dimensions, the stability of the ordered phase with respect to application of an infinitesimal random field. In the absence of a random field, the system at low temperature becomes a single domain of spins in, say, their "up" state. In the presence of the random field, it is possible that this single domain state breaks up into subdomains each oriented along the direction of the local random field. Therefore, in the Imry–Ma argument one compares the gain in random field energy of a domain of linear dimension R (if the spins locally orient themselves along the local random field) to the energy cost of making a domain wall surrounding this domain. Argue that the ordered phase is stable against such domain formation if $d > 2$.

7. This problem concerns a model which is useful for solid solutions of two atoms, say, copper and zinc, to make β brass, and also liquid mixtures such as phenol in water. In these cases, one has a phase diagram in the T-x plane where T is the temperature and x is the concentration of one of the species (Cu for the Cu–Zn system or phenol for the phenol–water system.) For simple systems one finds that at high temperatures a single phase of uniform concentration is thermodynamically stable, so that the two species are said to be completely *miscible.* As the temperature is lowered through a critical value, the system will spontaneously decompose into two coexisting phases, giving rise to a temperature-dependent "miscibility gap." One models such systems by a lattice model in which sites on the lattice are occupied, either by A atoms of by B atoms. If two A's are nearest neighbors, their interaction energy is ϵ_{AA}, if two are B's it is ϵ_{BB} and if the two are unlike, it is ϵ_{AB}. You are to study the partition function for a lattice in which every site is occupied, but the fraction of sites occupied by A atoms is x.

(**a**) Show that the partition function for this model can be exactly related to the Ising model partition providing $\epsilon_{AA} = \epsilon_{BB} = -\epsilon_{AB} < 0$.

(**b**) Using the phenomenology you know for the Ising model, construct (qualitatively) the phase diagram for this system in the T-x plane and discuss what happens when a system at some fixed overall concentration x has its temperature lowered toward $T = 0$. (If the system phase separates, discuss the relative abundance of the two phases.)

8. Use the fact that for self-avoiding walks in two spatial dimensions, it is believed that the exponents assume the exact values

$$x_{\mathrm{SAW}} = \frac{11}{32}, \qquad \nu_{\mathrm{SAW}} = \frac{3}{4}.$$

Use this information to tabulate the exact values of the critical exponents α, β, γ, ν, and η for the n-component Heisenberg model in two spatial dimensions in the limit $n \to 0$.

9. Verify the results for the averages given in Eq. (15.55).

References

M. Aizenman, D.J. Barsky, Sharpness of the phase transition in percolation models. Commun. Math. Phys. **108**, 489 (1987)

P.G. de Gennes, Exponents for the excluded volume problem as derived by the Wilson method. Phys. Lett. A **38**, 339 (1972)

C. Domb, Self-avoiding walks on lattices, in *Advances in Chemical Physics*, ed. by K.E. Shuler, vol. 15 (Wiley Online Library, 1969), pp. 229–259

P.J. Flory, Molecular size distribution in three dimensional polymers I–III. J. Am. Chem. Soc. **63**, 3083, 3091, 3096 (1941)

C.M. Fortuin, P.W. Kasteleyn, On the random-cluster model: I. Introduction and relation to other models. Physica (Utrecht) **57**, 536 (1972)

I.S. Gradshteyn, I.M. Ryzhik, *Table of Integrals, Series, and Products*, 5th edn. (Academic Press, 1994)

A.B. Harris, Effect of random defects on the critical behaviour of Ising models. J. Phys. C **7**, 1671 (1974)

F. Hausdorff, Dimension und äußeres Maß. Math. Ann. **79**, 157179 (1919)

Y. Imry, S.-K. Ma, Random-field instability of the ordered state of continuous symmetry. Phys. Rev. Lett. **35**, 1399 (1975)

P.W. Kasteleyn, C.M. Fortuin, Phase transitions in lattice systems with random local properties. J. Phys. Soc. Jpn. Suppl. **26**, 11 (1969)

B.B. Mandelbrot, *Fractals: Form, Chance, and Dimension*, Revised edn. (W. H. Freeman and Co., 1977)

B.M. McCoy, T.T. Wu, Random impurities as the cause of smooth specific heats near the critical temperature. Phys. Rev. Lett. **21**, 549 (1968)

G. Slade, Self-avoiding walks. Math. Intell. **16**, 29–35 (1994); The self-avoiding walk: a brief survey, in *Surveys in Stochastic Processes, Proceedings of the 33rd SPA Conference in Berlin, EMS Series of Congress Reports*, ed. by J. Blath, P. Imkeller, S. Roelly (2010), pp. 181–199

D. Stauffer, A. Aharony, *Introduction to Percolation Theory*, 2nd edn. (Taylor and Francis, 1992)

Chapter 16
Series Expansions

16.1 Introduction

Series expansions provide a method for obtaining useful, accurate, and analytic results for interacting many-body systems for which exact solutions are not available. The usefulness of the series expansion method is greatly enhanced by two important tools. The first is the availability of powerful computers, which allow automated computation of many more terms in a series than could possibly be achieved by hand, pencil, and paper. The second important tool involves mathematical methods for analyzing series, including Padé approximant methods and other powerful techniques. As a result, it is often possible to obtain essentially exact numerical results based on finite series expansions, even in the absence of exact solutions. For broad overviews of series expansions methods and their analysis, applied to problems in statistical mechanics, see Baker (1990) and Oitmaa et al. (2006).

In this chapter, we are going to discuss the techniques used to generate series expansions. These are usually expansions in powers of a coupling constant λ, such that the problem for $\lambda = 0$ is exactly soluble. We will discuss three cases where such expansions are useful.

The first case we consider is the classical monatomic nonideal gas, for which the partition function, Z, is

$$Z = \frac{1}{C_N} \int e^{-\beta \sum_i p_i^2/(2m)} \prod_i d\mathbf{p}_i \int \prod_{i<j} e^{-\beta V(r_{ij})} \prod d\mathbf{r}_i \, . \qquad (16.1)$$

We write $Z = Z_0 \hat{Z}$, where Z_0 is the partition function in the absence of interactions. Then

$$\hat{Z} = \frac{\int \prod_{i<j} e^{-\beta V(r_{ij})} \prod_i d\mathbf{r}_i}{\int \prod_i d\mathbf{r}_i} \, . \qquad (16.2)$$

A. J. Berlinsky and A. B. Harris, *Statistical Mechanics*, Graduate Texts in Physics,
https://doi.org/10.1007/978-3-030-28187-8_16

The purpose of this factorization is to isolate the effects of interactions into the factor $\hat{Z}$. Now if we think that we are near the ideal gas limit, then the factor $\exp[-\beta V(r_{ij})]$ can nearly be replaced by unity. So we write

$$\begin{aligned} e^{-\beta V(r_{ij})} &= 1 + \lambda[e^{-\beta V(r_{ij})} - 1] \\ &\equiv 1 + \lambda f(r_{ij}) , \end{aligned} \tag{16.3}$$

where we have introduced the expansion parameter λ, which will be set equal to one at the end of the calculation. Then the aim is to calculate the free energy by calculating $\ln \hat{Z}$ as a power series in λ. This will lead to corrections to the equation of state in powers of the density, and, by an outrageous extrapolation, one can also derive a simple model of the liquid–gas transition.

The second case we consider is the high-temperature expansion of a spin system on a lattice with short-range interactions. For instance, if we have an interaction, $\mathcal{H}_{ij}$, between nearest neighboring spins on a lattice which we write in the form

$$\mathcal{H}_{ij} = J h_{ij} , \tag{16.4}$$

where J is a characteristic energy and h_{ij} is a dimensionless interaction, then the partition function is of the form

$$Z = \mathrm{Tr}\, e^{-\beta J \sum_{\langle ij \rangle} h_{ij}} , \tag{16.5}$$

where $\langle ij \rangle$ indicates a sum over pairs of nearest neighbors. Our aim is to expand the free energy in powers of βJ. More generally, if $\mathbf{S}(i)$ is the spin vector at site i, then we want to similarly develop the susceptibility, χ, as a high-temperature expansion in the variable βJ, where we may write (in the disordered phase)

$$\chi = \frac{1}{NkT} \frac{\sum_{kl} \mathrm{Tr}\left(e^{-\beta J \sum_{\langle ij \rangle} h_{ij}} S_\alpha(k) S_\alpha(l) \right)}{\mathrm{Tr} e^{-\beta J \sum_{\langle ij \rangle} h_{ij}}} , \tag{16.6}$$

where S_α labels the Cartesian component and N is the total number of sites. Higher order derivatives of the free energy with respect to the order-parameter field are also subject to high-temperature expansions. For percolation, the analog of high temperature is low bond concentration, p. So for percolation, we may consider expansion of various quantities in powers of p. In these cases the objective is to generate a sufficiently large number of terms in, say, the susceptibility, to allow determination of the numerical value of the critical exponent describing the singular behavior near the order–disorder phase transition. Whereas in the nonideal gas expansion the practical limit is about five terms, for spin models the number of terms that can be evaluated is often about 15, and in special cases many more than that. So, accurate extrapolation into the critical regime is possible for models involving discrete variables on a lattice.

The third case is that of perturbation theory where we write

$$\mathcal{H} = \mathcal{H}_0 + \lambda \mathcal{V} \,. \tag{16.7}$$

Then we may consider the expansion of the ground state energy as a function of λ. Here, one might imagine that one has only to look up the appropriate formulas in any quantum mechanics text. The complication is that we are dealing with a many-body system whose energy is an extensive quantity. This circumstance raises special issues which are easily addressed using the techniques presented here for use at nonzero temperature. This will be explored briefly in an exercise.

16.2 Cumulant Expansion

Here, we present several general results which will have repeated application in this chapter. We begin by replacing the single coupling constant λ by the set of coupling constants $\{\lambda_{ij}\}$ where each pairwise interaction now has its own coupling constant. For instance, for the nonideal gas, we would write

$$\begin{aligned} e^{-\beta V(r_{ij})} &= 1 + \lambda_{ij}[e^{-\beta V(r_{ij})} - 1] \\ &\equiv 1 + \lambda_{ij} f(r_{ij}) \,. \end{aligned} \tag{16.8}$$

For the spin model, we write

$$\chi = \frac{1}{NkT} \frac{\sum_{kl} \mathrm{Tr}\left(e^{\sum_{\langle ij \rangle} \lambda_{ij} h_{ij}} S_\alpha(k) S_\alpha(l) \right)}{\mathrm{Tr} e^{\sum_{\langle ij \rangle} \lambda_{ij} h_{ij}}} \,. \tag{16.9}$$

Now we consider the multivariable expansion in powers of the λ's. We will only consider expansions for extensive functions and will focus on the thermodynamic limit. For concreteness we consider the expansion of the free energy, $\mathcal{F}$, in powers of the λ's. It is convenient to define and calculate quantities which are zero in the absence of interactions, so we discuss the series expansion of $F \equiv \mathcal{F}(\{\lambda_\alpha\}) - \mathcal{F}(\{\lambda_\alpha = 0\})$.

Consider the expansion

$$F(\{\lambda\}) = \sum_n F_n(\{\lambda\}) \,, \tag{16.10}$$

where

$$F_1(\{\lambda\}) = \sum_\alpha F(\lambda_\alpha) \tag{16.11a}$$

$$F_2(\{\lambda\}) = \sum_{\alpha<\beta} \Big(F(\lambda_\alpha, \lambda_\beta) - F(\lambda_\alpha) - F(\lambda_\beta) \Big) \tag{16.11b}$$

$$\begin{aligned} F_3(\{\lambda\}) = \sum_{\alpha<\beta<\gamma} \Big(& F(\lambda_\alpha, \lambda_\beta, \lambda_\gamma) - F(\lambda_\alpha, \lambda_\beta) - F(\lambda_\beta, \lambda_\gamma) - F(\lambda_\alpha, \lambda_\gamma) \\ & + F(\lambda_\alpha) + F(\lambda_\beta) + F(\lambda_\gamma) \Big) \end{aligned} \tag{16.11c}$$

and so forth. Here λ_α denotes one of the λ_{ij}'s, and all arguments of F which are not explicitly written are set equal to zero. So, for instance,

$$F(\lambda_\alpha) \equiv F(0, \ldots 0, \lambda_\alpha, 0, \ldots 0) \tag{16.12a}$$

$$F(\lambda_\alpha, \lambda_\beta) \equiv F(0, \ldots 0, \lambda_\alpha, 0, \ldots 0, \lambda_\beta, 0, \ldots 0) \tag{16.12b}$$

and so forth. The right-hand sides of Eqs. (16.11) define the nth order *cumulant* which we denote by the subscript "c":

$$F(\lambda_\alpha)_c \equiv F(\lambda_\alpha) \tag{16.13a}$$

$$F(\lambda_\alpha, \lambda_\beta)_c \equiv F(\lambda_\alpha, \lambda_\beta) - F(\lambda_\alpha) - F(\lambda_\beta) \tag{16.13b}$$

$$\begin{aligned} F(\lambda_\alpha, \lambda_\beta, \lambda_\gamma)_c \equiv\ & F(\lambda_\alpha, \lambda_\beta, \lambda_\gamma) - F(\lambda_\alpha, \lambda_\beta) - F(\lambda_\beta, \lambda_\gamma) \\ & - F(\lambda_\alpha, \lambda_\gamma) + F(\lambda_\alpha) + F(\lambda_\beta) + F(\lambda_\gamma). \end{aligned} \tag{16.13c}$$

Because $F(0) = 0$, we see that $F(\lambda_\alpha)_c$, when expanded in powers of λ_α (which are then all set equal to λ), contains terms of order λ^1 *and higher*. Likewise $F(\lambda_\alpha, \lambda_\beta)_c$ vanishes if *either* λ_α or λ_β vanishes. Therefore the multivariable expansion of $F(\lambda_\alpha, \lambda_\beta)_c$ contains only terms which are proportional to the product $\lambda_\alpha \lambda_\beta$. So F_2 contains only terms which are of order λ^2 *and higher*. In general, when $\lambda_\alpha = \lambda$ for all α, $F_n(\lambda)_c$ contains terms of order λ^n and *higher*.

The general form of Eq. (16.11) is

$$F_n(\{\lambda\}) = \sum_{\Gamma_n} F(\Gamma_n)_c \,, \tag{16.14}$$

where Γ_n denotes a set of n variables: $\Gamma_n = \lambda_{\alpha_1}, \lambda_{\alpha_2} \ldots \lambda_{\alpha_n}$ and we sum over all such sets, Γ_n, of n arguments. The cumulant $F(\Gamma_n)_c$ is given by

$$F(\Gamma_n)_c = F(\Gamma_n) + \sum_{\gamma \in \Gamma_n} F(\gamma)(-1)^{n-n(\gamma)} \,, \tag{16.15}$$

where γ is a subset containing $n(\gamma)$ of these variables. In, the summation over subsets of Γ_n, we do *not* include the set Γ_n itself as an allowable subset γ. (In mathematical nomenclature γ is a *proper* subset of Γ_n.) One sees that F_n contains contributions to F which are of order $\prod_\alpha \lambda_\alpha \to \lambda^n$ *and higher*. Note that F_n contains no terms of order lower than λ^n, but in contrast to the Taylor series expansion, the nth-order term contains contributions not only of nth order but also of *higher order* than nth. Thus, to evaluate the function correctly up to order λ^m requires keeping all contributions in each F_n for $n \leq m$ which are of order up to λ^m.

A very important alternative formula for the cumulant of $F(\Gamma_n)$ is a recursive one. Since the are no subtractions for $n = 1$, we write

$$F(\Gamma_1)_c = F(\Gamma_1) . \tag{16.16}$$

For larger values of n we write

$$F(\Gamma_n)_c = F(\Gamma_n) - \sum_{\gamma_m \in \Gamma_n : m < n} F(\gamma_m)_c , \tag{16.17}$$

where, since $m < n$, γ_m is a proper subset of Γ_n. It is left as an exercise to show that this recursive definition of the cumulant is the same as the explicit one given in Eq. (16.15). A big advantage of the recursive definition is that the subtractions at nth order are given in terms of cumulants which have previously been calculated at lower order. Furthermore, we shall see that most subsets have zero cumulants. As a result, the recursive equation has many less contributions to consider than is the case with Eq. (16.15). To understand the meaning of the cumulant operation one should realize that it isolates the terms which depend on it all the λ_{ij}'s of the diagram in question. Any term which does not depend on all the λ_{ij}'s of the diagram was counted at a lower order in the expansion and therefore must not be counted again.

The final remark we make is a trivial one, but one which has far-reaching consequences. Suppose that F is a separable function in the sense that

$$F(x_1, x_2, \ldots x_n) = f(x_1, x_2, \ldots x_k) + g(x_{k+1}, x_{k+2}, \ldots x_n) , \tag{16.18}$$

so that F, which depends on n coupling constants, is the sum of two functions of k and $n - k$ coupling constants which do not have any common arguments. The cumulant value of such a function is zero!!

$$\Big(f(x_1, x_2, \ldots x_k) + g(x_{k+1}, x_{k+2}, \ldots x_n) \Big)_c = 0 . \tag{16.19}$$

This result follows from the fact that the cumulant has to depend on all its variables, i.e.,

$$F(x_1, x_2, x_3, \ldots x_n)_c = \prod_{k=1}^n x_k \sum_{j_1, j_2, j_3 \ldots j_n = 0}^{\infty} a_{\{j\}} \prod_{k=1}^n x_k^{j_k} . \tag{16.20}$$

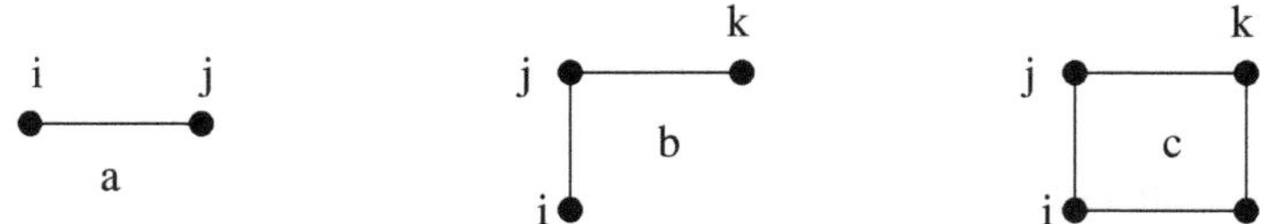

Fig. 16.1 Subset diagrams a, b, and c for the $F(\Gamma)_c$ whose cumulant contribution are given in Eq. (16.22a). For the specific case of the nonideal gas the corresponding contributions are written in Eqs. (16.29), (16.32), and (16.33), respectively

For what we have called a separable function, there are no such terms in its multivariable expansion, so Eq. (16.19) follows. A special case of this is that when Z factorizes, i.e., if

$$Z(x_1, x_2, \dots x_n) = X(x_1, x_2, \dots x_k) Y(x_{k+1}, x_{k+2}, \dots x_n) , \tag{16.21}$$

then the cumulant value of $\ln Z$ is zero. This means that, for the free energy or any of its derivatives (e.g., the susceptibility), the cumulant value of any contribution vanishes if it is of the form of Eq. (16.19).

In carrying out these calculations it is helpful to employ a diagrammatic representation, of which a number have been proposed in the literature. Here we introduce what we call "subset diagrams." For each set of coupling constants Γ, we will associate a diagram with the corresponding cumulant value $F(\Gamma)_c$. For the diagrams of Fig. 16.1 we have the following algebraic expressions:

$$F(\Gamma_a)_c = F(\lambda_{ij}) \tag{16.22a}$$

$$F(\Gamma_b)_c = F(\lambda_{ij}, \lambda_{jk}) - F(\lambda_{ij}) - F(\lambda_{jk}) \tag{16.22b}$$

$$\begin{aligned} F(\Gamma_c)_c = {} & F(\lambda_{ij}, \lambda_{jk}, \lambda_{kl}, \lambda_{li}) - F(\lambda_{ij}, \lambda_{jk}, \lambda_{kl})_c - F(\lambda_{ij}, \lambda_{kl}, \lambda_{li})_c \\ & - F(\lambda_{ij}, \lambda_{jk}, \lambda_{li})_c - F(\lambda_{jk}, \lambda_{kl}, \lambda_{li})_c - F(\lambda_{ij}, \lambda_{jk})_c \\ & - F(\lambda_{jk}, \lambda_{kl})_c - F(\lambda_{kl}, \lambda_{li})_c - F(\lambda_{li}, \lambda_{ij})_c \\ & - F(\lambda_{ij})_c - F(\lambda_{jk})_c - F(\lambda_{kl})_c - F(\lambda_{li})_c . \end{aligned} \tag{16.22c}$$

Note that for diagram c, we used the recursive definition of the cumulant. This representation is more compact because the subtractions involve only connected subdiagrams.

The description of separable functions above means that contributions from disconnected diagrams will vanish for any function which is separable in the sense of Eq. (16.18). In Exercise 5, the reader is asked to show explicitly that the cumulant value for a disconnected diagram does *not* vanish when F is a function which is not separable in this sense.

16.2.1 Summary

To summarize, we will construct the series expansion of any function $F(\{\lambda\})$ as the cumulant expansion of Eq. (16.10). Normally we use this equation for a system of N particles in the limit $N \to \infty$ by evaluating terms up to those involving Γ_n with n as large as one has the *sitzfleisch* to evaluate. For a finite system where F is a function of p coupling constants, Eq. (16.10) has a finite number of terms, where the largest value of n is p. Almost always we may restrict our attention to connected diagrams. As we saw previously, one can transform the Hamiltonian so that diagrams with one or more free ends give zero contribution. As we will see later in this chapter for certain problems it is possible to construct a series expansion using the much smaller set of *star* graphs which will be defined below. The method of obtaining a series expansion presented here is quite general and can be used for complicated interactions, such as dipolar interactions in a quantum spin system. However, in many cases, special properties of the model may give rise to important simplifications.

16.3 Nonideal Gas

Here, we derive perturbative corrections to the free energy or equation of state of a nearly ideal monatomic gas. In contrast to many texts, we calculate directly $F(T, V)$ for fixed N and avoid recourse to the grand partition function for fixed chemical potential. In addition, the "standard" derivation (Pathria 1996) involves resummation of a series with just the right combinatorial factors to produce the final result. This algebraic derivation obscures the fact that the connectedness property of the free energy (which plays a central role in our discussion) is the fundamental relation that leads to the simple result.

As indicated, for the nonideal gas we consider

$$\hat{Z}(\{\lambda\}) = \frac{\int \left(\prod_{i<j}[1 + \lambda_{ij} f(r_{ij}]\right) \prod_i d\mathbf{r}_i}{\int \prod_i d\mathbf{r}_i} . \tag{16.23}$$

We wish to expand the free energy in powers of the λ's. Of course, we will set each $\lambda = 1$ at the end of the calculation.

Note that in the absence of interactions $\hat{Z} = 1$ and $\hat{F}$, the perturbation to the free energy due to interactions, which is given by

$$\hat{F} = -kT \ln[\hat{Z}] , \tag{16.24}$$

vanishes when all the λ's are set equal to zero. So we can use Eq. (16.14) to calculate the expansion of the free energy in powers of λ.

We now look at explicit expressions for low order terms in order to get some intuition about the nature of this expansion. Start by defining the free energies for any set of nonzero coupling constants:

$$\begin{aligned}\hat{F}(\lambda_\alpha) &= -kT \ln \hat{Z}(\lambda_\alpha) \\ \hat{F}(\lambda_\alpha, \lambda_\beta) &= -kT \ln \hat{Z}(\lambda_\alpha, \lambda_\beta)\end{aligned} \tag{16.25}$$

and so forth, where λ_α denotes one of the λ_{ij}'s. We have

$$\hat{Z}(\lambda_{ij}) = \frac{\int \left(1 + \lambda_{ij}[e^{-\beta V(r_{ij})} - 1]\right) \prod_i d\mathbf{r}_i}{\int \prod_i d\mathbf{r}_i} . \tag{16.26}$$

The integrations over coordinates which do not explicitly appear in the integrand cancel, so that

$$\begin{aligned}\hat{Z}(\lambda_{ij}) &= V^{-2} \int d\mathbf{r}_i \int d\mathbf{r}_j \left(1 + \lambda_{ij}[e^{-\beta V(r_{ij})} - 1]\right) \\ &= 1 + \lambda_{ij} V^{-2} \int d\mathbf{r}_i \int d\mathbf{r}_j [e^{-\beta V(r_{ij})} - 1] .\end{aligned} \tag{16.27}$$

Introduction of a relative coordinate enables us to write this as

$$\hat{Z}(\lambda_{ij}) = 1 + \lambda_{ij} V^{-1} \int d\mathbf{r} [e^{-\beta V(r)} - 1] . \tag{16.28}$$

Thus the free energy when only a single λ_{ij} is nonzero is

$$\begin{aligned}-\frac{F(\lambda_{ij})}{kT} &= \ln\left(1 + \lambda_{ij} V^{-1} \int d\mathbf{r}[e^{-\beta V(r)} - 1]\right) \\ &= \lambda_{ij} V^{-1} \int d\mathbf{r}[e^{-\beta V(r)} - 1] \\ &\quad -\frac{1}{2}\lambda_{ij}^2 V^{-2} \left(\int d\mathbf{r}[e^{-\beta V(r)} - 1]\right)^2 + \cdots\end{aligned} \tag{16.29}$$

The sum of the terms in Eq. (16.14) for the free energy which depend on a single λ and which we denote F_1 is (with $\lambda_{ij} = \lambda$)

$$\begin{aligned}F_1 &= \sum_\alpha F(\lambda_\alpha)_c = \sum_{i<j} F(\lambda_{ij})_c \equiv \sum_{i<j} F(\lambda_{ij}) \\ &= \frac{1}{2} N(N-1) F(\lambda) .\end{aligned} \tag{16.30}$$

We want to determine the dependence of each term on N and V. Note, from Eq. (16.29), that the term of order λ is of order $N^2/V = N\rho$, where $\rho \equiv N/V$ is the particle density. So this term survives in the thermodynamic limit. The term of order λ^2 is of order $N^2/V^2 = \rho^2$. So this term (and similarly those of higher order in λ_{ij}) are negligible in the thermodynamic limit.

Now we substitute into this equation the power series expansion of the logarithm from Eq. (16.29). In the thermodynamic limit we have

$$F_1(\lambda = 1) = -\frac{1}{2}kTV\rho^2 \int [e^{-\beta V(r)} - 1]d\mathbf{r} . \qquad (16.31)$$

As we will see later, contributions F_2, F_3, etc., which depend on more than a single λ_{ij} are higher order in the density. One feature of this simple calculation which turns out to be general for the nonideal gas is the following. In Eq. (16.29), only the first term in the expansion of the logarithm survived in the thermodynamic limit. This happens generally in higher order.

16.3.1 Higher Order Terms

We now consider higher order terms in the cumulant expansion. It is very natural to introduce a diagrammatic representation of Eq. (16.14). Each term on the right-hand side of this equation will be represented by a diagram. Since each term in Eq. (16.14) corresponds to the free energy of a subsystem whose Hamiltonian consists of a subset of interactions V_{ij}, we represent such a term in the free energy by a diagram in which each interaction V_{ij} is represented by a line joining vertices i and j. With any such diagram we associate the *cumulant* value of the free energy. These correspond to the "subset diagrams," introduced above in Fig. 16.1.

Corresponding to diagram a of Fig. 16.1 we have Eq. (16.29), and for the diagram of Fig. 16.2 we have

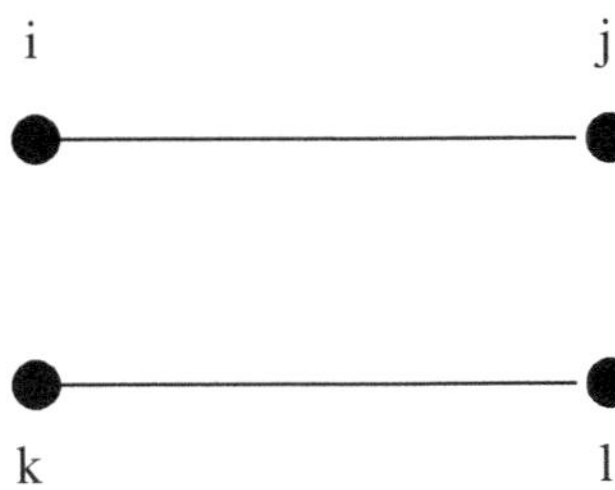

Fig. 16.2 A disconnected diagram

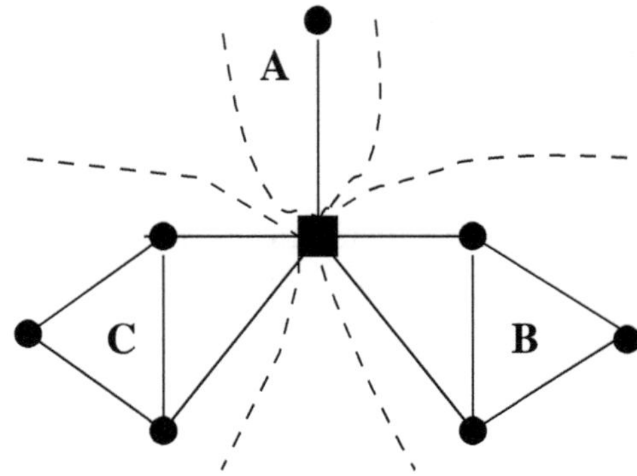

Fig. 16.3 A subset diagram with an articulation point (indicated by the filled square). When this point is removed, the diagram is decomposed into disconnected subdiagrams A, B, and C

$$\left(\hat{F}(\lambda_{ij},\lambda_{kl})\right)_c = -kT\left(\ln \hat{Z}(\lambda_{ij},\lambda_{kl})\right)_c$$
$$= -kT\left(\ln \hat{Z}(\lambda_{ij}) + \ln \hat{Z}(\lambda_{kl})\right)_c . \qquad (16.32)$$

But Eq. (16.19) tells us that the cumulant value of this separable function is zero.

Corresponding to diagram b of Fig. 16.1 we have

$$\left(F(\lambda_{ij},\lambda_{jk}\right)_c = -kT\left(\ln \hat{Z}(\lambda_{ij},\lambda_{jk})\right)_c . \qquad (16.33)$$

You are asked in Exercise 1 to show that this contribution is zero.

It should be clear that the partition function for a disconnected diagram (such as the diagram of Fig. 16.2) factorizes and therefore gives no contribution to the free energy series. We now consider the contribution to the free energy from a diagram with an *articulation point*. An articulation point is a point such that if it is removed, the diagram breaks into two or more disconnected subdiagrams. In diagram b of Fig. 16.1 vertex j is an articulation point. In Fig. 16.3 we show the case where there are three subdiagrams which become disconnected from one another when the articulation point is removed. Each of these subdiagrams is surrounded by a dashed line. We start our analysis of this diagram by considering the partition function $\hat{Z}$ associated with it. To carry out the integration over all coordinates, we introduce coordinates relative to that of the articulation point. So for the coordinates in subdiagram A we write

$$\mathbf{r}_{A,i} = \mathbf{r}_0 + \mathbf{u}_i , \qquad (16.34)$$

where $\mathbf{r}_0$ is the coordinate associated with the articulation point. Similarly for subdiagrams B and C we write

$$\mathbf{r}_{B,j} = \mathbf{r}_0 + \mathbf{v}_j , \qquad (16.35)$$
$$\mathbf{r}_{C,k} = \mathbf{r}_0 + \mathbf{w}_k . \qquad (16.36)$$

Note that when we perform the integrations required to evaluate $\hat{Z}$, the integrand depends only on the differences $\mathbf{r}_{ij}$ between particles. Thus the integrand does not depend on $\mathbf{r}_0$. Thus the integrals over the $\mathbf{u}$'s give a factor which is denoted $G_A(\{\lambda_A\})$ to indicate that it is a function only of λ's associated with the subdiagram A. Likewise the integrals over the $\mathbf{v}$'s and $\mathbf{w}$'s give the factors we similarly denote $G_B(\{\lambda_B\})$ and $G_C(\{\lambda_C\})$, so that from this diagram we get

$$\begin{aligned} Z(\{\lambda\}) &= \int d\mathbf{r}_0 \Big(\int \prod_{\alpha \in A} [1 + \lambda_\alpha f(u)] \prod_{i \in A} d\mathbf{u}_i \Big) \\ &\quad \times \Big(\int \prod_{\beta \in B} [1 + \lambda_\beta f(v)] \prod_{j \in B} d\mathbf{v}_j \Big) \\ &\quad \times \Big(\int \prod_{\gamma \in C} [1 + \lambda_\gamma f(w)] \prod_{k \in C} d\mathbf{w}_k \Big) \\ &= V G_A(\{\lambda_A\}) G_B(\{\lambda_B\}) G_C(\{\lambda_C\}) \, . \end{aligned} \tag{16.37}$$

In the first equality, the product over α indicates a product over factors $\lambda_\alpha f(u)$ corresponding to each line in the subdiagram A. Each such factor is a function of one or two u's, and similarly for the products over B and C. The product over i is over all relative coordinates associated with the subdiagram A and similarly for the products over j and k for subdiagrams B and C. The integration over $\mathbf{r}_0$ gives a factor of the volume. Because the partition function of any diagram with one or more articulation points factorizes, its logarithm (which gives the free energy) is separable and therefore its cumulant value is zero. Thus a diagram with one or more articulation points has zero cumulant free energy.

Accordingly, we need consider only contributions to $\hat{F}$ from *star graphs*, which are connected graphs with no articulation point. Let $\mathcal{S}$ be the set of λ's which connect the n_s points (and whose diagram is a star graph). Then we expand the product over $i < j$ in Eq. (16.23) to write

$$\begin{aligned} \hat{Z}(\mathcal{S}) &= 1 + \sum_{g \in \mathcal{S}} \int \Big(\prod_{\lambda \in g} \lambda_{ij} f(r_{ij}) \Big) \prod_{i \in g} \Big[\frac{d\mathbf{r}_i}{V} \Big] \Big) \\ &\equiv 1 + \sum_{g \in \mathcal{S}} \delta Y(g) \equiv 1 + Y(\mathcal{S}) \, , \end{aligned} \tag{16.38}$$

where the sum is over all nonempty sets, g, of variables λ_{ij} which are subsets of the set $\mathcal{S}$ of variables λ_{ij} for $\hat{Z}$ which is to be evaluated. (Here the sum over g includes $g = \mathcal{S}$.) Also the product over i is over all sites which appear in at least one $f(\mathbf{r})$. If $\mathcal{S}$ contains n_s variables λ_α, then the right-hand side of the first line of Eq. (16.38) has 2^{n_s} terms, the first of which (unity) comes from the empty subset and the last of which comes from $g = \mathcal{S}$. Thus the corresponding contribution to the free energy is

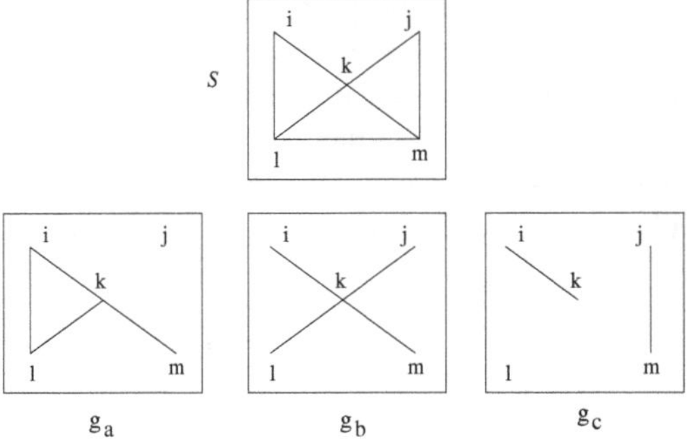

Fig. 16.4 Example of a set $\mathcal{S}$ of interactions $\{\lambda_{ij}\}$ as in Eq. (16.38). A few examples of subsets g are also shown

$$\begin{aligned} \hat{F}(\mathcal{S}) &= -kT[\ln \hat{Z}(\mathcal{S})]_c \\ &= -kT\left(Y(\mathcal{S}) - \frac{1}{2}Y(\mathcal{S})^2 + \cdots\right)_c , \end{aligned} \tag{16.39}$$

In Fig. 16.4, we show an example of a set $\mathcal{S}$ together with a few of the subsets g_i. To make this discussion more concrete we give the expressions for the corresponding contributions

$$\begin{aligned} \hat{Z}(\mathcal{S}) = \int \Big(&[1 + \lambda_{ik} f(\mathbf{r}_{ik})][1 + \lambda_{kj} f(\mathbf{r}_{kj})][1 + \lambda_{lk} f(\mathbf{r}_{lk})] \\ &\times [1 + \lambda_{km} f(\mathbf{r}_{km})][1 + \lambda_{lm} f(\mathbf{r}_{lm})][1 + \lambda_{il} f(\mathbf{r}_{il})][1 + \lambda_{jm} f(\mathbf{r}_{jm})]\Big) \\ &\times \frac{d\mathbf{r}_i}{V}\frac{d\mathbf{r}_j}{V}\frac{d\mathbf{r}_k}{V}\frac{d\mathbf{r}_l}{V}\frac{d\mathbf{r}_m}{V} . \end{aligned} \tag{16.40}$$

Associated with the diagrams shown in Fig. 16.4 are the algebraic expressions (for $\lambda_\alpha = 1$)

$$\delta Y(g_a) = \int f(\mathbf{r}_{ik}) f(\mathbf{r}_{il}) f(\mathbf{r}_{lk}) f(\mathbf{r}_{km}) \frac{d\mathbf{r}_i}{V}\frac{d\mathbf{r}_k}{V}\frac{d\mathbf{r}_l}{V}\frac{d\mathbf{r}_m}{V} \tag{16.41a}$$

$$\begin{aligned} \delta Y(g_b) = \int & f(\mathbf{r}_{ik}) f(\mathbf{r}_{kj}) f(\mathbf{r}_{lk}) f(\mathbf{r}_{km}) \\ &\times \frac{d\mathbf{r}_i}{V}\frac{d\mathbf{r}_j}{V}\frac{d\mathbf{r}_k}{V}\frac{d\mathbf{r}_l}{V}\frac{d\mathbf{r}_m}{V} \end{aligned} \tag{16.41b}$$

$$\delta Y(g_c) = \int f(\mathbf{r}_{ik}) f(\mathbf{r}_{jm}) \frac{d\mathbf{r}_i}{V}\frac{d\mathbf{r}_j}{V}\frac{d\mathbf{r}_k}{V}\frac{d\mathbf{r}_m}{V} . \tag{16.41c}$$

In the Appendix, we give the technical analysis which shows that terms proportional to $Y(\mathcal{S})^n$ with $n > 1$ do not contribute to the free energy in the thermodynamic limit. Assuming that result, we write

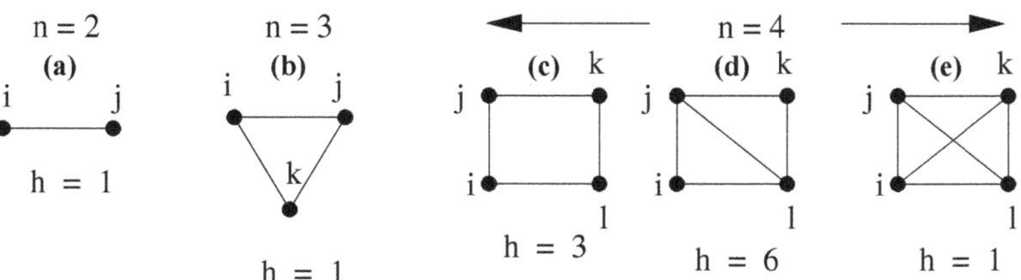

Fig. 16.5 Star graphs which contribute to $V^{-1}F$ at nth order (for n up to and including $n = 4$) in the particle density, where $n = N_s$, where N_s is the number of particles involved in the diagram. We also give the symmetry number h for each diagram

$$\hat{F}(\mathcal{S}) = -kTY(\mathcal{S})_C = -kT\int\prod_{\mathbf{f}\in\mathcal{S}} f(r_{ij})\prod_{i\in\mathcal{S}}\left[\frac{d\mathbf{r}_i}{V}\right], \tag{16.42}$$

where $\mathbf{f} \in \mathcal{S}$ means that the product is over all $f(\mathbf{r}_{ij})$'s such that λ_{ij} (which we set equal to unity) is in $\mathcal{S}$. To write this result we noted that the only term in $Y(\mathcal{S})$ which depends on all the λ's (and therefore the only term which contributes to the cumulant) is the one with $g = \mathcal{S}$.

We now switch from subset diagrams to what we call "V-diagrams" in which each line connecting particles i and j represents a single factor $f(r_{ij})$. (But we still have to integrate over $\mathbf{r}_i$ for each particle labeled i.) This switch is useful because now we have established that the only term in the subset diagram that we need to keep is the one in which each perturbation, $\lambda_{ij} f(\mathbf{r}_{ij})$, occurs once and only once. So the contribution to $\hat{F}$ from a subset diagram is just the contribution from that diagram interpreted as a V-diagram. Thus, for the diagram $\mathcal{S}$ shown in Fig. 16.4, we have (for $\lambda_\alpha = 1$)

$$\hat{F}(\mathcal{S}) = -kT\int f(\mathbf{r}_{ik})f(\mathbf{r}_{kj})f(\mathbf{r}_{lk})f(\mathbf{r}_{km})f(\mathbf{r}_{lm}) \\ \times f(\mathbf{r}_{il})f(\mathbf{r}_{jm})\frac{d\mathbf{r}_i}{V}\frac{d\mathbf{r}_j}{V}\frac{d\mathbf{r}_k}{V}\frac{d\mathbf{r}_l}{V}\frac{d\mathbf{r}_m}{V}. \tag{16.43}$$

According to our discussion we restrict the sum to star graphs, of which some low-order examples are shown in Fig. 16.5. Note that Eq. (16.42) gives the contribution for a fixed set of particle labels. But we need to sum over all possible ways to assign particle labels to a given topology Γ of diagrams involving $N_s(\Gamma)$ particles.

First, we consider how the sum over site indices i, j, k, etc depends on N, the total number of particles. Let $\hat{F}_N(\Gamma)$ be the contribution to the free energy of the system of N particles from all diagrams with the topology Γ. The dependence on N comes because there are $N!/([(N - N_s(\Gamma)]!N_s(\Gamma)!)$ ways to select the set of $N_s(\Gamma)$ indices out of the total number of indices of particles in the system. So

$$\hat{F}_N(\Gamma) = \frac{N!}{[N - N_s(\Gamma)]!N_s(\Gamma)!}F_{N_s(\Gamma)}(\Gamma). \tag{16.44}$$

Thus, the quantity we want is related to the value of $\hat{F}(\Gamma)$ calculated for a system for which the number of particles in the total system is the same as the number of particles involved in the diagram Γ.

Now we have to address the question of counting how many ways, $h(\Gamma)$, the indices on the diagram Γ should be permuted. We will illustrate the discussion by analyzing the contributions from the diagrams Γ_α of Fig. 16.5 for which α assumes the values a, b, c, d, and e, reading from left to right. We settle the counting question by going back to the algebraic expression for the free energy and seeing how many algebraic terms correspond to a given topology. For instance, for diagram a we consider the calculation of $\hat{F}_2(\Gamma_a)$, which is

$$\begin{aligned}\hat{F}_2(\Gamma_a) &\sim -kT\left(\ln \int [1+\lambda_{12} f(\mathbf{r}_{12})] \frac{d\mathbf{r}_1}{V}\frac{d\mathbf{r}_2}{V}\right)_c \\ &\sim -kT\lambda_{12}\left(\int f(\mathbf{r}_{12}) \frac{d\mathbf{r}_1}{V}\frac{d\mathbf{r}_2}{V}\right). \end{aligned} \tag{16.45}$$

There is clearly only one term which corresponds to the topology of diagram a, so we conclude that $h(a) = 1$. Similarly, for diagram b we consider the calculation of $\hat{F}_3(\Gamma_b)$, which is

$$\begin{aligned}\hat{F}_3(\Gamma_b) &\sim -kT\left(\ln \int [1+\lambda_{12} f(\mathbf{r}_{12})][1+\lambda_{13} f(\mathbf{r}_{13})]\right. \\ &\qquad \left.\times [1+\lambda_{23} f(\mathbf{r}_{23})] \frac{d\mathbf{r}_1}{V}\frac{d\mathbf{r}_2}{V}\frac{d\mathbf{r}_3}{V}\right)_c \\ &\sim -kT\lambda_{12}\lambda_{23}\lambda_{13}\left(\int f(\mathbf{r}_{12}) f(\mathbf{r}_{23}) f(\mathbf{r}_{13}) \frac{d\mathbf{r}_1}{V}\frac{d\mathbf{r}_2}{V}\frac{d\mathbf{r}_3}{V}\right). \end{aligned} \tag{16.46}$$

Again there is only one term which corresponds to the topology of diagram b, so we conclude that $h(b) = 1$.

Finally consider the diagrams involving four particles. These contributions come from the free energy of the four-particle system, which is

$$\begin{aligned}F_4 = -kT &\left\{\ln \int \Big([1+\lambda_{12} f(\mathbf{r}_{12})][1+\lambda_{13} f(\mathbf{r}_{13})][1+\lambda_{14} f(\mathbf{r}_{14})]\right. \\ &\times [1+\lambda_{23} f(\mathbf{r}_{23})][1+\lambda_{24} f(\mathbf{r}_{24})][1+\lambda_{34} f(\mathbf{r}_{34})]\Big) \\ &\left.\times \frac{d\mathbf{r}_1}{V}\frac{d\mathbf{r}_2}{V}\frac{d\mathbf{r}_3}{V}\frac{d\mathbf{r}_4}{V}\right\}_c . \end{aligned} \tag{16.47}$$

For diagram c of Fig. 16.5, we want to select from the expansion of F_4 those terms which involve four λ's which correspond to diagram c, namely we want to get a term of the form $\lambda_{ij}\lambda_{jk}\lambda_{kl}\lambda_{il}$. [Keep in mind that when we talk about expanding in powers of λ, we only keep terms which correspond to $Y(\mathcal{S})$ in Eq. (16.42). Also, we do not distinguish between λ_{ij} and λ_{ji}.] There are three such terms:

$$\lambda_{12}\lambda_{23}\lambda_{34}\lambda_{14}, \qquad \lambda_{13}\lambda_{14}\lambda_{23}\lambda_{24}, \, and \qquad \lambda_{12}\lambda_{13}\lambda_{34}\lambda_{24} \, , \tag{16.48}$$

so that $h(c) = 3$. An alternate way to come to this conclusion is to realize that diagram c (in the form of a square) is such that vertices i and k do not interact and vertices j and l do not interact. The number of ways to dividing four indices into two equivalent pairs is 3, and so, as above, we conclude that $h(c) = 3$. The factor $h(d)$ is the number of ways we can obtain a term having the form of diagram d from the expansion of F_4 in powers of the λ's. The easy way to get $h(d)$ is to note that we want to count the number of terms which have any five of the six λ's. There are clearly six ways to choose the five λ's, so $h(d) = 6$. A more geometric way to reach this conclusion is to ask how many ways we can pick the two indices (i and k) out of four which do not interact. This can be done in $4 \times 3/2 = 6$ ways, so again $h(d) = 6$. Finally, diagram e requires taking all terms in which all the λ's appear. There is only one such term, so $h(e) = 1$.

So the general result for the contribution to δF from a diagrams of a given topology Γ is

$$-\beta\hat{F}(\{\Gamma\}) = \frac{N!h(\Gamma)}{[N - N_s(\Gamma)]!N_s(\Gamma)!}\int\left(\prod_{f\in\Gamma} f(r_{ij})\prod_{i\in\Gamma}\left[\frac{d\mathbf{r}_i}{V}\right]\right) , \tag{16.49}$$

where $N_s(\Gamma)$ is the number of vertices in the diagram Γ and f is an interaction in the diagram Γ.

We now check the dependence of this contribution $-\beta\hat{F}(\{\Gamma\})$ on N and V. The combinatorial factor gives $N^{N_s(\Gamma)}$ since $N_s(\Gamma) \ll N$. Each $d\mathbf{r}$ carries a factor V^{-1}, giving a factor $V^{-N_s(\Gamma)}$. But the integration over the center-of-mass coordinate gives an extra factor of V. So we see that overall the contribution to $-\beta\hat{F}$ from this topology of diagram is of order $V(N/V)^{N_s(\Gamma)}$. As expected $-\beta\hat{F}$ is an extensive quantity in the thermodynamic limit and the contribution to $-V^{-1}\beta F(\{\Gamma\})$ is proportional to the $N_s(\Gamma)$ power of the particle density. This expansion in powers of the particle density is known as the "virial expansion."

16.3.2 Van der Waals Equation of State

We now proceed to an explicit evaluation of the leading correction due to particle interactions. It is

$$\hat{F} = -\frac{1}{2}NkT(N/V)\int[e^{-\beta V(r)} - 1]d\mathbf{r} \, . \tag{16.50}$$

The integral is of order the microscopic volume $v_0 \equiv \xi^3$, where ξ is the range over which the potential energy is less than kT. It is customary to introduce the macroscopic volume, called the *second virial coefficient*, denoted $B_2(T)$ and defined by

$$B_2(T) = \frac{1}{2}N_A \int [1 - e^{-\beta V(r)}]d\mathbf{r} , \tag{16.51}$$

where N_A is Avogadro's number. In terms of this quantity we may write the free energy per mole, f, as

$$f(V,T) = f_0(V,T) + RT B_2(T)/v , \tag{16.52}$$

where v is the molar volume and $R = N_A k$ is the gas constant. Then the pressure is obtained as

$$\begin{aligned} P &= -\left.\frac{\partial f}{\partial v}\right|_T \\ &= \frac{RT}{v}\left[1 + \frac{B_2(T)}{v} + \cdots\right], \end{aligned} \tag{16.53}$$

where $R = N_A k$. One model of interactions which is easy to use has a hard core for $r < r_0$ and van der Waals attraction for larger r, so that for $r > r_0$, $V(r) = -u(r_0/r)^6$. Then if we further say that $u \ll kT$, we have

$$B_2 = \frac{2\pi r_0^3}{3} N_A \left(1 - \frac{u}{kT}\right) . \tag{16.54}$$

This gives

$$P = \frac{RT}{v}\left[1 + \frac{2}{3}\pi r_0^3 (N_A/v) - \frac{2}{3}\pi r_0^3 (N_A u)/(vkT)\right], \tag{16.55}$$

or approximately

$$P + \frac{2}{3}\pi r_0^3 u (N_A/v)^2 = \frac{RT}{v}\left[1 - \frac{2}{3}\pi r_0^3 (N_A/v)\right]^{-1} , \tag{16.56}$$

which is

$$\left(P + \frac{a}{v^2}\right)[v - b] = RT , \tag{16.57}$$

where $a = \frac{2}{3}\pi r_0^3 u N_A^2$ and $b = \frac{2}{3}\pi r_0^3 N_A$. It can be shown that at leading order in the density, b represents correctly the excluded volume of hard spheres. Equation (16.57) is the van der Waals equation of state which can reasonably be used when $v \gg b$ and $v \gg a/(RT)$.

However, we will pretend that this equation holds for smaller values of v, and explore its consequences. First of all, note that different gases will have different values of the constants a and b, of course. The question now is can we rescale the variables so as that a and b do not appear explicitly? If so, then we have a universal equation of state. From Eq. (16.57) the factor $v - b$ indicates that we should define a reduced volume variable by $\hat{v} \equiv v/b$. Then if we do that the first factor becomes $P + (a/b^2)\hat{v}^{-2}$, which indicates that we should define a reduced pressure by $\hat{P} = Pb^2/a$. Then we have

$$\left(\hat{P} + \hat{v}^{-2}\right)(\hat{v} - 1) = RTa/b \,, \tag{16.58}$$

which suggests that we define $\hat{T} = RTa/b$ to get a universal equation of state. In fact, this is usually done by writing

$$P = P_c\hat{P} \,, \qquad v = v_c\hat{v} \,, \qquad T = T_c\hat{T} \,, \tag{16.59}$$

where

$$P_c = \frac{a}{27b^2} \,, \qquad v_c = 3b \,, \qquad T_c = \frac{8a}{27bR} \,. \tag{16.60}$$

Then the equation of state in terms of these reduced variables is

$$\left(\hat{P} + \frac{3}{\hat{v}^2}\right)(3\hat{v} - 1) = 8\hat{T}. \tag{16.61}$$

and is plotted in Fig. 16.6 for indicated values of reduced temperature. The critical point (where the liquid phase first appears) is the point at which both $\partial\hat{P}/\partial\hat{v}|_{\hat{T}}$ and $\partial^2\hat{P}/\partial\hat{v}^2|_{\hat{T}}$ vanish. These conditions, which determine the critical point, give

$$\begin{aligned} \frac{6}{\hat{v}^3} &= \frac{24\hat{T}}{(3\hat{v}-1)^2} \\ \frac{18}{\hat{v}^4} &= \frac{144\hat{T}}{(3\hat{v}-1)^3} \,. \end{aligned} \tag{16.62}$$

Divide the second equation by the first to give

$$\frac{3}{\hat{v}} = \frac{6}{3\hat{v}-1} \,, \tag{16.63}$$

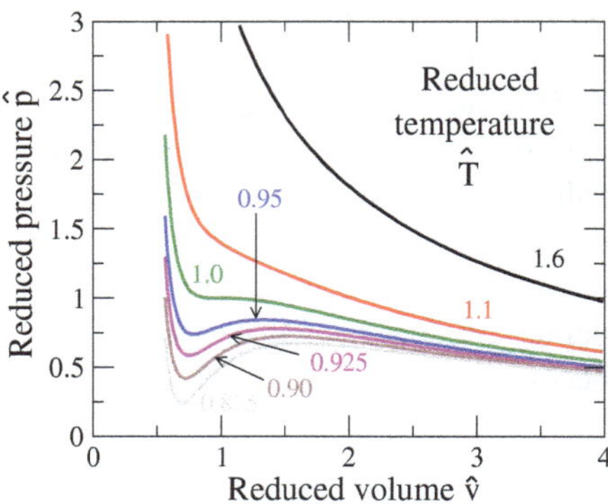

Fig. 16.6 The van der Waals equation of state of Eq. (16.61). Here we show isotherms whose reduced temperatures are indicated

so that $\hat{v}_c = 1$ is the value of $\hat{v}$ at the critical point. Then we find the values of $\hat{P}$ and $\hat{T}$ at the critical point to be $\hat{T}_c = \hat{P}_c = 1$. That is why the constants in the reduced variables in Eq. (16.59) were chosen as in Eq. (16.60). The analysis leads to the prediction that $Pv_c/RT_c = 3/8 = 0.375$. This result is not well satisfied experimentally. For many gases the value is around 0.28, and it can range between 0.23 (for water) to 0.31 (for ^{4}He).

Now what are we to make of a curve which has an instability in it? Thermodynamics requires that $-\partial P/\partial v|_T$ has to be positive. For $T < T_c$ there are regions of the curve where this is violated. In Fig. 16.7 is shown a schematic view of such a curve. Between V_B and V_C the microscopic condition for thermodynamic stability is violated, so these points can never be realized. Within this region the total free energy of the system can be lowered by breaking into a two phase system in which a fraction of the system is in a phase with volume, V_α, and the rest is in phase with volume, V_γ. In order to minimize the free energy, the volumes V_α and V_γ are determined by the "equal area" condition, namely, that the area under the straight line between points at V_α and V_γ is the same as the area under the hypothetical curve between V_α and V_γ.

One sees that even beyond the region of microscopic (or local) instability one has global instability that even though $-\partial p/\partial v|_T$ is positive, the system is nevertheless unstable relative to phase separation. However, since the system is not locally unstable, it is possible to superheat or supercool and get into this region of global but not local instability. (We already had such a discussion in connection with first order transitions.) Note the similarity between this phase transition and that of a magnet. If we vary the magnetic field H so as to cross the phase boundary at $H = 0$ below T_c, we see a discontinuous jump in the magnetization M. This jump becomes progressively smaller as we approach the critical temperature T_c. The same happens in the liquid–gas transition as described by the van der Waals equation. As T is raised toward T_c, the discontinuity in the molar volume is reduced and this jump vanishes exactly at T_c. It was this similarity which motivated studies of the Ising model in place of studies of the more algebraically complicated nonideal gas. Whereas development of the virial expansion has been taken up to order about 5, the high-temperature expansions for the Ising model have gone up to terms of order 40, at least, and maybe more, by

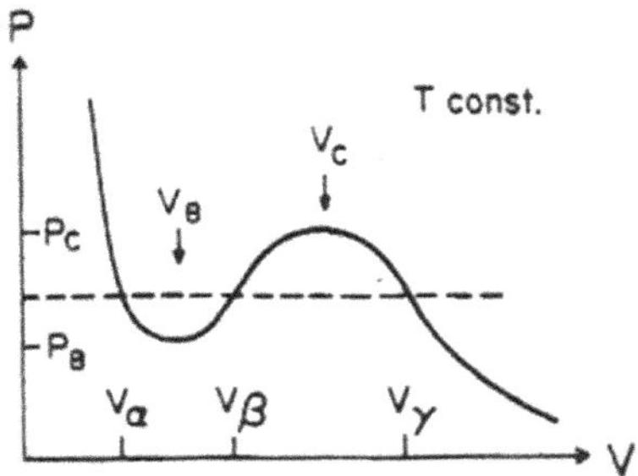

Fig. 16.7 Schematic representation for an isotherm for $T < T_c$ with a range of volume (between V_B and V_C) of local instability, where $\partial P/\partial V)_T > 0$. The equal area construction, which was the subject of Exercise 24 in Chap. 3, replaces this unphysical curve by one with a coexistence region between V_α and V_γ

now. Such studies enable one to effectively extrapolate into the critical region and make very accurate numerical predictions of critical exponents, which compare well with less accurate numerical results based on the renormalization group. (But the RG provides a conceptual framework and explanation of the physics of the critical point that series expansions can never provide.)

Although the equation of state developed by J. D. van der Waals was not completely and systematically justified, it nevertheless showed that a simple picture could reproduce much of the physics associated with the liquid–gas phase transition. It explained how the transition might arise from a microscopic model. It also explained that the transition was generically a discontinuous transition accompanied inevitably by hysteresis. It is therefore not surprising that for this work van der Waals was awarded the Nobel prize in Physics in 1910.

16.4 High-Temperature Expansions

In this section we discuss high-temperature expansions. We have already seen in Sect. 14.5 the use of an effective potential in order to eliminate diagrams in an expansion which have free ends. This so-called no-free-end method thereby leads to exact solutions for the Cayley tree on which there are no diagrams without free ends.

Here, we consider the high-temperature expansion for the two-point correlation function. Since we are not in the ordered phase, this function is proportional to the susceptibility. In particular, if ψ_i is a variable associated with site i such that $\mathrm{Tr}\psi_i = 0$, then we develop a high-temperature expansion for $\chi = \sum_{ij} \chi_{i,j}$, where

$$\chi_{ij} \equiv \langle \psi_i \psi_j \rangle = \frac{\mathrm{Tr} e^{-\beta \sum_{\langle ij \rangle} \mathcal{H}_{ij}} \psi_i \psi_j}{\mathrm{Tr} e^{-\beta \sum_{\langle ij \rangle} \mathcal{H}_{ij}}} , \qquad (16.64)$$

where we have assumed only nearest-neighbor interactions. For instance, for the Ising model we take $\psi_i = \sigma_i = \pm 1$, in which case $\chi_{ii} = 1$.

For the percolation problem, we may consider a low-concentration expansion for the pair-connectedness susceptibility, defined by

$$\chi_{ij} \equiv \langle \nu_{i,j} \rangle = \sum_{ij} \sum_{\mathcal{C}} P(\mathcal{C}) \nu_{i,j}(\mathcal{C}) \,, \tag{16.65}$$

where $\nu_{i,j}$ is unity if sites i and j are in the same cluster in the configuration $\mathcal{C}$ and is zero otherwise. The probability of having the configuration $\mathcal{C}$ is $p^{\text{occ}}(1-p)^{\text{vac}}$, where "occ" is the number of occupied bonds and "vac" the number of vacant bonds in the configuration $\mathcal{C}$. Here also $\chi_{ii} = 1$.

In the general development of series expansions, we represented each term in Eq. (16.14) by a *subset* diagram as follows. Each diagram corresponds to the cumulant value of the susceptibility calculated for a subsystem whose Hamiltonian includes the interactions indicated by the lines in the diagram. So we place a line between sites i and j for each variable λ_{ij}, and indicating by dots, the two sites i and j for which the susceptibility is being evaluated. Examples of such diagram are shown in Fig. 16.8 For instance, the susceptibility associated with diagram (a) of Fig. 16.8 is given in the notation of Eq. (16.4) by

$$\chi_{ij} = \frac{\text{Tr}e^{-\lambda_{ik}\mathcal{H}_{ik} - \lambda_{kj}\mathcal{H}_{kj}} \psi_i \psi_j}{\text{Tr}e^{-\lambda_{ik}\mathcal{H}_{ik} - \lambda_{kj}\mathcal{H}_{kj}}} \,. \tag{16.66}$$

The contribution to the series expansion for the susceptibility requires forming the cumulant of χ_{ij} which is best done using Eq. (16.17). Since χ_{ij} of Eq. (16.66) (for diagram a of Fig. 16.8) has no contributions from subdiagrams, the value given in Eq. (16.66) is also the cumulant value for this quantity.

Now consider the contribution to χ_{ij} from a diagram which is disconnected. If the diagram is disconnected, the contribution to the susceptibility is either zero (as it is for diagram b), or if it is nonzero (as for diagram c), it is not a function involving all the variables. In this case, the cumulant contribution (which is what contributes to the series expansion) vanishes. So we only need consider connected diagrams.

Now we consider a diagram having s sites and an articulation point at site k. (In diagram a of Fig. 16.8, point k is an articulation point, i.e., a point, which when

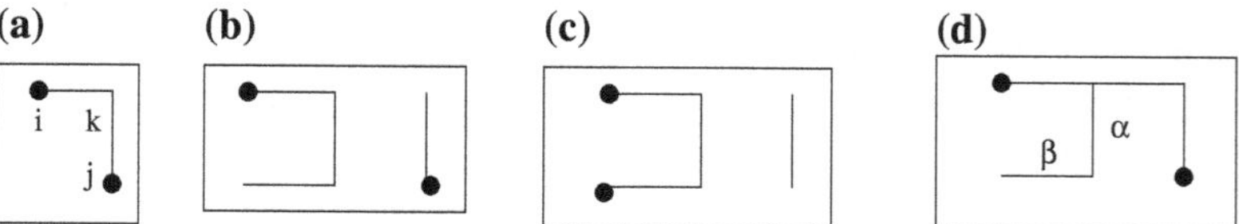

Fig. 16.8 Subset diagrams for the high-temperature expansion. Each line represents an interaction $\lambda_{ij}\mathcal{H}_{ij}$ and each dot represents a ψ_i. Diagrams b and c are disconnected. The cumulant value of diagram d is zero because the contribution to $\chi_{i,j}$ does not depend on the λ for bonds α and β

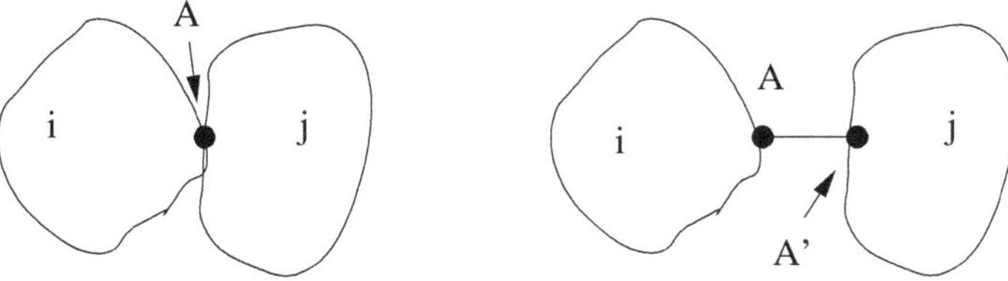

Fig. 16.9 Left: The two-point correlation function χ_{ij} between sites i and j for a diagram with an articulation point A. The subdivided diagram in which the articulation point A is replaced by a point A connected to the left subcluster and a point A' connected to the right subcluster. To reproduce the system shown at left we must force σ_A to be equal to $\sigma_{A'}$

removed causes the diagram to become disconnected.) We label sites so that when the articulation point is removed, sites $1, 2, \ldots, k-1$ are in one cluster and sites $k+1, k+2, \ldots, s$ are in the other cluster(s). The Hamiltonian which governs sites $1, 2, \ldots, k$ is called $\mathcal{H}_<$ and that which governs sites $k, k+1, \ldots\ s$ will be denoted $\mathcal{H}_>$. Now we will develop the *star graph* expansion, which holds for most classical systems. In particular, this expansion holds for systems for which the χ_{ij} for graphs with an articulation point at vertex k obeys the rules

$$\begin{aligned}
\chi_{ij} &= \frac{\mathrm{Tr}\psi_i \psi_j e^{-\beta\mathcal{H}_<}}{\mathrm{Tr}e^{-\beta\mathcal{H}_<}} \equiv A(i,j)\,, && 1 \le i < j \le k \text{ (a)} \\
\chi_{ij} &= \frac{\mathrm{Tr}\psi_i \psi_j e^{-\beta\mathcal{H}_>}}{\mathrm{Tr}e^{-\beta\mathcal{H}_>}} \equiv B(i,j)\,, && k \le i < j \le s \text{ (b)} \\
\chi_{ij} &= A(i,k)B(k,j) && i < k, \quad k < j\,. \text{ (c)}
\end{aligned} \tag{16.67}$$

Also we have assumed that $\chi_{ii} = 1$.

We now verify that the conditions of Eq. (16.67) are satisfied by the nearest-neighbor Ising model. We explicitly require that the external magnetic field is zero. Since Eqs. (16.67a) and (16.67b) are easy to verify, we will only explicitly verify Eq. (16.67c). For that purpose we evaluate χ_{ij} when the points i and j are on opposite sides of the articulation point, in which case we have the situation shown in Fig. 16.9 and we write

$$\chi_{ij} = \frac{\mathrm{Tr}e^{-\beta\mathcal{H}_<} e^{-\beta\mathcal{H}_>} \sigma_i \sigma_j}{\mathrm{Tr}e^{-\beta\mathcal{H}_<} e^{-\beta\mathcal{H}_>}}\,, \tag{16.68}$$

where $\mathcal{H}_<$ ($\mathcal{H}_>$) represent the interactions to the left (right) of the articulation point. Now consider the system shown in the right-hand panel. Only if we constrain σ_A and $\sigma_{A'}$ to be equal are the systems in the right-hand and left-hand panels equivalent. This constraint can be enforced by including the factor $(1 + \sigma_A \sigma_{A'})/2$. So we may write

$$\chi_{ij} = \frac{\text{Tr} e^{-\beta\mathcal{H}_<} e^{-\beta\mathcal{H}_>} \sigma_i \sigma_j (1 + \sigma_A \sigma_{A'})/2}{\text{Tr} e^{-\beta\mathcal{H}_<} e^{-\beta\mathcal{H}_>} (1 + \sigma_A \sigma_{A'})/2}$$
$$\equiv \frac{N}{D}, \qquad (16.69)$$

where now the system is that of the right-hand panel of Fig. 16.9 so that in $\mathcal{H}_>$ σ_A is replaced by $\sigma_{A'}$.

In the last line of Eq. (16.69) N (D) is the numerator (denominator) of the preceding line. Now the Hamiltonian for the left-hand and right-hand clusters are independent. Thus

$$N = \text{Tr} e^{-\beta\mathcal{H}_<} e^{-\beta\mathcal{H}_>} \frac{1}{2}\Big[\sigma_i\sigma_j + \sigma_i\sigma_j\sigma_A\sigma_{A'}\Big]$$
$$= \frac{1}{2}\Big[\text{Tr}_< e^{-\beta\mathcal{H}_<}\sigma_i\Big]\Big[\text{Tr}_> e^{-\beta\mathcal{H}_>}\sigma_j\Big]$$
$$+\frac{1}{2}\Big[\text{Tr}_< e^{-\beta\mathcal{H}_<}\sigma_i\sigma_A\Big]\Big[\text{Tr}_> e^{-\beta\mathcal{H}_>}\sigma_j\sigma_{A'}\Big], \qquad (16.70)$$

where $\text{Tr}_<$ ($\text{Tr}_>$) indicates a trace only over variables in the $<$ ($>$) subsystem. Since we do not have long-range order, $\text{Tr}_<\sigma_i \exp(-\beta\mathcal{H}_<) = 0$ and

$$N = \frac{1}{2}\Big[\text{Tr}_< e^{-\beta\mathcal{H}_<}\sigma_i\sigma_A\Big]\Big[\text{Tr}_> e^{-\beta\mathcal{H}_>}\sigma_j\sigma_{A'}\Big]. \qquad (16.71)$$

Similarly, we have

$$D = \text{Tr} e^{-\beta\mathcal{H}_<} e^{-\beta\mathcal{H}_>} \frac{1}{2}\Big[1 + \sigma_A\sigma_{A'}\Big]$$
$$= \frac{1}{2}\Big[\text{Tr}_< e^{-\beta\mathcal{H}_<}\Big]\Big[\text{Tr}_> e^{-\beta\mathcal{H}_>}\Big] + \frac{1}{2}\Big[\text{Tr}_< e^{-\beta\mathcal{H}_<}\sigma_A\Big]\Big[\text{Tr}_> e^{-\beta\mathcal{H}_>}\sigma_{A'}\Big]$$
$$= \frac{1}{2}\Big[\text{Tr}_< e^{-\beta\mathcal{H}_<}\Big]\Big[\text{Tr}_> e^{-\beta\mathcal{H}_>}\Big]. \qquad (16.72)$$

Thus

$$\chi_{ij} = \frac{\Big[\text{Tr}_< e^{-\beta\mathcal{H}_<}\sigma_i\sigma_A\Big]\Big[\text{Tr}_> e^{-\beta\mathcal{H}_>}\sigma_j\sigma_{A'}\Big]}{\Big[\text{Tr}_< e^{-\beta\mathcal{H}_<}\Big]\Big[\text{Tr}_> e^{-\beta\mathcal{H}_>}\Big]}. \qquad (16.73)$$

The right-hand side can be identified as being $\chi_{iA}\chi_{Aj}$, Q. E. D.

The relations of Eq. (16.67) also hold for the pair-connectedness susceptibility of the percolation problem. In general, this type of relation does *not* hold in the presence of a nonzero order-parameter field. Nor does this relation hold for quantum systems, for which the quantum wave function does not factorize into a single product of a wave function for sites with $i \leq k$ times a wave function for sites with $i \geq k$. But this condition is nevertheless very widely satisfied as exemplified by some of the exercises.

Under the above assumptions we see that for a subset diagram with an articulation point at vertex k we may write the matrix χ_{ij} as

$$\chi_{ij} = \mathcal{I}_{i,j} + A(i, j) + B(i, j) + A(i, k)B(k, j) + B(i, k)A(k, j) , \quad (16.74)$$

where $\mathcal{I}$ is the unit matrix and $A(i, j)$ and $B(i, j)$ are zero if either index is outside its range of definition in Eq. (16.67). In particular, the diagonal elements of the matrices **A** and **B** are defined to be zero. The diagonal elements of the matrix χ are taken into account by the unit matrix.

In matrix notation, we may write this as

$$\chi = \mathcal{I} + \mathbf{A} + \mathbf{B} + \mathbf{AB} + \mathbf{BA} , \quad (16.75)$$

because the matrices **A** and **B** only have the single site k in common.

16.4.1 Inverse Susceptibility

Now we consider the series expansion for the matrix inverse of χ. As we will see, the calculation of this quantity can be used to simplify the calculation of the susceptibility. As usual we want to develop a series for a quantity which vanishes in the absence of interactions. In the absence of interactions, $\chi_{ij} = \delta_{i,j}$. Accordingly, we define the quantity

$$M \equiv \sum_{ij} \left([\chi^{-1}]_{ij} - \delta_{i,j} \right) . \quad (16.76)$$

We now consider the series expansion for M. We apply the cumulant expansion, which requires us to sum over the cumulant contributions to M from all connected subset type diagrams, Γ, where Γ denotes some set of coupling constants. What we will now show is that if the conditions of Eq. (16.67) are fulfilled, then the cumulant contribution to M from any diagram Γ_A with an articulation point vanishes. To show that the cumulant of $M(\Gamma_A)$ vanishes, it suffices to show that this quantity is separable, i.e., all terms depend either on the variables of one side of the articulation point or on the variables of the other side of the articulation point but there are no terms which depend on both sets of variables. (If removal of the articulation point creates more than two connected subclusters, we arbitrarily consider one subcluster as being on one side of the articulation point and all the other subclusters to be on the other side of the articulation point.)

In particular, we will show that $M(\Gamma_A)$ has no terms which are functions of both the matrices **A** and **B**. To show this we first point out that by their ranges of indices that

$$\mathbf{ABA} = 0 = \mathbf{BAB} \,. \tag{16.77}$$

To see this consider

$$[\mathbf{ABA}]_{ij} = \sum_{l,m} A_{i,l} B_{l,m} A_{m,j} \,. \tag{16.78}$$

Here, the indices l and m must both belong simultaneously to the allowed set of indices for *both* **A** and **B**. So that requires that $l = m = k$, where k is the index of the articulation point. But $B_{k,k} = 0$ because we have defined these matrices to have zero diagonal elements. (The diagonal contributions to the susceptibility were specifically included by the unit matrix.) Thus we have proved the first part of Eq. (16.77). The proof of the second part is similar.

We will now show, using Eq. (16.77) that for a subset diagram Γ which has an articulation point, so that χ is given by Eq. (16.75), then

$$[\chi(\Gamma)]^{-1} = \mathcal{I} - \frac{\mathbf{A}}{\mathcal{I}+\mathbf{A}} - \frac{\mathbf{B}}{\mathcal{I}+\mathbf{B}} \,. \tag{16.79}$$

This relation implies that $M(\Gamma)$ is separable and hence that the cumulant contribution to $M(\Gamma)$ vanishes for diagrams which have an articulation point.

We now derive Eq. (16.79). Using Eq. (16.75) we write

$$\begin{aligned} [\chi(\Gamma)]^{-1} &= (\mathcal{I} + \mathbf{A} + \mathbf{B} + \mathbf{AB} + \mathbf{BA})^{-1} \\ &= ([\mathcal{I} + \mathbf{A}][\mathcal{I} + \mathbf{B}] + \mathbf{BA})^{-1} \,. \end{aligned} \tag{16.80}$$

But using Eq. (16.77) we have

$$\mathbf{BA} = [\mathcal{I} + \mathbf{A}]\mathbf{BA}[\mathcal{I} + \mathbf{B}] \tag{16.81}$$

so that

$$\begin{aligned} [\chi(\Gamma)]^{-1} &= \Big([\mathcal{I} + \mathbf{A}][\mathcal{I} + \mathbf{BA}][\mathcal{I} + \mathbf{B}]\Big)^{-1} \\ &= [\mathcal{I} + \mathbf{B}]^{-1}[\mathcal{I} + \mathbf{BA}]^{-1}[\mathcal{I} + \mathbf{A}]^{-1} \\ &= \left[\mathcal{I} - \frac{\mathbf{B}}{\mathcal{I}+\mathbf{B}}\right][\mathcal{I} - \mathbf{BA}]\left[\mathcal{I} - \frac{\mathbf{A}}{\mathcal{I}+\mathbf{A}}\right] . \end{aligned} \tag{16.82}$$

In the last step we used $\mathbf{BA}(\mathbf{BA})^n = 0$ for $n > 0$ from Eq. (16.77). Thus

$$[\chi(\Gamma)]^{-1} = \mathcal{I} - \frac{\mathbf{A}}{\mathcal{I}+\mathbf{A}} - \frac{\mathbf{B}}{\mathcal{I}+\mathbf{B}} + \frac{\mathcal{I}}{\mathcal{I}+\mathbf{B}}\mathbf{X}\frac{\mathcal{I}}{\mathcal{I}+\mathbf{A}}, \tag{16.83}$$

where

$$\begin{aligned}\mathbf{X} &= \mathbf{BA} - [\mathcal{I}+\mathbf{B}]\mathbf{BA}[\mathcal{I}+\mathbf{A}] + \mathbf{B}^2\mathbf{A}[\mathcal{I}+\mathbf{A}] \\ &\quad +[\mathcal{I}+\mathbf{B}]\mathbf{BA}^2 - \mathbf{B}^2\mathbf{A}^2 = 0\,.\end{aligned} \tag{16.84}$$

Thus we have established Eq. (16.79) and therefore that the contribution to $M(\Gamma)$ from a diagram Γ which has an articulation point is zero. So out of all diagrams, we need consider only the relatively small subset of diagrams which are star graphs.

Of course, we really want the susceptibility, and not its matrix inverse. Note that we are dealing with a translationally invariant system, i.e., one for which $\chi(i, j) = \chi(\mathbf{r}_i - \mathbf{r}_j)$. In this case, the general Fourier transform

$$\chi(\mathbf{q}, \mathbf{q}') \equiv (1/N)\sum_{i,j} e^{i(\mathbf{q}\cdot\mathbf{r}_i - \mathbf{q}'\cdot\mathbf{r}_j)}\chi_{ij} \tag{16.85}$$

is diagonal: i.e., it is nonzero only for $\mathbf{q} = \mathbf{q}'$. Therefore we define the Fourier transformed susceptibility, $\chi(\mathbf{q})$ via

$$\chi(\mathbf{q}) = \sum_i e^{i\mathbf{q}\cdot(\mathbf{r}_i - \mathbf{r}_j)}\chi_{ij}\,. \tag{16.86}$$

The inverse transform is

$$\chi_{ij} = \frac{1}{N}\sum_{\mathbf{q}} e^{-i\mathbf{q}\cdot(\mathbf{r}_i - \mathbf{r}_j)}\chi(\mathbf{q})\,. \tag{16.87}$$

Using Eq. (13.97) one can show that inverse susceptibility matrix is given by

$$[\chi^{-1}]_{ij} = \frac{1}{N}\sum_{\mathbf{q}} e^{-i\mathbf{q}\cdot(\mathbf{r}_i - \mathbf{r}_j)}\chi(\mathbf{q})^{-1}\,. \tag{16.88}$$

From this it follows that

$$N^{-1}\sum_{ij}[\chi^{-1}]_{ij} = \sum_{\mathbf{q}}\delta_{\mathbf{q},0}\chi(\mathbf{q})^{-1} = [\chi(\mathbf{q}=\mathbf{0})]^{-1} = [\sum_i \chi_{ij}]^{-1}\,. \tag{16.89}$$

In other words $\chi \equiv \sum_i \chi_{ij}$ satisfies

$$\chi^{-1} = \frac{1}{N}\sum_{ij}[\chi^{-1}]_{ij} = 1 + \frac{1}{N}M\,, \tag{16.90}$$

where M is the quantity introduced in Eq. (16.76).

In summary, one constructs an expansion for the quantity M of Eq. (16.76) as a sum over contributions from all star graphs, of which those with up to nine bonds on a simple cubic lattice are shown in Fig. 16.10. For each such star graph Γ with n vertices one calculates the n by n matrix $\chi_{i,j}$ (including here the diagonal elements $\chi_{i,i} = 1$). Then for that diagram Γ one inverts the n by n matrix χ, to evaluate $M(\Gamma)$. This is usually called the "bare" value of M. Now use the recursive definition of the cumulant in Eq. (16.17) to form the cumulant value of M. Finally, sum over all star graphs having up to m interactions in order to get the series result up to order λ^m. From the series expansion of M one can get the desired series expansion for χ using Eq. (16.90). We will illustrate this technique in a moment. However, even at this stage one can see a simplification for the Cayley tree. For such a structure the only star graph is the diagram consisting of a single bond. This means that the complete power series for M is given in closed form in terms of the value for a single bond!

16.5 Enumeration of Diagrams

In order to develop series expansions for lattice models, one needs to evaluate $W(\Gamma)$ the number of ways the subset diagram Γ can be placed on the lattice. Here, we assume that we may lump all diagrams with the same topology into a given topology class. For instance, if we were to tabulate $W(\Gamma)$ for a chain of three bonds on the hypercubic lattice, it is obviously important to consider whether or not the Hamiltonian depends on the orientation of the three bonds. If it does not, then all such chains, irrespective of their shape, can be counted as members of the same topology class. If one were dealing, say, with dipole–dipole interactions, then one would have to enumerate chains whose shapes give distinguishable properties separately. However, for models such as Ising models, Heisenberg models, Potts models, and the like, properties depend only on topology and we consider here only such models. We give results for $w(\Gamma) = W(\Gamma)/N$, which is the number of ways *per site* that a diagram topologically equivalent to Γ can be placed on the lattice. (In the literature, these are sometimes called the *weak embedding constants*). In Fig. 16.10 we show star graphs with up to 9 bonds which can occur on a hypercubic lattice in arbitrary spatial dimension, d and in Table 16.1 we give the associated value of $w(\Gamma)$. In Fig. 16.11, we give the additional data for those diagrams with no free ends which have articulation points. Data for diagrams with up 15 bonds are given in Harris and Meir (1840).

16.5.1 *Illustrative Construction of Series*

We illustrate the construction of a series expansion by calculating the susceptibility of the Ising model on a hypercubic lattice in d spatial dimensions correct up to order t^4, where $t \equiv \tanh[J/(kT)]$.

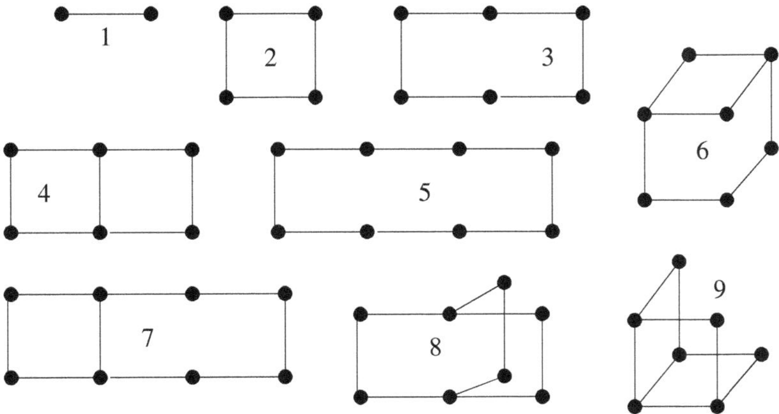

Fig. 16.10 Star graphs on hypercubic lattices with up to nine bonds

Table 16.1 Number of occurrences per lattice site $w(\Gamma)$ for subset diagrams of Figs. 16.10 and 16.11 for a hypercubic lattice in d dimensions

Diagram (Γ)	$w(\Gamma)$
1	d
2	$d(d-1)/2$
3	$(8d^3-21d^2+13d)/3$
4	$2d^3-5d^2+3d$
5	$(54d^4-262d^3+415d^2-207d)/2$
6	$4d^3-12d^2+8d$
7	$32d^4-144d^3+214d^2-102d$
8	$(4d^4-14d^3+14d^2-4d)/3$
9	$(4d^3-12d^2+8d)/3$
10	$2d^4-8d^3+11d^2-5d$
11	$4d^5-16d^4+24d^3-15d^2+4d$

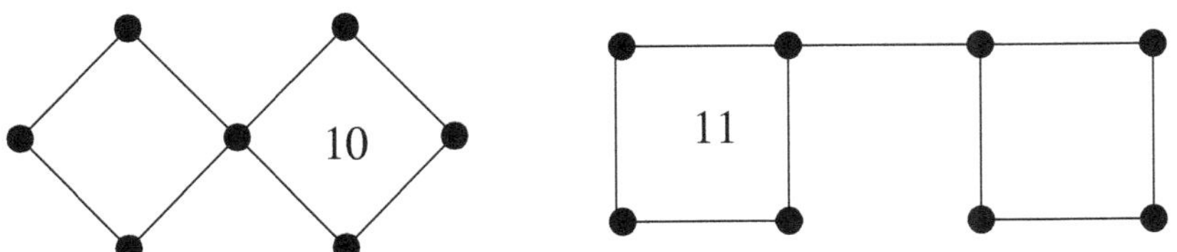

Fig. 16.11 Diagrams on hypercubic lattices with up to nine bonds which have no free ends but which are not star graphs

We have

$$\chi^{-1} = 1 + \sum_{\Gamma} w(\Gamma) M_c(\Gamma) , \tag{16.91}$$

where $w(\Gamma)$ is the number of occurrences *per lattice site* of the subset diagram of topology Γ and $M_c(\Gamma)$ is the cumulant value of M evaluated for the diagram Γ, where M is defined in Eq. (16.76).

To start consider the one bond star graph Γ_1 of Fig. 16.10. The susceptibility matrix for this diagram is

$$\chi(\Gamma_1) = \begin{bmatrix} 1 & t \\ t & 1 \end{bmatrix} . \tag{16.92}$$

We then have the matrix inverse of the susceptibility as

$$\chi^{-1}(\Gamma_1) = \frac{1}{1-t^2} \begin{bmatrix} 1 & -t \\ -t & 1 \end{bmatrix} , \tag{16.93}$$

from which we get

$$\begin{aligned} M(\Gamma_1) &= \sum_{ij} \left([\chi^{-1}(\Gamma_1)]_{ij} - \delta_{i,j} \right) \\ &= \frac{2(1-t)}{1-t^2} - 2 = -\frac{2t}{1+t} . \end{aligned} \tag{16.94}$$

Since this diagram has no subdiagrams, this is also the cumulant value. For a hypercubic lattice in d spatial dimensions, this diagram occurs $w(\Gamma_1) = d$ times per lattice site, so at this order we have

$$\chi^{-1} = 1 - \frac{2dt}{1+t} = \frac{1-(2d-1)t}{1+t} . \tag{16.95}$$

For the Cayley tree, there are no further star graphs, so this is the full answer (when $2d$ is replaced by the coordination number z) and it does agree with exact solutions such as Eq. (14.24).

Now we include the first correction away from the Cayley tree result, by including the contribution from Γ_2 of Fig. 16.10. Here the susceptibility matrix is

$$\chi(\Gamma_2) = \frac{1}{1+t^4} \begin{bmatrix} 1+t^4 & t+t^3 & 2t^2 & t+t^3 \\ t+t^3 & 1+t^4 & t+t^3 & 2t^2 \\ 2t^2 & t+t^3 & 1+t^4 & t+t^3 \\ t+t^3 & 2t^2 & t+t^3 & 1+t^4 \end{bmatrix} \tag{16.96}$$

and its inverse is

$$\chi^{-1}(\Gamma_2) = \frac{1+t^4}{(1-t^2)^2(1+t^2)} \begin{bmatrix} (1+t^2) & -t & 0 & -t \\ -t & (1+t^2) & -t & 0 \\ 0 & -t & (1+t^2) & -t \\ -t & 0 & -t & (1+t^2) \end{bmatrix} \tag{16.97}$$

from which we get

$$\begin{aligned} M(\Gamma_2) &= \sum_{ij} [([\chi^{-1}(\Gamma_2)]_{ij} - \delta_{i,j}) \\ &= -8\frac{t+t^2+t^3}{(1+t)^2(1+t^2)} \,. \end{aligned} \tag{16.98}$$

Now we need to implement the subtraction of cumulants from all lower order subdiagrams according to Eq. (16.17). In this case the only subdiagrams with nonzero cumulants are the four single-bond subdiagrams. Thus

$$\begin{aligned} M(\Gamma_2)_c &= M(\Gamma_2) - 4M(\Gamma_1)_c \\ &= -8\frac{t+t^2+t^3}{(1+t)^2(1+t^2)} + \frac{8t}{1+t} \\ &= \frac{8t^4}{(1+t)^2(1+t^2)} \,. \end{aligned} \tag{16.99}$$

(It is a partial check on our calculations that $M_c(\Gamma_2)$ has no terms of order lower than t^4.) Now the number of occurrences of a square on a d-dimensional hypercubic lattice is $w(\Gamma_2) = d(d-1)/2$, so to order t^4 we have

$$\chi^{-1} = \frac{1-(2d-1)t}{1+t} + 4t^4 d(d-1) \,. \tag{16.100}$$

This gives the series for χ as

$$\begin{aligned} \chi &= 1 + 2dt + 2d(2d-1)t^2 + 2d(2d-1)^2 t^3 \\ &\quad + 2dt^4\left(8d^3 - 12d^2 + 4d + 1\right) + \mathcal{O}(t^5) \,. \end{aligned} \tag{16.101}$$

16.6 Analysis of Series

As mentioned, for models of spins on lattices, the series evaluations have been carried to very high order, with the aim of determining, by extrapolation, the critical exponents at the order–disorder phase transition. (This assumes one is dealing with a continuous phase transition. If the transition is discontinuous, much less useful information about the phase transition is obtained from series expansions.) With a short series (i.e., one of the five or six terms, say), one may hope to get an estimate of the critical point which is more accurate than that from mean field theory. One way to locate the critical point is by constructing a series for a quantity which is known to diverge at the critical point. The susceptibility is usually such a quantity. Then, a common technique of analysis is to use *Padé Approximants*. Having n terms in a series in the coupling constant, say, $\lambda \equiv \beta J$ will allow one to approximate this series by a ratio of an rth degree polynomial $N_r(\lambda)$ divided by an sth degree polynomial $D_s(\lambda)$ where the coefficients in the polynomials are chosen so that the first $r+s$ terms of the power series expansion of $N_r(\lambda)/D_s(\lambda)$ reproduce those of the original series. This scheme requires that $r+s$ be less than or equal to n. The rational function $G_{r,s}(\lambda) \equiv N_r(\lambda)/D_s(\lambda)$ is called the $[r, s]$ *Padé Approximant* to the original series. Then the smallest value of λ for which $D(\lambda) = 0)$ is an estimate for the critical value of the coupling constant. Indeed, one may generate a family of estimates for different values of r and s such that their sum does not exceed n. The central value of these estimates then forms an approximation for the critical value of the coupling constant.

The above scheme is not optimal because we are trying to approximate a function with a noninteger critical exponent by a function which has integer poles and zeros. It would be better to try to approximate a function which has that kind of simple singularity. To do that we use the following approach. Since $\chi = 1$ for $\lambda \equiv \beta J = 0$, one can develop a power series for $\ln \chi(\lambda)$ from that of $\chi(\lambda)$. If χ has a divergence of the form $\chi \sim |\lambda - \lambda_c|^{-\gamma}$, then $d \ln \chi(\lambda)/d\lambda$ has a simple pole at $\lambda = \lambda_c$ with residue $-\gamma$. This scheme has the virtue that the approximant has the same local analytic structure as the function we wish to approximate. So, proceeding as above, we can construct Padé approximants for $d \ln \chi/d\lambda$. In this analysis the critical temperature is taken from the smallest argument at which $D_s(\lambda) = 0$. Then the residue of $N_r(\lambda)/D_s(\lambda)$ at this critical value of λ gives the value of γ. This type of analysis is referred to a "dlog Padé" analysis. Note that, when using the dlog Padé analysis of a star graph series, it is not necessary to evaluate the series for χ itself. Instead one can use the simpler series for χ^{-1} since $\ln \chi = -\ln[\chi^{-1}]$. More sophisticated analyses of series expansions are possible, but the dlog Padé method is a reasonable one.

16.7 Summary

A technique to generate series expansions for many-body systems is given by Eq. (16.10), where F_n (the contribution from subset diagrams containing n perturbative interactions) is given in terms of cumulants as written in Eq. (16.14). The cumulants are best obtained recursively via Eq. (16.17). For two-point correlation functions of classical systems a star graph expansion (in which appear only graphs with no articulation point) is often useful. For higher order susceptibilities of classical system, for which there may not be a star graph expansion, one may have recourse to the no-free-end method described in the chapter on the Cayley tree. For quantum systems, there appears to be no such simplification. Here we discuss only general methods. For any specific problem there may well be a special simplification peculiar to that problem. A popular method of analyzing such series is based on the use of Padé approximants. In particular, one fits the derivative of the logarithm of a divergent susceptibility to a ratio of polynomials. Then the location of the pole in this approximant gives an estimate for the location of the critical point and the residue gives the associated critical exponent.

16.8 Exercises

1. Show, by explicit evaluation, that the contribution from diagram b of Fig. 16.1, which is given in Eq. (16.33) is zero.

2. This problem concerns the resistance correlation function for percolation clusters. We define the resistive "susceptibility" χ_R via $\chi_R \equiv \sum_j \chi_R(i, j) = \sum_j \langle \nu_{ij} R_{ij} \rangle_p$, where ν_{ij} is the pair-connectedness function and R_{ij} is the resistance between sites i and j when a unit resistance is attributed to each occupied bond. (When the points i and j are in different clusters, $\nu_{ij} R_{ij}$ is interpreted to be zero.) It is desired to use a star graph expansion to get an expansion for χ_R. However, the resistance of a cluster with an articulation point does not obey the product rule of Eq. (16.67). Show that if we define

$$\chi(i, j; \lambda) \equiv \langle \nu_{ij} e^{-\lambda R_{ij}} \rangle_p , \tag{16.102}$$

then for this quantity Eq. (16.67) is satisfied. Evaluate the star graph expansion for $\chi(\lambda) \equiv \sum_j \chi(0, j; \lambda)$, where 0 labels the seed site. From this quantity obtain the result for the resistive susceptibility which is exact for the Cayley tree.

3. Obtain the expansion for the susceptibility of an Ising model correct up to order t^6 for a hypercubic lattice in d dimensions.

4. Write explicitly the entire series in Eq. (16.10) for a function of four variables, $F(\lambda_1, \lambda_2, \lambda_3, \lambda_4)$.

5. Recall that $\exp(\beta J \sigma_i \sigma_j)$, where $\sigma_i = \pm 1$, can be written as $\cosh(\beta J) + \sigma_i \sigma_j \sinh(\beta J)$. Thus the part of the partition function due to interactions can be written as

$$\hat{Z} = \prod_{\langle ij \rangle} \left(1 + \lambda_{ij} t \sigma_i \sigma_j \right) , \tag{16.103}$$

where $t = \tanh(\beta J)$. You are to calculate the cumulant value $(\hat{Z})_c$ for the diagram Γ shown in Fig. 16.12. Even though the diagram Γ is disconnected, $(\hat{Z})_c$ is not zero because $\hat{Z}$ is not separable in the sense of Eq. (16.18).

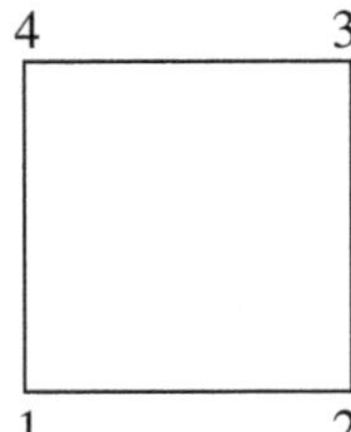

Fig. 16.12 The diagram Γ for which the calculations of Prob. 5 are to be done

To start you off, we will give the bare value, $\hat{Z}(\Gamma)$:

$$\hat{Z}(\Gamma) = \left(1 + \lambda_{12}\lambda_{23}\lambda_{34}\lambda_{41}t^4\right)\left(1 + \lambda_{56}\lambda_{67}\lambda_{78}\lambda_{85}t^4\right) .$$

6. Consider the one-dimensional chain of spins 1/2 governed by the Hamiltonian

$$\mathcal{H} = -H \sum_i \sigma_{i,z} + J \sum_i \sigma_{i,x}\sigma_{i+1,x} , \tag{16.104}$$

where $\sigma_{i,\alpha}$ is the α-component Pauli matrix for the ith spin. Obtain an expansion for the ground state energy for such a system of N spins in the limit $N \to \infty$ as a power series in J/H for small J correct up to and including order $(J/H)^4$. Do this by using Eq. (16.14) where the function F is identified for this problem as the ground state energy.

7. Show that Eq. (16.17) is a consequence of Eq. (16.15). Hint: do this by induction. Equation (16.17) is true for $n = 1$. Assume it to be true for $n \leq n_0$. Then show (with this assumption) that Eq. (16.17) holds for $n = n_0 + 1$.

8. (**a**) Show that Eq. (16.67) is obeyed by the classical Heisenberg model for which

$$\mathcal{H} = -J \sum_{\langle ij \rangle} \mathbf{S}_i \cdot \mathbf{S}_j , \tag{16.105}$$

where **S** is a classical unit vector and the susceptibility is $\chi(k, l) = \langle \mathbf{S}_k \cdot \mathbf{S}_l \rangle_T$.

(**b**) Obtain the high-temperature expansion for χ correct up to and including order $[\beta J]^4$.

9. Consider the quenched random Ising system with only nearest-neighbor interactions. For a given set of exchange constants $\{J_{ij}\}$) the Hamiltonian is

$$\mathcal{H}(\{J_{ij}\}) = -\sum_{\langle ij \rangle} J_{ij} S_i S_j ,$$

where $S_i = \pm 1$. Since we consider quenched randomness, one is supposed to calculate the free energy as a function of the J_{ij}'s and then average over the probability distribution of the J_{ij}'s. Thus the susceptibility is defined to be the configurationally averaged quantity

$$\chi \equiv \int \chi(\{J_{ij}\}) \prod_{\langle ij \rangle} P(J_{ij}) dJ_{ij} \, .$$

As indicated, it is assumed that each nearest-neighbor bond is an independent random variable governed by the same probability distribution function $P(J)$. Show that the susceptibility of this model obeys Eq. (16.67) and therefore that the star graph expansion can be used for the quenched random Ising model.

10. In Eq. (14.54), we developed an expansion in which only diagrams with no free ends contribute to the partition function and later on we determined the auxiliary function g_i correct up to order H^2.

(**a**) What diagram on the hypercubic lattice in d dimensions gives the lowest-order (in V) nonzero correction to the result for the Cayley tree.

(**b**) Evaluate this correction so as to obtain the analogous correction to the susceptibility. You should thereby reproduce the result of Eq. (16.101).

11(**a**) By constructing a high-temperature expansion for the susceptibility we were able to develop an estimate of its critical exponent γ. Superficially it might seem impossible to develop an analogous estimate for the order-parameter exponent β. However, consider the higher order susceptibility $\chi^{(3)}(T) \equiv -\partial^4 F(H,T)/\partial H^4)_T$ evaluated at $H = 0$ and define the critical exponent γ_3 by $\chi^{(3)}(T) \sim |T - T_c|^{-\gamma_3}$ for $T \to T_c$. Use a scaling argument to express the critical exponent γ_3 in terms of β and γ.

(**b**) Obtain the first three nonzero terms in the high-temperature expansion for $\chi^{(3)}(T)$ for a nearest-neighbor ferromagnet Ising model on a simple cubic lattice. (Part (a) indicates that an analysis using a much longer series will give an estimate for the value of β.)

16.9 Appendix: Analysis of Nonideal Gas Series

Here we discuss the terms in Eq. (16.39) involving $Y(\mathcal{S})^p$ for $p > 1$. To illustrate the discussion we consider the set $\mathcal{S}$ corresponding to the diagram of Fig. 16.13.

To make the diagrammatic analysis we note that a contribution involving $Y(\mathcal{S})^p$ is of the form

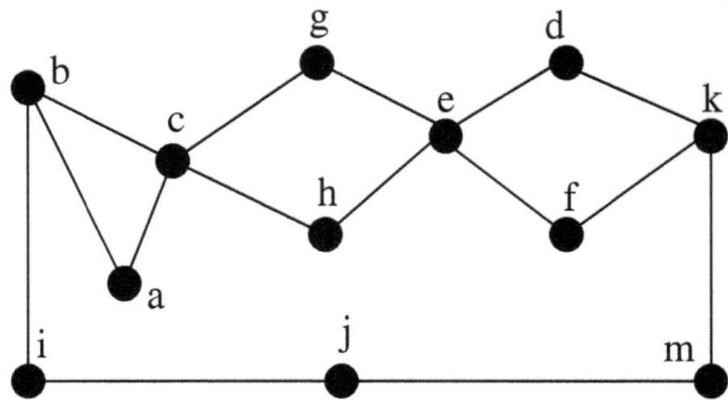

Fig. 16.13 Diagrammatic representation for an arbitrarily chosen $Y(\mathcal{S})_c$

$$\left(\prod_{f\in\Gamma_1} f(r_{ij}^{(1)}) \prod_{i\in\Gamma_1} \frac{d\mathbf{r}_i^{(1)}}{V}\right) \times \left(\prod_{f\in\Gamma_2} f(r_{ij}^{(2)}) \prod_{i\in\Gamma_2} \frac{d\mathbf{r}_i^{(2)}}{V}\right) \dots$$
$$\times \left(\prod_{f\in\Gamma_p} f(r_{ij}^{(p)}) \prod_{i\in\Gamma_p} \frac{d\mathbf{r}_i^{(p)}}{V}\right), \qquad (16.106)$$

where Γ_i is a subset (including Γ itself) of Γ. In $Y(\mathcal{S})^p$ we have to introduce integration variables $\mathbf{r}_i^{(k)}$ for $k = 1, 2, \dots p$. Maybe it helps to think of the integrals over the $\mathbf{r}^{(k)}$'s as giving rise to a diagram in the kth level. In this diagram, for the kth level we draw a solid line connecting points i and j for each factor of $f(r_{ij}^{(k)})$. The same lines and/or the same particle labels can appear in more than one level.

The only constraint is that the term is connected, which here means that all lines and vertices of $\mathcal{S}$ must appear in at least one level. In any given level the diagram may be disconnected. But if we connect points with the same site label which are in different levels, then the diagram must be connected to contribute. To be explicit, let us connect sites with the same particle label but which are in different levels by dashed lines as follows. If, say, site i appears in levels, $n_1 < n_2 < n_3 < n_4 < \cdots n_k$, then we connect site i in level n_1 to site i in level n_2, and also site i in level n_2 to site i in level n_3, and so forth up to making a connection between site i in level n_{k-1} to site i in level n_k. This means that what we have is a diagram like that shown above: on each level we have a collection of disconnected subdiagrams. But when the connections between points with the same site labels which are in different levels are taken into account, the diagram is connected. In addition, as we have already said, we only need consider diagrams that do not have an articulation point. This means that, except for diagrams having only a single level, the number of interlevel dashed lines has to be at least as large as the number of connected components (of which there may be several in any given level). In Fig. 16.14 we show a diagram which corresponds to the following contribution to $Y(\mathcal{S})^4$:

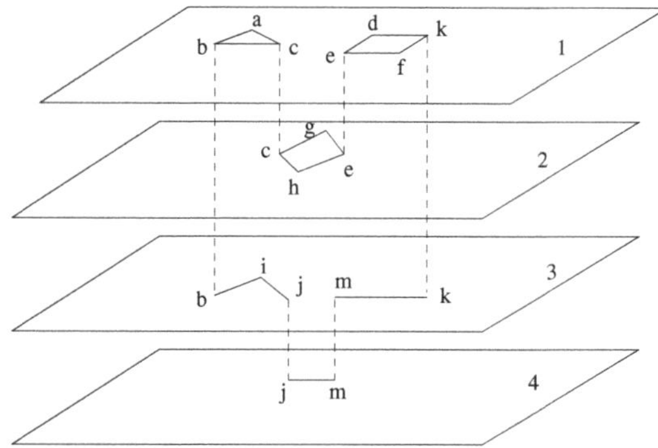

Fig. 16.14 Diagrammatic representation of a term arising from $Y(\mathcal{S})^4$, where $Y(\mathcal{S})$ is that shown in Fig. 16.13. At the right we label the levels as 1, 2, …4. Here there are $N_X = 6$ clusters (two clusters on level 1, one on level 2, two on level 3, and one on level 4), $N_{IL} = 6$ interlevel dashed lines, and $N_s = 12$ sites (a, b, c, …m)

$$\left(V^{-7} \int f(r_{ab}) f(r_{bc}) f(r_{ac}) f(r_{df}) f(r_{fk}) f(r_{ke}) f(r_{ed}) \times dr_a dr_b dr_c dr_d dr_e dr_f dr_k\right)$$
$$\times \left(V^{-4} \int f(r_{cg}) f(r_{ge}) f(r_{eh}) f(r_{hc}) dr_c dr_g d_h dr_e)\right)$$
$$\times \left(V^{-5} \int f(r_{bi}) f(r_{ij}) f(r_{mk}) dr_b dr_i dr_j dr_m dr_k\right)$$
$$\times \left(V^{-2} \int f(r_{jm}) dr_j dr_m\right) . \tag{16.107}$$

Now let us count the number of factors of N and V from such a diagram. The number of position coordinates integrated over is equal to the number of sites, N_s in $\mathcal{S}$ *plus* the number of interlevel lines, N_{IL}. Thereby we take account of the fact that a site (e.g., b) can be integrated over more than once (twice). Each such coordinate carries a factor $1/V$, so the powers of $1/V$ give

$$V^{-N_s - N_{IL}} . \tag{16.108}$$

From the combinatorial factor $N!/(N - N_s)!$ we get

$$N^{N_s} . \tag{16.109}$$

Finally we we integrate over the position variables in a given level we have one factor of V for the center-of-mass coordinate for each connected components within the level in question. If there are N_X such connected components, this gives a factor

$$V^{N_X} , \tag{16.110}$$

So in all we have

$$N^{N_X - N_{IL}} \tag{16.111}$$

times constants (for the symmetry factors) and powers of the density (to convert V to N). If we are speaking about the term involving $Y(\mathcal{S})^1$, then $N_{IL} = 0$ and $N_X = 1$ because we keep the term with a single connected component on one level. This gives an extensive contribution and it is the contribution we have already kept.

When there is more than one level, then N_{IL} has to be at least as large as N_X, because each connected subdiagram has to be connected to the rest of the diagram by at least *two* interlevel lines. If this were not so, the vertex at which one line entered the subdiagram would give an articulation point. If N_{IL} is at least as large as N_X, then by Eq. (16.111) we are talking about a contribution of order N^0 or smaller, which we may neglect in the thermodynamic limit. This argument shows that we only need to consider the term involving $Y(\mathcal{S})^1$ in Eq. (16.39).

References

G.A. Baker Jr., *Quantitative Theory of Critical Phenomena* (Academic Press, 1990)

A.B. Harris, Y. Meir, Recursive enumeration of clusters in general dimension on hypercubic lattices. Phys. Rev. A **36**, 1840 (1987)

J. Oitmaa, C. Hamer, W. Zheng, *Series Expansion Methods for Strongly Interacting Lattice Models* (Cambridge University Press, 2006)

R.K. Pathria, *Statistical Mechanics*, 2nd edn. (Butterworth-Heinemann, 1996)

Chapter 17
The Ising Model: Exact Solutions

This chapter further supports the case for the Ising spin as the Drosophila of statistical mechanics, that is the system that can be used to model virtually every interesting thermodynamic phenomenon and to test every theoretical method. Exact solutions are extremely rare in many-body physics. Here, we consider three of them: (1) the one-dimensional Ising model, (2) the one-dimensional Ising model in a transverse field, the simplest quantum spin system, and (3) the two-dimensional Ising model in zero field. The results are paradigms for a host of more complex systems and situations that cannot be solved exactly but which can be understood qualitatively and even quantitatively on the basis of simulations and asymptotic methods such as series expansions and the renormalization group.

17.1 The One-Dimensional Ising Model

The Ising model was first formulated by Lenz (1920). The one-dimensional case was solved by Lenz's student, Ernst Ising, in his 1924 Ph.D. thesis (Ising 1925). Ising's solution demonstrated that spontaneous magnetization does not occur in the one-dimensional model.

The Hamiltonian for the one-dimensional Ising model is simply

$$\mathcal{H} = -J\sum_i S_i S_{i+1} - H\sum_i S_i \ . \tag{17.1}$$

We first consider the case with $H = 0$ and N spins in a chain with free boundary conditions. Then the partition function is

$$Z_N = \text{Tr}e^{-\beta\mathcal{H}} = \sum_{S_1=\pm 1} \dots \sum_{S_N=\pm 1} e^{\beta J\sum_{i=1}^{N-1} S_i S_{i+1}} \ . \tag{17.2}$$

A. J. Berlinsky and A. B. Harris, *Statistical Mechanics*, Graduate Texts in Physics,
https://doi.org/10.1007/978-3-030-28187-8_17

We do the trace over S_N first

$$\sum_{S_N=\pm 1} e^{\beta J S_{N-1} S_N} = 2\cosh \beta J. \tag{17.3}$$

Then

$$Z_N = 2\cosh \beta J \sum_{S_1=\pm 1} \cdots \sum_{S_{N-1}=\pm 1} e^{\beta J \sum_{i=1}^{N-2} S_i S_{i+1}} . \tag{17.4}$$

Iterating, we obtain

$$Z_N = (2\cosh \beta J)^{N-2} Z_2 , \tag{17.5}$$

where

$$Z_2 = \sum_{S_1=\pm 1} \sum_{S_2=\pm 1} e^{\beta J S_1 S_2} = 4\cosh \beta J . \tag{17.6}$$

so that

$$Z_N = 2(2\cosh \beta J)^{N-1} . \tag{17.7}$$

The free energy per spin is then

$$f = -\frac{kT}{N} \ln Z_N = -\frac{kT}{N} \ln 2 - \frac{kT(N-1)}{N} \ln(2\cosh \beta J) \tag{17.8}$$

and

$$f \to -kT \ln(2\cosh \beta J) \quad \text{as } N \to \infty . \tag{17.9}$$

This result is perfectly analytic as a function of T. That is, there is no phase transition.

17.1.1 Transfer Matrix Solution

Consider $H \neq 0$ and periodic boundary conditions so that $S_{N+1} = S_1$. Then

$$\mathcal{H} = -\sum_{i=1}^{N} \Big[J S_i S_{i+1} + H(S_i + S_{i+1})/2 \Big] \tag{17.10}$$

and

$$\begin{aligned} Z_N &= \sum_{\{S_i\}} \prod_i e^{\beta [J S_i S_{i+1} + H(S_i + S_{i+1})/2]} \\ &= \sum_{\{S_i\}} P(S_1, S_2) P(S_2, S_3) \ldots P(S_N, S_1) \\ &= \mathrm{Tr}(P^N) , \end{aligned} \tag{17.11}$$

where the 2×2 matrix P has elements

$$\begin{aligned} P(1,1) &= e^{\beta(J+H)} \quad P(1,-1) = e^{-\beta J} \\ P(-1,1) &= e^{-\beta J} \quad P(-1,-1) = e^{\beta(J-H)} . \end{aligned} \tag{17.12}$$

Since the trace is invariant under unitary transformations, the easiest way to calculate $\mathrm{Tr}\, P^N$ is to diagonalize P and find its eigenvalues, λ_1 and λ_2. Then

$$Z_N = \lambda_1^N + \lambda_2^N . \tag{17.13}$$

Note that only the larger of the two eigenvalues is required for the limit $N \to \infty$. In that case, for $\lambda_1 > \lambda_2$,

$$\begin{aligned} f &= -\frac{kT}{N} \ln(\lambda_1^N + \lambda_2^N) \to -kT \ln \lambda_1 - \mathcal{O}\left[\frac{T}{N}\left(\frac{\lambda_2}{\lambda_1}\right)^N\right] \\ &= -kT \ln \lambda_1 , \end{aligned} \tag{17.14}$$

as $N \to \infty$.

Note that λ_1 and λ_2 satisfy the secular equation

$$\begin{aligned} \mathrm{Det}(P - \lambda I) &= (\lambda - e^{\beta(J+H)})(\lambda - e^{\beta(J-H)}) - e^{-2\beta J} \\ &= \lambda^2 - 2\lambda e^{\beta J} \cosh \beta H + 2 \sinh 2\beta J = 0 . \end{aligned} \tag{17.15}$$

The solutions are

$$\begin{aligned} \lambda_{1,2} &= e^{\beta J} \cosh \beta H \pm \sqrt{e^{2\beta J} \cosh^2 \beta H - 2 \sinh 2\beta J} \\ &= e^{\beta J} \cosh \beta H \pm \sqrt{e^{2\beta J} \sinh^2 \beta H + e^{-2\beta J}} . \end{aligned} \tag{17.16}$$

Then for $N \to \infty$,

$$f = -kT \ln\left[e^{\beta J} \cosh \beta H + \sqrt{e^{2\beta J} \sinh^2 \beta H + e^{-2\beta J}}\right] . \tag{17.17}$$

The magnetization per spin $m = -\partial f / \partial H$ is

$$\begin{aligned} m &= \frac{\left[e^{\beta J} \sinh \beta H + \frac{e^{2\beta J} \sinh \beta H \cosh \beta H}{\sqrt{e^{2\beta J} \sinh^2 \beta H + e^{-2\beta J}}}\right]}{e^{\beta J} \cosh \beta H + \sqrt{e^{2\beta J} \sinh^2 \beta H + e^{-2\beta J}}} \\ &= \frac{\sinh \beta H}{\sqrt{\sinh^2 \beta H + e^{-4\beta J}}} . \end{aligned} \tag{17.18}$$

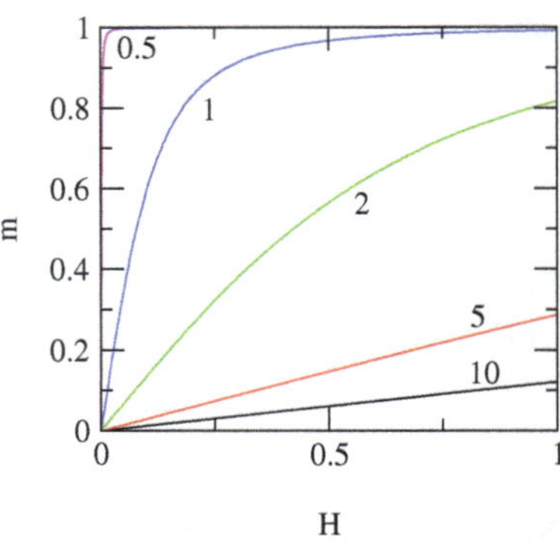

Fig. 17.1 m versus H curves of the classical Ising chain for the values of kT indicated with $J = 1$. We only show results for positive H. $m(-H) = -m(H)$

Figure 17.1 shows a plot of m versus H for various temperatures. Note that $m = 0$ for $H = 0$ at all temperatures, but that $m \approx 1$ for $\sinh \beta H > e^{-2\beta J}$, i.e., for $H > kTe^{-2\beta J}$ which can be very small at low T. Thus at low T, m versus H looks almost like a step function, going from -1 to $+1$ as H crosses zero from below.

17.1.2 Correlation Functions

Consider $H = 0$ for the open chain. We want to calculate $\langle S_i S_{i+n} \rangle$ for $1 \le n \le N - i$. (We could use our previous calculation for a Cayley tree, setting $z = 2$. However, for variety we carry out the calculation in a different way here.) For the present calculation we write

$$\mathcal{H}(J_1, \ldots, J_N) = \sum_{i=1}^{N-1} J_i S_i S_{i+1} . \tag{17.19}$$

Then

$$Z_N(J_1, \ldots, J_N) = 2 \prod_{i=1}^{N-1} (2 \cosh \beta J_i) , \tag{17.20}$$

where all of the J_i will be set equal to J at the end of the calculation.

Also note that, since $S_i^2 = 1$,

$$\begin{aligned}\langle S_i S_{i+n} \rangle &= Z_N^{-1} \sum_{\{S_i\}} (S_i S_{i+n}) e^{\beta \sum_j J_j S_j S_{j+1}} \\ &= Z_N^{-1} \sum_{\{S_i\}} (S_i S_{i+1})(S_{i+1} S_{i+2}) \\ &\qquad \ldots (S_{i+n-1} S_{i+n}) e^{\beta \sum_j J_j S_j S_{j+1}} .\end{aligned} \tag{17.21}$$

We can write this in terms of derivatives with respect to the J_k for $k = i, \ldots, i + n - 1$, evaluated for J_i equal to J.

$$\begin{aligned}
\langle S_i S_{i+n} \rangle &= Z_N^{-1} \beta^{-n} \left[\frac{\partial^n}{\partial J_i \partial J_{i+1} \ldots \partial J_{i+n-1}} Z_N(J_1, J_2, \ldots, J_{N-1}) \right]_J \\
&= Z_N^{-1} \left[Z_N(J_1, J_2, \ldots, J_{N-1}) \prod_{j=i}^{i+n-1} \frac{\sinh \beta J_j}{\cosh \beta J_j} \right]_J \\
&= \Big(\tanh \beta J \Big)^n .
\end{aligned} \tag{17.22}$$

Since $|\tanh x| < 1$, this function decays exponentially with increasing n at all T. At low T, $\tanh \beta J \approx 1 - 2e^{-2\beta J}$, so

$$\langle S_i S_{i+n} \rangle = e^{n \ln \tanh \beta J} \to e^{-n(2/e^{2\beta J})} \equiv e^{-n/\xi} \tag{17.23}$$

and the correlation length diverges as

$$\xi(T) \to \frac{e^{2\beta J}}{2} \quad \text{as } \beta \to \infty . \tag{17.24}$$

Thus, although there is no long-range order at any finite temperature, the correlation length diverges rapidly with decreasing T. The energy in the exponent, $2J$, is the energy cost of an interface between a region of up and down spins.

17.2 Ising Model in a Transverse Field

We can think of the Ising spin S_i as the z-component of a quantum spin 1/2, S_i^z, an operator that has eigenvalues $\pm 1/2$. As long as the Hamiltonian contains only operators, S_i^z, $i = 1, \ldots N$, then everything commutes, and the problem is effectively classical. To simplify the algebra in this section and eliminate a proliferation of factors of 2, we instead work with the Pauli operators, $\vec{\sigma}_i = 2\vec{S}_i$. The eigenvalues of σ_i^z are ± 1, exactly like those of the Ising operators, S_i.

For the standard case of ferromagnetic interactions between neighboring spins, the Hamiltonian is, as usual,

$$\mathcal{H} = -J \sum_{(i,j)} \sigma_i^z \sigma_j^z \qquad\qquad J > 0 , \tag{17.25}$$

which favors parallel spin alignment along one of the $\pm\hat{z}$ directions.

The situation changes dramatically if we apply a magnetic field along the x-direction so that

$$\mathcal{H} = -J \sum_{(i,j)} \sigma_i^z \sigma_j^z \; - H \sum_i \sigma_i^x . \tag{17.26}$$

The effect of the operator σ_i^x is to flip an "up" spin down and a "down" spin up. Thus the lowest energy eigenstate of the operator $-H\sigma_i^x$ for spin i is the symmetric linear superposition of "up" and "down", $\frac{1}{\sqrt{2}}[| \uparrow\rangle + | \downarrow\rangle]$, which has energy eigenvalue, $-H$.

This interaction competes with the effect of the spin–spin interaction which favors the states in which either all spins are $| \uparrow\rangle$ or all are $| \downarrow\rangle$. In this sense it is like the kinetic energy of the interacting particle problem, which competes with the potential energy. Typically, interactions favor static arrangements in which the particles are localized in a way that minimizes their potential energy, whereas the kinetic energy favors states in which the particles are delocalized. In quantum mechanics, this competition is reflected in the fact that the kinetic and potential energies do not commute. Similarly, for the Ising model in a transverse field, the spin–spin interaction and the transverse field term do not commute.

17.2.1 Mean-Field Theory ($T = 0$)

For $d > 1$ we expect to find that long-range order ($\langle \sigma_i^z \rangle$) is nonzero for H and T sufficiently small, as indicated in the left panel of Fig. 17.2. (For a mean-field treatment of this phase diagram, see Exercise 3.) For $d = 1$ we know that long-range order does not occur for nonzero T for $H = 0$. Accordingly, we now study the case of $T = 0$ for $d = 1$ to see if there is an order–disorder phase transition as a function of H. We will find such a transition, so that the phase diagram for $d = 1$ is that in the right panel of Fig. 17.2.

We now develop mean-field theory for $T = 0$. Consider an approximate solution of the mean-field type in which the ground state wave function is assumed to be a direct product of single-particle states. Since we want to allow competition between the state for $H = 0$ in which the spins lies along $\hat{z}$ and the state for $J = 0$ in which the spins lie along $\hat{x}$, we put each spin in the state $|\theta\rangle$, where

$$|\theta\rangle = \cos\frac{\theta}{2} | \uparrow\rangle + \sin\frac{\theta}{2} | \downarrow\rangle . \tag{17.27}$$

For this state

$$\begin{aligned} \langle \sigma^z \rangle &= \cos\theta \\ \langle \sigma^x \rangle &= \sin\theta \end{aligned} \tag{17.28}$$

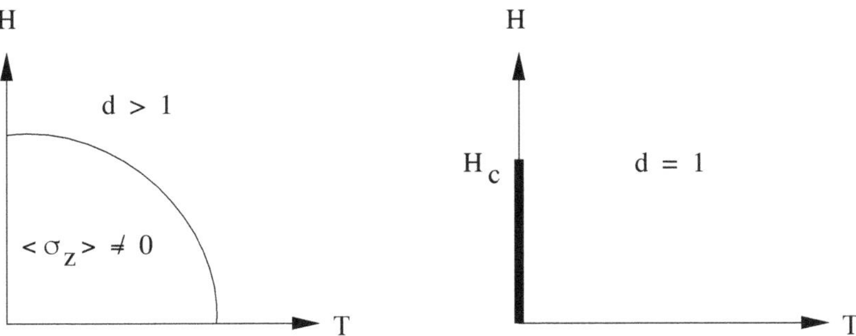

Fig. 17.2 Phase diagram for the Ising model in a transverse field H at temperature T for spatial dimension d greater than 1 (left) and $d = 1$ (right). For $d = 1$ long-range order only exists for $T = 0$

and the mean-field energy of the Hamiltonian of Eq. (17.26) is

$$\langle \mathcal{H} \rangle = -\frac{JzN}{2} \cos^2 \theta - HN \sin \theta \, , \tag{17.29}$$

where z is the coordination number. This result may be obtained less intuitively by taking the density matrix of the ith spin to be

$$\rho_i = \begin{bmatrix} \cos^2(\theta/2) & \cos(\theta/2)\sin(\theta/2) \\ \cos(\theta/2)\sin(\theta/2) & \sin^2(\theta/2) \end{bmatrix} . \tag{17.30}$$

Minimizing the energy with respect to θ, we find that

$$\begin{aligned} \sin\theta &= \frac{H}{Jz} \implies \frac{\langle \mathcal{H} \rangle}{N} = -\frac{Jz}{2} - \frac{H^2}{2Jz} \quad \text{for} \quad \frac{H}{Jz} < 1, \\ \sin\theta &= 1 \quad \implies \frac{\langle \mathcal{H} \rangle}{N} = -H \qquad\qquad \text{for} \quad \frac{H}{Jz} > 1 \, . \end{aligned} \tag{17.31}$$

Thus, there is a phase transition in which the spins start to develop a nonzero value of $\langle \sigma^z \rangle$ as the field is reduced below the critical value, $H_c = Jz$. We will obtain an exact solution for this system in $d = 1$ below.

17.2.2 Duality

As a prelude to finding the exact solution to this one-dimensional problem, we derive an exact mapping between weak and strong coupling regimes. Such a relation is an example of what is called "duality." To start we recall various relations obeyed by the spin operators, σ_i^α, $\alpha = x, y, z$.

1. The square of any spin operator is unity:

$$\left(\sigma_i^\alpha\right)^2 = I \; , \tag{17.32}$$

where I is a unit operator.
2. Spin operators for different sites commute:

$$\left[\sigma_i^\alpha, \sigma_j^\beta\right] = 0 \quad \text{for } i \neq j \; . \tag{17.33}$$

3. Spin operators on the same site anticommute:

$$\sigma_i^\alpha \sigma_i^\beta = -\sigma_i^\beta \sigma_i^\alpha \quad \text{for } \alpha \neq \beta \; . \tag{17.34}$$

4. Spin operators on the same site obey the angular momentum algebra:

$$\sigma_i^x \sigma_i^y = i\sigma_i^z \quad \text{and cyclic permutations} \; . \tag{17.35}$$

These four relations uniquely define the set of operators, $\vec{\sigma}_i$.

Next we consider a set of operators defined on the bonds between neighboring sites, the so-called "dual" lattice. We define

$$\tau_i^x = \sigma_i^z \sigma_{i+1}^z \tag{17.36a}$$

$$\tau_i^z = \prod_{j \le i} \sigma_j^x \; . \tag{17.36b}$$

The operator τ_i^x measures whether the spins on sites i and $i+1$ are parallel or antiparallel, while the operator τ_i^z flips all of the spins to the left of site $i+1$. Starting from a fully aligned state, τ_i^z would create a defect on the ith bond.

There is a problem with this transformation that has to do with boundary conditions. For example, for open boundary conditions, $i = 1, \ldots N$, there is one less bond than site, so that τ_N^x is not defined. For periodic boundary conditions, the problem is that $\tau_{N+1}^z \neq \tau_1^z$. We will sidestep this problem by assuming that N is sufficiently large that boundary effects of order $1/N$ can be ignored. Except for problems at the boundaries, it is straightforward to show that τ_i^x, τ_i^z, and their partners

$$\tau_i^y = i\tau_i^x \tau_i^z \tag{17.37}$$

obey the same algebra, Eqs. (17.32)–(17.35), as the $\{\sigma_i^\alpha\}$.

Using the transformations, (17.36a) and (17.36b), in the Hamiltonian, Eq. (17.26), we obtain

$$\mathcal{H} = -J \sum_i \tau_{i+1}^x - H \sum_i \tau_i^z \tau_{i+1}^z . \tag{17.38}$$

Comparing Eqs. (17.38) and (17.26) we can conclude that the energy eigenvalue spectrum of $\mathcal{H}$ is invariant under interchange of J and H. Of course the eigenfunction associated with a particular eigenvalue will, in general, be different when J and H are interchanged. However quantities such as the free energy depend only on a sum over energy eigenvalues. Therefore, we can write

$$F(H, J, T) = F(J, H, T), \tag{17.39}$$

or, if we define $f(H/J, T/J)$ by $F(H, J, T) = Jf(H/J, T/J)$, then we can write

$$Jf(H/J, T/J) = Hf(J/H, T/H) \tag{17.40a}$$
$$f(H/J, 0) = (H/J)f(J/H, 0)\,, \tag{17.40b}$$

where the last relation applies only at $T = 0$.

The mean-field calculation done above, suggests that there is a kind of phase transition for this system at $T = 0$ as the transverse field is reduced from some large value toward zero. The way that we use the duality symmetry and the relation (17.40b) above is to argue that, if there is a single phase transition point defined by some ratio of H to J, then it must map into itself under Eq. (17.40b). This will only happen for $H = J$, which is thus the expected critical field, H_c. Note that the value H_c that we obtain from this duality argument is half the result that we obtained from the mean-field calculation, $H_c = Jz$, where $z = 2$ for $d = 1$. This reduction of H_c due to fluctuations neglected in mean-field theory is analogous to the similar reduction in T_c which is found in thermal phase transition when fluctuations are taken into account.

17.2.3 Exact Diagonalization

The Hamiltonian of Eq. (17.26) can be diagonalized exactly by a transformation to fermion operators. However, first, we perform a rotation of the coordinate system in spin space around the y-axis which rotates x into z and z into $-x$. Then the Hamiltonian becomes

$$\mathcal{H} = -J\sum_{i=1}^{N-1} \tilde{\sigma}_i^x \tilde{\sigma}_{i+1}^x + V - H\sum_{i=1}^{N} \tilde{\sigma}_i^z\,, \tag{17.41}$$

where V is zero for free-end boundary conditions and $V = J\tilde{\sigma}_N^x\tilde{\sigma}_1^x$ for periodic boundary conditions. Henceforth, we assume periodic boundary conditions and consider a chain with an *even* number of sites. In this new coordinate system, we define the raising and lowering operators

$$\tilde{\sigma}_j^+ = \frac{\tilde{\sigma}_j^x + i\tilde{\sigma}_j^y}{2} \tag{17.42a}$$

$$\tilde{\sigma}_j^- = \frac{\tilde{\sigma}_j^x - i\tilde{\sigma}_j^y}{2} . \tag{17.42b}$$

Since a spin 1/2 can be raised or lowered at most once, they satisfy

$$\left(\tilde{\sigma}_j^+\right)^2 = \left(\tilde{\sigma}_j^-\right)^2 = 0 . \tag{17.43}$$

The Hamiltonian can then be written as

$$\mathcal{H} = -J \sum_{i=1}^{N} \left(\tilde{\sigma}_i^+ + \tilde{\sigma}_i^-\right)\left(\tilde{\sigma}_{i+1}^+ + \tilde{\sigma}_{i+1}^-\right) - H \sum_{i=1}^{N} \tilde{\sigma}_i^z , \tag{17.44}$$

where $\sigma_{N+1}^{\pm} = \sigma_1^{\pm}$.

The fact that a spin cannot be lowered twice suggests a connection to Fermi statistics. The squares of fermion operators are also zero. We can transform this problem to one of interacting, spinless fermions with the help of what is called a Jordan–Wigner transformation. The Jordan–Wigner transformation may be written in a number of different forms. Here we write it as

$$c_n = (-1)^n P_{n-1} \tilde{\sigma}_n^- \tag{17.45a}$$

$$c_n^+ = (-1)^n \tilde{\sigma}_n^+ P_{n-1} , \tag{17.45b}$$

where $P_0 = 1$ and $P_n = \prod_{i=1}^{n} \tilde{\sigma}_i^z$. These operators obey the identities

$$c_n^+ c_n = \tilde{\sigma}_n^+ \tilde{\sigma}_n^- = \frac{1}{2}\left(1 + \tilde{\sigma}_n^z\right) \tag{17.46a}$$

$$c_n c_n^+ = \tilde{\sigma}_n^- \tilde{\sigma}_n^+ = \frac{1}{2}\left(1 - \tilde{\sigma}_n^z\right) , \tag{17.46b}$$

so that

$$c_n^+ c_n + c_n c_n^+ = 1 . \tag{17.47}$$

Similarly, using the fact that different Pauli operators on the same site anticommute, one can show that

$$c_n c_m^+ + c_m^+ c_n = c_n c_m + c_m c_n = c_n^+ c_m^+ + c_m^+ c_n^+ = 0 \quad \text{for } n \neq m . \tag{17.48}$$

Thus the c_n and c_n^+ obey Fermi anticommutation relations.

Next we write the Ising Hamiltonian, Eq. (17.44) in terms of these new operators. Equations (17.46a) and (17.46b) allow us to write

$$\tilde{\sigma}_n^z = c_n^+ c_n - c_n c_n^+ = 2c_n^+ c_n - 1 . \tag{17.49}$$

Thus

$$\tilde{\sigma}_n^- = (-1)^n P_{n-1} c_n \quad (17.50a)$$
$$\tilde{\sigma}_n^+ = (-1)^n c_n^+ P_{n-1} \,. \quad (17.50b)$$

[If desired, P_i can be expressed in terms of the c's using Eq. (17.50b).] We also need the relations (for $1 \leq n < N$)

$$\tilde{\sigma}_n^+ \tilde{\sigma}_{n+1}^- = -c_n^+ P_{n-1} P_n c_{n+1} = c_n^+ c_{n+1} \quad (17.51a)$$
$$\tilde{\sigma}_n^+ \tilde{\sigma}_{n+1}^+ = -c_n^+ P_{n-1} c_{n+1}^+ P_n = c_n^+ c_{n+1} \,, \quad (17.51b)$$

which, together with their hermitian conjugate relations, allow us to evaluate the terms in Eq. (17.44) proportional to J. The result is

$$\mathcal{H} = HN \; - \; 2H \sum_{i=1}^{N} c_i^+ c_i \; - J \sum_{i=1}^{N-1} \big(c_i^+ - c_i\big)\big(c_{i+1}^+ + c_{i+1}\big) + V \,, \quad (17.52)$$

where

$$\begin{aligned} V &= -J(\tilde{\sigma}_N^+ + \tilde{\sigma}_N^-)(\tilde{\sigma}_1^+ + \tilde{\sigma}_1^-) = J(c_N^+ P_{N-1} + P_{N-1} c_N)(c_1^+ + c_1^-) \\ &= J P_{N-1}(c_N^+ + c_N)(c_1^+ + c_1^-) = J P_N (c_N^+ - c_N)(c_1^+ + c_1^-) \,. \end{aligned} \quad (17.53)$$

Note that P_N commutes with the Hamiltonian. Although the Hamiltonian does *not* conserve the number of fermions, eigenstates consist of a superposition of states each of which has an even number of fermions (these are called "even" states) or a superposition of states each of which has an odd number of fermions (these are called "odd" states). We see that for periodic boundary conditions, the eigenvalues are the eigenvalues of $\mathcal{H}_+$ when $P_N = +1$ (i.e., for even states) and are those of $\mathcal{H}_-$ when $P_N = -1$ (i.e., for odd states), where

$$\mathcal{H}_\pm = \mathcal{H}_0 \pm J(c_N^+ - c_N)(c_1^+ + c_1) \,, \quad (17.54)$$

where

$$\mathcal{H}_0 = HN \; - \; 2H \sum_{i=1}^{N} c_i^+ c_i \; - J \sum_{i=1}^{N-1} \big(c_i^+ - c_i\big)\big(c_{i+1}^+ + c_{i+1}\big) \,. \quad (17.55)$$

We also see from Eq. (17.52) that, if $J = 0$, then the ground state is one in which all sites are occupied by fermions, and the ground state energy is $-HN$. According to Eq. (17.49) occupied sites correspond to up spins.

For $J \neq 0$ the problem is more complicated. However it is also a very familiar-looking problem, a quadratic quantum Hamiltonian containing terms which create and destroy pairs of particles. It can be diagonalized by a Bogoliubov transformation of the kind that we used to solve the superconductivity problem. Of course, there are a number of differences between this problem and the one treated in Chap. 12.

1. First of all, Eq. (17.52) is the *exact* Hamiltonian for this problem. It is not some variational Hamiltonian that will be used to generate an approximation to the density matrix. The spectrum of $\mathcal{H}$ in Eq. (17.52) is the exact spectrum of the system, and a knowledge of this spectrum should allow us to calculate all thermodynamic properties.
2. The fermions in this problem are spinless. The pairing terms will create pairs of spinless particles with opposite momenta.
3. The magnetic field term looks like a chemical potential term. The J-interaction term looks like the sum of kinetic energy and pairing terms in the BCS Hamiltonian with a special relation between the magnitude of the kinetic energy and the gap.

The next step toward solving this problem is to Fourier transform so that the Hamiltonian is diagonal in Fourier wavevector k. For $P_N = -1$, the values of k which diagonalize the Hamiltonian are

$$\frac{k}{2\pi} = -\frac{N}{2N}, -\frac{N-2}{2N}, \ldots \frac{N-4}{2N}, \frac{N-2}{2N}, \tag{17.56}$$

whereas for $P_N = 1$ the values of k which diagonalize the Hamiltonian are

$$\frac{k}{2\pi} = -\frac{N-1}{2N}, -\frac{N-3}{2N}, \ldots \frac{2N-3}{2N}, \frac{2N-1}{2N}. \tag{17.57}$$

Eventually, sums over k will become integrals, in which case the distinction between even and odd values of k will become irrelevant. We write

$$c_n^+ = \frac{1}{\sqrt{N}} \sum_k c_k^+ e^{ikn}\,; \qquad c_n = \frac{1}{\sqrt{N}} \sum_k c_k e^{-ikn}. \tag{17.58}$$

Then the Hamiltonian becomes

$$\mathcal{H} = HN - 2\sum_k (H + J\cos k) c_k^+ c_k + iJ \sum_k \sin k \left(c_k^+ c_{-k}^+ + c_k c_{-k} \right). \tag{17.59}$$

This Hamiltonian couples states k and $-k$. The special cases $k = 0$ and $k = \pi$ can be dropped in the thermodynamic limit. In that case we write $\mathcal{H}$ as a sum over positive k as

$$\begin{aligned} \mathcal{H} = HN &- 2\sum_{k>0} (H + J\cos k)\left(c_k^+ c_k + c_{-k}^+ c_{-k}\right) \\ &+ 2iJ \sum_{k>0} \sin k \left(c_k^+ c_{-k}^+ + c_k c_{-k} \right). \end{aligned} \tag{17.60}$$

At this point, we can utilize the method that was used to diagonalize the BCS Hamiltonian in Eq. (12.4). For $k > 0$,

$$c_k = u_k \gamma_k - v_k^* \gamma_{-k}^+ \tag{17.61a}$$

$$c_{-k} = u_k \gamma_{-k} + v_k^* \gamma_k^+ \ . \tag{17.61b}$$

The analog of the gap is $2J \sin k \, e^{i\pi/2}$. Then

$$u_k = \tilde{u}_k e^{i\pi/4} \tag{17.62a}$$

$$v_k = \tilde{v}_k e^{-i\pi/4} \ , \tag{17.62b}$$

where

$$2\tilde{u}_k \tilde{v}_k = \frac{2J \sin k}{\epsilon_k} \tag{17.63a}$$

$$\tilde{u}_k^2 - \tilde{v}_k^2 = \frac{-2(H + J \cos k)}{\epsilon_k} \tag{17.63b}$$

and

$$\begin{aligned} \epsilon_k &= 2\sqrt{(H + J \cos k)^2 + (J \sin k)^2} \\ &= 2\sqrt{H^2 + J^2 + 2HJ \cos k} \ . \end{aligned} \tag{17.64}$$

Then the Hamiltonian becomes

$$\mathcal{H} = HN - \frac{1}{2} \sum_k \{2(H + J \cos k) + \epsilon_k\} + \sum_k \epsilon_k \, \gamma_k^+ \gamma_k \ . \tag{17.65}$$

The ground state will have $\langle c_0^+ c_0 \rangle = 1$, as long as $H + J > 0$, and all of the $\langle \gamma_k^+ \gamma_k \rangle$ will be zero. Therefore, the ground state energy is

$$\begin{aligned} E_0 &= HN - \frac{1}{2} \sum_k \{2(H + J \cos k) + \epsilon_k\} \\ &= - \sum_k \sqrt{H^2 + J^2 + 2HJ \cos k} \ , \end{aligned} \tag{17.66}$$

where the last step follows because the terms in H cancel and the sum over k of $\cos k$ is zero. Converting the sum over k to an integral, we obtain the analytic result for the ground state energy

$$E_0(H, J) = \frac{2N}{\pi} |H + J| \, E\left(\frac{4HJ}{(H + J)^2} \right) , \tag{17.67}$$

where $E(x)$ is the complete elliptic integral of the second kind. $E_0(H, J)$ has the expected property that $E_0(H, J) = E_0(J, H)$. The elliptic function is nonanalytic when its argument is equal to 1, i.e., for $H = J$, as expected from the duality argument. Furthermore, given that all of the excitation energies are symmetric under interchange of H and J (cf. Eq. 17.64), it is clear that Eq. (17.39) is satisfied everywhere in positive octant of the space of H, J, and T.

17.2.4 Finite Temperature and Order Parameters

For $T > 0$, the free energy is

$$\begin{aligned} F(H, J, T) &= E_0 - T \sum_k \ln\left(1 + e^{-\beta\epsilon_k}\right) \\ &= -T \sum_k \ln\left(2\cosh\frac{\beta\epsilon_k}{2}\right). \end{aligned} \tag{17.68}$$

As expected, this function is perfectly analytic, except at $T = 0$ (where F is equal to E_0) and $J = H$.

There are two thermal averages which are easy to calculate. The first is the average transverse magnetization in the original coordinate system of Eq. (17.26).

$$\begin{aligned} \langle \sigma_i^x \rangle &= -\frac{1}{N}\frac{\partial F(H, J)}{\partial H} \\ &= \frac{1}{2N}\sum_k \frac{\partial \epsilon_k}{\partial H} \tanh\left(\frac{\beta\epsilon_k}{2}\right). \end{aligned} \tag{17.69}$$

The second thermal average is the average longitudinal pairing

$$\begin{aligned} \langle \sigma_i^z \sigma_{i+1}^z \rangle &= -\frac{1}{N}\frac{\partial F(H, J)}{\partial J} \\ &= \frac{1}{2N}\sum_k \frac{\partial \epsilon_k}{\partial J} \tanh\left(\frac{\beta\epsilon_k}{2}\right). \end{aligned} \tag{17.70}$$

For $T \to 0$ the tanh's can be set equal to 1. Alternatively, at $T = 0$ the above averages can be calculated simply by substituting $E_0(H, J)$ for $F(H, J, T)$ in the above expressions. The graph of the resulting functions is plotted versus H/J in Fig. 17.3.

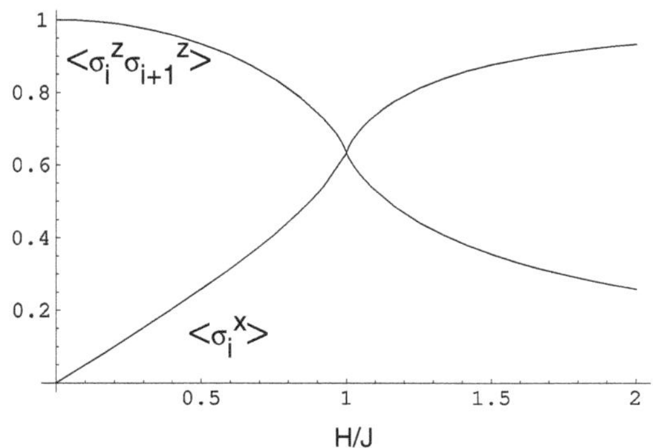

Fig. 17.3 Thermal averages for the Ising model in a transverse field at $T = 0$, plotted versus H/J, where H is the field and J is the ferromagnetic Ising coupling constant. The critical point is at $H/J = 1$

17.2.5 Correlation Functions

It is clear from Fig. 17.3 that neither $\langle\sigma_i^x\rangle$ nor $\langle\sigma_i^z\sigma_{i+1}^z\rangle$ is an order parameter in the classic sense used for order–disorder transitions where the order parameter is zero in the disordered state and nonzero in the ordered one. $\langle\sigma_i^x\rangle$, which couples linearly to the transverse field, is always nonzero, except when $H = 0$, in which case it is identically zero. Intuitively, it would seem that the relevant order parameter should be $\langle\sigma_i^z\rangle$ which we expect to be nonzero for $H < H_c = J$. The question is, does this average act like an order parameter and go to zero for $H \geq H_c$? To answer this question, we calculate the correlation function,

$$C_{zz}(m) = \frac{1}{N}\sum_n \langle\sigma_n^z\sigma_{n+m}^z\rangle\ . \tag{17.71}$$

Because we invoke periodic boundary conditions, the summand is independent of n and we henceforth take this into account. The magnitude of the order parameter can then be obtained from

$$|\langle\sigma^z\rangle| = \sqrt{\lim_{m\to\infty} C_{zz}(m)}\ . \tag{17.72}$$

Just as we did in Sect. 17.1.2, we can write the correlation function as

$$\begin{aligned} C_{zz}(m) &= \langle\left(\sigma_n^z\sigma_{n+1}^z\right)\left(\sigma_{n+1}^z\sigma_{n+2}^z\right)\dots\left(\sigma_{n+m-1}^z\sigma_{n+m}^z\right)\rangle \\ &= \langle\left(\tilde\sigma_n^x\tilde\sigma_{n+1}^x\right)\left(\tilde\sigma_{n+1}^x\tilde\sigma_{n+2}^x\right)\dots\left(\tilde\sigma_{n+m-1}^x\tilde\sigma_{n+m}^x\right)\rangle\ . \end{aligned} \tag{17.73}$$

Then using Eq. (17.50b) we write $C_{zz}(x)$ in terms of fermion operators as

$$C_{zz}(m) = \langle\left(c_n^+ - c_n\right)\left(c_{n+1}^+ + c_{n+1}\right)\dots\left(c_{n+m-1}^+ - c_{n+m-1}\right)\left(c_{n+m}^+ + c_{n+m}\right)\rangle\ . \tag{17.74}$$

The average can be evaluated with the help of Wick's theorem which applies to averages taken with respect to a quadratic Hamiltonian, such as we have here. For this purpose it is convenient to define new operators, linear in the fermion operators:

$$A_i = c_i^+ - c_i , \qquad B_i = c_i^+ + c_i \; . \tag{17.75}$$

In terms of these operators

$$C_{zz}(m) = \langle A_n B_{n+1} A_{n+1} B_{n+2} \dots A_{n+m-1} B_{n+m} \rangle \tag{17.76}$$

and Wick's theorem relates this average to a sum of products of all possible pairwise averages of the various A_r and B_r. The operators A_r and B_r have the useful property that

$$\langle A_j A_{j'} \rangle = -\delta_{j,j'} , \qquad \langle B_j B_{j'} \rangle = \delta_{j,j'} \; . \tag{17.77}$$

Since no pairs of A_j operators (or pairs of B_j operators) with the same site index occur in Eq. (17.76), the only pairwise averages that we need to consider are those of the form $\langle A_j B_{j'} \rangle = -\langle B_{j'} A_j \rangle$. The expansion, by Wick's theorem, of Eq. (17.76) can be written as a determinant,

$$C_{zz}(m) = \begin{vmatrix} \langle A_n B_{n+1} \rangle & \langle A_n B_{n+2} \rangle & \dots & \langle A_n B_{n+m} \rangle \\ \langle A_{n+1} B_{n+1} \rangle & \langle A_{n+1} B_{n+2} \rangle & \dots & \langle A_{n+1} B_{n+m} \rangle \\ \dots & \dots & \dots & \dots \\ \langle A_{n+m-1} B_{n+1} \rangle & \langle A_{n+m-1} B_{n+2} \rangle & \dots & \langle A_{n+m-1} B_{n+m} \rangle \end{vmatrix} \; . \tag{17.78}$$

Each term in the determinant is a product of m factors of the pair correlation functions, $\langle A_j B_{j'} \rangle$, with the appropriate plus or minus sign to keep track of the number of permutations required to return to the order of factors in Eq. (17.76). Of course, for the periodic boundary conditions that we are assuming, the averages, $\langle A_j B_{j'} \rangle$, only depend on the difference $j - j'$. If we define $G(m) = \langle A_n B_{n+m} \rangle$, then the correlation function becomes

$$C_{zz}(n) = \begin{vmatrix} G(1) & G(2) & \dots & G(n) \\ G(0) & G(1) & \dots & G(n-1) \\ \dots & \dots & \dots & \dots \\ G(2-n) & G(3-n) & \dots & G(1) \end{vmatrix} , \tag{17.79}$$

where, for example, $C_{zz}(1) = G(1)$ is the function, $\langle \sigma_i^z \sigma_{i+1}^z \rangle$ plotted in Fig. 17.3.

To proceed further, we need to calculate $G(n)$ for general n. We write

$$
\begin{aligned}
G(m) &= \langle (c_n^+ - c_n)(c_{n+m}^+ + c_{n+m}) \rangle \\
&= \langle c_n^+ c_{n+m} + c_{n+m}^+ c_n \rangle + \langle c_n^+ c_{n+m}^+ + c_{n+m} c_n \rangle - \delta_{m,0} \\
&= \frac{1}{N} \sum_k \sum_{k'} \Big\{ \left(e^{ikn} e^{-ik'(n+m)} + e^{ik(n+m)} e^{-ik'n} \right) \langle c_k^+ c_{k'} \rangle \\
&\quad + e^{ikn} e^{ik'(n+m)} \langle c_k^+ c_{k'}^+ \rangle + e^{-ik'(n+m)} e^{-ikn} \langle c_{k'} c_k \rangle \Big\} - \delta_{m,0} \, .
\end{aligned}
\tag{17.80}
$$

Using Eqs. (17.61b)–(17.64), we find that specializing to $T = 0$, where $\langle \gamma_k^+ \gamma_k \rangle = 0$, the function $G(m)$ is given by

$$
\begin{aligned}
G(m) &= \frac{1}{N} \sum_k \Big\{ \left(1 + \frac{2(H + J\cos k)}{\epsilon_k} \right) \cos(km) \\
&\quad + \left(\frac{2J \sin k}{\epsilon_k} \right) \sin(km) \Big\} - \delta_{m,0}
\end{aligned}
\tag{17.81a}
$$

$$
\begin{aligned}
&= \frac{1}{N} \sum_k \frac{H \cos(km)}{\sqrt{H^2 + J^2 + 2HJ\cos k}} \\
&\quad + \frac{1}{N} \sum_k \frac{J \cos[k(m-1)]}{\sqrt{H^2 + J^2 + 2HJ\cos k}} \, .
\end{aligned}
\tag{17.81b}
$$

The right-hand side of Eq. (17.79) is the determinant of what is called a Toeplitz matrix. This matrix is of the form

$$
\begin{pmatrix}
a_0 & a_1 & \dots & a_{x-1} \\
a_{-1} & a_0 & \dots & a_{x-2} \\
\dots & \dots & \dots & \dots \\
a_{1-x} & a_{2-x} & \dots & a_0
\end{pmatrix} ,
\tag{17.82}
$$

where $a_n = G(n+1)$ is a coefficient in the Fourier expansion of a function $f(k)$

$$
a_n = \frac{1}{2\pi} \int_{-\pi}^{\pi} e^{-ikn} f(k) dk \, .
\tag{17.83}
$$

[Note that, in the limit $N \to \infty$, the sum defining $G(m)$ is of this form.] For such determinants, a theorem due to Szego (1939), which was generalized by Kac (1954), allows the calculation of the limit as $x \to \infty$. This theorem states that the determinant, $D_m(f)$, satisfies

$$
\lim_{m \to \infty} \frac{D_m(f)}{Q(f)^{m+1}} = \exp\left(\sum_{n=1}^{\infty} n q_n q_{-n} \right) ,
\tag{17.84}
$$

where the q_n are coefficients in the Fourier expansion of $\ln f(k)$,

$$\ln f(k) = \sum_{n=-\infty}^{\infty} q_n e^{ink} \tag{17.85}$$

and

$$Q(f) = \exp \frac{1}{2\pi} \int_{-\pi}^{\pi} \ln f(k) dk \,. \tag{17.86}$$

Using the above definitions, we find, for the Ising model in a transverse field

$$f(k) = \frac{He^{-ik} + J}{\sqrt{H^2 + J^2 + 2HJ\cos k}} \,. \tag{17.87}$$

If we define $z = H/J$, then for $H < J$

$$\begin{aligned}
\ln f(k) &= \ln \frac{1 + ze^{-ik}}{\sqrt{1 + z^2 + z(e^{ik} + e^{-ik})}} \\
&= \ln \frac{1 + ze^{-ik}}{\sqrt{(1 + ze^{-ik})(1 + ze^{ik})}} \\
&= \frac{1}{2} \ln(1 + ze^{-ik}) - \frac{1}{2} \ln(1 + ze^{ik}) \\
&= -\frac{1}{2} \sum_{n=1}^{\infty} \frac{(-z)^n}{n} e^{-ikn} + \frac{1}{2} \sum_{n=1}^{\infty} \frac{(-z)^n}{n} e^{ikn}
\end{aligned} \tag{17.88}$$

and the values of the Fourier coefficients q_n are

$$q_n = \frac{(-z)^n}{2n} \tag{17.89a}$$

$$q_0 = 0 \tag{17.89b}$$

$$q_{-n} = -q_n \,. \tag{17.89c}$$

The fact that $q_0 = 0$ means that $Q(f) = 1$. So, using Szego's theorem, we find

$$\begin{aligned}
\lim_{x\to\infty} C(x) = \exp\left(-\frac{1}{4}\sum_{n=1}^{\infty} \frac{z^{2n}}{n}\right) &= e^{\frac{1}{4}\ln(1-z^2)} \\
&= \left(1 - z^2\right)^{\frac{1}{4}} \,.
\end{aligned} \tag{17.90}$$

Substituting H/J for z and using Eq. (17.72), we finally obtain the result

$$|\langle\sigma^z\rangle| = \left[1 - \left(\frac{H}{J}\right)^2\right]^{\frac{1}{8}} \,. \tag{17.91}$$

Thus as the field is increased so that H/J approaches 1, the z component of the magnetization falls to zero with critical exponent, $\beta = \frac{1}{8}$.

17.3 The Two-Dimensional Ising Model

As is mentioned in footnote 2 of Fig. 17.4, the partition function of the two-dimensional Ising model in zero field was first calculated by Onsager and published in 1944 (Onsager 1944). Onsager's solution was quite complicated, relying, as it did, on the author's intimate familiarity with the properties of elliptic functions. Onsager's original work stimulated a great deal of activity by many authors to extend the calculation to include correlation functions and nonzero magnetic field, and to provide alternative, hopefully simpler solutions. The 1964 review article by Schultz et al. (1964) is possibly the most straightforward reproduction of Onsager's result and we will give a brief description of this approach. It makes use of the transformation to fermion operators used above to solve the one-dimensional Ising model in a transverse field. A more complete exposition of the exact solution of the two-dimensional Ising model appears in many Statistical Mechanics text books, such as the ones by Huang (1987), Feynman (1972), and Plischke and Bergersen (1994), to name a few.

17.3.1 Exact Solution via the Transfer Matrix

In this subsection we will address the solution for the free energy of the 2D Ising model with different coupling constants J_1 and J_2 in the x and y directions. We will see how T_c depends on J_1 and J_2 and what critical exponents describe the phase transition.

The Hamiltonian of the 2D Ising model is

$$\mathcal{H} = -\sum_{m,n} \left\{ J_1 S_{m,n} S_{m+1,n} + J_2 S_{m,n} S_{m,n+1} \right\} , \tag{17.92}$$

where $S_{m,n}$ is the spin at $\mathbf{r} = (ma, na)$. The partition function is

$$Z = \sum_{\{S_{m,n}=\pm 1\}} \prod_{m,n} \exp\left(\beta J_1 S_{m,n} S_{m+1,n} + \beta J_2 S_{m,n} S_{m,n+1}\right) . \tag{17.93}$$

Here, we will display the steps which lead to the evaluation of Z in terms of a transfer matrix. Recall that in one dimension the transfer matrix is labeled by two indices, one for the state of the spin at site j and the other for the state of the spin at site $j+1$. Since each state can assume two values, this matrix is a 2×2 matrix, as we have seen. Here we construct the analogous transfer matrix which is labeled by two indices, one for the jth row of L spins (we assume L to be even) and the other

Fig. 17.4 Introduction to the paper, "Correlations and Spontaneous Magnetization of the Two-Dimensional Ising Model," by Montroll et al. (1963). The exponent on the right-hand side of Equation (0) is $\frac{1}{8}$

IN the days of Kepler and Galileo it was fashionable to announce a new scientific result through the circulation of a cryptogram which gave the author priority and his colleagues headaches. Onsager is one of the few moderns who operates in this tradition.

This paper concerns, among other things, the Onsager formula[1]

$$M = (1 - k^2)^{\frac{1}{8}}, \tag{0}$$

with

$$k = [\sinh(2J_1/kT)\sinh(2J_2/kT)]^{-1},$$

for the spontaneous magnetization of a two-dimensional Ising ferromagnet. This famous Onsager cryptogram required four years for its decipherment. It was first exposed to the public on 23 August 1948 on a blackboard at Cornell University on the occasion of a conference on phase transitions. Lazlo Tisza had just presented a paper on the General Theory of Phase Transitions. Gregory Wannier opened the discussion with a question concerning the compatability of the theory with some properties of the Ising model.[2] Onsager continued this discussion and then remarked that—incidentally the formula for the spontaneous magnetization of the two-dimensional model is just that given by (0). To tease a wider audience, the formula was again exhibited during the discussion which followed a paper by Rushbrooke at the first postwar IUPAP statistical mechanics meeting in Florence in 1948; it finally appeared in print as a discussion remark in reference (1). However, Onsager never published his derivation. The puzzle was finally solved by C. N. Yang[3] and its solution published in 1952.

Yang's analysis is very complicated. While many derivations of Onsager's thermodynamic formulas exist and are often presented in statistical mechanics courses, no new derivation of (0) appears in the literature nor is it considered to be appropriate as a classroom exercise.

---

* Part of this work was done while two of the authors (Elliott W. Montroll and John C. Ward) were visitors at the Brookhaven National Laboratories.

† On leave from the University of Adelaide, South Australia.

[1] L. Onsager, Nuovo Cimento, Suppl. 6, 261 (1949).

[2] Onsager's celebrated derivation of the partition function of the two-dimensional model is in Phys. Rev. 65, 117 (1944).

[3] C. N. Yang, Phys. Rev. 85, 808 (1952). See also C. H. Chang, Phys. Rev. 88, 1422 (1952), where the result for the rectangular lattice (different vertical and horizontal interactions) is obtained.

for the $j + 1$st row of L spins. Since a row has 2^L states, we are now dealing with a $2^L \times 2^L$ dimensional transfer matrix. The logical way to write this index which labels the state of a row is to give the states of spins 1, 2, ... L in that row. Since each individual spin has two states, there are 2^L possible states of a row. Thus, each of the two indices of the transfer matrix assume the form $\sigma_1^z, \sigma_2^z, \ldots \sigma_L^z$. Thus, the transfer matrix $\mathbf{V}$ associated with rows j and $j + 1$ has components

$$\Big[\mathbf{V}\Big]_{\sigma_1^z(j),\sigma_2^z(j),\ldots\sigma_L^z(j);\sigma_1^z(j+1),\sigma_2^z(j+1),\ldots\sigma_L^z(j+1)} \equiv \mathbf{V}_{\sigma(j);\sigma(j+1)} \,, \tag{17.94}$$

where $\sigma_m^z(j)$ defines the state of the mth spin in row j. Now if the system consists of M rows, we want

$$\sum_{\sigma(1),\sigma(2),\ldots\sigma(M)} V_{\sigma(1);\sigma(2)} V_{\sigma(2);\sigma(3)} \ldots V_{\sigma(M);\sigma(1)} = \mathrm{Tr}[\mathbf{V}]^M \tag{17.95}$$

to reproduce the partition function. In writing this we have assumed periodic boundary conditions so that row $M+1$ is in the same state as the first row. In that case the sum over indices is the trace of a product of M matrices, each of which is 2^L dimensional. We will reproduce the partition function if $V_{\sigma_j;\sigma_{j+1}}$ reproduces $e^{-\beta\{\frac{1}{2}E[\sigma(j)]+\frac{1}{2}E[\sigma(j+1)]+E[\sigma(j),\sigma(j+1)]\}}$, where $E[\boldsymbol{\sigma}(j)]$ is the energy of row j in the state $\boldsymbol{\sigma}(j)$ due to interactions within row j and $E[\boldsymbol{\sigma}(j),\boldsymbol{\sigma}(j+1)]$ is the energy of interaction between all spins in row j in the state $\boldsymbol{\sigma}(j)$ with their neighbors in row $j+1$ in the state $\boldsymbol{\sigma}(j+1)$. Thus we write

$$\mathbf{V} = \mathbf{V}_1\mathbf{V}_2\mathbf{V}_1 \,, \tag{17.96}$$

and we want $\mathbf{V}_1$ to be diagonal and reproduce the factor $e^{-\frac{1}{2}\beta E(j)}$ and $[\mathbf{V}_2]_{\sigma(j),\sigma(j+1)}$ should give the factor $e^{-\beta E[\sigma(j),\sigma(j+1)]}$. Thus we write

$$V_1 = e^{\frac{1}{2}J\sum_n \sigma_n^z\sigma_{n+1}^z} \,. \tag{17.97}$$

Next we write $e^{-\beta E(j,j+1)}$, which represents the interaction between one row and the next. To do this we note that the interaction energy between the kth spin in row j and the kth spin in row $j+1$ will give rise to a factor $e^{\beta J_2}$ if the two spins are in the same state and a factor $e^{-\beta J_2}$ if the two spins are in different states. These scenarios are encompassed by the operator

$$e^{\beta J_2}\mathcal{I} + \sigma_k^x e^{-\beta J_2} \,, \tag{17.98}$$

where $\mathcal{I}$ is the unit operator (which guarantees that the kth spin is in the same state in both rows), and σ_k^x, the Pauli matrix, flips the spin on going from one row to the next. We need this operator for each spin in the row, so

$$e^{-\beta E[\sigma(j),\sigma(j+1)]} = \left[\prod_n \left(e^{\beta J_2}\mathcal{I} + \sigma_n^x e^{-\beta J_2}\right)\right]_{\sigma(j);\sigma(j+1)} . \tag{17.99}$$

Thus, we write

$$Z = \mathrm{Tr}\Big([\mathbf{V}_1]\mathbf{V}_2[\mathbf{V}_1]\Big)^M \,, \tag{17.100}$$

where $\mathbf{V}_1$ is given in Eq. (17.97) and

$$[\mathbf{V}_2] = \prod_n \left(e^{\beta J_2} \mathcal{I} + \sigma_n^x e^{-\beta J_2} \right) . \tag{17.101}$$

A matrix of the form $C + D\sigma_x$ can always be put into the form $A \exp(B\sigma_x)$. So we may write

$$\mathbf{V}_2 = A^L e^{B \sum_n \sigma_n^x} , \tag{17.102}$$

where $A^2 = 2 \sinh(2\beta J_2)$ and $\tanh B = e^{-2\beta J_2}$. As before it is convenient to rotate coordinates so that $\sigma^x \to \tilde{\sigma}^z$ and $\sigma^z \to \tilde{\sigma}^x$. Also we introduce fermionic variables in the same way as before, so that

$$\mathbf{V}_2 = A^L e^{B \sum_{n=1}^{L} (2c_n^+ c_n - 1)} , \tag{17.103}$$

and

$$\begin{aligned} [\mathbf{V}_1] = \exp\Biggl(& \frac{1}{2} \beta J_1 \sum_{n=1}^{L-1} (c_n^+ - c_n)(c_{n+1}^+ + c_{n+1}) \\ & - \frac{1}{2} \beta J_1 P_L (c_L^+ - c_L)(c_1^+ + c_1) \Biggr) , \end{aligned} \tag{17.104}$$

where the last term ensures periodicity within the jth row and P_L was defined just after Eq. (17.45b). From the treatment of the 1D Ising model in a transverse field, we know that the effect of P_L is to create two sectors with different meshes of wavevectors. But the two meshes become identical when sums over wavevector become integrals. So we omit the corresponding details. Then, in terms of Fourier transformed fermionic variables, denoted $c^+(k)$ and $c(k)$ for k in the interval $[-\pi, \pi]$, we have

$$\mathbf{V}_2 = A^L e^{2B \sum_{k>0} [c(k)^+ c(k) + c(-k)^+ c(-k) - 1]} \tag{17.105}$$

and

$$\begin{aligned} [\mathbf{V}_1] = \exp\Biggl[\beta J_1 \sum_{k>0} \Bigl(& \cos k [c(k)^+ c(k) + c(-k)^+ c(-k)] \\ & - i \sin k [c(k)^+ c(-k)^+ + c(k) c(-k)] \Bigr) \Biggr] . \end{aligned} \tag{17.106}$$

Since bilinear operators at different wavevectors commute, we may write

$$\mathbf{V} = \prod_{k>0} \Big([\mathbf{V}_1(k)]\mathbf{V}_2(k)[\mathbf{V}_1(k)] \Big) \equiv \prod_{k>0} V(k) \, , \tag{17.107}$$

where

$$[\mathbf{V}_1(k)] = \exp\Big[\beta J_1 \Big(\cos k[c(k)^+ c(k) + c(-k)^+ c(-k)] - i \sin k[c(k)^+ c(-k)^+ + c(k)c(-k)] \Big) \Big] . \tag{17.108}$$

and

$$\mathbf{V}_2(k) = A^2 e^{2B[c(k)^+ c(k) + c(-k)^+ c(-k) - 1]} . \tag{17.109}$$

Thus

$$Z = \prod_{k>0} \Big(\mathrm{Tr}_k V(k)^M \Big) , \tag{17.110}$$

where Tr_k indicates a trace over the four states generated by $c(k)^+$ and $c(-k)^+$, namely, $|0\rangle$, $c(k)^+|0\rangle$, $|c(-k)^+|0\rangle$, and $c(k)^+ c(-k)^+|0\rangle$. Thus we may write

$$Z = \prod_{k>0} \Big(\lambda_1(k)^M + \lambda_2(k)^M + \lambda_3(k)^M + \lambda_4(k)^M \Big) , \tag{17.111}$$

where λ_k is an eigenvalue of $\mathbf{V}(k)$. As discussed by Schultz et al., in the thermodynamic limit only the largest eigenvalue need be kept. The determination of the four eigenvalues of $\mathbf{V}(k)$ is not difficult. The only term which does not conserve particles is one which creates or destroys pairs of particles. So $c(k)^+|0\rangle$ and $c(-k)^+|0\rangle$ are eigenvectors of $\mathbf{V}(k)$. The associated eigenvalues are

$$\lambda_1(k) = \lambda_2(k) = A^2 e^{2\beta J_1 \cos k} \equiv \lambda(k) . \tag{17.112}$$

We now determine λ_3 and λ_4 by considering the subspace $\mathcal{S}$ spanned by $c(k)^+ c(-k)^+|0\rangle$ and $|0\rangle$. We now introduce operators τ_y and τ_z which are Pauli matrices within this subspace $\mathcal{S}$. Then, *within this subspace*,

$$\begin{aligned} \mathbf{V}_1(k) &= \exp\Big[\beta J_1 \Big([\mathcal{I} + \tau_z] \cos k + \tau_y \sin k \Big) \Big] \\ &\equiv \exp(\beta J_1 \cos k) \exp(\hat{n}(k) \cdot \tau) \\ &\equiv \exp(\beta J_1 \cos k) \mathbf{v}_1(k) \end{aligned} \tag{17.113}$$

and

$$\mathbf{V}_2(k) = A^2 e^{2B\tau_z} \equiv A^2 \mathbf{v}_2(k) . \tag{17.114}$$

To simplify the algebra we note that

$$\begin{aligned}\lambda_3\lambda_4 &= \mathrm{Det}[\mathbf{V}_1(k)\mathbf{V}_2(k)\mathbf{V}_1(k)] \\ &= A^4 e^{4\beta J_1 \cos k}\left(\mathrm{Det}\mathbf{v}_1(k)\right)^2 \mathrm{Det}\mathbf{v}_2(k) ,\end{aligned} \tag{17.115}$$

where the determinants are taken within the subspace $\mathcal{S}$. But for any vector r one has $\mathrm{Det}[e^{\mathbf{r}\cdot\tau}] = 1$. We therefore see that

$$\lambda_3\lambda_4 = \lambda(k)^2 , \tag{17.116}$$

so that we may write

$$\lambda_{3,4} = \lambda(k)e^{\pm\epsilon(k)} . \tag{17.117}$$

We then have

$$2\cosh[\epsilon(k)] = \mathrm{Tr}_k \mathbf{V}(k) , \tag{17.118}$$

where the trace is over the two states of the subspace $\mathcal{S}$. In an exercise you are asked to show that this gives

$$\begin{aligned}\cosh\epsilon(k) &= \cosh(2\beta J_1)\cosh(2B) \\ &\quad + \sinh(2K_1)\sinh(2B)\cos q .\end{aligned} \tag{17.119}$$

It then follows that the free energy per spin, f, is

$$f = -kT\left[\ln(2\sinh 2B)^{1/2} + \frac{1}{4\pi}\int_{-\pi}^{\pi} \epsilon(k)dk\right] . \tag{17.120}$$

The free energy per spin of the 2D Ising model can be written in many different forms. Onsager's original result displays the symmetry of the 2D lattice. It is

$$\begin{aligned}f = -kT\Bigg\{\ln 2 + \frac{1}{2}\int_{-\pi}^{\pi}\frac{dk}{2\pi}\int_{-\pi}^{\pi}\frac{dk'}{2\pi}\ln[\cosh 2\beta J_1\cosh 2\beta J_2 \\ - \sinh 2\beta J_1 \cos k - \sinh 2\beta J_2 \cos k']\Bigg\} .\end{aligned} \tag{17.121}$$

(Complicated algebra (Huang 1987) is required to go from Eq. (17.120) to Onsager's result (Onsager 1944).) The integral in Eq. (17.121) is nonanalytic when

$$\cosh 2\beta_c J_1 \cosh 2\beta_c J_2 = \sinh 2\beta_c J_1 + \sinh 2\beta_c J_2 \tag{17.122}$$

because, for this value, $\beta = \beta_c$, the argument of the logarithm vanishes at $k = k' = 0$ where the cosines are stationary. This condition can also be written as

$$\sinh 2\beta_c J_1 \sinh 2\beta_c J_2 = 1. \tag{17.123}$$

Two special limiting cases for this expression are $J_1 = J_2$ and $J_2 \ll J_1$. For $J_1 = J_2 = J$, Eq. (17.123) gives

$$kT_c = \frac{2J}{\sinh^{-1} 1} = 2.26919\, J\ , \tag{17.124}$$

a classic result for the square Ising ferromagnet. For general J_2/J_1 one can solve Eq. (17.123) for J_2/J_1 as a function of T_c/J_1. The result is

$$\frac{J_2}{J_1} = \frac{kT_c}{2J_1} \sinh^{-1}\left[\frac{1}{\sinh(2J_1/kT_c)}\right] \tag{17.125a}$$

$$\approx \frac{kT_c}{J_1} e^{-2J_1/kT_c}, \qquad \text{for} \quad \frac{kT_c}{J_1} \ll 1\ . \tag{17.125b}$$

Roughly speaking, for $J_2/J_1 \ll 1$, $kT_c \approx 2J_1/\ln(J_1/J_2)$. In the limit $J_2 \to 0$, the system becomes one-dimensional and $T_c \to 0$. What is striking, however, is how

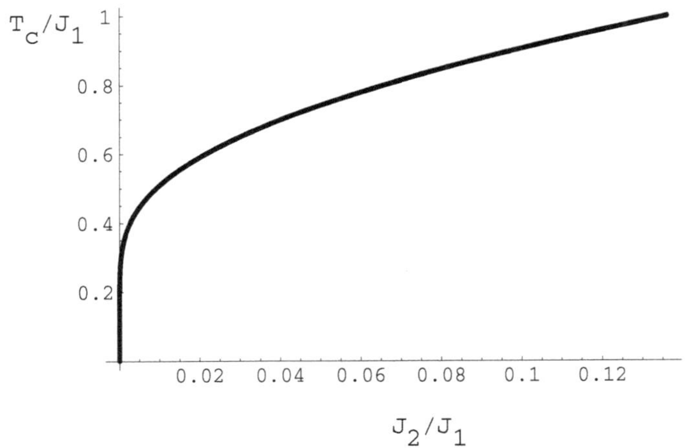

Fig. 17.5 Transition temperature, T_c/J_1, as a function of coupling constant anisotropy, J_2/J_1, for the x- and y-directions of the two-dimensional Ising ferromagnet

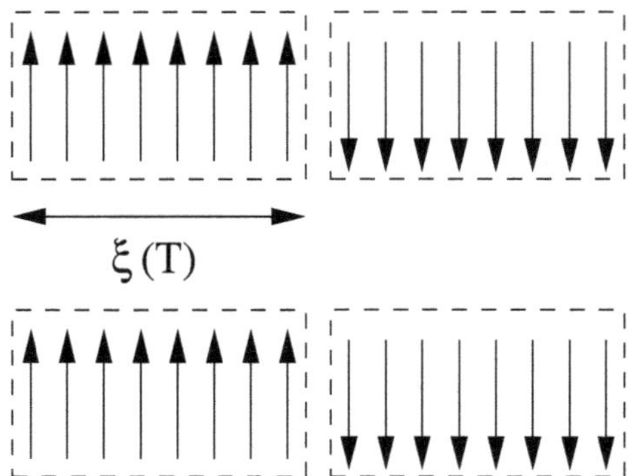

Fig. 17.6 Heuristic argument to obtain T_c for weak coupling $J_\perp$ between adjacent Ising chains. Here, $\xi(T)$ is the correlation length at temperature T for a chain with nearest-neighbor interaction J. The transition temperature is estimated to occur when the interaction energy between segment of length ξ becomes of order kT. This gives $kT_c = \xi(T)J_\perp$

quickly a nonzero T_c turns on, even for very small J_2/J_1. The behavior is illustrated in Fig. 17.5. It demonstrates how easily a finite temperature transition can occur for weakly coupled quasi-one-dimensional chains. In Fig. 17.6, we show the simple heuristic argument which gives an estimate for T_c in the case of weak interchain coupling, $J_\perp$. For a one-dimensional Ising chain we have seen the result $\xi(T) \sim e^{2J/kT}$. Then we estimate that

$$kT_c = J_\perp \xi(T) = J_\perp e^{2J/kT} , \tag{17.126}$$

which is the same as Eq. (17.125b).

Next we calculate the internal energy and specific heat for the isotropic system, i.e., $J_1 = J_2$. For this case the free energy is

$$\beta f = -\ln 2 - \frac{1}{2} \int_0^\pi \frac{dk}{\pi} \int_0^\pi \frac{dk'}{\pi} \ln \left[\cosh^2 2\beta J - \sinh 2\beta J (\cos k + \cos k') \right] \tag{17.127}$$

and the internal energy is

$$\begin{aligned} u &= \frac{\partial \beta f}{\partial \beta} \\ &= -\frac{J}{2} \int_0^\pi \frac{dk}{\pi} \int_0^\pi \frac{dk'}{\pi} \left(\frac{4\cosh 2\beta J \sinh 2\beta J - \sinh 2\beta J(\cos k + \cos k')}{\cosh^2 2\beta J - \sinh 2\beta J(\cos k + \cos k')} \right) \\ &= -\frac{J \cosh 2\beta J}{\sinh 2\beta J} \int_0^\pi \frac{dk}{\pi} \int_0^\pi \frac{dk'}{\pi} \\ &\quad \left[\frac{\sinh^2 2\beta J - 1}{\cosh^2 2\beta J - \sinh 2\beta J(\cos k + \cos k')} + 1 \right] . \end{aligned} \tag{17.128}$$

The double integral is done as follows:

$$\begin{aligned} I(a,b) &= \int_0^{\pi} \frac{dk}{\pi} \int_0^{\pi} \frac{dk'}{\pi} \frac{1}{a - b(\cos k + \cos k')} \\ &= \int_0^{\pi} \frac{dk}{\pi} \int_0^{\pi} \frac{dk'}{\pi} \int_0^{\infty} e^{-(a-b\cos k - bcosk')x} dx \\ &= \int_0^{\infty} e^{-ax} \left[I_0(bx)\right]^2 dx = \frac{2K\left(\frac{2b}{a}\right)}{\pi a}, \end{aligned} \tag{17.129}$$

where $K(x)$ is the complete elliptic integral of the first kind. This integral is defined as long as $|b/a| < 1/2$.

For the internal energy, we have $a = \cosh^2 2\beta J$ and $b = \sinh 2\beta J$. Then it is easy to show that $b/a < 1/2$ *except* at the critical point where $\sinh 2\beta J = 1$. Using the above identity, we obtain

$$u = -\frac{J \cosh 2\beta J}{\sinh 2\beta J}\left[1 + \frac{2(\sinh^2 2\beta J - 1)}{\pi \cosh^2 2\beta J} K\left(\frac{2\sinh 2\beta J}{\cosh^2 2\beta J}\right)\right]. \tag{17.130}$$

It is straightforward to differentiate the above expression to obtain the specific heat. It is left as an exercise for the reader to show that the specific heat diverges logarithmically at T_c so that the specific heat exponent $\alpha = 0$.

The calculation of the spontaneous magnetization as a function of temperature below T_c is similar in spirit to but more complicated than the calculation for the Ising chain in a transverse field. It was first done by Yang (1952) and is also done in the paper by Montroll et al. (1963). The result, as discussed in Fig. 17.4, is

$$M = (1 - k^2)^{\frac{1}{8}}, \tag{17.131a}$$

with

$$k = \left[\sinh 2\beta J_1 \sinh 2\beta J_2\right]^{-1}. \tag{17.131b}$$

17.4 Duality

We now give a brief discussion of duality for two-dimensional models. A duality transformation is one which maps the strong coupling regime of one model onto the weak coupling regime of another model. The model is said to be self-dual if it is the same as its dual. We will study duality first for the Ising model and then for the q state Potts model.

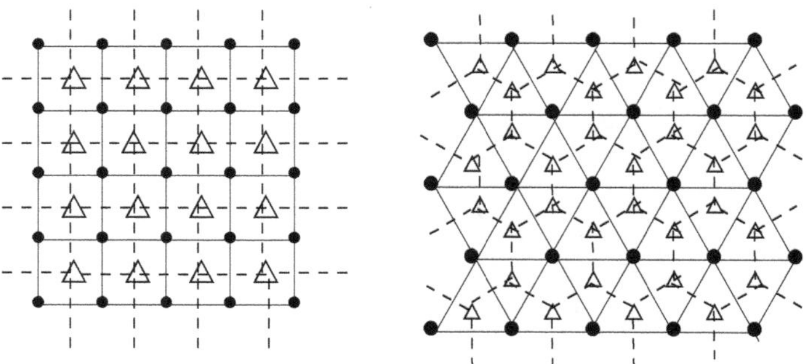

Fig. 17.7 Left: a square lattice with vertices indicated by filled circles with its dual lattice whose vertices are indicated by triangles. Since the dual lattice is also a square lattice, we see that the square lattice is self-dual. Right: a similar representation of a triangular lattice (filled circles) and its dual (triangles). In this case the dual lattice is a lattice of hexagons, called a honeycomb lattice

17.4.1 The Dual Lattice

For simplicity we confine the discussion to two-dimensional lattices. Since the question of boundary conditions introduces purely technical complications, we consider an infinitely large lattice without explicitly discussing its topology at infinity. We start with the direct lattice which consists of sites (or vertices) which are connected by nearest-neighbor bonds (or edges). These edges define plaquettes or faces. As illustrated in Fig. 17.7, the dual lattice consists of vertices placed inside each face, so the number of vertices in the dual lattice is equal to the number of faces in the direct lattice. We now define what we mean by bonds (or edges) in the dual lattice. We define edges on the dual lattice to be connections between those pairs of vertices of the dual lattice such that the associated pairs of faces of the original lattice have an edge in common. Thus each bond in the dual lattice crosses its unique partner bond in the direct lattice. The dual of the dual lattice is the original lattice. Typically the duality relation is one between a model on the original lattice and a similar model on the dual lattice.

17.4.2 2D Ising Model

We now consider the Ising model in two dimensions. We rewrite the partition function using the identity (for $S_i = \pm 1$):

$$\begin{aligned} e^{\beta J S_i S_j} &= \cosh(\beta J) + S_i S_j \sinh(\beta J) \\ &= \cosh(\beta J)[1 + t S_i S_j] \,, \end{aligned} \qquad (17.132)$$

where $t \equiv \tanh(\beta J)$. Thereby for a system of N spins, each of which is coupled to its z nearest neighbors, we get

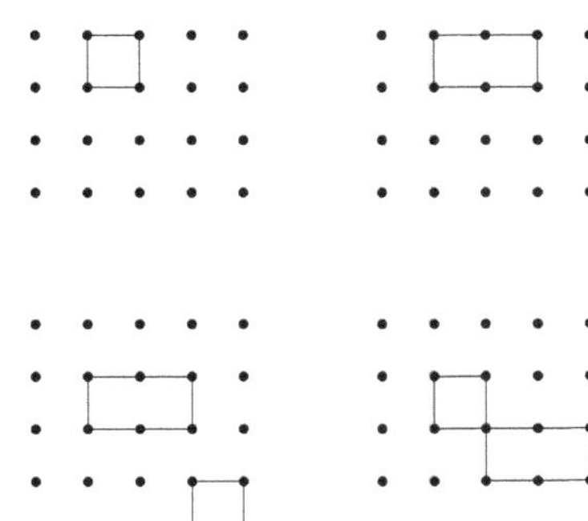

Fig. 17.8 Diagrams corresponding to the expansion of Z in powers of t. The only diagrams we show here are those which survive the trace over spin variables. Such diagrams consist of a union of one or more polygons. Two polygons may have a site in common, but each bond can be used either 0 or 1 times

$$Z = \cosh(\beta J)^{Nz/2} \sum_{S_1=\pm 1} \sum_{S_2=\pm 1} \cdots \sum_{S_N=\pm 1} \prod_{\langle ij \rangle} [1 + tS_iS_j] . \tag{17.133}$$

We now consider the expansion of Z in powers of t. The only terms in the expansion of the square bracket which survive the trace over spin variables are those in which each S_i appears an even number of times. If each term is represented graphically by associating a line connecting sites i and j for each factor tS_iS_j, we see that only diagrams consisting of one or more polygons contribute to the expansion of Z in powers of t. Such diagrams are shown in Fig. 17.8. The result is

$$Z(\beta J) = [\cosh(\beta J)]^{Nz/2} \sum_r n_r \tanh(\beta J)^r , \tag{17.134}$$

where n_r is the number of ways of forming a set of polygons using a total of r nearest-neighbor bonds. This expansion is a high-temperature expansion because $\tanh(\beta J) \ll 1$ for $\beta J \ll 1$.

In contrast we may also consider a low-temperature expansion. The idea here is to consider states in which almost all spins are, say, up. Such configurations are shown in Fig. 17.9. Then the partition function is

$$Z = e^{Nz\beta J/2} \sum e^{-\beta E_{\rm inter}} , \tag{17.135}$$

where the sum is over all configurations and $E_{\rm inter}$ is the interfacial energy. (We omit a factor of 2 in the partition function due to the degeneracy of the ground state, since this factor is irrelevant in the thermodynamic limit.) Since a bond between antiparallel spins has energy $+J$, we see that the energy cost of making a pair of spins be antiparallel is $2J$. The interfacial energy is related to the perimeter of the cluster. The general prescription for drawing the perimeter of a cluster of down spins is as follows: for each down spin, A, in the cluster which has an up spin neighbor, B, draw the unit length perpendicular bisector of this bond AB. The union of such perpendicular bisectors will define the bounding polygon of the cluster of down

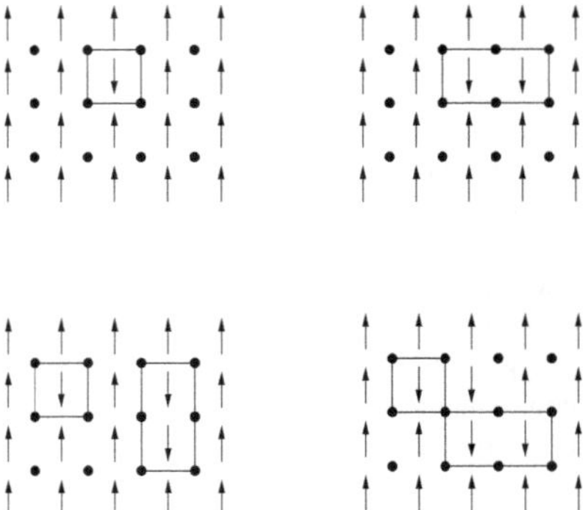

Fig. 17.9 Diagrams corresponding to the low-temperature expansion of Z. Here we count the interfacial energy attributed to clusters of down spins in a background of up spins

spins. Then the interfacial energy of this cluster (in units of $2J$) is the perimeter of the polygon. The lattice of possible endpoints of such perpendicular bisectors forms the dual lattice. Note that in Eq. (17.135) any configuration of clusters of down spins generates a unique set of one or more bounding polygons. Alternatively, any set of bounding polygons generates a unique set of clusters of down spins. Thus there is a one-to-one correspondence between clusters of down spins and bounding polygons on the dual lattice. What this means is that Eq. (17.135) may be written as

$$Z = e^{Nz\beta J/2} \sum n_r^* e^{-2r\beta J} . \tag{17.136}$$

where X^* denotes the quantity X evaluated on the dual lattice. From Eq. (17.134) we see that

$$\sum n_r x^r = Z(\tanh^{-1} x)(1 - x^2)^{Nz/4} . \tag{17.137}$$

Thus Eq. (17.136) becomes

$$Z(\beta J) = e^{Nz\beta J/2} Z^*[\tanh^{-1}(e^{-2\beta J})][1 - e^{-4\beta J}]^{Nz/4} . \tag{17.138}$$

(Here we used $Nz = N^* z^*$.) This is an exact relation between the partition function on the direct lattice at coupling constant βJ and the partition function on the dual lattice at coupling constant βK, where

$$\tanh(\beta K) = e^{-2\beta J} . \tag{17.139}$$

Note that Eq. (17.138) relates the high-temperature regime ($\beta J \ll 1$) of one model to its counterpart on the dual lattice in the low-temperature regime ($e^{-2\beta J} \ll 1$). In general the lattice and its dual are different lattices. However, some lattices are the same as their dual lattice, in which case the lattice is "self-dual." This is the case for the square lattice, as Fig. 17.7 shows. If we assume that the square lattice model has only one critical point, then we can locate the critical temperature by setting

$$\tanh[J/(kT_c)] = e^{-2J/(kT_c)} . \tag{17.140}$$

This gives $e^{2J/(kT_c)} = 1 + \sqrt{2}$, or $\sinh[2J/(kT_c)] = 1$.

The dual lattice to the triangular lattice is the honeycomb lattice. In this case, the duality relation relates models on two *different* lattices and hence would not seem to pin down T_c. However, by invoking the additional transformation, known as the *star-triangle transformation*, one can use duality to obtain exact results for the critical points for Ising models on these lattices.

17.4.3 2D q-State Potts Model

The duality transformation is not limited to Ising models. We now consider the q-state Potts model, whose Hamiltonian is

$$\mathcal{H} = -J \sum_{\langle ij \rangle} [q\delta_{s_i,s_j} - 1] , \tag{17.141}$$

where the Potts variable s_i associated with site i can assume the values $1, 2, \dots q$.

In Chap. 15 we developed the representation of the partition function, derived in Eqs. (15.7) to (15.13), relating the Potts model to the bond percolation problem, which forms the starting point for the present analysis:

$$Z = e^{N_B \beta J(q-1)} \sum_{\mathcal{C}} P(\mathcal{C}) q^{N_c(\mathcal{C})} , \tag{17.142}$$

where the sum is over all possible configurations of the N_B bonds. Here, a configuration is determined by specifying the state (occupied or unoccupied) of each of the N_B bonds in the lattice and

$$P(\mathcal{C}) = p^{\mathcal{N}(\mathcal{C})} (1-p)^{N_B - \mathcal{N}(\mathcal{C})} , \tag{17.143}$$

where $p = 1 - e^{-\beta J q}$ and $\mathcal{N}(\mathcal{C})$ is the number of occupied bonds and $N_c(\mathcal{C})$ the number of clusters of sites in the configuration $\mathcal{C}$. A cluster of sites is a group of sites which are connected by a sequence of occupied bonds (see Fig. 15.2). Thus

$$\begin{aligned} Z(p) &= [(1-p)e^{\beta J(q-1)}]^{N_B} \sum_{\mathcal{C}} [p/(1-p)]^{\mathcal{N}(\mathcal{C})} q^{N_c(\mathcal{C})} \\ &= (1-p)^{N_B/q} \sum_{\mathcal{C}} [p/(1-p)]^{\mathcal{N}(\mathcal{C})} q^{N_c(\mathcal{C})} . \end{aligned} \tag{17.144}$$

We now relate this result to a partition function on the dual lattice. For this purpose we make a one-to-one correspondence between a configuration $\mathcal{C}$ of occupied bonds on the direct lattice to a configuration $\mathcal{C}_D$ of occupied bonds on the dual lattice. Note that each bond of the direct lattice has its unique partner in the dual lattice. With each *occupied* bond of the direct lattice we associate an *unoccupied* bond of the dual lattice and *vice versa*. In this way, we make a one-to-one correspondence between any possible configuration $\mathcal{C}$ of the direct lattice and its associated configuration, $\mathcal{C}_D$ on the dual lattice. Since either the bond on the direct lattice is occupied or its partner on the dual lattice is occupied, we have that

$$\mathcal{N}(\mathcal{C}) + \mathcal{N}^*(\mathcal{C}_D) = N_B \ . \tag{17.145}$$

We now wish to relate the number of clusters on the direct lattice $N_c(\mathcal{C})$ and the number of clusters on the dual lattice $N_c^*(\mathcal{C}_D)$. When $\mathcal{C}$ has no occupied bonds, then $N_c(\mathcal{C}) = N$, the total number of sites in the direct lattice. For this configuration we have that $N_c^*(\mathcal{C}_D) = 1$, because when all bonds on the dual lattice are occupied, all sites of the dual lattice must be in the same cluster. For this configuration

$$N_c(\mathcal{C}) - N_c^*(\mathcal{C}_D) = N - \mathcal{N}(\mathcal{C}) - 1 \ , \tag{17.146}$$

because $N_c(\mathcal{C}) = N$, $N_c^*(\mathcal{C}_D) = 1$, and $\mathcal{N}(\mathcal{C}) = 0$. We will give an argument to show that this relation holds generally. We do this by considering the effect of adding occupied bonds to the initially unoccupied ($\mathcal{N} = 0$) lattice. If Δ indicates the change in a quantity (i.e., its final value minus its initial value), then, to establish Eq. (17.146), we need to show that when a bond is added to $\mathcal{C}$

$$\Delta N_c(\mathcal{C}) - \Delta N_c^*(\mathcal{C}_D) = -\Delta\mathcal{N} = -1 \ . \tag{17.147}$$

The argument for this is given in the caption to Fig. 17.10.

We use Eqs. (17.145) and (17.146) to write

$$\begin{aligned} Z(p) &= (1-p)^{N_B/q} \sum_{\mathcal{C}} [p/(1-p)]^{N_B - \mathcal{N}^*(\mathcal{C}_D)} q^{N_c^*(\mathcal{C}_D) + N - N_B + \mathcal{N}^*(\mathcal{C}_D) - 1} \\ &= [p(1-p)^{(1-q)/q}]^{N_B} q^{N-N_B-1} \\ &\quad \times \sum_{\mathcal{C}} [q(1-p)/p]^{\mathcal{N}^*(\mathcal{C}_D)} q^{N_c^*(\mathcal{C}_D)} \ . \end{aligned} \tag{17.148}$$

This result shows that the partition function at coupling constant $p = 1 - e^{-\beta Jq}$ is (apart from constants and analytic functions) identical to that on the dual lattice at coupling constant $p^* = 1 - e^{-\beta Kq}$, where

$$\frac{p^*}{1-p^*} = \frac{q(1-p)}{p} \ . \tag{17.149}$$

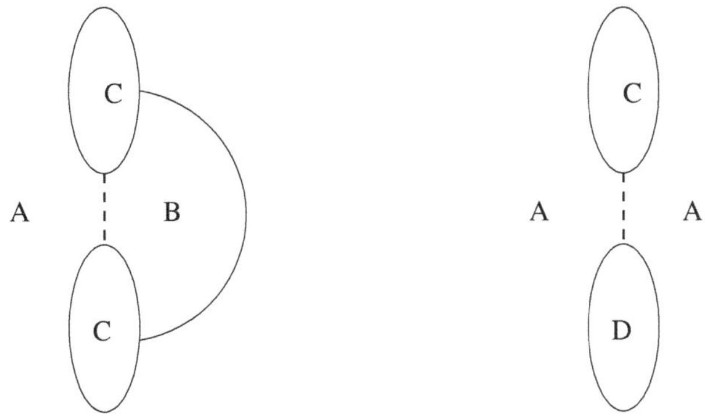

Fig. 17.10 The effect of adding a bond to $\mathcal{C}$, so that $\Delta\mathcal{N} = +1$. Here the bond added is the dashed line. Such a bond must connect two sites which may, before the bond is added, either be in the same cluster C (left) or in different clusters C and D (right). In the left panel, $\Delta N_c(\mathcal{C}) = 0$ (because the two sites were already in the same cluster C) and $\Delta N_c(\mathcal{C}_D) = +1$, because, by occupying the dashed bond on the direct lattice, we have made unoccupied the bond b_D on the dual lattice which intersects this bond. The bond b_D is a bond which was previously was needed to connect (on the dual lattice) the regions A and B on opposite sites of the bond added in the direct lattice. In the right panel, $\Delta N_c(\mathcal{C}) = -1$ because, by assumption the added bond connects previously unconnected clusters, C and D, on the direct lattice. But since the two clusters were not connected, it means that there has to exist a path of occupied bonds on the dual lattice which joins the regions on opposite sides of the bond added in the direct lattice, so that $\Delta N_C(\mathcal{C}_D) = 0$

When p is small (so that the direct lattice model is in its disordered phase), then the dual lattice model has p large, and is in the order phase regime. If there is a single-phase transition for the model on the (self-dual) square lattice, it must occur for $p^* = p$, or

$$e^{-\beta Jq} = \frac{1}{\sqrt{q}+1} . \tag{17.150}$$

For percolation $q = 1$, this gives $p_c = 1/2$, which is the known result for bond percolation on the square lattice. For $q = 2$ we reproduce the result obtained above for the Ising model.

17.4.4 Duality in Higher Dimensions

In three (or more) spatial dimensions duality is more complicated. In one dimension the dual of a bond is a point. In two dimensions the dual of a bond is a transverse bond. In three dimensions the dual of a bond is a perpendicular plaquette. A discussion of plaquette Hamiltonians is beyond the scope of this text.

Table 17.1 Critical exponents for q-state 2D Potts models

Model:	Percolation (q = 1)	Ising (q = 2)	3-State (q = 3)	4-State (q = 4)
α	−2/3	0	1/3	2/3
β	5/36	1/8	1/9	1/12
γ	43/18	7/4	13/9	7/6
δ	91/5	15	14	15
η	5/24	1/4	4/15	1/4
ν	4/3	1	5/6	2/3
Q[a]	0	1/2	4/5	1

[a]In two dimensions, conformal field theory yields exact values of the critical exponents. In that case there are only certain allowed universality classes, which are characterized by a "central charge," Q. Blöte et al. (1986)

17.5 Summary

Perhaps the most important aspect of Onsager's exact solution of the two-dimensional Ising model, was that it provided a rigorous example of a phase transition at which the various thermodynamic functions displayed nonanalyticity with nontrivial critical exponents. Since then, many exact solutions having nontrivial critical exponents have been found. We have listed some of those results for the 2D q-state Potts model in Table 17.1.

17.6 Exercises

1. Prove the identities given in Eqs. (17.77).

2. For the one-dimensional Ising model in a transverse field at zero temperature, show that the order parameter, $|\langle\sigma^z\rangle| = 0$ for $H > J$.

3. Carry out a mean-field analysis for the Ising model in a transverse field for nonzero temperature.

a. Show that nonzero long-range order, $\langle\sigma_i^z\rangle$, occurs for $H < H_c(T)$ and give an equation which determines (implicitly) $H_c(T)$.

b. Let $t \equiv [T_c(H) - T]/T_c(H)$. Give the dominant nonanalytic contributions to both $\langle\sigma_i^x\rangle$ and $\langle\sigma_i^z\rangle$ for small $|t|$.

4. Show that τ, given in Eqs. (17.36a), (17.36b), and (17.37), does obey Eqs. (17.32)–(17.35).

5a. Write down the transformation inverse to Eq. (17.58). Substitute this transformation into Eq. (17.59). Show that you thereby reproduce the two cases of Eq. (17.54) depending on whether the mesh of k's is given by Eq. (17.56) or by Eq. (17.57).

(**b**) Suppose you wanted to recover Eq. (17.41) with $V = 0$ so as to implement free-end boundary conditions. What mesh of k points would you use instead of either Eq. (17.56) or (17.57)?

6. Consider E_0/J as given in Eq. (17.66) as a function of the complex variable $z \equiv H/J$. You are to locate the singularities in this function which are on the *real* z-axis. Locate any such singularities and give the dominant nonanalytic form for E_0/J near the singularity. (OK—you can cheat and look up analytic properties of elliptic integrals, but it more stimulating to analyze the integral.)

7. Verify Eq. (17.119).

8. The dual of the dual lattice is the original lattice. Thus if you iterate the duality relation twice, you should get back the original coupling constant. Verify that this is the case for the q-state Potts model. (Since $q = 2$ is the Ising model, this exercise includes the Ising model as a special case.)

9. The dual of the dual lattice is the original lattice. Consider Eq. (17.146) when the original lattice is taken to be the dual lattice. The left-hand side of this equation is $N_c^*(\mathcal{C}_D) - N_c(\mathcal{C})$. Construct the right-hand side of this equation. It may not be obvious that the equation you have obtained is equivalent to Eq. (17.146). To establish this equivalence you may wish to invoke Euler's celebrated theorem for graphs (i.e., lattices) on a sphere: $F + V = E + 2$, where F is the number of faces, V the number of vertices, and E the number of edges.

References

H.W.J. Blöte, J.L. Cardy, M.P. Nightingale, Conformal invariance, the central charge, and universal finite-size amplitudes at criticality. Phys. Rev. Lett. **56**, 742 (1986)

R.P. Feynman, *Statistical Mechanics*; a set of lectures (1972) (Notes taken by R. Kikuchi and H.A. Fiveson, W.A. Benjamin)

K. Huang, *Statistical Mechanics*, 2nd edn. (Wiley, 1987)

E. Ising, Contribtions to the theory of ferromagnetism. Ph.D. thesis, and Z. Phys. **31**, 253 (1925)

M. Kac, Toeplitz matrices, transition kernels and a related problem in probability theory. Duke Math. J. **21**, 501 (1954)

W. Lenz, Contributions to the understanding of magnetic properties in solid bodies (in German). Physikalische Zeitschrift **21**, 613 (1920)

E.W. Montroll, R.B. Potts, J.C. Ward, Correlations and spontaneous magnetization of the two-dimensional ising model. J. Math. Phys. **4**, 308 (1963)

L. Onsager, Crystal statistics. I. A two-dimensional model with an order-disorder transition. Phys. Rev. **65**, 117 (1944)

M. Plischke, B. Bergersen, *Equilibrium Statistical Physics*, 2nd edn. (World Scientific, 1994)

T.D. Schultz, D.C. Mattis, E.H. Lieb, Two-dimensional ising model as a soluble problem of many fermions. Rev. Mod. Phys. **36**, 856 (1964)

G. Szego, *Orthogonal Polynomials*, American Mathematical Society, Vol. XXIII (Colloquium Publications, 1939)

C.N. Yang, The spontaneous magnetization of a two-dimensional ising model, Phys. Rev. **85**, 808 (1952). (For the rectangular lattice see C.H. Chang, The spontaneous magnetization of a two-dimensional rectangular ising model. Phys. Rev. **88**, 1422, 1952)

Chapter 18
Monte Carlo

18.1 Motivation

The Monte Carlo method is a technique for calculating statistical averages using a computer. The method is particularly suited to the calculation of thermal averages of properties of an interacting many-body system. It was originally formulated for the case of an interacting gas, but it is easily adapted for use in just about any problem in classical statistical mechanics. The extension to quantum systems is more problematic.

Metropolis et al. (1953) considered a gas of N particles in a box with periodic boundary conditions. The potential energy for this system is

$$V = \frac{1}{2} \sum_{i,j=1}^{N} v(r_{i,j}) \tag{18.1}$$

and the average of any function $\mathcal{O}$ of the particle positions is

$$\langle \mathcal{O} \rangle = \frac{\int \mathcal{O}(\{\vec{r}_i\}) e^{-\beta V} d\{\vec{r}_i\}}{\int e^{-\beta V} d\{\vec{r}_i\}}. \tag{18.2}$$

Thus, the calculation of $\langle \mathcal{O} \rangle$ requires the evaluation of an $N \times d$-dimensional integral for N particles in d dimensions.

The simplest Monte Carlo algorithm for evaluating this average involves turning the sum into an integral, evaluated at random points in configuration space. Then

$$\langle \mathcal{O} \rangle \to \frac{\sum_{i=1}^{M} \mathcal{O}_i e^{-\beta V_i}}{\sum_{i=1}^{M} e^{-\beta V_i}}, \text{ as } M \to \infty, \tag{18.3}$$

where $\mathcal{O}_i$ and V_i are evaluated at the point $\{\vec{r}_i\}$ in configuration space.

A. J. Berlinsky and A. B. Harris, *Statistical Mechanics*, Graduate Texts in Physics,
https://doi.org/10.1007/978-3-030-28187-8_18

This algorithm is inefficient because, for random points distributed uniformly in configuration space, the factor $e^{-\beta V_i}$ is almost always very small. A more general prescription for calculating the integral would involve sampling configuration space nonuniformly, using a probability distribution $p_i \equiv p(\{\vec{r}_i\})$. In this case, the average is given by

$$\langle \mathcal{O} \rangle = \lim_{M \to \infty} \frac{\sum_{i=1}^{M} \mathcal{O}_i p_i^{-1} e^{-\beta V_i}}{\sum_{i=1}^{M} p_i^{-1} e^{-\beta V_i}} \tag{18.4}$$

where the factors p_i^{-1} correct for the nonuniform sampling probability. This approach is called "importance sampling" because p_i weighs the relative "importance" of different regions of configuration space to the sum. By far the simplest and most useful prescription for importance sampling in statistical mechanics is

$$p_i = \frac{e^{-\beta V_i}}{\sum_{j=1}^{M} e^{-\beta V_j}}, \tag{18.5}$$

which leads to the simple result

$$\langle \mathcal{O} \rangle = \lim_{M \to \infty} \frac{1}{M} \sum_{i=1}^{M} \mathcal{O}_i. \tag{18.6}$$

18.2 The Method: How and Why It Works

The Big Question is then: How does one generate a set of random points which sample configuration space with such relative probabilities? The answer, which is the prescription invented by Metropolis and co-workers, is as follows:

Assume some initial configuration with a particle at the point $\vec{r} = (x, y, z)$. A Monte Carlo step begins with choosing another nearby point, $\vec{r}\,' = (x + \alpha\ell, y + \beta\ell, z + \gamma\ell)$ where α, β, γ are random numbers between -1 and 1, and ℓ is the maximum step size which is chosen to optimize the procedure. For this particular problem, one would expect ℓ to be of order or smaller than the average interparticle spacing.

Next one decides whether the particle should move from $\vec{r}$ to $\vec{r}\,'$ or remain at $\vec{r}$ according to the following scheme:

1. Let $\Delta E = V(\vec{r}\,') - V(\vec{r})$. If $\Delta E < 0$ the particle is moved to $\vec{r}\,'$.
2. If $\Delta E > 0$, one generates a random number $0 \leq x \leq 1$. If $x < e^{-\beta \Delta E}$, the particle is moved to $\vec{r}\,'$.
3. Otherwise the particle remains at $\vec{r}$.

Whatever the outcome, this constitutes a single Monte Carlo step for a particle. A measurement would typically be taken after N such steps, one MC step per particle.

Why does this scheme work?

1. It is "ergodic". Every particle can reach every point in the box.
2. After many steps, a long sequence of further steps will correspond to a Boltzmann distribution.

To understand this, we need to look more closely at the procedure. Typically, we start the system off in some random configuration which could be highly improbable. We then perform i MC moves per particle to allow the system to relax. Next we perform M more steps per particle, generating an "ensemble" of M configurations. We call this the "ith ensemble," where i simply labels the method used to prepare it. Thermal averages depend on the frequency with which (or number of times that) each state occurs in the ensemble. We call this frequency $\nu_r^{(i)}$ for state r. Then

$$\sum_r \nu_r^{(i)} = M \tag{18.7}$$

If i is large enough that the system loses all memory of its initial state, then $\{\nu_r^{(i)}\}$ approximates a Boltzmann distribution. That is

$$\nu_r^{(i)}/\nu_s^{(i)} = e^{-\beta(E_r - E_s)} \tag{18.8}$$

This happens because the Monte Carlo algorithm drives the system toward such a distribution.

Consider the situation where each state in the ensemble differs from its neighbors by the move of a single particle, i.e., by one Monte Carlo decision, and consider the transition from state j to state $j+1$ within this ensemble of M states.

Let P_{rs} be the probability of generating s from r (say by moving a particle from $\vec{r}$ to $\vec{r}\,'$) *before* deciding whether or not to accept the move. Then clearly

$$P_{rs} = P_{sr}. \tag{18.9}$$

Assume that $E_r > E_s$. Then the number of transitions from r to s that occur within the ensemble of M steps is

$$\nu_r^{(i)} P_{rs}$$

since this move is always accepted. The number of transitions from s to r is

$$\nu_s^{(i)} P_{sr} e^{-\beta(E_r - E_s)}$$

Thus, the contribution to $\nu_s^{(i)}$ due to transitions to or from the state r is

$$P_{rs}\left[\nu_r^{(i)} - \nu_s^{(i)} e^{-\beta(E_r - E_s)}\right]$$

and $\nu_s^{(i)}$ increases at the expense of $\nu_r^{(i)}$ if

$$\frac{\nu_r^{(i)}}{\nu_s^{(i)}} > e^{-\beta(E_r - E_s)} \tag{18.10}$$

The ratio $\nu_r^{(i)}/\nu_s^{(i)}$ is stationary with respect to transitions between the two states if

$$\frac{\nu_r^{(i)}}{\nu_s^{(i)}} = e^{-\beta(E_r - E_s)} \tag{18.11}$$

that is, if these frequencies follow a Boltzmann distribution.

These statements apply to the average behavior of the ensemble and hence are only meaningful when M is very large. They apply to all pairs of states which differ only by the position of one particle and for which $P_{rs} \neq 0$ However, since any state can be reached from any other state by a succession of moves, we conclude that the ensemble must approach a Boltzmann distribution.

All of the above can be applied to the Ising model. In fact, the algorithm seems even simpler. A Monte Carlo step involves trying to flip a single spin. If the energy to flip a spin is ΔE, then

1. If $\Delta E < 0$, the spin is flipped.
2. If $\Delta E > 0$, then we calculate a random number, $0 \leq x \leq 1$. If $x < e^{-\beta \Delta E}$, the spin is flipped.
3. Otherwise the spin is left unchanged. In each case the result is counted as a new configuration.

18.3 Example: The Ising Ferromagnet on a Square Lattice

Following the above prescription, we now examine the results of Monte Carlo simulations for the ferromagnetic Ising model. The actual results presented in this chapter were generated on laptop and desktop PCs of one of the authors. However, an excellent reference for this topic is the 1976 article by Landau (1976).

The raw output from the Monte Carlo simulation, for a given temperature and field, consists of time series, such as the total magnetization and total energy of the system after each cycle through the lattice. Consider the time series for the magnetization. At high temperatures, the distribution of magnetizations is, to a good approximation, a Gaussian centered at some average value with a mean square width proportional to the uniform susceptibility. As the temperature is lowered, this distribution broadens as the susceptibility increases. Of course, Monte Carlo simulations are simulations of finite systems for a finite number of timesteps. The trick is to try to infer the long-time behavior of the infinite system. In principle, one could approach this simply by simulating large enough systems for long enough times. However, we shall see that a more efficient approach is to study systems of varying sizes to see how their properties vary with increasing size.

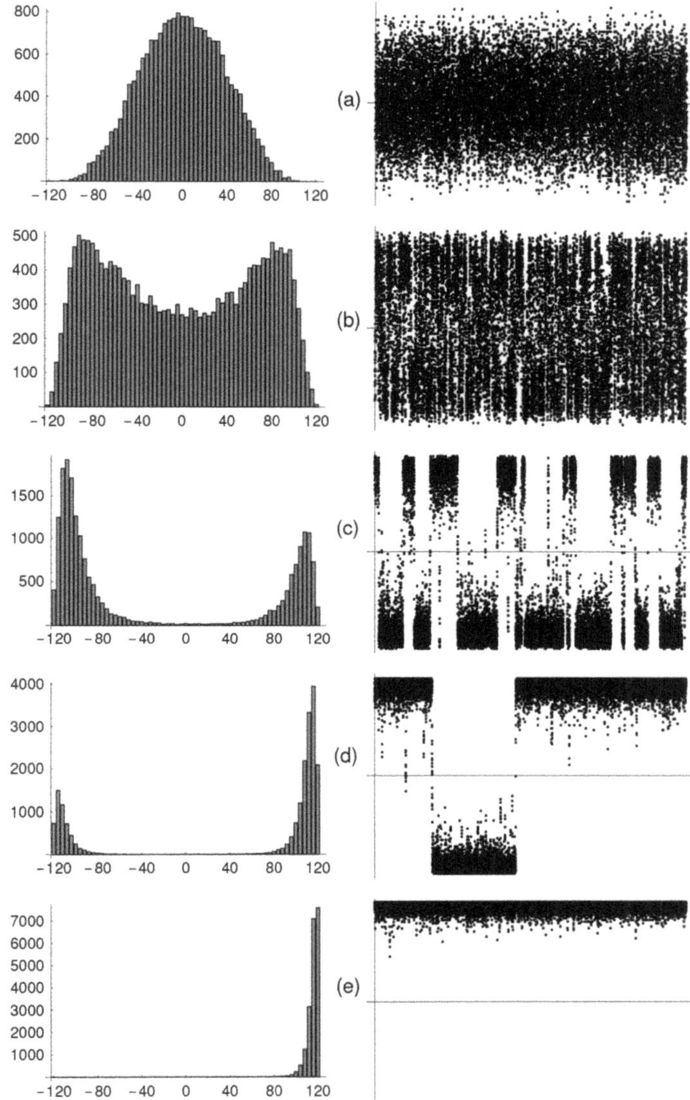

Fig. 18.1 Panels on the left are distributions of the total magnetization of an 11×11 lattice of ferromagnetically coupled Ising spins in zero field with periodic boundary conditions for 2×10^4 steps, each taken after one cycle through the lattice (1 MC step/site). The right hand panels are the time series from which the distributions are derived. The temperatures are **a** $T/J = 3.0$, **b** $T/J = 2.5$, **c** $T/J = 2.2$, **d** $T/J = 2.0$, **e** $T/J = 1.8$

We begin by studying one system of relatively small size as it is cooled from high to low temperature. Figure 18.1 shows simulation data for an 11×11 system of ferromagnetically coupled Ising spins with periodic boundary conditions on a square lattice in zero magnetic field. The left-hand panels are distributions of the total magnetization of the finite lattice, with each measurement taken after one sweep through the lattice (1 MC step/site). The right-hand panels are the time series from which the distributions are derived. They are plots of 20,000 successive values of the total magnetization of 121 spins. The temperatures range from $T/J = 3.0$ to $T/J = 1.8$. For reference, the mean field T_c is $4J$ and the exact T_c for the infinite system is around $2.269J$. (In this chapter, we set the Boltzmann constant, k equal to 1.) At the highest temperature, (a), $T/J = 3$, the system fluctuates rapidly and spends most of its time around $M = 0$. At $T/J = 2.5$, (b), the system spends somewhat less time around $M = 0$, and the distribution is double-peaked. At $T/J = 2.2$, (c), the system spends very little time around $M = 0$, and spends most of its time around $M = \pm L^2$, where L^2 is the number of spins. Note that the number of zero crossings is much less here than it was at higher temperature. Finally, for $T/J = 2.0$, (d) the magnetization changes sign only a few times in 20,000, and for $T/J = 1.8$, (e), it remains positive for the entire time.

What are we to infer from these measurements? We know that, for finite-size systems, there is no phase transition and no long-range ordered state. However, the simulations show that with decreasing temperature the system spends more and more time with most spins up or most spins down, and less and less time around magnetization zero. As the size of the system increases, these time scales grow longer. In the limit of infinite system size, below the critical temperature, the time to reverse the (infinite) total magnetization becomes infinite, and the system spends all of its time around one sign of the magnetization. This phenomenon is called "ergodicity breaking" because the system does not sample all energetically accessible states uniformly. Instead, it stays stuck at positive or negative magnetization.

How can we study this behavior when the most interesting phenomena correspond to infinite system sizes and times? We begin by looking at the temperature dependencies of various quantities as functions of temperature and system size. Consider two such quantities, the internal energy

$$E = \langle \mathcal{H} \rangle, \tag{18.12}$$

and the magnetization

$$M = \langle | \sum_i S_i | \rangle. \tag{18.13}$$

Of the two, the energy is by far the more boring. It is a well-defined, rather smooth function which measures nearest-neighbor correlations. Its behavior, for a range of system sizes, is shown in Fig. 18.2b. Each data point represents an average over 20,000 MC steps per spin.

By contrast, the average magnetization is almost pathological. For any given system size, for large enough averaging time, the average of M is zero. At low T, where sign reversals of M are rather rare, then it is reasonable to ignore them and simply average $|M|$. At higher temperatures, where the distribution is roughly Gaussian with zero mean, averaging $|M|$ clearly gives the wrong answer. However,

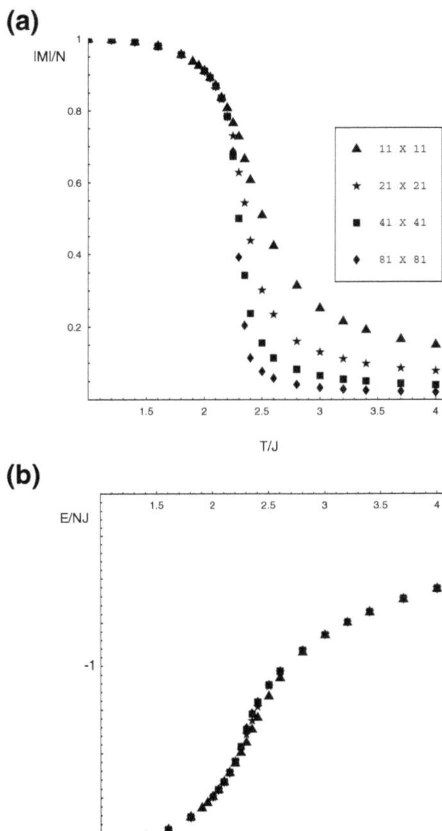

Fig. 18.2 **a** Magnetization versus temperature for $L \times L$ lattices. **b** Internal energy versus temperature for the same lattice sizes as in (**a**). Each data point represents an average over 20,000 MC steps per spin

this quantity approaches the correct result in the limit of infinite system size, even above T_c, because the width of the distribution of magnetizations scales like $\sqrt{N}$, while the magnetization itself scales like N. Therefore, $\langle |M|/N \rangle \to 0$, albeit slowly, above T_c as $N \to \infty$. Figure 18.2a shows the temperature dependence of $\langle |M|/N \rangle$ for different system sizes.

Neither the internal energy nor the magnetization are particularly well suited for locating the transition temperature, T_c or for characterizing the critical behavior around T_c. These properties are more clearly reflected in the fluctuation averages, the heat capacity,

$$C = \frac{\langle \mathcal{H}^2 \rangle - \langle \mathcal{H} \rangle^2}{T^2} = \frac{\partial U}{\partial T} \tag{18.14}$$

and the magnetic susceptibility,

$$\chi = \frac{\langle (\sum_i S_i)^2 \rangle - \langle \sum_i S_i \rangle^2}{T} \tag{18.15}$$

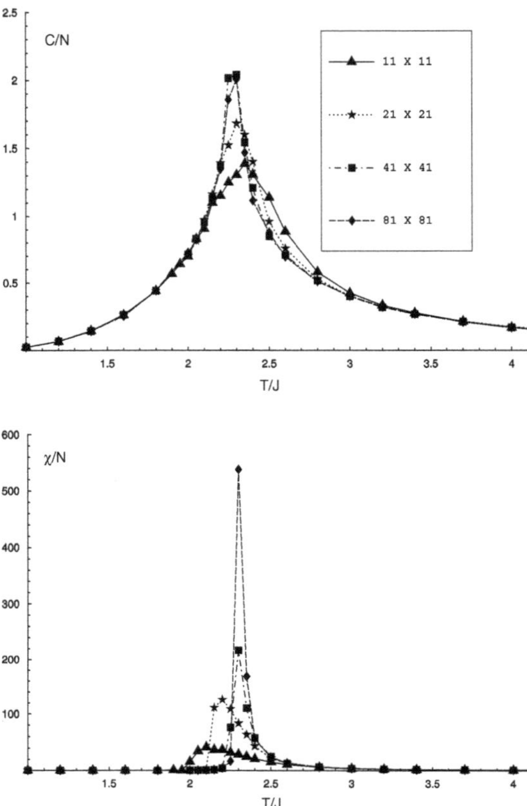

Fig. 18.3 Specific heat (upper figure) and magnetic susceptibility per spin (lower figure) for $L \times L$ lattices

Monte Carlo data for these two quantities for the simulations used to generate Fig. 18.2 are shown in Fig. 18.3.

From these results, we can draw a number of conclusions. First of all, away from the critical point where the fluctuations peak, the various thermodynamic quantities are relatively insensitive to system size, except for the high-temperature tail of $|M|/N$. However, above T_c, $\langle M \rangle/N$, the actual magnetization is exactly zero, independent of system size, so it is fair to say that the the dependence of these quantities on system size, for this system with periodic boundary conditions, is mainly significant close to T_c. The situation could be different if different boundary conditions were imposed. For example, open boundary conditions, in which spins on the boundaries saw fewer nearest neighbors, would lead to effects of order $\sqrt{N}$ in all thermodynamic quantities.

Figure 18.3 illustrates a weakness of the Monte Carlo approach, as presented so far. In order to explore the critical behavior close to T_c it is necessary to study larger and larger systems at closely spaced temperatures around T_c Even then, fluctuations in the MC results, make it difficult to fit the actual thermal behavior at T_c. We next consider a method that is particularly well suited to studying thermal behavior in the critical region.

18.4 The Histogram Method

In the prescription given so far, simulations of a system at a single temperature are used only to calculate thermal averages at that temperature. It turns out that this is a rather wasteful use of simulation results. To see why, consider an alternative approach to calculating the properties of a system of N spins. We define the quantities

$$S = \sum_{(i,j)} S_i S_j , \tag{18.16}$$

$$M = \sum_i S_i . \tag{18.17}$$

One can imagine enumerating all the states of an Ising system and sorting them according to values of S and M. Although possible, in principle, this procedure involves $2^N \approx 10^{3N/10}$ steps which is impractical for N larger than about 40–50. The result of such a calculation would be the function $\mathcal{N}_N(S, M)$, the number of states with zero field energy $E = -JS$ and magnetization M for a system of N spins. Knowledge of $\mathcal{N}_N(S, M)$ allows the calculation of thermal averages, as a function of $T = 1/\beta$ and H, of quantities that depend only on the energy and magnetization using the relation

$$\langle \mathcal{O} \rangle_{T,H} = \frac{\sum_{S,M} \mathcal{O}(S, M) \mathcal{N}_N(S, M) e^{\beta(JS+HM)}}{\sum_{S,M} \mathcal{N}_N(S, M) e^{\beta(JS+HM)}} \tag{18.18}$$

Alternatively, one can calculate by Monte Carlo simulation a histogram which is the distribution of occurrences of states with specific values of (S, M) at a given temperature $T_0 = 1/\beta_0$, in zero magnetic field.

$$\mathcal{P}_N(S, M; T_0) \propto \mathcal{N}_N(S, M) e^{\beta_0 JS}. \tag{18.19}$$

where the proportionality is only exact in the limit of infinite averaging time. Then, from Eq. (18.18), the average can be written as

$$\langle \mathcal{O} \rangle_{T,H} \approx \frac{\sum_{S,M} \mathcal{O}(S, M) \mathcal{P}_N(S, M; T_0) e^{(\beta-\beta_0)JS+\beta HM}}{\sum_{S,M} \mathcal{P}_N(S, M; T_0) e^{(\beta-\beta_0)JS+\beta HM}} \tag{18.20}$$

where again, the approximation becomes exact for infinite averaging times. Equation (18.20) suggests that, in principle, thermal averages at any temperature can be

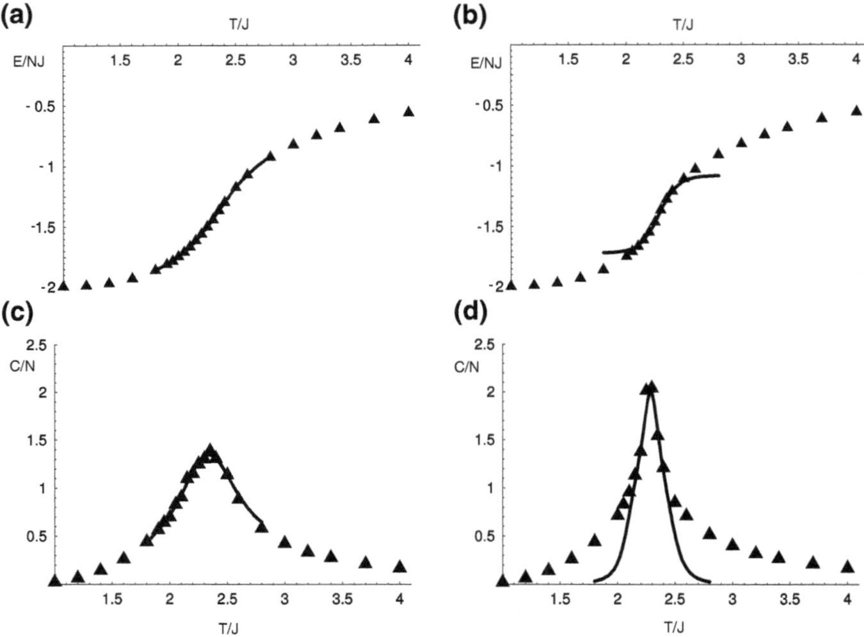

Fig. 18.4 Comparison of results of Monte Carlo simulations at discrete temperatures (triangles) to results of the histogram method based on a simulation at $T_c = 2.269J$ (solid lines). **a** and **c** are the internal energy and specific heat for an 11×11 lattice. **b** and **d** are the internal energy and specific heat for a 41×41 lattice

calculated using histograms computed at a single temperature T_0 in zero magnetic field.

As a practical matter, for finite averaging times, Eq. (18.20) cannot be used to compute averages far from $(T, H) = (T_0, 0)$ because the histogram generated by the finite Monte Carlo average will not contain accurate information about the density of states far away from the values of S and M that are most probable for temperature T_0. However, Eq. (18.20) does allow a smooth extrapolation of results to a region around the point $(T_0, 0)$. If T_0 is chosen to equal the infinite-system T_c then this method allows continuous results to be obtained in the critical region. The results of such a histogram calculation are shown by the solid lines in Fig. 18.4 and compared to simulations at discrete temperature points, denoted by triangles, for 11×11 and 41×41 spin systems.

The figure shows that the histogram results for the smaller system are accurate over a much wider temperature range than those for the larger system. Why is this? Consider for simplicity functions of the energy in zero magnetic field so that the energy histogram consists of a sum over all magnetizations. Then the distribution of energies at some temperature T_0 will be approximately Gaussian around the average energy with a mean square width that scales like the heat capacity. Since both the heat capacity and the mean energy scale like the system size, the ratio of the width

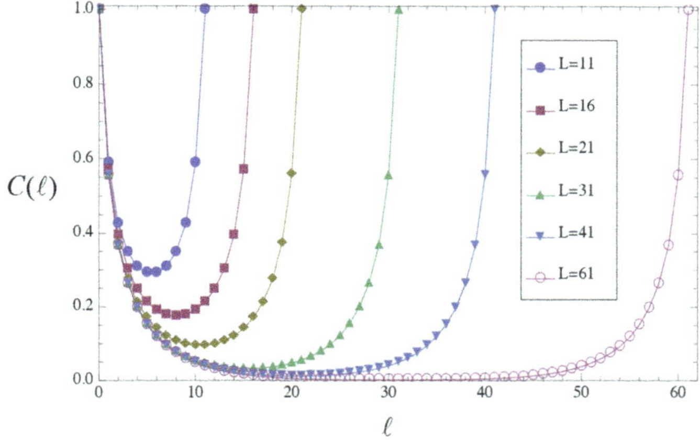

Fig. 18.5 Spin–spin correlation function for $T > T_c$, plotted as a function of the number of lattice spaces along the x or y direction for a range of system sizes, as indicated in the legend, and for $T - T_c = 0.1T_c$. The data show that the correlation length converges to a finite value of a few lattice spaces as the system size $L \to \infty$ and that the correlation function itself decays to zero. Data points are averages of 100,000 MC steps per site

to the mean energy scales like $1/\sqrt{N}$. Thus, particularly for larger systems, the distribution function generated in a finite averaging time at a single temperature will be inaccurate away from its most probable value because those regions will only rarely if ever be sampled. This means that a single histogram is only useful for extrapolation nearby the averaging temperature used in its generation. A useful generalization of the method is to generate multiple histograms and use histograms generated at nearby temperatures and fields to interpolate between them.

18.5 Correlation Functions

In addition to studying the thermodynamic functions directly, useful information can be learned by using Monte Carlo to calculate the spin–spin correlation functions

$$C(\vec{r}_i - \vec{r}_j) = \langle S_{\vec{r}_i} S_{\vec{r}_j} \rangle. \tag{18.21}$$

For the case where $\vec{r}_i - \vec{r}_j$ is a nearest-neighbor vector, this correlation function is proportional to the internal energy.

As we saw in Chap. 13, the generic, asymptotic behavior of the correlation function above T_c is given by Eq. (13.34) which is a product of a power-law decay, with exponent, $-(d - 2 + \eta)$, times an exponentially decaying function of r/ξ, where ξ is a temperature-dependent correlation length that diverges like $|T - T_c|^{-\nu}$ at T_c. On the other hand, below T_c as $r \to \infty$, the correlation function must approach a

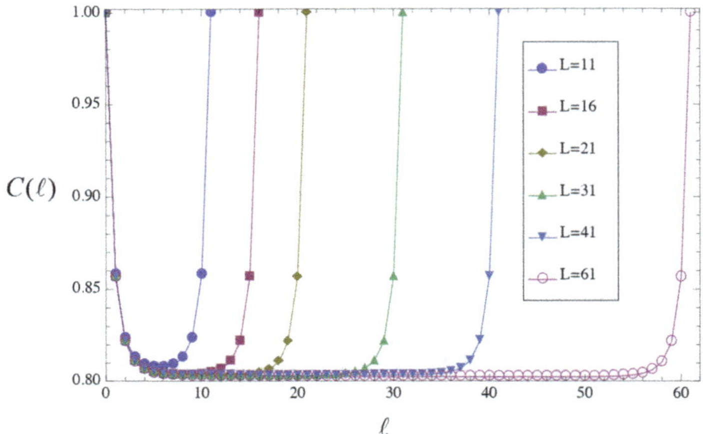

Fig. 18.6 Spin–spin correlation function for $T < T_c$, plotted as a function of the number of lattice spaces along the x or y direction for a range of system sizes, as indicated in the legend, and for $T_c - T = 0.1T_c$. Note that the vertical scale ranges from 0.8 to 1. The data show that the correlation function decays to a value of $m^2 \approx 0.8$ and that the correlation length for this decay is smaller than for $T = 1.1T_c$, shown in the previous figure. Data points here are also averages of 100,000 MC steps per site

value equal to the square of the average of a single spin because two spins far apart are uncorrelated, i.e., to the square of the site magnetization m. Thus, for $T < T_c$, $C(r)$ decays from 1 at $r = 0$ to m^2 for $r = \infty$. Of course, we can only Monte Carlo finite systems. If we do this for systems with periodic boundary conditions, then the correlation function exhibits a minimum half-way across the system. This behavior is demonstrated by the correlation functions shown in Figs. 18.5 and 18.6 for $|T - T_c| = 0.1T_c$ for a range of system sizes.

Figure 18.5 shows that the correlation function does, in fact, decay to zero for sufficiently large system size. Starting at 1 for $\ell = 0$, it decays to its minimum value at $\ell = L/2$ and then increases back to 1 at $\ell = L$. Furthermore, it exhibits a finite correlation length, which is a few lattice spacings for the temperature shown, again for sufficiently large system size. The corresponding case for $T < T_c$ is shown in Fig. 18.6. There are two main differences between this and the previous figure for $T > T_c$. First the vertical scale ranges only from 0.8 to 1 since the magnetization squared is just over 0.8 for a sufficiently large system at this temperature. Second, it is clear that the decay length is noticeably shorter for this case. This might seem surprising since the power-law exponent ν is the same above and below T_c. However, the amplitude of the power law is different for the two cases in a way that is similar to the behavior of the mean field amplitudes of the susceptibility which was discussed in Chap. 8 and displayed in Eq. (8.26). This is known from the exact solution and from the general behavior of scaling functions. Furthermore, Renormalization Group Theory will teach us that the ratio of amplitudes such as this, above and below T_c is "universal" (Aharony 1976; Aharony and Hohenberg 1976) in the sense that it

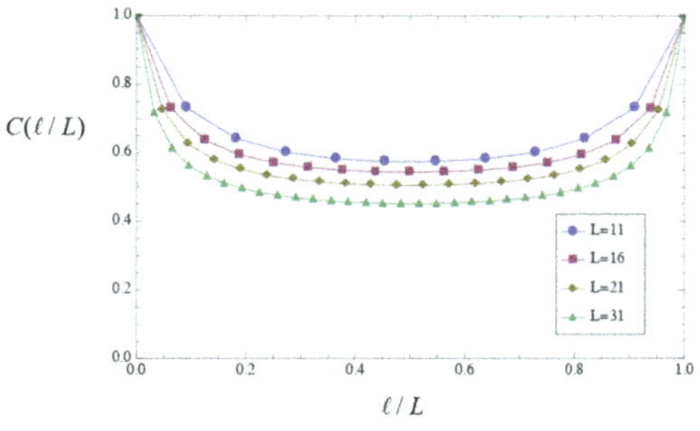

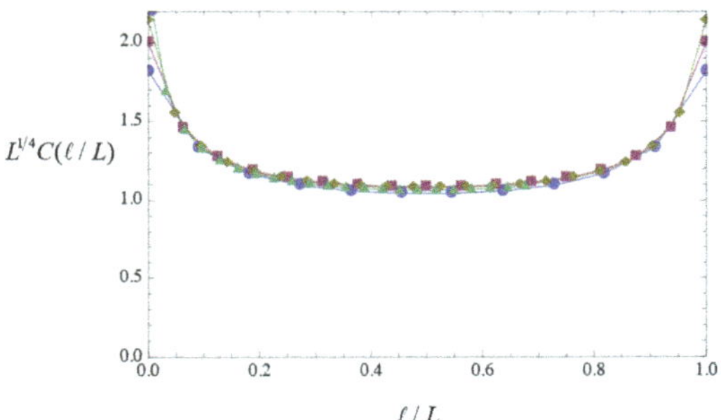

Fig. 18.7 (Upper graph) Spin–spin correlation function for $T = T_c$, plotted as a function of ℓ/L along x or y, where ℓ is the distance in lattice spaces and L is the system size, for a range of system sizes as indicated in the legend. (Lower graph) The same data scaled by $L^{1/4}$ which causes them to collapse onto a single curve. Data points here are also averages of 100,000 MC steps per site

does not depend on any model parameter (which in this case could only be J or equivalently T_c), but is a pure number depending only on the dimensionality of the lattice and the number of spin components. For the 2D Ising model, this universal amplitude ratio is known to be exactly $\frac{1}{2}$ (Privman et al. 1991) That is, the correlation length below T_c is exactly half that at the corresponding temperature above T_c.

The behavior of the correlation function at exactly $T = T_c$ is strikingly different from the two previous cases. This is shown in Fig. 18.7, where the correlation functions are plotted versus distance as a fraction of the system size. All the curves have essentially the same shape, regardless of system size, reflecting the fact that there is no length scale for the power-law decay at a critical point. In fact, after scaling with a factor of $L^{1/4}$, the data collapse onto a single curve. We discuss the significance of the exponent 1/4 below.

18.6 Finite-Size Scaling for the Correlation Function

We will see below, by analyzing the magnetization, that the scaling behavior of the spin–spin correlation function can be written as

$$C(\ell, \epsilon, L) = L^{-2\beta/\nu} G(\ell/L, \epsilon L^{1/\nu}), \tag{18.22}$$

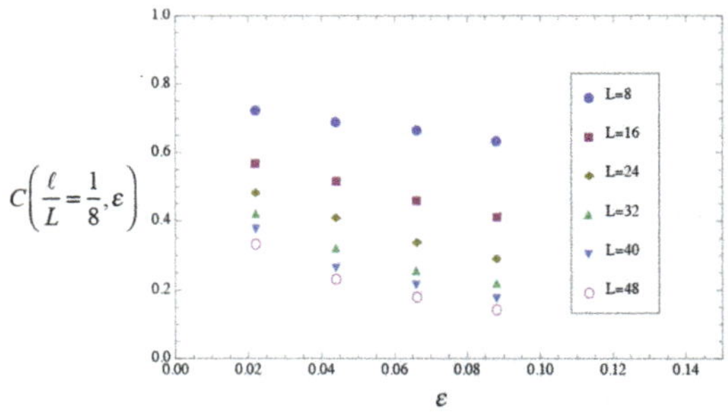

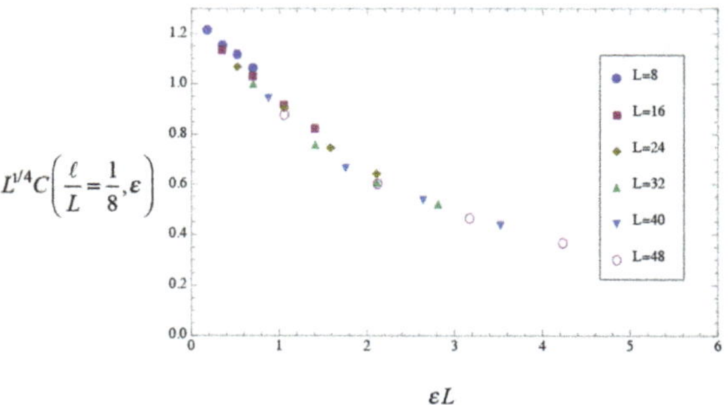

Fig. 18.8 (Upper graph) Spin–spin correlation function versus temperature for $\epsilon = (\mathrm{T} - \mathrm{T_c})/\mathrm{T_c} > 0$ and $\ell = L/8$ along x or y, where ℓ is the distance in lattice spaces and L is the system size, for a range of system sizes as indicated in the legend. (Lower graph) The same data scaled by $L^{1/4}$ plotted versus ϵL which causes them to collapse onto a single curve. Data points here are averages of 50,000 MC steps per site

where $\epsilon = (\mathrm{T} - \mathrm{T_c})/\mathrm{T_c}$. The first argument of G is the distance between the two spins in the correlator written as a fraction of the system size, and the second argument is essentially $(L/\xi)^{1/\nu}$. From Table 17.1, we know the exact 2D Ising exponents, $\beta = 1/8$ and $\nu = 1$, so the exponent $2\beta/\nu = 1/4$ which explains why the factor $L^{1/4}$ leads to the data collapse shown in Fig. 18.7 for $\epsilon = 0$.

We can test the full, two-parameter scaling of Eq. (18.22) by studying the temperature and size dependence of the correlation function for some fixed value of ℓ/L. An example is shown in Fig. 18.8 for the case of $\ell/L = 1/8$. In the upper panel of the figure, values of the correlation function at $\ell/L = 1/8$ are plotted versus ϵ for different system sizes. In the lower panel, the correlation function is scaled by $L^{1/4}$ and the horizontal axis is scaled by $L^{1/\nu} = L$. Through the magic of scaling, the points are rearranged both vertically and horizontally to fall approximately on a single curve. The scaling works well, particularly for larger values of L.

18.7 Finite-Size Scaling for the Magnetization

For the case of periodic boundary conditions, we expect to be able to write the magnetization per site, M in terms of a scaling function g as

$$M(L,\xi) = L^a g(L/\xi) \tag{18.23}$$

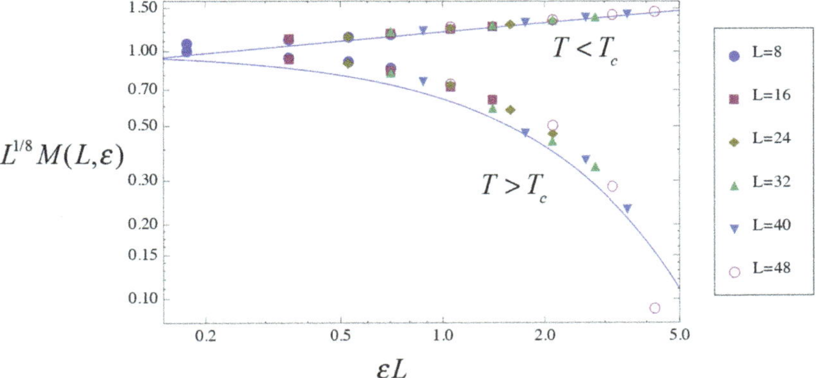

Fig. 18.9 Finite-size scaling plot for the magnetization versus scaled reduced temperature. The magnetization is derived for periodic boundary conditions from simulations of the spin–spin correlation function along x or y evaluated at a distance equal to half the sample size, L. For $T < T_c$ the solid line is $1.2x^{1/8}$. For $T > T_c$ the correlation function decays exponentially as discussed in the text. Each point is an average of 50000 MC steps per site

This is essentially a dimensional analysis argument. Since the linear dimension of the system, L, and the correlation length, ξ, are the only two lengths in the problem, g must be a function of their dimensionless ratio. Then the factor L^a takes care of any other possible L-dependence.

Alternatively, as we did for the correlation function, we can write Eq. (18.23) as

$$M(L, \epsilon) = L^a X^0(\epsilon L^{1/\nu}) \tag{18.24}$$

As $L \to \infty$, we know that $M \sim \epsilon^\beta$ for $T < T_c$ where $\beta = 1/8$ and $\nu = 1$ for the 2D Ising model. So it is necessary that $X^0(\epsilon L^{1/\nu}) \to \left(\epsilon L^{1/\nu}\right)^\beta$ in order that M have the correct ϵ dependence. Since the L-dependence must drop out as $L \to \infty$, a must satisfy $a = -\beta/\nu$ so that

$$M(L) = L^{-\beta/\nu} X^0(\epsilon L^{1/\nu}). \tag{18.25}$$

This result is the finite-size scaling form for the magnetization. Then the corresponding result for the spin–spin correlation function, which is essentially the magnetization–magnetization correlation function, has the prefactor, $L^{-2\beta/\nu}$ as was assumed above in Eq. (18.22).

If we define the square of the magnetization, M^2, as the value of the correlation function at a distance $L/2$, then, in the limit $L \to \infty$ for $\mathrm{T} < \mathrm{T_c}$, this should agree with the magnetization calculated using Eq. (18.13). There is a difference, however, between the two definitions for $\mathrm{T} > \mathrm{T_c}$. Equation (18.13) prescribes averaging the absolute value of the sum of L^2 spins over many Monte Carlo steps. Each sum will be of order $1/L$ per site and hence the magnetization should go to zero slowly, like $1/L$. However, as we have seen, calculating M^2 from the correlation function evaluated at $L/2$ gives a value proportional to $e^{-L/2\xi}$ so that $M(L/\xi) \sim e^{-L/4\xi}$. Thus, the two methods for simulating M can give quite different results for $T > T_c$, with the correlation function method giving much smaller results for large systems well above T_c.

Scaled results for the magnetization obtained from the correlaton function, $C(L/2, \epsilon, L)$ for a range of values of L and ϵ are shown in Fig. 18.9 for $T < T_c$ and $T > T_c$. The results for $T < T_c$ are well approximated by the power-law fit, $1.2(\epsilon L)^{1/8}$. For $T > T_c$, the solid line is the function $e^{-L/4\xi}$ where the approximate expression, Privman et al. (1991)

$$1/\xi(\epsilon > 0) \approx [2\ln(1+\sqrt{2}]\epsilon\ , \tag{18.26}$$

has been used.

The type of argument used above for finite-size scaling can also be used to study the dependence of thermodynamic functions on field and on $T - T_c$ near the critical point $T = T_c,\ \ H = 0$. In fact, this approach can be extended to analyze dynamic properties such as the conductivity near the critical point.

18.8 Summary

In this chapter, we have described the Monte Carlo method for the simulation of classical many-body thermal systems, and illustrated its use by studying various properties of the 2D Ising model. We have explained how Monte Carlo data can be utilized more efficiently through the histogram method, and we have looked in detail at the spin–spin correlation as a means of justifying and applying the finite-size scaling method. The student is encouraged to try out these methods as an Exercise.

18.9 Exercises

1. Do a Monte Carlo calculation for the nearest-neighbor, one-dimensional Ising ferromagnet of L spins with periodic boundary conditions.

Write your own computer program. Concentrate on calculating the spin–spin correlation function

$$f_n = \frac{1}{N}\sum_{i=1}^{N}\langle S_i S_{i+n}\rangle, \quad n = 0, \ldots, N-1$$

for various values of N

Given f_n as a function of T, generate plots of the internal energy and the magnetic susceptibility per spin versus temperature for different values of N. By plotting f_n

versus n, determine the correlation length ξ as a function of T. Compare it to exact results for the 1D Ising model

2. Perform similar calculations of the correlation function for the 2D Ising model,

$$C(\ell, m, T) = \frac{1}{L^2} \sum_{i,j=1}^{L} \langle S_{(i,j)} S_{(i+\ell,j+m)} \rangle. \tag{18.27}$$

3. Study the lattice Fourier transform of the spin–spin correlation functions for your simulations of the 1D and 2D Ising models. Examine how the shape of this "structure factor" behaves around $\mathbf{q} = 0$ as $T \to 0$ for the 1D case and for T around T_c for the 2D case.

References

A. Aharony, Dependence of Universal critical behaviour on symmetry and range of interaction, in *Phase Transitions and Critical Phenomena* vol. 6, ed. by C. Domb, M.S. Green (Academic Press, 1976)

A. Aharony, P.C. Hohenberg, Universal relations among thermodynamic critical amplitudes. Phys. Rev. B **13**, 3081 (1976)

D.P. Landau, Finite-size behavior of the Ising square lattice. Phys. Rev. B **13**, 2997 (1976)

N. Metropolis, A.W. Rosenbluth, M.N. Rosenbluth, A.H. Teller, E. Teller, Equation of state calculations by fast computing machines. J. Chem. Phys. **21**, 1087 (1953)

V. Privman, P. C. Hohenberg, A. Aharony, in *Phase Transitions and Critical Phenomena*, ed. by C. Domb, J.L. Lebowitz, vol. 14, Chap. 1 (Academic Press, 1991), 1–134 & 364–367

Chapter 19
Real Space Renormalization Group

The renormalization group technique is a method in which the free energy for a system with one degree of freedom per unit volume and coupling constants $(K_1, K_2, \ldots)$ is related to the free energy of a system with the same Hamiltonian but with only one degree of freedom per volume L^d with $L > 1$, and coupling constants $(K_1', K_2', \ldots)$, as shown in Fig. 13.1. The correlation length in the primed system is smaller by a factor of L than that of the unprimed system. In principle, one iterates this transformation until the correlation length becomes of order one lattice constant. At that point, one may treat the resulting Hamiltonian using mean-field theory. This program sounds nice, but the big question is how to implement it. We begin our discussion of the renormalization group approach by applying it to the one-dimensional Ising model.

19.1 One-Dimensional Ising Model

We write the Hamiltonian for the one-dimensional Ising model as

$$-\beta\mathcal{H} = K\sum_i S_i S_{i+1} + H\sum_i S_i + NA\,. \tag{19.1}$$

Note that we have included a constant, NA, in the Hamiltonian. In the initial Hamiltonian this constant is zero. We will see in a moment why it is necessary to include such a term. The partition function for N spins is

A. J. Berlinsky and A. B. Harris, *Statistical Mechanics*, Graduate Texts in Physics,
https://doi.org/10.1007/978-3-030-28187-8_19

$$\begin{aligned} Z_N &= e^{NA}\text{Tr}\prod_i e^{KS_iS_{i+1}+\frac{H}{2}(S_i+S_{i+1})} \\ &= e^{NA}\text{Tr}\, e^{KS_1S_2+\frac{H}{2}(S_1+S_2)}\, e^{KS_2S_3+\frac{H}{2}(S_2+S_3)}\dots \\ &= e^{NA}\text{Tr}\prod_{i\ \text{odd}} Z(S_i, S_{i+2})\,, \end{aligned} \tag{19.2}$$

where

$$\begin{aligned} Z(S_1, S_3) &= \sum_{S_2=\pm 1} e^{K(S_1S_2+S_2S_3)+\frac{H}{2}(S_1+2S_2+S_3)} \\ &= e^{H(S_1+S_3)/2}\, 2\cosh[K(S_1+S_3)+H]\,. \end{aligned} \tag{19.3}$$

Now, in the spirit of Kadanoff's block spin argument (Kadanoff 1966), we ask whether the $Z(S_1, S_3)$ can be written as the exponential of an Ising Hamiltonian for spins 1 and 3. We write

$$e^{2A}Z(S_1, S_3) = e^{A'}e^{K'S_1S_3+\frac{H'}{2}(S_1+S_3)}\,. \tag{19.4}$$

There are three independent $Z(S_1, S_3)$, because $Z(S_1, S_3) = Z(S_3, S_1)$, and there are three new coupling constants, A', K', and H'. They satisfy

$$\ln Z(++) = A' - 2A + K' + H' \tag{19.5a}$$
$$\ln Z(+-) = A' - 2A - K' \tag{19.5b}$$
$$\ln Z(--) = A' - 2A + K' - H'\,. \tag{19.5c}$$

Then

$$A' = 2A + \frac{1}{4}\ln[Z(++)Z(+-)^2Z(--)] \tag{19.6a}$$
$$K' = \frac{1}{4}\ln\left[\frac{Z(++)Z(--)}{Z(+-)^2}\right] \tag{19.6b}$$
$$H' = \frac{1}{2}\ln\left[\frac{Z(++)}{Z(--)}\right]\,, \tag{19.6c}$$

where

$$Z(++) = 2e^H\cosh[2K+H] \tag{19.7a}$$
$$Z(+-) = 2\cosh H \tag{19.7b}$$
$$Z(--) = 2e^{-H}\cosh[2K-H]\,. \tag{19.7c}$$

This gives us a complete set of "recursion relations" for A', K', and H' in terms of A, K, and H. The mapping is exact and can be used to generate an infinite sequence of coupling constants. We now see why it was necessary to include the constant NA in the Hamiltonian. Even though A is initially zero, the recursion relations take the coupling constants into the larger space in which A is nonzero. It is a general feature of the renormalization group that the recursion relations act in a space of coupling constants which is larger than the initial, or "bare," values of the coupling constants in the initial Hamiltonian we are considering. In the present case, the situation is simple: the larger space simply includes an additional constant term. As we will see later, more generally the space in which the recursion relations act includes infinitely many variables and we must devise some way of truncating this space to include only the most important coupling constants. The ϵ-expansion, treated in the next chapter, provides a controlled way of carrying out this truncation.

We now return to analyze the one-dimensional problem. What does the sequence generated by the recursion relations of Eqs. (19.6) and (19.7) look like? As an example, consider the case $H = 0$. Then

$$Z(+-) = 2, \qquad Z(++) = Z(--) = 2\cosh 2K \ . \tag{19.8}$$

This means that $H' = 0$, or, in other words, that $H = 0$ is preserved under this transformation. It also implies that

$$K' = \frac{1}{2}\ln\cosh 2K \ . \tag{19.9}$$

For large K, it is clear that

$$K' \approx K - \frac{1}{2}\ln 2 < K \ . \tag{19.10}$$

For $K \ll 1$, we can expand the right hand side of Eq. (19.9) to obtain

$$K' \approx K^2 \ll K \ . \tag{19.11}$$

In fact, it is straightforward to show that, for any value of K not equal to 0 or ∞, $K' < K$, so that the sequence of coupling constants $\{K,\ K^{(1)},\ K^{(2)}, \ldots\}$ flows monotonically to zero as is shown in Fig. 19.1. There are only two "fixed points", $K = 0$, corresponding to weak coupling or high temperature, and $K = \infty$, corresponding to strong coupling or $T = 0$. These two fixed points are distinguished by their stability relative to an infinitesimal displacement in the value of K. The fixed point at $K = 0$ is stable, whereas that at $K = \infty$ is unstable. As we will see in Fig. 19.3, a critical point is represented by a somewhat different type of unstable fixed point.

Next, we consider how to calculate the the free energy from the recursion relations. To do this, we write

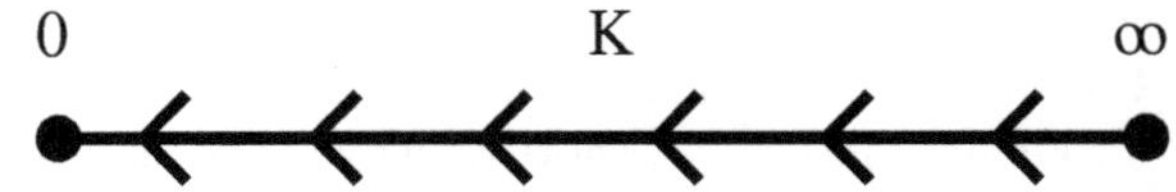

Fig. 19.1 Flow diagram for the one-dimensional Ising model in zero field

$$Z_N(A, K, H) = e^{NA} e^{Nf(K,H)} \tag{19.12a}$$
$$= Z_{N/2}(A', K', H') \tag{19.12b}$$
$$= e^{NA'/2} e^{Nf(K',H')/2} , \tag{19.12c}$$

where the function $f(K, H)$ is actually minus the free energy per site divided by the temperature, but, for simplicity, we will refer to it as the free energy. Taking the logarithm and dividing by N, we obtain

$$A + f(K, H) = \frac{1}{2}\left[A' + f(K', H')\right] . \tag{19.13}$$

In general, for the nth iteration, we can write

$$A + f(K, H) = \frac{1}{2^n}\left[A^{(n)} + f(K^{(n)}, H^{(n)})\right] . \tag{19.14}$$

We showed above that for $H = 0$ the coupling constant $K^{(n)}$ flows to zero. In fact, it is not difficult to show that $K^{(n)}$ flows to zero, even for nonzero H. So, for large n, $f\big(K^{(n)}, H^{(n)}\big) \to f(0, H^{(n)}) = \ln(2\cosh H^{(n)})$.

Next we calculate the free energy explicitly for $H = 0$ and $A = 0$. From Eqs. (19.6a), (19.6b) and (19.8), we can write

$$K^{(n+1)} = \frac{1}{2}\ln\cosh 2K^{(n)} \tag{19.15a}$$
$$A^{(n+1)} = 2A^{(n)} + \frac{1}{2}\ln\cosh 2K^{(n)} + \ln 2 . \tag{19.15b}$$

Using these relations, we find that

$$A^{(1)} = \frac{1}{2}\ln\cosh 2K + \ln 2 \tag{19.16a}$$
$$A^{(2)} = \ln\cosh 2K + \frac{1}{2}\ln\cosh(\ln\cosh 2K) + 3\ln 2 \tag{19.16b}$$
$$A^{(3)} = 2\ln\cosh 2K + \ln\cosh(\ln\cosh 2K) + \frac{1}{2}\ln\cosh(\ln\cosh(\ln\cosh 2K)) + 7\ln 2 . \tag{19.16c}$$

Extrapolating to $n = \infty$ and making use of Eq. (19.14) and the fact that $f(0, 0) = \ln 2$, we obtain a series solution for the free energy as a function of K

$$f(K,0) = \frac{1}{4}\Bigg[\ln\cosh 2K + \frac{1}{2}\ln\cosh(\ln\cosh 2K) + \frac{1}{4}\ln\cosh(\ln\cosh(\ln\cosh 2K)) + \cdots\Bigg] + \ln 2 \qquad (19.17a)$$

$$\equiv \frac{1}{4}\Bigg[\phi(2K) + \frac{1}{2}\phi^{(2)}(2K) + \frac{1}{4}\phi^{(3)}(2K) + \cdots\Bigg] + \ln 2\,, \qquad (19.17b)$$

where $\phi(u) = \ln\cosh u$, and $\phi^{(n)}(2K)$ denotes the n-fold nested function. It is not obvious that this result is identical to that we obtained previously for the 1d Ising model, namely

$$f(K,0) = \ln[\cosh K] + \ln 2\,. \qquad (19.18)$$

It is instructive to show that these two results are indeed equivalent. We start by obtaining an identity for $\psi(2u) \equiv \ln[1+\cosh(2u)] - \ln 2 = 2\phi(u)$:

$$\begin{aligned}
\psi(2u) &= \frac{1}{2}\ln[1+\cosh(2u)]^2 - \ln 2 \\
&= \frac{1}{2}\ln[1+2\cosh(2u)+\cosh^2(2u)] - \ln 2 \\
&= \frac{1}{2}\ln\Bigg\{\Bigg[\cosh(2u)\Bigg]\Bigg[2+\cosh(2u)+\frac{1}{\cosh(2u)}\Bigg]\Bigg\} - \ln 2 \\
&= \frac{1}{2}\ln\cosh(2u) - \frac{1}{2}\ln 2 + \frac{1}{2}\ln\Bigg[1+\frac{\cosh(2u)}{2}+\frac{1}{2\cosh(2u)}\Bigg] \\
&= \frac{1}{2}\ln\cosh(2u) - \frac{1}{2}\ln 2 + \frac{1}{2}\ln\Bigg[1+\cosh\Bigg(\ln\cosh(2u)\Bigg)\Bigg] \\
&= \frac{1}{2}\phi(2u) + \frac{1}{2}\psi[\ln\cosh(2u)] \\
&= \frac{1}{2}\phi(2u) + \frac{1}{2}\psi[\phi(2u)]\,.
\end{aligned} \qquad (19.19)$$

When we iterate this relation, we get

$$\psi(2u) = \frac{1}{2}\phi(2u) + \frac{1}{4}\phi[\phi(2u)] + \cdots\,. \qquad (19.20)$$

Since $\phi(u) = \psi(2u)/2$, we have

$$\phi(u) = \sum_{n=1}^{\infty} 2^{-n-1}\phi^{(n)}(2u)\,. \qquad (19.21)$$

Using this identity one sees that Eqs. (19.17b) and (19.18) are identical. Thus we have shown that the one-dimensional Ising model can be solved exactly by the approach which Kadanoff envisioned.

19.2 Two-Dimensional Ising Model

We would like to solve the two-dimensional Ising model by successively tracing over half the spins and then rewriting the resulting partition function as the exponential of a Hamiltonian of the same form. Not surprisingly, there are problems with this approach which we will now describe. We write the Hamiltonian for the two dimensional Ising model as

$$-\beta\mathcal{H} = \frac{K}{2}\sum_{i,j} S_{i,j}\left[S_{i+1,j} + S_{i,j+1} + S_{i-1,j} + S_{i,j-1}\right] + NA\ , \qquad (19.22)$$

where i and j are summed over integers. We can divide the square lattice up into two sublattices labeled by X's and O's as shown in Fig. 19.2. If we restrict the indices i, j to correspond to an X-site (by saying, for example that $i + j$ is an even integer), then the factor of 1/2 in $-\beta\mathcal{H}$ can be dropped.

If we trace over the $S_{i,j}$ on X-sites, then the partition function becomes

$$\begin{aligned} Z = e^{NA}\sum\dots\sum&\left[e^{K(S_X^1+S_X^2+S_X^3+S_X^4)} + e^{-K(S_X^1+S_X^2+S_X^3+S_X^4)}\right] \\ &\times\left[\text{similar factors involving 4 neighbors of every X site}\right], \end{aligned} \qquad (19.23)$$

where S_X^n is a spin on an O site which is the nth neighbor of spin X.

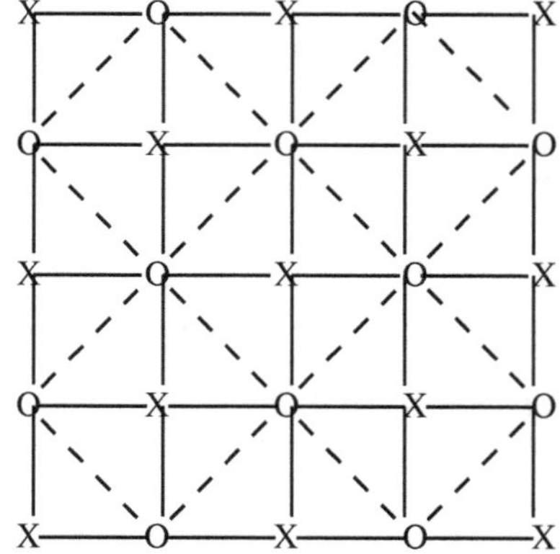

Fig. 19.2 Original lattice of sites (both X's and O's) with nearest neighbor bonds in heavy solid lines, which are mapped into a lattice of O's, with nearest neighbor bonds indicated by dashed lines. The correlation length (measured in lattice constants) is smaller in the renormalized lattice than that in the original lattice by a factor of $\sqrt{2}$

There are $N/2$ factors of the square brackets in the above equation We would like to write each factor in square brackets in the form

$$e^{2A}\left[e^{K(S_1+S_2+S_3+S_4)}+e^{-K(S_1+S_2+S_3+S_4)}\right]=e^{A'}e^{-\beta\mathcal{H}'(S_1,S_2,S_3,S_4)} . \tag{19.24}$$

Unfortunately, the form of $\mathcal{H}'$ which can satisfy this relation for all values of (S_1, S_2, S_3, S_4) is more complicated than that of the original Hamiltonian since it must involve, not only nearest neighbor, but also, second- neighbor and four-spin interactions. In other words, $\mathcal{H}'$ must be of the form

$$\begin{aligned} -\beta\mathcal{H}' = &\frac{1}{2}K_1(S_1S_2+S_2S_3+S_3S_4+S_4S_1) \\ &+K_2(S_1S_3+S_2S_4)+K_3S_1S_2S_3S_4 \end{aligned} \tag{19.25}$$

For this form of $\mathcal{H}'$, it is straightforward to solve for the four new coupling constants, A', K_1, K_2, and K_3, in terms of the original coupling constants, A and K. For the various configurations of (S_1, S_2, S_3, S_4), Eq. (19.24) yields

$$S_1=S_2=S_3=S_4 \rightarrow e^{2A}\left(e^{4K}+e^{-4K}\right)=e^{A'}e^{2K_1+2K_2+K_3} \tag{19.26a}$$

$$S_1S_2S_3S_4<0 \rightarrow e^{2A}\left(e^{2K}+e^{-2K}\right)=e^{A'}e^{-K_3} \tag{19.26b}$$

$$S_1=S_2=-S_3=-S_4 \rightarrow 2e^{2A}=e^{A'}e^{-2K_2+K_3} \tag{19.26c}$$

$$S_1=S_3=-S_2=-S_4 \rightarrow 2e^{2A}=e^{A'}e^{-2K_1+2K_2+K_3} . \tag{19.26d}$$

Equations (19.26c) and (19.26d) may be combined to give $K_1=2K_2$. Multiplying Eqs. (19.26b) and (19.26c) together gives

$$4e^{4A}\cosh 2K=e^{2A'}e^{-2K_2} \tag{19.27}$$

and multiplying Eqs. (19.26a) and (19.26b) together gives

$$4e^{4A}\cosh 2K\cosh 4K=e^{2A'}e^{6K_2} , \tag{19.28}$$

where we have used the fact that $K_1=2K_2$. Multiplying the cube of Eq. (19.27) by Eq. (19.28)

$$4^4e^{16A}\cosh^4 2K\cosh 4K=e^{8A'} \tag{19.29}$$

or

$$A'=2A+\frac{1}{2}\left[\ln 4\cosh 2K+\frac{1}{4}\ln\cosh 4K\right] . \tag{19.30}$$

Combining this result with Eq. (19.26b) gives the relation for K_3

$$K_3 = \frac{1}{8}\ln\cosh 4K - \frac{1}{2}\ln\cosh 2K\,. \tag{19.31}$$

Similarly, Eqs.(19.27) and (19.30) together with the fact that $K_1 = 2K_2$ give

$$K_2 = \frac{1}{8}\ln\cosh 4K \tag{19.32}$$

$$K_1 = \frac{1}{4}\ln\cosh 4K. \tag{19.33}$$

So we have succeeded in expressing the four new coupling constants, A', K_1, K_2, K_3, in terms of the two original coupling constants, A and K. However, it is clear that iterating this procedure would couple more and more spins in evermore complicated ways. Therefore this procedure, as it stands, is not manageable since it leads to a proliferation of coupling constants. Nevertheless, it is fair to ask whether this approach provides a basis for a useful approximation scheme.

Notice, for example, that for large K the coupling constants K_2 and K_3 are smaller than K_1. A somewhat extreme approximation would be to set K_2 and K_3 equal to zero, leaving the single recursion relation, Eq. (19.33), for the nearest neighbor coupling constant. Unfortunately, this recursion relation has the same property as the exact recursion relation that we derived for the one-dimensional Ising model. It flows smoothly from large to small coupling, and hence describes a system with no phase transition.

A better approximation, which also leads to a closed set of recursion relations, involves incorporating the effect of further-neighbor couplings into a renormalized nearest neighbor coupling. Physically, it is easy to see how this should work. Tracing out the spin at the X-site generates a second-neighbor interaction which is ferromagnetic ($K_2 > 0$). The effect of this extra ferromagnetic coupling is, to a good approximation, equivalent to an enhancement of the nearest neighbor coupling. (This is exactly the case in mean-field theory.) On the other hand, the four-spin coupling K_3 does not distinguish between local ferro- and antiferromagnetic order. This suggests the admittedly uncontrolled approximation of setting $K_3 = 0$ and defining a renormalized nearest neighbor coupling constant

$$K' = K_1 + K_2 = \frac{3}{8}\ln\cosh 4K. \tag{19.34}$$

In this form, the recursion relation can be iterated indefinitely. This is the key point: near criticality the correlation length is very large, which makes the calculation of the thermodynamic properties very difficult. A single iteration reduces the correlation length by a factor of L, but the correlation length is still large because L is a small factor. However, if we can iterate the recursion relations indefinitely, then, no matter how large the initial correlation function maybe, we can eventually, through iteration, relate the original problem to one at a small correlation length where simple techniques, like mean-field theory, work well. We will also see that many results, such as critical exponents, can be deduced from the recursion relations without any further calculations.

Equation (19.34) defines a flow diagram for the nearest neighbor coupling constant. When combined with Eq. (19.30) it defines a complete set of recursion relations for calculating the free energy as a function of the starting coupling constant K. Following the same steps that led up to Eq. (19.13), we find that the free energy per site obeys

$$\begin{aligned} f(K') &= 2f(K) - (A' - 2A) \\ &= 2f(K) - \ln\left[2\cosh^{1/2} 2K \cosh^{1/8} 4K\right] \end{aligned} \tag{19.35}$$

Consider the flow diagram defined by Eq. (19.34). For large K, we find that

$$K' = \frac{3}{2}K - \frac{3}{8}\ln 2 > K \tag{19.36}$$

For small K,

$$K' = 3K^2 < K. \tag{19.37}$$

Clearly then there is some value, $K = K' = K^*$, corresponding to a fixed point, as shown in Fig. 19.3. This point has the numerical value $K^* = 0.506981$, corresponding to $T_c = 1.97J$, which is a bit lower than the exact value of $2.27J$, perhaps because of the neglect of the four- spin term.

How do we interpret the above recursion relation and its associated flow diagram? For small coupling constant K, the system on large length scales looks like (i.e. flows to) a very weakly coupled system. This regime is what one would expect for a disordered phase. For large coupling constant, the system flows toward ever stronger coupling, as one would expect for an ordered phase. The fixed point defines the boundary between these two regimes. So the fixed point describes the system exactly at criticality. We expect that the nature of the renormalization group flows near the fixed point will give information on the critical exponents, or indeed, on the asymptotic form of the various critical properties. Near the critical point, K^*, the recursion relation is given approximately by its linearized form

$$K' = K^* + (K - K^*)\left.\frac{dK'}{dK}\right)_{K=K^*}, \tag{19.38}$$

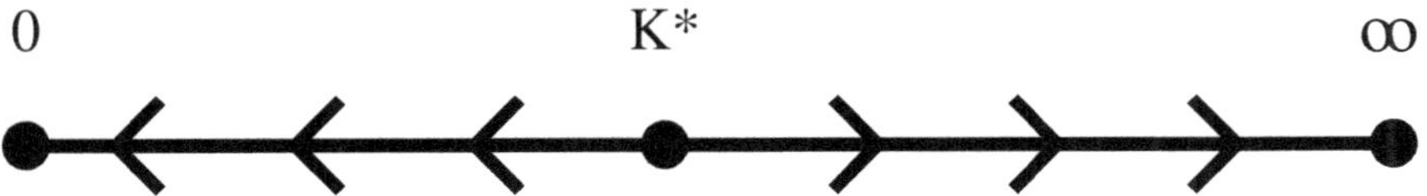

Fig. 19.3 Flow diagram for the two-dimensional Ising model in zero field

where

$$\left.\frac{dK'}{dK}\right)_{K=K^*} = \frac{3}{2}\tanh 4K^* = 1.4489\,. \tag{19.39}$$

Note that the transformation takes the original lattice of X's and O's into a the lattice of X's. This means that the length scale is changed by a factor of $\sqrt{2}$. If ξ' is the correlation length before transformation and ξ that after transformation, then we have the relation

$$\xi' = \xi/\sqrt{2}\,. \tag{19.40}$$

(Exactly, at the critical point this is satisfied by $\xi = \xi' = \infty$.) We suppose that

$$\xi \sim |T - T_c|^{-\nu} \sim |K - K^*|^{-\nu}\,, \tag{19.41}$$

where $K^* = J/(kT_c)$. Thus, Eq. (19.40) is

$$(K' - K^*)^{-\nu} = (K - K^*)^{-\nu}/\sqrt{2}\,. \tag{19.42}$$

Combining this with Eq. (19.38), we see that

$$\frac{dK'}{dK} = \left(\frac{1}{\sqrt{2}}\right)^{-1/\nu}\,, \tag{19.43}$$

or that

$$\begin{aligned}\nu &= \frac{\ln 2}{2\ln\left|\left.\frac{dK'}{dK}\right)_{K=K^*}\right|}\\ &= 0.935\,.\end{aligned} \tag{19.44}$$

This result is much larger than the mean-field value $\nu = 1/2$, and is not too different from the exact answer $\nu = 1$.

Alternatively, we can calculate the critical exponent α from the recursion relation for K. As a function of K, the second term in Eq. (19.35) is perfectly analytic at $K = K^*$. Therefore, we can separate off the singular part of the recursion relation, and write Eq. (19.35) as

$$f_s(K') = 2f_s(K). \tag{19.45}$$

Writing $f_s(K)$ in the scaling form

$$f_s(K) = a|K - K^*|^{2-\alpha}. \tag{19.46}$$

Equation (19.45) becomes

$$a|K' - K^*|^{2-\alpha} = 2a|K - K^*|^{2-\alpha}. \tag{19.47}$$

Then, using Eq. (19.38),

$$a\left|(K-K^*)\frac{dK'}{dK}\right)_{K=K^*}\bigg|^{2-\alpha} = 2a|K-K^*|^{2-\alpha}$$
$$\left|\frac{dK'}{dK}\right)_{K=K^*}\bigg|^{2-\alpha} = 2\,, \tag{19.48}$$

so that

$$\alpha = 2 - \frac{\ln 2}{\ln\left|\frac{dK'}{dK}\right)_{K=K^*}\big|}$$
$$= 0.131\,. \tag{19.49}$$

As expected, this result satisfies the so-called hyperscaling relation $\alpha = 2 - d\nu$. The exact result is, of course, $\alpha = 0$, so the exponent obtained from this approximate renormalization group calculation is not exactly correct. However, the approximate exponent is not bad. A power law divergence with an exponent of 0.131 is not so different from a log divergence.

So far, we have shown that there is a fixed point at a special value, K^*, of the coupling constant K. But we know that even if $K = K^*$ we can avoid the transition if $H \neq 0$. So in addition to flow away from the critical point in the coordinate K, we should also have flow away from the critical point in the coordinate H. So we now wish to include the effect of the magnetic field term, $-H\sum_{i,j} S_{i,j}$, in the Hamiltonian and we are interested in the limit of small H. Then Eq. (19.24) becomes

$$e^{2A}\left[e^{K(S_1+S_2+S_3+S_4)+H} + e^{-K(S_1+S_2+S_3+S_4)-H}\right] = e^{A'}e^{-\beta\mathcal{H}'(S_1,S_2,S_3,S_4)}\,. \tag{19.50}$$

For small H, this becomes

$$-\beta\mathcal{H}'(S_1,S_2,S_3,S_4;H) = -\beta\mathcal{H}'(S_1,S_2,S_3,S_4;H=0)$$
$$+H\tanh[K(S_1+S_2+S_3+S_4)]\,. \tag{19.51}$$

Now we set

$$\tanh[K(S_1+S_2+S_3+S_4)] = a(S_1+S_2+S_3+S_4)$$
$$+b(S_1S_2S_3+S_1S_2S_4+S_1S_3S_4+S_2S_3S_4)\,. \tag{19.52}$$

(This equation is true when the S's are restricted to be ± 1.) By considering the two special cases: (1) when all the S's have the same sign and (2) when three S's have one sign and one S has the other sign, we see that the coefficients a and b are determined by

$$\tanh(4K) = 4a + 4b \tag{19.53a}$$

$$\tanh(2K) = 4a - 2b \ , \tag{19.53b}$$

so that

$$a = \frac{1}{8}[\tanh(4K) + 2\tanh(2K)] \ , \tag{19.54a}$$

$$b = \frac{1}{8}[\tanh(4K) - 2\tanh(2K)] \ . \tag{19.54b}$$

As before, we see that although the starting Hamiltonian has no three-spin terms, this recursion relation introduces such three spin terms. It is clear that further iteration will introduce even higher spin terms. In the same spirit of simplification wherein we included second-neighbor interactions as being the same as first neighbor interactions we drop these three-spin terms. Here, we do not include them as having the same effect as single spin terms. The reason for this is that the three- spin terms do not favor a single direction of net magnetization. That is, fixing $S_1S_2S_3$ to be $+1$ does not lead to a net magnetization because it equally favors the four states $(+1, +1, +1)$, $(+1, -1, -1)$, $(-1, +1, -1)$, and $(-1, -1, +1)$, whose average magnetization is zero. So the three-spin term is not similar to the single spin term, and therefore the best approximation is to set $b = 0$ in Eq. (19.53b). The single spin, S_O at an O site gets contributions aHS_O induced by summing over S_X^n for its four neighbors with $n = 1, \ldots 4$. Thus, the recursion relation for H is

$$H' = H + 4aH \ , \tag{19.55}$$

so that

$$\left.\frac{dH'}{dH}\right)_{H=0,K=K^*} = 1 + \frac{1}{2}[\tanh(4K^*) + 2\tanh(2K^*)]$$
$$= 2.25 \ . \tag{19.56}$$

Indeed this means that under iteration, H grows. So, to be at the critical points does require that $H = 0$. Now we expect that

$$F(H,T) \sim t^{2-\alpha} g(H/t^{\Delta}) \ , \tag{19.57}$$

where $t = |T - T_c|/T_c$, Δ is related to the more familiar critical exponents via $\Delta = \beta + \gamma$, and g is some undetermined function. We can express this in terms of the correlation length as

$$F(H,T) \sim \xi^{-d\nu} h(H/\xi^{\Delta/\nu}) \ , \tag{19.58}$$

where h is an undetermined function. This implies that under renormalization we should have

$$\frac{H'}{H} = \left(\frac{\xi'}{\xi}\right)^{-\Delta/\nu} . \tag{19.59}$$

so that

$$\begin{aligned}\frac{\Delta}{\nu} &= \frac{\ln(H'/H)}{\ln(\xi/\xi')} \\ &= \frac{\left.\ln \frac{dH'}{dH}\right)_{H=0, K=K^*}}{\ln \sqrt{2}} \\ &= 2.34 .\end{aligned} \tag{19.60}$$

This result has some rather strange implications. We have

$$\Delta = \beta + \gamma = 2 - \alpha - \beta = d\nu - \beta , \tag{19.61}$$

so that

$$\beta = d\nu - \Delta . \tag{19.62}$$

Since the order parameter is finite, we see that Δ should not be larger than $d\nu$, which in this case is 2. The result $\Delta = d\nu$ would indicate a discontinuous (first order) transition in which the order parameter appears discontinuously below $T = T_c$. What, then, are we to make of our present result that Δ is larger than $d\nu$? It simply means that our approximations were too drastic. Indeed, if one extends the present scheme by allowing three-spin, four-spin, further-neighbor etc. interactions, one obtains the much better result that $\Delta < d\nu$ and in fact, one gets a reasonably good result for β, whose exact value is $1/8$. We will discuss such results in a later section.

To summarize what we have learned here: The renormalization scheme leads to recursion relations from which we get a thermal exponent (related to the correlation length exponent ν) and a magnetic exponent (related to the order parameter exponent β). From these two exponents, we can get all the other static critical exponents, as indicated in Eq. (13.68). Note that the lack of analyticity and noninteger exponents comes, not from any non-analyticity in the recursion relations, but rather from the slopes in the linearized recursion relations. The renormalization group mapping from one length scale to the next is perfectly analytic. The critical exponents are related to the details of the fixed point. We will discuss these points more systematically in the next section.

19.3 Formal Theory

Now that we have some feeling for how a renormalization group calculation works, we will look more closely at the formalism and its associated terminology.

We start by considering the "partition functional" of the spins, the object whose trace is the partition function.

$$Z(\mathcal{H}_N) = \mathrm{Tr} e^{-\beta\mathcal{H}_N\{\sigma\}} \,. \tag{19.63}$$

Here, the subscript N indicates the total number of spins in the Hamiltonian. In most of what follows, the "spins" could be any kind of degree of freedom.

We may write a general Hamiltonian for N spins as

$$\begin{aligned} -\beta\mathcal{H}_N(g,\{K\}) = Ng + H\sum_{\mathbf{r}}\sigma_{\mathbf{r}} + K_1\sum_{\mathbf{r},\delta_1}\sigma_{\mathbf{r}}\sigma_{\mathbf{r}+\delta_1} + K_2\sum_{\mathbf{r},\delta_2}\sigma_{\mathbf{r}}\sigma_{\mathbf{r}+\delta_2} \\ +K_3\sum_{1,2,3}\sigma_1\sigma_2\sigma_3 + K_4\sum_{1,2,3,4}\sigma_1\sigma_2\sigma_3\sigma_4 + \cdots \end{aligned} \tag{19.64}$$

where g is a constant, $\boldsymbol{\delta}_1$ and $\boldsymbol{\delta}_2$ are nearest and second-neighbor vectors, and terms with coupling constants K_3 and K_4 are three- and four-spin interactions. In principle, there could be many more such terms. However many there are, we can represent the set of coupling constants as

$$\{K\} = \{H, K_1, K_2, K_3, \ldots\}. \tag{19.65}$$

In any practical calculation, we will only be able to consider a small number of such coupling constants. Typically, our bare or starting Hamiltonian only has a few nonzero K_i. We will find that a renormalization group calculation can tell us which kinds of coupling constants are generated through renormalization and which ones remain zero. It can also tell us which ones are "relevant" because they affect the values of the critical exponents and which are not.

The renormalization group transformation is a transformation on the Hamiltonian which we may be viewed as a transformation on the array of coupling constants $\{K_i\}$. We write this transformation as

$$K_i' = R_i(\{K\})\,; \qquad g' = g'(g;\{K_i\})\,. \tag{19.66}$$

(The role of the constant g is obviously special: it does not enter the recursion relations for the K_i.) This transformation is constructed to leave the partition function invariant:

$$Z[\mathcal{H}_{N/L^d}(g',\{K'\})] = Z[\mathcal{H}_N(g,\{K\})]\,. \tag{19.67}$$

It is important to recognize that the recursion relations themselves, the $R_i(\{K\})$ and the function $g(\{K\})$ are assumed to be analytic functions of the coupling constants $\{K\}$. There is no reason to think that the transformation causing a change of scale by a *finite* factor is singular. Thermodynamic quantities can be singular, because they implicitly depend on infinitely iterating the recursion relations. We used the analyticity of the recursion relations earlier, in the calculation for the 2D Ising model, when we expanded K' in powers of $K - K^*$.

We define a fixed point, $\{K^*\}$ by the relation

$$K_i^* = R_i\{K^*\} \quad \forall\, i. \tag{19.68}$$

Of course, in general in a given space of coupling constants, there will be more than one such point.

If the recursion relations are analytic in the vicinity of a fixed point, we can linearize these relations around $\{K^*\}$. Then

$$K_i' = K_i^* + \sum_j \left(\frac{\partial R_i}{\partial K_j}\right)_{K^*} \left(K_j - K_j^*\right). \tag{19.69}$$

We define the coefficient matrix

$$\Lambda_{i,j} = \left(\frac{\partial R_i}{\partial K_j}\right)_{K^*} \tag{19.70}$$

Then

$$\left(K_i' - K_i^*\right) = \sum_j \Lambda_{i,j}\left(K_j - K_j^*\right). \tag{19.71}$$

The following trick will allow us to define a principal axis coordinate system in the space of coupling constants and to compute the rate of flow toward and away from fixed points. Let $\{\lambda_\alpha, \phi_i^\alpha\}$ be the eigenvalues and left eigenvectors of the matrix Λ, so that

$$\sum_i \phi_i^\alpha \Lambda_{i,j} = \lambda_\alpha \phi_j^\alpha \tag{19.72}$$

Then

$$\begin{aligned} u_\alpha' &\equiv \sum_i \phi_i^\alpha \left(K_i' - K_i^*\right) = \sum_{i,j} \phi_i^\alpha \Lambda_{i,j}\left(K_j - K_j^*\right) \\ &= \lambda_\alpha \sum_j \phi_j^\alpha \left(K_j - K_j^*\right) \equiv \lambda_\alpha u_\alpha \end{aligned} \tag{19.73}$$

The u_α are called the "scaling fields." They are linear combinations of displacements from the fixed point, corresponding to principal directions of the flows under the renormalization group transformation. The λ_α, which may be larger or smaller than one, measure the rate of flow away from (for $\lambda_\alpha > 1$) or toward (for $\lambda_\alpha < 1$) the fixed point. [Here and below, we assume that the λ_α are all positive.] We can see the way λ_α must depend on L, the compression factor for the length scale. Whether we compress in two steps or one, the result must be the same. So

$$\lambda_\alpha(L_1 L_2) = \lambda_\alpha(L_1)\lambda_\alpha(L_2) . \tag{19.74}$$

Set $L_2 = 1 + \delta$, where δ is an infinitesimal. Then Eq. (19.74) is

$$\lambda_\alpha(L_1) + L_1\delta\frac{d\lambda_\alpha}{dL_1} = \lambda_\alpha(L_1)\left[1 + \delta\frac{d\lambda_\alpha}{dL}\right)_{L=1}\Bigg] . \tag{19.75}$$

This has the solution $\lambda_\alpha(L) = L^{y_\alpha}$, where $y_\alpha = d\lambda_\alpha/dL)_{L=1}$. So the eigenvalues are characterized by their exponents y_α. The argument leading to Eq. (19.74) indicates furthermore that the linearized coefficient matrix $\Lambda(L)$ has to obey

$$\Lambda(L_1 L_2) = \Lambda(L_1)\Lambda(L_2) . \tag{19.76}$$

A symmetric matrix $\mathbf{M}$ can be written in terms of its eigenvalues $\{\lambda\}$ and its eigenvectors $\{|v\rangle\}$ as

$$\mathbf{M} = \sum_i |v_i\rangle\lambda_i\langle v_i| . \tag{19.77}$$

Under certain conditions this expansion also works for a nonsymmetric matrix like Λ, where now $|v_i\rangle$ is a right eigenvector and $\langle v_i|$ is a left eigenvector, with

$$\langle v_i|v_j\rangle = \delta_{i,j} . \tag{19.78}$$

In our case Eq. (19.77) is

$$\Lambda(L) = \sum_i |\phi_i\rangle L^{y_i}\langle\phi_i| . \tag{19.79}$$

Using Eq. (19.78) one can verify that Eq. (19.76) is satisfied. Of course, for this to be true it is essential that the left and right eigenvectors do *not* depend on L.

If $\lambda_\alpha > 1$, y_α is positive and $u'_\alpha > u_\alpha$ in the vicinity of the fixed point. In this case, u_α flows toward larger values, and the scaling field u_α is said to be "relevant (in the renormalization group sense)." If $y_\alpha < 0$, then u_α flows to zero and this field is said to be "irrelevant." If $y_\alpha = 0$ then the field u_α is called "marginal." Since u_α is defined in terms of the operator that it couples to (multiplies) in the Hamiltonian, these terms are also applied to the corresponding operators. For example, the operator which couples to a marginal field is called a marginal operator.

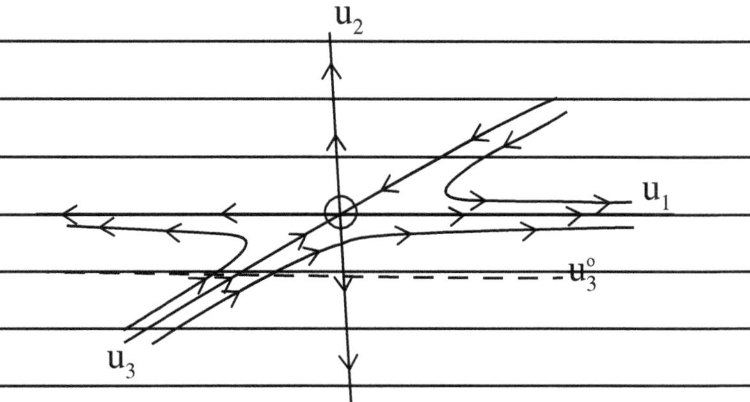

Fig. 19.4 Flow diagram near a fixed point

In the simplest case, which includes the 2D Ising model, there are two relevant fields, $u_1 = K_1 - K_1^*$ and $u_2 = H$, and the rest are irrelevant. In the larger space of coupling constants, $\{u_\alpha\}$, one can define an "attractive surface" or "critical surface" in field space by the condition $u_1 = u_2 = 0$. Any point on this surface moves toward K^* under the RG transformation, because all the other u_α flow to zero. So on the critical surface, the flow is toward the fixed point. What this means is that one has *universality* in the sense that the critical exponents do not depend on the values assigned to *irrelevant* variables because such irrelevant variables are renormalized to zero by the recursion relations. This picture is one of the triumphs of RG theory: it gives a clear physical explanation of universality classes. Near the critical surface, where one or both of u_1 and u_2 are nonzero but small, the flow eventually diverges from K^* as u_1 and/or u_2 becomes large. This situation is illustrated in Fig. 19.4, where the u_3 axis corresponds to $u_1 = u_2 = 0$.

So far we have indicated that the eigenvalues of the linearized recursion relations determine the critical exponents. But the eigenvector associated with the thermal eigenvalue also encodes interesting information. To see this look at the linearized equation for the critical surface (in this approximation the critical surface is a plane):

$$\begin{aligned} 0 = u_T &= \sum_\alpha \phi_i^T (K_\alpha - K_\alpha^*) \\ &= \sum_\alpha \phi_i^T [(J_\alpha / kT_c) - K_\alpha^*] . \end{aligned} \tag{19.80}$$

Thus T_c is determined by

$$kT_c(\{J_i\}) = \frac{\sum_\alpha \phi_\alpha^T J_\alpha}{\sum_\alpha \phi_\alpha^T K_\alpha^*} . \tag{19.81}$$

Let $T_c^{(0)}$ denote the value of T_c with only nearest neighbor interactions. It is given by

$$kT_c^{(0)} = \frac{\phi_{\text{nn}}^T J_{\text{nn}}}{\sum_\alpha \phi_\alpha^T K_\alpha^*} , \tag{19.82}$$

where the subscript nn indicates nearest neighbor. Then Eq. (19.81) may be written as

$$T_c(\{J_i\}) = T_c^{(0)} \left[1 + \sum_{\alpha \neq \text{nn}} (\phi_\alpha^T / \phi_{\text{nn}}^T)(J_\alpha / J_{\text{nn}}) \right] . \tag{19.83}$$

Thus from the thermal eigenvector we obtain an explicit prediction for the dependence of T_c on the addition of other than nearest neighbor interactions.

19.4 Finite Cluster RG

Recursion relations play a central role in renormalization group theory. They define the critical surfaces, which separate regions corresponding to different phases in coupling constant space; they have fixed points which control renormalization group flows; and, through their derivatives at fixed points, they determine the critical exponents. A big part of any renormalization group calculation is the derivation of the recursion relations. When the calculation cannot be done exactly, it is nevertheless worthwhile to consider methods for calculating approximate recursion relations. If the approximation is reasonable, the approximate recursion relations may capture the important physics of the problem.

In the early days of RG theory, one popular method for generating approximate recursion relations was the finite cluster method. In this method, a finite cluster, usually with periodic boundary conditions was transformed to a smaller cluster with renormalized coupling constants. In this section, we consider a particular case of a finite cluster RG calculation for the Ising model on a square lattice.

19.4.1 Formalism

Before presenting these calculations, it is appropriate to give a generalized description of the method. The initial Hamiltonian is taken to be a function of the initial spin variables σ_i for $i = 1, N$. We wish to define a transformed Hamiltonian in terms of a new set of spins $\{\mu\}$ on a lattice with a lattice constant which is larger by a factor of L. This means that, if $N_\sigma = N$ is the number of spins σ, then $N_\mu = N/L^d$ is the number of spins μ. The renormalization group transformation $T\{\mu, \sigma\}$ is defined by

$$Z'\{\mu\} = \sum_{\{\sigma\}} T\{\mu, \sigma\} Z\{\sigma\}. \tag{19.84}$$

$T\{\mu, \sigma\}$ is constrained by the condition that the value of the partition function is not changed by this transformation. That is that

$$Z = \sum_{\{\sigma\}} Z\{\sigma\} = \sum_{\{\mu\}} Z'\{\mu\}. \tag{19.85}$$

This condition is automatically satisfied if

$$\sum_{\{\mu\}} T\{\mu, \sigma\} = 1. \tag{19.86}$$

Of course, this condition is far from what is necessary to define the transformation $T\{\mu, \sigma\}$.

A useful form for $T\{\mu, \sigma\}$ is

$$T\{\mu, \sigma\} = \prod_{i=1}^{N_\mu} \frac{1}{2} \left[1 + \mu_i t\{\sigma\}\right]. \tag{19.87}$$

where $t\{\sigma\}$ is a function of the σ's which we might expect to depend only on those σ's in the cell around the so-called "cell spin", μ_i. Note that, for Ising spins, the condition on the trace of this transformation is satisfied because

$$\sum_{\mu_i = \pm 1} \frac{1}{2} \left[1 + \mu_i t\right] = \frac{1}{2}[1 + t] + \frac{1}{2}[1 - t] = 1. \tag{19.88}$$

As an example, consider the transformation which maps three spins, $\sigma_1, \sigma_2, \sigma_3$ into one spin μ. A commonly used transformation is the "majority rule", which has the form

$$T\{\mu; \sigma_1, \sigma_2, \sigma_3\} = \frac{1}{2}\left[1 + \frac{1}{2}(\sigma_1 + \sigma_2 + \sigma_3 - \sigma_1\sigma_2\sigma_3)\mu\right] \tag{19.89}$$

This transformation has the values

$$T = \frac{1}{2}[1 + \mu] \quad \text{if} \quad \sigma_1 + \sigma_2 + \sigma_3 > 0 \tag{19.90a}$$

$$T = \frac{1}{2}[1 - \mu] \quad \text{if} \quad \sigma_1 + \sigma_2 + \sigma_3 < 0. \tag{19.90b}$$

Therefore, this transformation has the property of assigning to μ the value $\text{sign}(\sigma_1 + \sigma_2 + \sigma_3)$. That is, the cell spin is assigned the value of the majority of spins in that cell. Such a transformation works well when the cell contains an odd number of spins. If the cell contains an even number of spins, then the states with zero total magnetization must be treated specially.

19.4.2 Application to the 2D Square Lattice

The example that we will consider involves a transformation from a system of 16 spins in 2D with periodic boundary conditions to one of four spins with periodic boundary conditions. However, much of the logic of the calculation is independent of the specific choice of cluster, but results instead from the underlying symmetries of the problem.

Consider a Hamiltonian for the four-spin system with periodic boundary conditions, involving one-, two-, three-, and four-spin interactions.

$$\begin{aligned}-\beta\mathcal{H} = 4A' + H'(S_1+S_2+S_3+S_4) + 2K_1'(S_1S_2+S_2S_3+S_3S_4+S_4S_1)\\ +4K_2'(S_1S_3+S_2S_4) + 4L'(S_2S_3S_4+S_1S_3S_4\\ +S_1S_2S_4 + +S_1S_2S_3) + 4MS_1S_2S_3S_4. \qquad (19.91)\end{aligned}$$

The possible energies, spin configurations and degeneracies for this four-spin Hamiltonian are as follows.

Energy +4A'	Spin configurations	Degeneracy
$4H' + 8K_1' + 8K_2' + 16L' + 4M'$	$\begin{matrix}+&+\\+&+\end{matrix}$	1
$2H' - 8L' - 4M'$	$\begin{matrix}+&-\\+&+\end{matrix}$, etc.	4
$-8K_2' + 4M'$	$\begin{matrix}+&+\\-&-\end{matrix}$, etc.	4
$-8K_1' + 8K_2' + 4M'$	$\begin{matrix}+&-\\-&+\end{matrix}$, etc.	2
$-2H' + 8L' - 4M'$	$\begin{matrix}-&+\\-&-\end{matrix}$, etc.	4
$-4H' - 8K_1' - 8K_2' - 16L' - 4M'$	$\begin{matrix}-&-\\-&-\end{matrix}$	1

The free energy corresponding to this Hamiltonian (for arbitrary numbers of spins) will have the symmetry property

$$F(-H, K_1, K_2, -L, M) = F(H, K_1, K_2, L, M). \qquad (19.92)$$

A renormalization group transformation that respects this symmetry will have the property that if

$$(H, K_1, K_2, L, M) \to (H', K_1', K_2', L', M') \qquad (19.93)$$

under a renormalization group transformation, then

$$(-H, K_1, K_2, -L, M) \to (-H', K_1', K_2', -L', M'). \tag{19.94}$$

In particular, this implies that

$$(0, K_1, K_2, 0, M) \to (0, K_1', K_2', 0, M'), \tag{19.95}$$

which means that the even-spin interactions, K_1, K_2, M form an "invariant subspace." Let us restrict our attention to this subspace. For this case, the energies, spin configurations and degeneracies are

Energy $+4A'$	Spin configurations	Degeneracy
$8K_1' + 8K_2' + 4M'$	$\begin{matrix}+&+\\+&+\end{matrix}, \begin{matrix}-&-\\-&-\end{matrix}$	2
$-4M'$	$\begin{matrix}+&-\\+&+\end{matrix}$, etc. and $\begin{matrix}-&+\\-&-\end{matrix}$, etc.	8
$-8K_2' + 4M'$	$\begin{matrix}+&+\\-&-\end{matrix}$, etc.	4
$-8K_1' + 8K_2' + 4M'$	$\begin{matrix}+&-\\-&+\end{matrix}$, etc.	2

From this table, it is easy to see that

1. $K_1 > 0$ favors a ferromagnetic state $\begin{matrix}+&+\\+&+\end{matrix}$ or $\begin{matrix}-&-\\-&-\end{matrix}$.
2. $K_1 < 0$ favors an antiferromagnetic state $\begin{matrix}+&-\\-&+\end{matrix}$.
3. $K_2 < 0$ favors rows or columns $\begin{matrix}+&-\\+&-\end{matrix}$ or $\begin{matrix}+&+\\-&-\end{matrix}$.
4. $M < 0$ favors states of the type $\begin{matrix}+&-\\+&+\end{matrix}$ and $\begin{matrix}-&+\\-&-\end{matrix}$.

Although the states favored are different, the free energy is invariant under reversal of the sign of the nearest coupling.

$$F(-K_1, K_2, M) = F(K_1, K_2, M) \tag{19.96}$$

The argument for this symmetry is based on the idea that the definition of the "up" direction can be reversed on one of the two sublattices in the antiferromagnetic state, making it look like a ferromagnetic state. This argument was discussed in Sect. 2.1.3 of Chap. 2. The key point here is that the new choice of coordinate system does not affect the form of the second-neighbor or four-spin interactions.

This kind of argument can be carried even further. If $K_1 = 0$, then it can be shown by reversing the coordinate system on every other row or column that

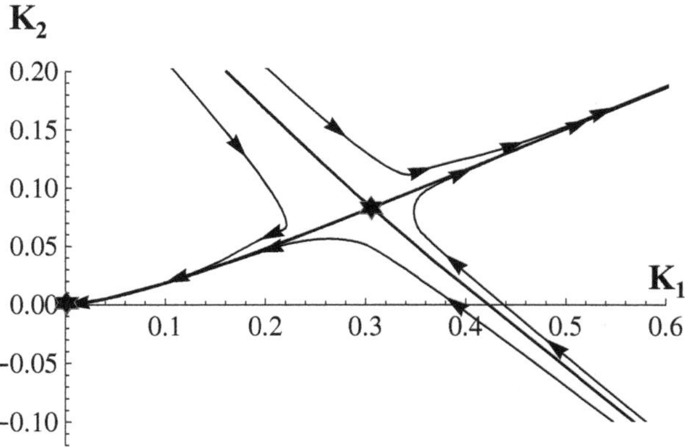

Fig. 19.5 Flow diagram near the unstable ferromagnetic fixed point in the plane $M = 0$ for the two-dimensional Ising model according to the real-space recursion relations of Refs. (Nauenberg and Nienhuis 1974), (Nienhuis and Nauenberg 1975). The critical line crosses the $K_2 = M = 0$ axis at $K_1 = 0.42$

$$F(0, -K_2, M) = F(0, K_2, M). \tag{19.97}$$

Finally, it can be shown, by reversing the coordinate system on one of the four sublattices, that

$$F(0, 0, -M) = F(0, 0, M). \tag{19.98}$$

The finite cluster RG calculation for a transformation from 16 to 4 spins with periodic boundary calculations was done by Nauenberg and Nienhuis in the mid-1970s and the results were published in a series of papers (Nauenberg and Nienhuis 1974), (Nienhuis and Nauenberg 1975). The renormalization group flow diagram in the positive quadrant of the $K_1 - K_2$ plane (with $M = 0$) is shown in Fig. 19.5. Critical surfaces are indicated by the solid lines, and the surrounding lines with arrows indicate the directions of flow. The $M = 0$ plane gives a reasonably accurate picture of the ferromagnetic and antiferromagnetic fixed points which actually lie at a small negative value of M^*. The actual locations are $(K_1^*, K_2^*, M^*) = (\pm 0.307, 0.084, -0.004)$. The entire flow diagram exhibits the $K_1 \rightarrow -K_1$ symmetry of Eq.(19.96). The eigenvalues associated with these fixed points have the values $\lambda_1 = 1.914$, $\lambda_2 = 0.248$, and $\lambda_3 = 0.137$. The value of K_1, on the critical surface for $K_2 = M = 0$ is $K_{1c}^* = 0.420$, which gives $T_c = 2.38$, slightly higher than the exact value of 2.27. The thermal exponent is $y_T = \ln \lambda_1 / \ln 2 = 0.937$. Then using $2 - \alpha = d/y_T$ We find $\alpha = -0.135$ so that the specific heat has a cusp rather than a divergence. Nevertheless, the thermodynamic functions calculated by Nauenberg and Nienhuis closely resemble the exact functions, as shown in Fig. 19.6.

Two additional and symmetrically located fixed points (not shown in Fig. 19.5) occur along the $K_1 = 0$ axis of the phase diagram. This is not completely unexpected since, for $K_1 = M = 0$ and $K_2 \neq 0$, the system consists of two interpenetrating Ising models, each with coupling constant K_2. Thus, one would expect fixed points to occur

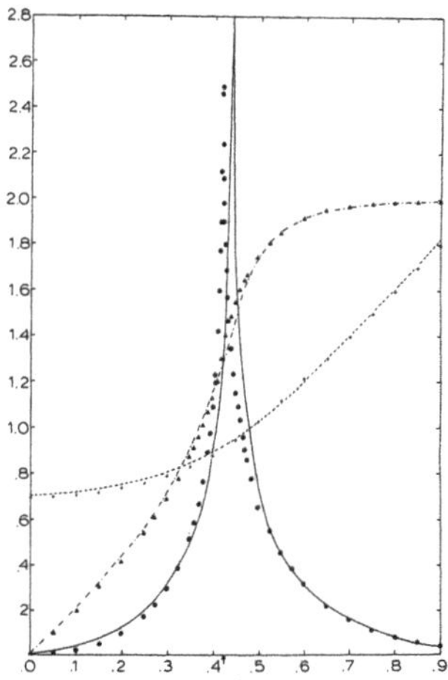

Fig. 19.6 Thermodynamic functions for the 2D Ising model from the real-space RG calculations of Refs. (Nauenberg and Nienhuis 1974), (Nienhuis and Nauenberg 1975)

at $(0, \pm K_2^*, 0)$. These two fixed points, at which both sublattices develop either ferro- or antiferromagnetic order, depending on the sign of K_2, are denoted SF and SAF respectively.

The model with $K_1 = 0$ is a symmetric case of the eight-vertex model, which was solved exactly by Baxter in 1971 (Baxter 1971). An interesting feature of the exact solution is that the critical exponents are not "universal." They depend on the values of the coupling constants. From the RG perspective, exponents which depend on the values of the coupling constants result from having a continuous line of fixed points. At every point along this line, the coupling constants go into themselves. This requires a kind of fine-tuning. For example, if we think of an unstable fixed point as a mountain pass, where the surface curves up along one direction and down along the perpendicular direction, then the analog of a line of fixed points would be a perfectly level ridge. As an aside, this situation corresponds to having a marginal operator, with eigenvalue 1. It seems intuitively clear that using approximate recursion relations, rather than exact ones, is likely to distort the surface of the ridge so that it is no longer exactly level, resulting in isolated fixed points rather than lines of fixed points. Another way of looking at this is that considerable care must be taken to preserve the internal symmetry of a problem so that subtle features, such as lines of fixed points are not lost due to approximations.

19.5 Summary

In this chapter, we have introduced the basic concepts and procedures of the renormalization group theory of phase transitions, starting from the exact solution of the 1D Ising model and proceeding through a sequence of approximate treatments of the 2D Ising model. We have defined and illustrated the use of recursion relations and explained how they can be used to derive critical exponents, and we have discussed the importance of preserving the symmetries of the Hamiltonian in order for approximate recursion relations to yield physical results. We have described the different kinds of fixed points, relevant, irrelevant, and marginal, and also explained the meaning of scaling fields and critical lines. What is missing so far is a prescription for deriving recursion relations which, if not exact, at least involves controlled approximations. In the next chapter, we will consider a method that provides such a controlled approximation that can be improved by going to successively higher orders in an expansion parameter.

19.6 Exercises

1. This exercise is easily done on a pocket calculator. Iterate the function $f(x) = \sin x$. On most calculators you can easily obtain $f^{*n}(x)$ by pressing the "sin x" button n times. Starting from some initial value of x to what fixed point does this sequence flow? Repeat for $f(x) = \cos(x)$. Repeat for $f(x) = \tanh(kx)$. Do the fixed points depend on the magnitude and/or sign of k? If so, give a graphical interpretation.

2. Perform a real-space cluster RG calculation, similar to the one for the square lattice, described in the text, for a triangular lattice. Choose a 9-spin cluster, as shown in Fig. 19.7, with periodic boundary conditions, where the cluster consists of three triangles of spins labeled, A, B, and C. Transform the three A spins into a single spin, using the majority rule and do the same for the B and C triangles. The starting Hamilton should contain a constant term, A, a field term, H, a nearest neighbor interaction, K, and a 3-spin interaction, P. Begin your solution by listing the sequence of steps that need to be performed in this calculation.

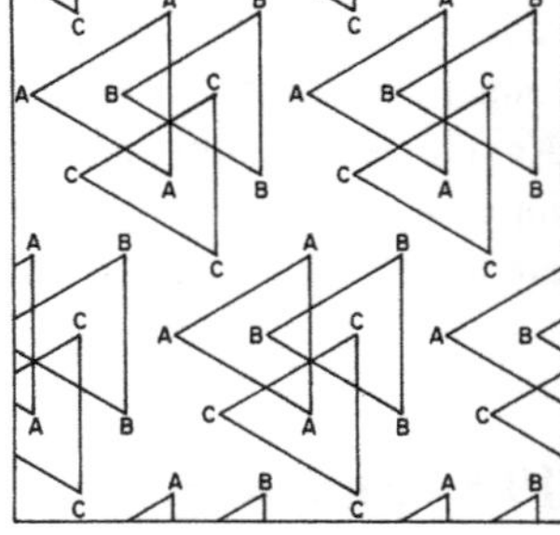

Fig. 19.7 Sites of the triangular lattice grouped into 9-spin clusters of interpenetrating 3-spin triangles. A real-space RG scheme can be defined to map spins on each triangle into a single spin using the majority rule. Figure from Schick et al. (1977)

References

R.J. Baxter, Eight-vertex model in lattice statistics. Phys. Rev. Lett. **26**, 832–3 (1971)

L.P. Kadanoff, Scaling laws for Ising models near T_c. Physics **2**, 263 (1966)

M. Nauenberg, B. Nienhuis, Critical surface for square Ising spin lattice. Phys. Rev. Lett. **33**, 944 and Renormalization-group approach to the solution of general Ising models, ibid. 1598 (1974)

B. Nienhuis, M. Nauenberg, Renormalization-group calculation for the equation of state of an Ising ferromagnet. Phys. Rev. B **11**, 4152 (1975)

M. Schick, J.S. Walker, M. Wortis, Phase diagram of the triangular Ising model: Renormalization-group calculation with application to adsorbed monolayers, Phys. Rev. B **16**, 2205 (1977)

Chapter 20
The Epsilon Expansion

20.1 Introduction

The renormalization group (RG) provides a prescription for computing the partition function by repeatedly integrating out the short length scale degrees of freedom. At each step, the problem is rewritten in its initial form yielding a set of recursion relations for the coupling constants at successive length scales. We have seen that, given the recursion relations, it is relatively straightforward to calculate the free energy and critical exponents. The challenge is calculating the recursion relations without making uncontrolled approximations.

In this chapter, we discuss the RG approach in a situation where the derivation of the recursion relations can be controlled, namely near the upper critical dimension, d_c, above which mean-field theory is asymptotically correct, even near the critical point. As we have seen, for the Ising model, $d_c = 4$, so that for this model the expansion we are about to describe is in the variable $\epsilon = 4 - d$. (More generally $\epsilon = d_c - d$, where d_c is the upper critical dimension.) In our discussion, we will assume that the reader has already digested the material in Chap. 13 on scaling and in the preceding chapter on the real-space RG. These contain many of the central concepts which the reader needs in order to deal with the present chapter.

20.2 Role of Spatial Dimension, d

It may be difficult for the reader to reconstruct the mindset in the phase-transition community before the RG was developed by Wilson (1971) (for which he was awarded the Nobel prize in physics in 1982). In those days, most people believed that the critical exponents smoothly approached their mean-field values as the spatial dimensionality d was taken to infinity. However, as we have seen, the Ginsburg criterion indicates that for the Ising model mean-field theory provides a correct description asymptotically close to the critical point as long as $d > 4$. This means that the crit-

A. J. Berlinsky and A. B. Harris, *Statistical Mechanics*, Graduate Texts in Physics,
https://doi.org/10.1007/978-3-030-28187-8_20

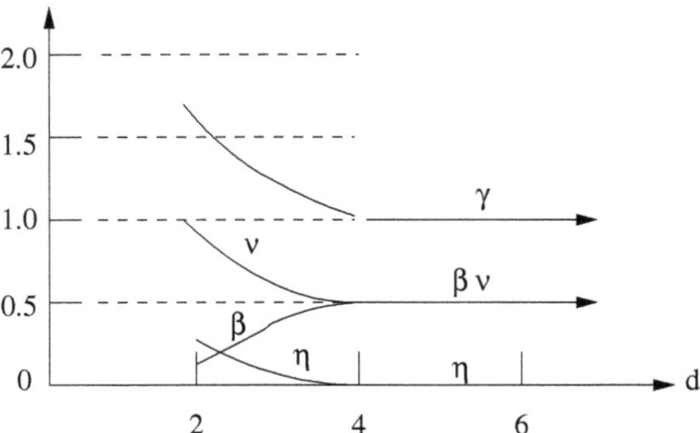

Fig. 20.1 Schematic plot of critical exponents of the Ising model versus continuous spatial dimension d

ical exponents do not depend on d for $d > 4$, where they assume their mean-field values. If we assume no discontinuities at $d_c = 4$, then the behavior of the critical exponents as a function of continuous dimension for $2 < d < \infty$ is that illustrated in Fig. 20.1. The idea of continuous dimension requires some comment. As we will see, the ϵ-expansion is developed by interpreting integrals which depend on d by simply substituting $d = 4 - \epsilon$ in the most obvious way. This analytic continuation from integer to continuous values of d is not a unique one. However, it happens that this extrapolation is exactly the same one as is implemented in series expansion when the diagram multiplicities are written as polynomials in d as as done in Table 16.1. This means that series results for critical exponents can be obtained for continuous d and appropriately compared to results from the RG.

For the reader's convenience, we repeat here the definitions of the various critical exponents:

$$\text{susceptibility} \quad \chi \sim t^{-\gamma} \tag{20.1a}$$

$$\text{specific heat} \quad C \sim t^{-\alpha} \tag{20.1b}$$

$$\text{order parameter} \quad M \sim t^{\beta} \tag{20.1c}$$

$$\text{correlation length} \quad \xi \sim t^{-\nu} \tag{20.1d}$$

$$\text{correlation function} \quad \langle \sigma(0)\sigma(\mathbf{r})\rangle_{T=T_c} \sim r^{-d+2-\eta}\,. \tag{20.1e}$$

$$\text{magnetic field} \quad H(T = T_c) \sim M^{\delta} \tag{20.1f}$$

$$\text{free energy density} \quad f(H, T) \sim t^{2-\alpha} f(H/t^{\Delta})\,, \tag{20.1g}$$

where $t \equiv (T - T_c)/T_c$ and Eqs. (20.1a)–(20.1e) apply for $H = 0$. Then we have

$$\gamma = (2 - \eta)\nu \tag{20.2a}$$

$$\alpha = 2 - d\nu \tag{20.2b}$$

$$\alpha + 2\beta + \gamma = 2 \tag{20.2c}$$

$$\delta = \frac{d + 2 - \eta}{d - 2 + \eta} \tag{20.2d}$$

$$\Delta = \beta + \gamma \ . \tag{20.2e}$$

Note that all critical exponents can be expressed in terms of η and ν which is referred to as "two exponent scaling." The above equations apply for $d_< < d \leq d_c$, where d_c is the upper critical dimension and $d_<$ is the lower critical dimension at which long-range order at any nonzero temperature is destroyed by thermal fluctuations. For $d > d_c$, d should be replaced by d_c.

In view of Fig. 20.1, it is reasonable to hope that $\epsilon \equiv 4 - d$ can be an expansion parameter. However, since $\epsilon > 0$ and $\epsilon < 0$ represent qualitatively different behavior, this expansion can at best be an asymptotic one. Nevertheless, it may yield reasonably accurate numerical values for critical exponents. In any case, the most important result from RG theory is the explanation it provides for the phenomenology of the critical point.

20.3 Qualitative Description of the RG ϵ-Expansion

20.3.1 Gaussian Variables

As mentioned, the plan is to integrate out repeatedly short-wavelength fluctuations. At each step in this process, we obtain a mapping from an initial Hamiltonian to a renormalized Hamiltonian which then becomes the initial Hamiltonian for the next step. As we will see, for d near 4 we can truncate the space of operators whose recursion relations we need to follow. In that case, we obtain recursion relations in the space of a manageably small number of variables.

If we were working with a Hamiltonian expressed in terms of spin operators, integrating out short-wavelength fluctuations would mean that we would have to trace over variables $S(\mathbf{q}) = \sum_i S_i e^{i\mathbf{q}\cdot\mathbf{r}_i}$ for large q while retaining variables with small q untraced over. We have not yet learned how to do this other than within the real-space RG. To proceed further toward this goal, we start from the Hamiltonian derived in Sect. 13.6 using the Hubbard–Stratonovich transformation. For a spatially uniform external magnetic field H, we can rewrite the partition function from Eq. (13.95b), setting all the $H_i = H$, as

$$Z(H, T) = A' \left[\prod_i \int_{-\infty}^{\infty} dy_i \right] e^{-\mathcal{H}(\{y_i\})} \ , \tag{20.3}$$

where we set $y_i = x_i + \beta H$. Then

$$\begin{aligned} \mathcal{H}(\{y_i\}) = & \frac{kT}{2} \sum_{ij} y_i [\mathbf{J}^{-1}]_{ij} y_j - \sum_i \ln\cosh(y_i) \\ & -[H/J(0)] \sum_i y_i + \frac{\beta}{2} N H^2 / J(0) \ , \end{aligned} \tag{20.4}$$

where

$$J(0) = \sum_j J_{ij}. \tag{20.5}$$

Expanding the $\ln \cosh(y_i)$ to order y^4 gives

$$\ln \cosh(y_i) = \frac{1}{2}y_i^2 - \frac{1}{12}y_i^4 + \dots . \tag{20.6}$$

Next, we introduce Fourier transformed variables via

$$x(\mathbf{q}) = \sum_i y_i e^{i\mathbf{q}\cdot\mathbf{r}_i} , \qquad y_i = \frac{1}{N}\sum_{\mathbf{q}} x(\mathbf{q})e^{-i\mathbf{q}\cdot\mathbf{r}_i} . \tag{20.7}$$

Following Eq. (13.100), we can write

$$\frac{kT}{2}\sum_{ij} y_i[\mathbf{J}^{-1}]_{ij}y_j = \frac{kT}{2N}\sum_{\mathbf{q}} x(\mathbf{q})x(-\mathbf{q})\left(\frac{1}{zJ} + cq^2\right) , \tag{20.8}$$

where $zJ = J(0)$ for nearest neighbor interactions.

Then the partition function can be written as

$$Z(H,T) = \int_{-\infty}^{\infty} dx(\mathbf{q}_1) \int_{-\infty}^{\infty} dx(\mathbf{q}_2) \dots e^{-\mathcal{H}[\{x(\mathbf{q})\}]} , \tag{20.9}$$

where

$$\mathcal{H}[\{x(\mathbf{q})\}] = -\tilde{H}x(0) + \frac{1}{2N}\sum_{\mathbf{q}}[r + \tilde{c}q^2]x(\mathbf{q})x(-\mathbf{q})$$
$$+ \frac{u}{N^3} \sum_{\mathbf{q}_1,\mathbf{q}_2,\mathbf{q},\mathbf{q}_4} x(\mathbf{q}_1)x(\mathbf{q}_2)x(\mathbf{q}_3)x(\mathbf{q}_4)\Delta(\mathbf{q}_1 + \mathbf{q}_2 + \mathbf{q}_3 + \mathbf{q}_4) , \tag{20.10}$$

$\mathcal{H}$ is called the Landau–Ginsburg–Wilson (LGW) free energy functional (which we will usually refer to as simply the "Hamiltonian"). In Eq. (20.10), $\tilde{H} = H/J(0) = H/zJ$, $r = (kT/zJ) - 1$, where the 1 comes from the first term in Eq. (20.6), $\tilde{c} = kTc$ and $u = 1/12$. Note that, in this model, the bare T_c is given by $kT_c = zJ$ or equivalently $r = 0$. Henceforth, we will drop the tildes in this expression.

The symbol, $\Delta(\mathbf{q})$ in Eq. (20.10) means that $\mathbf{q}$ is zero, modulo a reciprocal lattice vector. The sums can be converted into integrals with the prescription (when the lattice constant is taken to be unity) $N^{-1}\sum_{\mathbf{q}} \to (2\pi)^{-d}\int d^d\mathbf{q}$, so that

$$\mathcal{H}[\{x(\mathbf{q})\}] = -Hx(0) + \frac{1}{2}\int_0^{\Lambda} [r + cq^2]x(\mathbf{q})x(-\mathbf{q})\frac{d^d\mathbf{q}}{(2\pi)^d}$$

$$+(2\pi)^d u \int_0^\Lambda \frac{d^d\mathbf{q}_1}{(2\pi)^d} \int_0^\Lambda \frac{d^d\mathbf{q}_2}{(2\pi)^d} \int_0^\Lambda \frac{d^d\mathbf{q}_3}{(2\pi)^d} \int_0^\Lambda \frac{d^d\mathbf{q}_4}{(2\pi)^d} x(\mathbf{q}_1)x(\mathbf{q}_2)$$
$$\times x(\mathbf{q}_3)x(\mathbf{q}_4)\delta(\mathbf{q}_1+\mathbf{q}_2+\mathbf{q}_3+\mathbf{q}_4)+\dots . \tag{20.11}$$

Here we have made several simplifications and changes. Instead of carrying the integral over $\mathbf{q}$ over the first Brillouin zone, as we should, we have integrated over a sphere of the same volume having radius Λ. It is useful to think of Λ as a cutoff, and we can characterize quantities according to whether or not they depend on Λ. (We expect critical exponents, for example, to be independent of Λ, whereas the actual value of the transition temperature T_c will depend on Λ.) Note that we have dropped all terms for which a nonzero reciprocal lattice vector is allowed by $\Delta(\mathbf{q})$. This type of approximation will be fine as long as only long-wavelength (small q) degrees of freedom are important. Note also that, although we have included the coupling to the external field H, we have dropped an operator-independent term proportional to H^2 from Eq. (20.4) which does not affect the critical properties. Also, we have dropped quadratic terms involving higher powers of q than q^2, and terms involving products of more than four fields $x(\mathbf{q})$. We will later justify these truncations.

Equation (20.11) is the generic Hamiltonian we will consider in developing the ϵ-expansion, and it is characterized by the four parameters r, c, u, and H, whose initial values will be indicated by the subscript "0." Note: there is no universal agreement on whether the coefficient of the quartic term should be u (as we take here), $u/8$ (as Ma takes), or $u/24$ (as Aharony takes). Also, note that we write the quartic term so that it is symmetric in the four wave vectors. If one expresses $x(\mathbf{q}_4)$ as $x(-\mathbf{q}_1-\mathbf{q}_2-\mathbf{q}_3)$ inside an integral over $\mathbf{q}_1$, $\mathbf{q}_2$, and $\mathbf{q}_3$, then when the first three $\mathbf{q}$'s are inside the cutoff, the fourth one need not be inside the cutoff. It is preferable to use the symmetric form, as is done here. Finally, it is convenient to introduce some simplifying notation. We define

$$\int_{q=q_<}^{q=q_>} \equiv \int_{q_<}^{q_>} \frac{d^d\mathbf{q}}{(2\pi)^d} \tag{20.12}$$

so that

$$\mathcal{H}[\{x(\mathbf{q})\}] = -Hx(q=0) + \frac{1}{2}\int_{q=0}^{q=\Lambda} [r+cq^2]x(\mathbf{q})x(-\mathbf{q})$$
$$+(2\pi)^d u \int_{q_1=0}^{q_1=\Lambda} \int_{q_2=0}^{q_2=\Lambda} \int_{q_3=0}^{q_3=\Lambda} \int_{q_4=0}^{q_4=\Lambda} x(\mathbf{q}_1)x(\mathbf{q}_2)x(\mathbf{q}_3)x(\mathbf{q}_4)$$
$$\times\delta(\mathbf{q}_1+\mathbf{q}_2+\mathbf{q}_3+\mathbf{q}_4)+\cdots . \tag{20.13}$$

Equation (20.13) is referred to as the "ϕ^4-model," where ϕ^4 indicates the presence of a perturbation that is a product of four operators.

20.3.2 Qualitative Description of the RG

We now discuss the RG as applied to the ϕ^4-model. We will describe one step of renormalization as consisting of three types of transformations: (1) integrating out high momentum degrees of freedom, (2) rescaling the sphere of momentum integration, and (3) rescaling the size of the spin variable. Our discussion is quite similar (but much briefer) than that given by Ma (1976) pp. 163–218.

Integrating Out Large q Variables

We integrate out variables whose wavevector lies in the shell $\Lambda/b < q < \Lambda$, where the factor b is greater than 1. (In the preceding chapter, the analogous parameter was denoted L.) A convenient notation is helpful for describing this process. We divide $\mathcal{H}$ into three terms: $\mathcal{H}_<$ contains all terms in $\mathcal{H}$ which involve *only* $x(\mathbf{q})$'s with $q < \Lambda/b$, $\mathcal{H}_>$ contains all terms in $\mathcal{H}$ which involve *only* $x(\mathbf{q})$'s with $\Lambda/b < q < \Lambda$, and $\mathcal{H}_{\rm int}$ contains interaction terms, i.e. terms which depend simultaneously on both small-q and large-q variables. Specifically, one such term, with two small-q and two large-q variables, is

$$\delta\mathcal{H}_{\rm int} = 6(2\pi)^d u \int_{q_1=0}^{q_1=\Lambda/b} \int_{q_2=0}^{q_2=\Lambda/b} \int_{q_3=\Lambda/b}^{q_3=\Lambda} \int_{q_4=\Lambda/b}^{q_4=\Lambda} \times x(\mathbf{q}_1)x(\mathbf{q}_2)x(\mathbf{q}_3)x(\mathbf{q}_4)\delta(\mathbf{q}_1+\mathbf{q}_2+\mathbf{q}_3+\mathbf{q}_4) , \qquad (20.14)$$

where the factor 6 is the number of ways of selecting which two $\mathbf{q}$'s are less than Λ/b and which are greater than Λ/b.

We define a renormalized Hamiltonian, $\mathcal{H}'$ by integrating out variables with $\Lambda/b < q < \Lambda$, as follows:

$$e^{-\mathcal{H}'} = e^{-\mathcal{H}_<} \int_{-\infty}^{\infty} dx(\mathbf{q}_{>,1}) \int_{-\infty}^{\infty} dx(\mathbf{q}_{>,2}) \dots \int_{-\infty}^{\infty} dx(\mathbf{q}_{>,k}) \dots \times e^{-\mathcal{H}_>} e^{-\mathcal{H}_{\rm int}} , \qquad (20.15)$$

where $x(\mathbf{q}_{>,k})$ is a variable with $\Lambda/b < q < \Lambda$. Obviously, $\mathcal{H}'$ is defined so that when the integration over small-q variables is carried out, it will give exactly the same partition function as does $\mathcal{H}$. We shall see that, for small $\epsilon = 4 - d$, we can calculate $\mathcal{H}'$ perturbatively without too much difficulty.

Rescaling the Cutoff

To obtain the partition function from $\mathcal{H}'$ note that all integrations over the $\mathbf{q}$'s now have the cutoff Λ/b. So to make the problem identical to that before integrating out the high-q variables, we need to set $\mathbf{q} = \mathbf{q}'/b$, in which case, in terms of $\mathbf{q}'$, the functional integral of the $x(\mathbf{q})$'s will look the same as before. In other words, everywhere you saw $\mathbf{q}$, replace it by $\mathbf{q}'/b$, so that the cutoff remains Λ.

Rescaling the Spin Variables

Formally, it may seem that we have done enough to map the problem into itself. But roughly speaking, each spin variable, $x(\mathbf{q}'/b)$, describes the state of a block of b^d spins. Therefore, the values of the new spin variable which are statistically important ought to be larger than those of the initial spin variable. So for $\mathcal{H}'$ to reproduce the physics of $\mathcal{H}$, we ought to replace $x(\mathbf{q})$ by $g(b)x(\mathbf{q}'/b)$, where $g(b)$ takes account of the larger scale of the new block spin variable. We now discuss the form of $g(b)$. Consider the external field term, where we replace $-Hx(0)$ by $-Hx(0)g(b)$, which we may write as $-H'x(0)$, with

$$H' = Hg(b) . \tag{20.16}$$

This is actually the recursion relation for H. Compare this with what we expect from scaling for the free energy density. From Eq. (13.48)

$$\begin{aligned} f &= t^{2-\alpha} f(H/t^{\Delta}) = t^{2-\alpha} f(H/t^{\beta+\gamma}) \\ &= t^{2-\alpha} f[H\xi^{(\beta+\gamma)/\nu}] . \end{aligned} \tag{20.17}$$

This equation says that when ξ is reduced in one step of the RG by a factor of b, H should be increased by a factor of $b^{(\beta+\gamma)/\nu}$ which is usually written as $b^{(d+2-\eta)/2}$. Thus, the RG is accomplished by first integrating out the large-q degrees of freedom and then in the surviving small-q Hamiltonian [obtained as in Eq. (20.15)], one makes the replacement

$$\mathbf{q} \to \mathbf{q}'/b \tag{20.18a}$$

$$\begin{aligned} x(\mathbf{q}) &\to b^{(d+2-\eta)/2} x(\mathbf{q}'/b) \\ &= b^{(d+2-\eta)/2} x'(\mathbf{q}') . \end{aligned} \tag{20.18b}$$

With these replacements, the partition function calculated in terms of x and $\mathbf{q}$ will be the same as that calculated in terms of x' and $\mathbf{q}'$ (and thus the primes in the final expression can be dropped).

Expected Results

What do we expect to learn from the above RG procedure? First of all, the Hamiltonian near the critical point will have two variables which *grow* under renormalization. These are called *relevant* variables. We know that this must be the case, because to be at criticality we must fix *two* variables, one of which is the temperature, $T = T_c$, and the other which is the magnetic field H which is zero at the critical point. We identify the temperature-like variable to be r [since $r_0 \sim (T - T_c)$] and the field variable is obviously H. The variable u turns out to be *irrelevant*, but for $T = T_c$ and $H = 0$ the recursion relations take it to a fixed point value u^*. For $d > 4$, $u^* = 0$ because the Gaussian model and Mean-Field Theory work for $d > 4$, whereas for $d < 4$ we expect to have $u^* > 0$. Accordingly, it is not surprising that for $d < 4$, u^* is of order ϵ.

What happens to c, the coefficient of the gradient-squared term, is less obvious. Clearly, we do not want an additional relevant variable, so c can not grow indefinitely. Nor can c go to zero, because a theory with a zero gradient term would not have spatial correlations. Similarly, we would not want c to be an infinitesimal, e.g. of order ϵ to some power. As we will see, the recursion relations allow us to keep c constant, so that the usual procedure is to keep $c = 1$ at each step of renormalization. Thus, c is a *marginal* variable. Already one can see that the fact that the fixed point has two relevant variables explains why all critical exponents can be expressed in terms of two exponents, say, ν and η.

20.3.3 Gaussian Model

We start by treating the Gaussian model, which is the special case when $u = 0$, so that the Hamiltonian is quadratic. From Eq. (20.15), we see that the renormalized Hamiltonian, $\mathcal{H}'$, has two contributions: one from $\mathcal{H}_<$ and another which involves $\mathcal{H}_{\rm int}$. Since $\mathcal{H}_{\rm int}$ vanishes for the Gaussian model we have only to deal with the first contribution, which one may think of as a kinematic contribution and leads to "naive scaling." As we will see, this model is a useful one to understand because the ϕ^4-model in $4 - \epsilon$ dimensions differs only perturbatively from the Gaussian model.

We now carry out the renormalization of the two terms in the Hamiltonian, applying the prescription given in Eqs. (20.18a) and (20.18b). The first term is

$$
\begin{aligned}
r \int_0^\Lambda d^d\mathbf{q} x(\mathbf{q}) x(-\mathbf{q}) &\to r \int_0^{\Lambda/b} d^d\mathbf{q} x(\mathbf{q}) x(-\mathbf{q}) \\
&= r \int_0^\Lambda d^d(\mathbf{q}'/b) x(\mathbf{q}'/b) x(-\mathbf{q}'/b) \\
&\to r b^{-d} \int_0^\Lambda d^d\mathbf{q}' b^{d+2-\eta} x'(\mathbf{q}') x'(-\mathbf{q}') \\
&\to r b^{2-\eta} \int_0^\Lambda d^d\mathbf{q} x(\mathbf{q}) x(-\mathbf{q}) \,,
\end{aligned}
\tag{20.19}
$$

where the primes have been dropped in the last step and

$$
r' = b^{2-\eta} r \,. \tag{20.20}
$$

This shows that r is relevant, as long as $\eta < 2$. (As we will see, $\eta = 0$ for $u = 0$.)

Likewise, for the cq^2 term,

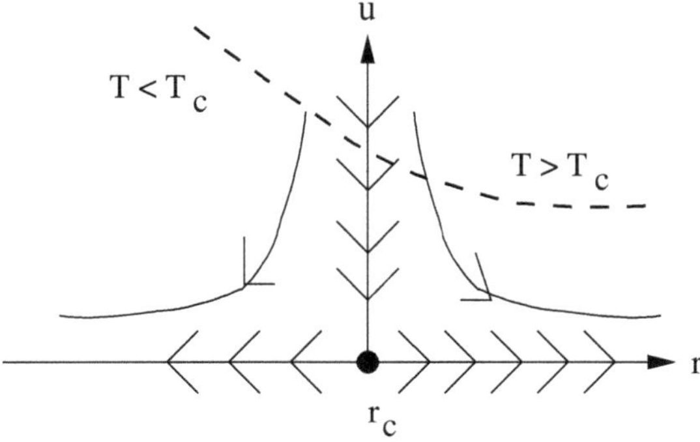

Fig. 20.2 Flow diagram for the LGW Hamiltonian for $d > 4$. Note that on this diagram (for $H = 0$) only r is relevant. The fixed point is at $r = u = 0$. The horizontal axis ($u = 0$) corresponds to the Gaussian model, and the flow for $u > 0$ obeys Eq. 20.24 for $\epsilon < 0$. The dashed line shows schematically the trajectory in the r-u plane which results when the temperature T is varied. (Changing T changes r and it also can have some effect on u, although obviously not causing u to change sign.) The phase transition occurs at the special value of temperature for which the dashed line intersects the flow to the critical point

$$\begin{aligned} c\int_0^{\Lambda} d^d\mathbf{q}q^2x(\mathbf{q})x(-\mathbf{q}) &\rightarrow c\int_0^{\Lambda/b} d^d\mathbf{q}q^2x(\mathbf{q})x(-\mathbf{q}) \\ &= c\int_0^{\Lambda} d^d(\mathbf{q}'/b)(q'/b)^2x(\mathbf{q}'/b)x(-\mathbf{q}'/b) \\ &\rightarrow cb^{-d-2}\int_0^{\Lambda} d^d\mathbf{q}'b^{d+2-\eta}(q')^2x'(\mathbf{q}')x'(-\mathbf{q}') \\ &\rightarrow cb^{-\eta}\int_0^{\Lambda} d^d\mathbf{q}q^2x(\mathbf{q})x(-\mathbf{q}) \ , \end{aligned} \tag{20.21}$$

which shows that the recursion relation for c is

$$c' = b^{-\eta}c \ . \tag{20.22}$$

As mentioned earlier, a theory in which c is renormalized to either 0 or infinity is not acceptable, because such a Hamiltonian would have unphysical spatial correlations. So $\eta = 0$ for the Gaussian model, and the constant, c is often set equal to 1. The flow diagram for the Gaussian model corresponds to the flow on the horizontal axis ($u = 0$) in Fig. 20.2.

20.3.4 Scaling of Higher Order Couplings

It is useful to study the naive scaling of other operators. Since we will consider models with $u \neq 0$, we first examine how u renormalizes, neglecting interaction effects. We have

$$
\begin{aligned}
& u \int_0^{\Lambda} d^d\mathbf{q}_1 \int_0^{\Lambda} d^d\mathbf{q}_2 \int_0^{\Lambda} d^d\mathbf{q}_3 \int_0^{\Lambda} d^d\mathbf{q}_4 \\
& \quad \times x(\mathbf{q}_1)x(\mathbf{q}_2)x(\mathbf{q}_3)x(\mathbf{q}_4)\delta(\mathbf{q}_1+\mathbf{q}_2+\mathbf{q}_3+\mathbf{q}_4) \\
& \to u \int_0^{\Lambda/b} d^d\mathbf{q}_1 \int_0^{\Lambda/b} d^d\mathbf{q}_2 \int_0^{\Lambda/b} d^d\mathbf{q}_3 \int_0^{\Lambda/b} d^d\mathbf{q}_4 \\
& \quad \times x(\mathbf{q}_1)x(\mathbf{q}_2)x(\mathbf{q}_3)x(\mathbf{q}_4)\delta(\mathbf{q}_1+\mathbf{q}_2+\mathbf{q}_3+\mathbf{q}_4) \\
& = u \int_0^{\Lambda} d^d(\mathbf{q}_1'/b) \int_0^{\Lambda} d^d(\mathbf{q}_2'/b) \int_0^{\Lambda} d^d(\mathbf{q}_3'/b) \int_0^{\Lambda} d^d(\mathbf{q}_4'/b) \\
& \quad \times x(\mathbf{q}_1'/b)x(\mathbf{q}_2'/b)x(\mathbf{q}_3'/b)x(\mathbf{q}_4)\delta(\mathbf{q}_1'/b+\mathbf{q}_2'/b+\mathbf{q}_3'/b+\mathbf{q}_4'/b) \\
& = b^{-3d} u \int_0^{\Lambda} d^d\mathbf{q}_1' \int_0^{\Lambda} d^d\mathbf{q}_2' \int_0^{\Lambda} d^d\mathbf{q}_3' \int_0^{\Lambda} d^d\mathbf{q}_4' \\
& \quad \times b^{2(d+2-\eta)} x'(\mathbf{q}_1')x'(\mathbf{q}_2')x'(\mathbf{q}_3')x'(\mathbf{q}_4)\delta(\mathbf{q}_1'+\mathbf{q}_2'+\mathbf{q}_3'+\mathbf{q}_4) \\
& = b^{4-d-2\eta}\, u \int_0^{\Lambda} d^d\mathbf{q}_1 \int_0^{\Lambda} d^d\mathbf{q}_2 \int_0^{\Lambda} d^d\mathbf{q}_3 \int_0^{\Lambda} d^d\mathbf{q}_4 \\
& \quad \times x(\mathbf{q}_1)x(\mathbf{q}_2)x(\mathbf{q}_3)x(\mathbf{q}_4)\delta(\mathbf{q}_1+\mathbf{q}_2+\mathbf{q}_3+\mathbf{q}_4)\,,
\end{aligned}
\tag{20.23}
$$

where we used $\delta(\mathbf{q}/b) = b^d\delta(\mathbf{q})$. Thus

$$u' = ub^{\epsilon-2\eta}\,. \tag{20.24}$$

This shows that u is weakly relevant or irrelevant depending on the sign of ϵ. (The flow for the case when $\epsilon < 0$ is shown in Fig. 20.2). As will become evident, the way u renormalizes plays a key role in critical phenomena.

Finally, let's look at a few strongly irrelevant operators. A term with 6 $x(\mathbf{q})$'s, it would have scale factors for the spin operators, giving $b^{3(d+2-\eta)}$. From the rescaling of the $\mathbf{q}$'s we would get b^{-5d}, so that in all, if we call the coefficient of such a term u_6 we would have

$$u_6' = b^{6-3\eta-2d}u_6\,, \tag{20.25}$$

which, at $d = 4$ is $u_6' = b^{-2}u_6$. Terms with higher numbers of $x(\mathbf{q})$'s are even more strongly irrelevant.

Next consider a term $eq^4x(\mathbf{q})x(-\mathbf{q})$. We have

$$e\int_0^\Lambda d^d\mathbf{q}q^4x(\mathbf{q})x(-\mathbf{q}) \to e\int_0^\Lambda d^d\mathbf{q}(q/b)^4b^{d+2-\eta}x(\mathbf{q}/b)x(-\mathbf{q}/b)$$
$$= eb^{-2-\eta}\int_0^\Lambda d^d\mathbf{q}q^4x(\mathbf{q})x(-\mathbf{q})\,, \tag{20.26}$$

so that e is strongly irrelevant. Obviously for each power of q one introduces in the coefficient, one obtains an addition power of $1/b$ in its recursion relation. Similarly, we may neglect the dependence of the coefficient u on any of its wave vectors $\mathbf{q}_1$, $\mathbf{q}_2$, etc. This reasoning shows why it was permissible to drop the terms we omitted from Eq. (20.13).

20.4 RG Calculations for the ϕ^4 Model

20.4.1 Preliminaries

We now carry out the RG procedure for the ϕ^4 Hamiltonian of Eq. (20.13). We need some notation to describe $\mathcal{H}_> + \mathcal{H}_{\rm int}$ in Eq. (20.15). We therefore introduce Feynman diagrams. Although we could present the calculations to leading order in ϵ without using diagrammatic techniques, we do so because these techniques are ubiquitous in papers on the ϵ-expansion. The formalism is designed to treat perturbations relative to a quadratic Hamiltonian, which in our case is

$$\mathcal{H}_{0,>} = \frac{1}{2}\int_{\Lambda/b}^\Lambda (r+cq^2)x(\mathbf{q})x(-\mathbf{q})\frac{d^d\mathbf{q}}{(2\pi)^d}\,. \tag{20.27}$$

Each perturbative term in the Hamiltonian is associated with a *vertex* in a diagram. From this vertex there emanate lines, each labeled by their wave vector $\mathbf{q}$. Lines directed outward from a vertex are associated with a factor $x(\mathbf{q})$ at that vertex and lines labeled $\mathbf{q}$ directed inwards are associated with a factor $x(-\mathbf{q})$. In the present context, we wish to distinguish between wave vectors in the shell $\Lambda/b < q < \Lambda$ which are integrated over and wave vectors with $q < \Lambda/b$ which will *not* be integrated over. The former are called *internal* lines and are represented by full lines, and the latter are called *external* lines and are represented by dashed lines. Here "integrated over" refers to the integration $\int_{-\infty}^\infty dx(\mathbf{q}_>)$ where $\Lambda/b < q_> < \Lambda$.

In the initial Hamiltonian, the various quartic contributions to $\mathcal{H}_> + \mathcal{H}_{\rm int}$ in Eq. (20.15) are

$$V^{(4)} = (2\pi)^d u\int_{\Lambda/b}^\Lambda \frac{d^d\mathbf{q}_1}{(2\pi)^d}\int_{\Lambda/b}^\Lambda \frac{d^d\mathbf{q}_2}{(2\pi)^d}\int_{\Lambda/b}^\Lambda \frac{d^d\mathbf{q}_3}{(2\pi)^d}\int_{\Lambda/b}^\Lambda \frac{d^d\mathbf{q}_4}{(2\pi)^d}$$

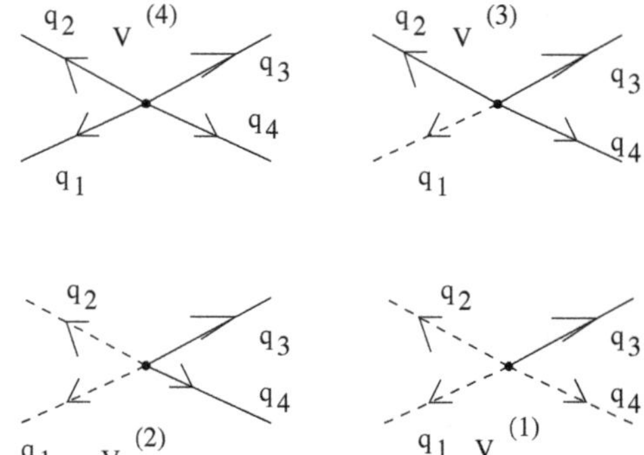

Fig. 20.3 Diagrammatic representation of the perturbations written in Eq. (20.28). Solid lines are to be labeled with wave vectors such that $\Lambda/b < q < \Lambda$ and dashed lines such that $q < \Lambda$. Each line represents a variable $x(\mathbf{q})$ and variables associated with solid lines are integrated over in each step of the RG

$$\times x(\mathbf{q}_1)x(\mathbf{q}_2)x(\mathbf{q}_3)x(\mathbf{q}_4)\delta(\mathbf{q}_1+\mathbf{q}_2+\mathbf{q}_3+\mathbf{q}_4) \quad (20.28a)$$

$$V^{(3)} = 4(2\pi)^d u \int_0^{\Lambda/b} \frac{d^d\mathbf{q}_1}{(2\pi)^d} \int_{\Lambda/b}^{\Lambda} \frac{d^d\mathbf{q}_2}{(2\pi)^d} \int_{\Lambda/b}^{\Lambda} \frac{d^d\mathbf{q}_3}{(2\pi)^d} \int_{\Lambda/b}^{\Lambda} \frac{d^d\mathbf{q}_4}{(2\pi)^d}$$
$$\times x(\mathbf{q}_1)x(\mathbf{q}_2)x(\mathbf{q}_3)x(\mathbf{q}_4)\delta(\mathbf{q}_1+\mathbf{q}_2+\mathbf{q}_3+\mathbf{q}_4) \quad (20.28b)$$

$$V^{(2)} = 6(2\pi)^d u \int_0^{\Lambda/b} \frac{d^d\mathbf{q}_1}{(2\pi)^d} \int_0^{\Lambda/b} \frac{d^d\mathbf{q}_2}{(2\pi)^d} \int_{\Lambda/b}^{\Lambda} \frac{d^d\mathbf{q}_3}{(2\pi)^d} \int_{\Lambda/b}^{\Lambda} \frac{d^d\mathbf{q}_4}{(2\pi)^d}$$
$$\times x(\mathbf{q}_1)x(\mathbf{q}_2)x(\mathbf{q}_3)x(\mathbf{q}_4)\delta(\mathbf{q}_1+\mathbf{q}_2+\mathbf{q}_3+\mathbf{q}_4) \quad (20.28c)$$

$$V^{(1)} = 4(2\pi)^d u \int_0^{\Lambda/b} \frac{d^d\mathbf{q}_1}{(2\pi)^d} \int_0^{\Lambda/b} \frac{d^d\mathbf{q}_2}{(2\pi)^d} \int_0^{\Lambda/b} \frac{d^d\mathbf{q}_3}{(2\pi)^d} \int_{\Lambda/b}^{\Lambda} \frac{d^d\mathbf{q}_4}{(2\pi)^d}$$
$$\times x(\mathbf{q}_1)x(\mathbf{q}_2)x(\mathbf{q}_3)x(\mathbf{q}_4)\delta(\mathbf{q}_1+\mathbf{q}_2+\mathbf{q}_3+\mathbf{q}_4)\,. \quad (20.28d)$$

The diagrammatic vertices representing these terms are shown in Fig. 20.3. We can rewrite Eq. (20.15) as

$$\mathcal{H}' = \mathcal{H}_< - \ln\left[\left\langle e^{-\sum_n V^{(n)}}\right\rangle\right], \quad (20.29)$$

where the average is taken with respect to $e^{-\mathcal{H}_{0,>}}$, with $\mathcal{H}_{0,>}$ given by Eq. (20.27). In principle, Eq. (20.29) requires evaluating averages of arbitrary powers of $V^{(n)}$. However, as will become apparent, since we are interested in the regime when u is of order ϵ, we will only need to consider diagrams with at most two $V^{(n)}$'s in the cumulant expansion of (20.29) which can be written as

$$\mathcal{H}' = \mathcal{H}_< + \sum_n \langle V^{(n)}\rangle + \frac{1}{2}\sum_{m,n}\langle V^{(n)}\rangle\langle V^{(m)}\rangle - \frac{1}{2}\sum_{n,m}\langle V^{(n)}V^{(m)}\rangle\,. \quad (20.30)$$

The evaluation of these averages is done in accord with Wick's theorem, which is reviewed in the Appendix at the end of this chapter. The result can be represented diagrammatically by connecting solid lines, forcing them to have the same $\mathbf{q}$ label, and then associating each such line, labeled by wave vector $\mathbf{q}$, with a so-called *propagator* $(r + cq^2)^{-1}$. (The lines so joined may emanate from either the same or different vertices.)

We will make frequent use of the identity,

$$\langle x(\mathbf{q})x(\mathbf{k})\rangle_0 = \frac{\delta(\mathbf{k}+\mathbf{q})(2\pi)^d}{r+cq^2} . \tag{20.31}$$

which may be derived with the help of the generating function

$$\mathcal{Z}(\{h(\mathbf{q})\}) = \int \mathcal{D}(\{x(\mathbf{q}\}) \exp\left(-\frac{1}{2}\int \left[(r+cq^2)x(\mathbf{q})x(-\mathbf{q}) - h(\mathbf{q})x(\mathbf{q})\right]\frac{d^d\mathbf{q}}{(2\pi)^d}\right) . \tag{20.32}$$

This derivation will be left as an exercise.

20.4.2 First Order in u

Next we consider the terms in Eq. (20.30) that are first order in u. We first consider

$$\begin{aligned}
\delta\mathcal{H}' &= \langle V^{(2)}\rangle_0 \\
&= 6(2\pi)^d u \int_0^{\Lambda/b} \frac{d\mathbf{q}_1}{(2\pi)^d}\int_0^{\Lambda/b} \frac{d\mathbf{q}_2}{(2\pi)^d}\int_{\Lambda/b}^{\Lambda} \frac{d\mathbf{q}_3}{(2\pi)^d}\int_{\Lambda/b}^{\Lambda} \frac{d\mathbf{q}_4}{(2\pi)^d} \\
&\quad \times x(\mathbf{q}_1)x(\mathbf{q}_2)\langle x(\mathbf{q}_3)x(\mathbf{q}_4)\rangle_0 \delta(\mathbf{q}_1+\mathbf{q}_2+\mathbf{q}_3+\mathbf{q}_4) \\
&= 6(2\pi)^d u \int_0^{\Lambda/b} \frac{d\mathbf{q}_1}{(2\pi)^d}\int_0^{\Lambda/b} \frac{d\mathbf{q}_2}{(2\pi)^d}\int_{\Lambda/b}^{\Lambda} \frac{d\mathbf{q}_3}{(2\pi)^d}\int_{\Lambda/b}^{\Lambda} \frac{d\mathbf{q}_4}{(2\pi)^d} \\
&\quad \times x(\mathbf{q}_1)x(\mathbf{q}_2)\frac{\delta(\mathbf{q}_3+\mathbf{q}_4)(2\pi)^d}{r+cq_3^2}\delta(\mathbf{q}_1+\mathbf{q}_2+\mathbf{q}_3+\mathbf{q}_4) \\
&= 6u\int_0^{\Lambda/b}\frac{d\mathbf{q}_1}{(2\pi)^d}x(\mathbf{q}_1)x(-\mathbf{q}_1)\int_{\Lambda/b}^{\Lambda}\frac{d\mathbf{q}_3}{(2\pi)^d(r+cq_3^2)} \\
&\equiv 6uJ(r)\int_0^{\Lambda/b}\frac{d\mathbf{q}_1}{(2\pi)^d}x(\mathbf{q}_1)x(-\mathbf{q}_1) ,
\end{aligned} \tag{20.33}$$

where

$$J(r) = \int_{\Lambda/b}^{\Lambda} \frac{d\mathbf{q}}{(2\pi)^d(r+cq^2)} . \tag{20.34}$$

This is the only contribution for $\mathcal{H}'$ at order u and its diagrammatic representation is shown in Fig. 20.4. The other first order terms either average to zero, for the cases

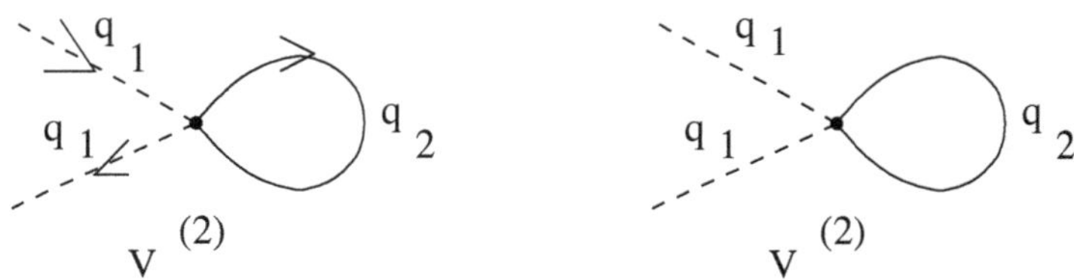

Fig. 20.4 Diagrammatic representation of the perturbation written in Eq. (20.33) (left) and with arrows removed (right)

of $V^{(1)}$ and $V^{(3)}$ or have no external lines and hence represent an additive constant, in the case of $V^{(4)}$.

We should point out that most authors do not assign directions to the interaction lines, as we have done. Indeed such arrows are not necessary for real fields. If one implements wave vector conservation, then in any given diagram without arrows on the lines, there is essentially only one way to assign arrows consistent with the momentum labels. So with a bit of practice, one can omit the arrows, as is customarily done, and as is illustrated in the right-hand panel of Fig. 20.4.

So at first order in u, we have the recursion relations,

$$r' = b^{2-\eta}\Big(r + 12uJ(r)\Big), \tag{20.35a}$$

$$u' = b^{\epsilon-2\eta}u, \tag{20.35b}$$

$$c' = b^{-\eta}c, \tag{20.35c}$$

$$H' = b^{(d+2-\eta)/2}H\ , \tag{20.35d}$$

where the $12u$ in Eq. (20.35a) is obtained from the previous $6u$ because the quadratic term in the Hamiltonian in Eq. (20.13) has a coefficient which is $r/2$ rather than just r. At this point, we see the following. As mentioned, we set $c' = c = 1$, so that, at this order $\eta = 0$. Then H and r are relevant, as expected, and, for $\epsilon > 0$, u is also relevant. But u should not be relevant because we can only have two relevant operators at the critical point. To resolve this problem, we have to extend the calculation to order u^2, which we do next. The point is that for $\epsilon > 0$ the point $u = 0$ must be unstable, as Eq. (20.35b) correctly indicates. Thus u must flow to a nonzero value (of order ϵ) where it is stable. This behavior can only occur when terms of order u^2 are included in the recursion relation.

20.4.3 Second Order in u

Accordingly we now consider the recursion relation for u at order u^2. Thus the term we want is one which has 4 external lines (so that it contributes to u') and two

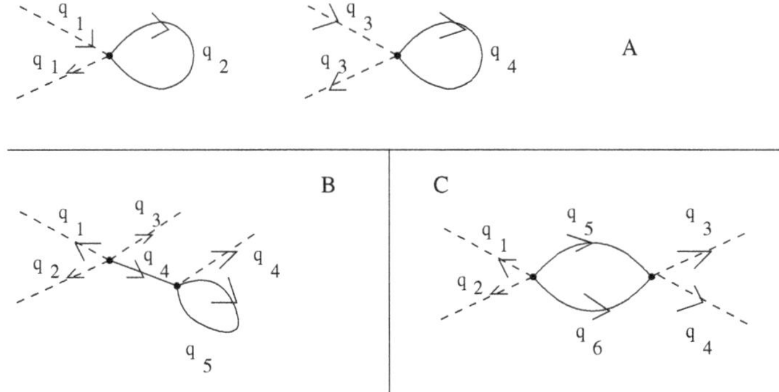

Fig. 20.5 Various contributions at order u^2. Diagram A is a disconnected diagram

internal lines. The only such diagrams are those shown in Fig. 20.5. The disconnected term shown in panel A vanishes because for such a diagram the last two terms in Eq. (20.30) cancel one another, as expected for cumulants. Also, remember that the q-dependence of u is irrelevant. So we evaluate the coefficient for the x^4 term in the limit when all wave vectors vanish. In this case, for the diagram in panel B $q_1 = q_2 = q_3 = q_4 = 0$, so that q_4 can not be in the shell $\Lambda/b < q_4 < \Lambda$. So the only nonzero contribution to u' at order u^2 comes from the diagram in panel C. For it we have to take account of the two ways to make this diagram, as explained in Fig. 20.6. So for the diagram of Fig. 20.6, Eq. (20.30) gives

$$\begin{aligned}
\delta\mathcal{H}' &= -2[(2\pi)^d(6u)]^2/2 \int_0^{\Lambda/b} \frac{d\mathbf{q}_1}{(2\pi)^d} \int_0^{\Lambda/b} \frac{d\mathbf{q}_2}{(2\pi)^d} \int_{\Lambda/b}^{\Lambda} \frac{d\mathbf{q}_5}{(2\pi)^d} \int_{\Lambda/b}^{\Lambda} \frac{d\mathbf{q}_6}{(2\pi)^d} \\
&\quad \times \int_{\Lambda/b}^{\Lambda} \frac{d\mathbf{q}_7}{(2\pi)^d} \int_{\Lambda/b}^{\Lambda} \frac{d\mathbf{q}_8}{(2\pi)^d} \int_0^{\Lambda/b} \frac{d\mathbf{q}_3}{(2\pi)^d} \int_0^{\Lambda/b} \frac{d\mathbf{q}_4}{(2\pi)^d} \\
&\quad \times x(\mathbf{q}_1)x(\mathbf{q}_2)x(\mathbf{q}_3)x(\mathbf{q}_4)\langle x(\mathbf{q}_5)x(\mathbf{q}_7)\rangle_0\langle x(\mathbf{q}_6)x(\mathbf{q}_8)\rangle_0 \\
&\quad \times \delta(\mathbf{q}_1+\mathbf{q}_2+\mathbf{q}_5+\mathbf{q}_6)\delta(\mathbf{q}_7+\mathbf{q}_8+\mathbf{q}_3+\mathbf{q}_4) \\
&\equiv -36u^2(2\pi)^d \int_0^{\Lambda/b} \frac{d\mathbf{q}_1}{(2\pi)^d} \int_0^{\Lambda/b} \frac{d\mathbf{q}_2}{(2\pi)^d} \int_0^{\Lambda/b} \frac{d\mathbf{q}_3}{(2\pi)^d} \int_0^{\Lambda/b} \frac{d\mathbf{q}_4}{(2\pi)^d} \\
&\quad \times M(r, \mathbf{q}_1+\mathbf{q}_2)x(\mathbf{q}_1)x(\mathbf{q}_2)x(\mathbf{q}_3)x(\mathbf{q}_4)\delta(\mathbf{q}_1+\mathbf{q}_2+\mathbf{q}_3+\mathbf{q}_4)\,, \qquad (20.36)
\end{aligned}$$

where

$$M(r, \mathbf{q}_1+\mathbf{q}_2) = \int_{\Lambda/b}^{\Lambda} \frac{d\mathbf{q}_5}{(2\pi)^d} \int_{\Lambda/b}^{\Lambda} d\mathbf{q}_6 \frac{\delta(\mathbf{q}_1+\mathbf{q}_2+\mathbf{q}_5+\mathbf{q}_6)}{(r+cq_5^2)(r+cq_6^2)}\,. \qquad (20.37)$$

Note that we integrate over the internal variables $x(\mathbf{q}_5)$, $x(\mathbf{q}_6)$, $x(\mathbf{q}_7)$, and $x(\mathbf{q}_8)$ to get (via Wick's theorem) the factor $\langle x(\mathbf{q}_5)x(\mathbf{q}_7)\rangle_0\langle x(\mathbf{q}_6)x(\mathbf{q}_8)\rangle_0$. The factor 2 from Fig. 20.6 is cancelled by the factor of 1/2 from expanding the exponential. As we have said, the dependence of M on the external wave vectors $\mathbf{q}_1$ and $\mathbf{q}_2$ is irrelevant.

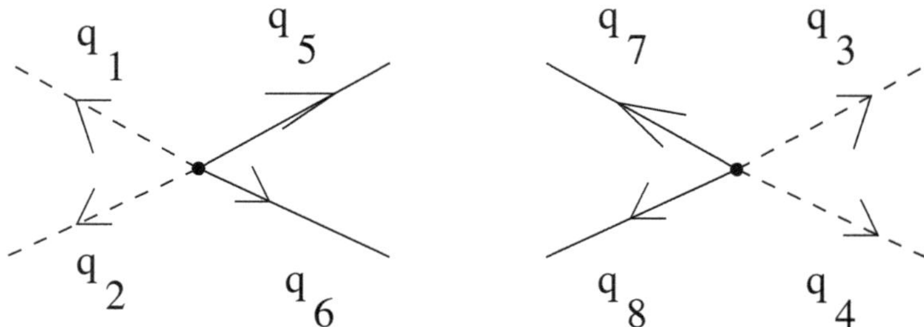

Fig. 20.6 Diagram C of Fig. 20.5 for the renormalization of u shown before the internal lines are connected in pairs. To make this diagram we can connect line $\mathbf{q}_5$ either to line $\mathbf{q}_7$ (so that $\mathbf{q}_5 + \mathbf{q}_7 = 0$) or to line $\mathbf{q}_8$ (so that $\mathbf{q}_5 + \mathbf{q}_8 = 0$). Thus diagram C has two identical contributions

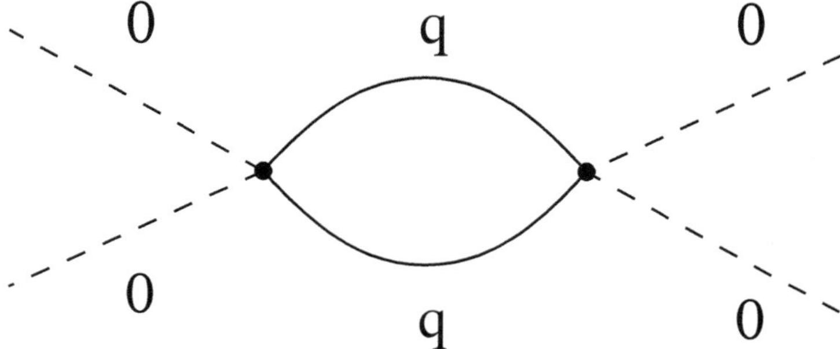

Fig. 20.7 Diagram for the renormalization of the zero wave vector component of u

Accordingly, we evaluate $M(r) \equiv M(r, 0)$ with all the external wave vectors zero, as represented in the diagram of Fig. 20.7. Then

$$M(r) = \int_{\Lambda/b}^{\Lambda} \frac{d\mathbf{q}_5}{(2\pi)^d} \frac{1}{(r + cq_5^2)^2} . \tag{20.38}$$

To evaluate the integrals $J(r)$ and $M(r)$ we need the element of volume in d dimensions. We set

$$\frac{d^d\mathbf{q}}{(2\pi)^d} = K_d q^{d-1} dq , \tag{20.39}$$

where $K_d = \Omega_d/(2\pi)^d$, where Ω_d, the area of the d-dimensional unit sphere, was given in Eq. (13.25), so that

$$K_d = \frac{2^{1-d}}{\pi^{d/2}\Gamma(d/2)} . \tag{20.40}$$

Note that $K_4 = 1/(8\pi^2)$. (To calculate critical exponents we do not actually need the explicit value of K_d.)

To the order in ϵ at which we work, we can evaluate the integrals for $d = 4$ and take r to be of order $u \sim \epsilon$. We write

$$\frac{1}{r + cq^2} = \frac{1}{cq^2} - \frac{r}{c^2 q^4} + \mathcal{O}(r^2) , \tag{20.41}$$

so that to order r we have

$$\begin{aligned} J(r) &= \frac{K_4}{c} \int_{\Lambda/b}^{\Lambda} \left(q - \frac{r}{cq} \right) dq \\ &= \frac{K_4 \Lambda^2}{2cb^2} (b^2 - 1) - \frac{K_4}{c^2} r \ln b . \end{aligned} \tag{20.42}$$

Similarly

$$\begin{aligned} M(0) &= \int_{\Lambda/b}^{\Lambda} \frac{K_4 q^3 dq}{c^2 q^4} \\ &= (K_4/c^2) \ln b . \end{aligned} \tag{20.43}$$

Thus the recursion relations are

$$r' = b^{2-\eta} \left(r + \frac{6K_4 \Lambda^2 (b^2 - 1) u}{cb^2} - 12 K_4 c^{-2} ur \ln b \right) \tag{20.44a}$$

$$u' = b^{\epsilon - 2\eta} \left(u - 36 \frac{K_4}{c^2} u^2 \ln b \right) \tag{20.44b}$$

$$c' = b^{-\eta} c \tag{20.44c}$$

$$H' = b^{(d+2-\eta)/2} H . \tag{20.44d}$$

At order ϵ we still have $\eta = 0$. But now we have a fixed point for $u = u^*$ whose value at order ϵ is given by

$$u^* = (1 + \epsilon \ln b) \left(u^* - 36 \frac{K_4}{c^2} (u^*)^2 \ln b \right) , \tag{20.45}$$

which gives

$$u^* = \frac{c^2 \epsilon}{36 K_4} . \tag{20.46}$$

To find r^* to order ϵ we analyze Eq. (20.44a) for $\eta = 0$, and we may omit the term in $ur \ln b$, which is higher order in ϵ. Then we get

$$\begin{aligned} r^* &= -6K_4\Lambda^2 u^*/c \\ &= -\frac{1}{6}c\Lambda^2\epsilon \,. \end{aligned} \tag{20.47}$$

Notice that r^* is negative at criticality. This means that $T - T_c^{(0)}$, where $T_c^{(0)}$ is the mean-field transition temperature, is negative at the transition. Thus, as one would expect, fluctuations cause the actual transition temperature to be less than its mean-field value.

Although the recursion relations involve b in a nontrivial way, it is noteworthy that the fixed point values of u and r are independent of b. (It would be totally unacceptable for r^* or u^* to depend on b.) Furthermore, although the value of b has been left unspecified, except that $b > 1$, it is often useful to consider infinitesimal scale transformations in which $0 < b - 1 << 1$, which lead to infinitesimal changes in the coupling constants, even if ϵ is not infinitesimal and hence to differential recursion relations for the coupling constants. This also serves to put expressions such as $b^\epsilon = 1 + \epsilon \ln b$, as used in Eq. (20.45), on a firmer footing.

The recursion relation for u is

$$\begin{aligned} u' - u^* &= (u - u^*)(du'/du)_{u=u^*} \\ &= (u - u^*)b^\epsilon\Big(1 - 72K_4(u^*/c)\ln b\Big) \\ &= (u - u^*)b^\epsilon\Big(1 - 2\epsilon \ln b\Big) \\ &\to b^{-\epsilon}(u - u^*) \,, \end{aligned} \tag{20.48}$$

so that, as we had hoped, u is irrelevant at the new non-Gaussian fixed point. This fixed point is stable (other than to the two relevant variables r and H). The recursion relation for r is

$$(r' - r^*) = (r - r^*)\frac{\partial r'}{\partial r}\Big)_{r=r^*,u=u^*} + (u - u^*)\frac{\partial r'}{\partial u}\Big)_{r=r^*,u=u^*} \,, \tag{20.49}$$

where

$$\begin{aligned} \frac{\partial r'}{\partial r}\Big)_{r=r^*,u=u^*} &= b^2\Big(1 - 12(K_4 u^*/c^2)\ln b\Big) \\ &= b^2[1 - \frac{1}{3}\epsilon \ln b] \\ &\to b^{2-\epsilon/3} \end{aligned} \tag{20.50}$$

and

$$\left.\frac{\partial r'}{\partial u}\right)_{r=r^*,u=u^*} = 6K_4\Lambda^2(b^2-1)/c - 12K_4b^2c^{-2}r^*\ln b$$
$$= 6K_4\Lambda^2(b^2-1)/c + 2K_4\Lambda^2b^2c^{-1}\epsilon\ln b\,. \qquad (20.51)$$

It is not consistent to keep the term here of order ϵ because we have omitted other terms in the recursion relation of this same order. So it is best to write

$$\left.\frac{\partial r'}{\partial u}\right)_{r=r^*,u=u^*} = 6[K_4\Lambda^2/c](b^2-1) + \mathcal{O}(\epsilon) \equiv X(b) + \mathcal{O}(\epsilon)\,. \qquad (20.52)$$

Thus

$$\begin{bmatrix} r'-r^* \\ u'-u^* \end{bmatrix} = \mathbf{R(b)} \begin{bmatrix} r'-r^* \\ u'-u^* \end{bmatrix}, \qquad (20.53)$$

where

$$\mathbf{R(b)} = \begin{bmatrix} b^{y_1} & X(b) \\ 0 & b^{y_2} \end{bmatrix}, \qquad (20.54)$$

where

$$y_1 = 2 - \frac{1}{3}\epsilon \equiv \frac{1}{\nu} \qquad (20.55a)$$
$$y_2 = -\epsilon\,. \qquad (20.55b)$$

Note that to get y_1 and y_2 correct to order ϵ it was necessary to calculate the diagonal elements of the matrix $\mathbf{R}$ correct up to order ϵ. To obtain the critical exponents to order ϵ we did not need to (and in fact we did not) calculate the off-diagonal element $X(b)$ correctly to order ϵ. Here the left eigenvectors are

$$\langle v_1| = (1, 6K_4\Lambda^2/c)\,, \qquad \langle v_2| = (0, 1) \qquad (20.56)$$

and the right eigenvectors are

$$|v_1\rangle = \begin{bmatrix} 1 \\ 0 \end{bmatrix}, \qquad |v_2\rangle = \begin{bmatrix} -6K_4\Lambda^2/c \\ 1 \end{bmatrix}. \qquad (20.57)$$

As required, these eigenvectors do not depend on b and the associated eigenvalues y_1 and y_2 do not depend on trivial constants (such as K_4, Λ, or c). Finally, we check that $\mathbf{R}$ has the form required by Eq. (19.79), according to which

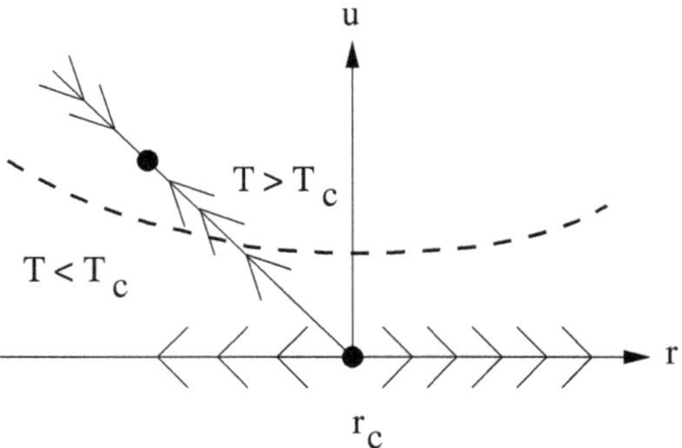

Fig. 20.8 Flow diagram for the ϕ^4-model. Note that on this diagram (for $H = 0$) only r is relevant. Here we show that the Gaussian fixed point ($r = u = 0$) is unstable, but the non-Gaussian fixed point with $u \neq 0$ is stable. As in Fig. 20.2 the dashed line shows schematically the trajectory in the r-u plane which results when the temperature T is varied. The phase transition occurs at the special value of temperature for which this trajectory intersects the line which flows into the non-Gaussian fixed point

$$\begin{aligned}\mathbf{R} &= b^{y_1}|v_1\rangle\langle v_1| + b^{y_2}|v_2\rangle\langle v_2| \\ &= b^{y_1}\begin{bmatrix} 1 & 64K_4\Lambda^2/c \\ 0 & 0 \end{bmatrix} + b^{y_2}\begin{bmatrix} 0 & -64K_4\Lambda^2/c \\ 0 & 1 \end{bmatrix} \\ &= \begin{bmatrix} b^{y_1} & [64K_4\Lambda^2/c][b^{y_1} - b^{y_2}] \\ 0 & b^{y_2} \end{bmatrix} . \end{aligned} \tag{20.58}$$

Comparison with Eq. (20.54) shows that the diagonal elements of $\mathbf{R}$ have the expected form. Also the off-diagonal matrix element is correct up to order ϵ^0, where $y_1 = 2$ and $y_2 = 0$. To get the correct form up to order ϵ would require keeping all terms in the recursion relations of order u^2.

The calculations we have presented so far, including some terms of order ϵ^2, illustrate most features of the RG. We see that the eigenvalues and eigenvectors of the RG transformation are independent of b, as they must be, but also do not depend on parameters like the cutoff, Λ. The fixed point values of the variables do depend on details like the cutoff. At order ϵ we found that u^* was independent of Λ. At higher order in ϵ, u^* does depend on Λ (Aharony 1976). This dependence on the details of the model indicates that the actual value of the transition temperature, for instance, is sensitive to the details of the model, whereas the critical exponents or critical eigenvectors are not sensitive to such details. The flow diagram is shown in Fig. 20.8.

20.4.4 Calculation of η

Thus far, low-order calculations in powers of $\epsilon = 4 - d$ have yielded $\eta = 0$, for both the Gaussian fixed point, $r_G = u_G = 0$, and for the non-Gaussian fixed point,

$r_{NG} = -c\Lambda^2\epsilon/6$, $u_{NG} = c^2\epsilon K_4^{-1}/36$. For the latter case, it was necessary to include diagrams of order u^2 to obtain the nontrivial renormalization of u. In light of the above, it is natural to ask what is the leading contribution to η in powers of ϵ and what diagrams must be included to obtain this contribution? We will now show that the leading correction to $\eta = 0$ is of order ϵ^2 and that to obtain this result one should consider all diagrams of order u^2, including some which were not required for the calculation of u_{NG}.

We start by considering a generalization of the ϕ^4 Hamiltonian by including a 6th order term. That is, we write

$$\begin{aligned}
\mathcal{H}[\{x(\mathbf{q})\}] &= \frac{1}{2}\int_{q=0}^{q=\Lambda}[r+cq^2]x(\mathbf{q})x(-\mathbf{q}) \\
&+ (2\pi)^d \int_{q_1=0}^{q_1=\Lambda}\int_{q_2=0}^{q_2=\Lambda}\int_{q_3=0}^{q_3=\Lambda}\int_{q_4=0}^{q_4=\Lambda} u_4(\mathbf{q}_1,\mathbf{q}_2,\mathbf{q}_3,\mathbf{q}_4) \\
&\quad \times x(\mathbf{q}_1)x(\mathbf{q}_2)x(\mathbf{q}_3)x(\mathbf{q}_4)\delta(\mathbf{q}_1+\mathbf{q}_2+\mathbf{q}_3+\mathbf{q}_4) \\
&+ (2\pi)^d \int_{q_1=0}^{q_1=\Lambda}\int_{q_2=0}^{q_2=\Lambda}\int_{q_3=0}^{q_3=\Lambda}\int_{q_4=0}^{q_4=\Lambda}\int_{q_5=0}^{q_5=\Lambda}\int_{q_6=0}^{q_6=\Lambda} u_6(\mathbf{q}_1,\mathbf{q}_2,\mathbf{q}_3,\mathbf{q}_4,\mathbf{q}_5,\mathbf{q}_6) \\
&\quad \times x(\mathbf{q}_1)x(\mathbf{q}_2)x(\mathbf{q}_3)x(\mathbf{q}_4)x(\mathbf{q}_5)x(\mathbf{q}_6)\delta(\mathbf{q}_1+\mathbf{q}_2+\mathbf{q}_3+\mathbf{q}_4+\mathbf{q}_5+\mathbf{q}_6) \, .
\end{aligned} \tag{20.59}$$

If we take as our initial Hamiltonian the ϕ^4 model of Eq. (20.13), then initially, $u_4(\mathbf{q}_1,\mathbf{q}_2,\mathbf{q}_3,\mathbf{q}_4) = u$ and $u_6(\mathbf{q}_1,\mathbf{q}_2,\mathbf{q}_3,\mathbf{q}_4,\mathbf{q}_5,\mathbf{q}_6) = 0$.

The leading contribution to u_6 is second order in the coupling constant u_4 and corresponds to the tree-diagram shown in Fig. 20.9 in which two quartic vertices are connected by a line with momentum $\mathbf{q}$ which is integrated over the range $\Lambda/b < q < \Lambda$.

Then the additional term in the Hamiltonian, generated by diagrams of the type shown in Fig. 20.9, corresponding to terms of the form $\langle V^{(1)}V^{(1)}\rangle$ in Eq. (20.30), is

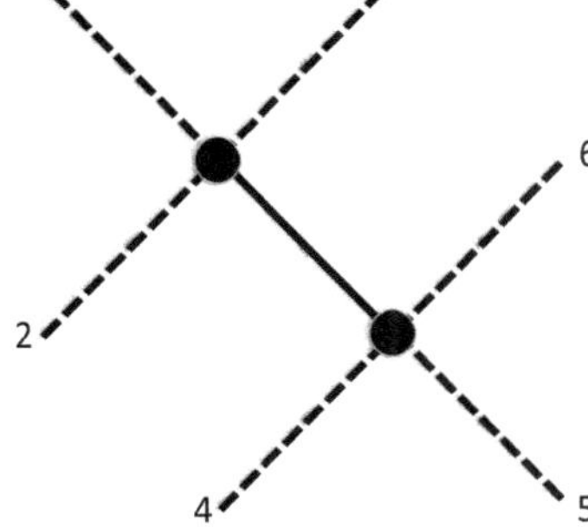

Fig. 20.9 The leading order contribution to the vertex corresponding to u_6 is generated by connecting two quartic vertices with a single line which is then integrated over the momentum shell, $\Lambda/b < q < \Lambda$

$$\delta\mathcal{H}_6 = -\frac{1}{2}4^2 u_4^2 (2\pi)^{2d} \int_{q_1=0}^{q_1=\Lambda} \int_{q_2=0}^{q_2=\Lambda} \int_{q_3=0}^{q_3=\Lambda} x(\mathbf{q}_1)x(\mathbf{q}_2)x(\mathbf{q}_3)$$
$$\times \int_{q_4=0}^{q_4=\Lambda} \int_{q_5=0}^{q_5=\Lambda} \int_{q_6=0}^{q_6=\Lambda} x(\mathbf{q}_4)x(\mathbf{q}_5)x(\mathbf{q}_6)$$
$$\times \int_{\Lambda/b}^{\Lambda} \frac{d\mathbf{q}}{(2\pi)^d (r + cq^2)} \delta(\mathbf{q}_1 + \mathbf{q}_2 + \mathbf{q}_3 + \mathbf{q})\delta(\mathbf{q}_4 + \mathbf{q}_5 + \mathbf{q}_6 - \mathbf{q}) \,, \tag{20.60}$$

where the factor 4^2 is the number of ways that the two quartic vertices can be connected.

In the calculation that follows, we take b to be only slightly larger than 1 by writing $b = e^{d\ell}$ where $d\ell$ represents an infinitesimal change in length scale. Then the change $\delta\mathcal{H}_6$ corresponds to a change in the coupling constant

$$\delta u_6 = -8u_4^2 \frac{\delta(|\mathbf{q}_1 + \mathbf{q}_2 + \mathbf{q}_3| - \Lambda)}{r + c\Lambda^2} \Lambda d\ell. \tag{20.61}$$

Up to this point, it has not been necessary to keep careful track of the momentum dependence of renormalized coupling constants. In Eqs. (20.44a) and (20.44b), the assumption was that the coupling constants r and u were independent of momentum, and, in the derivation of Eq. (20.44c) for c, it was simple to treat the momentum dependence explicitly. However, for calculating η it is important to keep track of the momentum dependencies of the coupling constants and this requires being a bit more explicit in our expressions for their recursion relations. For example, for the case of u_6, we should write

$$u_6(\{\mathbf{q}_i\}, b) = b^{-2}[u_6(\{\mathbf{q}_i/b\}, 1) + \delta u_6(\{\mathbf{q}_i\})], \tag{20.62}$$

where $i = 1, 2, 3$ and we have used Eq. (20.25) for the scaling behavior of u_6. Setting $b = e^{d\ell}$ and expanding to first order in $d\ell$, we derive the differential recursion relation

$$\frac{du_6(\{\mathbf{q}_i\})}{d\ell} = -2u_6(\{\mathbf{q}_i\}) - \sum_i \mathbf{q}_i \cdot \nabla_i u_6(\{\mathbf{q}_i\}) + \frac{1}{d\ell}\delta u_6(\{\mathbf{q}_i\}), \tag{20.63}$$

The fixed point, u_6^*, is determined from

$$\frac{du_6(\{\mathbf{q}_i\})}{d\ell} = 0. \tag{20.64}$$

Defining $w = |\mathbf{q}_1 + \mathbf{q}_2 + \mathbf{q}_3|^2$ and, noting that $u_6(w) = w^{-1}\hat{u}_6(w)$ is a function of w because $\delta u_6(\{\mathbf{q}_i\})$depends only on w, we can write the fixed point equation as

$$0 = -2w^{-1}\hat{u}_6(w) - \sum_i \mathbf{q}_i \cdot \nabla_i w \frac{\partial}{\partial w}\left(w^{-1}\hat{u}_6(w)\right) - \frac{8u_4^2}{c\Lambda}\delta(\sqrt{w} - \Lambda), \quad (20.65)$$

where in the last term, we have neglected r, which is of order ϵ at the fixed point, compared to $c\Lambda^2$. Noting that

$$\sum_i \mathbf{q}_i \cdot \nabla_i w = 2w \ , \quad (20.66)$$

we have

$$0 = -2w^{-1}\hat{u}_6(w) - 2w\left(-w^{-2}\hat{u}_6(w) + w^{-1}\frac{\partial \hat{u}_6(w)}{\partial w}\right) - \frac{8u_4^2}{c\Lambda}2\Lambda\delta(w - \Lambda^2),$$
$$= -2\frac{\partial \hat{u}_6(w)}{\partial w} - \frac{16u_4^2}{c}\delta(w - \Lambda^2). \quad (20.67)$$

Integrating with respect to w, we find

$$\hat{u}_6(w) = -\frac{8u_4^2}{c}\int_0^w dw'\delta(w' - \Lambda^2) = -\frac{8u_4^2}{c}\theta(w - \Lambda^2) \ , \quad (20.68)$$

where $\theta(x)$ is the unit step function, so that, using $u_6(w) = w^{-1}\hat{u}_6(w)$, the 6th order coupling constant at the fixed point is

$$u_6^*(w) = -8(u_4^*)^2\frac{\theta(w - \Lambda^2)}{cw}. \quad (20.69)$$

Next, we take a closer look at the perturbation to the quartic coupling from Eq. (20.36) which we rewrite to focus on the previously neglected dependence on the external momenta.

$$\delta\mathcal{H}_4 = -36u_4^2(2\pi)^d \int_0^{\Lambda/b}\frac{d\mathbf{q}_1}{(2\pi)^d}\int_0^{\Lambda/b}\frac{d\mathbf{q}_2}{(2\pi)^d}\int_0^{\Lambda/b}\frac{d\mathbf{q}_3}{(2\pi)^d}\int_0^{\Lambda/b}\frac{d\mathbf{q}_4}{(2\pi)^d}$$
$$\times M(r, \mathbf{q}_1 + \mathbf{q}_2)x(\mathbf{q}_1)x(\mathbf{q}_2)x(\mathbf{q}_3)x(\mathbf{q}_4)\delta(\mathbf{q}_1 + \mathbf{q}_2 + \mathbf{q}_3 + \mathbf{q}_4), \quad (20.70)$$

so that, using Eq. (20.37), the perturbation to u_4 is

$$\delta u_4 = -36u_4^2\int_{\Lambda/b}^{\Lambda}\frac{d\mathbf{q}_5}{(2\pi)^d}\int_{\Lambda/b}^{\Lambda}d\mathbf{q}_6\frac{\delta(\mathbf{q}_1 + \mathbf{q}_2 + \mathbf{q}_5 + \mathbf{q}_6)}{(r + cq_5^2)(r + cq_6^2)} \ . \quad (20.71)$$

Again taking $b = e^{d\ell}$, we can perform the integral over $\mathbf{q}_6$ obtaining

$$\delta u_4 = -36u_4^2 \int_{\Lambda e^{-d\ell}}^{\Lambda} \frac{d\mathbf{q}_5}{(2\pi)^d} \frac{\theta(|\mathbf{q}_1 + \mathbf{q}_2 + \mathbf{q}_5| - \Lambda e^{-d\ell})\theta(\Lambda - |\mathbf{q}_1 + \mathbf{q}_2 + \mathbf{q}_5|)}{(r + cq_5^2)(r + c|\mathbf{q}_1 + \mathbf{q}_2 + \mathbf{q}_5|^2)}, \tag{20.72}$$

where the $\mathbf{q}_5$ integral is restricted to the momentum shell of thickness, $\Lambda d\ell$, and the theta-functions impose an additional restriction that depends on the sum of external momenta, $\mathbf{q}_1 + \mathbf{q}_2$.

We said in the last subsection that the dependence of the perturbation on nonzero external momentum is not "relevant." It is instructive to explore the meaning of that statement by evaluating the leading momentum dependence of δu_4 from Eq. (20.72). Close to the fixed point, we can neglect $r << c\Lambda^2$, and we can also perform the radial part of the $\mathbf{q}_5$ integral while setting $\mathbf{q}_5 = \Lambda e^{-d\ell/2}\hat{\mathbf{q}}$ (so that $\mathbf{q}_5$ is inside the shell). Then

$$\delta u_4 = -\frac{36u_4^2}{c^2} d\ell \int \frac{d\Omega_q}{(2\pi)^4} \theta(|\mathbf{q}_1 + \mathbf{q}_2 + \Lambda e^{-d\ell/2}\hat{\mathbf{q}}| - \Lambda e^{-d\ell})\theta(\Lambda - |\mathbf{q}_1 + \mathbf{q}_2 + \Lambda e^{-d\ell/2}\hat{\mathbf{q}}|). \tag{20.73}$$

Consider the limit in which $|\mathbf{q}_1 + \mathbf{q}_2| << \Lambda d\ell$. Then the θ-functions are both 1, and we find

$$\delta u_4 = -\frac{36u_4^2}{c^2} K_4 d\ell \tag{20.74}$$

consistent with the result used in Eq. (20.44b). However, for the region, $\Lambda d\ell < |\mathbf{q}_1 + \mathbf{q}_2| < 2\Lambda$, the product of θ-functions selects out an annular region of the shell with a volume proportional to $\Lambda^d d\ell^2$, which is negligible compared to the result for $\mathbf{q}_1 + \mathbf{q}_2 = 0$. For external momenta in this region of nonzero $|\mathbf{q}_1 + \mathbf{q}_2|$, the renormalization of u_4 is zero, which means that the calculation of the renormalization of u in the previous section was incomplete, since the renormalized u_4 must be an analytic function of the external momenta. Fortunately, as we will now show, this is where the u_6 term comes to the rescue and provides the rest of the momentum-dependent renormalization of u_4.

The contribution of the Hamiltonian of Eq. (20.60), corresponding to the diagram of Fig. 20.9, to the renormalization of u_4 is obtained by connecting two of the six external momenta and integrating the wavevector of the resulting propagator over the momentum shell. There are two distinct ways in which this can be done. The first corresponds to connecting any two of the external lines $(1, 2, 3)$ or $(4, 5, 6)$. The second is to connect one line from $(1, 2, 3)$ to one from $(4, 5, 6)$. The two possibilities are illustrated in Fig. 20.10. In calculating these diagrams, we use the fixed point function, $u_6^*(w)$ where $w = |\mathbf{q}_1 + \mathbf{q}_2 + \mathbf{q}_3|^2$ is the squared magnitude of the sum of the momenta of the lines entering the diagram. $u_6^*(w)$ is given by Eq. (20.69) which is zero for $w < \Lambda^2$. For the first (left-hand) diagram in Fig. 20.10, $u_6^*(w) = 0$ since $\mathbf{q}_1 = -\mathbf{q}_2 = \mathbf{q}$, and $\mathbf{q}_3$ lies inside the momentum shell.

Diagrams of the second type in Fig. 20.10, of which there are nine, contribute an amount $\delta u_4'$ where, defining $\mathbf{s} = \mathbf{q}_1 + \mathbf{q}_2$,

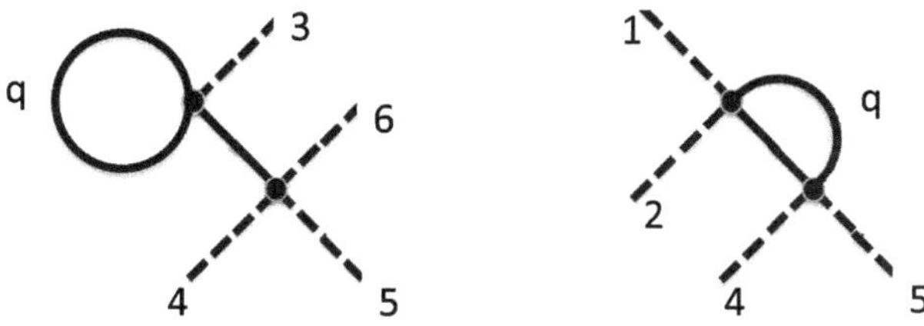

Fig. 20.10 Diagrams from $\delta\mathcal{H}_6$ which contribute to the renormalization of u_4. The wavevector $\mathbf{q}$ is integrated over the momentum shell

$$\delta u_4'(s) = -\frac{72(u_4)^2}{c}\int_{\Lambda/b}^{\Lambda}\frac{d\mathbf{q}}{(2\pi)^d}\frac{\theta(|\mathbf{s}+\mathbf{q}|^2-\Lambda^2)}{(r+cq^2)|\mathbf{s}+\mathbf{q}|^2}. \tag{20.75}$$

Unlike δu_4 of Eq. (20.74), $\delta u_4'(0)$ is zero because $\Lambda/b < q < \Lambda$.

Neglecting r and writing $\mathbf{q} = \Lambda\hat{\mathbf{q}}$, we can evaluate the integral with $s \neq 0$ for d=4, using, $d\Omega_q = \sin^2\psi d\psi \sin\theta d\theta d\phi$. Then

$$\delta u_4'(s) = -\frac{72(u_4)^2}{c^2}\frac{4\pi d\ell}{(2\pi)^4}\Lambda^2\int_0^{\pi}\frac{\theta(s^2+2s\Lambda\cos\psi)}{s^2+2s\Lambda\cos\psi+\Lambda^2}\sin^2\psi\, d\psi \tag{20.76}$$

We can replace the upper limit of the ψ integral by the value of ψ for which the value of the θ-function vanishes, which is

$$\psi_0(s) = \cos^{-1}\left(-\frac{s}{2\Lambda}\right). \tag{20.77}$$

Then

$$\delta u_4'(s) = -\frac{72(u_4)^2}{c^2}\frac{4\pi d\ell}{(2\pi)^4}\Lambda^2\int_0^{\psi_0(s)}\frac{\sin^2\psi\, d\psi}{s^2+2s\Lambda\cos\psi+\Lambda^2}. \tag{20.78}$$

Equation (20.76) was derived assuming that $s > 0$ so that the θ-function can be satisfied. However, the reader can check that, for Eq. (20.78) in the limit, $s \to 0$, $\delta u_4'(s) \to \delta u_4(0)$, where $\delta u_4(0)$ is given by the expression on the right-hand side of Eq. (20.74) (where $K_4 = 1/(8\pi^2)$), and hence the expression in Eq. (20.78) captures the full behavior for $s \geq 0$ and we can drop the prime. Furthermore, if we define the function

$$f(s) \equiv \frac{4}{\pi}\Lambda^2\int_0^{\psi_0(s)}\frac{\sin^2\psi\, d\psi}{s^2+2s\Lambda\cos\psi+\Lambda^2}, \tag{20.79}$$

then $f(0) = 1$, and

$$\delta u_4(s) = \delta u_4(0) f(s). \tag{20.80}$$

Next, as we did in Eqs. (20.62) and (20.63), we derive the differential recursion relation (also called the beta-function) for $u_4(\{\mathbf{q}_i\})$.

$$u_4(\{\mathbf{q}_i\}, e^{d\ell}) = e^{\epsilon d\ell}[u_4(\{\mathbf{q}_i e^{-d\ell}\}, 1) + \delta u_4(\{\mathbf{q}_i\})], \quad (20.81)$$

$$\frac{du_4(\{\mathbf{q}_i\})}{d\ell} = \epsilon u_4(\{\mathbf{q}_i\}) - \sum_i \mathbf{q}_i \cdot \nabla_i u_4(\{\mathbf{q}_i\}) + \frac{1}{d\ell}\delta u_4(\{\mathbf{q}_i\}), \quad (20.82)$$

where $i = 1, 2$. Treating $u_4(\{\mathbf{q}_i\})$ as a function of $s = \sqrt{|\mathbf{q}_1 + \mathbf{q}_2|^2}$, we have

$$\sum_i \mathbf{q}_i \cdot \nabla_i u_4(\{\mathbf{q}_i\}) = s\frac{\partial u_4}{\partial s}, \quad (20.83)$$

so that, at the fixed point,

$$0 = \epsilon u_4^*(s) - s\frac{\partial u_4^*(s)}{\partial s} - \frac{36 u_4^*(0)^2}{c^2} K_4 f(s). \quad (20.84)$$

For $s = 0$, this gives the now familiar result,

$$u_4^*(0) = \frac{c^2\epsilon}{36K_4}. \quad (20.85)$$

Defining $\tilde{u}_4^*(s) = u_4^*(s) - u_4^*(0)$, we then have

$$0 = \epsilon \tilde{u}_4^*(s) - s\frac{\partial \tilde{u}_4^*(s)}{\partial s} - \frac{36 u_4^*(0)^2}{c^2} K_4\left(f(s) - 1\right). \quad (20.86)$$

Equation (20.86) has a consistent solution for $\tilde{u}_4^*(s) \sim \epsilon^2$, in which case we can drop the first term which is of order ϵ^3. Defining $\Delta f(s) = f(s) - 1$, we obtain

$$\tilde{u}_4^*(s) = -\frac{c^2\epsilon^2}{36K_4}\int_0^s \frac{\Delta f(s')}{s'} ds'. \quad (20.87)$$

So we can write the fourth-order fixed point Hamiltonian as

$$\mathcal{H}_4^*[\{x(\mathbf{q})\}] = (2\pi)^d \int_{q_1=0}^{q_1=\Lambda}\int_{q_2=0}^{q_2=\Lambda}\int_{q_3=0}^{q_3=\Lambda}\int_{q_4=0}^{q_4=\Lambda} u_4^*(|\mathbf{q}_1 + \mathbf{q}_2|)$$
$$\times x(\mathbf{q}_1)x(\mathbf{q}_2)x(\mathbf{q}_3)x(\mathbf{q}_4)\delta(\mathbf{q}_1 + \mathbf{q}_2 + \mathbf{q}_3 + \mathbf{q}_4), \quad (20.88)$$

and, adding together the different ways that a pair of the $x(\mathbf{q}_i)$ can be combined into a propagator and integrated over the momentum shell, we find the perturbation to the quadratic Hamiltonian to be

$$\delta\mathcal{H}_2^*[\{x(\mathbf{q})\}] = \frac{1}{2}\int_0^{q=\Lambda} \frac{d\mathbf{q}}{(2\pi)^d} x(\mathbf{q})x(-\mathbf{q})\, 2\int_{\Lambda/b}^{\Lambda} \frac{2u_4^*(0) + 4u_4^*(s)}{r + ck^2}\frac{d\mathbf{k}}{(2\pi)^d}, \quad (20.89)$$

where $s = |\mathbf{q} + \mathbf{k}|$, and the factors, 1/2 and 2, have been inserted to put $\delta\mathcal{H}_2^*$ into the form of the quadratic term in Eq. (20.59).

We are looking for the momentum-dependent part of $\delta\mathcal{H}_2^*$ which will give the leading correction to η. To find this, we first expand $u_4^*(s)$ to order q^2.

$$u_4^*(s) = u_4^*(|\mathbf{k}|) + \mathbf{q} \cdot \nabla_{q'} u_4^*(|\mathbf{k} + \mathbf{q}'|)\Big|_{q'=0} + \frac{1}{2}(\mathbf{q} \cdot \nabla_{q'})(\mathbf{q} \cdot \nabla_{q'}) u_4^*(|\mathbf{k} + \mathbf{q}'|)\Big|_{q'=0}. \tag{20.90}$$

The first term contributes to the renormalization of r, and the second term contributes zero to $\delta\mathcal{H}_2^*$ because of the angular integration over $\mathbf{k}$ in Eq. (20.89). The q^2 term is

$$\begin{aligned} &\frac{1}{2}(\mathbf{q} \cdot \nabla_{q'})(\mathbf{q} \cdot \nabla_{q'}) u_4^*(|\mathbf{k} + \mathbf{q}'|)\Big|_{q'=0} \\ &\qquad = \frac{1}{2}\left(\frac{q^2}{k} - \frac{(\mathbf{q}\cdot\mathbf{k})^2}{k^3}\right)\frac{du_4^*(k)}{dk} + \frac{1}{2}\frac{(\mathbf{q}\cdot\mathbf{k})^2}{k^2}\frac{d^2u_4^*(k)}{dk^2}. \end{aligned} \tag{20.91}$$

For calculating the integral in Eq. (20.89), we can replace this by its angular average over $\mathbf{k}$, in which case we can write

$$\begin{aligned} &\left\langle \frac{1}{2}(\mathbf{q} \cdot \nabla_{q'})(\mathbf{q} \cdot \nabla_{q'}) u_4^*(|\mathbf{k} + \mathbf{q}'|)\Big|_{q'=0} \right\rangle_{\hat{\mathbf{k}}} \\ &\qquad = \frac{q^2}{8}\left(\frac{3}{k}\frac{du_4^*(k)}{dk} + \frac{d^2u_4^*(k)}{dk^2}\right), \end{aligned} \tag{20.92}$$

where we have used the fact that $\langle(\mathbf{q}\cdot\mathbf{k})^2\rangle_{\hat{\mathbf{k}}} = q^2k^2/4$ in $d = 4$ dimensions. If we define δc as the contribution of $\delta\mathcal{H}_2^*$ to the renormalization of the coefficient of q^2, we can write

$$\begin{aligned} \delta c &= 8\frac{1}{(2\pi)^d}\int_{\Lambda/b}^{\Lambda}\frac{1}{8}\left(\frac{3}{k}\frac{du_4^*(k)}{dk} + \frac{d^2u_4^*(k)}{dk^2}\right)\frac{d\mathbf{k}}{r + ck^2}, \\ &= \frac{K_4}{c}\Lambda^2 d\ell\left(\frac{3}{k}\frac{du_4^*(k)}{dk} + \frac{d^2u_4^*(k)}{dk^2}\right)_{k=\Lambda}, \end{aligned} \tag{20.93}$$

where we have, once again, set $r = 0$.

The momentum-dependent part of $u_4^*(k)$ is $\tilde{u}_4^*(k)$, defined in Eq. (20.87). Then

$$\delta c = -\frac{c\epsilon^2}{36}\Lambda^2 d\ell\left(\frac{2\Delta f(\Lambda)}{\Lambda^2} + \frac{1}{\Lambda}\frac{df(k)}{dk}\Big|_{k=\Lambda}\right). \tag{20.94}$$

So we need

$$2\Delta f(\Lambda) = \frac{4}{\pi}\int_0^{2\pi/3}\frac{\sin^2\psi\, d\psi}{1 + \cos\psi} - 2 = -\frac{2\sqrt{3}}{\pi} + \frac{2}{3} \tag{20.95}$$

where we have used Eq. (20.79) and $\cos^{-1}(-1/2) = 2\pi/3$, and

$$\begin{aligned}\frac{df(k)}{dk} &= \frac{\partial f}{\partial \psi_0}\frac{\partial \psi_0(k)}{\partial k} - \frac{4\Lambda^2}{\pi}\int_0^{\psi_0(k)} \frac{2(k+\Lambda\cos\psi)\sin^2\psi}{(k^2+2k\Lambda\cos\psi+\Lambda^2)^2} \\ &= \frac{4\Lambda^2}{\pi}\left[\frac{\sin^2\psi_0\frac{1}{\sqrt{4\Lambda^2-k^2}}}{(k^2+2k\Lambda\cos\psi_0+\Lambda^2)} - \int_0^{\psi_0(k)} \frac{2(k+\Lambda\cos\psi)\sin^2\psi\, d\psi}{(k^2+2k\Lambda\cos\psi+\Lambda^2)^2}\right].\end{aligned} \tag{20.96}$$

Setting $k = \Lambda$, $\cos\psi_0 = -1/2$, $\sin\psi_0 = \sqrt{3}/2$, we find

$$\Lambda\frac{df(k)}{dk}\bigg|_{k=\Lambda} = \frac{2\sqrt{3}}{\pi} - \frac{4}{3} \tag{20.97}$$

Substituting these results into Eq. (20.94), gives

$$\delta c = \frac{c\epsilon^2}{54}d\ell. \tag{20.98}$$

So the recursion relation for c, for which we previously had Eq. (20.44c), becomes

$$\begin{aligned}c' &= e^{-\eta d\ell}\,(c+\delta c) \\ \frac{dc}{d\ell} &= c\left(-\eta+\frac{\epsilon^2}{54}\right),\end{aligned} \tag{20.99}$$

so that c remains fixed, as it must, for

$$\eta = \frac{\epsilon^2}{54}. \tag{20.100}$$

We have gone into considerable detail for this calculation because it illustrates aspects of momentum-shell RG, which are not included in other textbooks and which are difficult to find in the literature. Furthermore, we find much of the discussion in the literature to be confusing and occasionally misleading. An example is the fact that the standard ϵ-expansion derivation of the fourth-order coupling at the fixed point is valid only for zero external momentum. We would also comment that the use of a sharp momentum cutoff does not seem to have caused any problem in the calculation of η above. We also note that the discussion of the calculation of η in the classic text by Ma (1976) appears to contain errors, specifically with regard to the evaluation of the momentum integral that determines the dependence of the coupling constant c on the renormalization factor b. A detailed calculation of η, similar in spirit to the one given above, can be found in Wegner and Houghton (1973).

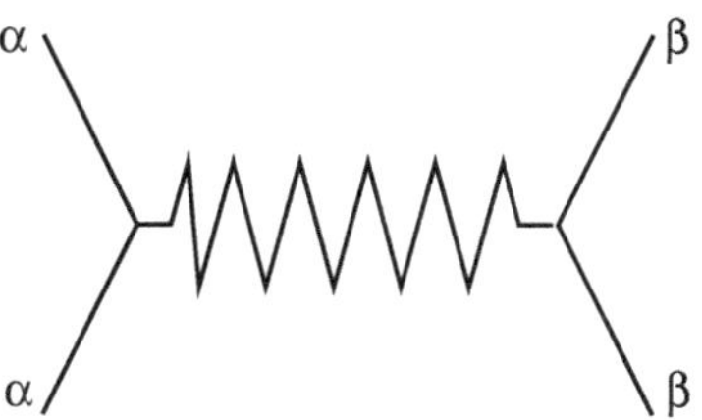

Fig. 20.11 Quartic Perturbation in the n-component classical Heisenberg model. The component labels are summed from 1 to n

20.4.5 Generalization to the n-Component Model and Higher Order in ϵ Terms

The calculations described in the previous sections may be generalized to the n-component model as follows. In the quadratic term of the Hamiltonian we replace $x(\mathbf{q})x(-\mathbf{q})$ by

$$\sum_{\alpha=1}^{n} x_\alpha(\mathbf{q})x_\alpha(-\mathbf{q}) \equiv \mathbf{x}(\mathbf{q}) \cdot \mathbf{x}(-\mathbf{q}) \, . \tag{20.101}$$

In the quartic term we replace $ux(\mathbf{q}_1)x(\mathbf{q}_2)x(\mathbf{q}_3)x(\mathbf{q}_4)$ by

$$u \, \mathbf{x}(\mathbf{q}_1) \cdot \mathbf{x}(\mathbf{q}_2) \, \mathbf{x}(\mathbf{q}_3) \cdot \mathbf{x}(\mathbf{q}_4) \, . \tag{20.102}$$

Now the fourth-order perturbation is represented diagrammatically as in Fig. 20.11. The renormalization diagrams for r, u, and c are shown in Fig. 20.12.

The calculations described above were intended to illustrate the ϵ-expansion method by direct computation of the leading order terms. With sufficient effort, the calculations can be extended to higher order. This was first done by Wilson who evaluated the ϵ-expansions for η and γ to order ϵ^3 and ϵ^2 respectively using a Feynman diagram expansion. Later the series were extended to higher order. The exponents, η and γ of the n-component model, to order ϵ^4 and ϵ^3, as calculated by Brezin and co-workers (1973) are as follows.

$$\begin{aligned} \eta = \frac{\epsilon^2}{2}\frac{(n+2)}{(n+8)^2}\Bigg\{ 1 + \epsilon\left[\frac{6(3n+14)}{(n+8)^2} - \frac{1}{4}\right] + \epsilon^2\Bigg[\frac{45(3n+14)^2}{(n+8)^4} \\ - \frac{8(3n^2+53n+160)}{(n+8)^3} + \frac{(-5n^2+234n+1076)}{16(n+8)^2} \\ - \frac{24(5n+22)}{(n+8)^3}\zeta(3)\Bigg]\Bigg\} + \mathcal{O}(\epsilon^5) \end{aligned} \tag{20.103a}$$

$$\gamma = 1 + \frac{\epsilon}{2}\frac{(n+2)}{(n+8)} + \epsilon^2\frac{(n+2)(n^2+22n+52)}{4(n+8)^3}$$

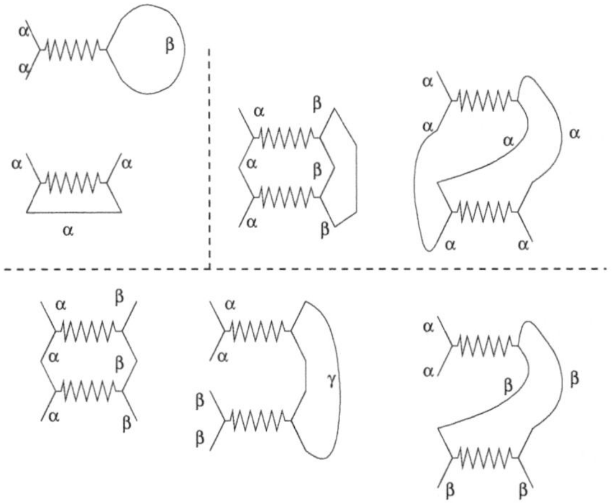

Fig. 20.12 Diagram for the renormalization of r (upper left), c (upper right), and u (bottom). The counting of these diagrams is left as an exercise for the reader. To start you off, we mention that the counting factor for the r diagrams, which was six for the Ising ($n = 1$) model, is here $2n + 4$, where n comes from the sum over β

$$+\epsilon^3(n+2)\left[\frac{55n^2+268n+424}{2(n+8)^5}+\frac{3(n+2)(n+3)}{(n+8)^4}+\frac{(n+2)^2}{8(n+8)^3}\right.$$
$$\left.-\frac{6(5n+22)}{(n+8)^4}\zeta(3)\right]+\mathcal{O}(\epsilon^4) \tag{20.103b}$$

where $\zeta(3) = 1.20206$ is a Riemann zeta function. These higher order expansions, like the lowest order expansions described above, are based on perturbation expansions in powers of the quartic coupling constant u. Such expansions are known to be divergent, and the expansion of the exponents in powers of ϵ is an asymptotic expansion. Techniques for analyzing and extrapolating asymptotic expansions, such as Borel summation, have been applied to the ϵ-expansion to obtain more accurate values of the exponents.

20.5 Universality Classes

An important idea in critical phenomena is the concept of a "universality class." The idea is that the critical exponents only depend on certain universal features such as the dimensionality and symmetry of the system and the type and number of components of the spins or spin-like degrees of freedom. This idea is exemplified by the expansions above in powers of $\epsilon = 4 - d$, with coefficients depending on n, the number of components of the spin. This idea has many important consequences. First of all, in the ϵ-expansion we have studied only what happens in the neighborhood of the fixed points. In particular, we analyzed the recursion relations assuming r and u were both of order ϵ. In real systems, it is unreasonable to believe that an appropriate model has starting values r_0 and u_0 which are small. What we know is that the fixed points we investigate will be relevant if the parameters are in the basin of attraction for that fixed point. It is assumed that this basin of attraction is *not* infinitesimal, so that what happens when r and u are of order ϵ is indicative of behavior at physically important values of these parameters. Furthermore, we extrapolate this to include the hope that the theory applies for d significantly away from $d = 4$, say, down to $d = 2$. We presume that the numerical values may not be well given by the ϵ-expansion, but that the exponents will vary smoothly and still satisfy the various scaling relations. Indeed a great triumph of the RG is that it provides a calculation whose result is that the number of relevant variables for Ising-like systems is two. Furthermore, although we did not explore this here, the RG provides a means of calculating scaling functions (and not just exponents) which describe the asymptotic critical properties. One finds that many dimensionless ratios (called amplitude ratios) are also universal (Aharony and Hohenberg 1976). For instance, if one writes

$$\chi(T, H = 0) \sim A_{\pm}|T - T_c|^{-\gamma} \quad \text{for } T \to T_c^{\pm} , \tag{20.104}$$

then the RG tells us that A_+ and A_- are *not* universal, but depend on the details of the system. However, the ratio A_+/A_- *is* universal. One can form many such dimensionless ratios and their values are found to vary much more significantly from one model to the next than do the critical exponents. So their study can provide a sensitive test of RG calculations.

However, the situation is not so simple as it may seem at first glance. For example, at the beginning of this book, we discussed the fact that the phase diagrams for the liquid–gas system and the ferromagnetic Ising model are similar and that the critical exponents are, for all practical purposes, identical. This similarity arises from the fact that the two systems are in the same universality class. What does this mean? One simple way of looking at it is to note that the liquid–gas system is well modeled by the lattice gas model which is a variation on the ferromagnetic Ising model. Therefore the two systems are in the same universality class because they are both basically Ising models.

On the other hand, in Chap. 10, we looked at the lattice gas with repulsive interactions on a triangular lattice and showed that the transition to the $\sqrt{3} \times \sqrt{3}$ ordered structure is closely analogous to the ordering transition of the ferromagnetic, three-state Potts model. The two-dimensional three-state Potts model is a solved problem with exponents different from those of the two-dimensional Ising model. It is the theoretical expectation, and there is good experimental evidence to back it up, that the exponents of the $\sqrt{3} \times \sqrt{3}$ lattice gas transitions will be the same as those for the three-state Potts model. Clearly, the operative principle here is that the number of components and symmetry of the order parameter, which develops at the transition (together with the spatial dimensionality) determine the universality class and the exponents.

This is beginning to sound like a real theory. However, it does not account for one very simple case, namely, tri-critical behavior. When the nonuniversal quartic term in the free energy changes sign, the transition goes from continuous to first order. Right at the boundary, the exponents corresponding to the tri-critical point are different from the critical exponents for the second-order transition. This suggests that the form of the Landau expansion, which also involves the spatial dimension and the number of components and symmetry of the critical order parameter, is the basis for defining a universality class, and this is indeed part of the story. The one remaining missing ingredient is spatial fluctuations.

Wilson's renormalization group theory is based on a free energy functional with quadratic and quartic terms. The quadratic part involves the coupling constant r which can be positive or negative. It also has a term proportional to q^2, which controls spatial fluctuations. In real space, this term corresponds to the square of a gradient. In a lattice model it comes from nearest (or further) neighbor interactions. For the Ising or n-component models, we expect the coupling constant u of the quartic term to be positive. If, for some reason u were negative, it would be necessary to consider sixth-order or higher order couplings. Furthermore, there are some interesting systems in which the gradient-squared q^2 terms are zero. In this case, the next order terms that control spatial fluctuations would be of order q^4, corresponding to the square of the local curvature. It is clear from the RG calculations in this chapter, that the absence of terms quadratic in q would drastically change the calculation of the exponents.

All of the above discussion can be incorporated into the statement that the universality class is defined by the form of the Landau–Ginzburg–Wilson (LGW) Hamiltonian, the analog of $\mathcal{H}$ in Eq. (20.10). The form of the LGW Hamiltonian depends on spatial dimensionality and on the symmetry and dimensionality of the order parameter. It also depends on the form of the gradient couplings and on other coupling constants. Of course, the LGW Hamiltonian can contain extra terms corresponding to irrelevant operators, terms whose coupling constants will flow to zero under renormalization. We have seen that such terms do not affect the values of the critical exponents. The exponents are determined by the unstable fixed point which controls the phase transition. The structure of this exponent is, in turn, determined by the form of the LGW Hamiltonian for the problem under consideration.

20.6 Summary

Mean-field theory is a valid description of continuous phase transitions for $d > d_c$, where d_c is the upper critical dimension. For the Ising model (where the nonquadratic term in the Landau expansion is quartic) $d_c = 4$. For percolation or for the spin-glass problem, $d_c = 6$. In either case, the fixed point which describes the critical point separates from the Gaussian fixed point as d is reduced below d_c. Following Wilson, one, therefore, expands in the parameter $\epsilon \equiv d_c - d$ and at the fixed point the nonquadratic potentials are of order ϵ and the critical exponents approach their mean-field values smoothly as $\epsilon \to 0$. We described the renormalization group (RG) approach whereby one recursively maps the problem onto itself by integrating out the short-wavelength degrees of freedom and rescales the spin magnitudes, so that renormalization group iterations lead to a "flow" in parameter space. For the Ising model, we expect that to reach the critical point requires fixing the temperature so that $T = T_c$ and the magnetic field so that $H = 0$. We, therefore, expect that under iteration there will be precisely two corresponding relevant parameters which grow under RG iteration. The other variables are irrelevant in that they renormalize to zero under iteration. The fact that a complete space of irrelevant parameters corresponds to a single fixed point explains beautifully the concept of universality, which states that the critical behavior of models depends only on the symmetries and dimensionality of space. We illustrate here the construction of the recursion relations within the ϵ-expansion. It goes without saying that there is a vast literature in which almost all conceivable fixed point structures have been investigated.

20.7 Exercises

1. Construct the figure for the critical exponents for percolation, analogous to that of Fig. 20.1 for the Ising model. Assume that the figure for percolation has the same structure as for the Ising model (namely that at high dimension the exponents are constant with their mean-field values for $d > d_c$ and below d_c they develop a dependence on dimensionality. Thereby deduce d_c for percolation. The solution to this homework problem (to find d_c for percolation) was done by G. Toulouse (see Harris et al. 1975) before the development of the field theory for percolation.

2. Using results from Sect. 13.6, verify Eqs. (20.4) and (20.5).

3. Show that the diagrams of Fig. 20.5b vanish identically because of momentum conservation.

4. Derive the identity, Eq. (20.31), using the generating function defined in Eq. (20.32).

5. Using the diagrams of Fig. 20.12 obtain ν correct to first order in ϵ for the n-component Heisenberg model, following the steps in Sect. 20.4.3.

6. Following the steps in Sect. 20.4.4, derive the leading order (ϵ^2) contribution to η for the n-component model.

20.8 Appendix: Wick's Theorem

We have already seen Wick's theorem for the average of a product of operators taken with respect to the density matrix of a Hamiltonian quadratic in creation and annihilation operators. The conclusion was that the average of such a product is equal to the sum over all ways of making pairwise contractions into products of averages of two operators. Exactly the same result applies here with respect to Gaussian variables.

Real-Valued Quadratic Forms

We introduce the notation

$$\tilde{\mathbf{s}}\mathbf{A}\mathbf{s} = \sum_{m,n} s_m A_{m,n} s_n \tag{20.105a}$$

$$\tilde{\rho}\mathbf{s} = \sum_m \rho_m s_m \ , \tag{20.105b}$$

where $\mathbf{A}$ is a symmetric $N \times N$ positive definite matrix. Then we have the identity

$$\begin{aligned}\int e^{-\frac{1}{2}\tilde{\mathbf{s}}\mathbf{A}\mathbf{s}+\tilde{\rho}\mathbf{s}} \prod_n ds_n &= \frac{(2\pi)^{N/2}}{\sqrt{\mathrm{Det}\mathbf{A}}} e^{\frac{1}{2}\tilde{\rho}[\mathbf{A}]^{-1}\rho} \\ &\equiv C e^{\frac{1}{2}\tilde{\rho}[\mathbf{A}]^{-1}\rho} \ ,\end{aligned} \tag{20.106}$$

in terms of the inverse matrix $[\mathbf{A}]^{-1}$.

Define the integrals

$$I(m_1, m_2, \ldots, m_k) = \int \left(s_{m_1} s_{m_2} \ldots s_{m_k}\right) e^{-\frac{1}{2}\tilde{\mathbf{s}}\mathbf{A}\mathbf{s}} \prod_n ds_n \ . \tag{20.107}$$

To calculate $I(m_1, m_2, \ldots, m_k)$ consider

$$\begin{aligned}&\int \left(s_{m_1} s_{m_2} \ldots s_{m_k}\right) e^{-\frac{1}{2}\tilde{\mathbf{s}}\mathbf{A}\mathbf{s}+\tilde{\rho}\mathbf{s}} \prod_n ds_n \\ &= C \frac{\partial}{\partial \rho_{m_1}} \cdots \frac{\partial}{\partial \rho_{m_k}} e^{-\frac{1}{2}\tilde{\rho}[\mathbf{A}]^{-1}\rho} \ .\end{aligned} \tag{20.108}$$

To obtain $I(m_1, m_2, \ldots, m_k)$ one simply takes the limit $[\rho_m \to 0$ for all $m]$ after differentiating. If k is odd, the result is zero because the exponent is a quadratic form. If k is even, we only need to consider terms of order ρ^k. Expanding the exponential, we obtain

$$\begin{aligned}\langle s_{m_1} s_{m_2} \ldots s_{m_k} \rangle &\equiv \frac{I(m_1, m_2, \ldots, m_k)}{\int e^{-\frac{1}{2}\tilde{\mathbf{s}}\mathbf{A}\mathbf{s}} \prod_n ds_n} \\ &= \lim_{\rho_m \to 0} \frac{\partial}{\partial \rho_{m_1}} \cdots \frac{\partial}{\partial \rho_{m_k}} \left[\frac{1}{2}\tilde{\rho}[\mathbf{A}]^{-1}\rho\right]^{\frac{k}{2}} \frac{1}{\left(\frac{k}{2}\right)!} \ ,\end{aligned} \tag{20.109}$$

where $(k/2)!$ comes from the expansion of the exponential.

To illustrate, consider the case $k = 4$ which involves

$$\langle s_i s_j s_k s_l \rangle = \frac{\partial}{\partial \rho_i} \frac{\partial}{\partial \rho_j} \frac{\partial}{\partial \rho_k} \frac{\partial}{\partial \rho_l} \frac{1}{8} \sum_{rstu} [\mathbf{A}]^{-1}_{rs} [\mathbf{A}]^{-1}_{tu} \rho_r \rho_s \rho_t \rho_u \ . \tag{20.110}$$

This gives 4!=24 terms corresponding to the 4! ways of making the set of indices r, s, t, and u be identical to the set i, j, k, and l. However, there are only 3 independent terms:

$$T_1 \equiv [\mathbf{A}]^{-1}_{ij}[\mathbf{A}]^{-1}_{kl} \,, \quad T_2 \equiv [\mathbf{A}]^{-1}_{ik}[\mathbf{A}]^{-1}_{jl} \,, \quad T_3 \equiv [\mathbf{A}]^{-1}_{il}[\mathbf{A}]^{-1}_{jk} \,. \tag{20.111}$$

The number (out of 24) ways one can get the first term is found as follows. We have four choices for which of r, s, t, or u we set equal to i. Then the partner index (out of the set r, s, t, u) must be equal to j. There are then remaining (out of r, s, t, u) two choices for k. So in all, each of the three terms in Eq. (20.111) is realized 8 times. Thus

$$\langle s_i s_j s_k s_l \rangle = [\mathbf{A}]^{-1}_{ij}[\mathbf{A}]^{-1}_{kl} + [\mathbf{A}]^{-1}_{ik}[\mathbf{A}]^{-1}_{jl} + [\mathbf{A}]^{-1}_{il}[\mathbf{A}]^{-1}_{jk} \,. \tag{20.112}$$

We see then that the general result is

$$\langle s_{m_1} s_{m_2} \dots s_{m_{2k}} \rangle = \sum_P \prod_{i=1}^{k} \langle s_{k_i} s_{l_i} \rangle \,, \tag{20.113}$$

where the product is over the k averages of pairs of s variables. The sum is over the $(2k-1)(2k-3)\dots 1$ different ways of dividing the indices in to k sets of pairs. This formula also holds when some of the m's are equal, as is illustrated in an exercise.

Complex-Valued Quadratic Forms

We are actually interested in quantities of the form

$$\begin{aligned} I = & \int_{-\infty}^{\infty} dx(\mathbf{q}_1) \int_{-\infty}^{\infty} dx(\mathbf{q}_2) \dots \int_{-\infty}^{\infty} dx(\mathbf{q}_k) \\ & \times e^{-\frac{1}{2}\int_q \frac{1}{2}(r+q^2)x(\mathbf{q})x(-\mathbf{q})}[x_{\mathbf{q}_1} x_{\mathbf{q}_2} \dots x_{\mathbf{q}_2 k}] \\ \equiv & \int \mathcal{D}x(\mathbf{q}) e^{-\frac{1}{2}\int_q \frac{1}{2}(r+q^2)x(\mathbf{q})x(-\mathbf{q})}[x_{\mathbf{q}_1} x_{\mathbf{q}_2} \dots x_{\mathbf{q}_2 k}] \,. \end{aligned} \tag{20.114}$$

We discuss very briefly how this is to be interpreted. The point is that $x(\mathbf{q})$ is really a complex valued quantity, which we write, in terms of real variables, as

$$x(\mathbf{q}) = [x_c(\mathbf{q}) \pm i x_s(\mathbf{q})]/\sqrt{2} \,, \tag{20.115}$$

where the + sign is taken for $\mathbf{q}$ "positive" and the − sign for $\mathbf{q}$ "negative." [Here "positive" or "negative" may be defined by the sign of the first component of the vector $\mathbf{q}$.] Thus, what we really have is

$$I = \int \mathcal{D}x_c(\mathbf{q}) \int \mathcal{D}x_s(\mathbf{q}) e^{-\frac{1}{2}\int_{q>0} \frac{1}{2}(r+q^2)[x_c(\mathbf{q})^2 + x_s(\mathbf{q})^2]}$$
$$\times \left(\frac{[x_c(\mathbf{q}_1) + i\eta_1 x_s(\mathbf{q}_1)]}{\sqrt{2}} \right) \left(\frac{[x_c(\mathbf{q}_2) + i\eta_2 x_s(\mathbf{q}_2)]}{\sqrt{2}} \right) \cdots$$
$$\times \left(\frac{[x_c(\mathbf{q}_{2k}) + i\eta_{2k} x_s(\mathbf{q}_{2k})]}{\sqrt{2}} \right), \tag{20.116}$$

where η_i is the sign of the original wave vector $\mathbf{q}_i$. When one makes a pairwise contraction one gets a factor $\langle x_c(\mathbf{q}) x_c(\mathbf{q}') \rangle (1 - \eta_i \eta_j)/2$ (since x_c^2 and x_s^2 have the same averages). This means that a pairwise average vanishes unless $\mathbf{q} + \mathbf{q}' = 0$. Thus, we have shown that Wick's theorem applies for complex-valued quadratic forms just as it does with real-valued forms, except that we have wavevector conservation, so that we have a factor $\delta(\mathbf{q}_1 + \mathbf{q}_2)$ and we eventually obtain Eq. (20.31) of the text.

References

A. Aharony, *Dependence of Universal Critical Behaviour on Symmetry and Range of Interaction* in *Phase Transitions and Critical Phenomena* vol. 6, ed. by C. Domb, M. Green (Academic Press, London, 1976)

A. Aharony, P.C. Hohenberg, Universal relations among thermodynamic critical amplitudes. Phys. Rev. B **13**, 3081 (1976)

E. Brezin, J.C. LeGuillou, J. Zinn-Justin, B.G. Nickel, Higher order contributions to critical exponents. Phys. Lett. A **44**, 227 (1973)

A.B. Harris, T.C. Lubensky, W.K. Holcomb, C. Dasgupta, Renormalization-group approach to percolation problems. Phys. Rev. Lett. 35, 327 (1975)

S.-K. Ma, *Modern Theory of Critical Phenomena* (Benjamin/Cummings, 1976)

F.J. Wegner, A. Houghton, Renormalization group equation for critical phenomena. Phys. Rev. A **8**, 401 (1973)

K.G. Wilson, Renormalization group and critical phenomena. i. renormalization group and the kadanoff scaling picture. Phys. Rev. B **4**, 3174; Renormalization group and critical phenomena. II. Phase-space cell analysis of critical behavior. ibid., 3184 (1971)

Chapter 21
Kosterlitz-Thouless Physics

21.1 Introduction

In this chapter, we will study how certain kinds of order are destroyed by thermal excitations. We will focus initially on crystals in different dimensions. The simplest excitations are the linear ones, phonons. However, there is another important class of excitations which have nontrivial topological properties, called dislocations, that can also play important roles in the destruction of order, particularly for the special case of two dimensions. Let us start, however, with phonons in d dimension and see where that leads.

21.2 Phonons in Crystal Lattices

We saw earlier, in Sect. 7.5, that the low-lying elementary excitations of a solid are its small amplitude, harmonic "normal modes", whose frequencies increase linearly from zero in proportion to their wave vector as given in Eq. (7.103). These are called acoustic modes or sound waves. The exact form of Eq. (7.103) is only approximate. For a crystal of atoms or molecules, the spectrum, in fact, bends over and becomes flat at the Brillouin zone boundary, but for present purposes this approximate form is adequate. There may also be optical modes, with frequencies that generally lie above the acoustic frequencies, that we will ignore in the following discussion. Phonons describe small displacements, which we call $\vec{u}(\vec{r})$, of the atom or molecule whose equilibrium position is $\vec{r}$. To calculate $\vec{u}(\vec{r})$ and its correlation functions which we will need to understand order in crystals, we first transform to boson creation and destruction operators. The Hamiltonian for the simple harmonic oscillator was given in Eq. (5.31), where x and p obey canonical commutation relations. Equation (5.31) is diagonalized by the transformation,

A. J. Berlinsky and A. B. Harris, *Statistical Mechanics*, Graduate Texts in Physics,
https://doi.org/10.1007/978-3-030-28187-8_21

$$x = \sqrt{\frac{\hbar}{2m\omega}}(a^\dagger + a) \tag{21.1a}$$

$$p = i\sqrt{\frac{\hbar m\omega}{2}}(a^\dagger - a) \tag{21.1b}$$

where a and $a^\dagger$ obey $[a, a^\dagger] = 1$. For a crystal of identical atoms, the displacement $\vec{u}(\vec{r})$ is the sum of the displacements due to each normal mode, which we write as

$$\vec{u}(\vec{r}) = \frac{1}{\sqrt{N}}\sum_{\vec{k},\alpha}\vec{U}_{\vec{k},\alpha}(\vec{r}), \tag{21.2}$$

where α is a polarization index for a normal mode with wave vector $\vec{k}$.

As for the single oscillator, the displacements due to each normal mode can be written in terms of boson operators. However, now these operators and the frequencies and polarization vectors associated with each mode are labeled by $\vec{k}$ and α. Thus, we write

$$\vec{U}_{\vec{k},\alpha}(\vec{r}) = \sqrt{\frac{\hbar}{2m\omega_{\vec{k},\alpha}}}(a^\dagger_{\vec{k},\alpha}e^{i\vec{k}\cdot\vec{r}} + a_{\vec{k},\alpha}e^{-i\vec{k}\cdot\vec{r}})\vec{\eta}_{\vec{k},\alpha} \tag{21.3}$$

where the boson operators obey $[a_{\vec{k},\alpha}, a^\dagger_{\vec{q},\beta}] = \delta_{\vec{k},\vec{q}}\delta_{\alpha,\beta}$ and $\vec{\eta}_{\vec{k},\alpha}$ is a unit vector describing the polarization of the mode. For example, if the mode is purely longitudinal, $\vec{\eta}_{\vec{k},\alpha}$ will be parallel to $\vec{k}$, while if it is purely transverse it will point in a direction perpendicular to $\vec{k}$. Because of crystal anisotropy, $\vec{\eta}_{\vec{k},\alpha}$ need not be purely longitudinal or transverse.

Equations (21.2) and (21.3) allow us to calculate a quantity that is a measure of whether a stable crystal can form and which also provides some indication of when it might melt. That quantity is $\langle|\vec{u}(\vec{r})|^2\rangle$ which is the thermally averaged mean-square amplitude of the motion of the atom at site $\vec{r}$, which can be written as

$$\langle(\vec{u}(\vec{r}))^2\rangle = \frac{1}{N}\sum_{\vec{k},\alpha}\frac{\hbar}{m\omega_{\vec{k},\alpha}}\left(n(\hbar\omega_{\vec{k},\alpha}/k_BT) + \frac{1}{2}\right), \tag{21.4}$$

where

$$n(\hbar\omega_{\vec{k},\alpha}/k_BT) = \langle a^\dagger_{\vec{k},\alpha}a_{\vec{k},\alpha}\rangle \tag{21.5}$$

is the mean phonon occupation number or, equivalently, the Bose factor for zero chemical potential,

$$n(x) = \frac{1}{e^x - 1}. \tag{21.6}$$

To get some idea of whether the crystal will be stable at all, even at $T = 0$, we examine

$$\langle(\vec{u}(\vec{r}))^2\rangle_0 = \frac{1}{N}\sum_{\vec{k},\alpha}\frac{\hbar}{2m\omega_{\vec{k},\alpha}}. \tag{21.7}$$

For d dimensions, $\alpha = 1, \ldots d$, and we will assume for simplicity that $\omega_{\vec{k},\alpha}$ is the same for all d polarizations. Converting the sum to an integral and using the simplified expression, Eq. (7.103) for the phonon frequency, we can write

$$\langle(\vec{u}(\vec{r}))^2\rangle_0 = \frac{d\hbar}{2mc}\left\langle\frac{1}{k}\right\rangle_{BZ} \tag{21.8}$$

where the factor d comes from the number of polarizations, and the average on the RHS is an average over the d-dimensional Brillouin zone (BZ). If we replace the BZ integration by a radial integration over the equivalent "volume," with a radius of π/a_D, then, using

$$\langle\ldots\rangle_{BZ} \equiv \frac{\int_{\pi/L}^{\pi/a_D}\ldots k^{d-1}dk}{\int_0^{\pi/a_D}k^{d-1}dk} = d\left(\frac{a_D}{\pi}\right)^d\int_{\pi/L}^{\pi/a_D}\ldots k^{d-1}dk \tag{21.9}$$

the average of $1/k$ can be written as

$$\left\langle\frac{1}{k}\right\rangle_{BZ} = \frac{d}{d-1}\frac{a_D}{\pi}, \qquad d \geq 2, \tag{21.10a}$$

$$= \frac{a_D}{\pi}\ln(L/a_D), \quad d = 1. \tag{21.10b}$$

In the first equation, valid for $d \geq 2$, the lower limit of the k-space integration has been taken to be zero, since the integral converges at its lower limit. However, the integral does not converge at $k = 0$ for $d = 1$, so it is necessary to explicitly use the wavevector of the lowest energy excitation which is π/L. This gives a logarithmic divergence for $L \to \infty$, implying that one-dimensional crystals are unstable even at $T = 0$. Note that this is different from the case of the 1D Ising model, which is effectively ordered at $T = 0$ as can be seen from Fig. 17.1. For $d > 1$,

$$\langle(\vec{u}(\vec{r}))^2\rangle_0 = \frac{d^2}{d-1}\frac{\hbar a_D}{2\pi mc} \tag{21.11}$$

Next we consider thermal effects, which requires calculating

$$\delta\langle(\vec{u}(\vec{r})^2\rangle_T \equiv \langle(\vec{u}(\vec{r}))^2\rangle_T - \langle(\vec{u}(\vec{r}))^2\rangle_0 = \frac{d\hbar}{mc}\left\langle\frac{n(\hbar ck/k_BT)}{k}\right\rangle_{BZ} \tag{21.12}$$

Then, changing integration variables to $x = \hbar ck/k_B T$, we can write

$$\delta\langle(\vec{u}(\vec{r}))^2\rangle_T = \frac{d^2\hbar^2}{mk_B}\frac{T^2}{\Theta_D^3}\int_{x_m}^{x_M}\frac{x^{d-2}}{e^x - 1}dx \tag{21.13}$$

where the integral goes from a low-energy cut-off up to $x_M = \hbar c\pi/(a_D k_B T) = \Theta_D/T$, where Θ_D is the Debye temperature. We write the low-energy cut-off as $x_m = a_D\Theta_D/(LT)$. It is immediately clear that the x^{d-2} in the integrand is going to lead to problems in $d = 2$, and we will return to that in a moment. However, for $d = 3$ Eq. (21.13) gives

$$\delta\langle(\vec{u}(\vec{r}))^2\rangle_T = \frac{9\hbar^2}{mk_B}\frac{T^2}{\Theta_D^3}\int_0^{\frac{\Theta_D}{T}}\frac{x}{e^x - 1}dx, \quad d = 3, \tag{21.14a}$$

$$= \frac{3\pi^2\hbar^2}{2mk_B}\frac{T^2}{\Theta_D^3}, \quad T << \Theta_D \tag{21.14b}$$

$$= \frac{9\hbar^2}{mk_B}\frac{T}{\Theta_D^2}, \quad T >> \Theta_D \tag{21.14c}$$

This last result, Eq. (21.14c), allows us to define a criterion for melting, called the Lindeman melting formula, which is based on the idea that, once the root mean-square thermal amplitude grows to some fraction, f, of a lattice spacing, a, then the crystal melts at the corresponding temperature, T_M. If we write this condition as

$$\delta\langle(\vec{u}(\vec{r}))^2\rangle_{T_M} = \frac{9\hbar^2}{mk_B}\frac{T_M}{\Theta_D^2} = f^2a^2 \tag{21.15}$$

Then

$$f = \frac{3\hbar}{a\Theta_D}\sqrt{\frac{T_M}{mk_B}} \tag{21.16}$$

If a is given in Å, m in atomic mass units, and T_M and Θ_D in Kelvin, then this can be written as

$$f = \frac{20.9}{a\Theta_D}\sqrt{\frac{T_M}{m}} \tag{21.17}$$

Values of a, m, T_M, and Θ_D are tabulated in those units for the elements on the inside front cover of *Solid State Physics* by Ashcroft and Mermin (Ashcroft and Mermin 1976). It is found empirically that $f \approx 0.1$ for most solids. Variations in the value of f may result from differences in lattice structure and in the type of thermally created defects which ultimately destroy crystalline order.

The case of $d = 2$ is less well behaved.

$$\delta\langle(\vec{u}(\vec{r}))^2\rangle_T = \frac{4\hbar^2}{mk_B}\frac{T^2}{\Theta_D^3}\int_{x_m}^{x_M}\frac{dx}{e^x - 1}. \tag{21.18}$$

The integral in Eq. (21.18) is elementary, and the result can be written as

$$\delta\langle(\vec{u}(\vec{r}))^2\rangle_T = \frac{4\hbar^2}{mk_B}\frac{T}{\Theta_D^2}\left[\ln\left(\frac{L}{a_D}\right)+\ln\left(\frac{T}{\Theta_D}\right)+\ln\left(1-e^{-\Theta_D/T}\right)\right],\ d=2. \tag{21.19}$$

The $\ln L$ divergence (in the limit $L \to \infty$) strongly suggests that 2D crystals are not stable for $T > 0$. However, it is worthwhile noting that the large amplitude thermal motion results from very-long-wavelength phonons, which means that, locally, the lattice could still look well ordered. In fact, to understand the full richness of what actually happens in 2D crystals, it is necessary to look more closely at the nature and size-dependence of correlation functions that describe their motions and to include, not only harmonic phonons, but also nonlinear crystal defects such as dislocations and disclinations.

21.3 2D Harmonic Crystals

A crystal is a long-range-ordered state of matter. In fact, it exhibits two kinds of order. First the crystal axes must pick out directions in space, and, second, the atoms must form periodic arrays, rows of atoms or molecules positioned along the crystal axes, each in registry with its neighboring rows. The equilibrium positions of atoms in a crystal are defined by lattice vectors. The subject of lattice vectors and lattice Fourier transforms was discussed earlier in Sect. 10.3. We write a general lattice vector, which is a sum of integral multiples of lattice basis vectors as $\vec{R}_n$. Then, for a static crystal, the number density of atoms as a function of position, $\vec{r}$, is

$$\rho_0(\vec{r}) = \sum_{n=1}^{N}\delta(\vec{r}-\vec{R}_n) \tag{21.20}$$

for a lattice of N sites. For a harmonic crystal, this becomes

$$\rho(\vec{r}) = \sum_{n=1}^{N}\delta(\vec{r}-\vec{R}_n-\vec{u}(\vec{R}_n)), \tag{21.21}$$

where $\vec{u}(\vec{R}_n)$ is the displacement of the atom associated with site n due to lattice vibrations.

The Fourier transform of the density is

$$\rho(\vec{q}) = \frac{1}{N}\sum_{n=1}^{N}e^{i\vec{q}\cdot\vec{R}_n}e^{i\vec{q}\cdot\vec{u}(\vec{R}_n)}. \tag{21.22}$$

The thermal average of $\rho(\vec{q})$ can be written as

$$\langle \rho(\vec{q})\rangle = \langle e^{i\vec{q}\cdot\vec{u}(\vec{R}_n)}\rangle \frac{1}{N}\sum_{n=1}^{N} e^{i\vec{q}\cdot\vec{R}_n} = \langle e^{i\vec{q}\cdot\vec{u}(\vec{R}_n)}\rangle \Delta(\vec{q}), \tag{21.23}$$

where $\Delta(\vec{q})$, which was defined in Eq. (10.51), is equal to 1 if $\vec{q}$ is a reciprocal lattice vector (RLV), and is zero otherwise. Thus for $\vec{Q}$ a RLV, the prefactor, $\langle e^{i\vec{Q}\cdot\vec{u}(\vec{R}_n)}\rangle$, which is independent of $\vec{R}_n$ is an order parameter for density order with wave vector $\vec{Q}$. If $\vec{u}(\vec{R}_n)$ was constrained to be zero (for example, for infinitely massive atoms with perfect crystalline order), then $\langle e^{i\vec{Q}\cdot\vec{u}(\vec{R}_n)}\rangle$ would equal 1 for all $\vec{Q}$. However, quantum zero-point and thermal vibrations reduce the value of this parameter.

According to a standard result for harmonic phonons (Ashcroft and Mermin 1976, Appendix N),

$$\langle e^{i\vec{Q}\cdot\vec{u}(\vec{R}_n)}\rangle = e^{-\langle(\vec{Q}\cdot\vec{u})^2\rangle/2} \equiv e^{-W(\vec{Q})} \tag{21.24}$$

where $e^{-2W(\vec{Q})}$ is called the Debye–Waller factor. For three dimensions, the RMS displacement amplitude, $\sqrt{\langle|u|^2\rangle}$ is a small fraction of a lattice spacing below the melting temperature, and so the Debye–Waller factor is close to one, at least for the smallest RLVs, and falls rapidly to zero for the larger ones. Such a nonzero Debye–Waller factor reflects long-range crystalline order.

The situation for two dimensions is very different since, according to Eq. (21.19), the RMS displacement diverges logarithmically with system size for $T > 0$. Just keeping the leading $\ln L$ term gives

$$W(\vec{Q}) \approx \frac{Q^2}{4}\frac{4\hbar^2}{mk_B}\frac{T}{\Theta_D^2}\ln\left(\frac{L}{a_D}\right) \tag{21.25}$$

Then

$$\langle \rho(\vec{Q})\rangle = e^{-W(\vec{Q})} \sim L^{-T_Q T/\Theta_D^2}, \tag{21.26}$$

where $T_Q = \hbar^2 Q^2/mk_B$. So $\langle\rho(\vec{Q})\rangle$ vanishes for $L \to \infty$ and $T > 0$, that is, the order parameter for crystalline order vanishes for $d = 2$, which is consistent with a rigorous result due to Mermin (1968).

Given that the order parameter vanishes, the next logical question to ask is how translational correlations fall to zero as a function of distance. This information is contained in the order–parameter–order–parameter correlation function. which we write as

$$\begin{aligned} C_{\vec{Q}}(\vec{R}_n - \vec{R}_0) &= \langle e^{i\vec{Q}\cdot(\vec{R}_n + \vec{u}(\vec{R}_n) - \vec{R}_0 - \vec{u}(\vec{R}_0))}\rangle \\ C_{\vec{Q}}(\vec{R}_n) &= \langle e^{i\vec{Q}\cdot(\vec{u}(\vec{R}_n) - \vec{u}(0))}\rangle \\ C_{\vec{Q}}(\vec{R}_n) &= e^{-\langle[\vec{Q}\cdot(\vec{u}(\vec{R}_n) - \vec{u}(0))]^2\rangle/2}, . \end{aligned} \tag{21.27}$$

In the second line, we have set $\vec{R}_0 = 0$, and in the third line, we have again used the identity for phonon operators, Eq. (21.24).

We define

$$w(\vec{q}, \vec{R}) = \frac{1}{2}\langle [\vec{q} \cdot (\vec{u}(\vec{R}_n) - \vec{u}(0))]^2 \rangle. \tag{21.28}$$

Then transforming to phonon operators and performing the thermal averages, we obtain

$$w(\vec{q}, \vec{R}) = \frac{1}{N}\sum_{\vec{k}} \frac{\hbar q^2}{2m\omega_{\vec{k}}}\left(2n(\hbar\omega_{\vec{k}}/k_B T) + 1\right)\left(1 - \cos\vec{k}\cdot\vec{R}\right), \tag{21.29}$$

We can evaluate $w(\vec{q}, \vec{R})$ in Eq. (21.29) approximately for finite temperature by replacing the factor $\left(2n(\hbar\omega_{\vec{k}}/k_B T) + 1\right)$ by its high temperature limit, $2k_B T/\hbar\omega_{\vec{k}}$ so that

$$w(\vec{q}, \vec{R}) \approx \frac{1}{N}\sum_{\vec{k}} \frac{q^2 k_B T}{m\omega_{\vec{k}}^2}\left(1 - \cos\vec{k}\cdot\vec{R}\right). \tag{21.30}$$

Next we replace the $\vec{k}$ sum by an integral over a circle of radius k_M, where $\pi k_M^2 = 4\pi^2$ (setting the lattice constant equal to 1 gives $k_M = 2\sqrt{\pi}$), and we make the Debye approximation, $\omega_{\vec{k}} = ck$. The angular average of $\cos\vec{k}\cdot\vec{R}$ is the Bessel function, $J_0(kR)$. Then

$$w(\vec{q}, \vec{R}) \approx \frac{2}{k_M^2}\frac{q^2 k_B T}{mc^2}\int_0^{k_M} \frac{(1 - J_0(kR))}{k} dk, \tag{21.31}$$

where $\hbar c k_M = k_B \Theta_D$ in terms of our earlier notation. Then, we can write Eq. (21.31) as

$$w(\vec{q}, \vec{R}) \approx 2\frac{T_q T}{\Theta_D^2}\int_0^{k_M R} \frac{(1 - J_0(x))}{x} dx, \tag{21.32}$$

where $T_q = \hbar^2 q^2/mk_B$. Integrating by parts,

$$\int_0^{k_M R} \frac{(1 - J_0(x))}{x} dx = \ln(x)\left(1 - J_0(x)\right)\Big|_0^{k_M R} - \int_0^{k_M R} \ln(x) J_1(x) dx \tag{21.33}$$

The second integral can be approximated by its value in the limit $k_M R \to \infty$,

$$-\int_0^{\infty} \ln(x) J_1(x) dx = \ln\frac{\gamma}{2}, \tag{21.34}$$

where $\ln\gamma = 0.577216\ldots$ is Euler's constant. Then, dropping the term proportional to $J_0(k_M R)$ which is negligible for large $k_M R$, we obtain

$$\int_0^{k_M R} \frac{(1 - J_0(x))}{x} dx \approx \ln\frac{\gamma k_M R}{2}. \tag{21.35}$$

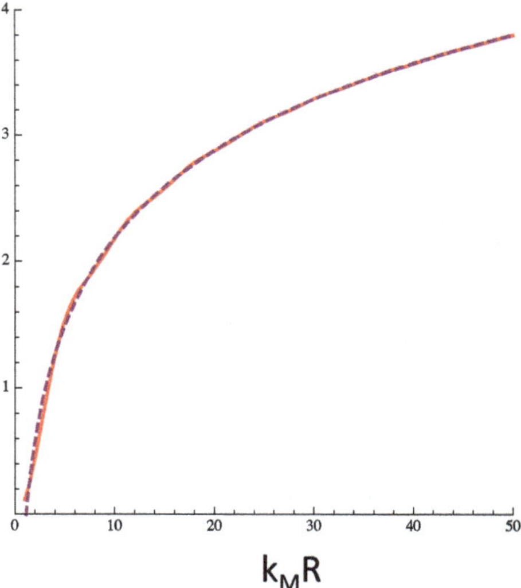

Fig. 21.1 Comparison of the left-hand (red solid line) and right-hand (black dashed line) sides of Eq. (21.35) for $1 \leq k_M R \leq 50$

This turns out to be a very good approximation as can be seen from the comparison of the dashed and solid lines in Fig. 21.1. Our final result is

$$w(\vec{q}, \vec{R}) \approx 2\frac{T_q T}{\Theta_D^2} \ln \frac{\gamma k_M R}{2}, \tag{21.36}$$

which then gives the translational order parameter correlation function in 2D as

$$C_{\vec{Q}}(\vec{R}) \approx e^{-2\frac{T_Q T}{\Theta_D^2} \ln \frac{\gamma k_M R}{2}} = \left(\frac{\gamma k_M R}{2}\right)^{-2\frac{T_Q T}{\Theta_D^2}}. \tag{21.37}$$

This remarkable result (Imry and Gunther 1971) shows that the order parameter correlation function for a 2D harmonic crystal decays as a power law, which we have learned is the distinctive property of a critical point. However, for this problem, critical behavior is not confined to a single temperature point. Instead critical behavior appears to persist for all temperatures with an exponent that increases linearly with T. We will see shortly that there are well-defined temperatures above which so-called quasi-long-range order disappears for harmonic crystals, one RLV at a time, with the order for the smallest RLVs disappearing at the highest temperatures. Note also that, because of the factor T_Q, the exponent is different for each RLV, being larger for larger values of Q.

In order to obtain a clearer physical picture of what this all means, we next examine the density–density correlation function,

$$\langle \rho(\vec{q})\rho(-\vec{q})\rangle = \frac{1}{N}\sum_{n=0}^{N-1} \langle e^{i\vec{q}\cdot(\vec{u}(\vec{R}_n)-\vec{u}(\vec{R}_0))}\rangle e^{i\vec{q}\cdot(\vec{R}_n-\vec{R}_0)}, \tag{21.38}$$

which is the function that describes X-ray scattering for momentum transfer $\vec{q}$. The order parameter for 3D crystals is the weight of the δ-function at Bragg peaks, each of which is labeled by wavevector $\vec{Q}$. Note that $\langle \rho(\vec{q})\rho(-\vec{q})\rangle$ is defined at all points $\vec{q}$ in reciprocal space—not just at RLVs. It can be written in terms of the function, $w(\vec{q}, \vec{R}_n)$ as

$$\langle \rho(\vec{q})\rho(-\vec{q})\rangle = \frac{1}{N}\sum_{n=0}^{N-1} e^{-w(\vec{q},\vec{R}_n)} e^{i\vec{q}\cdot\vec{R}_n}. \tag{21.39}$$

At this point we could proceed to make approximations, as we did above, to complete the calculation analytically. Instead, we first evaluate Eqs. (21.29) and (21.39) numerically, using *Mathematica*, in order to see what to expect. The only minor change that we make is to work with a phonon spectrum that has the periodicity of the reciprocal lattice by setting

$$\omega_{\vec{k}} = \omega_D \sqrt{\sin^2(k_x/2) + \sin^2(k_y/2)} \tag{21.40}$$

We continue to use the convention that $\hbar\omega_D = k_B\Theta_D$, and we can determine the value of the constant, $\hbar/(2m\omega_D)$, by using it to fix a particular value of the squared amplitude of the zero-point motion, $\langle |u|^2\rangle_0$, and so the behavior of the density–density correlation function is determined by $\langle |u|^2\rangle_0$ in units of the lattice constant squared, T/Θ_D, and, of course, the value of $N = L^2$, the size of the 2D crystal. Results for N times the correlation function are shown in Fig. 21.2 for $L = 50$, $\langle |u|^2\rangle_0 = 0.0025$, and for $T/\Theta_D = 0, 1, 2$.

The figure shows a number of interesting features: (1) There is a δ-function of weight N for $q = 0$ at all temperatures. (2) For $T = 0$ (the blue dots), the weight of the δ-function peaks at reciprocal lattice points $q_x/2\pi = 1, 2, 3, 4$ falls off slowly with increasing q, and there is a smooth background between Bragg points with an amplitude that increases with increasing q. The discrete δ-function-like peaks at Bragg points are indicative of the existence of long-range translational order at $T = 0$. (3) For $T > 0$, although it is perhaps less obvious, the peaks evolve into smooth functions of $|q_x - 2\pi n|$ around each Bragg point that will grow sharp in the limit $L \to \infty$. (4) Finally we note that, for large enough values of $T_Q T/\Theta_D^2$, the peaks disappear. This can be seen for $T/\Theta_D = 2$ (the yellow diamonds) at $q_x/2\pi = 4$.

Next we perform an approximate analytical calculation of the correlation function of Eq. (21.39). To begin, we specialize to the case where $\vec{q} = \vec{Q}$ is a reciprocal lattice vector so that all the phase factors are equal to 1. We replace the sum over lattice sites, $\vec{R}_n$, by an integral over R and write the density–density correlation function of Eq. (21.39) as

$$D(\vec{Q}) \equiv \sum_{n=1}^{N} e^{-w(\vec{Q},\vec{R}_n)} \to 2\pi \int_1^L e^{-2a_Q \ln(\gamma R/2)} R dR, \tag{21.41}$$

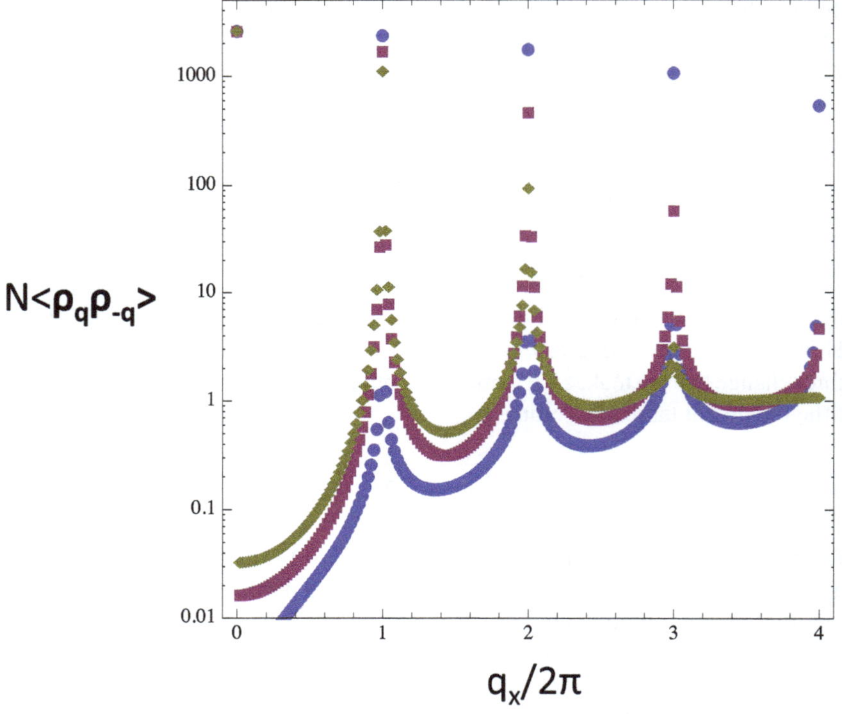

Fig. 21.2 Density–density correlation function for three different temperatures, $T = 0$ (blue circles), $T = \Theta_D$ (red squares), and $T = 2\Theta_D$ (yellow diamonds), plotted on a log scale, versus $q_x/2\pi$ for a 50×50 lattice with periodic boundary conditions and using the Debye–Waller factor given in Eq. (21.29). The value of the dimensional coefficients of q^2 has been chosen to correspond to an rms zero-point amplitude of 0.05 lattice spacings

where $a_Q = T_Q T/\Theta_D^2$ and $\pi L^2 = N$. Clearly $D(\vec{Q})$ depends only on the magnitude, Q. Then

$$D(Q) = 2\pi \left(\frac{2}{\gamma}\right)^{2a_Q} \int_1^L R^{1-2a_Q} dR.$$
$$= \pi \left(\frac{2}{\gamma}\right)^{a_Q} \frac{(N/\pi)^{1-a_Q} - 1}{1 - a_Q}. \tag{21.42}$$

or, for N large,

$$D(Q) = \left(\frac{2}{\pi\gamma}\right)^{a_Q} \frac{N^{1-a_Q}}{1 - a_Q}, \quad a_Q < 1, \tag{21.43a}$$
$$= \pi \left(\frac{2}{\gamma}\right)^{a_Q} \frac{1}{a_Q - 1}, \quad a_Q > 1, \tag{21.43b}$$

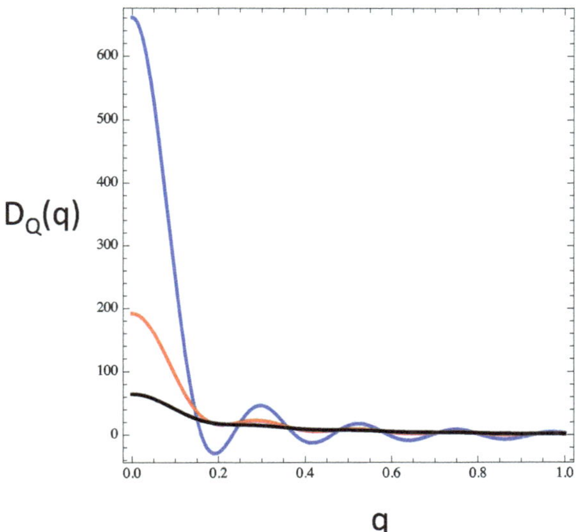

Fig. 21.3 Analytic result for the density–density correlation function of Eq. (21.45) for three values of a_Q, 1/4, 1/2 and 3/4 and for a lattice of 2500 sites

Equations (21.43a) show that the quasi-Bragg peaks remain large for $T > 0$, with magnitudes proportional to a fractional power of N, as long as $a_Q < 1$. However, at high enough temperature when $a_Q > 1$, the peak at reciprocal lattice vector $\vec{Q}$ disappears. To compare this prediction to the results of Fig. 21.2, we use the relation, $T_Q/\Theta_D = Q^2\langle u^2\rangle/2$. Then, for $\langle u^2\rangle = 0.0025$, $Q = 8\pi$ and $T = 2\Theta_D$, as is the case for the yellow diamond at the Bragg point $q_x/2\pi = 4$ on the right side of the figure, $a_Q = 1.58$, while, for the red square dot, which corresponds to $T = \Theta_D$, $a_Q = 0.79$ which is slightly less than 1. There is a slight gap between the red square and the background at this Bragg point, but there is no gap for the yellow diamond.

To complete this calculation, we consider the behavior of $D(\vec{Q} + \vec{q})$ where the assumption is that $\vec{q}$ is small in magnitude compared to π. Then we can write

$$D(\vec{Q} + \vec{q}) \approx \sum_{n=1}^{N} e^{-w(\vec{Q},\vec{R}_n)} e^{i\vec{q}\cdot\vec{R}_n} \to 2\pi \int_1^L e^{-2a_Q \ln(\gamma R/2)} J_0(qR) R dR, \tag{21.44}$$

where we have ignored the dependence of $a_{|\vec{Q}+\vec{q}|}$ on $\vec{q}$. We can write this as

$$D_Q(q) = 2\pi \left(\frac{2}{\gamma}\right)^{2a_Q} \int_1^L R^{1-2a_Q} J_0(qR) dR \tag{21.45}$$

Representative curves for $D_Q(q)$ are shown in Fig. 21.3 for a sample with $N = 2500$ and for three values of a_Q, $a_Q = 1/4,\ 1/2,\ 3/4$. As expected, the height of the peak at $q = 0$ falls rapidly with increasing a_Q. Note that the width of the peak at $q = 0$ is roughly π/L.

For $q > 1/L$, the peaks can be viewed as power-law divergences. To see this, we write (21.45) as

$$D_Q(q) = 2\pi \left(\frac{2}{\gamma}\right)^{2a_Q} \frac{1}{q^{2-2a_Q}} \int_0^\infty x^{1-2a_Q} J_0(x)dx$$
$$= \frac{4\pi}{\gamma^{2a_Q}} \frac{\Gamma(1-a_Q)}{\Gamma(a_Q)} \frac{1}{q^{2-2a_Q}}, \; q >> 1/L, \; 1/4 < a_Q < 1. \quad (21.46)$$

To summarize this section, we have seen that 2D harmonic crystals are quite remarkable. They exhibit no long-range translational order, but their order parameter correlation function decays like a power law, $R^{-\eta}$, where η is linear in T. Whereas a 3D crystal exhibits δ-function peaks in its X-ray diffraction pattern, 2D harmonic crystals exhibit power-law divergences, which saturate at peak heights that are proportional to a fractional power of the system size. With increasing temperature, the quasi-Bragg peaks disappear at temperatures that vary inversely with Q^2, that is, higher order Bragg peaks disappear before lower order ones. In fact, if we define the melting temperature, T_M of a 2D harmonic crystal as the temperature where the Bragg peak corresponding to the smallest RLV, $Q = 2\pi$ disappears, then we can relate the melting temperature to the mean-square zero-point amplitude by writing

$$\frac{T_{2\pi} T_M}{\Theta_D^2} = 1 \rightarrow T_M = \frac{\Theta_D}{2\pi^2 \langle u^2 \rangle_0}. \quad (21.47)$$

In three dimensions, the Lindemann melting criterion can be written as $\langle u^2 \rangle_{T_M} \approx 0.01$. However, our result, Eq. (21.47) suggests that, even if the mean-square amplitude at $T = 0$ were large enough melt a 3D crystal, the 2D harmonic crystal would only "melt", that is, its smallest-Q Bragg peaks disappear, at $T_M \approx 5\Theta_D$! This suggests that 2D quasi-long-range order is remarkably stable. However, what it really means is that the harmonic theory is unable to describe the actual mechanism by which 2D order is lost.

In the next section, we will look at how quasi-long-range order is destroyed by the proliferation of topological defects. In the case of 2D crystals, these defects are dislocations in the crystal lattice. However there are other equivalent systems, including the Ising model with $1/R^2$ interactions which disorders by the proliferation of domain walls and the 2D XY model which disorders via the unbinding of vortex–antivortex pairs. All of these and more fall under the general heading of Kosterlitz–Thouless physics, to which important contributions have also been made by others including particularly Anderson, Berezinskii, Halperin, Nelson, and Young. (For a recent, comprehensive review of this topic, see Kosterlitz (2016.)

21.4 Disordering by Topological Defects

We saw in Chap. 17, by direct calculation, that the 1D Ising model with nearest neighbor interactions is disordered for $T > 0$. That result can also be inferred from an argument due to Peierls and also Landau and Lifshitz. The form of the argument presented here, which applies to both short- and long-range interactions, is due to Thouless (1969).

Thouless considered an Ising chain of length N with a Hamiltonian of the form

$$H = -\frac{1}{2}\sum_{m=2}^{N}\sum_{n=1}^{m-1} J(m-n)S_m S_n, \tag{21.48}$$

where the $S_n, S_m = \pm 1$ are Ising spins, and $J(m-n)$ is the interaction between the spins at sites n and m. The form of H is flexible enough to accommodate short-range interactions, such at the nearest neighbor Hamiltonian for which $J(1) = J$ and all other $J(n)$ are zero, or very-long-range interactions, such as $J(n) = J/N$ for $1 \le n \le N$. A particular model, which was of great interest at the time, is the case, $J(m-n) = J/(m-n)^2$ which lies on the borderline between short- and long-range interactions as we shall see in a moment.

In order to address the question of whether long-range order is possible for such a system at $T > 0$, consider states with net magnetization μN, where $0 < \mu < 1$. For a statistical ensemble of such states, the magnetization would typically be distributed approximately uniformly along the length of the sample. For such a state, we can construct a new state with a magnetization of roughly $\mu(N-2L)$ by flipping the first L spins. The energy cost of doing this is

$$\Delta E = \sum_{m=L+1}^{N}\sum_{n=1}^{L} J(m-n)S_m S_n, \tag{21.49}$$

which is $\pm J$ for nearest neighbor interactions and which, in general, is independent of system size for a short-range interaction. Since the value of L can range from 1 to $N-1$, one can associate an entropy with this process of $\Delta S = k_B \ln(N-1)$, so that the difference in free energy between this ensemble of states and the ensemble from which they were generated is approximately

$$\Delta F = \sum_{m=L+1}^{N}\sum_{n=1}^{L} J(m-n)S_m S_n - k_B T \ln N, \tag{21.50}$$

For N sufficiently large and for short-range interactions, this suggests that the magnetization will decay in a series of such transitions until it becomes randomized with zero average value. On the other hand, for long-range interaction, where $\Delta E/J$ is large compared to $\ln N$, the magnetization will not be destroyed by this mechanism.

The argument above is useful and correct in its main conclusions but is somewhat lacking in rigor. In particular, the energy that enters in the free energy should be the average over the ensemble of states under study. General Ising states are defined by the arrangement of up and down spins, but they can equally well be defined by the number and location of "domain walls" between regions of up and down spins. This is particularly useful at low T, where the density of domain walls is low. The free energy considered above is that of a system with total magnetization μN. Such a state might have as few as one domain wall, with $(1+\mu)N/2$ up spins on one side and $(1-\mu)N/2$ on the other, or several or many domain walls, positioned so that $N_{\mathrm{up}} - N_{\mathrm{down}} = \mu N$.

The process of flipping the first L spins can increase or decrease the energy, but, to the extent that the magnetization extends across the system, it will usually decrease the magnetization. For nearest neighbor interactions, the energy is lowered by J if the bond to the right of site L contains a domain wall and increases by J otherwise. If the number of domain walls is small, then $\langle \Delta E \rangle > 0$. However, in any case, $\Delta E = \pm J$ is independent of N, and so the free energy for the average of ΔE over this ensemble of states is reduced by the addition of a single domain wall which, on average, decreases the magnetization as described above.

These domain walls have certain interesting properties. Consider a fully aligned state. Then flipping a block of L neighboring spins in the interior of the chain, creates two domain walls separated by a distance L. One can think of these two as a "wall" and an "anti-wall" since, they can "move" toward each other, decreasing L, and annihilate when L goes to zero. Similarly, one or both of the domain walls can move to the end of the open chain and disappear. For a chain with periodic boundary conditions, a ring, domain walls can only be created or destroyed in pairs.

For the purpose of this chapter on Kosterlitz–Thouless physics, it is relevant to consider the energy of a single domain wall in the interior of a long chain for the case of an inverse square interaction. Using Eq. (21.49), we can approximate this domain wall energy as

$$\begin{aligned} \Delta E_{DW} &= J \int_{L+1}^{N} dy_1 \int_{1}^{L} dy_2 \frac{1}{(y_1 - y_2)^2} \\ &= J \ln \frac{L(N-L)}{N-1} \\ \Delta E_{DW} &\approx J \ln(N/4), \end{aligned} \tag{21.51}$$

where the last result holds as long as the domain wall is well away from the chain ends. This falls neatly at the boundary between even longer range interactions, for which thermal excitation of domain walls cannot destroy the order, and shorter range interactions for which the system cannot order.

Since the inverse square system can apparently order at low T, where the energy to create a domain wall is of the same order, $\ln N$, but larger than $k_B T S$, long-range order, i.e., a nonzero magnetization, μN, should occur. If $|\mu| < 1$ for $T > 0$, then there will be a renormalizaton of the domain wall energy which we can write as

$$\Delta E_{DW} \approx J\mu^2 \ln N. \tag{21.52}$$

The condition that the free energy to create a domain wall be positive, for the case of nonzero magnetization, can then be written as

$$\mu^2 > \frac{T}{J}. \tag{21.53}$$

Alternatively, above the temperature where ΔE_{DW} changes sign, we expect that $\mu^2 = 0$. Such reasoning led Thouless to predict that the magnetization jumps, discontinuously, at some temperature, T_c, which we would expect to obey $T_c \lesssim J$, from $\mu = 0$ to a nonzero value, $|\mu| \geq T_c/J$. This prediction turns out to be correct, although it took almost 20 years for it to be proven rigorously (Aizenman et al. 1988).

It is straightforward to calculate the energy for the case where a region of spins of length L is flipped in an otherwise perfectly ordered chain. This calculation, which is left as an exercise, leads to an effective attractive interaction, between a domain wall and its "anti-wall" partner, that goes like $-J \ln L$. Thus a dilute system of domain walls in an Ising chain with $1/(m-n)^2$ spin–spin interactions can be viewed as a system of particles interacting via a $\ln R$ potential.

21.5 Related Problems

In order to understand how this problem fits together with the class of problems that are the subject of this chapter, it is useful to understand why the Ising ferromagnet with inverse square interactions was "trending" in the late 1960s. This was mainly due to the current hot topic of the day, the Kondo Problem (Kondo 1964), which had been attacked by, but had so-far defeated, most of the world's best condensed matter theorists. The Kondo Effect is the low-temperature resistivity minimum observed in metals containing a low density of magnetic impurities. In 1964, Kondo used perturbation theory to identify a $\ln(T_K/T)$ divergence in the resistivity, where T_K is the so-called Kondo temperature, that could account for the minimum, but which also implied an unphysical divergence (hence the Kondo Problem) at $T = 0$. Eventually significant progress was made by P.W. Anderson and his collaborators at Bell Labs (Anderson et al. 1970; Anderson and Yuval 1971), who, using a path integral formulation, managed to map the zero-space-plus-one-time, $(0+1)$- dimension Kondo Problem (i.e., the problem of a magnetic point impurity in a metal) onto the problem of a classical 1D Ising chain with $1/R^2$ interactions. This led to further progress on the Ising problem by Anderson's group which developed a so-called "Poor Man's ScalingTheory," an early version of the renormalization group method, to address the Ising and consequently also the Kondo problem. Subsequently, this scaling RG method was applied to related 2D problems (Kosterlitz and Thouless 1973), and later Wilson (1975) derived his own renormalization group method for the Kondo Problem.

Before launching into an examination of RG theory for the Kosterlitz–Thouless transition, it is worthwhile to review the systems that exhibit such behavior. As mentioned above, the $0 + 1$-dimensional Kondo problem and the 1D classical $1/r^2$ Ising model fall into this class. In 2 dimensions, we looked earlier at the 2D harmonic crystal and concluded that some new physics must intervene to destroy translational order. The new physics is that of dislocations.

A dislocation in 2D can be thought of as an extra row of atoms, inserted into the crystal from the edge and extending only part-way across. The position where the extra line of atoms ends is the dislocation core. A distinctive property of a dislocation can be described as follows. Normally, in a crystal, for example, in a square lattice, if you move n steps along x and then m steps along y and then $-n$ steps along x and $-m$ steps along y, the path describes a closed rectangle. However, if the path encloses a dislocation core, then the last step does not end at the starting point. The additional step required to get back to the starting point is called the Burgers vector. The dislocation creates a strain field in the crystal which falls off slowly, like $1/R$, and the square of the strain field, which is an elastic energy, integrated over the entire crystal, diverges logarithmically with the size of the crystal. In this way, an isolated dislocation is a bit like a single domain wall in the $1/r^2$ Ising ferromagnet. Extending the analogy, one can imagine removing L neighboring atoms from a row in the crystal. The effect is to create two dislocations, with equal and opposite Burgers vectors, whose cores are separated by a distance L. It is relatively straightforward to show that an effective attractive interaction exists between these two dislocations that goes like $\ln L$.

A similar but even simpler picture arises for the case of the 2D planar (XY) ferromagnet where, in the ground state, all spins are parallel and low-energy excitations from the ground state are harmonic spin waves. A vortex in an XY magnet corresponds to a state in which the spins wind by $\pm 2\pi$ around any closed path that encircles the vortex core once. The sign of the winding is determined by whether the spins are arranged clockwise or counterclockwise around the core. Far from the core, spins in any given region are nearly parallel, but, as one moves in a large circle around a single vortex core, that direction rotates by $\pm 2\pi$. The energy associated with an isolated vortex diverges logarithmically with system size and there is an effective inter-vortex interaction that goes like $-q_i q_j \ln R_{ij}$ where R_{ij} is the distance between the cores of vortices i and j, and $q_i, q_j = \pm 1$ represent the directions that the spins wind around each of the vortices.

The XY model is analogous to a neutral superfluid with the spin direction corresponding to the local phase of the superfluid order parameter. Gradients in that phase describe supercurrents, and vortices in a 2D superfluid film consist of supercurrent circulating around the vortex core with a velocity that falls off inversely with distance from the core. Again there are positive and negative vortices with log interactions between them. The situation is somewhat different for 2D charged superfluids, i.e., superconductors, because the interaction between superconducting vortices, although logarithmic at short distances, $\xi < R < \lambda$, falls off exponentially for $R > \lambda$. Here, ξ is the superconducting coherence length which is also the size of a vortex core, and λ is the superconducting penetration length. However, it is also the case that for a thin superconducting film, screening by supercurrents is not very effective, and so the effective penetration length can be very large, i.e., larger than the sample size, and thus Kosterlitz–Thouless physics can also be observed in thin superconducting films.

These models are all similar to the 2D Coulomb gas, which can be thought of as a gas of line charges that are perpendicular to the 2D layer, leading to log interactions between them. Kosterlitz and Thouless noted that the disordered phase of any of these 2D models is similar in nature to that of a plasma of equal numbers of such positive and negative charges, while the low-temperature, ordered phase is analogous to one in which + and − charges are bound into neutral molecules, forming an insulating but polarizable state. Thus, solving the problem of the statistical mechanics of the neutral 2D Coulomb gas also amounts to solving all these problems.

21.6 The 2D XY Model

In this section, we will perform a sequence of transformations on the 2D XY model and eventually make an approximation that preserves the basic symmetry of the Hamiltonian and leads to a form that is better suited to treatment by renormalization group methods. In some ways, this will be reminiscent of the transformations that we used in Sect. 13.6 to transform the Ising model into a field theory which was then studied using the ϵ-expansion in Chap. 20. We begin with the 2DXY Hamiltonian, which we write as

$$-\beta\mathcal{H} = K \sum_{(\vec{r},\vec{r}')} \left[\cos\left(\theta(\vec{r}) - \theta(\vec{r}')\right) - 1\right], \tag{21.54}$$

where $(\vec{r},\vec{r}')$ is summed over the $2N$ bonds of a square lattice of N sites with PBCs. We follow the general approach of José et al. (1978).

21.6.1 *The Villain Approximation*

In the partition function, Z, the potential function, $V(\theta) = K[\cos\theta - 1]$, enters in an exponential which is, of course, also a periodic function of θ and hence can be written as a Fourier sum

$$e^{V(\theta)} = \sum_{n=-\infty}^{\infty} e^{in\theta} e^{V_n}, \tag{21.55}$$

where we have written the Fourier coefficient as an exponential. Then

$$e^{V_n} = \frac{1}{2\pi}\int_0^{2\pi} e^{-in\theta}e^{V(\theta)} = e^{-K}I_n(K), \tag{21.56}$$

where $I_n(K)$ is a modified Bessel function of the first kind. In the limit of small and large K, e^{V_n} has asymptotic behavior

$$e^{V_n} \approx \frac{(K/2)^n}{n!}, \quad K << 1. \tag{21.57a}$$

$$e^{V_n} \approx \frac{e^{-n^2/2K}}{\sqrt{2\pi K}}, \quad K >> 1. \tag{21.57b}$$

Unfortunately, for K large which is relevant for any quasi-long-range-ordered state, the Fourier sum in Eq. (21.55) converges slowly. A useful device for finding a more rapidly converging expression is the so-called Poisson summation formula which can be easily understood from the theory of lattice sums. The formula is

$$\sum_{n=-\infty}^{\infty} \delta(\phi - n) = \sum_{m=-\infty}^{\infty} e^{-2\pi i m\phi}. \tag{21.58}$$

Multiplying both sides of (21.58) by $e^{i\phi\theta}e^{V_\phi}$ and integrating over ϕ, we obtain

$$e^{V(\theta)} = \sum_{m=-\infty}^{\infty}\int_{-\infty}^{\infty} d\phi e^{i\phi(\theta-2\pi m)}e^{V_\phi}. \tag{21.59}$$

Then, substituting the large K expression, Eq. (21.57b) for e^{V_ϕ} and performing the Gaussian integral, we obtain

$$e^{V(\theta)} \approx \sum_{m=-\infty}^{\infty} e^{-\frac{K}{2}(\theta-2\pi m)^2}. \tag{21.60}$$

The approximation of using the large K approximation, Eq. (21.57b), for the Fourier coefficient, e^{V_ϕ}, is due to Villain (1975), and the model potential function of Eq. (21.60) is called the Villain model. José et al. note that, although (21.60) is an accurate representation of $V(\theta) = K[\cos\theta - 1]$ for K large, it can be used over the entire range of K if K is replaced by $K_V = Kf(K)$ where

$$\begin{aligned} f(K) &\to 1 \qquad \text{for } K \to \infty, \\ f(K) &\to \frac{1}{2K\ln(2/K)} \quad \text{for } K \to 0. \end{aligned} \tag{21.61}$$

21.6.2 Duality Transformation

Now, going back to Eq. (21.55), the partition function corresponding to the Hamiltonian of Eq. (21.54) can be written as

$$Z = \left(\prod_{\vec{r}} \int_0^{2\pi} \frac{d\theta(\vec{r})}{2\pi} \right) \prod_{(\vec{r},\vec{r}\,')} \sum_{n(\vec{r},\vec{r}\,')} e^{in(\vec{r},\vec{r}\,')[\theta(\vec{r})-\theta(\vec{r}\,')]} e^{V_{n(\vec{r},\vec{r}\,')}}, \tag{21.62}$$

where each $n(\vec{r}, \vec{r}\,')$ is associated with the bond connecting site $\vec{r}$ to site $\vec{r}\,'$ and is summed over integer values between $-\infty$ and ∞, and we chose the bond labels, $(\vec{r}, \vec{r}\,')$ so that site $\vec{r}$ lies either to the left of or below site $\vec{r}\,'$. It is now straightforward to perform the integrals over $\theta(\vec{r})$. A single such integral may be written as

$$\begin{aligned} \int_0^{2\pi} \frac{d\theta(\vec{r})}{2\pi} e^{i[n(\vec{r},\vec{r}+\hat{x})+n(\vec{r},\vec{r}+\hat{y})-n(\vec{r}-\hat{x},\vec{r})-n(\vec{r}-\hat{y},\vec{r})]\theta(\vec{r})} \\ = \delta_{n(\vec{r},\vec{r}+\hat{x})+n(\vec{r},\vec{r}+\hat{y}),n(\vec{r}-\hat{x},\vec{r})+n(\vec{r}-\hat{y},\vec{r})}. \end{aligned} \tag{21.63}$$

Equation (21.63) amounts to a zero divergence condition requiring that the sum of the integers on the bonds to the left and below site $\vec{r}$ equals the sum of those on the bonds to the right and above. For a square lattice of N sites, there are N such conditions and $2N$ bond variables. We can satisfy the conditions on the $n(\vec{r}, \vec{r}\,')$ exactly by defining integer variables on the sites of the "dual" lattice, which is the lattice formed by the centers of each square of the original lattice. If we call the sites of the dual lattice $\vec{R}(\vec{r}) = \vec{r} + (\hat{x} + \hat{y})/2$ and define N integers $n(\vec{R})$ on the dual lattice, then we can write the original bond integers in terms of the site integers of the dual lattice as

$$n(\vec{r}, \vec{r} + \hat{x}) = n\big(\vec{R}(\vec{r})\big) - n\big(\vec{R}(\vec{r}) - \hat{y}\big), \tag{21.64a}$$

$$n(\vec{r}, \vec{r} + \hat{y}) = n\big(\vec{R}(\vec{r})\big) - n\big(\vec{R}(\vec{r}) - \hat{x}\big). \tag{21.64b}$$

When the bond integers are written in terms of these new site integers, the conditions of Eq. (21.63) are satisfied exactly, and the partition function can now be written as

$$Z = \sum_{\{n(\vec{R})\}} e^{\sum_{(\vec{R},\vec{R}')} V_{n(\vec{R})-n(\vec{R}')}}. \tag{21.65}$$

Unfortunately, like Eq. (21.55), the $n(\vec{R})$ sums in this expression are not easy to calculate and do not converge particularly quickly, and so we again use the Poisson summation formula to reexpress the result as

$$Z = \sum_{\{m(\vec{R})\}} \left(\prod_{\vec{r}} \int_{-\infty}^{\infty} d\phi(\vec{r}) \right) \exp\left(\sum_{(\vec{R},\vec{R}')} V_{\phi(\vec{R})-\phi(\vec{R}')} + \sum_{\vec{R}} 2\pi i \phi(\vec{R}) m(\vec{R}) \right) \tag{21.66}$$

21.6.3 Generalized Villain Model

We now specialize to the Villain approximation for the 2DXY model by writing

$$e^{V_\phi} = \frac{1}{\sqrt{2\pi K}} e^{-\phi^2/2K}. \tag{21.67}$$

Then Eq. (21.66) becomes

$$Z = \sum_{\{m(\vec{R})\}} \left(\prod_{\vec{R}} \int_{-\infty}^{\infty} \frac{d\phi(\vec{R})}{2\pi K} \right) \exp\left(-\frac{1}{2K} \sum_{(\vec{R},\vec{R}')} \left(\phi(\vec{R}) - \phi(\vec{R}')\right)^2 + 2\pi i \sum_{\vec{R}} \phi(\vec{R}) m(\vec{R}) \right) \tag{21.68}$$

We note that the coefficient of $\left(\phi(\vec{R}) - \phi(\vec{R}')\right)^2$ in the above equation is $1/2K$, rather than $K/2$. This is what one expects for the interaction on the dual lattice, as we saw in Chap. 17. However, for the Hamiltonian of Eq. (21.68), it is straightforward to recover the strong-coupling form by rescaling each ϕ by a factor of K. Equation (21.68) then becomes

$$Z = \sum_{\{m(\vec{R})\}} \left(\prod_{\vec{R}} \int_{-\infty}^{\infty} \frac{d\phi(\vec{R})}{2\pi} \right) \exp\left(-\frac{K}{2} \sum_{(\vec{R},\vec{R}')} \left(\phi(\vec{R}) - \phi(\vec{R}')\right)^2 + 2\pi i K \sum_{\vec{R}} \phi(\vec{R}) m(\vec{R}) \right) \tag{21.69}$$

Equation (21.69) has the attractive feature that the integrals over $\phi(\vec{R})$ can be evaluated in terms of Gaussian integrals. Also, as we will see in a moment, (21.69) is essentially the partition function for spin waves and vortices of strength $m(\vec{R}) = 0, \pm 1, \ldots$.

21.6.4 Spin Waves and Vortices

The exponent in Eq. (21.69) is of the form, $-\beta\mathcal{H}$, where

$$\beta\mathcal{H} = \frac{K}{2} \sum_{(\vec{R},\vec{R}')} \left(\phi(\vec{R}) - \phi(\vec{R}')\right)^2 - 2\pi i K \sum_{\vec{R}} \phi(\vec{R}) m(\vec{R}). \tag{21.70}$$

We can write $(\vec{R}, \vec{R}') \Rightarrow \vec{R}, \hat{\delta}$ with $\hat{\delta}$ summed over $\hat{\delta} = \hat{x}, \hat{y}$ for the square lattice. Then, writing, the $\phi(\vec{R})$ and $m(\vec{R})$ in terms of their lattice Fourier transforms,

$$\phi_{\vec{k}} = \frac{1}{\sqrt{N}} \sum_{\vec{R}} \phi(\vec{R}) e^{-i\vec{k}\cdot\vec{R}} \leftrightarrow \phi(\vec{R}) = \frac{1}{\sqrt{N}} \sum_{\vec{k}} \phi_{\vec{k}} e^{i\vec{k}\cdot\vec{R}}, \qquad (21.71a)$$

$$m_{\vec{k}} = \frac{1}{\sqrt{N}} \sum_{\vec{R}} m(\vec{R}) e^{-i\vec{k}\cdot\vec{R}} \leftrightarrow m(\vec{R}) = \frac{1}{\sqrt{N}} \sum_{\vec{k}} m_{\vec{k}} e^{i\vec{k}\cdot\vec{R}}. \qquad (21.71b)$$

Note that, since both $\phi(\vec{R})$ and $m(\vec{R})$ are real, we have $\phi_{-\vec{k}} = \phi^*_{\vec{k}}$ and $m_{-\vec{k}} = m^*_{\vec{k}}$

Then $\beta\mathcal{H}$ becomes

$$\beta\mathcal{H} = \frac{K}{2} \sum_{\vec{k}} \phi_{\vec{k}}\phi_{-\vec{k}} \left[|1 - e^{ik_x}|^2 + |1 - e^{ik_y}|^2 \right] - 2\pi i K \sum_{\vec{k}} \phi_{\vec{k}} m_{-\vec{k}}. \qquad (21.72)$$

The quantity in square brackets can be written as

$$[\dots] = 4 \left(\sin^2 \frac{k_x}{2} + \sin^2 \frac{k_y}{2} \right) \equiv G^{-1}(\vec{k}), \qquad (21.73)$$

which also defines the square lattice Green's function, $G(\vec{k})$ through its inverse $G^{-1}(\vec{k})$. Note that $G^{-1}(\vec{k}) \approx k^2$ and $G(\vec{k}) \approx 1/k^2$ for small k. So $\beta\mathcal{H}$ can be written as

$$\beta\mathcal{H} = \frac{K}{2} \sum_{\vec{k}} \left[G^{-1}(\vec{k})\phi_{\vec{k}}\phi_{-\vec{k}} - 2\pi i \left(\phi_{\vec{k}} m_{-\vec{k}} + \phi_{-\vec{k}} m_{\vec{k}} \right) \right]. \qquad (21.74)$$

The quantity in square brackets is clearly a quadratic form in the $\phi_{\vec{k}}$, and hence we expect to be able to do the ϕ-integrals in the partition function as Gaussian integrals. To see how this works, we first complete the square. Then

$$\begin{aligned}\beta\mathcal{H} &= \frac{K}{2} \sum_{\vec{k}} G^{-1}(\vec{k}) \left(\phi_{\vec{k}} - i2\pi G(\vec{k}) m_{\vec{k}} \right) \left(\phi_{-\vec{k}} - i2\pi G(\vec{k}) m_{-\vec{k}} \right) \\ &\qquad + 2\pi^2 K \sum_{\vec{k}} G(\vec{k}) m_{\vec{k}} m_{-\vec{k}} \\ &= \beta\mathcal{H}_{SW} + \beta\mathcal{H}_{VV}, \end{aligned} \qquad (21.75)$$

where we have called the first sum, $\beta\mathcal{H}_{SW}$, the spin-wave Hamiltonian and the second, $\beta\mathcal{H}_{VV}$, the vortex Hamiltonian. The spin-wave Hamiltonian will give the usual spin-wave (classical harmonic) partition function for 2D. We note that the fact that the Hamiltonian is a sum of decoupled spin-wave and vortex terms is a special property of the Villain approximation.

The vortex Hamiltonian has the property that, because $G(\vec{k})$ is infinite for $\vec{k} = 0$, states with a nonzero $m_{\vec{k}=0}$ are forbidden. This means that

$$\sum_{\vec{R}} m(\vec{R}) = 0. \qquad (21.76)$$

If we think of $m(\vec{R})$ as the vortex "charge" on site $\vec{R}$, then Eq. (21.76) is a condition that the total vortex charge is zero, i.e., that the overall system is charge neutral. Performing the inverse Fourier transform back to position space, $\beta\mathcal{H}_{VV}$ becomes

$$\beta\mathcal{H}_{VV} = 2\pi^2 K \sum_{\vec{R},\vec{R}'} G(\vec{R}-\vec{R}')m(\vec{R})m(\vec{R}'), \tag{21.77}$$

where

$$G(\vec{R}-\vec{R}') = \frac{1}{N}\sum_{\vec{k}\neq 0} \frac{e^{i\vec{k}\cdot(\vec{R}-\vec{R}')}}{4\left(\sin^2\frac{k_x}{2}+\sin^2\frac{k_y}{2}\right)}. \tag{21.78}$$

We can evaluate $G(\vec{R})$ approximately, replacing the sum by an integral, as

$$\begin{aligned} G(\vec{R}) &\approx \frac{1}{(2\pi)^2}\int_0^{2\pi} d\theta \int_{2\sqrt{\pi}/L}^{2\sqrt{\pi}} kdk\frac{e^{i\vec{k}\cdot\vec{R}}}{k^2} \\ &= \frac{1}{2\pi}\int_{2\sqrt{\pi}R/L}^{2\sqrt{\pi}R} \frac{J_0(x)}{x}dx \quad \text{for } \vec{R}\neq 0 \\ G(\vec{R}) &\approx -\frac{1}{2\pi}\ln R - \frac{1}{2\pi}\ln\gamma\sqrt{\pi} + \frac{1}{2\pi}\ln L, \qquad (21.79\text{a}) \\ G(0) &\approx \frac{1}{2\pi}\ln L, \qquad (21.79\text{b}) \end{aligned}$$

where L is the linear dimension of the system, and, as before, $\ln\gamma = 0.577216\ldots$ is Euler's constant. So $G(\vec{R})$ contains a term $\frac{1}{2\pi}\ln L$ for every value of $\vec{R}$ including $\vec{R}=0$, and the contribution of this term to $\beta\mathcal{H}_{VV}$ vanishes due to charge neutrality.

$$\pi K \ln L \sum_{\vec{R},\vec{R}'} m(\vec{R})m(\vec{R}') = \pi K \ln L \left(\sum_{\vec{R}} m(\vec{R})\right)^2 = 0 \tag{21.80}$$

Consequently, we can subtract $\frac{1}{2\pi}\ln L$ from $G(\vec{R})$, which implies, after this subtraction, that $G(0)=0$.

Substituting the Green's function for $\vec{R}\neq 0$ (after subtracting the $\frac{1}{2\pi}\ln L$ term) into Eq. (21.77), we obtain the vortex Hamiltonian as

$$\beta\mathcal{H}_{VV} = -\pi K \sum_{\vec{R}\neq\vec{R}'} m(\vec{R})m(\vec{R}')\ln(|\vec{R}-\vec{R}'|) + \pi K \ln\gamma\sqrt{\pi}\sum_{\vec{R}} m(\vec{R})^2, \tag{21.81}$$

where the second term was obtained after again invoking charge neutrality. It represents the energy of the vortex cores. The final result for the vortex Hamiltonian may be written as

$$\beta\mathcal{H}_{VV} = -\pi K \sum_{\vec{R}\neq\vec{R}'} m(\vec{R})m(\vec{R}')\ln(|\vec{R}-\vec{R}'|) - \ln y_0 \sum_{\vec{R}} m(\vec{R})^2, \quad (21.82)$$

where

$$y_0 = e^{-\pi K \ln \gamma\sqrt{\pi}}. \quad (21.83)$$

y_0 may be thought of as the vortex fugacity, $\exp(-\beta\mu)$, where μ is the chemical potential of a vortex or equivalently as $\exp(-\beta E_c)$ where $E_c = \pi J \ln \gamma\sqrt{\pi} \approx 3.6J$ is the energy of a vortex core.

21.6.5 The K-T Transition

You are asked in an exercise to calculate the energy of a single vortex for a 2D XY system with linear dimension L. The result is

$$E_V = \pi K \ln L. \quad (21.84)$$

The entropy for placing a single vortex on a lattice of $N = L^2$ sites is $2\ln L$ and hence the free energy for adding a single vortex vanishes when

$$\frac{1}{K} = \frac{\pi}{2}, \quad (21.85)$$

or, equivalently, $k_B T_0 = \pi J/2$. As discussed above for the case of the 1D Ising model with $1/r^2$ interactions, such an argument strongly suggests that a transition occurs, but may not provide an accurate value for T_c.

Kosterlitz and Thouless (1972, 1973) derived an alternative approach to estimating when the stability of the low-temperature state is disrupted. They first calculated the average size of a pair of unit vortices with opposite charge as

$$\langle r^2\rangle = \frac{\int_a^\infty r^2 \exp(-2\pi K \ln(r/a))rdr}{\int_a^\infty \exp(-2\pi K \ln(r/a))rdr} = a^2\frac{\pi K - 1}{\pi K - 2}, \quad (21.86)$$

where the 2 in $2\pi K$ comes from the double sum over $\vec{R}$ and $\vec{R}'$ in Eq. (21.82). They then calculated the average distance, d, between neutral pairs from the density of pairs, n_{pairs}, in a low-density approximation,

$$n_{\text{pairs}} = \frac{1}{A}\sum_{\vec{R}\neq\vec{R}'} \exp\left[-2\pi K \ln(\vec{R}-\vec{R}'|/a) + 2\ln y_0\right] + \mathcal{O}(y_0^4), \quad (21.87)$$

where $A = L^2a^2$ is the area of the system. Changing from the double sum to L^2 times an integral,

$$n_{\text{pairs}} \equiv \frac{1}{d^2} = \frac{2\pi y_0^2}{a^4} \int_a^{\sqrt{A/\pi}} \exp(-2\pi K \ln(r/a)) r dr = \frac{\pi y_0^2}{a^2} \frac{1}{(\pi K - 1)}, \tag{21.88}$$

where y_0^2 is the fugacity for a pair of vortices, $y_0 \ll 1$, and $\sqrt{A/\pi}$ is large enough that the upper limit of the integral can be taken to ∞. Then the ratio of the mean-square radius of a pair to the area per pair is

$$\frac{\langle r^2 \rangle}{d^2} = \frac{\pi y_0^2}{(\pi K - 2)}. \tag{21.89}$$

The physical picture is that, at a certain temperature, when $1/K$ is slightly less than $\pi/2$, the mean-square pair radius grows larger than the distance between pairs so that the neutral pairs overlap and become indistinguishable from a disordered plasma. At lower temperatures, individual pairs are smaller and act as molecular dipoles. $y_0 \ll 1$ represents a dilute gas of such dipoles.

Smaller dipoles will screen the interaction between the charges of larger dipoles, and so the effective interaction between charges in each pair depends on the density of other pairs which act as a polarizable medium. Note that we expect this effect to be less important for small pairs and more important for large pairs. If we represent the effect of the polarizability due to a finite density of bound pairs on the effective interaction by a dielectric constant, ϵ, then we expect ϵ to be a function of r. If we write the electric force of one positive unit charge at the origin on its negative partner at $\vec{r}$ as

$$\vec{F}(\vec{r}) = -\frac{2\pi K}{\epsilon(r) r} \hat{r}. \tag{21.90}$$

Then the work required to separate two oppositely charged particles by a distance R,

$$U_{\text{eff}}(R) = 2\pi K \int_a^R \frac{dr}{\epsilon(r) r} = 2\pi \int_a^R \frac{K(r) dr}{r} \equiv 2\pi U(R) \ln(R/a) \tag{21.91}$$

is the effective potential for the two charges. We have also defined two new functions, $K(r) = K/\epsilon(r)$, and $U(R)$. Note that we expect $\epsilon(a) = 1$ since there is negligible screening for nearest neighbor pairs, and so $K(a) = K$.

We can derive a self-consistent theory for this system by using U_{eff} to calculate $\epsilon(r)$. For a two-dimensional medium containing a density n of dipoles with moments $\vec{p} = q\vec{r}$ at temperature $1/\beta$ in an electric field of strength $\vec{E}$, the polarization density of the medium is $\vec{P} = n\beta p^2 \vec{E}/2$, and the dielectric constant (again for 2D) is increased by an amount,

$$\epsilon = 1 + 2\pi \frac{P}{E} \Rightarrow \epsilon - 1 = 2\pi\beta n p^2/2 = \pi\beta q^2 r^2 n. \tag{21.92}$$

For our problem, the size of the dipoles is distributed over a range of values of r. In considering the interaction that binds a large molecule of size R, we need to consider that the region interior to the molecule might contain smaller dipoles of sizes ranging from a to nearly R which would modify the effective potential that binds the molecule. Noting that for the vortex problem the charge is given by $\beta q^2 = 2\pi K$, and, using the same approach as in Eq. (21.88), we write the distance-dependent dielectric constant $\epsilon(r)$ as

$$\epsilon(r) = \frac{K}{K(r)} = 1 + 2\pi^2 K \langle r^2 n_{\rm pair}(r) \rangle$$
$$= 1 + \frac{4\pi^3 K y_0^2}{a^4} \int_a^r r^2 e^{-U_{\rm eff}(r)} r dr, \tag{21.93}$$

or

$$K(r)^{-1} = K(a)^{-1} + 4\pi^3 y_0^2 \int_a^r \left(\frac{r'}{a}\right)^{3-2\pi U(r')} dr'/a. \tag{21.94}$$

One would like to solve this equation for $K(r)$. However, the presence in the exponent on the right-hand side of Eq. (21.94), of the function $U(r')$ which is defined in Eq. (21.91), makes the solution cumbersome. This led Kosterlitz and Thouless (1973) to make an approximation which, although convenient and reasonably accurate, was later shown to be unnecessary. By replacing $U(r')$ by $K(r')$, they were able to solve the equation self-consistently and obtain useful results. Subsequently, Kosterlitz (1974) solved this same problem by a completely different method, using Anderson's poor man's scaling renormalization group approach (Anderson et al. 1970). The resulting recursion relations were found not to agree perfectly with those of the self-consistent method described above. A year later, Young (1978) showed that it was possible to solve the self-consistent problem without making the approximation $U(r') \to K(r')$ and the results agreed exactly with Kosterlitz's RG solution, demonstrating that the self-consistent mean-field approach works well for this 2D problem with long-range interactions. Since we have come this far, we will complete the original K-T calculation making use of Young's approach.

Young defined a distance-dependent fugacity function, $y(r)$, by

$$y(r)^2 \equiv y_0^2 \left(\frac{r}{a}\right)^{4-2\pi U(r)}, \tag{21.95a}$$

$$y(r)^2 = y(a)^2 \exp\left[4\ln(r/a) - 2\pi \int_{\ln a}^{\ln r} K(r') d\ln r'\right], \tag{21.95b}$$

where $y(a) = y_0$, and in the second line we have used the definition of $U(r)$ from Eq. (21.91). Then Eq. (21.94) can be rewritten in terms of $y(r)$ as

$$K(r)^{-1} = K(a)^{-1} + 4\pi^3 \int_{\ln a}^{\ln r} y(r')^2 d\ln r'. \tag{21.96}$$

Differentiating Eqs. (21.95b) and (21.96) with respect to $\ln r$, we obtain the differential recursion relations,

$$\frac{dK(r)^{-1}}{d\ln r} = 4\pi^3 y(r)^2, \tag{21.97a}$$

$$\frac{dy(r)}{d\ln r} = \big(2 - \pi K(r)\big)y(r). \tag{21.97b}$$

If we write r as ae^{ℓ}, then $d\ln r = d\ell$, so that Eqs. (21.97) become

$$\frac{dK(\ell)^{-1}}{d\ell} = 4\pi^3 y(\ell)^2, \tag{21.98a}$$

$$\frac{dy(\ell)}{d\ell} = \big(2 - \pi K(\ell)\big)y(\ell). \tag{21.98b}$$

Equation (21.98) is easy to integrate. Substituting one power of $y(\ell)$ from the second relation into the first gives

$$\big(2 - \pi K(\ell)\big)\frac{dK(\ell)^{-1}}{d\ell} = 2\pi^3\frac{dy(\ell)^2}{d\ell}, \tag{21.99}$$

which implies that

$$\frac{2}{K(\ell)} + \pi \ln K(\ell) - 2\pi^3 y(\ell)^2 = \frac{2}{K(0)} + \pi \ln K(0) - 2\pi^3 y(0)^2 \tag{21.100}$$

is independent of ℓ.

Furthermore, it was already clear from Eq. (21.98) that the point, $K(\ell) = 2/\pi,\ y(\ell) = 0$ is special. It is a fixed point since, from the second relation, if $K(\ell) = 2/\pi$, then $y(\ell)$ does not change, and, from the first, if $y(\ell) = 0$, then $K(\ell)$ does not change. We can scale and shift Eq. (21.100) so that $K(\ell)$ appears only in the combination, $\pi K(\ell)/2$, and we can also define a temperature-like variable, $t(\ell)$, by

$$\frac{2}{\pi K(\ell)} \equiv 1 + t(\ell). \tag{21.101}$$

where t is of the form $(T - T_c)/T_c$. Then Eq. (21.100) becomes

$$1 + t(\ell) - \ln\big(1 + t(\ell)\big) - 2\pi^2 y(\ell)^2 = 1 + C(t(0), y(0)), \tag{21.102a}$$

$$C(t(0), y(0)) = t(0) - \ln\big(1 + t(0)\big) - 2\pi^2 y(0)^2, \tag{21.102b}$$

$$\approx \frac{1}{2}\big(t(0)^2 - 4\pi^2 y(0)^2\big) + \mathcal{O}(t(0)^3), \tag{21.102c}$$

where we have defined the function $C(t(0), y(0))$ that controls the shape of the trajectories of $\big(t(\ell), y(\ell)\big)$, and, in the third line, approximated $C(t(0), y(0))$ for the case, $t(0) \ll 1$. (Remember that $t(0)$ and $y(0)$ are the "initial values" of $t(\ell)$ and $y(\ell)$ and are determined by the bare coupling constant, $K_0 = K(0)$ and the fugacity y_0.)

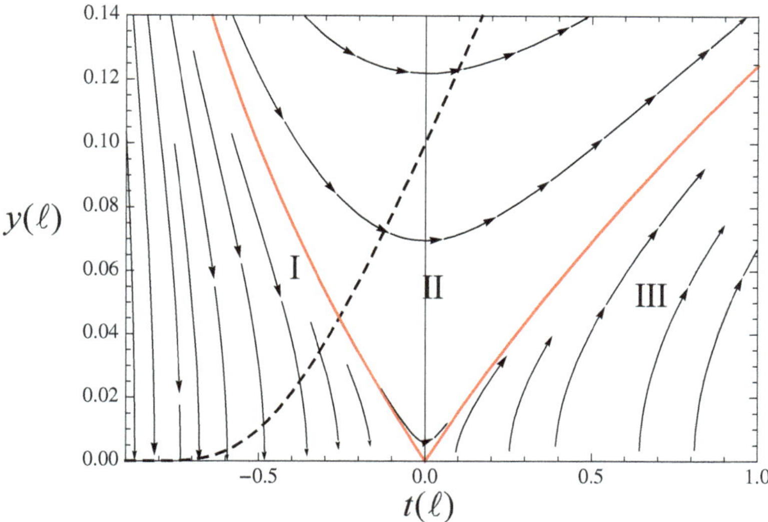

Fig. 21.4 Flow diagram corresponding to the differential recursion relations in Eq. (21.103). The solid (red) curves define three regions. In region **I**, the fugacity, $y(\ell)$ flows to zero, while $t(\ell) < 0$ flows to a nonzero constant. In regions **II** and **III**, both $y(\ell)$ and $t(\ell)$ flow to infinity, although, in region **II**, $y(\ell)$ passes though a positive minimum value at $t(\ell) = 0$. The dashed black curve represents the initial conditions, $y(0) = \exp(-(2E_c/\pi)/(1 + t(0)))$, where E_c is given below Eq. (21.83) and we have also used Eq. (21.101)

This is all best understood with the help of the flow diagram for the recursion relations, which we rewrite for $t(\ell)$ and $y(\ell)$ as

$$\frac{dt(\ell)}{d\ell} = 8\pi^2 y(\ell)^2, \tag{21.103a}$$

$$\frac{dy(\ell)}{d\ell} = \frac{2t(\ell)}{1+t(\ell)} y(\ell). \tag{21.103b}$$

The flows corresponding to these recursion relations are shown in Fig. 21.4. The solid red lines in the figure, separating regions I, II, and III, are critical lines. They are described by the equation, $C(t(0), y(0)) = 0$. Initial values, $(t(0), y(0))$ on the left-hand line (where $t(0) < 0$) flow to the fixed point, (0, 0), while initial values on the right-hand line (where $t(0) > 0$) flow to infinity. Region I is a low-temperature, ordered region. In this region, the fugacity flows to zero and $t(\ell)$ flows to a nonzero, negative value. Regions II and III correspond to disordered states. In both regions, $y(\ell)$ and $t(\ell)$) both flow to infinity. The dashed black line represents a possible set of initial conditions $(y(0), K(0))$ as described in the figure caption.

21.6.6 Disconituity of the Order Parameter at T_c

Next, we will examine the nature and temperature dependence of the "order" in region I. For this problem of the neutral 2D Coulomb gas, order means that there are at most a small number of tightly bound molecules of ± 1 charges. Then the dielectric constant is essentially equal to 1. As noted in Eq. (21.92), the dielectric constant is one plus 2π times the polarizability, where the polarizability is small when the temperature and y_0 are small, but increases with increasing temperature and y_0. Thus, a suitable order parameter for this problem would be $1/\epsilon_\infty$ where ϵ_∞ is the effective dielectric constant for widely separated test charges (or, equivalently, for very large $\pm$ pairs). In the notation that we are using

$$\frac{1}{\epsilon_\infty} = \frac{1 + t(0)}{1 + t(\infty)}. \tag{21.104}$$

We can calculate $1/\epsilon_\infty$ as a function of the temperature, $1 + t(0)$, and the fugacity $y(0)$. (Note that $y(\infty)$ is zero in region I.) In fact, for regions II and III, $1/\epsilon_\infty = 0$ since, for initial conditions in those regions, $t(\infty)$ is infinite. A particularly interesting case is when $t(0)$ and $y(0)$ lie on the critical line separating regions I and II. In that case, $t(\infty) = 0$ for any initial conditions on the critical line, and hence

$$\frac{1}{\epsilon_\infty} = 1 + t_c(y(0)) \text{ for } t(0) = t_c(y(0)), \tag{21.105}$$

where $t_c(y(0))$ is the critical temperature as a function of fugacity. This is a remarkable result. At the transition temperature, the order parameter jumps, discontinuously with decreasing temperature, from zero to a nonzero value which is the ratio of the transition temperature to the fixed point temperature.

To get a more complete picture of how this works, it is useful to see some plots of $1/\epsilon_\infty$ versus temperature for different fugacities as shown in Fig. 21.5. For purposes of illustration in Fig. 21.5, we have written the fugacity as $y_0(t) = \exp[-(2E_c/\pi)/(1+t)]$, where we treat the vortex core energy, E_c as a parameter. The assumed values of E_c are given in the caption of Fig. 21.5.

The curves in Fig. 21.5 show that $1/\epsilon_\infty$ falls to $1 + t_c(y(0)) > 0$ at the critical temperature, $1 + t_c(y(0))$, with what appears to be infinite slope. We can check that the slope is, in fact, infinite by examining solutions to Eq. (21.102) in the limit $\ell \to \infty$, which we write as

$$T(t_0, y_0) - \ln T(t_0, y_0) = 1 + C(t_0, y_0), \tag{21.106}$$

where $T(t_0, y_0) = 1 + t(\infty)$ for initial values $(t(0), y(0)) = (t_0, y_0)$, and we have used the fact that $y(\infty) = 0$ in the ordered region.

The solution to Eq. (21.106) can be written in terms of the Lambert W-function (Corless et al. 1996), which is the solution to the equation, $z = W(z) \exp W(z)$ and which is called the ProductLog in *Mathematica*. It is

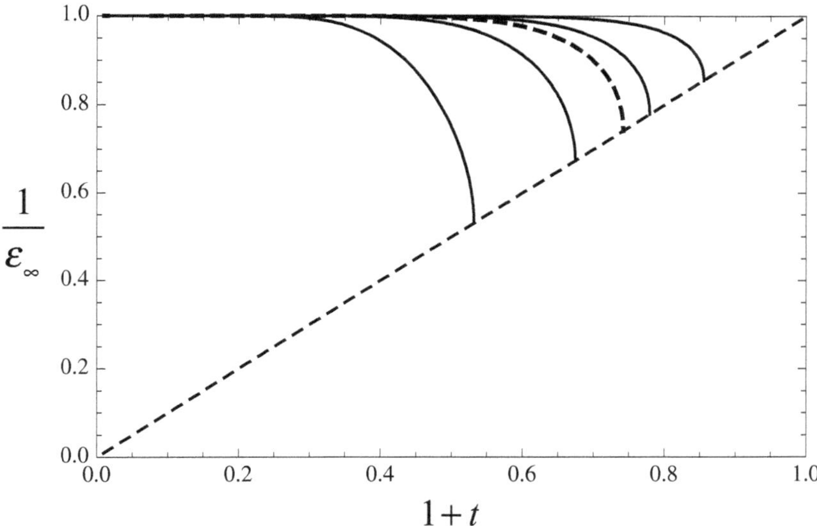

Fig. 21.5 Order parameter, $1/\epsilon_\infty$, for the neutral Coulomb gas as defined in Eq. (21.104), plotted versus temperature for several values of the fugacity. The fugacity for each curve is given by $y_0(t) = \exp(-(2E_c/\pi/(1+t))$. (See Fig. 21.4 for an example.) Curves are for $E_c = 2, 3, \pi\ \ln(\gamma\sqrt{\pi}) \approx$ 3.6, 4 and 5, with larger values of E_c corresponding to higher transition temperatures

$$T(t, y) = -W\big(-e^{-1-C(t,y)}\big). \tag{21.107}$$

Expanding the W-function in powers of C gives

$$T(C) = 1 - \sqrt{2C} + \frac{2C}{3} - \frac{C^{3/2}}{9\sqrt{2}} - \frac{2C^2}{135} + \mathcal{O}(C^{5/2}). \tag{21.108}$$

So the leading correction to $T(0) = 1$ is $-\sqrt{2C}$. Since, close to the critical line, C is linear in $|t_0 - t_c(y_0)|$ as can be seen from the last of Eqs. (21.102), it is indeed the case that the curves in Fig. 21.5 start off with a square root edge, implying infinite slope at the transition.

The physics of Fig. 21.5 extends to other related problems. For example, Anderson and Yuval (1971) showed that the temperature dependence of the squared magnetization, $m(T/T_c)^2$ for the ferromagnetic Ising model with $1/r^2$ interactions is essentially identical to that of $1/\epsilon_\infty$. For that problem, the fugacity for domain walls is controlled by a nearest neighbor coupling parameter that can be varied to make domain wall formation easier or more difficult. Their result also supported the conjecture of Thouless that the magnetization drops discontinuously to zero, for which a rigorous proof was provided later by Aizenman et al. (1988).

Similar physics occurs in thin superfluid ^{4}He films where the superfluid phases, $\phi(\vec{r})$, of neighboring volume elements play the role of XY spins in the 2D XY model

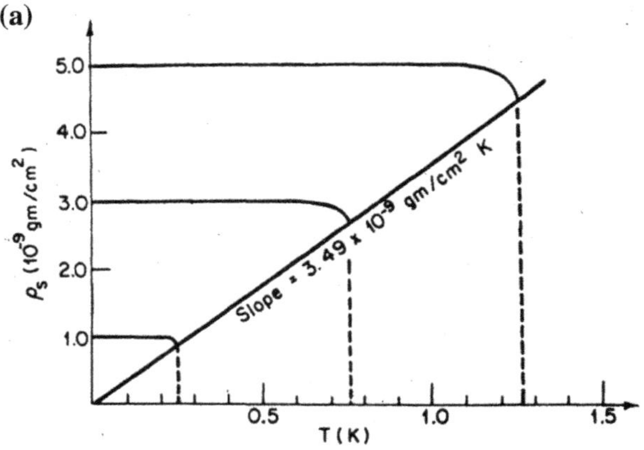

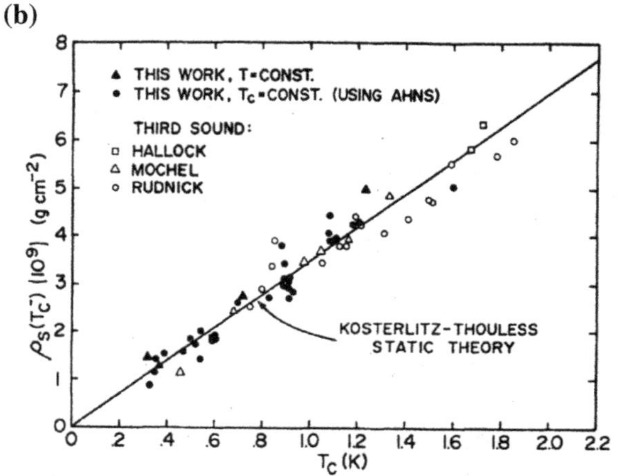

Fig. 21.6 **a** Superfluid density for three different thickness films as a function of temperature according to the Kosterlitz–Thouless static theory, and **b** results of torsional oscillator measurements and third-sound results for the discontinuous superfluid density jump $\rho_s(T_c)$ as a function of T_c. The solid line is the Kosterlitz–Thouless static theory. Both figures are from Bishop and Reppy (1980)

and the analog of $1/\epsilon_\infty$ is the superfluid density, ρ_s, which has units of mass per unit area. An attractive feature of thin superfluid ^{4}He films is that both the superfluid density and T_c scale linearly with the thickness of the mobile layer of liquid. (There is a solid layer at the substrate which does not participate in the superfluid.) The Kosterlitz–Thouless theory for this system leads to the theoretical prediction that

$$\frac{\rho_s(T_c)}{T_c} = \frac{2m^2 k_B}{\pi \hbar^2} = 3.491 \times 10^{-9} \frac{\text{g}}{\text{cm}^2\text{K}}. \tag{21.109}$$

Bishop and Reppy (1980) measured the superfluid density as a function of temperature, using a torsional oscillator method. In that method, the frequency and damping of the torsional oscillator vary with the amount of superfluid mass that decouples from the oscillator and the dissipation due to the motion of vortices in the superfluid.

Predictions and results (corrected for finite frequency effects) for $\rho_s(T)$ as a function of T and T_c for different film thicknesses are shown in Fig. 21.6.

21.6.7 Correlation Length

The physical picture used to calculate the dielectric constant, $\epsilon(\ell)$, suggests defining a correlation length associated with the distribution of sizes of bound, neutral pairs of ± 1 charges. In the insulating state, one expects the pairs to be relatively tightly bound, with the probability of large pairs falling off rapidly above a characteristic size, $\xi(\ell^*) = ae^{\ell^*}$. We argued that $\epsilon(0) = 1$ because there is no room for bound pairs to screen charges one lattice space apart. As ℓ increases, the effective dielectric constant for two test charges a distance ae^{ℓ} apart should saturate for $\ell > \ell^*$. To see how this works, we calculate $\epsilon(\ell)$ for $(t(0), y(0)) = (t_0, y_0)$ close to the critical point.

We want to calculate

$$\epsilon(\ell) = \frac{1 + t(\ell)}{1 + t_0}. \tag{21.110}$$

Using Eq. (21.102a) to solve for $y(\ell)^2$ and then substituting into Eq. (21.103a), we obtain a differential equation for $t(\ell)$ as follows:

$$\frac{dt(\ell)}{d\ell} = 4\left[t(\ell) - \ln\left(1 + t(\ell)\right) - C(t_0, y_0)\right]. \tag{21.111}$$

We are interested in solutions to Eq. (21.111) close to the critical point, $t = y = 0$, so we assume that $|t_0| \ll 1$. Furthermore, we know from Fig. 21.4 that $t(\ell) > t_0$ in region I. In addition, we can use Eq. (21.102c) and the fact that $C(t_0, y_0) = 0$ on the critical line to define $t_c(y_0) = -2\pi y_0$. The situation is illustrated in Fig. 21.7 where the flow is shown for two initial conditions, $(t_0, y_0) = (-0.15, 0.01)$ and $(t_0', y_0) = (-0.075, 0.01)$. The distinctive difference between these two curves is that curve A flows to a limiting value, $t(\infty) < t_c(y_0)$ while, for curve B, $t(\infty) > t_c(y_0)$. The boundary between these two limiting behaviors can be calculated using Eq. (21.108) which we write as

$$t(\infty) = -\sqrt{2C} \tag{21.112}$$

where we have used Eq. (21.102c),

$$C(t_0, y_0) \approx \frac{1}{2}\left(t_0^2 - t_c(y_0)^2\right). \tag{21.113}$$

We can then obtain the condition for the boundary of the "true" critical region as

$$t(\infty) = t_c(y_0) \rightarrow t_0 = \sqrt{2}t_c(y_0). \tag{21.114}$$

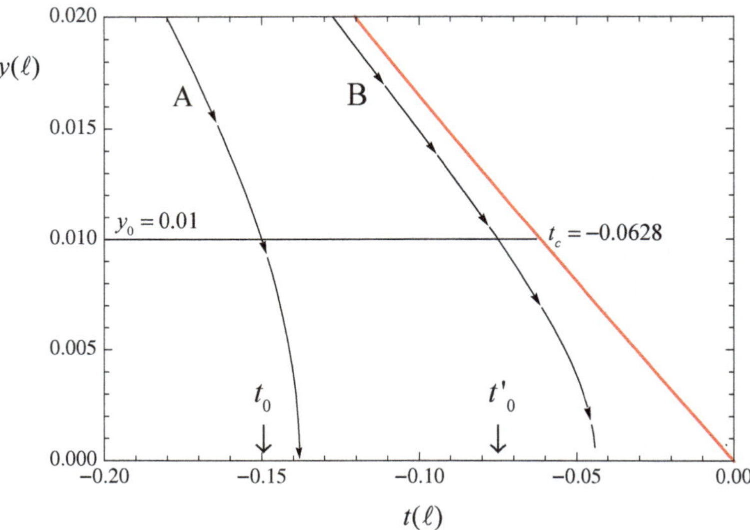

Fig. 21.7 Flows for two initial conditions. For curve A, $(t_0, y_0) = (-0.15, 0.01)$ and, for curve B, $(t_0', y_0) = (-0.075, 0.01)$. For both cases, $t_c = 2\pi y_0 = -0.0628$. The end point for curve A, where $y(\infty) = 0$, is less than t_c, while the end point for curve B is greater than t_c, i.e., closer to the critical point value of zero

We now return to the problem of solving Eq. (21.111) in the critical region. Expanding the logarithm on the right-hand side of Eq. (21.111) to order $t(\ell)^2$ and using the approximate expression for $C(t_0, y_0)$, Eq. (21.111) becomes

$$\frac{dt(\ell)}{d\ell} = 2\left[t(\ell)^2 - 2C\right], \tag{21.115}$$

which has the solution

$$\begin{aligned} 2\ell &= \int_{t_0}^{t(\ell)} \frac{dt}{t^2 - 2C} \\ &= \frac{1}{\sqrt{2C}}\left[\tanh^{-1}\frac{t_0}{\sqrt{2C}} - \tanh^{-1}\frac{t(\ell)}{\sqrt{2C}}\right], \end{aligned} \tag{21.116}$$

or, solving for $t(\ell)$,

$$t(\ell) = -\sqrt{2C}\left[\frac{t_0 - \sqrt{2C}\tanh(\sqrt{8C}\ell)}{t_0\tanh(\sqrt{8C}\ell) - \sqrt{2C}}\right]. \tag{21.117}$$

Equation (21.117) has the expected property that, for sufficiently large ℓ, the tanh's go to 1, as does the quantity in square brackets, so that $t(\infty) \to -\sqrt{2C}$. The argument of the tanh allows us to define the value $\ell^* = 1/\sqrt{8C}$ where $t(\ell)$ and hence the dielectric constant, $\epsilon(\ell)$, stop changing, which means that, beyond the length scale, $\xi(\sqrt{2C}) = ae^{\ell^*}$, the distribution of bound pair sizes becomes constant. Thus the correlation length is

$$\xi(\sqrt{2C}) = a \exp \frac{1}{2\sqrt{2C}} \tag{21.118a}$$

$$\approx a \exp \frac{1}{2|t_0|} \qquad \text{for } t_0^2 \gg t_c(y_0)^2, \tag{21.118b}$$

$$\approx a \exp \frac{1}{2\sqrt{2t_0(t_0 - t_c(y_0))}} \quad \text{for } |t_0 - t_c(y_0)| \ll |t_0|. \tag{21.118c}$$

Note that the behavior of the correlation length at $t_0 = t_c(y_0)$ in Eq. (21.118c) is of the form of an essential singularity. This will be reflected in all the thermodynamic properties at $t_c(y_0)$.

21.6.8 Free Energy and Specific Heat

In this section, we calculate the free energy approximately in a low-temperature, low-density (of vortices) limit using the self-consistent mean-field theory described above. We begin by writing the partition function for the Hamiltonian of Eq. (21.82).

$$Z = \sum_{\{m(\vec{R}_i)\}'} \exp\left[\pi K_0 \sum_{\vec{R}\neq\vec{R}'} m(\vec{R})m(\vec{R}') \ln(|\vec{R} - \vec{R}'|) + \ln y_0 \sum_{\vec{R}} m(\vec{R})^2\right], \tag{21.119}$$

where the set, $\{m(\vec{R}_i)\}'$, satisfies the neutrality condition, Eq. (21.76).

We would like to evaluate Z approximately in the limit of large K_0 (low temperature) and small y_0 where we can restrict the values of the $m(\vec{R}_i)$ to be 0 or ± 1 and the number of nonzero $m(\vec{R}_i)$ is small. Under these conditions, we can define the number of vortices as $2n$ where

$$2n = \sum_{\vec{R}} m(\vec{R})^2 \tag{21.120}$$

The partition function can then be rewritten as a sum over the number of vortices as

$$Z = \sum_{n=0}^{\infty} y_0^{2n} Z_{2n}, \tag{21.121}$$

where

$$Z_0 = 1, \tag{21.122a}$$

$$Z_2 = N \sum_{\vec{R} \neq 0} e^{-2\pi K_0 \ln |\vec{R}|}. \tag{21.122b}$$

Things start to get more complicated with Z_4 which involves two positive and two negative vortices.

$$Z_4 = \frac{1}{(2!)^2} \sum_{\{\vec{R}_i^{\pm}\}'} e^{\mathcal{H}_4(\vec{R}_1^+, \vec{R}_2^+, \vec{R}_1^-, \vec{R}_2^-)}, \tag{21.123}$$

where the prime means that no two charges can occupy the same site and the factor of $1/(2!)^2$ corrects for the fact that interchanging the positions of two vortices with the same charge does not generate a distinct configuration. If we define $V(\vec{R}, \vec{R}') = -2\pi K_0 \ln(|\vec{R} - \vec{R}'|)$, then we can write $\mathcal{H}_4$ as

$$\begin{aligned} \mathcal{H}_4(\vec{R}_1^+, \vec{R}_2^+, \vec{R}_1^-, \vec{R}_2^-) = \ & V(\vec{R}_1^+, \vec{R}_1^-) + V(\vec{R}_2^+, \vec{R}_2^-) \\ & + V(\vec{R}_1^+, \vec{R}_2^-) + V(\vec{R}_1^-, \vec{R}_2^+) \\ & - V(\vec{R}_1^+, \vec{R}_2^+) - V(\vec{R}_1^-, \vec{R}_2^-). \end{aligned} \tag{21.124}$$

The Independent Pair Approximation

Now we make an approximation that corresponds to our picture of relatively closely bound, neutral pairs separated by distances much larger than their size. If $\vec{R}_1^+$ and $\vec{R}_1^-$ are the coordinates of the vortices in one pair, and $\vec{R}_2^+$ and $\vec{R}_2^-$ are the coordinates for the other pair, then the last 4 terms in Eq. (21.124) are relatively negligible, since they correspond to the interaction of two well-separated neutral pairs. However, there is another equivalent region in the configuration sum of Eq. (21.123) where the pair of particles $\vec{R}_1^+$ and $\vec{R}_2^-$ are close together but well separated from the particles at $\vec{R}_2^+$ and $\vec{R}_1^-$ which also form a neutral pair. This means we can write

$$Z_4 \approx \frac{2!}{(2!)^2} (Z_2)^2, \tag{21.125}$$

where the factor of 2! in the numerator corresponds to the two distinct regions of configuration space corresponding to two bound neutral pairs. Equation (21.125) has been written in this way to suggest the fact that an even more tedious line of reasoning will show that the general term, Z_{2n}, can be approximated, in this independent pair picture, by

$$Z_{2n} \approx \frac{1}{n!} (Z_2)^n, \tag{21.126}$$

Then the full partition function is

$$Z_{ip} = \sum_{n=0}^{\infty} \frac{y_0^{2n} Z_2^n}{n!} = e^{y_0^2 Z_2}, \tag{21.127}$$

This result, as discussed by Solla and Riedel (1981), implies a free energy per site of

$$f_{ip} = \frac{1}{N} \ln Z_{ip} = y_0^2 \sum_{\vec{R} \neq 0} e^{-2\pi K_0 \ln |\vec{R}|} = \frac{\pi y_0^2}{\pi K_0 - 1}, \tag{21.128}$$

which, for unit lattice constant ($a = 1$), is identical to the result for n_{pairs} given in Eq. (21.88). The free energy can be used to calculate the specific heat from the identity

$$c = K_0^2 \frac{\partial^2 f}{\partial K_0^2}. \tag{21.129}$$

The result found using the independent pair approximation for the free energy, f_{ip}, turns out to be very close, at low temperatures, to that of the more accurate renormalization group (or self-consistent mean field) solution that we will discuss next.

Self-Consistent Free Energy and Specific Heat

The approximation made to obtain Z_{ip} in Eq. (21.127) is a bit unsatisfying since it neglects all the effects of the finite background density of polarizable pairs that we calculated earlier. However, we know how to introduce those effects in an approximate but consistent way. The prescription is to replace $K_0 \ln(r/a)$ by $U(r) \ln(r/a)$ as defined in Eq. (21.91) and then express $U(r)$ in terms of $y(r)^2$ using Eq. (21.95a). Physically, this amounts to replacing the bare log interaction by a distance-dependent screened interaction due to the background density of pairs.

Thus, we define a self-consistent free energy

$$f_{sc} = 2\pi y_0^2 \int_a^{\infty} \frac{r dr}{a^2} e^{-2\pi U(r) \ln(r/a)} \tag{21.130a}$$

$$= 2\pi y_0^2 \int_1^{\infty} \left(\frac{r}{a}\right)^{1-2\pi U(r) \ln(r/a)} d(r/a) \tag{21.130b}$$

$$= 2\pi y_0^2 \int_1^{\infty} \left(\frac{a}{r}\right)^3 \left(\frac{r}{a}\right)^{4-2\pi U(r) \ln(r/a)} d(r/a) \tag{21.130c}$$

$$= 2\pi \int_0^{\infty} \left(\frac{a}{r}\right)^2 y(r)^2 d \ln(r/a). \tag{21.130d}$$

If we make the usual substitution, $r = ae^{\ell}$, then $d \ln(r/a) = d\ell$, and

$$f_{sc} = 2\pi \int_0^{\infty} e^{-2\ell} y(\ell)^2 d\ell. \tag{21.131}$$

In order to calculate $f_{sc}(K_0, y_0)$, one simply integrates Eq. (21.98) to obtain $y(\ell)$ for given initial values, (K_0, y_0), and then performs the rapidly converging integral, Eq. (21.131). We should emphasize that this treatment and result applies only in the region below t_c where $y(\ell)$ flows to zero. Above t_c in regions II and III of Fig. 21.4, a different model must be used for the fully ionized "plasma" of vortices. Whereas, in the low- temperature insulating phase, it was correct to use the dielectric function derived from Eq. (21.92) to describe the screening of the vortex–antivortex interaction, above t_c one should use a theory such as Debye–Huckel theory to describe the effective interaction in the conducting phase. We will not treat the ionized phase in this book, and will instead refer the reader to relevant treatments in the literature (Kosterlitz 2016 and references therein).

The result for the free energy in Eq. (21.131) can be reformulated to show the scaling behavior and flow of $f_{sc}(\ell)$. In the following discussion, we will drop the subscript and write the free energy as $f(\ell) = f(K(\ell), y(\ell))$. Dividing the integral of Eq. (21.131) into two parts, we write

$$f(K_0, y_0) = 2\pi \int_0^{\ell} e^{-2\ell'} y(\ell')^2 d\ell' + 2\pi \int_{\ell}^{\infty} e^{-2\ell'} y(\ell')^2 d\ell'. \tag{21.132}$$

The lower limit of the second integral can be shifted to zero giving

$$f(K_0, y_0) = 2\pi \int_0^{\ell} e^{-2\ell'} y(\ell')^2 d\ell' + 2\pi e^{-2\ell} \int_0^{\infty} e^{-2\ell'} y(\ell + \ell')^2 d\ell' \tag{21.133a}$$

$$f(K_0, y_0) = 2\pi \int_0^{\ell} e^{-2\ell'} y(\ell')^2 d\ell' + e^{-2\ell} f(K(\ell), y(\ell)). \tag{21.133b}$$

This recursion relation for the free energy is of the same type as that implicit in the recursion relation for the partition function in Eq. (19.67).

Differentiating Eq. (21.133b) with respect to ℓ gives the differential recursion relation for the free energy,

$$\frac{df(\ell)}{d\ell} = 2f(\ell) - 2\pi y(\ell)^2, \tag{21.134}$$

which supplements Eq. (21.98), providing a complete set of RG recursion relations for this problem.

Results for the specific heat for a slightly modified version of the recursion relations (21.98) and (21.134) were calculated by Solla and Riedel (1981) for initial conditions, $y_0 = \exp\left(-\frac{1}{2}\pi^2 K_0\right)$. They also calculated the low-temperature specific heat using the independent pair model of Eq. (21.128), and the high-temperature specific heat using the Debye–Huckel approximation which becomes exact in the high temperature limit. Their results are shown in Fig. 21.8. Note that the initial conditions used for their calculation intersect the critical line at $y^* = 0.15$, giving

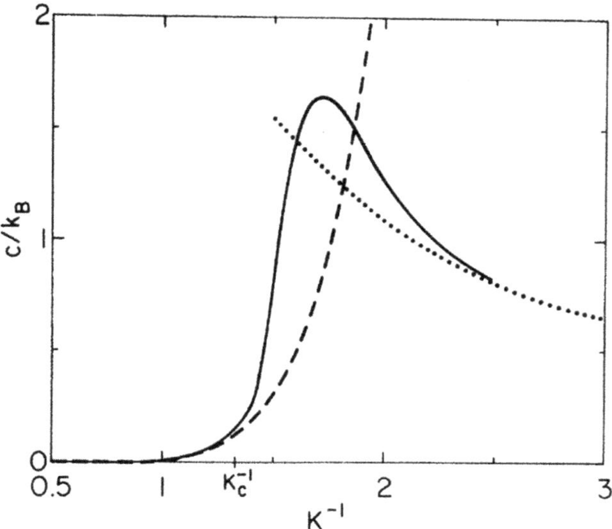

Fig. 21.8 Specific heat, calculated by Solla and Riedel (1981), for the 2D Coulomb gas from an approximate renormalization group approach (full curve) and comparison with exact low- and high-temperature results (dashed and dotted curves, respectively). The transition occurs at $K_c^{-1} = 1.33$. The asymptotic behavior at high temperatures is $c = 1/2$

a critical temperature of $K_c^{-1} = \frac{\pi}{2}(1 + t_c) = 1.33$, or $t_c = -0.1533$ in our notation. Note also that there is an essential singularity in the specific heat at t_c which is, however, unobservably weak.

21.7 Conclusion

A great deal more could be said about the physics of the Kosterlitz–Thouless transition for which Kosterlitz and Thouless received the 2016 Nobel Prize "for theoretical discoveries of topological phase transitions and topological phases of matter," along with Duncan Haldane. Fortunately, there is a relatively recent 2016 review of work on this subject by Kosterlitz which includes more than 400 references to the literature. Hopefully, the material in this chapter provides a useful introduction and guide to anyone who wishes to pursue further study of this fascinating subject.

21.8 Exercises

1. Calculate the numerical values of the fraction, f, in Eqs. (21.16) and (21.17), using data for for Al, Cu, and W.

2. Perform a calculation, similar to the one that gave Eq. (21.51), for a configuration of an Ising chain of N spins in which all spins are up except for the spins on sites from

$L_1 + 1$ to $L_2 = L_1 + L$ which are down and where both L_1 and L_2 are well away from the chain ends. Explain why your result corresponds to an attractive interaction of the form, $-J \ln L$ between the domain wall at L_1 and its "anti-wall" partner at L_2.

3. The goal of this exercise is to calculate the energy of a vortex in a 2D square lattice, nearest neighbor XY model, using the Hamiltonian in Eq. (21.54). For the sake of definiteness, assume that the vortex is centered at the origin, and that the spin at site, (m, n) points along $\theta = \tan^{-1}(n/m) + \pi/2$.

(a) Draw a picture of the spin configuration in a region around the origin.

(b) How does the spin at the origin contribute to the total energy if its direction is taken to be θ_0?

(c) For spins far enough away from the origin, derive an approximate expression for the energy of the interaction of a spin at site $\vec{r} = (m, n)$ with its nearest neighbors along $+\hat{x}$ and $+\hat{y}$.

(d) Defining the energy density as the sum of the two energies derived in part (c), integrate this approximate expression around the vortex core and from $|\vec{r}| = 1$ up to L to find the vortex energy.

4. **(a)** Use a program such as *Mathematica* or MATLAB to verify Eq. (21.57b) numerically, or, if you know your Bessel functions well, try deriving it analytically.

(b) Check the result, (21.60) by substituting Eq. (21.57b) into Eq. (21.59) and performing the Gaussian integral.

5. Solla and Riedel derive the density of vortices from the expression

$$n_v = \frac{\partial f}{\partial \ln y_0} = \frac{1}{N} \sum_{\vec{R}} \langle m(\vec{R})^2 \rangle. \tag{21.135}$$

Using this relation show that the density of vortices derived using Eq. (21.128) is consistent with the result of Eq. (21.88).

References

M. Aizenman, J.M. Chayes, L. Chayes, C.M. Newman, Discontinuity of the magnetization in one-dimensional $1/|x - y|^2$ Ising and Potts models. J. Stat. Phys. **50**, 140 (1988)

N.W. Ashcroft, N.D. Mermin, *Solid State Physics* (Brooks Cole, 1976)

P.W. Anderson, G. Yuval, D.R. Hamann, Exact results in the Kondo problem. II. Scaling theory, qualitatively correct solution, and some new results on one-dimensional classical statistical models. Phys. Rev. **B1**, 4464 (1970)

P.W. Anderson, G. Yuval, Some numerical results on the Kondo problem and the inverse square one-dimensional Ising model. J. Phys. C Solid St. Phys. **4**, 607 (1971)

D.J. Bishop, J.D. Reppy, Study of the superfluid transition in two-dimensional 4He films. Phys. Rev. B **22**, 5171 (1980)

R.M. Corless, G.H. Gonnet, D.E.G. Hare, D.J. Jeffrey, D.E. Knuth, On the Lambert W function. Adv. Comp. Math. **5**, 329 (1996)

Y. Imry, L. Gunther, Fluctuations and physical properties of the two-dimensional crystal lattice. Phys. Rev. B **3**, 3939 (1971)

J. José, L.P. Kadanoff, S. Kirkpatrick, D.R. Nelson, Renormalization, vortices, and symmetry-breaking perturbations in the two-dimensional planar model. Phys. Rev. **B16**, 1217 (1977); (Erratum) Phys. Rev. **B17** 1477 (1978)

J. Kondo, Resistance minimum in dilute magnetic alloys. Prog. Theor. Phys. **32**, 37 (1964)

J.M. Kosterlitz, The critical properties of the two-dimensional xy model. J. Phys. C Solid State Phys. **7**, 1046 (1974)

J.M. Kosterlitz, Kosterlitz Thouless physics: a review of key issues. Rep. Prog. Phys. **79**, 026001 (2016)

J.M. Kosterlitz, D.J. Thouless, Long range order and metastability in two-dimensional solids and superfluids. J. Phys. C Solid State Phys. **5**, L124 (1972)

J.M. Kosterlitz, D.J. Thouless, Ordering, metastability and phase transitions in two-dimensional systems. J. Phys. C Solid State Phys. **6**, 1181 (1973)

N.D. Mermin, Crystalline order in two dimensions. Phys. Rev. 176 (1968)

S.A. Solla, E.K. Riedel, Vortex excitations and specific heat of the planar model in two dimensions. Phys. Rev. B **23**(6008), 12 (1981)

D.J. Thouless, Long range order in one-dimensional Ising systems. Phys. Rev. **187**, 732 (1969)

J. Villain, Theory of one-dimensional and two-dimensional magnets with an easy magnetization plane: planar classical two-dimensional magnet. J. Phys. **36**, 581 (1975)

K.G. Wilson, The renormalization group: critical phenomena and the Kondo problem. Rev. Mod. Phys. **47**, 773 (1975)

A.P. Young, On the theory of the phase transition in the two-dimensional planar spin model. J. Phys. C Solid State Phys. **11**, L453 (1978)

Index

A. J. Berlinsky and A. B. Harris, *Statistical Mechanics*, Graduate Texts in Physics,
https://doi.org/10.1007/978-3-030-28187-8

R

S

T

GPSR Compliance
The European Union's (EU) General Product Safety Regulation (GPSR) is a set of rules that requires consumer products to be safe and our obligations to ensure this.

If you have any concerns about our products, you can contact us on

ProductSafety@springernature.com

In case Publisher is established outside the EU, the EU authorized representative is:

Springer Nature Customer Service Center GmbH
Europaplatz 3
69115 Heidelberg, Germany

www.ingramcontent.com/pod-product-compliance
Ingram Content Group UK Ltd.
Pitfield, Milton Keynes, MK11 3LW, UK
UKHW021832270726
14058UKWH00001B/105

9783030281892